Cathey Carter

Communication Systems

Commu-nication Systems,

Second Edition

SIMON HAYKIN

McMaster University

John Wiley & Sons

New York / Chichester / Brisbane / Toronto / Singapore

Library of Congress Cataloging in Publication Data:

Haykin, Simon S.
Communication systems.

(Wiley series in management, ISSN 0271-6046)
Includes bibliographical references and index.
1. Telecommunication. 2. Signal theory (Telecommunication) I. Title. II. Series.
TK5101.H37 1983 621.38'0413 82-17593
ISBN 0-471-09691-1

In loving memory of my loving wife Vera,
and to Michael and Juanne.

Preface

This book is devoted to the study of principles of communication theory as applied to the transmission of information, with equal emphasis given to analog and digital communications. The focus is on basic issues, relating theory to practice wherever possible, and an effort has been made to present the material in a logical and interesting manner. Numerous examples, worked out in detail, have been included to help the reader develop an intuitive grasp of the theory under discussion. Each chapter begins with introductory remarks and (except for Chapter 1) ends with a set of problems that is intended not only to help readers test their understanding of the material covered in the chapter, but also challenge them to extend this material. For the interested reader, the text includes footnotes that provide historical references and suggestions for further reading.

Chapter 1 is an introduction to communication theory and a review of some basic concepts.

In Chapter 2 we review two classical methods for frequency analysis: the Fourier series and the Fourier transform. We present a unified treatment of the spectral density and correlation functions of energy signals and power signals. We also review the time-domain and frequency-domain descriptions of signal transmission through linear filters and channels. Next, we discuss the Hilbert transform, which is usedul in describing narrow-band signals and in implementing certain modulation techniques. We develop a procedure for evaluating the response of a band-pass filter or channel to a band-pass signal applied to the input. This procedure not only simplifies the analysis of band-pass systems but also provides the basis for their simulation on a digital computer. The chapter concludes with a discussion of phase and group delay.

Chapter 3 is devoted to amplitude modulation. Specifically, we develop the mathematical descriptions and the spectral characteristics of full amplitude modulation, double-sideband suppressed-carrier modulation, quadrature-carrier multiplexing or quadrature-amplitude modulation, single-sideband modulation, and vestigial side-band modulation. We discuss devices for their generation and demodulation. We describe frequency-division multiplexing for the transmission of a plurality of different message sources over a common channel.

In Chapter 4 we study frequency modulation. The bandwidth occupied by a frequency-modulated wave is given special attention. We describe devices for the generation and demodulation of frequency-modulated waves. The use of a phase-locked loop for demodulating an FM wave is given detailed attention. We describe a procedure for the use of computer simulation to evaluate the effects of passing a frequency-modulated wave through a linear filter. The chapter also includes a discussion of the effects of channel nonlinearities on FM system performance.

In Chapter 5 we briefly review probability theory. We discuss the issue of stationarity, and the mean, correlation, and covariance functions of a random process. We discuss time averages and the related issue of ergodicity. We describe the effects of passing a random process through a linear filter. Next, we develop the concept of power spectral density and discuss its properties. We discuss the properties of a Gaussian process. The properties of narrow-band noise and its representations are also given detailed attention. We derive the probability distribution for the envelope of a sine-wave plus narrow-band noise, a result which is used later in the book.

Chapter 6 is devoted to analyzing the effects of white Gaussian noise on continuous-wave modulation. In particular, we define signal-to-noise ratios for assessing the performance of such systems in the presence of additive noise. Then, we study in detail the noise performances of various types of amplitude-modulation and frequency-modulation systems. Particular attention is given to the threshold effect that can arise in these systems under certain operating conditions.

In Chapter 7 we present a detailed treatment of the sampling process and related practical issues. We discuss the sampling of band-pass signals and the issue of recovering a random message process from its samples. Then we discuss time-division multiplexing, and the characteristics of pulse-amplitude modulation, pulse-duration modulation, pulse-position modulation, and devices for their generation and demodulation.

In Chapter 8 we describe pulse-code modulation (PCM). We evaluate the effects of transmission noise and quantizing noise on the performance of PCM. We define a quantitative measure of information and use it to derive the Hartley-Shannon law for the channel capacity of an idealized white band-limited Gaussian channel. Then, we examine the channel capacity of a PCM system that is quantizing noise-limited and compare the result with the ideal communication system. We describe the characteristics of differential pulse-code modulation, delta modulation, and the use of adaptive schemes to improve their performances. We also look at digital multiplexers for the time-division multiplexing of a plurality of digital streams. The T1 carrier system developed by the Bell system is given special attention.

Chapter 9 is devoted to the issues involved in baseband data transmission. We describe baseband shaping as a means of combating the effects of intersymbol interference. We next present a procedure for the optimum design of transmitting and receiving filters in the combined presence of intersymbol interference and noise. We describe correlative coding or partial-response signaling that uses intersymbol interference in a controlled manner to improve the capability of a data transmission system. We describe adaptive equalization in order to realize the full capability of a telephone channel. We then describe the eye pattern for assessing the performance of a data transmission system.

Finally, in Chapter 10 we study band-pass data transmission. In particular, we use the geometric representation of signals as the basis of this study. We study the coherent detection of a known signal in the presence of additive white Gaussian noise, and so derive the correlation receiver and matched filter receiver. We next describe the use of coherent binary phase-shift keying and frequency-shift keying for the transmission of data. We describe coherent quadriphase-shift keying and minimum-shift keying as examples of quadrature-carrier multiplexing. We derive formulas for the average probability of symbol error and the spectral characteristics of these binary and quadrature signaling techniques, and describe devices for their generation and detection. We then discuss the issue of detecting a sinewave of unknown phase in the presence of noise. We derive in detail the formulas for the average probability of symbol error for the noncoherent detection of binary FSK signals, and differential phase-shift keying. We present highlights of coherent M-ary PSK and FSK techniques. We describe how digital computer simulation may be used to study the effects of intersymbol interference in band-pass data transmission techniques. The chapter ends with a discussion of maximum-likelihood estimation and the phase-locked loop as an example of this parameter estimation procedure.

The book is essentially self-contained, and suited for a one-academic year or one-semester course on communication theory. It is suitable for electrical engineering students in their final year. It is expected that the reader has a knowledge of electronics, circuit theory, and probability theory. The makeup of the material for the course may be determined only by the background of the students and the interests of the teacher involved. However, the material covered in the book is broad enough to satisfy a variety of backgrounds and interests, thereby allowing considerable flexibility in making up the course material. As an aid to the teacher of the course, a detailed solutions manual for all the problems in the book is available from the publisher.

Simon Haykin

Acknowledgments

I would like to express my deep gratitude to Dr. S. Pasupathy, University of Toronto, for reading the manuscript and for his numerous suggestions that have had a major influence on this second edition of the book. I am also grateful to Dr. R. deBuda, Canadian General Electric, Toronto; Dr. C. R. Carter, Dr. D. P. Taylor, and Dr. K. M. Wong, McMaster University; and Dr. S. B. Kesler, Drexel University, for their help. I have also benefited from reviews of the first edition of the book by Dr. M. Sablatash and Dr. J. H. Roberts. The numerous suggestions made by two anonymous reviewers have been very helpful in finalizing the manuscript.

I am indebted to the American Telephone and Telegraph Company for their permission to reproduce the following two diagrams:

(a) Figure 5.33 from the paper: S. O. Rice, "Mathematical analysis of random noise," *Bell System Technical Journal*, vol. 24, pp. 46–156, Jan. 1945.
(b) Figure 8.2 from the paper: W. R. Bennett, "Spectra of quantized signals," *Bell System Technical Journal*, vol. 27, pp. 446–472, July 1948.

Finally, the guidance and encouragement received from my editor, Mr. Merrill G. Floyd, are particularly appreciated. I am also grateful to other staff members of John Wiley & Sons for their help in the production of this book.

Contents

Chapter 1 Introduction

1.1 Classifications of Signals 4
1.2 Fourier Analysis of Signals 6
1.3 Model of a Communication System 7

Chapter 2 Representation of Signals and Systems

2.1 Fourier Series 12
2.2 Fourier Transform 17
2.3 Properties of the Fourier Transform 23
2.4 Numerical Computation of the Fourier Transform 40
2.5 Dirac Delta Function 42
2.6 Fourier Transforms of Periodic Signals 49
2.7 Spectral Density 51
2.8 Autocorrelation Function 54
2.9 Cross-Correlation Functions 58
2.10 Transmission of Signals Through Linear Systems 60
2.11 Ideal Low-Pass Filters 70
2.12 Hilbert Transform 74
2.13 Pre-Envelope 79
2.14 Band-Pass Signals 80
2.15 Band-Pass Systems 85
2.16 Phase and Group Delay 91
Problems 93

Chapter 3 Amplitude Modulation

3.1 Amplitude Modulation (AM) 114
3.2 Generation of AM Waves 119
3.3 Demodulation of AM Waves 123
3.4 Double-Sideband Suppressed-Carrier Modulation (DSBSC) 125
3.5 Generation of DSBSC Waves 127

3.6 Coherent Detection of DSBSC Waves 130
3.7 Quadrature-Carrier Multiplexing 135
3.8 Single-Sideband Modulation (SSB) 137
3.9 Generation of SSB Waves 141
3.10 Demodulation of SSB Waves 146
3.11 Vestigial Sideband Modulation (VSB) 149
3.12 Discussion 156
3.13 Frequency Translation 158
3.14 Frequency-Division Multiplexing (FDM) 160
Problems 163

Chapter 4 Angle Modulation

4.1 Basic Definitions: Phase Modulation (PM) and Frequency Modulation (FM) 180
4.2 Single-Tone Frequency Modulation 183
4.3 Narrow-Band Frequency Modulation 185
4.4 Wide-Band Frequency Modulation 187
4.5 Multitone FM Waves 190
4.6 Transmission Bandwidth of FM Waves 194
4.7 Generation of FM Waves 197
4.8 Demodulation of FM Waves 202
4.9 Response of Linear Filters to FM Waves 216
4.10 Nonlinear Effects in FM Systems 217
Problems 219

Chapter 5 Random Processes

5.1 Probability 230
5.2 Random Variables 233
5.3 Random Processes 240
5.4 Stationarity 243
5.5 Mean, Correlation, and Covariance Functions 244
5.6 Time Averages and Ergodicity 250
5.7 Transmission of a Random Process Through a Linear Filter 254
5.8 Spectral Density 256
5.9 Gaussian Process 270
5.10 Noise 275
5.11 Narrow-Band Noise 283
5.12 Envelope of Sine-Wave Plus Narrow-Band Noise 296
Problems 299

Chapter 6 Noise in CW Modulation

6.1 AM Receivers 318
6.2 Signal-To-Noise Ratios (SNR) 320
6.3 Signal-To-Noise Ratios for Coherent Reception With DSBSC Modulation 322
6.4 Signal-To-Noise Ratios for Coherent Reception With SSB Modulation 325
6.5 Noise in AM Receivers Using Envelope Detection 328
6.6 FM Receivers 332
6.7 Noise in FM Reception 335
6.8 FM Threshold Effect 341
6.9 Pre-emphasis and De-emphasis in FM 348
6.10 Discussion 352
Problems 354

Chapter 7 Pulse-Analog Modulation

7.1 Sampling Theorem 364
7.2 Sampling of Band-Pass Signals 371
7.3 Practical Aspects of Sampling 376
7.4 Reconstruction of a Message Process from Its Samples 382
7.5 Time-Division Multiplexing (TDM) 384
7.6 Pulse-Amplitude Modulation (PAM) 385
7.7 Pulse-Time Modulation 389
Problems 397

Chapter 8 Pulse-Digital Modulation

8.1 Elements of Pulse-Code Modulation (PCM) 408
8.2 Noise in PCM Systems 420
8.3 Measure of Information 429
8.4 Channel Capacity 433
8.5 Channel Capacity of a PCM System 436
8.6 Differential Pulse-Code Modulation (DPCM) 439
8.7 Delta Modulation (DM) 445
8.8 Discussion 449
8.9 Adaptive Digital Waveform Coding Schemes 451
8.10 Digital Multiplexers 454
Problems 458

Chapter 9 Baseband Data Transmission

9.1 Elements of Baseband Binary PAM Systems 466

9.2 Baseband Shaping 469
9.3 Optimum Transmitting and Receiving Filters for Noise Immunity 473
9.4 Correlative Coding 476
9.5 Baseband M-ary PAM Systems 485
9.6 Adaptive Equalization 487
9.7 Eye Pattern 496
Problems 498

Chapter 10 Band-Pass Data Transmission

10.1 A Model of Band-Pass Data Transmission Systems 506
10.2 Gram–Schmidt Orthogonalization Procedure 510
10.3 Geometric Interpretation of Signals 517
10.4 Response of a Bank of Correlators to Noisy Input 519
10.5 Coherent Detection of Signals in the Presence of Noise 523
10.6 Correlation Receiver 528
10.7 Matched Filter Receiver 530
10.8 Coherent Binary Signaling Techniques 540
10.9 Coherent Binary PSK 541
10.10 Coherent Binary FSK 544
10.11 Spectral Properties of Binary PSK and FSK Signals 549
10.12 Coherent Quadrature-Signaling Techniques 552
10.13 Quadriphase-Shift Keying (QPSK) 553
10.14 Minimum Shift Keying (MSK) 561
10.15 Spectral Properties of QPSK and MSK Signals 573
10.16 M-ary Signaling Techniques 575
10.17 Detection of Signals with Random Phase in the Presence of Noise 578
10.18 Noncoherent Detection of Binary FSK Signals 581
10.19 Differential Phase-Shift Keying (DPSK) 585
10.20 Discussion 593
10.21 Effect of Intersymbol Interference 596
10.22 Maximum Likelihood Estimation 599
Problems 603

Appendix 1 Continuous Probability Distributions 616
Appendix 2 Error Function 619
Appendix 3 Noise Figure 623
Appendix 4 Bessel Functions 628
Appendix 5 Schwarz's Inequality 634
Appendix 6 Mathematical Tables 636

Glossary 643

Index 647

chapter 1 INTRODUCTION

Today, *communication* enters our daily lives in so many different ways that it is easy to overlook the multitude of its facets.* Aside from the telephones at our

* For essays on communications and other related disciplines (e.g., electronics, computers, radar, radio astronomy, satellites), see C. F. J. Overhage (Editor), *The Age of Electronics*, (McGraw-Hill, 1962). In particular, see the chapter on "Communications," by L. V. Berkner, pp. 35–50.

hands, and the radios and televisions in our living rooms, our newspapers are assembled by rapid communications from every corner of the globe. Communication provides the senses for ships on the high seas, aircraft in flight, and rockets and satellites in space. Communication keeps the weather forecaster informed of conditions measured by a multitude of sensors. The list of applications involving the use of communication in one way or another goes on.

In the most fundamental sense, communication involves implicitly the transmission of *information* from one point to another through a succession of processes, as described below:

1. The generation of a thought pattern or image in the mind of an originator.
2. The description of that image, with a certain measure of precision, by a set of aural or visual symbols.
3. The encoding of these symbols in a form that is suitable for transmission over a medium of interest.
4. The transmission of the encoded symbols to the desired destination.
5. The decoding and reproduction of the initial symbols.
6. The recreation of the original image, with a definable degradation in quality, in the mind of a recipient; the degradation is caused by imperfections in the system.

There are, of course, many other forms of communication that do not directly involve the human mind in real time. For example, in computer communications, human decisions may enter only in setting up the programs or commands for the computer, or in monitoring the results.

In this book we present a study of communication systems. Our approach will be from a *systems viewpoint*, emphasizing mathematical descriptions, representations, and processing of electrical signals which characterize such systems.

1.1 CLASSIFICATIONS OF SIGNALS

For our purposes, we define a *signal* as a single-valued function of time that conveys information. Consequently, for every instant of time (the independent variable) there is a unique value of the function (the dependent variable). This value may be a real number, in which case we have a *real-valued signal*, or it may be a complex number, in which case we have a *complex-valued signal*. In any case, the independent variable (namely, time) is real-valued.

For a given situation, we find that the most useful method of signal representation hinges on the particular type of signal being considered. Depending on the feature of interest, we may distinguish four different classes of signals:

(1) Periodic Signals, Nonperiodic Signals

A *periodic signal* $g(t)$ is a function that satisfies the condition

$$g(t)=g(t+T_0), \qquad (1.1)$$

for all t, where t denotes time and T_0 is a constant. The smallest value of T_0 that satisfies this condition is called the *period* of $g(t)$. Accordingly, the period T_0 defines the duration of one complete cycle of $g(t)$.

Any signal for which there is no value of T_0 to satisfy the condition of Eq. (1.1) is called a *nonperiodic* or *aperiodic signal.*

(2) Deterministic Signals, Random Signals

A *deterministic signal* is a signal about which there is no uncertainty with respect to its value at any time. Accordingly, we find that deterministic signals may be modeled as completely specified functions of time.

On the other hand, a *random signal* is a signal about which there is some degree of uncertainty before it actually occurs. Such a signal may be viewed as belonging to a collection or ensemble of signals, with each signal in the collection being different. Random signals are considered in Chapter 5.

(3) Energy Signals, Power Signals

In electrical systems, a signal may represent a voltage or a current. Consider a voltage $v(t)$ developed across a resistor R, producing a current $i(t)$. The *instantaneous power* dissipated in this resistor is defined by

$$p=\frac{|v(t)|^2}{R} \tag{1.2}$$

or, equivalently,

$$p=R|i(t)|^2 \tag{1.3}$$

In both cases, the instantaneous power p is proportional to the squared amplitude of the signal. Furthermore, for a resistor R equal to 1 ohm, we see that Eqs. (1.2) and (1.3) take on the same mathematical form. Accordingly, in signal analysis it is customary to work with a 1-ohm resistor, so that, regardless of whether a given signal $g(t)$ represents a voltage or a current, we may express the instantaneous power associated with the signal as

$$p=|g(t)|^2 \tag{1.4}$$

Based on this convention, we define the *total energy* of a signal $g(t)$ as

$$\begin{aligned} E &= \lim_{T\to\infty} \int_{-T}^{T} |g(t)|^2 \, dt \\ &= \int_{-\infty}^{\infty} |g(t)|^2 \, dt \end{aligned} \tag{1.5}$$

and its *average power* as

$$P=\lim_{T\to\infty} \frac{1}{2T} \int_{-T}^{T} |g(t)|^2 \, dt \tag{1.6}$$

We say that the signal $g(t)$ is an *energy signal* if and only if the total energy of the signal satisfies the condition

$$0 < E < \infty$$

We say that the signal $g(t)$ is a *power signal* if and only if the average power of the signal satisfies the condition

$$0 < P < \infty$$

The energy and power classifications of signals are mutually exclusive. In particular, an energy signal has zero average power, whereas a power signal has infinite energy. Also, it is of interest to note that, usually, periodic signals and random signals are power signals, and signals that are both deterministic and nonperiodic are energy signals.

(4) Analog Signals, Digital Signals

An *analog signal* is a continuous function of time, with the amplitude being continuous as well. Analog signals arise when a physical waveform such as an acoustic or a light wave is converted into an electrical signal. The conversion is effected by means of a *transducer*; examples include the microphone, which converts sound pressure variations into corresponding voltage or current variations, and the photoelectric cell, which does the same for light intensity variations.

On the other hand, a *discrete-time signal* is defined only at discrete times. Thus, in this case the independent variable takes on only discrete values, which are usually uniformly spaced. Consequently, discrete-time signals are described as sequences of samples whose amplitudes may take on a continuum of values. When each sample of a discrete-time signal is *quantized* (i.e., its amplitude is only allowed to take on a finite set of discrete values) and then *coded*, the resulting signal is referred to as a *digital signal*. The output of a digital computer is an example of a digital signal. Naturally, an analog signal may be converted into digital form by *sampling* it in time, then quantizing and coding it. The sampling process is described in Chapter 7, and the quantizing process and encoding in Chapter 8.

1.2 FOURIER ANALYSIS OF SIGNALS

In theory, there are many possible methods for the representation of signals. In practice, however, we find that *Fourier analysis*, involving the resolution of signals into sinusoidal components, overshadows all other methods in usefulness.* Basically, this is a consequence of the well-known fact that the response of a system to a sine-wave input is another sine-wave of the same frequency (but with a different phase and amplitude) under two conditions:

* For a historical account of the concept of frequency, see: J. M. Manley, "The concept of frequency in linear system analysis," *IEEE Communications Society Magazine*, pp. 26–35, Vol. 20, Jan. 1982.

1. The system is *linear* in that it obeys the *principle of superposition*. That is, if $y_1(t)$ and $y_2(t)$ denote the responses of a system to the inputs $x_1(t)$ and $x_2(t)$, respectively, the system is linear if the response to the composite input $a_1x_1(t) + a_2x_2(t)$ is equal to $a_1y_1(t)+a_2y_2(t)$, where a_1 and a_2 are arbitrary constants.
2. The system is *time-invariant*. That is, if $y(t)$ is the response of a system to the input $x(t)$, the system is time-invariant if the response to the time-shifted input $x(t-t_0)$ is equal to $y(t-t_0)$, where t_0 is constant.

There are several methods of Fourier analysis available for the representation of signals. The particular version that is used in practice depends on the type of signal being considered. For example, if the signal is periodic, then the logical choice is to use the *Fourier series* to represent the signal in terms of a set of harmonically related sine-waves. On the other hand, if the signal is an energy signal, then it is customary to use the *Fourier transform* to represent the signal. By using the Fourier series or the Fourier transform, we obtain the *frequency-domain description* or *spectrum* of the signal, by means of which we are often able to discern important characteristics of the signal in a way that may otherwise be difficult.

In Chapter 2 we review the Fourier series, and then present a detailed treatment of the Fourier transform, its properties, and its use in the frequency analysis of signals and linear systems.

1.3 MODEL OF A COMMUNICATION SYSTEM

The purpose of a communication system is to transmit *information-bearing signals* from a *source*, located at one point in space, to a *user destination*, located at another point. As a rule, the message produced by the source is not electrical in nature. Accordingly, an input transducer is used to convert the message generated by the source into a time-varying electrical signal called the *message signal*. By using another transducer at the receiver, the original message is recreated at the user destination.

Figure 1.1 shows the block diagram of a communication system. The system consists of three major parts: (1) *transmitter*, (2) *communication channel*, and (3)

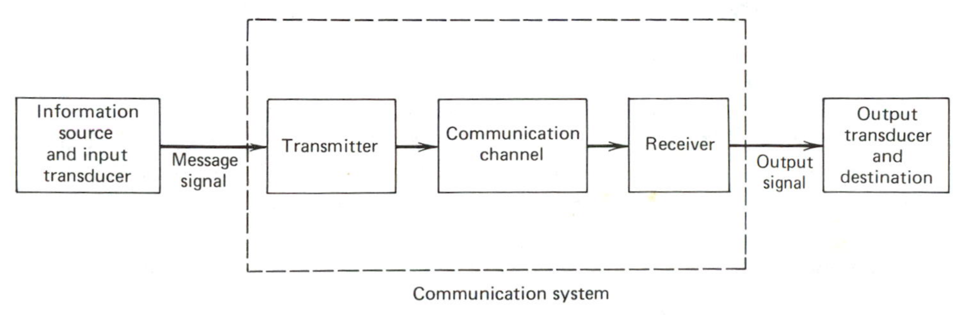

Figure 1.1 Model of an electrical communication system.

receiver. The main purpose of the transmitter is to modify the message signal into a form suitable for transmission over the channel. This modification is achieved by means of a process known as *modulation*, which involves varying some parameter of a *carrier wave* (e.g., the amplitude, frequency, or phase of a sinusoidal wave) in accordance with the message signal.

The communication channel may be a transmission line (as in telephony and telegraphy), an optical fiber (as in optical communications), or merely free space in which the signal is radiated as an electromagnetic wave (as in radio and television broadcasting). In propagating through the channel, the transmitted signal is distorted due to nonlinearities and/or imperfections in the frequency response of the channel. Other sources of degradation are *noise* and *interference* picked up by the signal during the course of transmission through the channel. Noise and distortion constitute two basic problems in the design of communication systems. Usually, the transmitter and receiver are carefully designed so as to minimize the effects of noise and distortion on the quality of reception.

The main purpose of the receiver is to recreate the original message signal from the degraded version of the transmitted signal after propagation through the channel. This recreation is accomplished by using a process known as *demodulation*, which is the reverse of the modulation process used in the transmitter. However, owing to the unavoidable presence of noise and distortion in the received signal, we find that the receiver cannot recreate the original message signal exactly. The resulting degradation in overall system performance is influenced by the type of modulation scheme used. Specifically, we find that some modulation schemes are less sensitive to the effects of noise and distortion than others.

In any communication system, there are two primary communication resources to be employed, namely, *transmitted power* and *channel bandwidth*. A general system design objective is to use these two resources as efficiently as possible. In most communication channels, one resource may be considered more important than the other. We may therefore classify communication channels as *power-limited* or *band-limited*. For example, the telephone circuit is a typical band-limited channel, whereas a space communication link or a satellite channel is typically power-limited.

For the case when the spectrum of a message signal extends down to zero or low frequencies, we define the bandwidth of the signal as that upper frequency above which the spectral content of the signal is negligible and, therefore, unnecessary for transmitting the pertinent information. For example, the average voice spectrum extends well beyond 10 kHz, though most of the energy is concentrated in the range of 100 to 600 Hz, and a band from 300 to 3,400 Hz gives good articulation. Accordingly, we find that telephone circuits that respond well to this latter range of frequencies give quite satisfactory commercial telephone service.

The major part of this book is devoted to a detailed treatment of various modulation schemes for the transmission of analog and digital signals, emphasizing their bandwidth properties and the effects of noise on their performance.

chapter 2 REPRESENTATION OF SIGNALS AND SYSTEMS

In this chapter, we begin our study of signal analysis by reviewing the Fourier series by which we are able to represent a periodic signal as an infinite sum of sine-wave components. Next, we develop the Fourier transform, which performs a similar role in the analysis of nonperiodic signals. The Fourier transform is more general in application than the Fourier series.* A motivation for using the Fourier series or the Fourier transform is to obtain the *spectrum* of a given signal, which describes the frequency content of the signal.

* The origin of the theory of Fourier series and Fourier transform is to be found in: J. B. J. Fourier, *The Analytical Theory of Heat*, translated by A. Freeman (Cambridge University Press, 1878).

2.1 FOURIER SERIES

Let $g_p(t)$ denote a periodic signal with period T_0. By using a *Fourier series expansion* of this signal, we are able to resolve the signal into an infinite sum of sine and cosine terms. This expansion may be expressed in the form:

$$g_p(t) = a_0 + 2 \sum_{n=1}^{\infty} \left[a_n \cos\left(\frac{2\pi nt}{T_0}\right) + b_n \sin\left(\frac{2\pi nt}{T_0}\right) \right] \tag{2.1}$$

where the coefficients a_n and b_n represent the unknown amplitudes of the cosine and sine terms, respectively. The quantity n/T_0 represents the nth harmonic of the fundamental frequency $1/T_0$. Each of the terms $\cos(2\pi nt/T_0)$ and $\sin(2\pi nt/T_0)$ is called a *basis function*. These basis functions form an *orthogonal set* over the interval T_0 in that they satisfy the following set of relations:

$$\int_{-T_0/2}^{T_0/2} \cos\left(\frac{2\pi mt}{T_0}\right) \cos\left(\frac{2\pi nt}{T_0}\right) dt = \begin{cases} T_0/2, & m = n \\ 0, & m \neq n \end{cases} \tag{2.2}$$

$$\int_{-T_0/2}^{T_0/2} \cos\left(\frac{2\pi mt}{T_0}\right) \sin\left(\frac{2\pi nt}{T_0}\right) dt = 0 \qquad \text{for all } m \text{ and } n \tag{2.3}$$

$$\int_{-T_0/2}^{T_0/2} \sin\left(\frac{2\pi mt}{T_0}\right) \sin\left(\frac{2\pi nt}{T_0}\right) dt = \begin{cases} T_0/2, & m = n \\ 0, & m \neq n \end{cases} \tag{2.4}$$

To determine the coefficient a_0, we integrate both sides of Eq. (2.1) over a complete period. We thus find that a_0 is the *mean value* of the periodic signal $g_p(t)$ over one period, as shown by the *time average*

$$a_0 = \frac{1}{T_0} \int_{-T_0/2}^{T_0/2} g_p(t) dt \tag{2.5}$$

To determine the coefficient a_n, we multiply both sides of Eq. (2.1) by $\cos(2\pi nt/T_0)$ and integrate over the interval $-T_0/2$ to $T_0/2$. Then, using Eqs. (2.2) and (2.3), we find that

$$a_n = \frac{1}{T_0} \int_{-T_0/2}^{T_0/2} g_p(t) \cos\left(\frac{2\pi nt}{T_0}\right) dt, \qquad n = 1, 2, \ldots \tag{2.6}$$

Similarly, we find that

$$b_n = \frac{1}{T_0} \int_{-T_0/2}^{T_0/2} g_p(t) \sin\left(\frac{2\pi nt}{T_0}\right) dt, \qquad n = 1, 2, \ldots \tag{2.7}$$

To apply the Fourier series representation of Eq. (2.1), it is sufficient that the function $g_p(t)$ satisfies the following conditions:

1. The function $g_p(t)$ is single-valued within the interval T_0.
2. The function $g_p(t)$ has at most a finite number of discontinuities in the interval T_0.
3. The function $g_p(t)$ has a finite number of maxima and minima in the interval T_0.

4. The function $g_p(t)$ is absolutely integrable, that is,

$$\int_{-T_0/2}^{T_0/2} |g_p(t)|dt < \infty.$$

These conditions are known as *Dirichlet's conditions.** They are satisfied by the periodic signals usually encountered in communication systems.

Complex Exponential Fourier Series

The Fourier series of Eq. (2.1) can be put into a much simpler and more elegant form with the use of complex exponentials. We do this by substituting in Eq. (2.1) the exponential form for the cosine and sine, namely:

$$\cos\left(\frac{2\pi nt}{T_0}\right)=\frac{1}{2}\left[\exp\left(\frac{j2\pi nt}{T_0}\right)+\exp\left(-\frac{j2\pi nt}{T_0}\right)\right]$$

$$\sin\left(\frac{2\pi nt}{T_0}\right)=\frac{1}{2j}\left[\exp\left(\frac{j2\pi nt}{T_0}\right)-\exp\left(-\frac{j2\pi nt}{T_0}\right)\right]$$

We thus obtain

$$g_p(t)=a_0+\sum_{n=1}^{\infty}\left[(a_n-jb_n)\exp\left(\frac{j2\pi nt}{T_0}\right)+(a_n+jb_n)\exp\left(-\frac{j2\pi nt}{T_0}\right)\right] \tag{2.8}$$

Let c_n denote a complex coefficient related to a_n and b_n by

$$c_n=\begin{cases} a_n-jb_n, & n>0 \\ a_0, & n=0 \\ a_n+jb_n, & n<0 \end{cases} \tag{2.9}$$

Then, we may simplify Eq. (2.8) as follows:

$$g_p(t)=\sum_{n=-\infty}^{\infty} c_n \exp\left(\frac{j2\pi nt}{T_0}\right) \tag{2.10}$$

where

$$c_n=\frac{1}{T_0}\int_{-T_0/2}^{T_0/2} g_p(t)\exp\left(-\frac{j2\pi nt}{T_0}\right)dt, \qquad n=0, \pm 1, \pm 2, \ldots \tag{2.11}$$

The series expansion of Eq. (2.10) is referred to as the *complex exponential Fourier series*. The c_n are called the *complex Fourier coefficients*. Equation (2.11) states that, given a periodic signal $g_p(t)$, we may determine the complete set of complex Fourier coefficients. On the other hand, Eq. (2.10) states that, given this set of values, we may reconstruct the original periodic signal exactly.

* The purpose of a good deal of theory dealing with the Fourier series has been to show that the series associated with a periodic function $g_p(t)$ does in fact converge. The rigorous development of this issue was initiated by Dirichlet in 1829. For a detailed discussion of the convergence of Fourier series, see R. V. Churchill, *Fourier Series and Boundary Value Problems*, Second Edition, Chapter 4 (McGraw-Hill, 1963).

According to this representation, a periodic signal contains all frequencies (both positive and negative) that are harmonically related to the fundamental. The presence of negative frequencies is simply a result of the fact that the mathematical model of the signal as described by Eq. (2.10) requires the use of negative frequencies. Indeed, this representation also requires the use of complex-valued basis functions, namely, $\exp(j2\pi nt/T_0)$, which have no physical meaning either. The reason for using complex-valued basis functions and negative frequency components is merely to provide a compact mathematical description of a periodic signal, which is well-suited for both theoretical and practical work.

Discrete Spectrum

The representation of a periodic signal by a Fourier series is equivalent to resolution of the signal into its various harmonic components. Thus, using the complex exponential Fourier series, we find that a periodic signal $g_p(t)$ with period T_0 has components of frequencies 0, $\pm f_0$, $\pm 2f_0$, $\pm 3f_0, \ldots$, and so forth, where $f_0 = 1/T_0$ is the fundamental frequency. That is, while the signal $g_p(t)$ exists in the time domain, we may say that its frequency-domain description consists of components of frequencies, 0, $\pm f_0$, $\pm 2f_0, \ldots$, called the *spectrum*. If we specify the periodic signal $g_p(t)$, we can determine its spectrum; conversely, if we specify the spectrum, we can determine the corresponding signal. This means that a periodic signal $g_p(t)$ can be specified in two equivalent ways: (1) the time-domain representation where $g_p(t)$ is defined as a function of time, and (2) the frequency-domain representation where the signal is defined in terms of its spectrum. While the two descriptions are separate aspects of a given phenomenon, they are not independent of each other, but are related, as Fourier theory shows.

In general, the Fourier coefficient c_n is a complex number, and so we may express it in the form:

$$c_n = |c_n| \exp[j \arg(c_n)] \tag{2.12}$$

The $|c_n|$ defines the amplitude of the nth harmonic component of the periodic signal $g_p(t)$, so that a plot of $|c_n|$ versus frequency yields the *discrete amplitude spectrum* of the signal. A plot of $\arg(c_n)$ versus frequency yields the *discrete phase spectrum* of the signal. We refer to the spectrum as a *discrete spectrum* because both the amplitude and phase of c_n have nonzero values only for discrete frequencies that are integer (both positive and negative) multiples of the fundamental frequency.

For a real-valued periodic function $g_p(t)$, we find, from the definition of the Fourier coefficient c_n given by Eq. (2.11), that

$$c_{-n} = c_n^* \tag{2.13}$$

where c_n^* is the complex conjugate of c_n. We therefore have

$$|c_{-n}| = |c_n| \tag{2.14}$$

and

$$\arg(c_{-n}) = -\arg(c_n) \tag{2.15}$$

That is, the amplitude spectrum of a real-valued periodic signal is symmetrical (an even function of n) and the phase spectrum is antisymmetrical (an odd function of n) about the vertical axis passing through the origin.

Example 1 Periodic pulse train

Consider a periodic train of rectangular pulses of duration T and period T_0, as shown in Fig. 2.1. For convenience the origin has been chosen to coincide with the center of the pulse. This signal may be described analytically over one period, $-(T_0/2) \leqslant t \leqslant (T_0/2)$, as follows

$$g_p(t) = \begin{cases} A, & -\dfrac{T}{2} \leqslant t \leqslant \dfrac{T}{2} \\ 0, & \text{for the remainder of the period.} \end{cases} \tag{2.16}$$

Using Eq. (2.11) to evaluate the complex Fourier coefficient c_n, we get

$$\begin{aligned} c_n &= \frac{1}{T_0} \int_{-T/2}^{T/2} A \exp\left(-\frac{j2\pi nt}{T_0} \right) dt \\ &= \frac{A}{n\pi} \sin\left(\frac{n\pi T}{T_0} \right), \qquad n = 0, \pm 1, \pm 2, \ldots \end{aligned} \tag{2.17}$$

To simplify notation in the above and subsequent results, we will use the so-called *sinc function* defined by

$$\text{sinc}(\lambda) = \frac{\sin(\pi\lambda)}{\pi\lambda} \tag{2.18}$$

where λ is the independent variable. The sinc function plays an important role in communication theory. As shown in Fig. 2.2, it has its maximum value of unity at $\lambda = 0$, and approaches zero as λ approaches infinity, oscillating through positive and negative values. It goes through zero at $\lambda = \pm 1, \pm 2, \ldots$, and so on. Thus, in terms of the sinc function we may rewrite Eq. (2.17) as follows

$$c_n = \frac{TA}{T_0} \text{sinc}\left(\frac{nT}{T_0} \right) \tag{2.19}$$

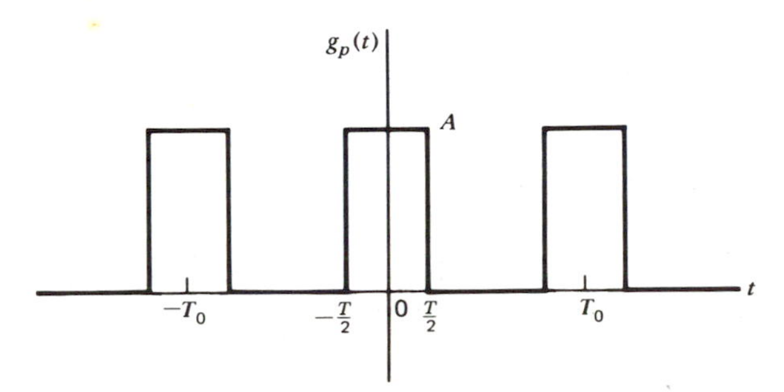

Figure 2.1 Periodic train of rectangular pulses of amplitude A, duration T, and period T₀.

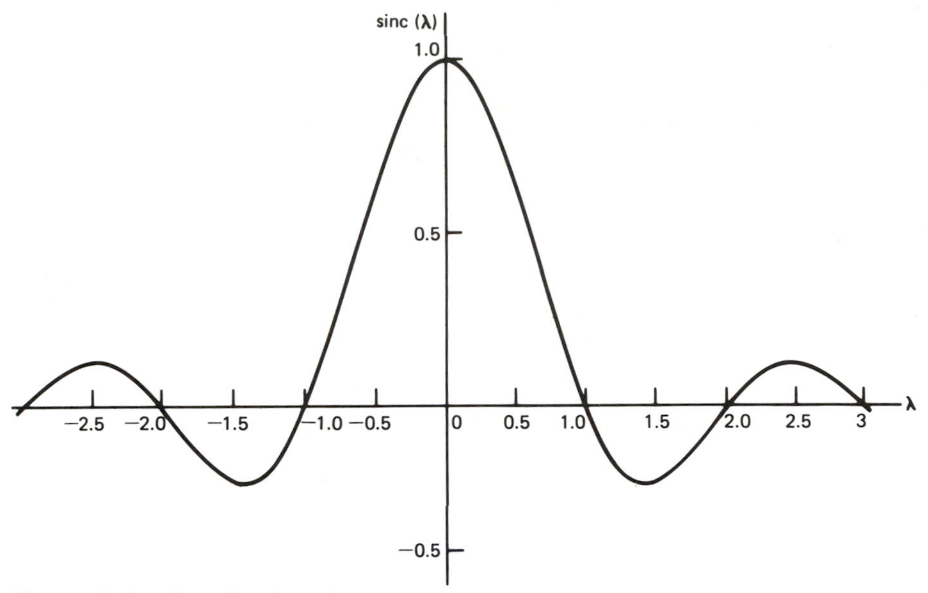

Figure 2.2 The sinc function.

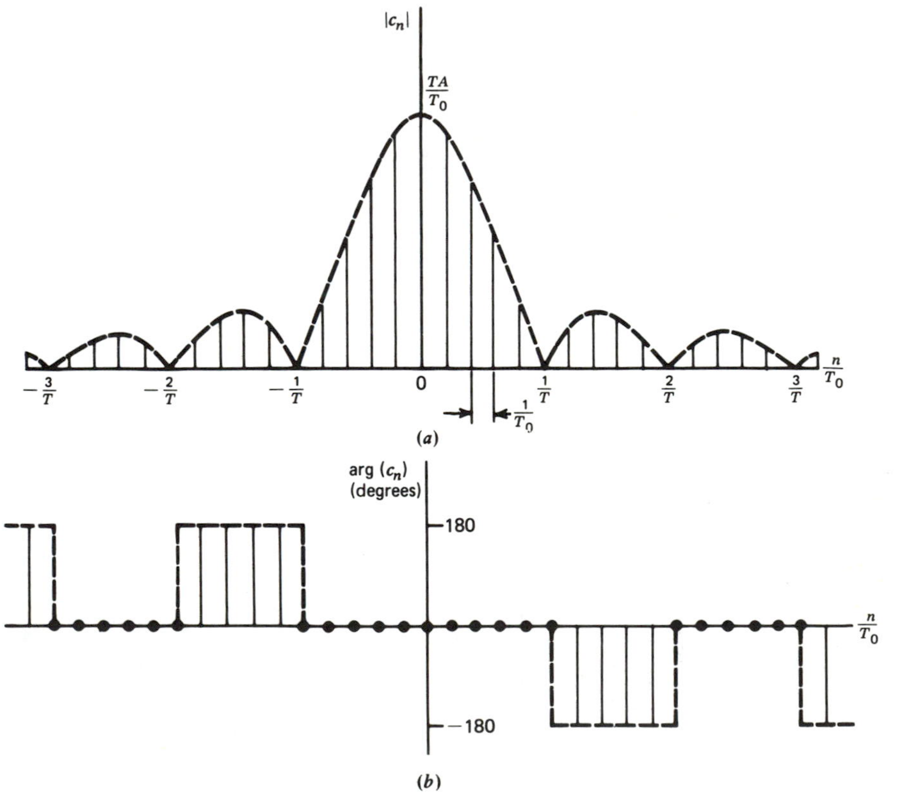

Figure 2.3 Discrete spectrum of a periodic train of rectangular pulses for a duty cycle $T/T_0 = 0.2$. (a) Amplitude spectrum. (b) Phase spectrum.

where n has discrete values only. In Fig. 2.3 we have plotted the amplitude spectrum $|c_n|$ and phase spectrum $\arg(c_n)$ versus the discrete frequency n/T_0 for a *duty cycle* T/T_0 equal to 0.2. We see that:

1. The line spacing in the amplitude spectrum in Fig. 2.3(*a*) is determined by the period T_0.
2. The envelope of the amplitude spectrum is determined by the pulse amplitude A, pulse duration T, and duty cycle T/T_0.
3. Zero-crossings occur in the envelope of the amplitude spectrum at frequencies that are multiples of $1/T$.
4. The phase spectrum takes on the values $0°$ and $\pm 180°$, depending on the polarity of $\text{sinc}(nT/T_0)$; in Fig. 2.3(*b*) we have used both $180°$ and $-180°$ to preserve antisymmetry.

2.2 FOURIER TRANSFORM

In the previous sections we used the Fourier series to represent a periodic signal. We now wish to develop a similar representation for a signal $g(t)$ that is nonperiodic, in terms of exponential signals. In order to do this, we first construct a periodic function $g_p(t)$ of period T_0 in such a way that $g(t)$ defines one cycle of this periodic function, as illustrated in Fig. 2.4. In the limit we let the period T_0 become infinitely large, so that we may write

$$g(t) = \lim_{T_0 \to \infty} g_p(t) \tag{2.20}$$

Representing the periodic function $g_p(t)$ in terms of the complex exponential form of the Fourier series, we have

$$g_p(t) = \sum_{n=-\infty}^{\infty} c_n \exp\left(\frac{j2\pi nt}{T_0}\right) \tag{2.21}$$

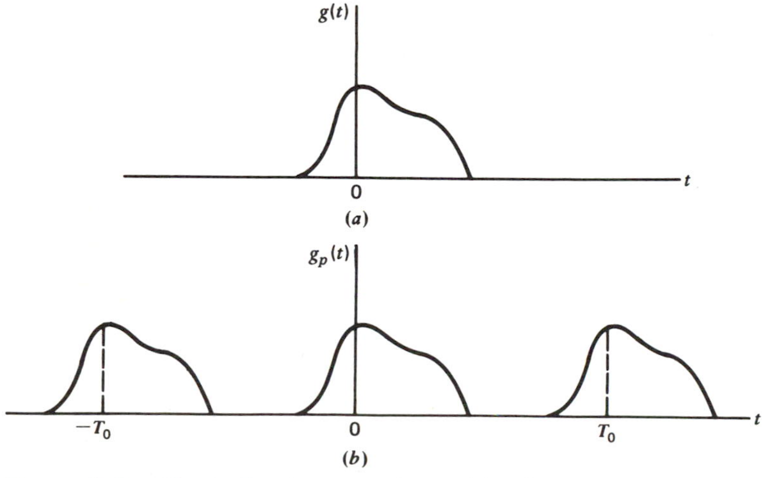

Figure 2.4 Illustrating the use of an arbitrarily defined function of time to construct a periodic waveform. *(a)* Arbitrarily defined function of time $g(t)$. *(b)* Periodic waveform $g_p(t)$ based on $g(t)$.

where

$$c_n = \frac{1}{T_0} \int_{-T_0/2}^{T_0/2} g_p(t) \exp\left(-\frac{j2\pi nt}{T_0}\right) dt \tag{2.22}$$

Define,

$$\Delta f = \frac{1}{T_0},$$

$$f_n = \frac{n}{T_0},$$

and

$$G(f_n) = c_n T_0$$

Thus, making this change of notation in the Fourier series representation of $g_p(t)$, given by Eqs. (2.21) and (2.22), we get the following relations for the interval $-(T_0/2) \leqslant t \leqslant (T_0/2)$,

$$g_p(t) = \sum_{n=-\infty}^{\infty} G(f_n) \exp(j2\pi f_n t) \Delta f \tag{2.23}$$

where

$$G(f_n) = \int_{-T_0/2}^{T_0/2} g_p(t) \exp(-j2\pi f_n t) dt \tag{2.24}$$

Suppose we now let the period T_0 approach infinity or, equivalently, its reciprocal Δf approach zero. Then we find that, in the limit, the discrete frequency f_n approaches the continuous frequency variable f, and the discrete sum in Eq. (2.23) becomes an integral defining the area under a continuous function of frequency f, namely, $G(f)\exp(j2\pi ft)$. Also, as T_0 approaches infinity, the function $g_p(t)$ approaches $g(t)$. Therefore, in the limit, Eqs. (2.23) and (2.24) become, respectively,

$$g(t) = \int_{-\infty}^{\infty} G(f) \exp(j2\pi ft) df \tag{2.25}$$

where

$$G(f) = \int_{-\infty}^{\infty} g(t) \exp(-j2\pi ft) dt \tag{2.26}$$

We have thus achieved our aim of representing an arbitrarily defined signal $g(t)$ in terms of exponential functions over the entire interval $(-\infty < t < \infty)$. Note that in Eqs. (2.25) and (2.26) we have used a lowercase letter to denote the time function and an uppercase letter to denote the corresponding frequency function.

Equation (2.26) states that, given a time function $g(t)$, we can determine a new function $G(f)$ of the frequency variable f. Equation (2.25) states that, given this new or transformed function $G(f)$, we can recover the original time function $g(t)$. Thus, since from $g(t)$ we can define the function $G(f)$ and from $G(f)$ we can reconstruct $g(t)$, the time function is also specified by $G(f)$. The function $G(f)$ can be thought of as a transformed version of $g(t)$ and is referred to as the *Fourier transform* of $g(t)$. The time function $g(t)$ is similarly referred to as the *inverse Fourier transform* of $G(f)$. The functions $g(t)$ and $G(f)$ are said to constitute a *Fourier transform pair*, and one is called the *mate* of the other.*

For a signal $g(t)$ to be Fourier transformable, it is sufficient that $g(t)$ satisfies Dirichlet's conditions:

1. The function $g(t)$ is single-valued, with a finite number of maxima and minima and a finite number of discontinuities in any finite time interval.
2. The function $g(t)$ is absolutely integrable, that is,

$$\int_{-\infty}^{\infty} |g(t)| dt < \infty$$

These conditions include all energy signals, for which we have †

$$\int_{-\infty}^{\infty} |g(t)|^2 \, dt < \infty$$

For convenience, we will use the symbol $\rightleftharpoons$ to indicate a Fourier transform pair, that is,

$$g(t) \rightleftharpoons G(f)$$

We will also use the symbol $F[\;]$ to indicate a Fourier transform operation; that is,

$$F[g(t)] = G(f)$$

and the symbol $F^{-1}[\;]$ to indicate an inverse Fourier transformation; that is,

$$F^{-1}[G(f)] = g(t)$$

Continuous Spectrum

By using the Fourier transform operation, a pulse signal $g(t)$ of finite energy is expressed as a continuous sum of exponential functions with frequencies in the interval $-\infty$ to ∞. The amplitude of a component of frequency f is infinitesimal, but it is proportional to $G(f)$, where $G(f)$ is the Fourier transform of $g(t)$. At any

* See Table A6.2 of Appendix 6 for a short list of Fourier transform pairs. For a more extensive list, see G. A. Campbell and R. M. Foster, *Fourier Integrals for Practical Applications*, (Van Nostrand, 1948).

† If the function $g(t)$ is such that the value of $\int_{-\infty}^{\infty} |g(t)|^2 \, dt$ is defined and finite, then the Fourier transform $G(f)$ of the function $g(t)$ exists and

$$\lim_{A \to \infty} \left[\int_{-\infty}^{\infty} |g(t) - \int_{-A}^{A} G(f) \exp(j2\pi ft) df|^2 \, dt \right] = 0$$

This result is known as *Plancherel's theorem*. For a proof, see E. C. Titchmarsh, *Introduction to the Theory of Fourier Integrals*, Second Edition, p. 69 (Oxford University Press, 1948). N. Wiener, *The Fourier Integral and Certain of Its Applications*, pp. 46–71 (Dover Publications, 1958).

frequency f, the exponential function $\exp(j2\pi ft)$ is weighted by the factor $G(f)df$, which is the contribution of $G(f)$ in an infinitesimal interval df centered at the frequency f. Thus we may express the function $g(t)$ in terms of the continuous sum of such infinitesimal components, as shown by

$$g(t) = \int_{-\infty}^{\infty} G(f)\exp(j2\pi ft)df$$

The Fourier transformation provides us with a tool to resolve a given signal $g(t)$ into its complex exponential components occupying the entire frequency interval from $-\infty$ to ∞. In particular, the Fourier transform $G(f)$ of the signal defines the frequency-domain representation of the signal in that it specifies relative amplitudes of the various frequency components of the signal. We may equivalently define the signal in terms of its time-domain representation by specifying the function $g(t)$ at each instant of time t. The signal is uniquely defined by either representation.

In general, the Fourier transform $G(f)$ is a complex function of frequency f, so that we may express it in the form

$$G(f) = |G(f)|\exp[j\theta(f)] \tag{2.27}$$

where $|G(f)|$ is the *continuous amplitude spectrum* of $g(t)$, and $\theta(f)$ is the *continuous phase spectrum* of $g(t)$. The spectrum is referred to as a *continuous spectrum* because both the amplitude and phase of $G(f)$ are defined for all frequencies. For a real-valued function $g(t)$, we have

$$G(f) = G^*(-f)$$

Therefore, it follows that if $g(t)$ is a real-valued function of time, then

$$|G(-f)| = |G(f)| \tag{2.28}$$

and

$$\theta(-f) = -\theta(f) \tag{2.29}$$

That is, the amplitude spectrum $|G(f)|$ of a real-valued signal is an even function of f, and the phase spectrum $\theta(f)$ is an odd function of f.

Example 2 Rectangular pulse

Consider a *rectangular pulse* of duration T and amplitude A, as shown in Fig. 2.5(a). To define this pulse mathematically in a convenient form, we will use the following notation*

$$\text{rect}(t) = \begin{cases} 1, & -\frac{1}{2} < t < \frac{1}{2} \\ 0, & |t| > \frac{1}{2} \end{cases} \tag{2.30}$$

which stands for a *rectangular function* of unit amplitude and unit duration centered at $t = 0$.

* P. M. Woodward, *Probability and Information Theory, with Applications to Radar*, Second Edition, p. 29 (Pergamon Press, 1964).

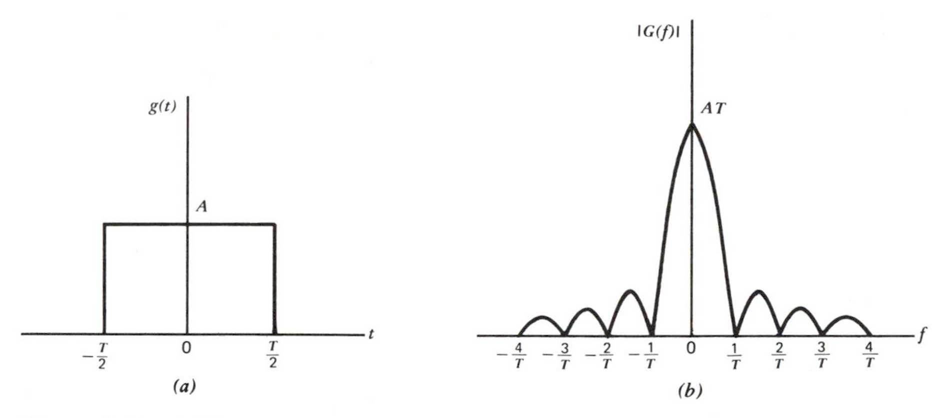

Figure 2.5 *(a)* Rectangular pulse. *(b)* Amplitude spectrum.

Then, in terms of this function, we may express the rectangular pulse of Fig. 2.5(*a*) simply as follows:

$$g(t) = A \operatorname{rect}\left(\frac{t}{T}\right) \tag{2.31}$$

The Fourier transform of this rectangular pulse is given by

$$\begin{aligned} G(f) &= \int_{-T/2}^{T/2} A \exp(-j2\pi f t)\,dt \\ &= AT\left[\frac{\sin(\pi f T)}{\pi f T}\right] \\ &= AT \operatorname{sinc}(fT) \end{aligned} \tag{2.32}$$

We thus have the Fourier transform pair:

$$A \operatorname{rect}\left(\frac{t}{T}\right) \rightleftharpoons AT \operatorname{sinc}(fT) \tag{2.33}$$

The amplitude spectrum $|G(f)|$ is shown plotted in Fig. 2.5(*b*). The first zero-crossing of the spectrum occurs at $f = \pm 1/T$. As the pulse duration T is decreased, this first zero-crossing moves up in frequency and, conversely, as the pulse duration T is increased, the first zero-crossing moves toward the origin.

This example shows that the relationship between the time-domain and frequency-domain descriptions of a signal is an *inverse* one. That is, a pulse, narrow in time, has a significant frequency description over a wide range of frequencies, and vice versa.

Example 3 Exponential pulse

A truncated form of decaying *exponential pulse* is shown in Fig. 2.6(*a*). We may define this pulse mathematically in a convenient form by using the *unit step function*:

$$u(t) = \begin{cases} 1, & t > 0 \\ \frac{1}{2}, & t = 0 \\ 0, & t < 0 \end{cases} \tag{2.34}$$

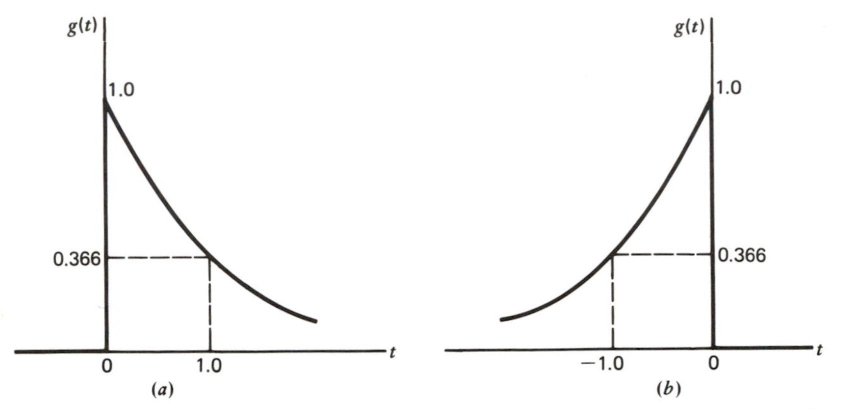

Figure 2.6 *(a)* Decaying exponential pulse. *(b)* Rising exponential pulse.

We may then express the exponential pulse of Fig. 2.6(*a*) as

$$g(t) = \exp(-t)u(t) \tag{2.35}$$

The Fourier transform of this pulse is

$$\begin{aligned} G(f) &= \int_0^\infty \exp(-t)\exp(-j2\pi ft)\,dt \\ &= \int_0^\infty \exp[-t(1+j2\pi f)]\,dt \\ &= \frac{1}{1+j2\pi f} \end{aligned} \tag{2.36}$$

Thus, combining Eqs. (2.35) and (2.36), we obtain the Fourier transform pair:

$$\exp(-t)u(t) \rightleftharpoons \frac{1}{1+j2\pi f} \tag{2.37}$$

A truncated rising exponential pulse is shown in Fig. 2.6(*b*), which is defined by

$$g(t) = \exp(t)u(-t) \tag{2.38}$$

Note that $u(-t)$ is equal to unity for $t<0$, one-half at $t=0$, and zero for $t>0$. The Fourier transform of this pulse is

$$\begin{aligned} G(f) &= \int_{-\infty}^0 \exp(t)\exp(-j2\pi ft)\,dt \\ &= \int_{-\infty}^0 \exp[t(1-j2\pi f)]\,dt \\ &= \frac{1}{1-j2\pi f} \end{aligned} \tag{2.39}$$

We thus have the Fourier transform pair:

$$\exp(t)u(-t) \rightleftharpoons \frac{1}{1-j2\pi f} \tag{2.40}$$

From the Fourier-transform pairs of Eqs. (2.37) and (2.40), we see that truncated decaying and rising exponential pulses have the same amplitude spectrum, but the phase spectrum of the one is the negative of that of the other.

2.3 PROPERTIES OF THE FOURIER TRANSFORM

It is useful to have a feeling for the relationship between a function $g(t)$ and its Fourier transform $G(f)$, and for the effect that various operations on the function $g(t)$ have on the transform $G(f)$. This may be achieved by examining certain properties of the Fourier transform. In this section we describe 12 of these properties, which we will prove, one by one. These properties are summarized in Table A6.1 of Appendix 6.

Property 1 Linearity (Superposition)

Let $g_1(t) \rightleftharpoons G_1(f)$ *and* $g_2(t) \rightleftharpoons G_2(f)$. *Then for all constants a and b, we have*

$$ag_1(t)+bg_2(t) \rightleftharpoons aG_1(f)+bG_2(f) \tag{2.41}$$

The proof of this property follows simply from the linearity of the integrals defining $G(f)$ and $g(t)$.

Example 4 Double Exponential Pulse

Consider a *double exponential pulse* defined by (see Fig. 2.7)

$$\begin{aligned} g(t) &= \begin{cases} \exp(-t), & t>0 \\ 1, & t=0 \\ \exp(t), & t<0 \end{cases} \\ &= \exp(-|t|) \end{aligned} \tag{2.42}$$

This pulse may be viewed as the sum of a truncated decaying exponential pulse and a truncated rising exponential pulse. Therefore, using the linearity property and the Fourier-transform pairs of Eqs. (2.37) and (2.40), we find that the Fourier transform of the double exponential pulse of Fig. 2.7 is as follows

$$\begin{aligned} G(f) &= \frac{1}{1+j2\pi f}+\frac{1}{1-j2\pi f} \\ &= \frac{2}{1+(2\pi f)^2} \end{aligned}$$

We thus have the Fourier transform pair

$$\exp(-|t|) \rightleftharpoons \frac{2}{1+(2\pi f)^2} \tag{2.43}$$

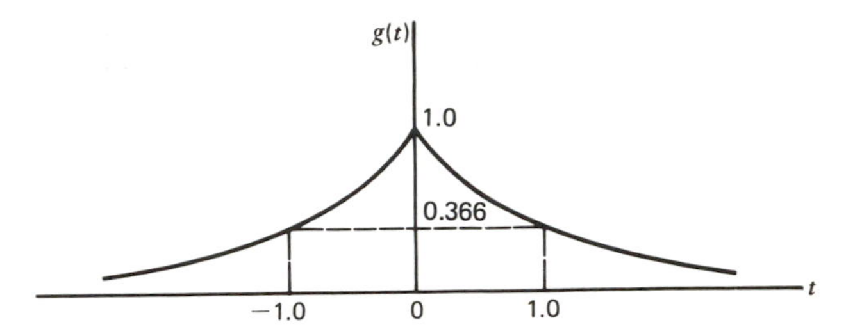

Figure 2.7 Double exponential pulse.

Property 2 Time scaling

Let $g(t) \rightleftharpoons G(f)$. *Then,*

$$g(at) \rightleftharpoons \frac{1}{|a|} G\left(\frac{f}{a}\right). \tag{2.44}$$

To prove this property, we note that

$$F(g(at)] = \int_{-\infty}^{\infty} g(at) \exp(-j2\pi ft)\, dt$$

Set $\tau = at$. There are two cases that can arise, depending on whether the scaling factor a is positive or negative. If $a > 0$, we get

$$\begin{aligned} F[g(at)] &= \frac{1}{a} \int_{-\infty}^{\infty} g(\tau) \exp\left[-j2\pi \left(\frac{f}{a}\right) \tau \right] d\tau \\ &= \frac{1}{a} G\left(\frac{f}{a}\right) \end{aligned}$$

On the other hand, if $a < 0$, the limits of integration are interchanged so that we have the multiplying factor $-(1/a)$ or, equivalently, $1/|a|$. This completes the proof of Eq. (2.44).

Note that the function $g(at)$ represents $g(t)$ compressed in time by a factor a, whereas the function $G(f/a)$ represents $G(f)$ expanded in frequency by the same factor a. Thus the scaling property states that the compression of a function $g(t)$ in the time domain is equivalent to the expansion of its Fourier transform $G(f)$ in the frequency domain by the same factor, or vice versa.

Example 5 Exponential pulse (continued)

Consider an exponential pulse defined by

$$g(t)=\begin{cases}\exp(-at), & t>0,\\ \frac{1}{2}, & t=0,\\ 0, & t<0\end{cases}$$

where $a>0$. This pulse is scaled in time by the factor a, compared to the exponential pulse of Eq. (2.35). Therefore, applying the time-scaling property of the Fourier transform to the transform pair of Eq. (2.37), we get

$$\begin{aligned}G(f)&=\frac{1}{a(1+j2\pi f/a)}\\ &=\frac{1}{a+j2\pi f}\end{aligned}$$

Property 3 Duality

If $g(t)\rightleftharpoons G(f)$, *then*

$$G(t)\rightleftharpoons g(-f). \tag{2.45}$$

This property follows from the relation defining the inverse Fourier transform by writing it in the form:

$$g(-t)=\int_{-\infty}^{\infty} G(f)\exp(-j2\pi ft)df$$

and then interchanging t and f.

Example 6 Sinc pulse

Consider a signal $g(t)$ in the form of a sinc function, as shown by

$$g(t)=A\ \mathrm{sinc}(2Wt) \tag{2.46}$$

To evaluate the Fourier transform of this function, we apply the duality and time-scaling properties to the Fourier transform pair of Eq. (2.33). Then, recognizing that the rectangular function is an even function, we obtain the following result:

$$A\ \mathrm{sinc}(2Wt)\rightleftharpoons\frac{A}{2W}\ \mathrm{rect}\left(\frac{f}{2W}\right) \tag{2.47}$$

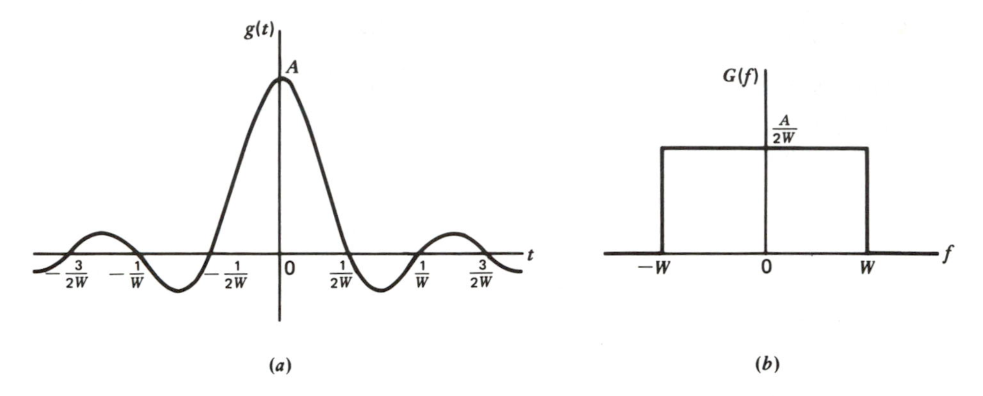

Figure 2.8 *(a)* Sinc pulse $g(t)$. *(b)* Fourier transform $G(f)$.

which is shown illustrated in Fig. 2.8. We thus see that the Fourier transform of a sinc pulse is zero for $|f|>W$. Note also that the sinc pulse itself is only asymptotically limited in time.

Example 7

Consider next the problem of evaluating the inverse Fourier transform of the one-sided frequency function $G(f)$ of Fig. 2.9, defined by

$$G(f)=\begin{cases}\exp(-f), & f>0\\ 0, & f<0\end{cases} \tag{2.48}$$

Applying the duality property to the Fourier transform pair of Eq. (2.40), we get the desired

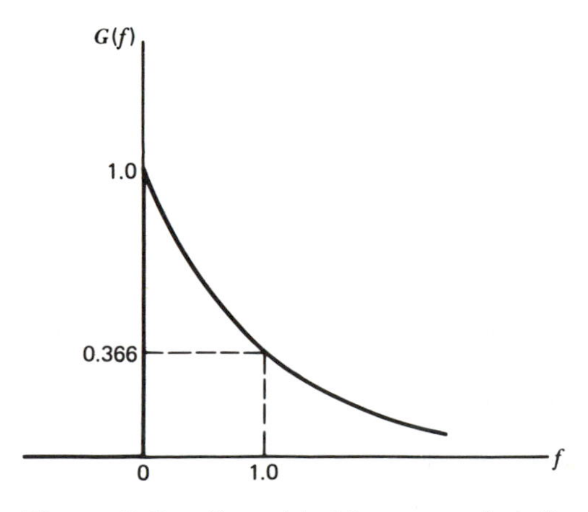

Figure 2.9 One-sided frequency function.

result:

$$g(t)=\frac{1}{1-j2\pi t}$$
$$=\frac{1}{1+(2\pi t)^2}+\frac{j2\pi t}{1+(2\pi t)^2} \tag{2.49}$$

Note that, in this example, the time function $g(t)$ is complex-valued, with a real as well as imaginary part. This is a direct consequence of the one-sided nature of the Fourier transform $G(f)$, as defined in Eq. (2.48).

Property 4 Time shifting

If $g(t)\rightleftharpoons G(f)$, *then*

$$g(t-t_0)\rightleftharpoons G(f)\exp(-j2\pi f t_0) \tag{2.50}$$

To prove this property, we take the Fourier transform of $g(t-t_0)$ and then set $\tau=t-t_0$ to obtain:

$$F[g(t-t_0)]=\exp(-j2\pi f t_0)\int_{-\infty}^{\infty} g(\tau)\exp(-j2\pi f\tau)d\tau$$
$$=\exp(-j2\pi f t_0)G(f)$$

The time-shifting property states that if a function $g(t)$ is shifted in the positive direction by an amount t_0, the effect is equivalent to multiplying its Fourier transform $G(f)$ by the factor $\exp(-j2\pi f t_0)$. This means that the amplitude of $G(f)$ is unaffected by the time shift but its phase is changed by the amount $-2\pi f t_0$.

Example 8 Rectangular pulse (continued)

Consider the rectangular pulse $g_a(t)$ of Fig. 2.10(*a*), which starts at $t=0$ and terminates at $t=T$. This pulse is obtained by shifting the rectangular pulse of Fig. 2.5 to the right by $T/2$ seconds. Therefore, applying the time-shifting property to the Fourier transform pair of

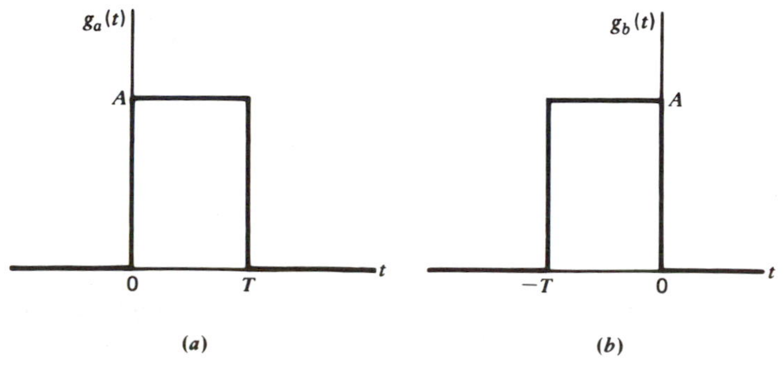

Figure 2.10 Time-shifted versions of a rectangular pulse.

Eq. (2.33), we find that the Fourier transform of the rectangular pulse of Fig. 2.10(*a*) is

$$G_a(f) = AT\ \mathrm{sinc}(fT)\exp(-j\pi fT)$$

Similarly, we find that the Fourier transform of the rectangular pulse $g_b(t)$ shown in Fig. 2.10(*b*) is

$$G_b(f) = AT\ \mathrm{sinc}(fT)\exp(j\pi fT)$$

Property 5 Frequency shifting

If $g(t) \rightleftharpoons G(f)$, *then*

$$\exp(j2\pi f_c t)g(t) \rightleftharpoons G(f-f_c) \tag{2.51}$$

where f_c *is a real constant.*

This property follows from the fact that

$$F[\exp(j2\pi f_c t)g(t)] = \int_{-\infty}^{\infty} g(t)\exp[-j2\pi t(f-f_c)]dt$$
$$= G(f-f_c)$$

That is, multiplication of a function $g(t)$ by the factor $\exp(j2\pi f_c t)$ is equivalent to shifting its Fourier transform $G(f)$ in the positive direction by the amount f_c. This property is called the *modulation theorem*, because a shift of the range of frequencies in a signal is accomplished by using modulation. Note the duality between the time-shifting and frequency-shifting operations.

Example 9 Radio frequency (RF) pulse

Consider the pulse signal $g(t)$ shown in Fig. 2.11(*a*) which consists of a sinusoidal wave of amplitude A and frequency f_c, extending in duration from $t=-T/2$ to $t=T/2$. This signal is sometimes referred to as an *RF pulse* when the frequency f_c falls in the radio-frequency band. The signal $g(t)$ of Fig. 2.11(*a*) may be expressed mathematically as follows:

$$g(t) = A\ \mathrm{rect}\left(\frac{t}{T}\right)\cos(2\pi f_c t) \tag{2.52}$$

To find the Fourier transform of this signal, we note that

$$\cos(2\pi f_c t) = \tfrac{1}{2}[\exp(j2\pi f_c t) + \exp(-j2\pi f_c t)]$$

Therefore, applying the frequency-shifting property to the Fourier transform pair of Eq. (2.33), we get the desired result

$$G(f) = \frac{AT}{2}\{\mathrm{sinc}[T(f-f_c)] + \mathrm{sinc}[T(f+f_c)]\} \tag{2.53}$$

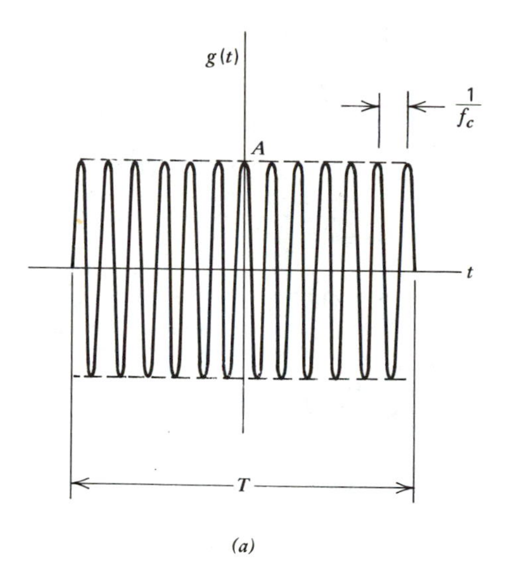

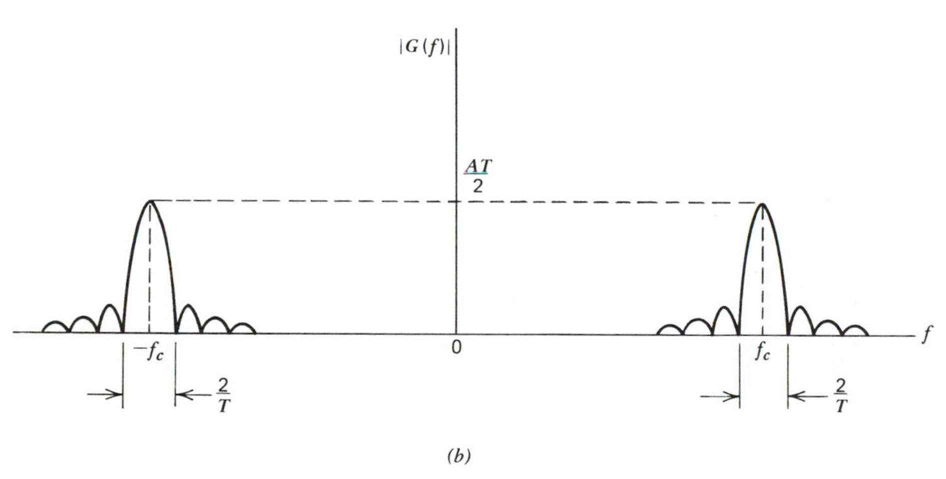

Figure 2.11 *(a)* RF pulse. *(b)* Amplitude spectrum.

In the special case of $f_c T \gg 1$, we may use the approximate result

$$G(f) \simeq \begin{cases} \frac{AT}{2} \operatorname{sinc}[T(f-f_c)], & f>0 \\ \frac{AT}{2} \operatorname{sinc}[T(f+f_c)], & f<0 \end{cases} \tag{2.54}$$

The amplitude spectrum of the RF pulse is shown in Fig. 2.11(*b*). This diagram, in relation to Fig. 2.5(*b*), clearly illustrates the frequency-shifting property of the Fourier transform.

Example 10 Damped sinusoidal wave

Consider next an exponentially damped sinusoidal wave defined by (see Fig. 2.12):

$$g(t)=\exp(-t)\sin(2\pi f_c t)u(t) \tag{2.55}$$

In this case, we note that

$$\sin(2\pi f_c t)=\frac{1}{2j}\left[\exp(j2\pi f_c t)-\exp(-j2\pi f_c t)\right]$$

Therefore, applying the frequency-shifting property to the Fourier transform pair of Eq. (2.37), we find that the Fourier transform of the damped sinusoidal wave of Fig. 2.12 is

$$\begin{aligned} G(f) &= \frac{1}{2j}\left[\frac{1}{1+j2\pi(f-f_c)}-\frac{1}{1+j2\pi(f+f_c)}\right] \\ &= \frac{2\pi f_c}{(1+j2\pi f)^2+(2\pi f_c)^2} \end{aligned} \tag{2.56}$$

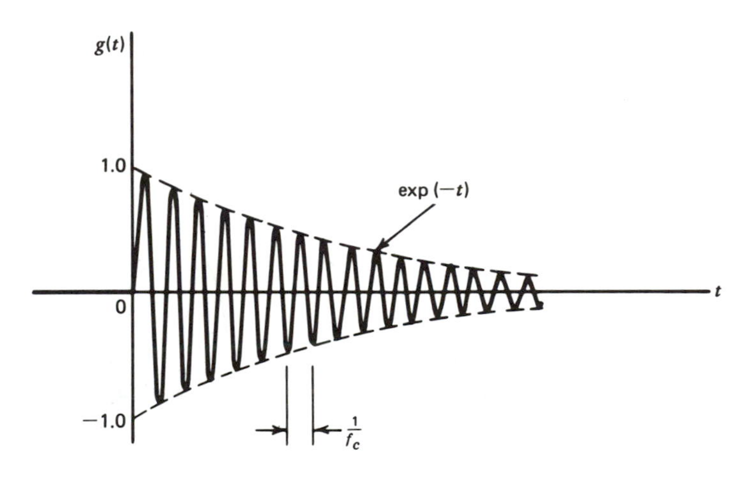

Figure 2.12 Damped sinusoidal wave.

Property 6 Area under g(t)

If $g(t)\rightleftharpoons G(f)$, *then*

$$\int_{-\infty}^{\infty} g(t)dt = G(0) \tag{2.57}$$

That is, the area under a function $g(t)$ *is equal to the value of its Fourier transform* $G(f)$ *at* $f=0$.

This result is obtained simply by putting $f=0$ in the formula defining the Fourier transform of the function $g(t)$.

Example 11 Sinc pulse (continued)

Consider again the sinc pulse of Fig. 2.8(*a*), which is defined by the Fourier transform pair:

$$A\ \mathrm{sinc}(2Wt) \rightleftharpoons \frac{A}{2W}\ \mathrm{rect}\left(\frac{f}{2W}\right)$$

For $f=0$, we have $G(0)=A/2W$. Therefore, using Eq. (2.57), we obtain

$$\int_{-\infty}^{\infty} A\ \mathrm{sinc}(2Wt)dt = \frac{A}{2W}$$

Cancelling out the common factor A and replacing $2Wt$ by t, we find that the total area under the sinc pulse $\mathrm{sinc}(t)$ is unity, as shown by

$$\int_{-\infty}^{\infty} \mathrm{sinc}(t)dt = 1 \tag{2.58}$$

Property 7 Area under G(f)

If $g(t) \rightleftharpoons G(f)$, *then*

$$g(0) = \int_{-\infty}^{\infty} G(f)dt \tag{2.59}$$

That is, the value of a function $g(t)$ *at* $t=0$ *is equal to the area under its Fourier transform* $G(f)$.

This result is obtained simply by putting $t=0$ in the formula defining the inverse Fourier transform of $G(f)$.

Example 12 Exponential pulse (continued)

Consider the exponential pulse $g(t)$ of Fig. 2.6(*a*). Recalling from Eq. (2.35) that $g(t)=\exp(-t)u(t)$, and noting that the unit step function $u(t)=\frac{1}{2}$ at $t=0$, we see that $g(0)=\frac{1}{2}$. Therefore, applying Eq. (2.59) to the Fourier transform pair of Eq. (2.37), we get

$$\int_{-\infty}^{\infty} \frac{df}{1+j2\pi f} = \frac{1}{2}$$

Property 8 Differentiation in the time domain

Let $g(t) \rightleftharpoons G(f)$, *and assume that the first derivative of* $g(t)$ *is Fourier transformable. Then*

$$\frac{d}{dt} g(t) \rightleftharpoons j2\pi f\, G(f) \tag{2.60}$$

That is, differentiation of a time function $g(t)$ *has the effect of multiplying its Fourier transform* $G(f)$ *by the factor* $j2\pi f$.

This result is obtained simply by taking the first derivative of both sides of the relation defining the inverse Fourier transform of $G(f)$, and then interchanging the operations of integration and differentiation.

We may generalize Eq. (2.60) as follows:

$$\frac{d^n}{dt^n} g(t) \rightleftharpoons (j2\pi f)^n G(f) \tag{2.61}$$

Example 13 Gaussian pulse

In this example we wish to use the differentiation property of the Fourier transform to derive the particular form of a pulse signal that has the same form as its own Fourier transform.

Let $g(t)$ denote the pulse expressed as a function of time, and $G(f)$ its Fourier transform. We note that by differentiating the formula for the Fourier transform $G(f)$ with respect to f, we have

$$-j2\pi t g(t) \rightleftharpoons \frac{d}{df} G(f) \tag{2.62}$$

which expresses the effect of differentiation in the frequency domain. From Eqs. (2.60) and (2.62) we thus deduce that if

$$\frac{d}{dt} g(t) = -2\pi t g(t) \tag{2.63}$$

then

$$\frac{d}{df} G(f) = -2\pi f G(f) \tag{2.64}$$

which means that the pulse signal and its transform are the same function. In other words, provided that the pulse signal $g(t)$ satisfies the differential equation (2.63), then $G(f)=g(f)$, where $g(f)$ is obtained from $g(t)$ by substituting f for t. Solving Eq. (2.63) for $g(t)$, we obtain

$$g(t) = \exp(-\pi t^2) \tag{2.65}$$

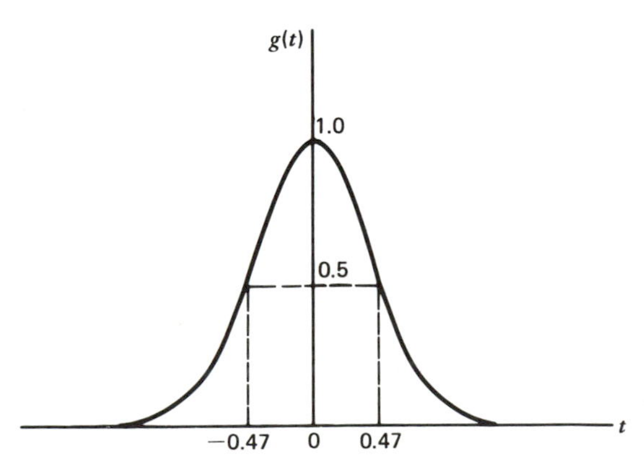

Figure 2.13 Gaussian pulse.

The pulse defined by Eq. (2.65) is called a *Gaussian pulse*, the name being derived from the similarity of the function to the Gaussian probability density function of probability theory (see Section 5.9). It is shown plotted in Fig. 2.13. By applying Eq. (2.57), we find that the area under this Gaussian pulse is unity, as shown by

$$\int_{-\infty}^{\infty} \exp(-\pi t^2)dt = 1 \tag{2.66}$$

When the central ordinate and the area under the curve of a pulse are both unity, as in the case of the Gaussian pulse of Eq. (2.65), we say that the pulse is *normalized*. We conclude therefore that the normalized Gaussian pulse is its own Fourier transform, as shown by

$$\exp(-\pi t^2) \rightleftharpoons \exp(-\pi f^2) \tag{2.67}$$

Property 9 Integration in the time domain

Let $g(t) \rightleftharpoons G(f)$. Then, provided $G(0)=0$, we have

$$\int_{-\infty}^{t} g(\tau)d\tau \rightleftharpoons \frac{1}{j2\pi f} G(f) \tag{2.68}$$

That is, integration of a time function $g(t)$ has the effect of dividing its Fourier transform $G(f)$ by the factor $j2\pi f$, assuming that $G(0)$ is zero.

This result is obtained by expressing $g(t)$ as

$$g(t) = \frac{d}{dt}\left[\int_{-\infty}^{t} g(\tau)d\tau\right]$$

and then applying the time differentiation property of the Fourier transform to obtain

$$G(f) = j2\pi f \left\{F\left[\int_{-\infty}^{t} g(\tau)d\tau\right]\right\}$$

from which Eq. (2.68) follows readily.

It is a straightforward matter to generalize Eq. (2.68) to multiple integration; however, the notation becomes rather cumbersome.

If $G(0)$ is nonzero, then the definite integral of $g(t)$ has a Fourier transform that includes a Dirac delta function or impulse $\delta(f)$ at the origin, as shown by:

$$\int_{-\infty}^{t} g(\tau)d\tau \rightleftharpoons \frac{1}{j2\pi f} G(f) + \frac{G(0)}{2}\delta(f) \tag{2.69}$$

The proof of this general relation is deferred until we have established the properties of a delta function (see Problem 2.12).

Example 14 Triangular pulse

Consider the *doublet pulse* $g_1(t)$ shown in Fig. 2.14(*a*). By integrating this pulse with respect to time, we obtain the *triangular pulse* $g_2(t)$ shown in Fig. 2.14(*b*). We note that the doublet

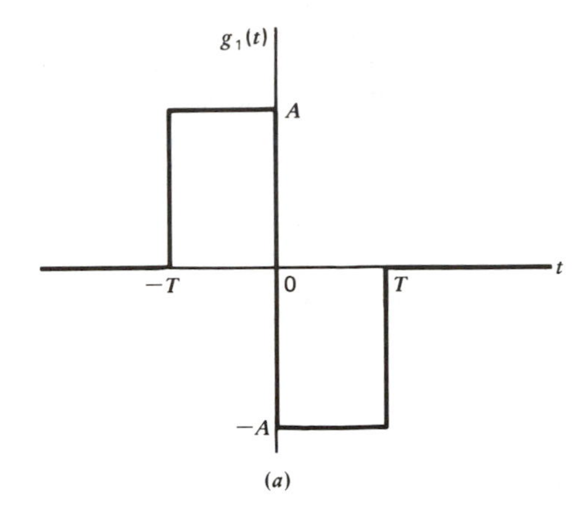

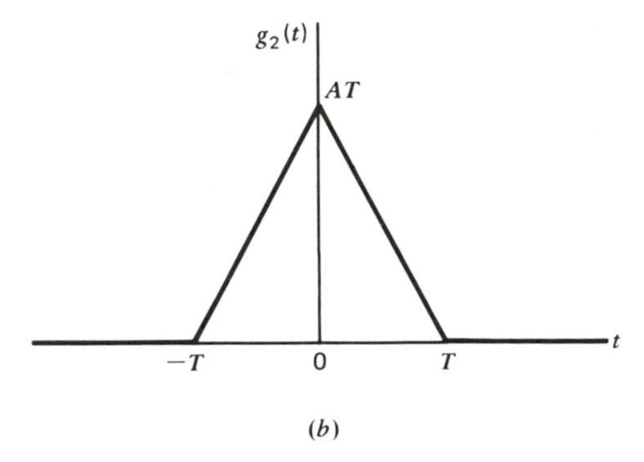

Figure 2.14 *(a)* Doublet pulse $g_1(t)$. *(b)* Triangular pulse $g_2(t)$ obtained by integrating $g_1(t)$.

pulse $g_1(t)$ consists of two rectangular pulses: one of amplitude A, defined for the interval $-T \leqslant t \leqslant 0$; the other of amplitude $-A$, defined for the interval $0 \leqslant t \leqslant T$. Therefore, using the results of Example 8, we find that the Fourier transform $G_1(f)$ of the doublet pulse $g_1(t)$ of Fig. 2.14(*a*) is given by

$$\begin{aligned} G_1(f) &= AT \operatorname{sinc}(fT)[\exp(j\pi fT) - \exp(-j\pi fT)] \\ &= 2jAT \operatorname{sinc}(fT)\sin(\pi fT) \end{aligned} \tag{2.70}$$

We further note that $G_1(0)$ is zero. Hence, using Eqs. (2.68) and (2.70), we find that the Fourier transform $G_2(f)$ of the triangular pulse $g_2(t)$ of Fig. 2.14(*b*) is given by

$$\begin{aligned} G_2(f) &= \frac{1}{j2\pi f} G_1(f) \\ &= AT \frac{\sin(\pi fT)}{\pi f} \operatorname{sinc}(fT) \\ &= AT^2 \operatorname{sinc}^2(fT) \end{aligned} \tag{2.71}$$

Property 10 Conjugate functions

If $g(t) \rightleftharpoons G(f)$, then for a complex-valued time function $g(t)$ we have

$$g^*(t) \rightleftharpoons G^*(-f), \tag{2.72}$$

where the asterisk denotes the complex conjugate operation.

To prove this property, we know from the inverse Fourier transform that

$$g(t) = \int_{-\infty}^{\infty} G(f) \exp(j2\pi ft)\, df$$

Taking the complex conjugates of both sides:

$$g^*(t) = \int_{-\infty}^{\infty} G^*(f) \exp(-j2\pi ft)\, df$$

Next, replacing f with $-f$:

$$g^*(t) = -\int_{\infty}^{-\infty} G^*(-f) \exp(j2\pi ft)\, df$$

$$= \int_{-\infty}^{\infty} G^*(-f) \exp(j2\pi ft)\, df$$

That is, $g^*(t)$ is the inverse Fourier transform of $G^*(-f)$, which is the desired result.

Example 15 Real and imaginary parts of a time function.

Expressing a complex-valued function $g(t)$ in terms of its real and imaginary parts, we may write

$$g(t) = \text{Re}[g(t)] + j\,\text{Im}[g(t)] \tag{2.73}$$

where Re denotes "the real part of" and Im denotes the "imaginary part of." The complex conjugate of $g(t)$ is

$$g^*(t) = \text{Re}[g(t)] - j\,\text{Im}[g(t)] \tag{2.74}$$

Adding Eqs. (2.73) and (2.74):

$$\text{Re}[g(t)] = \tfrac{1}{2}[g(t) + g^*(t)], \tag{2.75}$$

and subtracting them;

$$\text{Im}[g(t)] = \frac{1}{2j}[g(t) - g^*(t)] \tag{2.76}$$

Therefore, applying Property 10, we obtain the following two Fourier transform pairs:

$$\text{Re}[g(t)] \rightleftharpoons \frac{1}{2}[G(f) + G^*(-f)] \tag{2.77}$$

$$\text{Im}[g(t)] \rightleftharpoons \frac{1}{2j}[G(f) - G^*(-f)] \tag{2.78}$$

From Eq. (2.78), it is apparent that in the case of a real-valued time function $g(t)$, we have $G(f) = G^*(-f)$, that is, $G(f)$ exhibits *conjugate symmetry*.

Property 11 Multiplication in the time domain

Let $g_1(t) \rightleftharpoons G_1(f)$ *and* $g_2(t) \rightleftharpoons G_2(f)$. *Then*

$$g_1(t)g_2(t) \rightleftharpoons \int_{-\infty}^{\infty} G_1(\lambda)G_2(f-\lambda)d\lambda \tag{2.79}$$

To prove this property, we first denote the Fourier transform of the product $g_1(t)g_2(t)$ by $G_{12}(f)$, so that we may write

$$g_1(t)g_2(t) \rightleftharpoons G_{12}(f)$$

where

$$G_{12}(f) = \int_{-\infty}^{\infty} g_1(t)g_2(t)\exp(-j2\pi ft)dt$$

For $g_2(t)$, we next substitute the inverse Fourier transform

$$g_2(t) = \int_{-\infty}^{\infty} G_2(f')\exp(j2\pi f't)df'$$

in the integral defining $G_{12}(f)$ to obtain

$$G_{12}(f) = \int_{-\infty}^{\infty}\int_{-\infty}^{\infty} g_1(t)G_2(f')\exp[-j2\pi(f-f')t]df'dt$$

Define $\lambda = f - f'$. Then, interchanging the order of integration, we obtain

$$G_{12}(f) = \int_{-\infty}^{\infty} d\lambda\, G_2(f-\lambda) \int_{-\infty}^{\infty} g_1(t)\exp(-j2\pi\lambda t)dt$$

The inner integral is recognized simply as $G_1(\lambda)$ and so we may write

$$G_{12}(f) = \int_{-\infty}^{\infty} G_1(\lambda)G_2(f-\lambda)d\lambda$$

which is the desired result. This integral is known as the *convolution integral* expressed in the frequency domain, and the function $G_{12}(f)$ is referred to as the *convolution* of $G_1(f)$ and $G_2(f)$. We conclude that *the multiplication of two signals in the time domain is transformed into the convolution of their individual Fourier transforms in the frequency domain.* This property is known as the *multiplication theorem.*

In a discussion of convolution, the following shorthand notation is frequently used:

$$G_{12}(f) = G_1(f) \otimes G_2(f)$$

Note that convolution is commutative, that is,

$$G_{12}(f) = G_{21}(f)$$

or

$$G_1(f) \otimes G_2(f) = G_2(f) \otimes G_1(f)$$

Example 16 Truncated sinc pulse

Consider the truncation of the sinc pulse sinc($2Wt$), so that the resulting signal $g(t)$ is zero outside the interval $-(T/2)\leqslant t\leqslant(T/2)$, as shown in Fig. 2.15($a$). This signal may be expressed as the product of a sinc pulse and a rectangular pulse, as shown by

$$g(t)=\text{sinc}(2Wt)\text{rect}\left(\frac{t}{T}\right) \tag{2.80}$$

The Fourier transform of the sinc pulse sinc($2Wt$) is equal to $(1/2W)\text{rect}(f/2W)$, whereas that of the rectangular pulse rect(t/T) is equal to $T\,\text{sinc}(fT)$. Therefore, using Eq. (2.79), we find that the Fourier transform of the truncated sinc pulse $g(t)$ is given by

$$\begin{aligned} G(f) &= \frac{T}{2W}\int_{-\infty}^{\infty} \text{rect}\left(\frac{\lambda}{2W}\right)\text{sinc}[(f-\lambda)T]d\lambda \\ &= \frac{T}{2W}\int_{-W}^{W} \text{sinc}[(f-\lambda)T]d\lambda \\ &= \frac{T}{2W}\int_{-W}^{W} \frac{\sin[\pi(f-\lambda)T]}{\pi(f-\lambda)T}d\lambda \end{aligned} \tag{2.81}$$

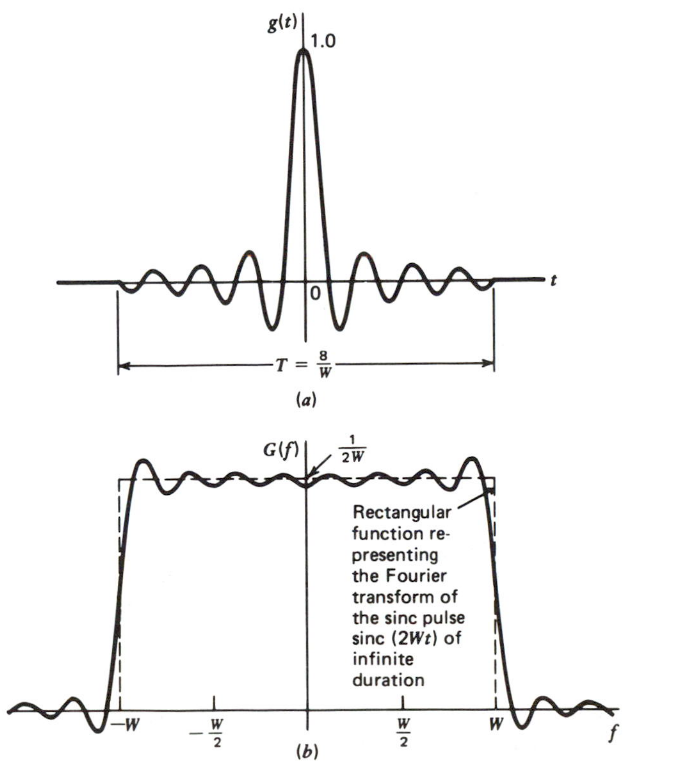

Figure 2.15 Illustrating the Gibbs phenomenon. *(a)* A truncated sinc function *g(t)*. *(b)* Fourier transform *G(f)*.

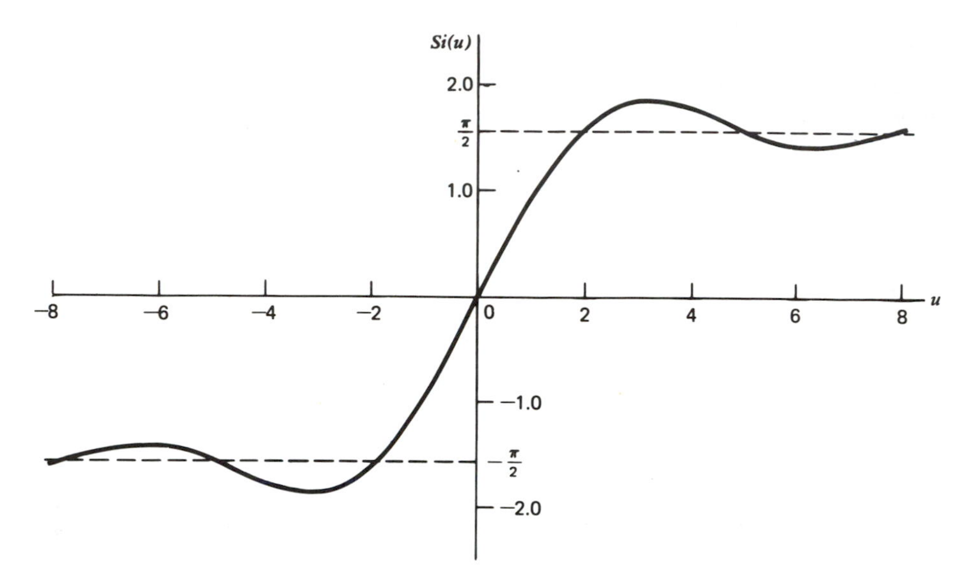

Figure 2.16 The sine integral.

The integral of the function sin x/x from zero up to some upper limit is called the *sine integral*, which is defined as follows

$$\mathrm{Si}(u) = \int_0^u \frac{\sin x}{x}\, dx \tag{2.82}$$

The sine integral $\mathrm{Si}(u)$ cannot be integrated in closed form in terms of elementary functions, but it can be integrated in a power series.* It is shown plotted in Fig. 2.16. We see that: (1) the sine integral $\mathrm{Si}(u)$ is odd symmetrical about $u=0$; (2) it has its maxima and minima at multiples of π; and (3) it approaches the limiting value $\pi/2$ for large values of u.

Thus substituting $x=\pi(f-\lambda)T$ in Eq. (2.81), we find that the Fourier transform $G(f)$ of the truncated sinc pulse may be expressed conveniently in terms of the sine integral as follows:

$$G(f) = \frac{1}{2\pi W}\left[\mathrm{Si}(\pi W T - \pi f T) + \mathrm{Si}(\pi W T + \pi f T)\right] \tag{2.83}$$

This relation is shown plotted in Fig. 2.15(*b*) for the case when $T=8/W$. We see that $G(f)$ approximates the Fourier transform of a sinc pulse $\mathrm{sinc}(2Wt)$ of infinite duration in an oscillatory fashion, with a maximum deviation of about 9 percent. Furthermore, for a given value of W, as the pulse duration T is increased, the ripples in the vicinities of the discontinuities of the rectangular function show a proportionality increased rate of oscillation versus the frequency f, whereas their amplitudes relative to the magnitude of the discontinuity remain the same. This effect is an example of *Gibb's phenomenon* in Fourier transforms.†

* For tables of values of the sine integral, see M. Abramowitz and I. A. Stegun, *Handbook of Mathematical Functions*, pp. 238–242 (Dover Publications, 1965). S. Goldman, *Frequency Analysis, Modulation, and Noise*, pp. 76–79 (McGraw,Hill, 1948).

† Gibb's phenomenon is also observed in Fourier series. For a detailed discussion of this phenomenon, see E. A. Guillemin, *The Mathematics of Circuit Analysis*, pp. 485–496 (Wiley, 1949).

Property 12 Convolution in the time domain

Let $g_1(t) \rightleftharpoons G_1(f)$ and $g_2(t) \rightleftharpoons G_2(f)$. Then

$$\int_{-\infty}^{\infty} g_1(\tau) g_2(t-\tau) d\tau \rightleftharpoons G_1(f) G_2(f) \tag{2.84}$$

This result follows directly by combining Property 3 (duality) and Property 11 (time-domain multiplication). We may thus state that *the convolution of two signals in the time domain is transformed into the multiplication of their individual Fourier transforms in the frequency domain.* This property is known as the *convolution theorem.* Its use permits us to exchange a convolution operation for a transform multiplication, an operation that is ordinarily easier to manipulate.

Using the shorthand notation for convolution, we may rewrite Eq. (2.84) in the form

$$g_1(t) \otimes g_2(t) \rightleftharpoons G_1(f) G_2(f) \tag{2.85}$$

where the symbol $\otimes$ denotes convolution.

Example 17 Derivative of a Convolution Integral

Let $g_{12}(t)$ denote the result of convolving two signals $g_1(t)$ and $g_2(t)$. Then the derivative of $g_{12}(t)$ is equal to the convolution of $g_1(t)$ with the derivative of $g_2(t)$, or vice versa. That is, if

$$g_{12}(t) = g_1(t) \otimes g_2(t)$$

then

$$\frac{d}{dt} g_{12}(t) = g_1(t) \otimes \left[\frac{d}{dt} g_2(t) \right]$$

and also

$$\frac{d}{dt} g_{12}(t) = \left[\frac{d}{dt} g_1(t) \right] \otimes g_2(t)$$

To prove this result, we use the differentiation property [i.e., Eq. (2.60)] in conjunction with the convolution property [i.e., Eq. (2.85)], obtaining

$$\frac{d}{dt} [g_1(t) \otimes g_2(t)] \rightleftharpoons j2\pi f [G_1(f) G_2(f)]$$

Associating the factor $j2\pi f$ with $G_1(f)$, we may write

$$\left[\frac{d}{dt} g_1(t) \right] \otimes g_2(t) \rightleftharpoons [j2\pi f G_1(f)] G_2(f)$$

which shows that

$$\frac{d}{dt} [g_1(t) \otimes g_2(t)] = \left[\frac{d}{dt} g_1(t) \right] \otimes g_2(t) \tag{2.86}$$

Next, associating the factor $j2\pi f$ with $G_2(f)$, we get

$$g_1(t) \otimes \left[\frac{d}{dt} g_2(t)\right] = G_1(f)[j2\pi f G_2(f)]$$

which shows that

$$\frac{d}{dt}[g_1(t) \otimes g_2(t)] = g_1(t) \otimes \left[\frac{d}{dt} g_2(t)\right] \tag{2.87}$$

2.4 NUMERICAL COMPUTATION OF THE FOURIER TRANSFORM

In this section we briefly describe a procedure for the computation of the Fourier transform, which is particularly well-suited for use on a digital computer. We assume that the given signal $g(t)$ is of finite duration. The procedure involves first, the *uniform sampling* of $g(t)$ to obtain a finite sequence of samples denoted by $g(0)$, $g(T_s)$, $g(2T_s), \ldots, g(NT_s - T_s)$, where T_s is the *sampling period* and N is the number of samples.* For a correct representation of the signal, the *sampling rate* $1/T_s$ must be equal to or greater than twice the highest frequency component of the signal; this issue will be discussed in detail in Section 7.1. For the purpose of our present discussion, it is adequate to assume that this requirement has been satisfied. It is possible, of course, that the signal initially may be in the form of a sequence of samples. In any event, for this sequence of samples, we may define a *discrete Fourier transform* denoted by $\{G(kF_s)\}$, which consists of another sequence of N samples separated in frequency by F_s hertz, as shown by

$$G(kF_s) = T_s \sum_{n=0}^{N-1} g(nT_s) \exp\left(-j\frac{2\pi}{N}kn\right), \qquad k = 0, 1, 2, \ldots, N-1 \tag{2.88}$$

Equation (2.88) is precisely the formula that would be obtained by using the trapezoidal rule for approximating the integral which defines the Fourier transform of the given signal $g(t)$. The difference between the actual Fourier transform and the sequence $\{G(kF_s)\}$ obtained from Eq. (2.88) gives the integration error evaluated for $f = kF_s$. The parameters T_s and F_s are related by

$$T_s F_s = \frac{1}{N} \tag{2.89}$$

To derive the inverse relationship expressing the sequence $\{g(nT_s)\}$ in terms of the discrete spectrum $\{G(kF_s)\}$, we multiply both sides of Eq. (2.88) by $\exp(j2\pi km/N)$ and sum over k, obtaining

$$\sum_{k=0}^{N-1} G(kF_s) \exp\left(j\frac{2\pi}{N}km\right) = T_s \sum_{k=0}^{N-1} \sum_{n=0}^{N-1} g(nT_s) \exp\left[j\frac{2\pi}{N}k(m-n)\right] \tag{2.90}$$

* In the literature, we often find that the sequence of samples is written as $g(0), g(1), g(2), \ldots, g(N-1)$, omitting the sampling period T_s from the notation. Yet another notation that is also used is as follows: $g_0, g_1, g_2, \ldots, g_{N-1}$.

Interchanging the order of summation on the right-hand side of Eq. (2.90), and using the fact that

$$\sum_{k=0}^{N-1} \exp\left[j\frac{2\pi}{N}k(m-n)\right] = \begin{cases} N, & m=n \\ 0, & \text{otherwise} \end{cases} \tag{2.91}$$

we get

$$\sum_{k=0}^{N-1} G(kF_s)\exp\left(j\frac{2\pi}{N}km\right) = NT_s g(mT_s) \tag{2.92}$$

Next, substituting the index n for m and rearranging the terms in Eq. (2.92), we get the desired relation

$$g(nT_s) = \frac{1}{NT_s}\sum_{k=0}^{N-1} G(kF_s)\exp\left(j\frac{2\pi}{N}kn\right), \qquad n=0, 1, \ldots, N-1 \tag{2.93}$$

which defines the *inverse discrete Fourier transform*. Here again, it is of interest to note that Eq. (2.93) is precisely the formula that would be obtained by using the trapezoidal rule for approximating the integral that defines the inverse Fourier transform.

The discrete Fourier transform, as defined in Eq. (2.88), has properties that are analogous to those of the continuous Fourier transform.* Some of these properties are discussed in Problem 2.20.

An important feature of the discrete Fourier transform is that the signal $\{g(nT_s)\}$ and its spectrum $\{G(kF_s)\}$ are both in discrete form. Furthermore, they are both periodic, with the period of either one consisting of a finite number of samples N. That is,

$$g(nT_s) = g(nT_s + NT_s) \tag{2.94}$$

and

$$G(kF_s) = G(kF_s + NF_s) \tag{2.95}$$

We thus find that the numerical computation of the discrete Fourier transform is well-suited for a digital computer or special-purpose digital processor. Indeed, it is this feature that makes the discrete Fourier transform so eminently useful in practice for spectral analysis and for the simulation of filters on digital computers. This is all the more so by virtue of the availability of an algorithm known as the *fast Fourier transform algorithm* (FFT), which provides a highly efficient procedure for computing the discrete Fourier transform of a finite-duration sequence. This algorithm takes advantage of the fact that the calculation of the coefficients of the discrete Fourier transform may be carried out in an iterative manner, thereby

* For a discussion of the relationship between the discrete and continuous Fourier transforms, see E. O. Brigham, *The Fast Fourier Transform*, pp. 99–108 (Prentice-Hall, 1974).

resulting in a considerable savings of computation time.* To compute the discrete Fourier transform of a sequence of N samples using the FFT algorithm, we require, in general, $N \log_2 N$ complex additions and $N \log_2 N$ complex multiplications. On the other hand, by using Eq. (2.88) to compute directly the discrete Fourier transform, we see that for each of the N possible values of k, we require $(N-1)$ complex additions and N complex multiplications, so that the direct computation of the discrete Fourier transform of N samples requires a total of $N(N-1)$ complex additions and N^2 complex multiplications. Accordingly, by using the FFT algorithm, the number of arithmetic operations is reduced by a factor of $N/\log_2 N$, which represents a considerable savings in computation effort for large N. For example, with $N=1024$, we reduce the computation effort by about two orders of magnitude by using the FFT algorithm. Indeed, it is this kind of improvement that also makes it possible to use special-purpose digital processors for the hardware implementation of the FFT algorithm.

2.5 DIRAC DELTA FUNCTION

Strictly speaking, the theory of the Fourier transform, as described in Sections 2.2 and 2.3, is applicable only to time functions that satisfy the Dirichlet conditions. Such functions include energy signals. However, it would be highly desirable to extend this theory in two ways:

1. To combine the Fourier series and Fourier transform into a unified theory, so that the Fourier series may be treated as a special case of the Fourier transform.
2. To include power signals in the list of signals to which we may apply the Fourier transform.

It turns out that both of these objectives can be met through the "proper use" of the *Dirac delta function* or *unit impulse*.†

The Dirac delta function, denoted by $\delta(t)$, is defined as having zero amplitude everywhere except at $t=0$ where it is infinitely large in such a way that it contains unit area under its curve; that is,‡

$$\delta(t)=0, \qquad t\neq 0 \tag{2.96}$$

* For a detailed discussion of the discrete Fourier transform and its properties, and for a description of the different forms of the fast Fourier transform algorithm, see A. V. Oppenheim and R. W. Schafer, *Digital Signal Processing*, Chapters 3 and 6 (Prentice-Hall, 1975). L. R. Rabiner and B. Gold, *Theory and Application of Digital Signal Processing*, Chapters 2 and 6 (Prentice-Hall, 1975). E. O. Brigham, *op. cit.*

† For a detailed treatment of the delta function, see R. Bracewell, *The Fourier Transform and Its Applications*, pp. 69–97 (McGraw-Hill, 1965). A. Papoulis, *The Fourier Integral and Its Applications*, pp. 35–52 and pp. 269–282 (McGraw-Hill, 1962). M. J. Lighthill, *An Introduction to Fourier Analysis and Generalized Functions* (Cambridge University Press, 1959). S. J. Mason and H. J. Zimmermann, *Electronic Circuits, Signals, and Systems*, pp. 310–318 (Wiley, 1960).

‡ The notation $\delta(t)$, which was first introduced into quantum mechanics by Dirac, is now in general use; see P. A. M. Dirac, *The Principles of Quantum Mechanics*, Third Edition (Oxford University Press, 1947).

and

$$\int_{-\infty}^{\infty} \delta(t)dt = 1 \tag{2.97}$$

Consider the integral of the product of $\delta(t)$ and any time function $g(t)$ that is continuous at $t=0$. As a consequence of the two conditions defining the delta function $\delta(t)$, we may write

$$\int_{-\infty}^{\infty} g(t)\delta(t)dt = g(0) \tag{2.98}$$

We refer to this statement as the *sifting property* of the delta function, since the operation on $g(t)$ indicated on the left-hand side of Eq. (2.98) sifts out a single value of $g(t)$, namely, $g(0)$. Equation (2.98) may also be used as the basis for defining a delta function.

It is clear that we may also write

$$\int_{-\infty}^{\infty} g(t)\delta(t-t_0)dt = g(t_0) \tag{2.99}$$

Noting that the delta function $\delta(t)$ is an even function of t (see Problem 2.21), we may rewrite Eq. (2.99) in a way emphasizing the resemblance to the convolution integral, as follows:

$$\int_{-\infty}^{\infty} g(\tau)\delta(t-\tau)d\tau = g(t) \tag{2.100}$$

or

$$g(t) \otimes \delta(t) = g(t) \tag{2.101}$$

That is, the convolution of any function with the delta function leaves that function unchanged. We refer to this statement as the *replication property* of the delta function.

The Fourier transform of the delta function is given by

$$F[\delta(t)] = \int_{-\infty}^{\infty} \delta(t)\exp(-j2\pi ft)dt$$

Using the sifting property of the delta function and noting that $\exp(-j2\pi ft)$ is equal to unity at $t=0$, we obtain

$$F[\delta(t)] = 1 \tag{2.102}$$

We thus have the Fourier transform pair:

$$\delta(t) \rightleftharpoons 1 \tag{2.103}$$

This relation states that the spectrum of the delta function $\delta(t)$ extends uniformly over the entire frequency interval from $-\infty$ to ∞, as shown in Fig. 2.17.

It is important to realize that the Fourier transform pair of Eq. (2.103) exists only in a limiting sense. The point is that no function in the ordinary sense has the

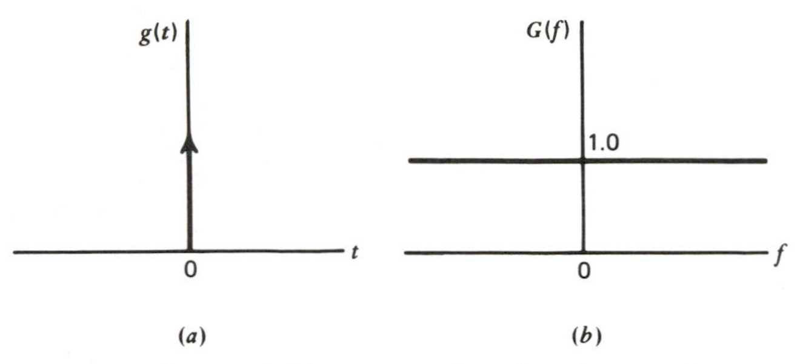

Figure 2.17 *(a)* Dirac delta function $\delta(t)$. *(b)* Spectrum of $\delta(t)$.

two properties of Eqs. (2.96) and (2.97) or the equivalent sifting property of Eq. (2.98). However, we can imagine a sequence of functions which have progressively taller and thinner peaks at $t=0$, with the area under the curve remaining equal to unity, whereas the value of the function tends to zero at every point, except $t=0$ where it tends to infinity. That is, we may view the delta function as the limiting form of a pulse of unit area as the duration of the pulse approaches zero. It is immaterial what sort of pulse shape is used. For example, we may use a Gaussian pulse, rectangular pulse, or sinc pulse. In Example 18, we consider the case of a Gaussian pulse; for the case of a rectangular pulse or sinc pulse, see Problem 2.22.

Example 18 The delta function as a limiting form of the Gaussian pulse

Consider a Gaussian pulse of unit area, defined by

$$g(t)=\frac{1}{\tau}\exp\left(-\frac{\pi t^2}{\tau^2}\right) \tag{2.104}$$

where τ is a variable parameter. The Gaussian function $g(t)$ has two useful properties: (1) its derivatives are all continuous and (2) it dies away more rapidly than any power of t. The delta function $\delta(t)$ is obtained by taking the limit $\tau\to 0$. The Gaussian pulse then becomes infinitely narrow in duration and infinitely large in amplitude, and yet its area remains finite and fixed at unity. Figure 2.18(*a*) illustrates the sequence of such pulses as the parameter τ varies.

The Gaussian pulse $g(t)$, defined above, is the same as the normalized Gaussian pulse $\exp(-\pi t^2)$ previously considered in Example 13, except that it is expanded in time by the factor τ and compressed in amplitude by the same factor. Therefore, applying the linearity and time-scaling properties of the Fourier transform to the transform pair of Eq. (2.67), we find that the Fourier transform of the Gaussian pulse $g(t)$ defined by Eq. (2.104) is also Gaussian, as shown by

$$G(f)=\exp(-\pi\tau^2 f^2) \tag{2.105}$$

Figure 2.18(*b*) illustrates the effect of varying the parameter τ on the spectrum of the Gaussian pulse $g(t)$. Thus, putting $\tau=0$, we find, as expected, that the Fourier transform of the delta function is unity.

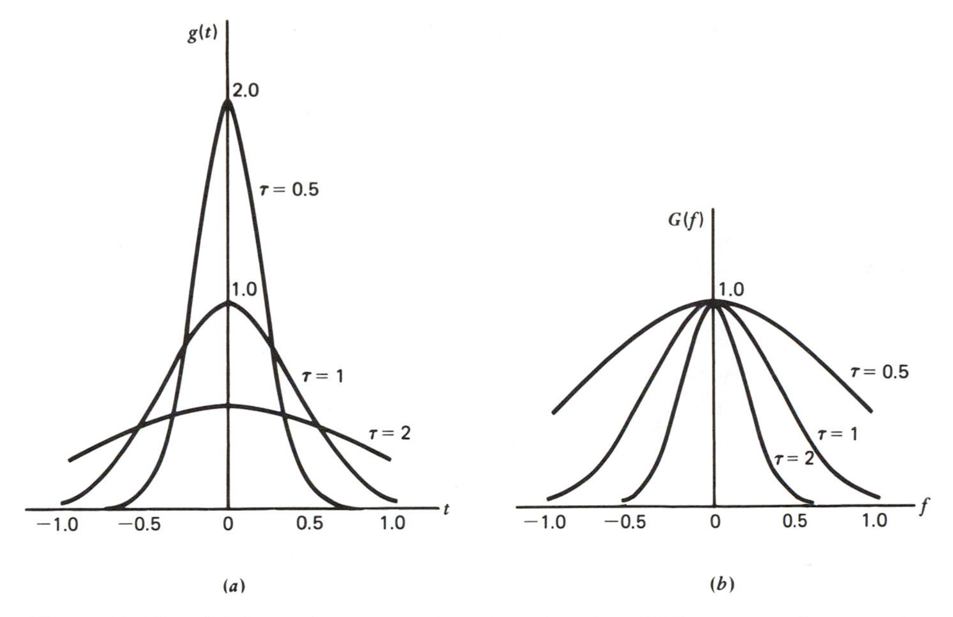

Figure 2.18 *(a)* Gaussian pulse of varying duration. *(b)* Corresponding spectrum.

Applications of the Delta Function

1. dc Signal

By applying the duality property to the Fourier transform pair of Eq. (2.103), and noting that the delta function is an even function, we obtain

$$1 \rightleftharpoons \delta(f) \tag{2.106}$$

Equation (2.106) states that a *dc signal* is transformed in the frequency domain into a delta function $\delta(f)$ occurring at zero frequency, as shown in Fig. 2.19. Of course, this result is intuitively satisfying.

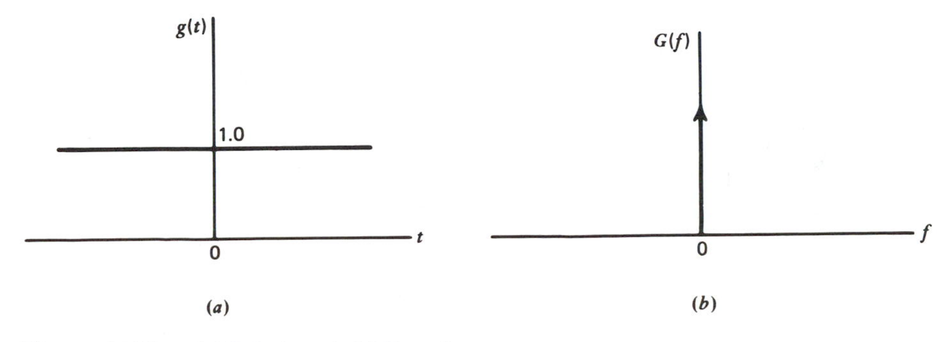

Figure 2.19 *(a)* DC signal. *(b)* Spectrum.

From Eq. (2.106) we also deduce the useful relation

$$\int_{-\infty}^{\infty} \exp(-j2\pi ft)dt = \delta(f) \tag{2.107}$$

2. Complex Exponential Function

Next, by applying the frequency-shifting property to Eq. (2.106), we obtain the Fourier transform pair

$$\exp(j2\pi f_c t) \rightleftharpoons \delta(f - f_c) \tag{2.108}$$

for a complex exponential function of frequency f_c. Equation (2.108) states that the complex exponential function $\exp(j2\pi f_c t)$ is transformed in the frequency domain into a delta function $\delta(f - f_c)$ occurring at $f = f_c$.

3. Sinusoidal Functions

Consider next the problem of evaluating the Fourier transform of the cosine function $\cos(2\pi f_c t)$. We first note that

$$\cos(2\pi f_c t) = \tfrac{1}{2}[\exp(j2\pi f_c t) + \exp(-j2\pi f_c t)]$$

Therefore, using Eq. (2.108), we find that the cosine function $\cos(2\pi f_c t)$ is represented by the Fourier transform pair

$$\cos(2\pi f_c t) \rightleftharpoons \tfrac{1}{2}[\delta(f - f_c) + \delta(f + f_c)] \tag{2.109}$$

In other words, the spectrum of the cosine function $\cos(2\pi f_c t)$ consists of a pair of delta functions occurring at $f = \pm f_c$, each of which is weighted by the factor $\frac{1}{2}$, as shown in Fig. 2.20.

Similarly, we may show that the sine function $\sin(2\pi f_c t)$ is represented by the

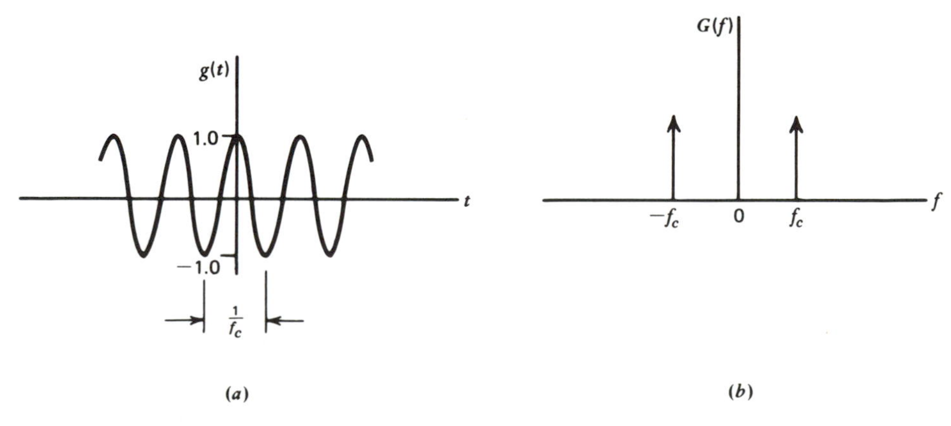

Figure 2.20 *(a)* Cosine function. *(b)* Spectrum.

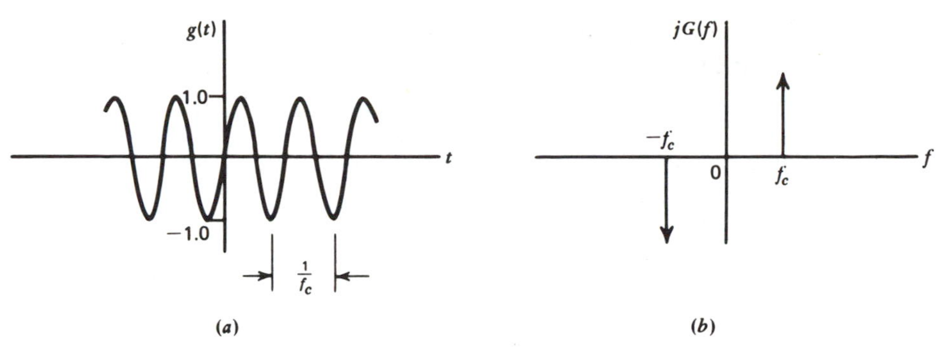

Figure 2.21 *(a)* Sine function. *(b)* Spectrum.

Fourier transform pair

$$\sin(2\pi f_c t) \rightleftharpoons \frac{1}{2j}[\delta(f-f_c)-\delta(f+f_c)] \tag{2.110}$$

which is illustrated in Fig. 2.21.

4. Unit Step Function

By applying the integration property [in particular, Eq. (2.69)] to the Fourier transform pair of Eq. (2.103), we get the result

$$\int_{-\infty}^{t} \delta(\tau)d\tau \rightleftharpoons \frac{1}{j2\pi f}+\frac{\delta(f)}{2} \tag{2.111}$$

We observe that the integral on the left-hand side of this Fourier transform pair is equal to zero for all negative values of t, one-half when $t=0$, and one for all positive values of t. In other words, it is equal to the unit step function, as shown by

$$u(t)=\int_{-\infty}^{t} \delta(\tau)d\tau \tag{2.112}$$

Therefore, the unit step function is represented by the Fourier transform pair

$$u(t) \rightleftharpoons \frac{1}{j2\pi f}+\frac{1}{2}\delta(f) \tag{2.113}$$

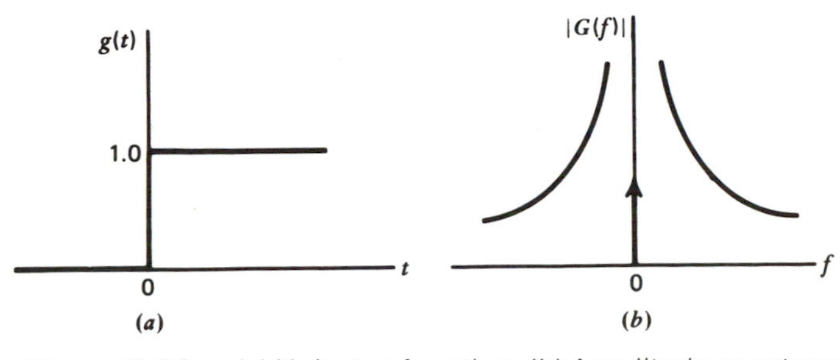

Figure 2.22 *(a)* Unit step function. *(b)* Amplitude spectrum.

This means that the spectrum of the unit step function contains a delta function at zero frequency, as illustrated in Fig. 2.22.

5. Signum Function

The *signum function*, denoted by sgn(t), is defined as follows

$$\operatorname{sgn}(t)=\begin{cases} 1, & t>0 \\ 0, & t=0 \\ -1, & t<0 \end{cases}$$

We may thus relate the signum function to the unit step function by

$$\operatorname{sgn}(t)=2u(t)-1 \tag{2.114}$$

Therefore, using Eqs. (2.106), (2.113), and (2.114), we obtain the Fourier transform pair (see Fig. 2.23):

$$\operatorname{sgn}(t)\rightleftharpoons\frac{1}{j\pi f} \tag{2.115}$$

Another useful Fourier transform pair, involving a signum function defined in the frequency domain, is obtained by applying Property 3 (duality) to Eq. (2.115) and so we obtain the following result:

$$\frac{1}{\pi t}\rightleftharpoons -j\operatorname{sgn}(f) \tag{2.116}$$

where the signum function sgn(f) is defined by

$$\operatorname{sgn}(f)=\begin{cases} 1, & f>0 \\ 0, & f=0 \\ -1, & f<0 \end{cases} \tag{2.117}$$

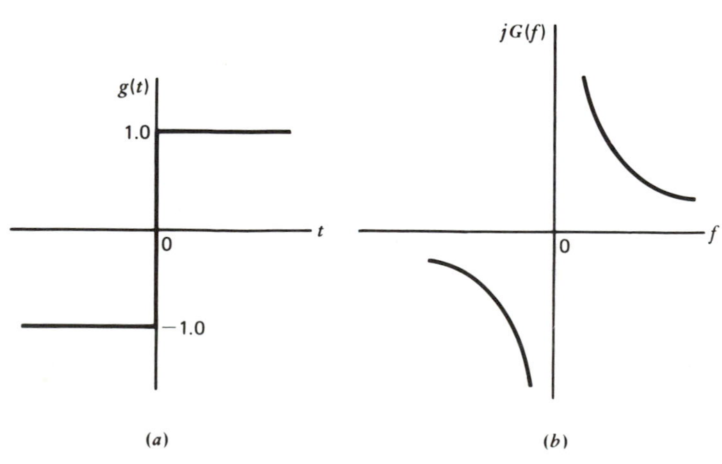

Figure 2.23 *(a)* Signum function. *(b)* Spectrum.

2.6 FOURIER TRANSFORMS OF PERIODIC SIGNALS

From Section 2.1 we recall that by using the Fourier series, a periodic signal $g_p(t)$ can be represented as a sum of complex exponentials. Also we know that, in a limiting sense, we can define Fourier transforms of complex exponentials. Therefore, it seems reasonable that a periodic signal can be represented in terms of a Fourier transform, provided that this transform is permitted to include delta functions.

Consider a periodic signal $g_p(t)$ of period T_0. We can represent $g_p(t)$ in terms of the complex exponential Fourier series as in Eq. (2.10), which is reproduced here for convenience,

$$g_p(t)=\sum_{n=-\infty}^{\infty} c_n \exp\left(\frac{j2\pi nt}{T_0}\right) \tag{2.118}$$

where c_n is the complex Fourier coefficient defined by

$$c_n=\frac{1}{T_0}\int_{-T_0/2}^{T_0/2} g_p(t)\exp\left(-\frac{j2\pi nt}{T_0}\right)dt \tag{2.119}$$

Let $g(t)$ be a pulse-like function, which equals $g_p(t)$ over one period and is zero elsewhere; that is,

$$g(t)=\begin{cases} g_p(t), & -\dfrac{T_0}{2}\leqslant t\leqslant\dfrac{T_0}{2} \\ 0, & \text{elsewhere} \end{cases} \tag{2.120}$$

The periodic signal $g_p(t)$ may be expressed in terms of the function $g(t)$ as an infinite summation, as shown by

$$g_p(t)=\sum_{m=-\infty}^{\infty} g(t-mT_0) \tag{2.121}$$

Based on this representation, we may view $g(t)$ as a *generating function*, which generates the periodic signal $g_p(t)$.

The function $g(t)$ is Fourier transformable. Accordingly, we may rewrite Eq. (2.119) as follows

$$\begin{aligned} c_n&=\frac{1}{T_0}\int_{-\infty}^{\infty} g(t)\exp\left(-\frac{j2\pi nt}{T_0}\right)dt \\ &=\frac{1}{T_0}G\left(\frac{n}{T_0}\right) \end{aligned} \tag{2.122}$$

where $G(n/T_0)$ is the Fourier transform of $g(t)$, evaluated at the frequency n/T_0. We may thus rewrite Eq. (2.118) as

$$g_p(t)=\frac{1}{T_0}\sum_{n=-\infty}^{\infty} G\left(\frac{n}{T_0}\right)\exp\left(\frac{j2\pi nt}{T_0}\right) \tag{2.123}$$

or, equivalently,

$$\sum_{m=-\infty}^{\infty} g(t-mT_0)=\frac{1}{T_0}\sum_{n=-\infty}^{\infty} G\left(\frac{n}{T_0}\right)\exp\left(\frac{j2\pi nt}{T_0}\right) \tag{2.124}$$

Equation (2.124) is one form of *Poisson's sum formula.*

Finally, using Eq. (2.108), which defines the Fourier transform of a complex exponential function, and Eq. (2.124), we deduce the following Fourier transform pair for a periodic signal $g_p(t)$ with a generating function $g(t)$ and period T_0:

$$g_p(t)=\sum_{m=-\infty}^{\infty} g(t-mT_0)\rightleftharpoons\frac{1}{T_0}\sum_{n=-\infty}^{\infty} G\left(\frac{n}{T_0}\right)\delta\left(f-\frac{n}{T_0}\right) \tag{2.125}$$

This relation simply states that the Fourier transform of a periodic signal consists of delta functions occurring at integer multiples of the fundamental frequency $1/T_0$, including the origin, and that each delta function is weighted by a factor equal to the pertinent value of $G(n/T_0)$. Indeed, this relation merely provides an alternative way to display the frequency content of a periodic signal $g_p(t)$.

It is of interest to observe that the function $g(t)$, constituting one period of the periodic signal $g_p(t)$, has a continuous spectrum defined by $G(f)$. On the other hand, the periodic signal $g_p(t)$ itself has a discrete spectrum. We conclude, therefore, that periodicity in the time domain has the effect of making discrete the frequency-domain description or spectrum of the signal at integer multiples of the reciprocal of the period.

Example 19 Ideal sampling function

An *ideal sampling function*, or *Dirac comb*, consists of an infinite sequence of uniformly spaced delta functions, as shown in Fig. 2.24(*a*). We will denote this waveform by

$$\delta_{T_0}(t)=\sum_{m=-\infty}^{\infty} \delta(t-mT_0) \tag{2.126}$$

We observe that the generating function $g(t)$ for the ideal sampling function $\delta_{T_0}(t)$ consists simply of the delta function $\delta(t)$. Therefore, $G(f)=1$, so that

$$G\left(\frac{n}{T_0}\right)=1, \qquad \text{for all } n \tag{2.127}$$

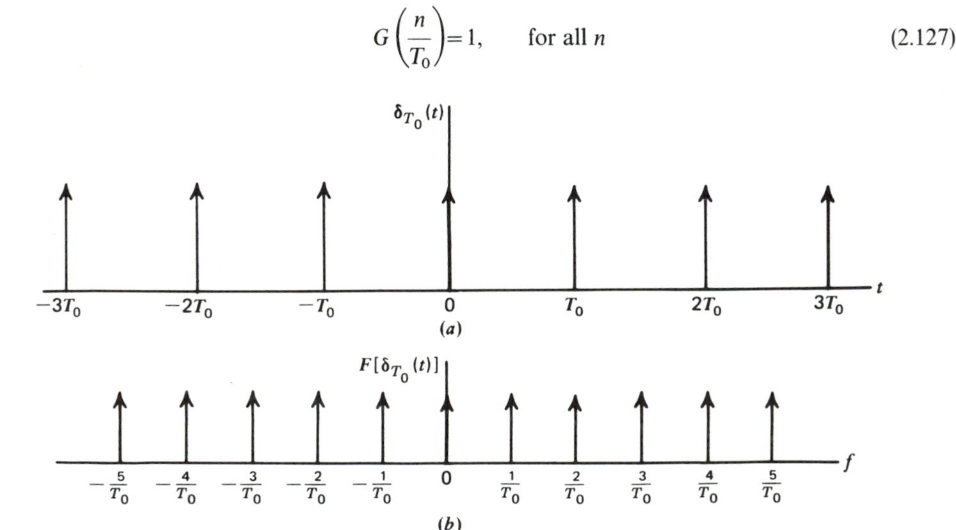

Figure 2.24 *(a)* Dirac comb. *(b)* Spectrum.

Thus the use of Eq. (2.125) yields the result

$$\sum_{m=-\infty}^{\infty} \delta(t-mT_0) \rightleftharpoons \frac{1}{T_0} \sum_{n=-\infty}^{\infty} \delta\left(f-\frac{n}{T_0}\right) \tag{2.128}$$

Equation (2.128) states that the Fourier transform of a periodic train of delta functions, spaced T_0 seconds apart, consists of another set of delta functions weighted by the factor $1/T_0$ and regularly spaced $1/T_0$ Hz apart along the frequency axis as in Fig. 2.24(*b*). In the special case of the period T_0 equal to 1 second, a periodic train of delta functions is, like a Gaussian pulse, its own transform.

We also deduce from Poisson's sum formula, Eq. (2.124), the following useful relation

$$\sum_{m=-\infty}^{\infty} \delta(t-mT_0) = \frac{1}{T_0} \sum_{n=-\infty}^{\infty} \exp\left(\frac{j2\pi nt}{T_0}\right)$$

The dual of this relation is as follows

$$\sum_{m=-\infty}^{\infty} \exp(j2\pi mfT_0) = \frac{1}{T_0} \sum_{n=-\infty}^{\infty} \delta\left(f-\frac{n}{T_0}\right) \tag{2.129}$$

2.7 SPECTRAL DENSITY

In this section we develop a procedure for calculating the energy or power of a signal by using the Fourier transform of the signal. In so doing, we introduce the notion of *spectral density*. Depending on whether the signal of interest is an energy signal or a power signal, we find that the total area under the spectral density curve plotted as a function of frequency is equal to the total energy or average power of the signal, respectively. We may therefore distinguish two distinct forms of spectral density, namely, *energy spectral density* and *power spectral density*.

Energy Spectral Density

Consider an energy signal $g(t)$ defined over the entire interval $-\infty < t < \infty$, and let its Fourier transform be denoted by $G(f)$. The signal $g(t)$ may be complex-valued. The total energy of the signal is defined by Eq. (1.5), which is reproduced here for convenience,

$$E = \int_{-\infty}^{\infty} |g(t)|^2 \, dt \tag{2.130}$$

The squared function $|g(t)|^2$ may be expressed as the product of two time functions, namely, $g(t)$ and its complex conjugate $g^*(t)$. The Fourier transform of $g^*(t)$ is equal to $G^*(-f)$, by virtue of Property 10 (complex conjugation). Then, applying property 11 (the multiplication theorem), or more specifically, applying Eq. (2.79) to the product $g(t)g^*(t)$ and evaluating the result for $f=0$, we obtain the relation

$$\int_{-\infty}^{\infty} g(t)g^*(t)dt = \int_{-\infty}^{\infty} G(\lambda)G^*(\lambda)d\lambda$$

Replacing λ with f in the right-hand side of this relation, and noting that $|G(f)|^2 = G(f)G^*(f)$, we may write

$$E=\int_{-\infty}^{\infty}|g(t)|^2\,dt=\int_{-\infty}^{\infty}|G(f)|^2\,df \tag{2.131}$$

This result is known as *Rayleigh's energy theorem*. To apply the theorem, we need to know only the amplitude spectrum $|G(f)|$ of the signal.

Let $\Psi_g(f)$ denote the squared amplitude spectrum of the signal $g(t)$, as shown by

$$\Psi_g(f)=|G(f)|^2 \tag{2.132}$$

Accordingly, we may express the energy of the signal $g(t)$ in terms of $\Psi_g(f)$ as follows

$$E=\int_{-\infty}^{\infty}\Psi_g(f)df \tag{2.133}$$

The quantity $\Psi_g(f)$ is referred to as the *energy spectral density* of the signal $g(t)$. To explain this, suppose that $g(t)$ denotes the voltage of a source connected across a 1-ohm resistor. Then the quantity

$$\int_{-\infty}^{\infty}|g(t)|^2\,dt$$

equals the total energy E delivered by the source. According to Rayleigh's theorem, this energy equals the total area under the $\Psi_g(f)$ curve. It follows therefore that the function $\Psi_g(f)$ is a measure of the density of the energy contained in $g(t)$ in joules per hertz. Note that since the amplitude spectrum $|G(f)|$ is an even function of f, the energy spectral density is symmetrical about the vertical axis passing through the origin.

Example 20 Sinc pulse (continued)

Consider again the sinc pulse $A\,\text{sinc}(2Wt)$. The energy of this pulse equals

$$E=A^2\int_{-\infty}^{\infty}\text{sinc}^2(2Wt)dt \tag{2.134}$$

The integral in the right-hand side of Eq. (2.134) is rather difficult to evaluate. However, noting that the Fourier transform of the sinc pulse $A\,\text{sinc}(2Wt)$ is equal to $(A/2W)\text{rect}(f/2W)$, and applying Rayleigh's energy theorem, we readily obtain the desired result, as shown by

$$\begin{aligned}E&=\left(\frac{A}{2W}\right)^2\int_{-\infty}^{\infty}\text{rect}^2\left(\frac{f}{2W}\right)df\\&=\left(\frac{A}{2W}\right)^2\int_{-W}^{W}df\\&=\frac{A^2}{2W}\end{aligned} \tag{2.135}$$

From Eqs. (2.134) and (2.135) we also deduce the following useful result

$$\int_{-\infty}^{\infty}\text{sinc}^2(2Wt)dt=\frac{1}{2W}$$

Replacing $2Wt$ by t, we thus find that the total area under the curve of $\text{sinc}^2\, t$ equals one, as shown by

$$\int_{-\infty}^{\infty} \text{sinc}^2\, t\, dt = 1$$

Power Spectral Density

The average power P of a so-called power signal $g(t)$ (voltage or current) dissipated in a 1-ohm load resistor is defined by Eq. (1.6), which is reproduced here for convenience,

$$P = \lim_{T\to\infty} \frac{1}{2T} \int_{-T}^{T} |g(t)|^2\, dt \tag{2.136}$$

Now, a periodic signal $g_p(t)$ is a special case of a power signal. With $g(t)$ in Eq. (2.136) replaced by $g_p(t)$, the integrand becomes periodic and we may take the time average over a single period of $g_p(t)$. Accordingly, the expression for the average power of a periodic signal $g_p(t)$ of period T_0 equals

$$P = \frac{1}{T_0} \int_{-T_0/2}^{T_0/2} |g_p(t)|^2\, dt \tag{2.137}$$

Replacing $g_p(t)$ by the complex Fourier series representation shown in Eq. (2.123), we get

$$P = \frac{1}{T_0^2} \int_{-T_0/2}^{T_0/2} g_p^*(t) \sum_{n=-\infty}^{\infty} G\left(\frac{n}{T_0}\right) \exp\left(\frac{j2\pi nt}{T_0}\right) dt \tag{2.138}$$

where $G(n/T_0)$ is equal to the Fourier transform of the generating function $g(t)$ evaluated at the frequency n/T_0. Since, by definition, $g(t)$ is equal to $g_p(t)$ over one period and zero elsewhere, we may rewrite Eq. (2.138) as follows

$$P = \frac{1}{T_0^2} \int_{-\infty}^{\infty} g^*(t) \sum_{n=-\infty}^{\infty} G\left(\frac{n}{T_0}\right) \exp\left(\frac{j2\pi nt}{T_0}\right) dt$$

Interchanging the order of summation and integration, we get

$$\begin{aligned} P &= \frac{1}{T_0^2} \sum_{n=-\infty}^{\infty} G\left(\frac{n}{T_0}\right) \int_{-\infty}^{\infty} g^*(t) \exp\left(\frac{j2\pi nt}{T_0}\right) dt \\ &= \frac{1}{T_0^2} \sum_{n=-\infty}^{\infty} G\left(\frac{n}{T_0}\right) G^*\left(\frac{n}{T_0}\right) \\ &= \frac{1}{T_0^2} \sum_{n=-\infty}^{\infty} \left| G\left(\frac{n}{T_0}\right) \right|^2 \end{aligned} \tag{2.139}$$

This relation is known as *Parseval's power theorem.* It states that the average power of a periodic signal $g_p(t)$ is equal to the sum of the squared amplitudes of all the harmonic components of the signal $g_p(t)$. Note that the theorem requires knowledge of the amplitude spectrum only.

By analogy with Eq. (2.133) defining the total area under the curve of the energy spectral density $\Psi_g(f)$ of an energy signal, we may define the *power spectral density* of a periodic signal $g_p(t)$ as that function of frequency the total area under which yields the average power of the signal when it is developed across a 1-ohm resistor. Using $S_{gp}(f)$ to denote this power spectral density, we may thus write

$$P = \int_{-\infty}^{\infty} S_{gp}(f)df \tag{2.140}$$

From Eqs. (2.139) and (2.140) we therefore deduce that the power spectral density of a periodic signal consists of a succession of weighted delta functions, as shown by

$$S_{gp}(f) = \frac{1}{T_0^2} \sum_{n=-\infty}^{\infty} \left| G\left(\frac{n}{T_0}\right) \right|^2 \delta\left(f - \frac{n}{T_0}\right) \tag{2.141}$$

According to Eq. (2.141), the power spectral density of a periodic signal is a discrete function of frequency. In Chapter 5, we evaluate the power spectral density of a random signal (which is another example of a power signal) and show that in this case it is a continuous function of frequency.

2.8 AUTOCORRELATION FUNCTION

The process of *autocorrelation* provides a measure of the similarity, or coherence, between a given signal and a replica of the signal delayed by a variable amount. It is defined in such a way that it is a function of this delay variable. However, in defining the autocorrelation function of a signal, we have to pay special attention to the type of signal being considered. In this section we develop definitions for the autocorrelation function of energy signals and periodic signals; the case of random signals is considered in Chapter 5.

Autocorrelation Function of Energy Signals

Consider an energy signal $g(t)$ that may be complex-valued. The autocorrelation of such a signal involves integrating the product of $g(t)$ and a delayed version of its complex conjugate, as shown by

$$R_g(\tau) = \int_{-\infty}^{\infty} g(t)g^*(t-\tau)dt \tag{2.142}$$

In this definition the time delay τ plays the role of a *scanning* or *searching parameter*. On the other hand, the physical time t plays the role of a dummy variable and disappears in the integration, so that the autocorrelation is a function of τ. In Eq. (2.142), the complex-conjugated version of the signal is shifted by the amount τ in the positive direction. Equivalently, we may shift the signal $g(t)$ itself by the amount τ in the negative direction, as shown by

$$R_g(\tau) = \int_{-\infty}^{\infty} g(t+\tau)g^*(t)dt \tag{2.143}$$

Note that when the signal $g(t)$ is complex-valued, the autocorrelation function $R_g(\tau)$ is complex-valued as well.

The autocorrelation function of an energy signal has the following properties:

Property 1

The autocorrelation function exhibits conjugate symmetry, as shown by

$$R_g(\tau)=R_g^*(-\tau) \tag{2.144}$$

That is, the real part of $R_g(\tau)$ is an even function of τ, whereas the imaginary part is an odd function of τ.

This property follows directly from Eq. (2.142) or (2.143).

Property 2

The value of the autocorrelation function at the origin is equal to the energy of the signal, that is

$$R_g(0)=\int_{-\infty}^{\infty} |g(t)|^2\, dt \tag{2.145}$$

This is obtained by putting $\tau=0$ in Eq. (2.142) or (2.143).

Property 3

The maximum value of the autocorrelation function $R_g(\tau)$ occurs at the origin.

As τ is increased, the similarity between the signal and its time-delayed version is reduced. Hence, it is logical to find that the autocorrelation function is reduced as τ is increased, so that we may write (see Problem 2.37)

$$|R_g(\tau)|\leqslant R_g(0), \qquad \text{for all } \tau \tag{2.146}$$

For a signal $g(t)$ of finite energy, $R_g(0)$ is finite, and so from Eq. (2.146) it follows that $R_g(\tau)$ is also finite for all values of τ.

Property 4

The autocorrelation function and energy spectral density form a Fourier transform pair; that is,

$$R_g(\tau)\rightleftharpoons\Psi_g(f). \tag{2.147}$$

To prove this property, we replace the dummy variable of integration, t, in Eq. (2.143) with a new variable $-\sigma$, obtaining

$$R_g(\tau)=\int_{-\infty}^{\infty} g(t-\sigma)g^*(-\sigma)d\sigma \tag{2.148}$$

Comparing this expression with the definition of convolution, we see that the autocorrelation function $R_g(\tau)$ can be obtained by convolving $g(t)$ with $g^*(-t)$. Let $G(f)$ denote the Fourier transform of $g(t)$. From Property 2 (time scaling) and Property 10 (complex conjugation), we find that the Fourier transform of $g^*(-t)$ is equal to $G^*(f)$. Moreover, from Property 12 (the convolution theorem), we recall that the convolution of two functions in the time domain is transformed into the multiplication of their respective Fourier transforms. It follows therefore that

$$R_g(\tau) \rightleftharpoons G(f)G^*(f)$$

However, by definition, the energy spectral density, $\Psi_g(f)$, is equal to $|G(f)|^2$ or, equivalently, $G(f)G^*(f)$. Hence, the Fourier transform pair of Eq. (2.147) is proved.

We may thus express the autocorrelation function $R_g(\tau)$ of the energy signal $g(t)$ as the inverse Fourier transform of the energy spectral density $\Psi_g(t)$ of the signal, as shown by

$$R_g(\tau) = \int_{-\infty}^{\infty} \Psi_g(f) \exp(j2\pi f\tau)df \tag{2.149}$$

Equivalently, we may express the energy spectral density $\Psi_g(f)$ as the Fourier transform of the autocorrelation function $R_g(\tau)$, as shown by

$$\Psi_g(f) = \int_{-\infty}^{\infty} R_g(\tau) \exp(-j2\pi f\tau)d\tau \tag{2.150}$$

Note that according to Eq. (2.149) or (2.150) we only retain information concerning the amplitude of the Fourier transform $G(f)$ of the signal $g(t)$. The information contained in the phase of $G(f)$ is completely lost in evaluating the autocorrelation function. This means that two different signals, whose Fourier transforms have the same amplitude but different phase angles, have the same autocorrelation function. Therefore, for a given function $g(t)$, there is a unique autocorrelation function $R_g(\tau)$. The converse of this statement, however, is not necessarily true.

Autocorrelation Function of Periodic Signals

Consider next a power signal $g(t)$. The autocorrelation function of such a signal is defined as the time average

$$R_g(\tau) = \lim_{T \to \infty} \frac{1}{2T} \int_{-T}^{T} g(t) g^*(t-\tau)dt \tag{2.151}$$

When the delay variable $\tau = 0$, we have $R_g(0) = P$, where P is the average power of the signal, as defined in Eq. (2.136).

When the power signal $g(t)$ is periodic, the integrand in Eq. (2.151) is also periodic and the time average may be taken over one period. Thus we may express the autocorrelation function of a periodic signal $g_p(t)$ of period T_0 as follows

$$R_{gp}(\tau)=\frac{1}{T_0}\int_{-T_0/2}^{T_0/2} g_p(t)g_p^*(t-\tau)dt \tag{2.152}$$

The autocorrelation function $R_{gp}(\tau)$ of a periodic signal has properties similar to those of an energy signal; specifically, we have

Property 1

The autocorrelation function exhibits conjugate symmetry; that is,

$$R_{gp}(\tau)=R_{gp}^*(-\tau) \tag{2.153}$$

Property 2

The value of the autocorrelation function at the origin is equal to the average power of the signal; that is,

$$R_{gp}(0)=\frac{1}{T_0}\int_{-T_0/2}^{T_0/2} |g_p(t)|^2 dt \tag{2.154}$$

Property 3

The maximum value of the autocorrelation function occurs at the origin; that is,

$$R_{gp}(\tau)\leqslant R_{gp}(0), \qquad \text{for all } \tau \tag{2.155}$$

Property 4

The autocorrelation function and power spectral density form a Fourier transform pair; that is,

$$R_{gp}(\tau)\rightleftharpoons S_{gp}(f) \tag{2.156}$$

We prove this property by making use of the generating function $g(t)$, which is defined as being equal to $g_p(t)$ over one period and zero elsewhere. Therefore, substituting Eqs. (2.120) and (2.121) in (2.152), we get

$$R_{gp}(\tau)=\frac{1}{T_0}\int_{-\infty}^{\infty} g(t)\sum_{m=-\infty}^{\infty} g^*(t-\tau-mT_0)dt$$

Interchanging the order of summation and integration, we get

$$\begin{aligned}R_{gp}(\tau)&=\frac{1}{T_0}\sum_{m=-\infty}^{\infty}\int_{-\infty}^{\infty} g(t)g^*(t-\tau-mT_0)dt\\ &=\frac{1}{T_0}\sum_{m=-\infty}^{\infty} R_g(\tau+mT_0)\end{aligned} \tag{2.157}$$

where $R_g(\tau+mT_0)$ is the autocorrelation function of the generating function $g(t)$ for a lag of $\tau+mT_0$. Next, taking the Fourier transform of both sides of Eq. (2.157), we get

$$F[R_{gp}(\tau)]=\frac{1}{T_0}\sum_{m=-\infty}^{\infty}F[R_g(\tau+mT_0)]$$
$$=\frac{1}{T_0}\sum_{m=-\infty}^{\infty}|G(f)|^2\exp(j2\pi mfT_0)$$
$$=\frac{1}{T_0}|G(f)|^2\sum_{m=-\infty}^{\infty}\exp(j2\pi mfT_0) \tag{2.158}$$

Finally, substituting Eq. (2.129) in (2.158), we get

$$F[R_{gp}(\tau)]=\frac{1}{T_0^2}|G(f)|^2\sum_{n=-\infty}^{\infty}\delta\left(f-\frac{n}{T_0}\right)$$
$$=\frac{1}{T_0^2}\sum_{n=-\infty}^{\infty}\left|G\left(\frac{n}{T_0}\right)\right|^2\delta\left(f-\frac{n}{T_0}\right)$$

which is recognized as the power spectral density of the periodic signal $g_p(t)$ [see Eq. (2.141)]. The Fourier transform pair of Eq. (2.156) is therefore proved.

From the Fourier transform pair of Eq. (2.156), we deduce that, for a periodic signal,

$$S_{gp}(f)=\int_{-\infty}^{\infty}R_{gp}(\tau)\exp(-j2\pi f\tau)d\tau \tag{2.159}$$

and conversely,

$$R_{gp}(\tau)=\int_{-\infty}^{\infty}S_{gp}(f)\exp(j2\pi f\tau)df \tag{2.160}$$

Equations (2.159) and (2.160) are known as the *Wiener–Khintchine relations.* In Chapter 5, we will show that the Wiener–Khintchine relations also apply to random signals.

Property 5

The autocorrelation function is periodic with the same period as the periodic signal itself; that is,

$$R_{gp}(\tau)=R_{gp}(\tau\pm nT_0), \qquad n=1, 2, \ldots \tag{2.161}$$

This property follows directly from Eq. (2.157)

2.9 CROSS-CORRELATION FUNCTIONS

The autocorrelation function provides a measure of the similarity between a signal and its time-delayed version. In a similar way, we may use the *cross-correlation function* as a measure of the similarity between a signal and the time-delayed version of a second signal. In subsequent chapters, we will see that both auto-correlation and cross-correlation functions are powerful signal processing tools in analytic and practical work.

The formula for the cross-correlation function depends on whether the pair of signals being considered are energy signals or power signals. We begin by first considering the case of energy signals. We then go on to consider periodic signals as a special case of power signals. The case of random signals is considered in Chapter 5.

Cross-Correlation of Energy Signals

Let $g_1(t)$ and $g_2(t)$ denote a pair of complex-valued signals of finite energy. The cross-correlation function of this pair of signals is defined by

$$R_{12}(\tau)=\int_{-\infty}^{\infty} g_1(t)g_2^*(t-\tau)dt \qquad (2.162)$$

We see that if the two signals $g_1(t)$ and $g_2(t)$ are somewhat similar, then the cross-correlation function $R_{12}(\tau)$ will be finite over some range of τ, thereby providing a quantitive measure of the similarity, or coherence, between them. The energy signals $g_1(t)$ and $g_2(t)$ are said to be *orthogonal* over the entire time interval if $R_{12}(0)$ is zero, that is, if

$$\int_{-\infty}^{\infty} g_1(t)g_2^*(t)dt=0 \qquad (2.163)$$

Equation (2.162) defines one possible value for the cross-correlation function for a specified value of the delay variable τ. We may define a second cross-correlation function for the energy signals $g_1(t)$ and $g_2(t)$ as follows

$$R_{21}(\tau)=\int_{-\infty}^{\infty} g_2(t)g_1^*(t-\tau)dt \qquad (2.164)$$

From the definitions of the cross-correlation functions $R_{12}(\tau)$ and $R_{21}(\tau)$ given above, we obtain the fundamental relationship (see Problem 2.45)

$$R_{12}(\tau)=R_{21}^*(-\tau) \qquad (2.165)$$

Equation (2.165) indicates that unlike convolution, correlation is not in general commutative, that is, $R_{12}(\tau)\neq R_{21}(\tau)$.

Another important property of cross-correlation is shown by the Fourier transform pair

$$R_{12}(\tau)\rightleftharpoons G_1(f)G_2^*(f) \qquad (2.166)$$

This relation is known as the *correlation theorem*. Its proof follows a procedure similar to that used for deriving Eq. (2.147). *The correlation theorem states that the cross-correlation of two energy signals corresponds to the multiplication of the Fourier transform of one signal by the complex conjugate of the Fourier transform of the other.*

Cross-Correlation of Periodic Signals

When $g_1(t)$ and $g_2(t)$ are power signals, then both signals have infinite energy, and the cross-correlation functions defined above do not exist. In such a situation, we use an average form of correlation to define the cross-correlation function $R_{12}(\tau)$, as shown by

$$R_{12}(\tau)=\lim_{T\to\infty}\frac{1}{2T}\int_{-T}^{T}g_1(t)g_2^*(t-\tau)dt \tag{2.167}$$

In a similar way we may define the second cross-correlation function $R_{21}(\tau)$. The pair of power signals $g_1(t)$ and $g_2(t)$ are orthogonal over the entire time interval if

$$\lim_{T\to\infty}\frac{1}{2T}\int_{-T}^{T}g_1(t)g_2^*(t)dt=0 \tag{2.168}$$

If the pair of power signals being correlated are periodic with the same period, Eq. (2.167) takes on the following special form

$$R_{12}(\tau)=\frac{1}{T_0}\int_{-T_0/2}^{T_0/2}g_{p1}(t)g_{p2}^*(t-\tau)dt \tag{2.169}$$

where $g_{p1}(t)$ and $g_{p2}(t)$ denote the periodic signals, and T_0 is their period. The resulting cross-correlation function is also periodic, with a period equal to T_0.

Another property of the cross-correlation function of a pair of periodic signals $g_{p1}(t)$ and $g_{p2}(t)$ is described by the Fourier transform pair (see Problem 2.48)

$$R_{12}(\tau)\rightleftharpoons\frac{1}{T_0^2}\sum_{n=-\infty}^{\infty}G_1\left(\frac{n}{T_0}\right)G_2^*\left(\frac{n}{T_0}\right)\delta\left(f-\frac{n}{T_0}\right) \tag{2.170}$$

where $G_1(n/T_0)$ and $G_2(n/T_0)$ are the Fourier transforms of the generating functions for $g_{p1}(t)$ and $g_{p2}(t)$, respectively, with both of them evaluated at a frequency equal to n/T_0.

2.10 TRANSMISSION OF SIGNALS THROUGH LINEAR SYSTEMS

A *system* refers to any physical device that produces an output signal in response to an input signal. It is customary to refer to the input signal as *excitation* and to the output signal as *response*. In a *linear* system the *principle of superposition* holds; namely, the response of a linear system to a number of excitations applied simultaneously is equal to the sum of the responses of the system when each excitation is applied individually. In this section we are concerned with a study of linear systems, with particular reference to *filters* and *communication channels* operating in their linear region. A filter refers to a frequency-selective device that is used to limit the spectrum of a signal to some band of frequencies. A channel refers to a transmission medium that connects the transmitter and receiver parts of a communication system. We wish to evaluate the effects of transmitting signals through linear filters and communication channels. This evaluation may be carried out in two ways, depending on the description adopted for the filter or channel. That is, we may use time-domain or frequency-domain ideas, as described below.

Time Response

In the time domain, a linear system is described in terms of its *impulse response, which is defined as the response of the system (with zero initial conditions) to a unit impulse or delta function* $\delta(t)$ *applied to the input of the system.* If the system is *time-invariant*, then the shape of the impulse response is the same no matter when the unit impulse is applied to the system. Thus, assuming that the unit impulse or delta function is applied at time $t=0$, we may denote the impulse response of a linear time-invariant system by $h(t)$. Let this system be subjected to an arbitrary excitation $x(t)$, as in Fig. 2.25(*a*). To determine the response $y(t)$ of the system, we begin by first approximating $x(t)$ by a staircase function composed of narrow rectangular pulses, each of duration $\Delta\tau$, as shown in Fig. 2.25(*b*). Clearly the approximation becomes better for smaller $\Delta\tau$. As $\Delta\tau$ approaches zero, each pulse approaches, in the limit, a delta function weighted by a factor equal to the height of the pulse times $\Delta\tau$. Consider a typical pulse, shown shaded in Fig. 2.25(*b*), which

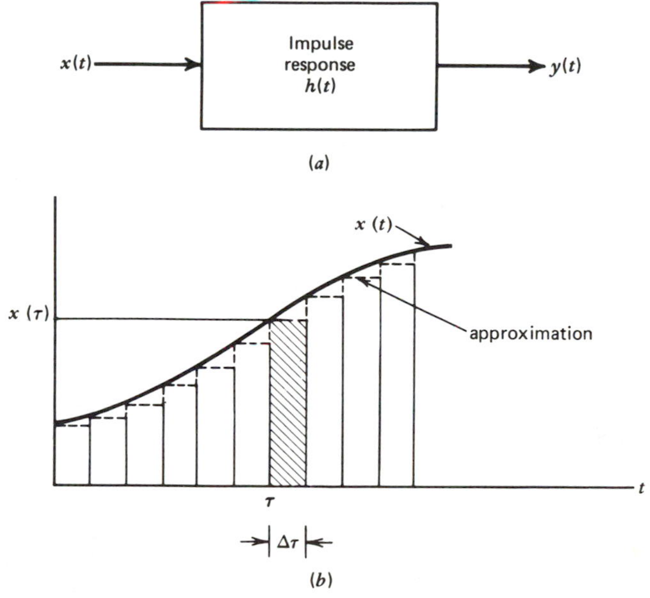

Figure 2.25 *(a)* Linear system. *(b)* Approximation of input *x(t)*.

occurs at $t=\tau$. This pulse has an area equal to $x(\tau)\Delta\tau$. By definition, the response of the system to a unit impulse or delta function $\delta(t)$, occurring at $t=0$, is $h(t)$. It follows, therefore, that the response of the system to a delta function, weighted by the factor $x(\tau)\Delta\tau$ and occurring at $t=\tau$, must be $x(\tau)h(t-\tau)\Delta\tau$. To find the total response $y(t)$ at some time t, we apply the principle of superposition. Thus, summing the various infinitesimal responses due to the various input pulses, we obtain in the limit, as $\Delta\tau$ approaches zero,

$$y(t)=\int_{-\infty}^{\infty} x(\tau)h(t-\tau)d\tau \tag{2.171}$$

This relation is called the *convolution intergral.*

In Eq. (2.171), three different time scales are involved: *excitation time* τ, *response time* t, and *system-memory time* $t-\tau$. This relation is the basis of time-domain analysis of linear time-invariant systems. It states that the present value of the response of a linear time-invariant system is a weighted integral over the past history of the input signal, weighted according to the impulse response of the system. Thus the impulse response acts as a *memory function* for the system.

In Eq. (2.171), the excitation $x(t)$ is convolved with the impulse response $h(t)$ to produce the response $y(t)$. Since convolution is commutative, it follows that we may also write

$$y(t)=\int_{-\infty}^{\infty} h(\tau)x(t-\tau)d\tau \tag{2.172}$$

where $h(t)$ is convolved with $x(t)$.

Example 21 Tapped-delay-line filter

Consider a linear time-invariant filter with impulse response $h(t)$. We assume that:

1. The impulse response $h(t)=0$ for $t<0$.
2. The impulse response of the filter is of finite duration, so that we may write $h(t)=0$ for $t\geqslant T_f$.

Then we may express the filter output $y(t)$ produced in response to the input $x(t)$ as follows:

$$y(t)=\int_{0}^{T_f} h(\tau)x(t-\tau)d\tau \tag{2.173}$$

Let the input $x(t)$, impulse response $h(t)$, and output $y(t)$ be *uniformly sampled* at the rate $1/\Delta\tau$ samples per second, so that we may put

$$t=n\Delta\tau,$$

and

$$\tau=k\Delta\tau,$$

where k and n are integers, and $\Delta\tau$ is the *sampling period*. Assuming that $\Delta\tau$ is small enough for the product $h(\tau)x(t-\tau)$ to remain essentially constant for $k\Delta\tau\leqslant\tau\leqslant(k+1)\Delta\tau$ for all values of k

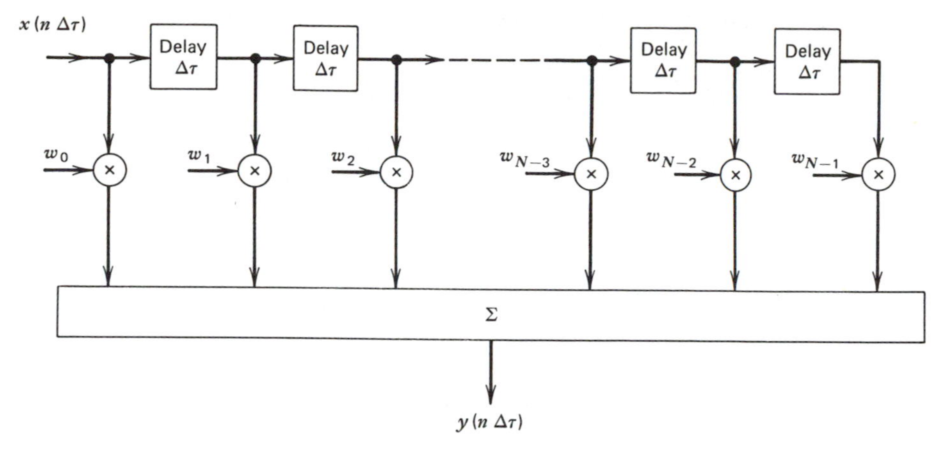

Figure 2.26 Tapped-delay-line filter.

and t of interest, we may approximate Eq. (2.173) by the *convolution sum*:

$$y(n\Delta\tau)=\sum_{k=0}^{N-1} h(k\Delta\tau)x(n\Delta\tau-k\Delta\tau)\Delta\tau \tag{2.174}$$

where $N\Delta\tau=T_f$. Defining

$$w_k=h(k\Delta\tau)\Delta\tau$$

we may rewrite Eq. (2.174) as

$$y(n\Delta\tau)=\sum_{k=0}^{N-1} w_k x(n\Delta\tau-k\Delta\tau) \tag{2.175}$$

Equation (2.175) may be realized using the circuit shown in Fig. 2.26, which consists of a set of *delay elements* (each producing a delay of $\Delta\tau$ seconds), a set of *multipliers* connected to the *delay-line taps*, a corresponding set of *weights* applied to the multipliers, and a *summer* for adding the multiplier outputs. This circuit is known as a *tapped-delay-line filter* or *transversal filter*. Note that in Fig. 2.26 the tap-spacing or basic increment of delay is equal to the sampling period of the input sequence $\{x(n\Delta\tau)\}$.

When a tapped-delay-line filter is implemented using digital hardware, it is commonly referred to as a *finite-duration impulse response* (*FIR*) *digital filter*. The required delay is provided by means of a *shift register*, with the basic increment of delay, $\Delta\tau$, equal to the clock period. An important feature of a digital filter is that it is programmable, thereby offering a high degree of flexibility in design.*

Causality and Stability

A system is said to be *causal* if it does not respond before the excitation is applied. For a linear time-invariant system to be causal, it is clear that the impulse response

* For a detailed treatment of the theory and design of digital filters, see A. V. Oppenheim and R. W. Schafer, *op. cit.* and L. R. Rabiner and B. Gold, *op. cit.*

$h(t)$ must vanish for negative time. That is, the necessary and sufficient condition for causality is

$$h(t)=0, \qquad t<0 \tag{2.176}$$

Clearly, for a system operating in *real time* to be physically realizable, it must be causal. However, there are many applications in which the signal to be processed is available in stored form; in these situations the system can be noncausal and yet physically realizable.

The system is said to be *stable* if the output signal is bounded for all bounded input signals. Let the input signal $x(t)$ be bounded, as shown by

$$|x(t)| \leqslant M \tag{2.177}$$

where M is a positive real finite number. Substituting Eq. (2.177) in (2.172), we get

$$|y(t)| \leqslant M \int_{-\infty}^{\infty} |h(\tau)| d\tau$$

It follows therefore that for a linear time-invariant system to be stable, the impulse response $h(t)$ must be absolutely integrable. That is, the necessary and sufficient condition for stability is

$$\int_{-\infty}^{\infty} |h(t)| dt < \infty \tag{2.178}$$

Frequency Response

Consider a linear time-invariant system of impulse response $h(t)$ driven by a complex exponential input of unit amplitude and frequency f, that is,

$$x(t)=\exp(j2\pi ft) \tag{2.179}$$

Using Eq. (2.172), the response of the system is obtained as

$$\begin{aligned} y(t) &= \int_{-\infty}^{\infty} h(\tau)\exp[j2\pi f(t-\tau)]d\tau \\ &= \exp(j2\pi ft) \int_{-\infty}^{\infty} h(\tau)\exp(-j2\pi f\tau)d\tau \end{aligned} \tag{2.180}$$

Define

$$H(f)=\int_{-\infty}^{\infty} h(\tau)\exp(-j2\pi f\tau)d\tau \tag{2.181}$$

Then we may rewrite Eq. (2.180) in the form

$$y(t)=H(f)\exp(j2\pi ft) \tag{2.182}$$

The response of a linear time-invariant system to a complex exponential function of frequency f is, therefore, the same complex exponential function multiplied by a constant coefficient $H(f)$. The quantity $H(f)$ is called the *transfer function* of the system. The transfer function $H(f)$ and impulse response $h(t)$ form a Fourier transform pair, as shown by the pair of relations:

$$H(f)=\int_{-\infty}^{\infty} h(t)\exp(-j2\pi ft)dt \tag{2.183}$$

and

$$h(t)=\int_{-\infty}^{\infty} H(f)\exp(j2\pi ft)df \tag{2.184}$$

An alternative definition of the transfer function may be deduced by dividing Eq. (2.182) by (2.179) to obtain

$$H(f)=\left.\frac{y(t)}{x(t)}\right|_{x(t)=\exp(j2\pi ft)} \tag{2.185}$$

Consider next an arbitrary signal $x(t)$ applied to the system. The signal $x(t)$ may be expressed in terms of its Fourier transform as

$$x(t)=\int_{-\infty}^{\infty} X(f)\exp(j2\pi ft)df \tag{2.186}$$

or, equivalently, in the limiting form

$$x(t)=\lim_{\substack{\Delta f\to 0\\ f=k\Delta f}} \sum_{k=-\infty}^{\infty} X(f)\exp(j2\pi ft)\Delta f \tag{2.187}$$

That is, the input signal $x(t)$ may be viewed as a superposition of complex exponentials of incremental amplitude. Because the system is linear, the response to this superposition of complex exponential inputs is

$$\begin{aligned} y(t) &= \lim_{\substack{\Delta f\to 0\\ f=k\Delta f}} \sum_{k=-\infty}^{\infty} H(f)X(f)\exp(j2\pi ft)\Delta f \\ &= \int_{-\infty}^{\infty} H(f)X(f)\exp(j2\pi ft)df \end{aligned} \tag{2.188}$$

The Fourier transform of the output is therefore

$$Y(f)=H(f)X(f) \tag{2.189}$$

A linear time-invariant system may thus be described quite simply in the frequency domain by noting that the Fourier transform of the output is equal to the product of the transfer function of the system and the Fourier transform of the input.

The result of Eq. (2.189) may, of course, be deduced directly by recognizing that the response $y(t)$ of a linear time-invariant system of impulse response $h(t)$ to an arbitrary input $x(t)$ is obtained by convolving $x(t)$ with $h(t)$, or vice versa, and by the fact that the convolution of a pair of time functions is transformed into the multiplication of their Fourier transforms. The derivation above is presented primarily to develop an understanding of why the Fourier representation of a time function as a superposition of complex exponentials is so useful in analyzing the behavior of linear time-invariant systems.

The transfer function $H(f)$ is a characteristic property of a linear time-invariant system. It is, in general, a complex quantity, so that we may express it in the form

$$H(f)=|H(f)|\exp[j\beta(f)] \tag{2.190}$$

where $|H(f)|$ is called the *amplitude response*, and $\beta(f)$ the *phase* or *phase angle*. In the case of a linear system with a real-valued impulse response $h(t)$, the transfer function $H(f)$ exhibits conjugate symmetry, which means that

$$|H(f)|=|H(-f)| \tag{2.191}$$

and

$$\beta(f)=-\beta(-f) \tag{2.192}$$

That is, the amplitude response $|H(f)|$ is an even function of frequency, whereas the phase $\beta(f)$ is an odd function of frequency.

In some applications it is preferable to work with the logarithm of $H(f)$ rather than with $H(f)$ itself. Define

$$\ln H(f)=\alpha(f)+j\beta(f) \tag{2.193}$$

where

$$\alpha(f)=\ln|H(f)| \tag{2.194}$$

The function $\alpha(f)$ is called the *gain* of the system. It is measured in *nepers*, whereas $\beta(f)$ is measured in *radians*. Equation (2.193) indicates that the gain $\alpha(f)$ and phase $\beta(f)$ are the real and imaginary parts of the logarithm of the transfer function $H(f)$, respectively. The gain may also be expressed in *decibels* (dB) by using the definition

$$\alpha'(f)=20\log_{10}|H(f)| \tag{2.195}$$

The two gain functions $\alpha(f)$ and $\alpha'(f)$ are related by

$$\alpha'(f)=8.69\alpha(f) \tag{2.196}$$

That is, 1 neper is equal to 8.69 dB.

As a means of specifying the constancy of the amplitude response $|H(f)|$ or gain $\alpha(f)$ of a system, we use a parameter called *bandwidth* of the system. In the case of a low-pass system, the bandwidth is customarily defined as the frequency at which the amplitude response $|H(f)|$ is $(1/\sqrt{2})$ times its value at zero frequency or, equivalently, the frequency at which the gain $\alpha'(f)$ drops by 3 dB below its value at zero frequency, as illustrated in Fig. 2.27(*a*). In the case of a band-pass system, the bandwidth is defined as the range of frequencies over which the amplitude response $|H(f)|$ remains within $(1/\sqrt{2})$ times its value at the mid-band frequency, as illustrated in Fig. 2.27(*b*). It will be shown later that the constancy of the amplitude response or gain of a system is one of the conditions necessary for the transmission of a signal with no distortion.

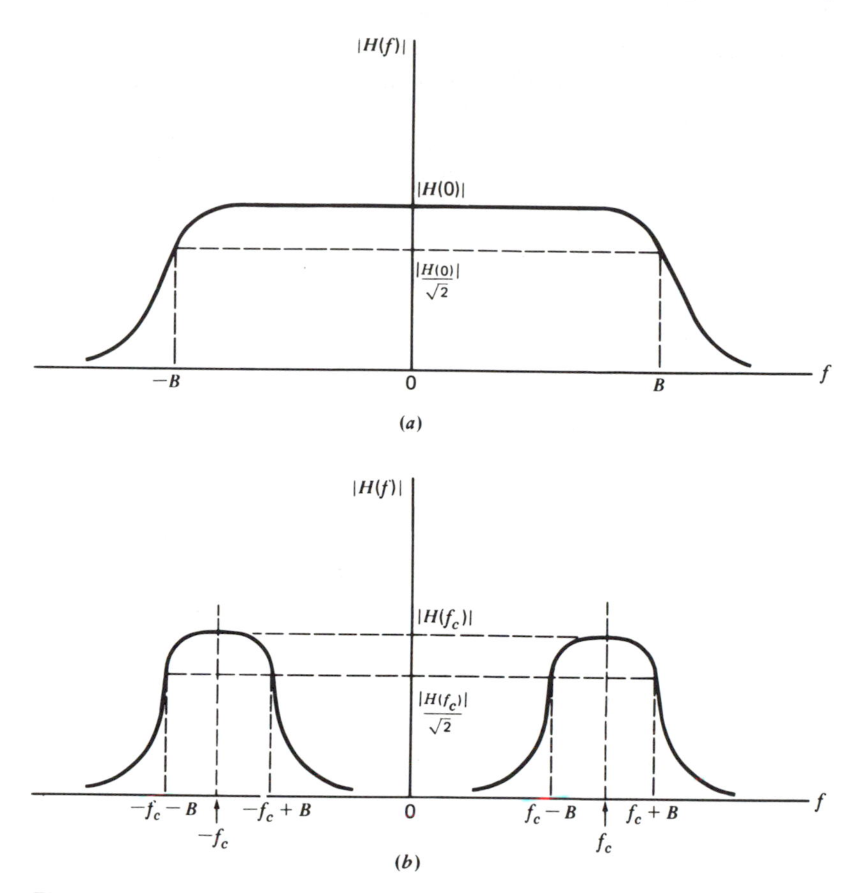

Figure 2.27 Illustrating the definition of system bandwidth. *(a)* Low-pass system. *(b)* Band-pass system.

Relation Among the Input and Output Energy Spectral Densities

Returning to the input–output relation of Eq. (2.189), we find that by taking the squared amplitude of both sides of this equation, the result is

$$|Y(f)|^2 = |H(f)|^2 |X(f)|^2 \tag{2.197}$$

Equivalently, we may write

$$\Psi_y(f) = |H(f)|^2 \Psi_x(f) \tag{2.198}$$

where $\Psi_y(y) = |Y(f)|^2$ and $\Psi_x(f) = |X(f)|^2$. The quantities $\Psi_y(f)$ and $\Psi_x(f)$ denote the energy spectral densities of the response (output), $y(t)$, and the excitation (input), $x(t)$, respectively. Equation (2.198) indicates that the output energy spectral density of a linear filter is obtained by multiplying the squared amplitude response of the filter by the input energy spectral density. That is, the relation between the input and output energy spectral densities of a linear filter is determined only by the amplitude response of the filter.

Equation (2.198) applies to energy signals. In a similar way, we may show that, in the case of power signals, the output power spectral density of a linear filter is equal to the input power spectral density multiplied by the squared amplitude response of the filter.

Paley-Wiener Criterion

A necessary and sufficient condition for a function $\alpha(f)$ to be the gain of a causal filter is the convergence of the integral

$$\int_{-\infty}^{\infty} \frac{|\alpha(f)|}{1+f^2}\, df < \infty \tag{2.199}$$

This condition is known as the *Paley–Wiener criterion.** It states that provided the gain $\alpha(f)$ satisfies the condition of Eq. (2.199), then we may associate with this gain a suitable phase $\beta(f)$, such that the resulting filter has a causal impulse response that is zero for negative time. In other words, the Paley–Wiener criterion is the frequency-domain equivalent of the causality requirement. According to Eq. (2.199), a system with a realizable gain characteristic may have infinite attenuation for a discrete set of frequencies, but it cannot have infinite attenuation over a band of frequencies. It is important to note, however, that for the Paley-Wiener criterion to be valid, the amplitude response $|H(f)|$ must be square-integrable, that is,

$$\int_{-\infty}^{\infty} |H(f)|^2 df < \infty \tag{2.200}$$

Distortionless Transmission

By *distortionless transmission* we mean that the output signal of a communication channel is an exact replica of the input signal, except for a possible change of amplitude and a constant time delay. We may therefore say that a signal $x(t)$ is transmitted through the channel without distortion if the output signal $y(t)$ is defined by

$$y(t) = Kx(t - t_0) \tag{2.201}$$

where the constant K accounts for the change in amplitude and the constant t_0 accounts for the delay in transmission.

Let $X(f)$ and $Y(f)$ denote the Fourier transforms of $x(t)$ and $y(t)$, respectively. Then, applying the Fourier transform to Eq. (2.201) and using the time-shifting

* R. E. A. C. Paley and N. Wiener, "Fourier transforms in the complex domain," American Mathematical Society Colloquium Publication, vol. 19, pp. 16 and 17, 1934. See, also, A. Papoulis, *The Fourier Integral and Its Applications*, pp. 215–217 (McGraw-Hill, 1962).

property of the Fourier transform, we get

$$Y(f) = KX(f)\exp(-j2\pi f t_0) \tag{2.202}$$

The transfer function of a distortionless channel is therefore

$$\begin{aligned} H(f) &= \frac{Y(f)}{X(f)} \\ &= K \exp(-j2\pi f t_0) \end{aligned} \tag{2.203}$$

A more general expression for distortionless transmission is

$$H(f) = K \exp[j(-2\pi f t_0 \pm n\pi)] \tag{2.204}$$

where n is an integer including zero. If n is even, we have $\exp(\pm jn\pi) = 1$, whereas if n is odd, $\exp(\pm jn\pi) = -1$, so that the condition for distortionless transmission is basically unchanged.

Equation (2.204) indicates that in order to achieve distortionless transmission through a channel, the transfer function of the channel must satisfy two conditions:

1. The amplitude response $|H(f)|$ is constant for all frequencies, as shown by

$$|H(f)| = K \tag{2.205}$$

2. The phase $\beta(f)$ is linear with frequency, passing through zero or a multiple of π at zero frequency, as shown by

$$\beta(f) = -2\pi f t_0 \pm n\pi \tag{2.206}$$

If, however, the spectrum of the transmitted signal is limited to a band of frequencies, then Eqs. (2.205) and (2.206) need to be satisfied by the channel only for that band of frequencies. This is illustrated in parts (*a*) and (*b*) of Figs. 2.28 and 2.29 for low-pass and band-pass channels, respectively. Note that in the band-pass case, the phase $\beta(f)$ may have any value at the mid-band frequency f_c, but at zero frequency the phase $\beta(f)$ must be a multiple of π.

In practice, the conditions for distortionless transmission, as described above, can only be satisfied approximately. That is to say, there is always a certain amount of distortion present in the output signal. The channel is then said to be *dispersive*. We may distinguish two forms of signal distortion in a dispersive channel:

1. When the amplitude response $|H(f)|$ of the channel is not constant with frequency inside the frequency band of interest, the frequency components of the input signal are transmitted with different amounts of gain or attenuation. This effect is called *amplitude distortion*. The most common cause of amplitude distortion is excess gain or attenuation at one or both ends of the frequency band of interest.
2. The second form of distortion arises when the phase response $\beta(f)$ of the channel is not linear with frequency. Then if the input signal is divided into a set of components, each one of which occupies a narrow band of frequencies,

we find that each of them is subject to a different delay in passing through the system, with the result that the output signal has a different waveform from the input. This form of distortion is called *phase* or *delay distortion*. We will have more to say on this issue in Section 2.16.

2.11 IDEAL LOW-PASS FILTERS

As previously mentioned, a *filter* is a frequency-selective device that is used to limit the spectrum of a signal to some specified band of frequencies. Its frequency response is characterized by a *passband* and a *stopband*. The frequencies inside the passband are transmitted with little or no distortion, whereas those in the stopband are rejected. The filter may be of the *low-pass*, *high-pass*, *band-pass*, or *band-stop* type, depending on whether it transmits low, high, intermediate, or all but intermediate frequencies, respectively. We have already encountered examples of low-pass and band-pass systems in Figs. 2.28 and 2.29.

In this section we study the time response of the *ideal low-pass filter* which transmits, without any distortion, all frequencies inside the passband and completely rejects all frequencies inside the stopband, as illustrated in Fig. 2.28. The transfer

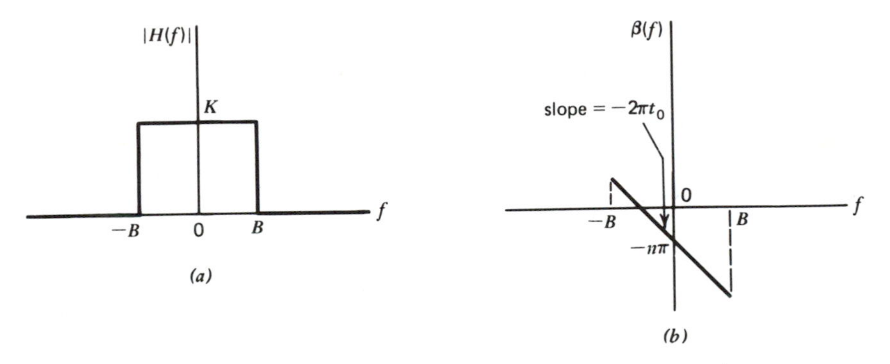

Figure 2.28 Frequency response of ideal low-pass channel.

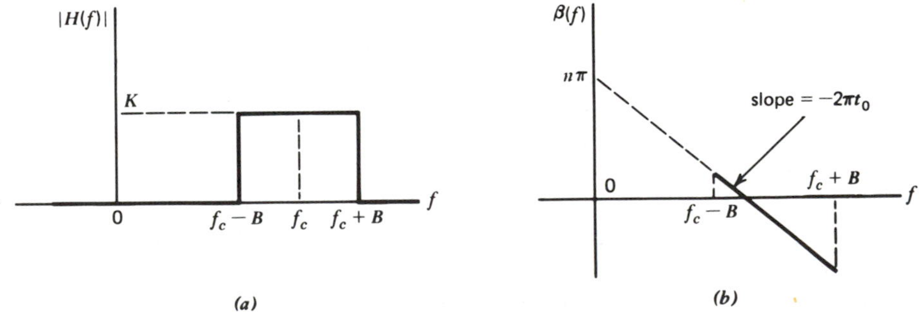

Figure 2.29 Frequency response of ideal band-pass channel (only the response for positive frequencies is shown).

function of an ideal low-pass filter is therefore defined by

$$H(f)=\begin{cases}\exp(-j2\pi f t_0), & -B\leqslant f\leqslant B\\ 0, & |f|>B\end{cases} \tag{2.207}$$

where, for convenience, we have set $K=1$ and $n=0$. The parameter B defines the bandwidth of the filter. The ideal low-pass filter is, of course, noncausal because it violates the Paley–Wiener criterion. This observation may also be confirmed by examining the impulse response $h(t)$. Thus, by evaluating the inverse Fourier transform of the transfer function of Eq. (2.207), we get

$$h(t)=\int_{-B}^{B}\exp[j2\pi f(t-t_0)]df \tag{2.208}$$

where the limits of integration have been reduced to the frequency band inside which $H(f)$ does not vanish. Equation (2.208) is readily integrated, yielding

$$\begin{aligned}h(t)&=\frac{\sin[2\pi B(t-t_0)]}{\pi(t-t_0)}\\ &=2B\,\mathrm{sinc}[2B(t-t_0)]\end{aligned} \tag{2.209}$$

This impulse response has a peak amplitude of $2B$ centered on time t_0, as shown in Fig. 2.30. The duration of the main lobe of the impulse response is $1/B$, and the build-up time from the zero at the beginning of the main lobe to the peak value is $1/2B$. We see from Fig. 2.30 that, for any finite value of t_0, there is some response from the filter before the time $t=0$ at which the unit impulse is applied to the input, confirming that the ideal low-pass filter is noncausal. However, in

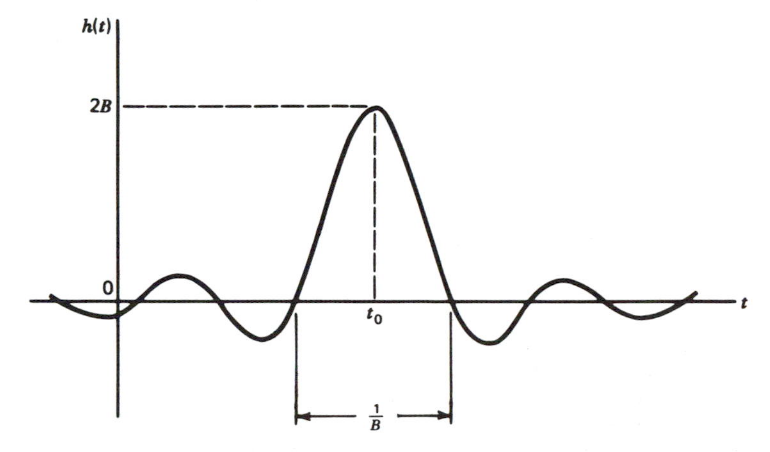

Figure 2.30 Impulse response of ideal low-pass filter.

spite of its noncausality, the ideal low-pass filter serves as a useful standard against which the response of causal filters may be measured.*

Example 22 Pulse response of ideal low-pass filter

Consider a rectangular pulse $x(t)$ of unit amplitude and duration T, which is applied to an ideal low-pass filter of bandwidth B. The problem is to determine the response $y(t)$ of the filter.

The impulse response $h(t)$ of the filter is defined by Eq. (2.209). Its response is therefore given by the convolution integral

$$\begin{aligned} y(t) &= \int_{-\infty}^{\infty} x(\tau)h(t-\tau)d\tau \\ &= 2B \int_{-T/2}^{T/2} \frac{\sin[2\pi B(t-t_0-\tau)]}{2\pi B(t-t_0-\tau)} d\tau \end{aligned} \tag{2.210}$$

Define

$$\lambda = 2\pi B(t-t_0-\tau)$$

Then, changing the integration variable from τ to λ, we may rewrite Eq. (2.210) as

$$\begin{aligned} y(t) &= \frac{1}{\pi} \int_{2\pi B(t-t_0-T/2)}^{2\pi B(t-t_0+T/2)} \frac{\sin \lambda}{\lambda} d\lambda \\ &= \frac{1}{\pi} \left[\int_0^{2\pi B(t-t_0+T/2)} \frac{\sin \lambda}{\lambda} d\lambda - \int_0^{2\pi B(t-t_0-T/2)} \frac{\sin \lambda}{\lambda} d\lambda \right] \\ &= \frac{1}{\pi} \{\mathrm{Si}[2\pi B(t-t_0+T/2) - \mathrm{Si}[2\pi B(t-t_0-T/2)]\} \end{aligned} \tag{2.211}$$

where the sine integral is defined by

$$\mathrm{Si}(u) = \int_0^u \frac{\sin \lambda}{\lambda} d\lambda$$

In Fig. 2.31 we have plotted the response $y(t)$ for three different values of the filter bandwidth B, assuming that t_0 is zero. We see that, in each case, the output is symmetrical about $t=0$. We further observe that the shape of the output is markedly dependent on the filter bandwidth B. In particular, we note the following:

1. When B is small compared with $1/T$, the output is a grossly distorted version of the input, as in Fig. 2.31(*c*).
2. When $B=1/T$, as in Fig. 2.31(*b*), the output is recognizable; however, the rise and fall times of the output are significant compared with the input pulse duration T.

* There are various ways of approximating the frequency response of an ideal filter, with the approximation being relizable in terms of physical elements. The choice of a particular filter design depends on the nature of the application in question. For a detailed discussion of the approximation problem and synthesis procedures used in filter design, see M. E. Van Valkenberg, *Introduction to Modern Network Synthesis* (Wiley, 1960). L. Weinberg, *Network Analysis and Synthesis* (McGraw-Hill, 1962). E. A. Guillemin, *Synthesis of Passive Networks* (Wiley, 1957). A. J. Zverev, *Handbook of Filter Synthesis* (Wiley, 1967).

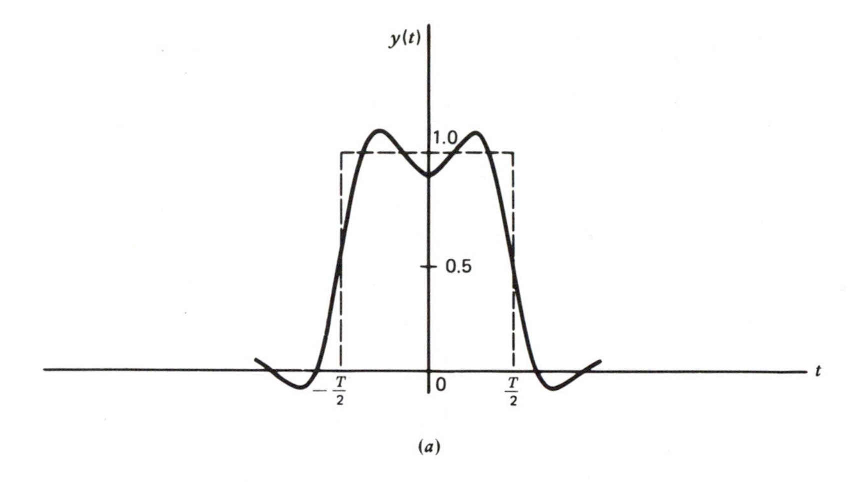

(a)

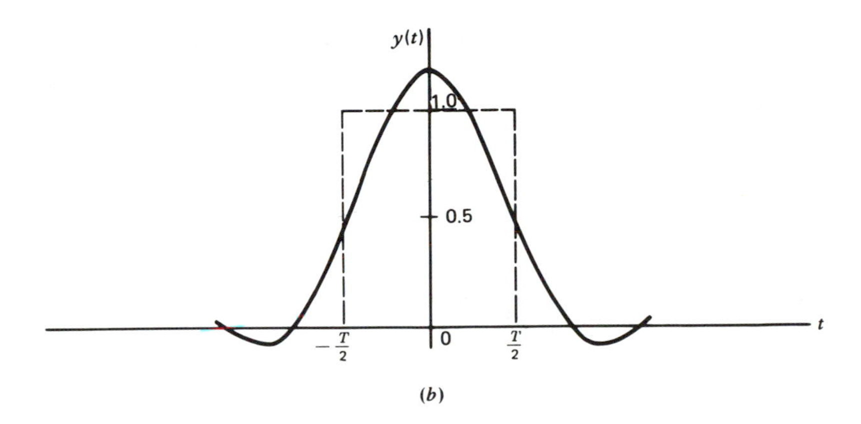

(b)

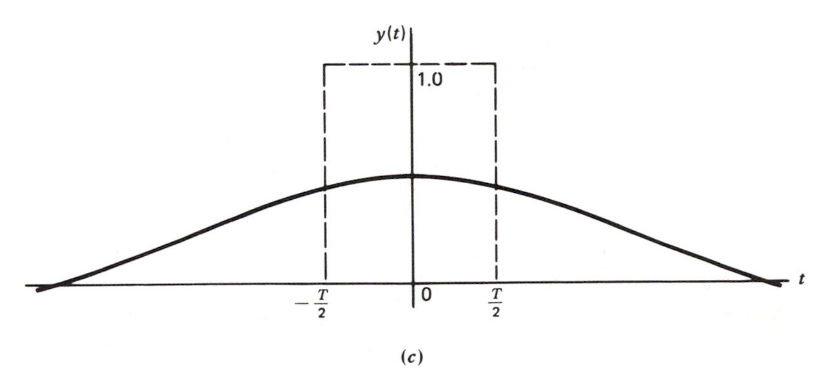

(c)

Figure 2.31 Pulse response of ideal low-pass filter for varying filter bandwidth. *(a)* $B = 2/T$. *(b)* $B = 1/T$. *(c)* $B = 1/4T$. The dotted rectangle in parts *(a)*, *(b)* and *(c)* of the figure represents the input signal.

3. When B is large compared with $1/T$, as in Fig. 2.31(a), the output has approximately the same duration as the input. However, it differs from the input in two major respects. First, the output, unlike the input, has nonzero rise and fall times that are inversely proportional to the filter bandwidth. Second, the output exhibits *ringing* at both the leading and trailing edges.

2.12 HILBERT TRANSFORM

The Fourier transform is particularly useful for evaluating the frequency content of an energy signal or, in a limiting sense, that of a power signal. As such, it provides a means of analyzing and designing frequency selective filters for the separation of signals on the basis of their frequency contents. Another method of separating signals is based on *phase selectivity*, which uses phase shifts between the pertinent signals to achieve the desired separation. The simplest phase shift is that of 180 degrees, which is merely a polarity reversal in the case of a sinusoidal signal. Shifting the phase angles of all components of a given signal wave by 180 degrees requires the use of an *ideal transformer*. Another phase shift of interest is that of ± 90 degrees. In particular, when the phase angles of all components of a given signal wave are shifted by ± 90 degrees, the resulting function of time is known as the *Hilbert transform* of the signal.

Consider a signal $g(t)$ with Fourier transform $G(f)$. The *Hilbert transform* of $g(t)$, which we shall denote by $\hat{g}(t)$, is defined by*

$$\hat{g}(t) = \frac{1}{\pi} \int_{-\infty}^{\infty} \frac{g(\tau)}{t-\tau}\, d\tau \tag{2.212}$$

Clearly, the Hilbert transformation of $g(t)$ is a linear operation. The *inverse Hilbert transform*, by means of which the original signal $g(t)$ is recovered from $\hat{g}(t)$, is defined by

$$g(t) = -\frac{1}{\pi} \int_{-\infty}^{\infty} \frac{\hat{g}(\tau)}{t-\tau}\, d\tau \tag{2.213}$$

The functions $g(t)$ and $\hat{g}(t)$ are said to constitute a *Hilbert-transform pair.*†

We note from the definition of the Hilbert transform, that $\hat{g}(t)$ may be interpreted as the convolution of $g(t)$ with the time function $1/(\pi t)$. We also know from the convolution theorem that the convolution of two functions in the time domain is transformed into the multiplication of their Fourier transforms in the frequency

* The integral in Eq. (2.212) is an improper integral in that the integrand has a singularity at $\tau = t$. In order to avoid this singularity, the integration must be carried out in a symmetrical manner about the point $\tau = t$. For this purpose, we use the definition

$$P \int_{-\infty}^{\infty} \frac{g(\tau)}{t-\tau}\, d\tau = \lim_{\varepsilon \to 0} \left[\int_{-\infty}^{t-\varepsilon} \frac{g(\tau)}{t-\tau}\, d\tau + \int_{t+\varepsilon}^{\infty} \frac{g(\tau)}{t-\tau}\, d\tau \right]$$

where the symbol P denotes *Cauchy's principal value of the integral*. For notational simplicity, the symbol P has been omitted from Eqs. (2.212) and (2.213).

† A short table of Hilbert-transform pairs is given in Table A6.3 in Appendix 6.

domain. The Fourier transform of $1/(\pi t)$ is equal to $-j\ \mathrm{sgn}(f)$, where $\mathrm{sgn}(f)$ is the signum function defined by (see Section 2.5)

$$\mathrm{sgn}(f)=\begin{cases} 1, & f>0 \\ 0, & f=0 \\ -1, & f<0 \end{cases} \tag{2.214}$$

It follows therefore that the Fourier transform $\hat{G}(f)$ of $\hat{g}(t)$ is given by

$$\hat{G}(f)=-j\ \mathrm{sgn}(f)G(f) \tag{2.215}$$

Equation (2.215) states that given a signal $g(t)$, we may obtain its Hilbert transform $\hat{g}(t)$ by passing $g(t)$ through a linear two-port device the transfer function of which is equal to $-j\ \mathrm{sgn}(f)$. This device may be considered as one that produces a phase shift of -90 degrees for all positive frequencies of the input signal and $+90$ degrees for all negative frequencies, as in Fig. 2.32. The amplitudes of all frequency components in the signal, however, are unaffected by transmission through the device. Such an ideal device is referred to as a *Hilbert transformer*.

The Hilbert transform has several important applications, which include the following:

1. It can be used to realize phase selectivity in the generation of a special kind of modulation known as *single sideband modulation*. We shall have more to say about this application in Chapter 3.
2. It provides the mathematical basis for the representation of band-pass signals. This application is discussed in Section 2.14.

The Hilbert transform, as defined above, applies to any signal that is Fourier transformable. Accordingly, it may be applied to energy signals as well as power signals.

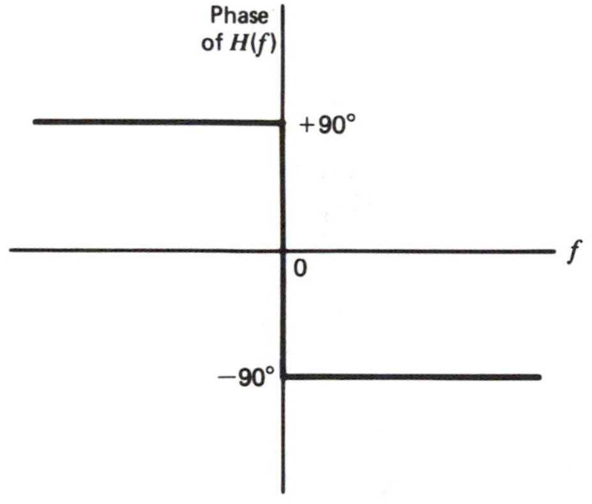

Figure 2.32 Phase characteristic of linear two-port device for obtaining the Hilbert transform of a signal.

Example 23 Sinusoidal Functions

Consider the cosine function

$$g(t)=\cos(2\pi f_c t) \tag{2.216}$$

whose Fourier transform is

$$G(f)=\frac{1}{2}\left[\delta(f-f_c)+\delta(f+f_c)\right] \tag{2.217}$$

Using Eqs. (2.215) and (2.217), we get

$$\begin{aligned}\hat{G}(f)&=-j\,\mathrm{sgn}(f)G(f)\\&=-\frac{j}{2}\left[\delta(f-f_c)+\delta(f+f_c)\right]\mathrm{sgn}(f)\\&=\frac{1}{2j}\left[\delta(f-f_c)-\delta(f+f_c)\right],\end{aligned}$$

which represents the Fourier transform of the sine function $\sin(2\pi f_c t)$. The Hilbert transform of the cosine function of Eq. (2.216) is therefore given by

$$\hat{g}(t)=\sin(2\pi f_c t) \tag{2.218}$$

In a similar way, we find that the sine function

$$g(t)=\sin(2\pi f_c t)$$

has its Hilbert transform given by

$$\hat{g}(t)=-\cos(2\pi f_c t)$$

Example 24 Product of low-pass and high-pass signals of non-overlapping spectra

Consider a signal $s(t)$ defined by

$$s(t)=c(t)m(t) \tag{2.219}$$

where $m(t)$ is a low-pass signal whose Fourier transform $M(f)$ vanishes for $|f|>W$, and $c(t)$ is a high-pass signal whose Fourier transform $C(f)$ vanishes for $|f|<W$. The problem is to determine the Hilbert transform of the product $s(t)=c(t)m(t)$.

Let $S(f)$ denote the Fourier transform of $s(t)$. Since $s(t)$ is equal to the product of $c(t)$ and $m(t)$, it follows that $S(f)$ is equal to the convolution of $C(f)$ and $M(f)$, as shown by

$$S(f)=\int_{-\infty}^{\infty} M(\lambda)C(f-\lambda)\,d\lambda$$

Let $\hat{s}(t)$ denote the Hilbert transform of $s(t)$, and $\hat{S}(f)$ the Fourier transform of $\hat{s}(t)$. Hence,

$$\hat{S}(f)=-j\,\mathrm{sgn}(f)S(f)$$

Using the formula for the inverse Fourier transform, we may then write

$$\begin{aligned}\hat{s}(t) &= \int_{-\infty}^{\infty} \hat{S}(f)\exp(j2\pi ft)df \\ &= \int_{-\infty}^{\infty} -j\,\mathrm{sgn}(f)S(f)\exp(j2\pi ft)df \\ &= \int_{-\infty}^{\infty}\int_{-\infty}^{\infty} -j\,\mathrm{sgn}(f)M(\lambda)C(f-\lambda)\exp(j2\pi ft)df\,d\lambda\end{aligned}$$

Let $\xi = f - \lambda$. Then, we may rewrite this relation as

$$\hat{s}(t) = -j\int_{-\infty}^{\infty}\int_{-\infty}^{\infty} \mathrm{sgn}(\xi+\lambda)M(\lambda)C(\xi)\exp[j2\pi(\xi+\lambda)t]d\xi\,d\lambda \tag{2.220}$$

However, by definition, the product $M(\lambda)C(\xi)$ is nonzero only for $|\lambda| < W$ and $|\xi| > W$. With the values of ξ and λ restricted in this manner, we find that $\mathrm{sgn}(\xi+\lambda) = \mathrm{sgn}(\xi)$. We may thus rewrite Eq. (2.220) in the form

$$\hat{s}(t) = \int_{-\infty}^{\infty} M(\lambda)\exp(j2\pi\lambda t)d\lambda \int_{-\infty}^{\infty} [-j\,\mathrm{sgn}(\xi)C(\xi)]\exp(j2\pi\xi t)d\xi \tag{2.221}$$

We recognize the first integral on the right-hand side of Eq. (2.221) as $m(t)$, and the second integral as $\hat{c}(t)$, the Hilbert transform of $c(t)$, and so we have the desired result

$$\hat{s}(t) = \hat{c}(t)m(t) \tag{2.222}$$

Equation (2.222) shows that the rule for evaluating the Hilbert transform of the product of non-overlapping low-pass and high-pass signals is to multiply the low-pass signal by the Hilbert transform of the high-pass signal.

Properties of the Hilbert Transform

The Hilbert transform has a number of useful properties, some of which are listed below. To derive these properties, we make use of Eq. (2.215), which defines the relationship between the Fourier transform of a signal $g(t)$ and that of its Hilbert transform $\hat{g}(t)$.

Property 1

A signal $g(t)$ and its Hilbert transform $\hat{g}(t)$ have the same spectral density.

To prove this property, we observe that the Fourier transform of $\hat{g}(t)$ is equal to $-j\,\mathrm{sgn}(f)$ times the Fourier transform of $g(t)$, and since the magnitude of $-j\,\mathrm{sgn}(f)$ is equal to one for all f, then $g(t)$ and $\hat{g}(t)$ will both have the same amplitude spectrum. It follows, therefore, that $g(t)$ and $\hat{g}(t)$ have the same energy spectral density if $g(t)$ is an energy signal, and the same power spectral density if it is a power signal.

As corollaries to this property, we may state that: (1) if a signal $g(t)$ is band-limited, then its Hilbert transform $\hat{g}(t)$ will also be band-limited, and (2) the signal $g(t)$ and its Hilbert transform $\hat{g}(t)$ have the same energy if $g(t)$ is an energy signal, and the same average power if it is a power signal.

Property 2

A signal $g(t)$ and its Hilbert transform $\hat{g}(t)$ have the same autocorrelation function.

This property is a direct consequence of Property 1 because the autocorrelation function and the spectral density form a Fourier transform pair.

Property 3

A signal $g(t)$ and its Hilbert transform $\hat{g}(t)$ are orthogonal.

To prove this property for the case of an energy signal $g(t)$, we first use Eqs. (2.162) and (2.166) to evaluate the cross-correlation function of the signal $g(t)$ and its Hilbert transform $\hat{g}(t)$ for a delay $\tau=0$. The result is [see part (f) of Problem 2.12]

$$\begin{aligned}\int_{-\infty}^{\infty} g^*(t)\hat{g}(t)dt &= \int_{-\infty}^{\infty} G^*(f)\hat{G}(f)df \\ &= \int_{-\infty}^{\infty} G^*(f)[-j \operatorname{sgn}(f)G(f)]df \\ &= -j\int_{-\infty}^{\infty} \operatorname{sgn}(f)|G(f)|^2 df \end{aligned} \tag{2.223}$$

The integrand in the right-hand side of Eq. (2.223) is an odd function of f, being the product of the odd function $\operatorname{sgn}(f)$ and the even function $|G(f)|^2$. Hence, the integral is zero, yielding the final result

$$\int_{-\infty}^{\infty} g^*(t)\hat{g}(t)dt = 0 \tag{2.224}$$

This shows that the energy signal $g(t)$ and its Hilbert transform $\hat{g}(t)$ are orthogonal.

Similarly, we may show that a power signal $g(t)$ and its Hilbert transform $\hat{g}(t)$ are also orthogonal, as shown by

$$\lim_{T\to\infty} \frac{1}{2T}\int_{-T}^{T} g^*(t)\hat{g}(t)dt = 0 \tag{2.225}$$

Property 4

If $\hat{g}(t)$ is the Hilbert transform of $g(t)$, then the Hilbert transform of $\hat{g}(t)$ is $-g(t)$.

To prove this property, we note that the process of Hilbert transformation is equivalent to passing $g(t)$ through a linear device with a transfer function equal to $-j\operatorname{sgn}(f)$. A double Hilbert transformation is therefore equivalent to passing $g(t)$ through a cascade of two such devices. The overall transfer function of such a cascade is equal to

$$[-j \operatorname{sgn}(f)]^2 = -1, \qquad \text{for all } f$$

The resulting output is thus $-g(t)$; that is, the Hilbert transform of $\hat{g}(t)$ is equal to $-g(t)$.

2.13 PRE-ENVELOPE

Consider a real-valued signal $g(t)$. We define the *pre-envelope* of the signal $g(t)$ as the complex-valued function*

$$g_+(t) = g(t) + j\hat{g}(t) \tag{2.226}$$

where $\hat{g}(t)$ is the Hilbert transform of $g(t)$. We note that the given signal $g(t)$ is the real part of the pre-envelope $g_+(t)$, and the Hilbert transform of the signal is the imaginary part of the pre-envelope. Just as the use of phasors simplifies manipulations of alternating currents and voltages, so we find that the pre-envelope is useful in handling band-pass signals and systems. The reason for the name "pre-envelope" is explained in Section 2.14.

One of the important features of the pre-envelope $g_+(t)$ is the behavior of its Fourier transform. Let $G_+(f)$ denote the Fourier transform of $g_+(t)$. Then, we may write

$$G_+(f) = G(f) + j[-j\,\mathrm{sgn}(f)]G(f)$$

Using the definition of the signum function sgn(f), we therefore have

$$G_+(f) = \begin{cases} 2G(f), & f > 0 \\ G(0), & f = 0 \\ 0, & f < 0 \end{cases} \tag{2.227}$$

where $G(0)$ is the value of $G(f)$ at frequency $f = 0$. This means that the pre-envelope of a signal has no frequency content (i.e., its Fourier transform vanishes) for all negative frequencies, as illustrated in Fig. 2.33.

For a given signal $g(t)$, we may determine its pre-envelope $g_+(t)$ in one of two equivalent ways:

1. We determine the Hilbert transform $\hat{g}(t)$ of the signal $g(t)$, and then use Eq. (2.226) to compute the pre-envelope $g_+(t)$.
2. We determine the Fourier transform $G(f)$ of the signal $g(t)$, use Eq. (2.227) to determine $G_+(f)$, and then evaluate the inverse Fourier transform of $G_+(f)$ to obtain

$$g_+(t) = 2\int_0^\infty G(f)\exp(j2\pi ft)\,df \tag{2.228}$$

* The complex representation of an arbitrary signal was first described by Gabor: D. Gabor, "Theory of communication," *J. IEE* (London), vol. 93, Part III, pp. 429–441, Nov. 1946. Gabor used the term "analytic signal." The term "pre-envelope" was used in the papers: R. Arens, "Complex processes for envelopes of normal noise," *IRE Trans. Information Theory*, vol. IT-3, pp. 204–207, Sept. 1957. J. Dungundji, "Envelopes and pre-envelopes of real wave-forms," *IRE Trans. Information Theory*, Vol. IT-4, pp. 53–57, Mar. 1958. For a review of the different envelopes, see S. O. Rice, "Envelopes of narrow-band signals," *Proc. IEEE*, vol. 70, pp. 692–699, July 1982.

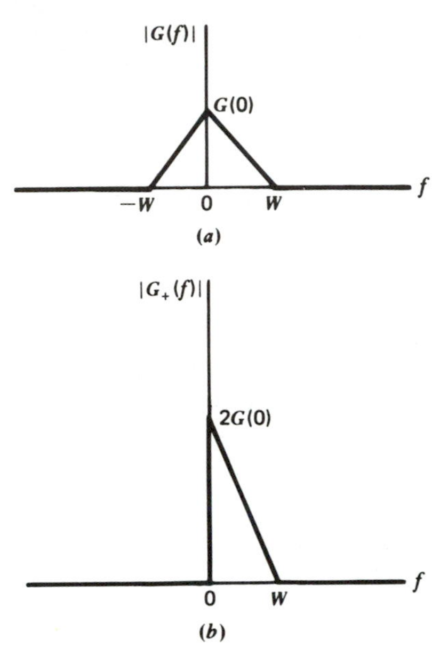

Figure 2.33 *(a)* Amplitude spectrum of low-pass signal *g(t)*. *(b)* Amplitude spectrum of pre-envelope $g_+(t)$.

For a particular signal $g(t)$ or Fourier transform $G(f)$, one way may be better than the other.

For the purpose of illustration in Fig. 2.33, we have used a *low-pass signal* with its spectrum limited to the band $-W \leqslant f \leqslant W$ and centered at the origin. Nevertheless, it should be emphasized that the pre-envelope can be defined for any signal that possesses a spectrum.

2.14 BAND-PASS SIGNALS

We say that a signal $g(t)$ is a *band-pass signal* if its Fourier transform $G(f)$ is non-negligible only in a band of frequencies of total extent $2W$, say, centered about some frequency $\pm f_c$. This is illustrated in Fig. 2.34(*a*). We refer to f_c as the *carrier frequency*. In the majority of communication signals, we find that the bandwidth $2W$ is small compared with f_c, and so we refer to such a signal as a *narrow-band signal*. However, a precise statement about how small the bandwidth must be in order for the signal to be considered narrow-band is not necessary for our present discussion.

Let the pre-envelope of a narrow-band signal $g(t)$, with its Fourier transform $G(f)$ centered about some frequency $\pm f_c$, be expressed in the form

$$g_+(t) = \tilde{g}(t)\exp(j2\pi f_c t) \tag{2.229}$$

We refer to $\tilde{g}(t)$ as the *complex envelope* of the signal. Equation (2.229) may be

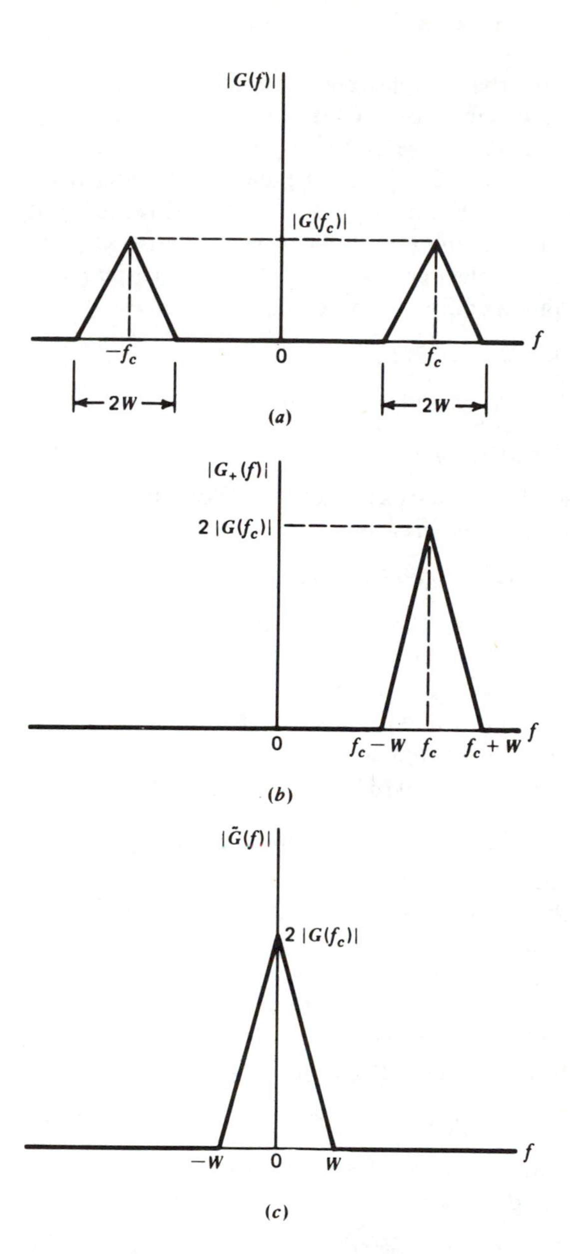

Figure 2.34 *(a)* Amplitude spectrum of band-pass signal $g(t)$. *(b)* Amplitude spectrum of pre-envelope $g_+(t)$. *(c)* Amplitude spectrum of complex envelope $\tilde{g}(t)$.

viewed as the basis of a definition for the complex envelope $\tilde{g}(t)$ in terms of the pre-envelope $g_+(t)$. We note that the spectrum of $g_+(t)$ is limited to the frequency band $f_c - W \leqslant f \leqslant f_c + W$, as in Fig. 2.34(*b*). Therefore, applying the frequency-shifting property of the Fourier transform to Eq. (2.229), we find that the spectrum of the complex envelope $\tilde{g}(t)$ is limited to the band $-W \leqslant f \leqslant W$ and centered at the origin, as in Fig. 2.34(*c*). That is, the complex envelope $\tilde{g}(t)$ is a low-pass signal.

By definition, the given signal $g(t)$ is the real part of the pre-envelope $g_+(t)$. We may thus express $g(t)$ in terms of the complex envelope $\tilde{g}(t)$ as follows

$$g(t) = \mathrm{Re}[\tilde{g}(t)\exp(j2\pi f_c t)] \tag{2.230}$$

In general, $\tilde{g}(t)$ is a complex-valued quantity, and so we may express it in the form

$$\tilde{g}(t) = g_c(t) + jg_s(t) \tag{2.231}$$

where $g_c(t)$ and $g_s(t)$ are both real-valued low-pass functions. We may therefore express the band-pass signal $g(t)$ in the *canonical form*:

$$g(t) = g_c(t)\cos(2\pi f_c t) - g_s(t)\sin(2\pi f_c t) \tag{2.232}$$

We refer to $g_c(t)$ as the *in-phase component* of the band-pass signal $g(t)$ and to $g_s(t)$ as the *quadrature component* of the signal, both with respect to the *carrier* $\cos(2\pi f_c t)$. The complex envelope $\tilde{g}(t)$ may thus be pictured as a *time-varying phasor* at the origin of the $g_c g_s$-plane. The end of the phasor moves about in the plane, and at the same time the plane rotates with an angular velocity equal to $2\pi f_c$ radians per second. The given signal $g(t)$ is the projection of this time-varying phasor on a fixed line.

Both $g_c(t)$ and $g_s(t)$ are limited to the band $-W \leqslant f \leqslant W$. Hence, except for scaling factors, they may be derived from the band-pass signal $g(t)$ by using the scheme shown in Fig. 2.35(*a*), where both low-pass filters are identical, each having a bandwidth equal to W (see Problem 2.62). To reconstruct $g(t)$ from its in-phase and quadrature components, we may use the scheme shown in Fig. 2.35(*b*). The two schemes shown in Fig. 2.35 are basic to the study of all *linear modulation systems*, as will be shown in subsequent chapters of the book.

Alternatively, we may express the complex envelope $\tilde{g}(t)$ in the form

$$\tilde{g}(t) = a(t)\exp[j\phi(t)] \tag{2.233}$$

where $a(t)$ and $\phi(t)$ are both real-valued low-pass functions. Based on this representation, the band-pass signal $g(t)$ is defined by

$$g(t) = a(t)\cos[2\pi f_c t + \phi(t)] \tag{2.234}$$

We refer to $a(t)$ as the *natural envelope* or the *envelope* of the band-pass signal $g(t)$ and to $\phi(t)$ as the *phase* of the signal.

It is therefore apparent that whether we represent the band-pass signal $g(t)$ in terms of its in-phase and quadrature components or in terms of its envelope and phase, the information content of the signal $g(t)$ is completely represented by the complex envelope $\tilde{g}(t)$. The particular virtue of using the complex envelope $\tilde{g}(t)$ to represent the band-pass signal is an analytical one, and will become evident in the next section.

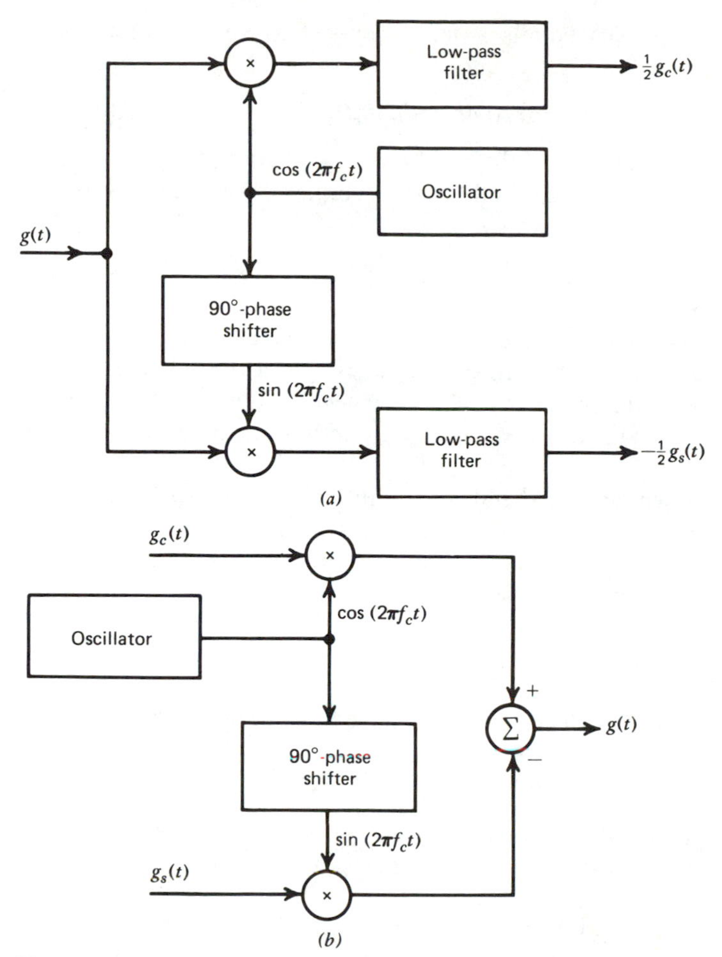

Figure 2.35 *(a)* Scheme for deriving the in-phase and quadrature components of a band-pass signal. *(b)* Scheme for reconstructing the band-pass signal from its in-phase and quadrature components.

The distinctions among the three different envelopes of a band-pass signal $g(t)$ should be carefully noted. We may summarize their definitions as follows:

1. The pre-envelope $g_+(t)$ is defined by

$$g_+(t) = g(t) + j\hat{g}(t)$$

where $\hat{g}(t)$ is the Hilbert transform of the signal $g(t)$. According to this representation, $\hat{g}(t)$ may be viewed as the quadratic function of $g(t)$. Correspondingly, in the frequency domain we have

$$G_+(f) = \begin{cases} 2G(f), & f > 0 \\ 0, & f < 0 \end{cases}$$

2. The complex envelope $\tilde{g}(t)$ equals a frequency-shifted version of the pre-envelope $g_+(t)$, as shown by

$$\tilde{g}(t) = g_+(t)\exp(-j2\pi f_c t)$$

where f_c is the carrier frequency of the band-pass signal $g(t)$.

3. The envelope $a(t)$ equals the magnitude of the complex envelope $\tilde{g}(t)$ and that of the pre-envelope $g_+(t)$, as shown by

$$a(t) = |\tilde{g}(t)| = |g_+(t)|$$

Note that for a band-pass signal $g(t)$, the pre-envelope $g_+(t)$ is a complex band-pass signal whose value depends on the carrier frequency f_c. On the other hand, the envelope $a(t)$ is always a real low-pass signal and, in general, the complex envelope $\tilde{g}(t)$ is a complex low-pass signal, whose values are independent of the choice of the carrier frequency f_c.

In Section 3.3 it is shown that the signal $a(t)$ results from envelope detection (i.e., rectification and low-pass filtering) of the band-pass signal $g(t)$. For this reason we call $a(t)$ the envelope of $g(t)$, and the complex signals $\tilde{g}(t)$ and $g_+(t)$ the complex envelope and pre-envelope of $g(t)$, respectively.

Example 25 RF Pulse (continued)

Suppose we wish to determine the different envelopes of the RF pulse defined by

$$g(t) = A \operatorname{rect}\left(\frac{t}{T}\right)\cos(2\pi f_c t)$$

We assume that $f_c T \gg 1$, so that the RF pulse $g(t)$ may be considered narrow-band.

From Example 9, we recall that the Fourier transform of $g(t)$ is given by

$$G(f) = \begin{cases} \dfrac{AT}{2}\operatorname{sinc}[T(f - f_c)], & f > 0 \\ \dfrac{AT}{2}\operatorname{sinc}[T(f + f_c)], & f < 0 \end{cases}$$

Taking the inverse Fourier transform of $G_+(f)$, we obtain the pre-envelope

$$g_+(t) = A \operatorname{rect}\left(\frac{t}{T}\right)\exp(j2\pi f_c t)$$

Correspondingly, the complex envelope equals

$$\tilde{g}(t) = A \operatorname{rect}\left(\frac{t}{T}\right)$$

and the envelope equals

$$a(t) = |\tilde{g}(t)| = A \operatorname{rect}\left(\frac{t}{T}\right) \tag{2.235}$$

which is intuitively satisfying. Note that in this example the complex envelope is real-valued and has the same value as the envelope.

2.15 BAND-PASS SYSTEMS

In this section we wish to develop a procedure for handling the analysis of band-pass systems. Specifically, we wish to show that the analysis of band-pass systems can be greatly simplified by establishing an analog (or, more precisely, an isomorphism) between low-pass and band-pass systems. This analogy is based on the use of the Hilbert transform for the representation of band-pass signals.

Consider first a narrow-band signal $x(t)$, with its Fourier transform denoted by $X(f)$. We assume that the spectrum of the signal $x(t)$ is limited to frequencies within $\pm W$ Hz of the carrier frequency f_c with $W < f_c$. Let this signal be represented in terms of its in-phase and quadrature components as follows:

$$x(t) = x_c(t)\cos(2\pi f_c t) - x_s(t)\sin(2\pi f_c t) \tag{2.236}$$

where $x_c(t)$ is the in-phase component and $x_s(t)$ is the quadrature component. Then, using $\tilde{x}(t)$ to denote the complex envelope of $x(t)$, we may write

$$\tilde{x}(t) = x_c(t) + jx_s(t) \tag{2.237}$$

Let the signal $x(t)$ be applied to a linear time-invariant band-pass system with impulse response $h(t)$ and transfer function $H(f)$. We assume that the frequency response of the system is limited to frequencies within $\pm B$ of the carrier frequency f_c. The system bandwidth $2B$ is usually narrower than or equal to the input signal bandwidth $2W$. We wish to represent the band-pass impulse response $h(t)$ in terms of two quadrature components, designated as $h_c(t)$ and $h_s(t)$. Thus, by analogy to the representation of band-pass signals, we may express $h(t)$ in the form

$$h(t) = 2h_c(t)\cos(2\pi f_c t) - 2h_s(t)\sin(2\pi f_c t) \tag{2.238}$$

where the factor 2 has been introduced for convenience in subsequent analysis. Defining the *complex impulse response* of the band-pass system as

$$\tilde{h}(t) = h_c(t) + jh_s(t), \tag{2.239}$$

we have the complex representation

$$h(t) = \text{Re}[2\tilde{h}(t)\exp(j2\pi f_c t)] \tag{2.240}$$

Note that $h_c(t)$, $h_s(t)$, and $\tilde{h}(t)$ are all low-pass functions limited to the frequency band $-B \leqslant f \leqslant B$.

We may determine the complex impulse response $\tilde{h}(t)$ in terms of the quadrature components $h_c(t)$ and $h_s(t)$ of the band-pass impulse response $h(t)$ by using Eq. (2.239). Alternatively, we may determine it from the band-pass transfer function $H(f)$ in the following way. We first note from Eq. (2.240) that

$$h(t) = \tilde{h}(t)\exp(j2\pi f_c t) + \tilde{h}^*(t)\exp(-j2\pi f_c t) \tag{2.241}$$

where $\tilde{h}^*(t)$ is the complex conjugate of $\tilde{h}(t)$. Therefore, applying the Fourier transform to Eq. (2.241), and using the complex-conjugation property of the Fourier transform, we get

$$H(f) = \tilde{H}(f - f_c) + \tilde{H}^*(-f - f_c) \tag{2.242}$$

where $\tilde{H}(f)$ is the Fourier transform of $\tilde{h}(t)$. Equation (2.242) satisfies the requirement that $H^*(f)=H(-f)$ for a real impulse response $h(t)$. Since $\tilde{H}(f)$ represents a low-pass transfer function limited to $|f|\leqslant B$, with $B<f_c$, we deduce from Eq. (2.242) that

$$\tilde{H}(f-f_c)=H(f), \qquad f>0 \tag{2.243}$$

Equation (2.243) indicates that for a specified band-pass transfer function $H(f)$, we may determine $\tilde{H}(f)$ by taking the part of $H(f)$ corresponding to positive frequencies, and then shifting it to the origin. To determine the complex impulse response $\tilde{h}(t)$, we take the inverse Fourier transform of $\tilde{H}(f)$, as shown by

$$\tilde{h}(t)=\int_{-\infty}^{\infty}\tilde{H}(f)\exp(j2\pi ft)df \tag{2.244}$$

The representations described above for band-pass signals and systems provide the basis of an efficient method for determining the output of a band-pass system driven by a band-pass signal. We assume that the spectrum of the input signal $x(t)$ and the transfer function $H(f)$ of the system are both centered around the same frequency f_c.* Let $y(t)$ denote the output signal of the system. It is clear that $y(t)$ is also a band-pass signal, so that we may represent it in terms of its low-pass complex envelope $\tilde{y}(t)$, as follows

$$y(t)=\mathrm{Re}[\tilde{y}(t)\exp(j2\pi f_c t)] \tag{2.245}$$

The output signal $y(t)$ is related to the input signal $x(t)$ and impulse response $h(t)$ of the system by the convolution integral

$$y(t)=\int_{-\infty}^{\infty}h(\tau)x(t-\tau)d\tau \tag{2.246}$$

In terms of pre-envelopes, we have $h(t)=2\,\mathrm{Re}[h_+(t)]$ and $x(t)=\mathrm{Re}[x_+(t)]$. We may therefore rewrite Eq. (2.246) in terms of the pre-envelopes $x_+(t)$ and $h_+(t)$ as follows

$$y(t)=2\int_{-\infty}^{\infty}\mathrm{Re}[h_+(\tau)]\mathrm{Re}[x_+(t-\tau)]d\tau \tag{2.247}$$

However, a basic property of pre-envelopes is described by the relation (see Problem 2.60)

$$\int_{-\infty}^{\infty}\mathrm{Re}[h_+(t)]\mathrm{Re}[x_+(t)]dt=\tfrac{1}{2}\mathrm{Re}\left[\int_{-\infty}^{\infty}h_+(t)x_+^*(t)dt\right] \tag{2.248}$$

Hence, applying this relation to Eq. (2.247), and then using the relationship

* In practice, there is no need to consider a situation in which the carrier frequency of the input signal is not aligned with the mid-band frequency of the band-pass system, since we have considerable freedom in choosing the carrier or mid-band frequency. Thus, changing the carrier frequency of the input signal by an amount Δf_c, say, simply corresponds to absorbing (or removing) the factor $\exp(\pm j2\pi\Delta f_c t)$ in the complex envelope of the input signal or the complex impulse response of the band-pass system.

between the pre-envelope and complex envelope of a band-pass function, we get

$$y(t) = \text{Re}\left[\int_{-\infty}^{\infty} h_{+}(\tau)x_{+}(t-\tau)d\tau\right]$$

$$= \text{Re}\left[\int_{-\infty}^{\infty} \tilde{h}(\tau)\exp(j2\pi f_c\tau)\tilde{x}(t-\tau)\exp(j2\pi f_c(t-\tau))d\tau\right]$$

$$= \text{Re}\left[\exp(j2\pi f_c t)\int_{-\infty}^{\infty} \tilde{h}(\tau)\tilde{x}(t-\tau)d\tau\right] \tag{2.249}$$

Thus, comparing the right-hand sides of Eqs. (2.245) and (2.249), we deduce that for a large enough carrier frequency f_c, the complex envelope $\tilde{y}(t)$ of the output signal is related to the complex envelope $\tilde{x}(t)$ of the input signal and the complex impulse response $\tilde{h}(t)$ of the band-pass system as follows

$$\tilde{y}(t) = \int_{-\infty}^{\infty} \tilde{h}(\tau)\tilde{x}(t-\tau)d\tau, \tag{2.250}$$

or, using the shorthand notation for convolution,

$$\tilde{y}(t) = \tilde{h}(t) \otimes \tilde{x}(t) \tag{2.251}$$

In other words, the complex envelope $\tilde{y}(t)$ of the output signal of a band-pass system is obtained by convolving the complex impulse response $\tilde{h}(t)$ of the system with the complex envelope $\tilde{x}(t)$ of the input band-pass signal. Equation (2.251) is the result of the isomorphism, for convolution, between band-pass functions and the corresponding low-pass functions. The scaling factor 2 was introduced in Eq. (2.238) so that Eq. (2.251) would have a familiar form.

The significance of this result is that in dealing with band-pass signals and systems, we need to deal only with the low-pass functions $\tilde{x}(t)$, $\tilde{y}(t)$, and $\tilde{h}(t)$, representing the excitation, the response, and the system, respectively. That is, the analysis of a band-pass system, which is complicated by the presence of the multiplying factor $\exp(j2\pi f_c t)$, is replaced by an equivalent but simpler low-pass analysis that completely retains the essence of the filtering process. This procedure is illustrated schematically in Fig. 2.36.

The complex envelope $\tilde{x}(t)$ of the input band-pass signal and the complex impulse response $\tilde{h}(t)$ of the band-pass system are defined in terms of their respective in-phase and quadrature components by Eqs. (2.237) and (2.239), respectively. Substituting these relations in Eq. (2.251), we get

$$\tilde{y}(t) = [h_c(t) + jh_s(t)] \otimes [x_c(t) + jx_s(t)] \tag{2.252}$$

Because convolution is distributive, we may rewrite Eq. (2.252) in the form

$$\tilde{y}(t) = [h_c(t) \otimes x_c(t) - h_s(t) \otimes x_s(t)] + j[h_s(t) \otimes x_c(t) + h_c(t) \otimes x_s(t)] \tag{2.253}$$

Defining the complex envelope $\tilde{y}(t)$ of the response in terms of its in-phase and quadrature components as

$$\tilde{y}(t) = y_c(t) + jy_s(t), \tag{2.254}$$

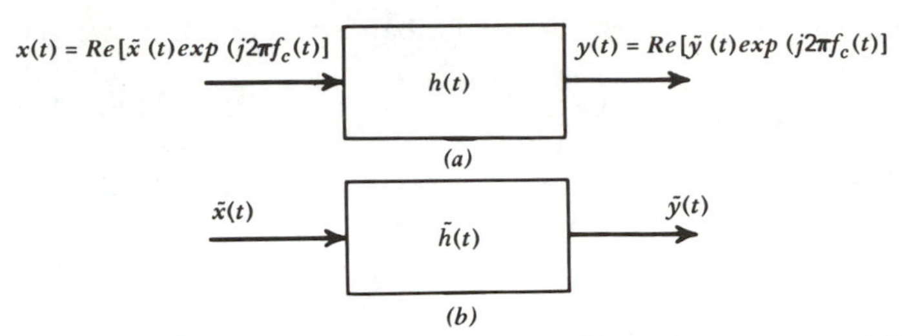

Figure 2.36 *(a)* Narrow-band filter of impulse response *h(t)* with narrow-band input signal *x(t)*. *(b)* Equivalent low-pass filter of complex impulse response $\tilde{h}(t)$ with complex low-pass input $\tilde{x}(t)$.

we have for the in-phase component $y_c(t)$ the relation

$$y_c(t) = h_c(t) \otimes x_c(t) - h_s(t) \otimes x_s(t), \tag{2.255}$$

and for the quadrature component $y_s(t)$ the relation

$$y_s(t) = h_s(t) \otimes x_c(t) + h_c(t) \otimes x_s(t) \tag{2.256}$$

Thus, for the purpose of evaluating the in-phase and quadrature components of the complex envelope $\tilde{y}(t)$ of the system output, we may use the *low-pass equivalent model* shown in Fig. 2.37. All the signals and impulse responses shown in

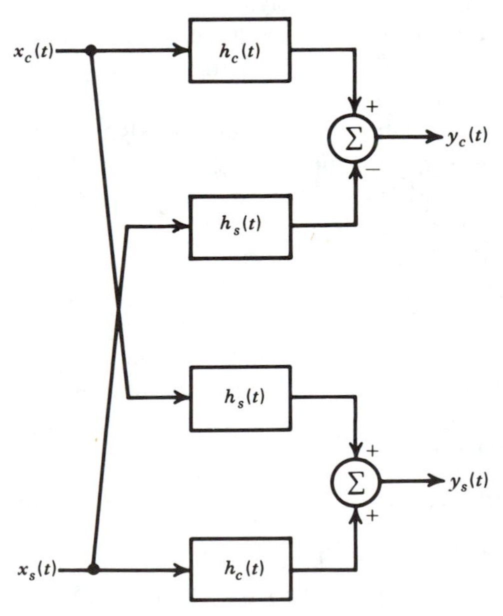

Figure 2.37 Block diagram illustrating the relationships between the in-phase and quadrature components of the response of a band-pass filter and those of the input signal.

this model are low-pass functions. Accordingly, this equivalent model provides a practical basis for the simulation of band-pass filters or communication channels on a digital computer. In carrying out this simulation, it is customary to use the fast Fourier transform algorithm (briefly discussed in Section 2.4) to perform the pertinent convolution operations on the computer, because it usually saves considerable computation time.

To sum up, the procedure for evaluating the response of a band-pass system to an input band-pass signal is as follows:

1. The input band-pass signal $x(t)$ is replaced by its complex envelope $\tilde{x}(t)$, which is related to $x(t)$ by
$$x(t) = \text{Re}[\tilde{x}(t)\exp(j2\pi f_c t)]$$
2. The band-pass system, with impulse response $h(t)$, is replaced by a low-pass analog, which is characterized by a complex impulse response $\tilde{h}(t)$ related to $h(t)$ by
$$h(t) = \text{Re}[2\tilde{h}(t)\exp(j2\pi f_c t)]$$
3. The complex envelope $\tilde{y}(t)$ of the output band-pass signal $y(t)$ is obtained by convolving $\tilde{h}(t)$ with $\tilde{x}(t)$, as shown by
$$\tilde{y}(t) = \tilde{h}(t) \otimes \tilde{x}(t)$$
4. The desired output $y(t)$ is derived from $\tilde{y}(t)$ by using the relation
$$y(t) = \text{Re}[\tilde{y}(t)\exp(j2\pi f_c t)]$$

Example 26 Response of an ideal band-pass filter to a pulsed RF wave

Consider an ideal band-pass filter of mid-band frequency f_c, the amplitude response of which is band-limited to $f_c - B \leqslant |f| \leqslant f_c + B$, as in Fig. 2.38(*a*), with $f_c > B$. We wish to determine the response of this filter to an RF pulse of duration T and frequency f_c defined by [see Fig. 2.39(*a*)].

$$x(t) = A \text{ rect}\left(\frac{t}{T}\right)\cos(2\pi f_c t) \tag{2.257}$$

where $f_c T \gg 1$.

Retaining the positive frequency part of the transfer function $H(f)$, defined in Fig. 2.38(*a*), and then shifting it to the origin, we find that the transfer function $\tilde{H}(f)$ of the low-pass equivalent filter is given by [see Fig. 2.38(*b*)]

$$\tilde{H}(f) = \begin{cases} \exp(-j2\pi f t_0), & -B < f < B \\ 0, & |f| > B \end{cases} \tag{2.258}$$

The complex impulse response in this example has only a real component, as shown by

$$\tilde{h}(t) = 2B \text{ sinc}[2B(t - t_0)] \tag{2.259}$$

From Example 25, we recall that the complex envelope $\tilde{x}(t)$ of the input RF pulse also has

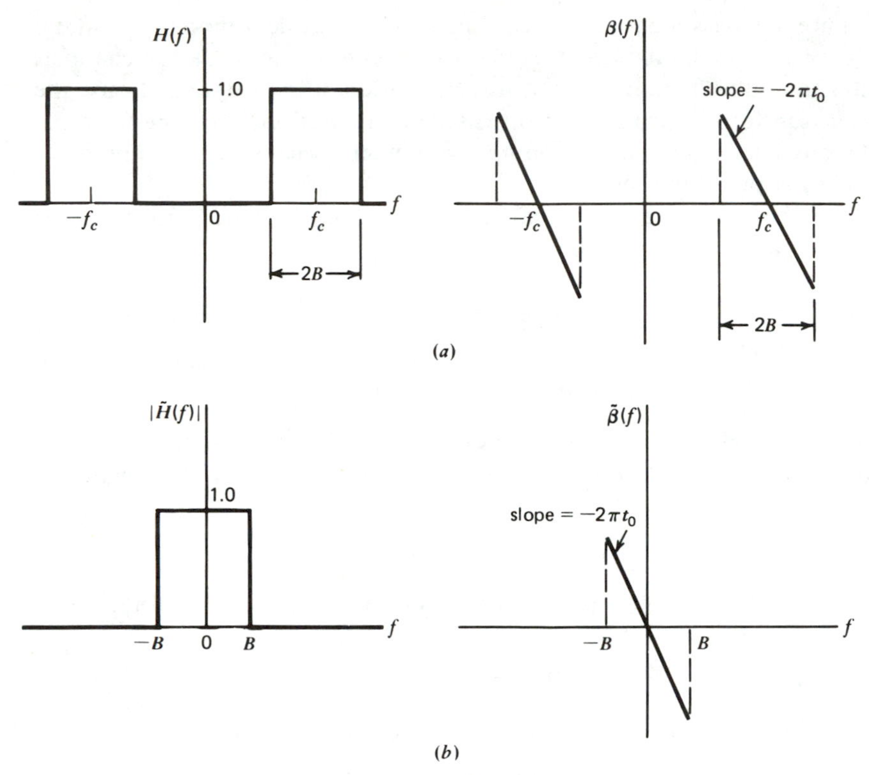

Figure 2.38 *(a)* Amplitude response *H(f)* and phase response *β(f)* of an ideal band-pass filter. *(b)* Corresponding components of complex transfer function $\tilde{H}(f)$.

only a real component, as shown by [see Fig. 2.39(*b*)]:

$$\tilde{x}(t) = A\,\mathrm{rect}\left(\frac{t}{T}\right) \tag{2.260}$$

The complex envelope $\tilde{y}(t)$ of the filter output is obtained by convolving the $\tilde{h}(t)$ of Eq. (2.259) with the $\tilde{x}(t)$ of Eq. (2.260). This convolution is exactly the same as the low-pass filtering operation that we studied in Example 22. Thus, using Eq. (2.211), we may write

$$\tilde{y}(t) = \frac{A}{\pi}\left\{\mathrm{Si}\left[2\pi B\left(t + \frac{T}{2} - t_0\right)\right] - \mathrm{Si}\left[2\pi B\left(t - \frac{T}{2} - t_0\right)\right]\right\} \tag{2.261}$$

As expected, the complex envelope $\tilde{y}(t)$ of the output has only a real component. Accordingly, from Eqs. (2.245) and (2.261), the output is obtained as

$$y(t) = \frac{A}{\pi}\left\{\mathrm{Si}\left[2\pi B\left(t + \frac{T}{2} - t_0\right)\right] - \mathrm{Si}\left[2\pi B\left(t - \frac{T}{2} - t_0\right)\right]\right\}\cos(2\pi f_c t) \tag{2.262}$$

which is the desired result. Equation (2.262) is shown sketched in Fig. 2.39(*c*) for the case when $B = 2/T$.

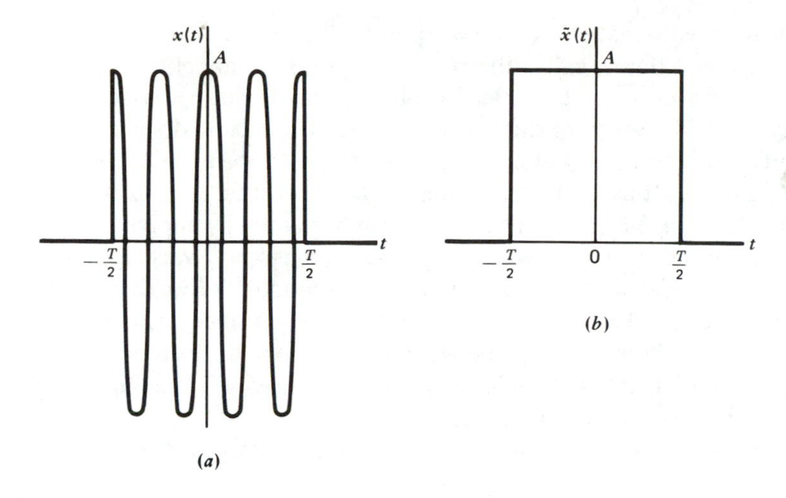

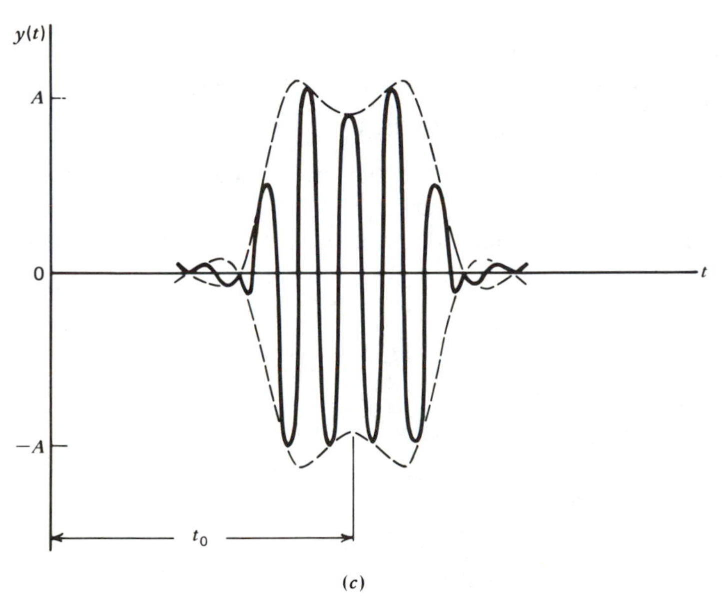

Figure 2.39 Illustrating the response of an ideal band-pass filter to RF pulse input. *(a)* RF pulse input $x(t)$. *(b)* Complex envelope $\tilde{x}(t)$ of RF pulse. *(c)* Response $y(t)$.

2.16 PHASE AND GROUP DELAY

Suppose a steady sinusoidal signal at frequency f_c is transmitted through a *dispersive* channel which has a total phase-shift of $\beta(f_c)$ radians at that frequency. By using two phasors to represent the input signal and the received signal, we see that the received signal phasor lags the input signal phasor by $\beta(f_c)$ radians. The time taken by the received signal phasor to sweep out this phase lag is simply equal to $\beta(f_c)/2\pi f_c$ seconds. This time is called the *phase delay* of the channel.

It is important, however, to realize that the phase delay is not necessarily the true signal delay. This follows from the fact that a steady sinusoidal signal does not carry information, and so it would be incorrect to deduce from the above reasoning that the phase delay is the true signal delay. In actual fact, as we will see in subsequent chapters, information can be transmitted only by applying some appropriate change to the sinusoidal wave. Suppose then a slowly varying signal is multiplied by a sinusoidal wave, so that the resulting modulated wave consists of a narrow group of frequencies. When this modulated wave is transmitted through the channel, we find that there is a delay between the envelope of the input signal and that of the received signal. This delay is called the *envelope* or *group delay* of the channel, and represents the true signal delay.

Assume that the dispersive channel is described by the transfer function

$$H(f)=K\exp[j\beta(f)] \tag{2.263}$$

where the amplitude K is a constant and the phase $\beta(f)$ is a nonlinear function of frequency. The input signal $x(t)$ consists of a narrow-band signal defined by

$$x(t)=x_c(t)\cos(2\pi f_c t)$$

where $x_c(t)$ is a low-pass function with its spectrum limited to the frequency interval $|f|\leqslant W$. We assume that $f_c \gg W$. By expanding the phase $\beta(f)$ in a Taylor series about the point $f=f_c$, and retaining only the first two terms, we may approximate $\beta(f)$ as

$$\beta(f)\simeq\beta(f_c)+(f-f_c)\left.\frac{\partial\beta(f)}{\partial f}\right|_{f=f_c} \tag{2.264}$$

Define

$$\tau_p=-\frac{\beta(f_c)}{2\pi f_c} \tag{2.265}$$

and

$$\tau_g=-\frac{1}{2\pi}\left.\frac{\partial\beta(f)}{\partial f}\right|_{f=f_c} \tag{2.266}$$

Then we may rewrite Eq. (2.264) in the form

$$\beta(f)\simeq-2\pi f_c\tau_p-2\pi(f-f_c)\tau_g \tag{2.267}$$

Correspondingly, the transfer function of the channel takes the form

$$H(f)\simeq K\exp[-j2\pi f_c\tau_p-j2\pi(f-f_c)\tau_g] \tag{2.268}$$

Following the procedure described in Section 2.15, in particular, using Eq. (2.243), we may replace the channel described by $H(f)$ by an equivalent low-pass filter whose transfer function equals

$$\tilde{H}(f)\simeq K\exp(-j2\pi f_c\tau_p-j2\pi f\tau_g) \tag{2.269}$$

Similarly, we may replace the input narrow-band signal $x(t)$ by its low-pass complex envelope $\tilde{x}(t)$ which equals

$$\tilde{x}(t)=x_c(t) \tag{2.270}$$

The Fourier transform of $\tilde{x}(t)$ is simply

$$\tilde{X}(f)=X_c(f) \tag{2.271}$$

where $X_c(f)$ is the Fourier transform of $x_c(t)$. Therefore, the Fourier transform of the complex envelope of the received signal is given by

$$\begin{aligned}\tilde{Y}(f)&=\tilde{H}(f)\tilde{X}(f)\\ &\simeq K\exp(-j2\pi f_c\tau_p)\exp(-j2\pi f\tau_g)X_c(f)\end{aligned} \tag{2.272}$$

We note that the multiplying factor $K\exp(-j2\pi f_c\tau_p)$ is a constant. We also note, from the time-shifting property of the Fourier transform, that the term $\exp(-j2\pi f\tau_g)X_c(f)$ represents the Fourier transform of the delayed signal $x_c(t-\tau_g)$. Accordingly, the complex envelope of the received signal equals

$$\tilde{y}(t)\simeq K\exp(-j2\pi f_c\tau_p)x_c(t-\tau_g) \tag{2.273}$$

Finally, we find that the received signal is itself given by

$$\begin{aligned}y(t)&=\text{Re}[\tilde{y}(t)\exp(j2\pi f_c t)]\\ &=Kx_c(t-\tau_g)\cos[2\pi f_c(t-\tau_p)]\end{aligned} \tag{2.274}$$

Equation (2.274) shows that, as a result of transmission through the channel, two delay effects occur:

1. The sinusoidal carrier wave $\cos(2\pi f_c t)$ is delayed by τ_p seconds; hence τ_p represents the phase delay. Sometimes, τ_p is also referred to as the *carrier delay*.
2. The envelope $x_c(t)$ is delayed by τ_g seconds; hence, τ_g represents the envelope or group delay. Note that τ_g is related to the slope of the phase $\beta(f)$, measured at $f=f_c$, as in Eq. (2.266).

Note also that when the phase response $\beta(f)$ is linear with frequency, and $\beta(0)=0$, the phase delay and group delay assume a common value. An example of this condition is illustrated in part (c) of Fig. 2.39.

Problems

Problem 2.1 A signal that is sometimes used in communication systems is a *raised cosine pulse*. Figure P2.1 shows a signal $g_p(t)$ that is a periodic sequence of these pulses with equal spacing between them. Show that the first three terms in the Fourier series expansion of $g_p(t)$ are as follows:

$$g_p(t)=\tfrac{1}{2}+\frac{8}{3\pi}\cos(\pi t)+\tfrac{1}{2}\cos(2\pi t)+\cdots$$

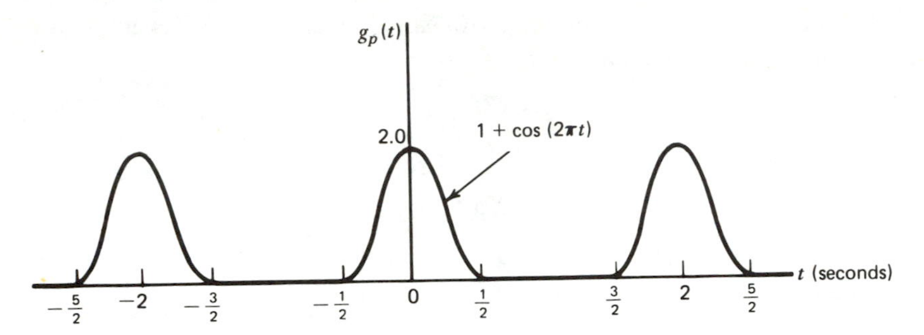

Figure P2.1

Problem 2.2 Evaluate the amplitude spectrum of the periodic pulsed RF waveform shown in Fig. P2.2, assuming that $f_c T_0 \gg 1$.

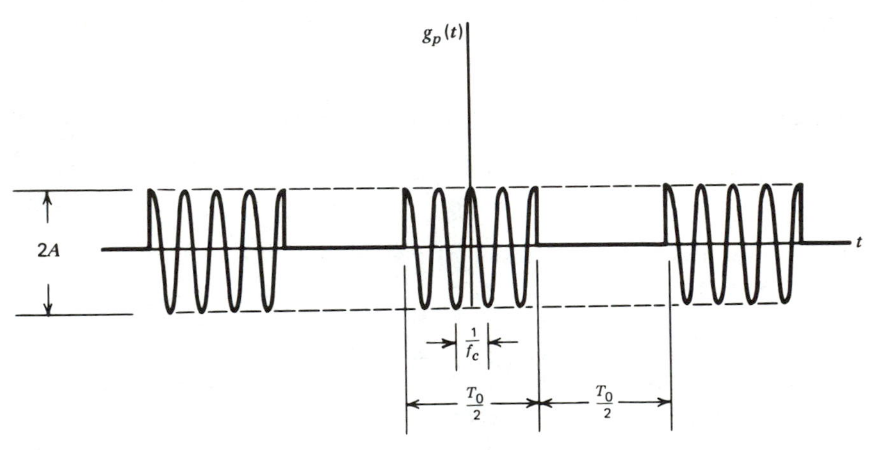

Figure P2.2

Problem 2.3 Prove the following properties of the Fourier series:

(a) If the periodic function $g_p(t)$ is even, that is,

$$g_p(-t) = g_p(t),$$

then the Fourier coefficient c_n is purely real and an even function of n.

(b) If $g_p(t)$ is odd, that is,

$$g_p(-t) = -g_p(t),$$

then c_n is purely imaginary and an odd function of n.

(c) If $g_p(t)$ has half-wave symmetry, that is,

$$g_p(t \pm \tfrac{1}{2}T_0) = -g_p(t),$$

where T_0 is the period of $g_p(t)$, then the Fourier series of such a signal consists of only odd-order terms.

Problem 2.4 Consider a set of n real-valued functions $\phi_1(t), \phi_2(t), \ldots, \phi_n(t)$ which are orthogonal to one another over an interval 0 to T; that is,

$$\int_0^T \phi_i(t)\phi_j(t)dt = 0, \qquad i \neq j$$

They are also normalized to have unit energy; that is,

$$\int_0^T \phi_j^2(t)dt = 1, \qquad j = 1, 2, \ldots, n$$

Let an arbitrary function $g(t)$ be approximated over an interval 0 to T by a linear combination of these n mutually orthogonal functions:

$$g(t) \simeq \sum_{k=1}^{n} g_k\phi_k(t)$$

Define the mean-square error of the approximation as

$$\mathscr{E} = \frac{1}{T}\int_0^T \left[g(t) - \sum_{k=1}^{n} g_k\phi_k(t)\right]^2 dt$$

(a) For the best approximation, that is, the one that will minimize the mean-square error $\mathscr{E}$, show that the coefficients of the expansion must be chosen as follows:

$$g_k = \int_0^T g(t)\phi_k(t)dt, \qquad k = 1, 2, \ldots, n$$

(b) Determine the minimum value of the mean-square error. What happens to this error as the number of terms n in the expansion is made infinity?

Problem 2.5 Determine the Fourier transform of the signal $g(t)$ consisting of three rectangular pulses, as shown in Fig. P2.3. Sketch the amplitude spectrum of this signal for the case when $T \ll T_0$.

Hint: Consider a rectangular pulse of amplitude A and duration T, and use the linearity and time-shifting properties of the Fourier transform.

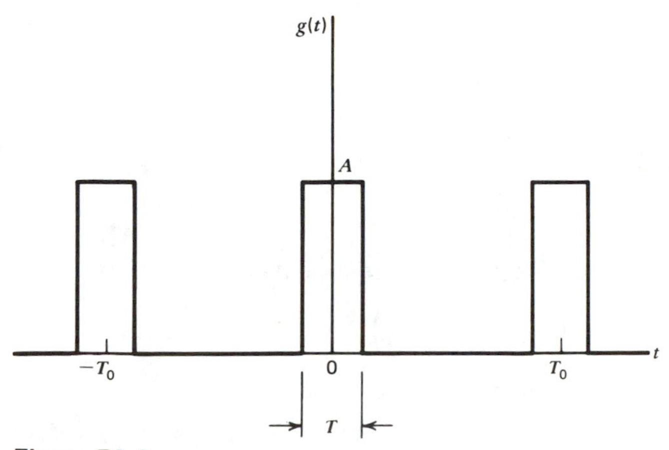

Figure P2.3

Problem 2.6

(a) Find the Fourier transform of the half-cosine pulse shown in Fig. P2.4(*a*).
(b) Apply the time-shifting property to the result obtained in part (a) to evaluate the spectrum of the half-sine pulse shown in Fig. P2.4(*b*).
(c) What is the spectrum of a half-sine pulse having a duration equal to aT?
(d) What is the spectrum of the negative half-sine pulse shown in Fig. P2.4(*c*)?
(e) Find the spectrum of the single sine pulse shown in Fig. P2.4(*d*).

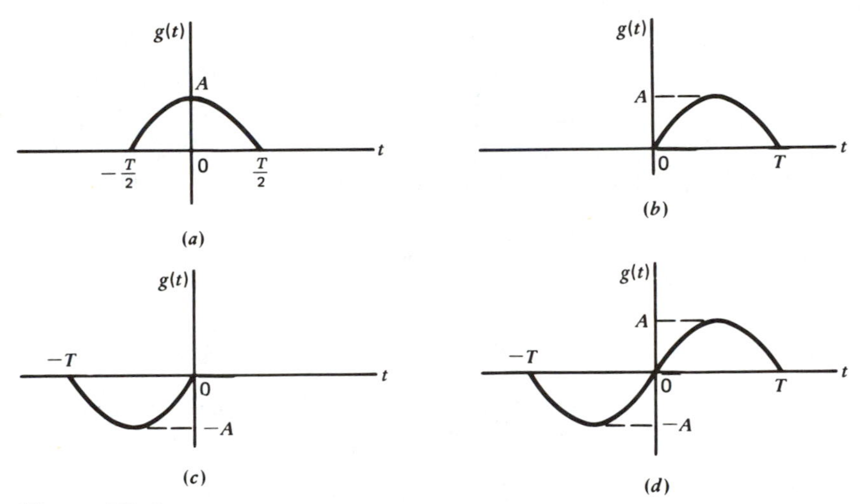

Figure P2.4

Problem 2.7 Any function $g(t)$ can be split unambiguously into an *even part* and an *odd part*, as shown by

$$g(t)=g_e(t)+g_o(t)$$

The even part is defined by

$$g_e(t)=\tfrac{1}{2}[g(t)+g(-t)]$$

and the odd part is defined by

$$g_o(t)=\tfrac{1}{2}[g(t)-g(-t)]$$

(a) Evaluate the even and odd parts of a rectangular pulse defined by

$$g(t)=A\ \mathrm{rect}\left(\frac{t}{T}-\frac{1}{2}\right)$$

(b) What are the Fourier transforms of these two parts of the pulse?

Problem 2.8 Determine the inverse Fourier transform of the frequency function $G(f)$ defined by the amplitude and phase spectra shown in Fig. P2.5.

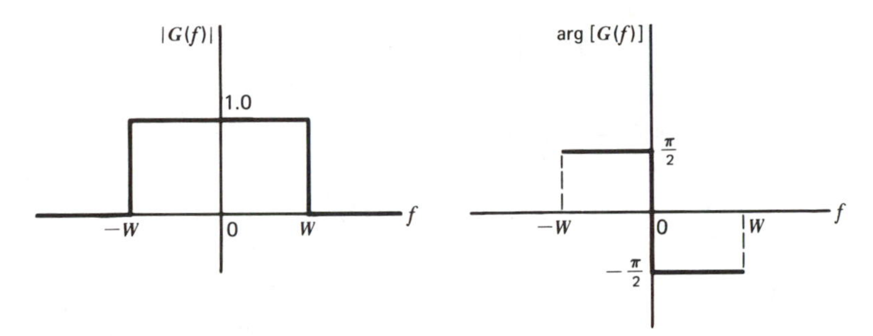

Figure P2.5

Problem 2.9 Assume the availability of a device that is capable of computing the Fourier transform of an energy signal $g(t)$ used as input. Explain the modifications that will have to be made to the input and output signals of such a device, so that it may also be used to compute the inverse Fourier transform of the quantity $G(f)$, where $g(t) \rightleftharpoons G(f)$.

Problem 2.10 A sinusoidal wave of amplitude A and frequency f_c is switched on abruptly for an odd number of cycles, denoted by N, and it is then switched off abruptly. Determine the spectrum of this signal for the following two cases:

(a) The sinusoidal wave is switched on and off at the zero-crossings.
(b) The sinusoidal wave is switched on and off when its amplitude is a maximum.

Sketch and compare the resulting amplitude spectra for $N=3$.

Problem 2.11 The following expression may be viewed as an approximate representation of a rectangular pulse with a finite rise time:

$$g(t)=\frac{1}{\tau}\int_{t-T}^{t+T} \exp\left(-\frac{\pi u^2}{\tau^2}\right)du$$

where it is assumed that $T \gg \tau$. Determine the Fourier transform of $g(t)$. What happens to this transform when we allow τ to become zero?

Hint: Express $g(t)$ as the superposition of two signals, one corresponding to integration from $t-T$ to 0, and the other from 0 to $t+T$.

Problem 2.12 The Fourier transform of a signal $g(t)$ is denoted by $G(f)$. Prove the following properties of the Fourier transform:

(a) If a real signal $g(t)$ is an even function of time t, the Fourier transform $G(f)$ is real. If a real signal $g(t)$ is an odd function of time t, $G(f)$ is imaginary.

(b)
$$\int_{-\infty}^{t} g(\tau)d(\tau) \rightleftharpoons \frac{1}{j2\pi f}G(f)+\frac{G(0)}{2}\delta(f)$$

Hint: Express the integral $\int_{-\infty}^{t} g(\tau)d\tau$ as the convolution of $g(t)$ and the unit step function $u(t)$.

(c)
$$t^n g(t) \rightleftharpoons \left(\frac{j}{2\pi}\right)^n G^{(n)}(f)$$

where $G^{(n)}(f)$ is the nth derivative of $G(f)$ with respect to f.

(d)
$$\int_{-\infty}^{\infty} t^n g(t)dt = \left(\frac{j}{2\pi}\right)^n G^{(n)}(0)$$

(e)
$$g_1(t)g_2^*(t) \rightleftharpoons \int_{-\infty}^{\infty} G_1(\lambda)G_2^*(\lambda - f)d\lambda$$

(f)
$$\int_{-\infty}^{\infty} g_1(t)g_2^*(t)dt = \int_{-\infty}^{\infty} G_1(f)G_2^*(f)df$$

Problem 2.13

(a) Show that the Fourier transform $G(f)$ of a signal $g(t)$ is bounded by the following inequalities:

$$|G(f)| \leqslant \int_{-\infty}^{\infty} |g(t)|dt,$$

$$|j2\pi f\, G(f)| \leqslant \int_{-\infty}^{\infty} \left|\frac{dg(t)}{dt}\right| dt,$$

and

$$|(j2\pi f)^2 G(f)| \leqslant \int_{-\infty}^{\infty} \left|\frac{d^2 g(t)}{dt^2}\right| dt$$

where it is assumed that the first and second derivatives of $g(t)$ exist.

Hint: Use a special form of Schwarz's inequality given by Eq. (A5.16) of Appendix 5.

(b) Construct these three bounds for the triangular pulse shown in Fig. P2.6 and compare your results with the actual amplitude spectrum of the pulse.

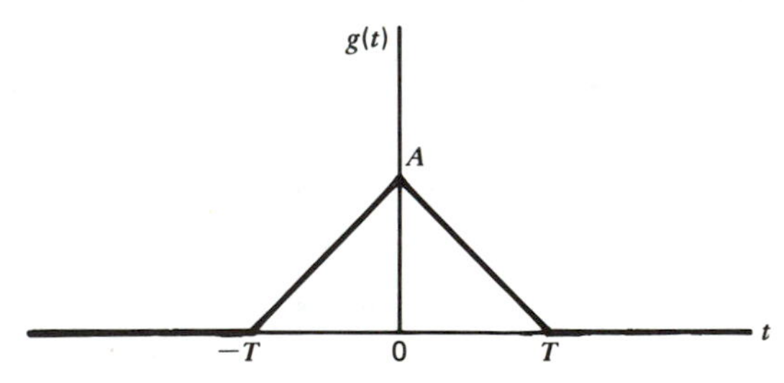

Figure P2.6

Problem 2.14 Prove the following properties of the convolution process:

(a) The commutative property:

$$g_1(t) \otimes g_2(t) = g_2(t) \otimes g_1(t)$$

(b) The associative property:

$$g_1(t) \otimes [g_2(t) \otimes g_3(t)] = [g_1(t) \otimes g_2(t)] \otimes g_3(t)$$

(c) The distributive property:

$$g_1(t) \otimes [g_2(t) + g_3(t)] = g_1(t) \otimes g_2(t) + g_1(t) \otimes g_3(t)$$

Problem 2.15 A signal $g_1(t)$ is defined by

$$g_1(t)=\exp(-\alpha t)u(t)$$

where $\alpha>0$.

(a) Find the function $g_2(t)$ obtained by convolving $g_1(t)$ with itself.
(b) Find the Fourier transform of $g_2(t)$.

Problem 2.16 A signal $x(t)$ of finite energy is applied to a square-law device whose output $y(t)$ is defined by

$$y(t)=x^2(t)$$

The spectrum of $x(t)$ is limited to the frequency interval $-W\leqslant f\leqslant W$. Hence, show that the spectrum of $y(t)$ is limited to $-2W\leqslant f\leqslant 2W$.

Hint: Express $y(t)$ as the product of $x(t)$ by $x(t)$.

Problem 2.17 Explain why it is not possible for a signal that is strictly limited to a band of frequencies to be simultaneously limited to a finite time duration, and vice versa. Use specific examples to illustrate your explanation.

Problem 2.18 Evaluate the discrete Fourier transform $\{G(kF_s)\}$ of a periodic sequence of samples $\{g(nT_s)\}$ defined by (for one period):

$$g(nT_s)=a^{-nT_s}, \qquad 0\leqslant n\leqslant N-1$$

Plot the amplitude spectrum $|G(kF_s)|$ and phase spectrum $\arg[G(kF_s)]$ for $N=8$ and $a=2$.

Problem 2.19 Evaluate the discrete Fourier transform of a sequence of samples $\{g(nT_s)\}$ defined by (for one period):

$$g(nT_s)=\begin{cases}1, & 0\leqslant n\leqslant \dfrac{N}{2}-1\\[2ex] 0, & \dfrac{N}{2}\leqslant n\leqslant N-1\end{cases}$$

Plot the amplitude spectrum $|G(kF_s)|$ and phase spectrum $\arg[G(kF_s)]$ for $N=10$.

Problem 2.20 Consider a sequence of samples denoted by $g(0), g(T_s), g(2T_s), \ldots, g(NT_s-T_s)$. The discrete Fourier transform of this sequence is defined by

$$G(kF_s)=T_s\sum_{n=0}^{N-1} g(nT_s)\exp\left(-j\frac{2\pi}{N}kn\right), \qquad n=0, 1, 2, \ldots, N-1$$

The inverse discrete Fourier transform is defined by

$$g(nT_s)=\frac{1}{NT_s}\sum_{k=0}^{N-1} G(kF_s)\exp\left(j\frac{2\pi}{N}kn\right), \qquad k=0, 1, 2, \ldots, N-1$$

The T_s and F_s are related by

$$T_sF_s=\frac{1}{N}$$

Prove the following properties of the discrete Fourier transform:

(a) Both $\{g(nT_s)\}$ and $\{G(kF_s)\}$ are periodic with a period equal to N samples.
(b) If $\{G(kF_s)\}$ is the discrete Fourier transform of $\{g(nT_s)\}$, then the discrete Fourier transform of $\{g(nT_s - n_0 T_s)\}$ is $\{G(kF_s)\exp[-j(2\pi/N)n_0 k]\}$.
(c) Let $\{g_1(nT_s)\}$ and $\{g_2(nT_s)\}$ be two periodic sequences of period N, with their discrete Fourier transforms denoted by $\{G_1(kF_s)\}$ and $\{G_2(kF_s)\}$. The *periodic convolution* of $g_1(nT_s)$ and $g_2(nT_s)$ is defined by

$$g_{12}(nT_s) = T_s \sum_{l=0}^{N-1} g_1(lT_s) g_2(nT_s - lT_s)$$

Then,

$$G_{12}(kF_s) = G_1(kF_s) G_2(kF_s)$$

where $\{G_{12}(kF_s)\}$ is the discrete Fourier transform of $\{g_{12}(nT_s)\}$.

Problem 2.21 Prove the following properties of the delta function $\delta(t)$:

(a) The delta function is an even function of time, that is,

$$\delta(t) = \delta(-t)$$

(b) The effect of scaling the argument of the delta function by a constant a may be described as follows

$$\delta(at) = \frac{1}{|a|}\delta(t)$$

(c) The effect of multiplying the delta function by a time function $g(t)$ may be described by

$$g(t)\delta(t - t_0) = g(t_0)\delta(t - t_0)$$

Problem 2.22 Evaluate the Fourier transform of the delta function by considering it as the limiting form of: (1) a rectangular pulse of unit area, and (2) a sinc pulse of unit area.

Problem 2.23 The Fourier transform $G(f)$ of a signal $g(t)$ is defined by

$$G(f) = \begin{cases} 1, & f > 0 \\ \frac{1}{2}, & f = 0 \\ 0, & f < 0 \end{cases}$$

Determine $g(t)$.

Problem 2.24 Consider a pulse-like function $g(t)$ that consists of a small number of straight-line segments. Suppose that this function is differentiated with respect to time t twice so as to generate a sequence of weighted delta functions, as shown by

$$\frac{d^2 g(t)}{dt^2} = \sum_i k_i \delta(t - t_i)$$

where the k_i are related to the slopes of the straight-line segments.

(a) Given the values of the k_i and the t_i, show that the Fourier transform of $g(t)$ is given by

$$G(f) = -\frac{1}{4\pi^2 f^2} \sum_i k_i \exp(-j2\pi f t_i)$$

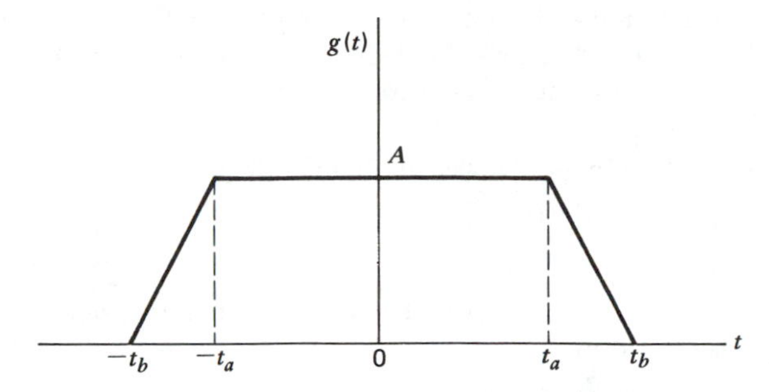

Figure P2.7

(b) Using this procedure, show that the Fourier transform of the trapezoidal pulse shown in Fig. P2.7 is given by

$$G(f)=\frac{A}{\pi^2 f^2(t_b-t_a)}\sin[\pi f(t_b-t_a)]\sin[\pi f(t_b+t_a)]$$

Problem 2.25 Using the procedure described in part (a) of Problem 2.24, show that the periodic train of triangular pulses depicted in Fig. P2.8 may be expanded in the form of a complex Fourier series as follows:

$$g_p(t)=\frac{AT}{T_0}\sum_{n=-\infty}^{\infty}\operatorname{sinc}^2\left(\frac{nT}{T_0}\right)\exp\left(\frac{j2\pi nt}{T_0}\right)$$

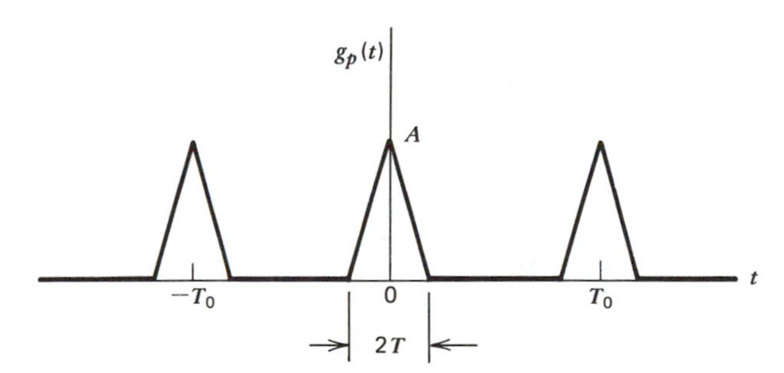

Figure P2.8

Problem 2.26 Evaluate the power spectral density of the periodic pulsed RF waveform shown in Fig. P2.2.

Problem 2.27 Show that the two different pulses defined in parts (*a*) and (*b*) of Fig. P2.4 have the same energy spectral density:

$$\Psi_g(f)=\frac{4A^2T^2\cos^2(\pi Tf)}{\pi^2(4T^2f^2-1)^2}$$

Problem 2.28 The *duration* of a signal provides a measure for describing the signal as a function of time. The *bandwidth* of the signal provides a measure for describing its frequency content. There is no unique set of definitions for the duration and bandwidth. However, regardless of the definitions used for the duration and bandwidth of a signal, we find that their product is always a constant. The choice of a particular set of definitions merely changes the value of this constant. Problems 2.28 and 2.29 are intended to explore these issues.

(a) The *root mean-square (rms) bandwidth* of a low-pass signal $g(t)$ of finite energy is defined by

$$W_{\mathrm{rms}} = \left[\frac{\int_{-\infty}^{\infty} f^2 |G(f)|^2 \, df}{\int_{-\infty}^{\infty} |G(f)|^2 \, df}\right]^{1/2}$$

where $|G(f)|^2$ is the energy spectral density of the signal. The corresponding *root mean-square (rms) duration* of the signal is defined by

$$T_{\mathrm{rms}} = \left[\frac{\int_{-\infty}^{\infty} t^2 |g(t)|^2 \, dt}{\int_{-\infty}^{\infty} |g(t)|^2 \, dt}\right]^{1/2}$$

Using these definitions, show that

$$T_{\mathrm{rms}} W_{\mathrm{rms}} \geqslant \frac{1}{4\pi}$$

Assume that $|g(t)| \to 0$ faster than $1/\sqrt{|t|}$ as $|t| \to \infty$.

(b) Consider a Gaussian pulse defined by

$$g(t) = \exp(-\pi t^2)$$

Show that, for this signal, the equality

$$T_{\mathrm{rms}} W_{\mathrm{rms}} = \frac{1}{4\pi}$$

can be reached.

Hint: Use Schwarz's inequality [see Eq. (A5.9) of Appendix 5]

$$\left\{\int_{-\infty}^{\infty} [g_1^*(t) g_2(t) + g_1(t) g_2^*(t)] dt\right\}^2 \leqslant 4 \int_{-\infty}^{\infty} |g_1(t)|^2 \, dt \int_{-\infty}^{\infty} |g_2(t)|^2 \, dt$$

in which we have

$$g_1(t) = t g(t)$$

and

$$g_2(t) = \frac{dg(t)}{dt}$$

Problem 2.29 Consider an energy signal $g(t)$ with an amplitude spectrum $|G(f)|$ that is defined for all frequencies from $-\infty$ to ∞, symmetric about $f=0$, and with its maximum value at $f=0$. The *equivalent rectangular bandwidth* of this signal is defined by

$$W_{\mathrm{eq}} = \frac{\int_{-\infty}^{\infty} |G(f)|^2 \, df}{2|G(0)|^2}$$

The corresponding *equivalent duration* of the signal is defined by

$$T_{\mathrm{eq}} = \frac{[\int_{-\infty}^{\infty} |g(t)| dt]^2}{\int_{-\infty}^{\infty} |g(t)|^2 \, dt}$$

(a) Show that the *time–bandwidth product* of the signal satisfies the condition: $T_{\mathrm{eq}} W_{\mathrm{eq}} \geqslant 0.5$.

(b) Show that, for a rectangular pulse of duration T centered about the origin, this condition holds with the equality sign.

Problem 2.30 Determine and sketch the autocorrelation functions of the following exponential pulses:

(a) $g(t)=\exp(-at)u(t)$
(b) $g(t)=\exp(-a|t|)$
(c) $g(t)=\exp(-at)u(t)-\exp(at)u(-t)$

Problem 2.31 Determine and sketch the autocorrelation function of a Gaussian pulse defined by

$$g(t)=\frac{1}{t_0}\exp\left(-\frac{\pi t^2}{t_0^2}\right),$$

Problem 2.32 Determine the autocorrelation function of the sinc pulse $A\,\text{sinc}(2Wt)$, and sketch it.

Problem 2.33 The Fourier transform of a signal is defined by $|\text{sinc}(f)|$. Show that the autocorrelation function of this signal is triangular in form.

Problem 2.34 Specify two distinctly different pulse signals that have exactly the same autocorrelation function.

Problem 2.35 Consider a signal $g(t)$ defined by

$$g(t)=A_0+A_1\cos(2\pi f_1 t+\theta)+A_2\cos(2\pi f_2 t+\theta)$$

(a) Determine the autocorrelation function $R_g(\tau)$ of this signal.
(b) What is the value of $R_g(0)$?
(c) Has any information about $g(t)$ been lost in obtaining the autocorrelation function?

Problem 2.36 Let $G(f)$ denote the Fourier transform of a real-valued signal $g(t)$, and $R_g(\tau)$ its autocorrelation function. Show that

$$\int_{-\infty}^{\infty}\left[\frac{dR_g(\tau)}{d\tau}\right]^2 d\tau=4\pi^2\int_{-\infty}^{\infty} f^2|G(f)|^4\,df$$

Problem 2.37 Consider an energy signal $g(t)$ with its autocorrelation function denoted by $R_g(\tau)$. Using Schwarz's inequality, show that

$$|R_g(\tau)|\leqslant R_g(0)$$

For a definition of Schwarz's inequality, see Appendix 5.

Problem 2.38 A signal $g(t)$ has its Fourier transform defined by $G(f)$, energy spectral density by $\Psi_g(f)$, and autocorrelation function by $R_g(\tau)$. The operations of spectral analysis involving this signal may be described in symbolic form as follows:

$$\begin{array}{ccc} g(t) & \rightleftharpoons & G(f) \\ \downarrow & & \downarrow \\ R_g(\tau) & \rightleftharpoons & \Psi_g(f) \end{array}$$

(a) Explain why the operations indicated by the vertical paths are irreversible.

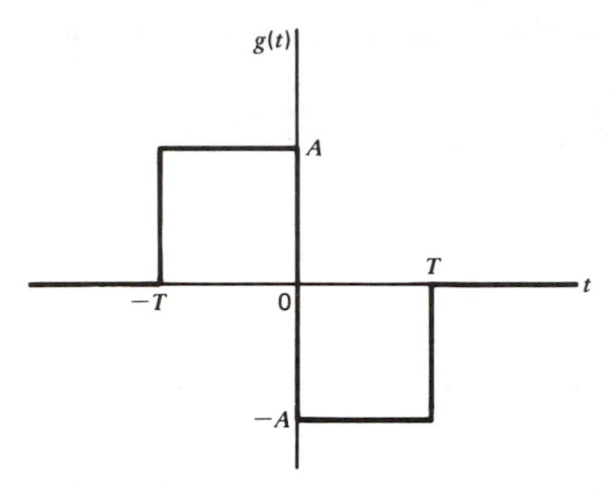

Figure P2.9

(b) For the doublet pulse shown in Fig. P2.9, determine and sketch the energy spectral density $\Psi_g(f)$ using the route $g(t) \rightarrow G(f) \rightarrow \Psi_g(f)$.

(c) For this pulse, show that the alternative route $g(t) \rightarrow R_g(\tau) \rightarrow \Psi_g(f)$ yields exactly the same result as in part (b).

Problem 2.39

(a) Let $R_g(\tau)$ denote the autocorrelation function of an energy signal $g(t)$. Show that

$$R_g^{(m+n)}(\tau) = (-1)^n \int_{-\infty}^{\infty} g^{(m)}(t)[g^{(n)}(t-\tau)]^* \, dt$$

where $g^{(m)}(t)$ denotes the mth derivative; similarly, for $g^{(n)}(t)$ and $R_g^{(m+n)}(\tau)$. The asterisk denotes complex conjugation.

(b) Use this relation to evaluate the autocorrelation function of the signal shown in Fig. P2.10.

Hint: Use $m=1$ and $n=0$ to evaluate $R_g^{(1)}(\tau)$, and then integrate with respect to τ.

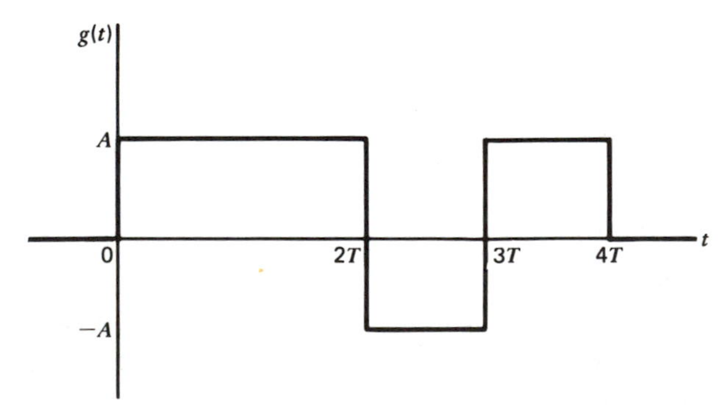

Figure 2.10

Problem 2.40

(a) Use the relation given in part (a) of Problem 2.39 to evaluate the autocorrelation function of the triplet pulse shown in Fig. P2.11. Sketch its dependence on the delay variable τ.

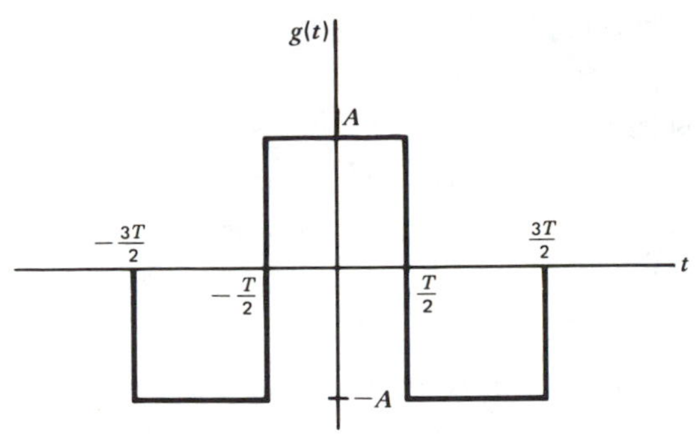

Figure 2.11

(b) Check your result in part (a), using the formula for the autocorrelation function of an energy signal.

Problem 2.41 Determine the cross-correlation function $R_{12}(\tau)$ of the pair of rectangular pulses shown in Fig. P2.12, and sketch it. What is the value of $R_{21}(\tau)$?

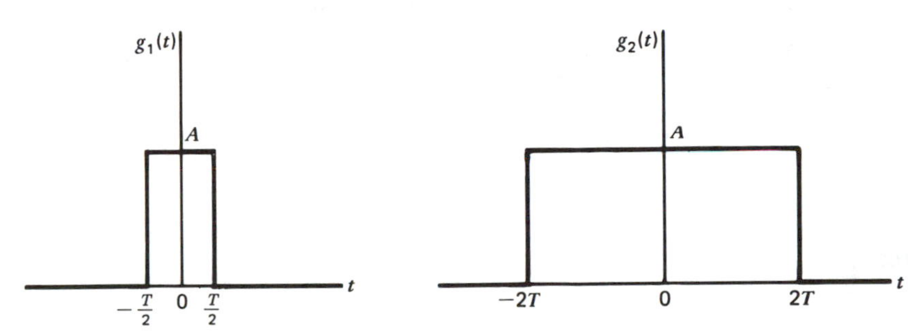

Figure P2.12

Problem 2.42 Determine the cross-correlation function $R_{12}(\tau)$ of the rectangular pulse $g_1(t)$ and triplet pulse $g_2(t)$ shown in Fig. P2.13, and sketch it. What is the value of $R_{21}(\tau)$? Are these signals orthogonal? Why?

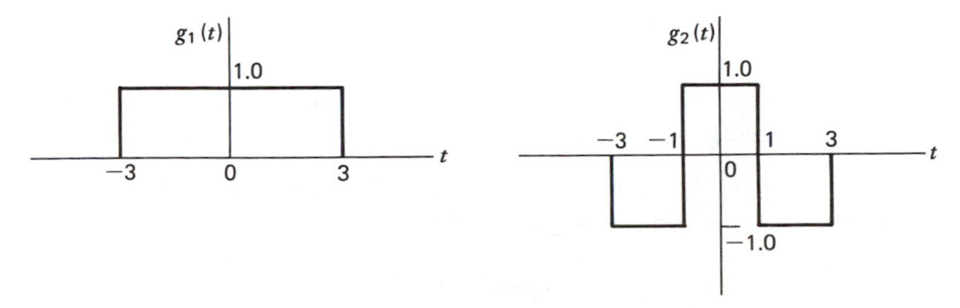

Figure P2.13

Problem 2.43 Determine the cross-correlation function $R_{12}(\tau)$ of the two signals $g_1(t)$ and $g_2(t)$ defined by

$$g_1(t)=\begin{cases} A\cos(2\pi f_1 t+\theta_1), & 0\leqslant t\leqslant T \\ 0, & \text{elsewhere} \end{cases}$$

$$g_2(t)=\begin{cases} A\cos(2\pi f_2 t+\theta_2), & 0\leqslant t\leqslant T \\ 0, & \text{elsewhere} \end{cases}$$

How does varying the frequency difference $|f_1-f_2|$ affect this cross-correlation function? Assume that $|f_1-f_2|$ is small compared with $(f_1+f_2)/2$.

Problem 2.44

(a) Let $R_{12}(\tau)$ denote the cross-correlation function of two energy signals $g_1(t)$ and $g_2(t)$. Show that

$$R_{12}^{(m+n)}(\tau)=(-1)^n\int_{-\infty}^{\infty} g_1^{(m)}(t)[g_2^{(n)}(t-\tau)]^*\,dt$$

where $g_1^{(m)}(t)$ denotes the mth derivative of $g_1(t)$; similarly, for $g_2^n(t)$ and $R_{12}^{(m+n)}(\tau)$.

(b) Use this relation to evaluate the cross-correlation function $R_{12}(\tau)$ of the rectangular pulse $g_1(t)$ and triplet pulse $g_2(t)$ shown in Fig. P2.13.

Hint: Use $m=1$ and $n=0$ to evaluate $R_{12}^{(1)}(\tau)$, and then integrate with respect to τ.

Problem 2.45 Let $R_{12}(\tau)$ and $R_{21}(\tau)$ denote the cross-correlation functions of the energy signals $g_1(t)$ and $g_2(t)$.

(a) Show that the total area under $R_{12}(\tau)$ is defined by

$$\int_{-\infty}^{\infty} R_{12}(\tau)d\tau=\left[\int_{-\infty}^{\infty} g_1(t)dt\right]\left[\int_{-\infty}^{\infty} g_2(t)dt\right]^*$$

(b) Show that

$$R_{12}(\tau)=R_{21}^*(-\tau)$$

Problem 2.46 Consider two energy signals $g_1(t)$ and $g_2(t)$, which may be complex-valued. Show that

$$\int_{-\infty}^{\infty} |g_1(t)-g_2(t)|^2\,dt=R_{11}(0)+R_{22}(0)-2\,\mathrm{Re}[R_{12}(0)]$$

where $R_{11}(\tau)$ and $R_{22}(\tau)$ are the autocorrelation functions of $g_1(t)$ and $g_2(t)$, respectively, and $R_{12}(\tau)$ is their cross-correlation function.

Problem 2.47 Consider two signals $g_1(t)$ and $g_2(t)$. These two signals are delayed by amounts equal to t_1 and t_2 seconds, respectively. Show that the time delays are additive in convolving the pair of delayed signals, whereas they are subtractive in cross-correlating them.

Problem 2.48 Consider two periodic signals $g_{p1}(t)$ and $g_{p2}(t)$, both of period T_0. Show that the cross-correlation function $R_{12}(\tau)$ of this pair of signals satisfies the Fourier Transform pair:

$$R_{12}(\tau)\rightleftharpoons\frac{1}{T_0^2}\sum_{n=-\infty}^{\infty} G_1\left(\frac{n}{T_0}\right)G_2^*\left(\frac{n}{T_0}\right)\delta\left(f-\frac{n}{T_0}\right)$$

where $G_1(n/T_0)$ and $G_2(n/T_0)$ are the Fourier transforms of the two generating functions for the periodic signals $g_{p1}(t)$ and $g_{p2}(t)$, respectively, evaluated at the frequency $f=n/T_0$.

Problem 2.49 A periodic signal $x_p(t)$ of period T_0 is applied to a linear time-invariant filter of impulse response $h(t)$. Use the complex Fourier series representation of $x_p(t)$ and the convolution integral to evaluate the response of the filter.

Problem 2.50 Let $x(t)$ and $y(t)$ be the input and output signals of a linear time-invariant filter. Using Rayleigh's energy theorem, show that if the filter is stable and the input signal $x(t)$ has finite energy, then the output signal $y(t)$ also has finite energy. That is, given that

$$\int_{-\infty}^{\infty} |x(t)|^2 \, dt < \infty,$$

then show that

$$\int_{-\infty}^{\infty} |y(t)|^2 \, dt < \infty$$

Problem 2.51 Evaluate the transfer function of a linear system represented by the block diagram shown in Fig. P2.14.

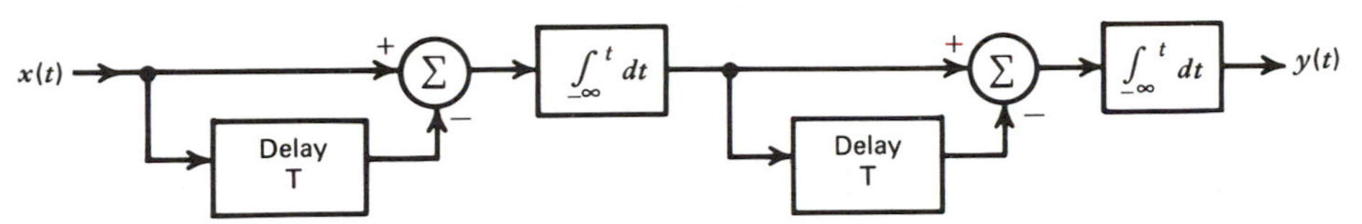

Figure P2.14

Problem 2.52

(a) Determine the overall amplitude response of the cascade connection shown in P2.15, consisting of N identical stages, each with a time constant RC equal to τ_0.
(b) Show that as N approaches infinity, the amplitude response of the cascade connection approaches the Gaussian function $\exp(-\frac{1}{2}f^2T^2)$, where for each value of N, the time constant τ_0 is selected so that

$$\tau_0^2 = \frac{T^2}{4\pi^2 N}$$

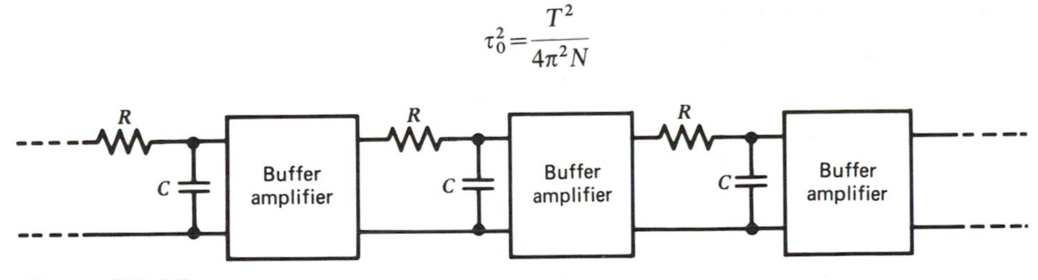

Figure P2.15

Problem 2.53 Suppose that, for a given signal $x(t)$, the integrated value of the signal over an interval T is required, as shown by

$$y(t) = \int_{t-T}^{t} x(\tau) d\tau$$

(a) Show that $y(t)$ can be obtained by transmitting the input signal $x(t)$ through a filter with its transfer function given by

$$H(f) = T \operatorname{sinc}(fT) \exp(-j\pi fT)$$

(b) An adequate approximation to this transfer function is obtained by using a low-pass filter with a bandwidth equal to $1/T$, passband amplitude response T, and delay $T/2$. Assuming this low-pass filter to be ideal, determine the filter output at time $t=T$ due to a unit step function applied to the filter at $t=0$. and compare the result with the corresponding output of the ideal integrator. Note that $\mathrm{Si}(\pi)=1.85$ and $\mathrm{Si}(\infty)=\pi/2$.

Problem 2.54

(a) Consider a signal $g(t)$ limited to the band $-B \leqslant f \leqslant B$. This signal is applied to a low-pass filter with an amplitude response $|H(f)|$ and linear phase, as in Fig. P2.16(a). Determine the resulting filter output.

(b) Suppose that the filter has a constant amplitude response but nonlinear phase, as in Fig. P2.16(b). Determine the resulting output. Assume that the constant b_1 is small enough to justify using the approximation:

$$\exp\left[jb_1 \sin\left(\frac{\pi f}{B}\right)\right] \simeq 1 + jb_1 \sin\left(\frac{\pi f}{B}\right)$$

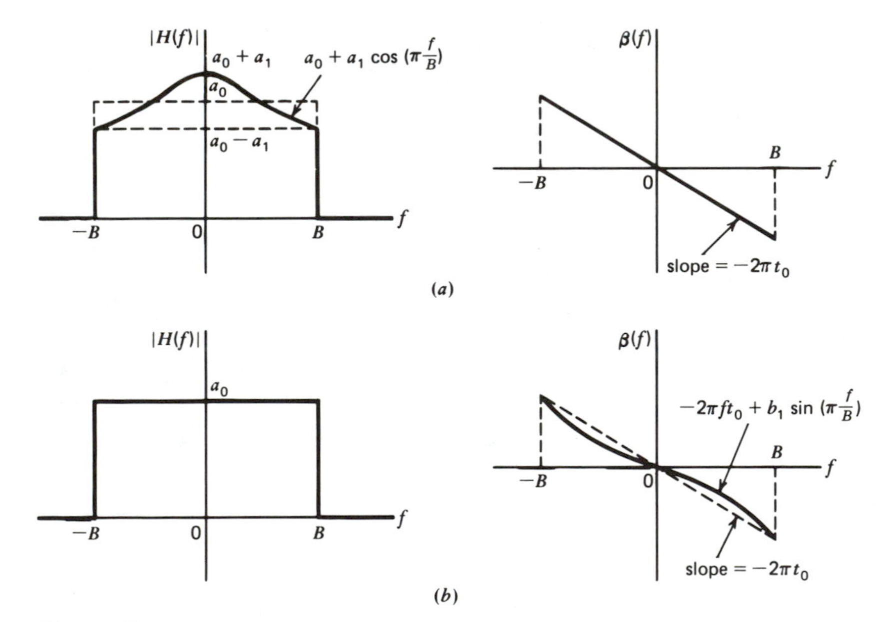

Figure P2.16

Problem 2.55 A tapped-delay-line filter consists of N weights, where N is odd. It is symmetric with respect to the center tap, that is, the weights satisfy the condition

$$w_n = w_{N-1-n} \qquad 0 \leqslant n \leqslant N-1$$

(a) Find the amplitude response of the filter.

(b) Show that this filter has a linear phase response.

Problem 2.56 Let $\hat{g}(t)$ denote the Hilbert transform of $g(t)$. Derive the following set of Hilbert transform pairs:

$g(t)$	$\hat{g}(t)$
$\dfrac{\sin t}{t}$	$\dfrac{1-\cos t}{t}$
$\text{rect}(t)$	$-\dfrac{1}{\pi}\ln\left\lvert\dfrac{t-\frac{1}{2}}{t+\frac{1}{2}}\right\rvert$
$\delta(t)$	$\dfrac{1}{\pi t}$
$\dfrac{1}{1+t^2}$	$\dfrac{t}{1+t^2}$

Problem 2.57 Let $\hat{g}(t)$ denote the Hilbert transform of a real-valued signal $g(t)$. Show that the cross-correlation functions of $g(t)$ and $\hat{g}(t)$ are given by

$$R_{g\hat{g}}(\tau) = -\hat{R}_g(\tau)$$
$$R_{\hat{g}g}(\tau) = \hat{R}_g(\tau)$$

where $\hat{R}_g(\tau)$ is the Hilbert transform of the autocorrelation function of $g(t)$.

Problem 2.58 Show that the complex envelope of the sum of two narrow-band signals (with the same center frequency) is equal to the sum of their individual complex envelopes.

Problem 2.59 Determine the pre-envelope $g_+(t)$ corresponding to each of the following two signals:

(a) $g(t) = \text{sinc}(t)$
(b) $g(t) = [1 + k\cos(2\pi f_m(t)]\cos(2\pi f_c t)$

Problem 2.60

(a) Consider two signals $g_1(t)$ and $g_2(t)$ whose pre-envelopes are denoted by $g_{1+}(t)$ and $g_{2+}(t)$, respectively. Show that

$$\int_{-\infty}^{\infty} \text{Re}[g_{1+}(t)]\text{Re}[g_{2+}(t)]dt = \tfrac{1}{2}\,\text{Re}\left[\int_{-\infty}^{\infty} g_{1+}(t)g_{2+}^*(t)dt\right]$$

How is this relation modified if $g_2(t)$ is replaced by $g_2(-t)$?

(b) Assuming that $g(t)$ is a narrow-band signal with complex envelope $\tilde{g}(t)$ and carrier frequency f_c, use the result of part (a) to show that

$$\int_{-\infty}^{\infty} g^2(t)dt = \frac{1}{2}\int_{-\infty}^{\infty} |\tilde{g}(t)|^2\,dt$$

Problem 2.61 Let a narrow-band signal $g(t)$ be expressed in the form:

$$g(t) = g_c(t)\cos(2\pi f_c t) - g_s(t)\sin(2\pi f_c t)$$

Using $G_+(f)$ to denote the Fourier transform of the pre-envelope of $g(t)$, show that the Fourier transforms of the in-phase component $g_c(t)$ and quadrature component $g_s(t)$ are given by, respectively,

$$G_c(f) = \tfrac{1}{2}[G_+(f+f_c) + G_+^*(-f+f_c)]$$

$$G_s(f)=\frac{1}{2j}[G_+(f+f_c)-G_+^*(-f+f_c)]$$

where the asterisk denotes complex conjugation.

Problem 2.62 The block diagram of Fig. 2.35(*a*) illustrates a method for extracting the in-phase component $g_c(t)$ and quadrature component $g_s(t)$ of a narrow-band signal $g(t)$. Given that the spectrum of $g(t)$ is limited to the interval $f_c-W \leq |f| \leq f_c+W$, demonstrate the validity of this method. Hence, show that

$$G_c(f)=\begin{cases} G(f-f_c)+G(f+f_c), & -W \leq f \leq W \\ 0, & \text{elsewhere} \end{cases}$$

and

$$G_s(f)=\begin{cases} j[G(f-f_c)-G(f+f_c)], & -W \leq f \leq W \\ 0, & \text{elsewhere} \end{cases}$$

where $G_c(f)$, $G_s(f)$, and $G(f)$ are the Fourier transforms of $g_c(t)$, $g_s(t)$, and $g(t)$, respectively.

Problem 2.63 Explain what happens to the low-pass equivalent model of Fig. 2.37 when the amplitude response of the corresponding band-pass filter has even symmetry and the phase response has odd symmetry with respect to the mid-band frequency f_c.

Problem 2.64 Consider an ideal band-pass filter with center frequency f_c and bandwidth $2B$, as defined in Fig. P2.17. The carrier wave $A\cos(2\pi f_0 t)$ is suddenly applied to this filter at time $t=0$. Assuming that $|f_c-f_0|$ is large compared to the bandwidth $2B$, determine the response of the filter.

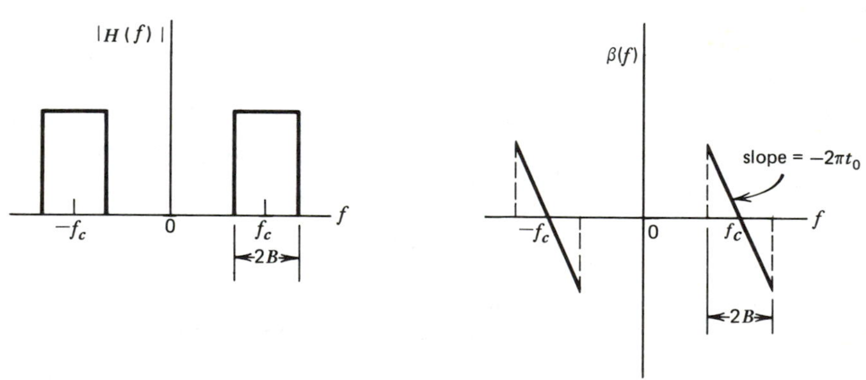

Figure P2.17

Problem 2.65 The rectangular RF pulse

$$x(t)=\begin{cases} A\cos(2\pi f_c t), & 0 \leq t \leq T \\ 0, & \text{elsewhere} \end{cases}$$

is applied to a linear filter with impulse response

$$h(t)=x(T-t)$$

Assume that the frequency f_c equals a large integer multiple of $1/T$. Determine the response of the filter, and sketch it.

Problem 2.66 Consider an integral of the form defined by

$$\tilde{U}(f)=\int_{-\infty}^{\infty} a(t)\exp[j\mu(t, f)]dt$$

where

$$\mu(t, f)=\theta(t)-2\pi ft$$

and $\tilde{u}(t)$, the inverse Fourier transform of $\tilde{U}(f)$, is defined by

$$\tilde{u}(t)=a(t)\exp[j\theta(t)]$$

Assume that the envelope $a(t)$ varies only slowly with t, whereas the factors $\cos[\mu(t, f)]$ and $\sin[\mu(t, f)]$ go through a large number of periods within the range of integration. According to *the principle of stationary phase*, in an integral of this type, $\tilde{U}(f)$ will have its greatest values when

$$\frac{\partial}{\partial t}[\mu(t, f)]=0$$

(a) Using the first three terms in a Taylor series expansion of $\mu(t, f_k)$, show that by applying the principle of stationary phase, we may approximate the value of $\tilde{U}(f)$ at $f=f_k$ as

$$\tilde{U}(f_k)\simeq\sqrt{\frac{2\pi j}{u''(t_k, f_k)}}\, a(t_k)\exp[j\mu(t_k, f_k)]$$

where t_k denotes the value of time t for which the first derivative of $\mu(t, f_k)$ is zero, and $\mu''(t_k, f_k)$ denotes the second derivative of $\mu(t, f_k)$ with respect to time, evaluated at $t=t_k$. Note that

$$\int_{-\infty}^{\infty} \exp(jy^2)dy=\sqrt{\pi j}$$

(b) The complex envelope of a pulse $u(t)$ is defined by

$$\tilde{u}(t)=\begin{cases}\exp(j\pi kt^2), & -\frac{T}{2}\leqslant t\leqslant\frac{T}{2}\\ 0, & \text{elsewhere}\end{cases}$$

where k is a positive constant having the dimensions of hertz squared. Assuming that k is large, use the approximate formula for $\tilde{U}(f)$ in part (a) to show that the spectrum of $\tilde{u}(t)$ is given approximately by

$$\tilde{U}(f)=\begin{cases}\frac{1}{\sqrt{k}}\exp\left(-j\frac{\pi f^2}{k}+j\frac{\pi}{4}\right), & -\frac{kT}{2}\leqslant f\leqslant\frac{kT}{2}\\ 0, & \text{elsewhere}\end{cases}$$

chapter 3 AMPLITUDE MODULATION

The purpose of a communication system is to transmit *information-bearing signals* or *baseband signals* through a communication channel separating the transmitter from the receiver. The term *baseband* is used to designate the band of frequencies representing the original signal as delivered by a source of information. The efficient utilization of the communication channel requires a shift of the range of baseband frequencies into other frequency ranges suitable for transmission, and a corresponding shift back to the original frequency range after reception. For example, a radio system must operate with frequencies of 30 kHz and upward, whereas the baseband signal usually contains frequencies in the audio frequency range, and so some form of frequency-band shifting must be used for the system to operate satisfactorily. A shift of the range of frequencies in a signal is accomplished by using *modulation* which is defined as *the process by which some characteristic of a carrier is varied in accordance with a modulating wave.** The baseband signal is referred to as the *modulating wave*, and the result of the modulation process is referred to as the *modulated wave*. At the receiving end of the communication system, we usually require the original baseband signal or modulating wave to be restored. This is accomplished by using a process known as *demodulation*, which is the reverse of the modulation process.

In this chapter and the next one we will study two families of *continuous-wave* (CW) modulation systems, namely, *amplitude modulation* and *angle modulation* systems, in which the amplitude and angle of a sinusoidal carrier wave are varied in accordance with the information-bearing signal to be transmitted, respectively. Amplitude modulation is studied in this chapter, whereas angle modulation is studied in Chapter 4.

3.1 AMPLITUDE MODULATION (AM)

Consider a sinusoidal carrier wave $c(t)$ defined by

$$c(t) = A_c \cos(2\pi f_c t) \tag{3.1}$$

where A_c is the carrier amplitude and f_c is the carrier frequency. For convenience, we have assumed that the phase of the carrier wave is zero in Eq. (3.1). Let $m(t)$ denote the baseband signal which carries the specification of the message. The carrier wave $c(t)$ is independent of $m(t)$. *Amplitude modulation (AM) is defined as a process in which the amplitude of the carrier wave $c(t)$ is varied about a mean value, linearly with the baseband signal $m(t)$.* An amplitude-modulated (AM) wave may thus be described as a function of time in the form

$$s(t) = A_c[1 + k_a m(t)]\cos(2\pi f_c t) \tag{3.2}$$

where k_a is a constant called the *amplitude sensitivity* of the modulator.

Part (*a*) of Fig. 3.1 shows a baseband signal $m(t)$, and parts (*b*) and (*c*) show the corresponding AM wave $s(t)$ for two values of amplitude sensitivity k_a and a carrier amplitude $A_c = 1$ volt. We observe that the *envelope* of $s(t)$ has the same shape as the baseband signal $m(t)$ provided two requirements are satisfied:

1. The amplitude of $k_a m(t)$ is always less than unity, that is,

$$|k_a m(t)| < 1, \quad \text{for all } t \tag{3.3}$$

This condition is illustrated in Fig. 3.1(*b*). It ensures that the function $1 + k_a m(t)$

* *IEEE Standard Dictionary of Electrical and Electronics Terms*, p. 351 (Wiley-Interscience, 1972).

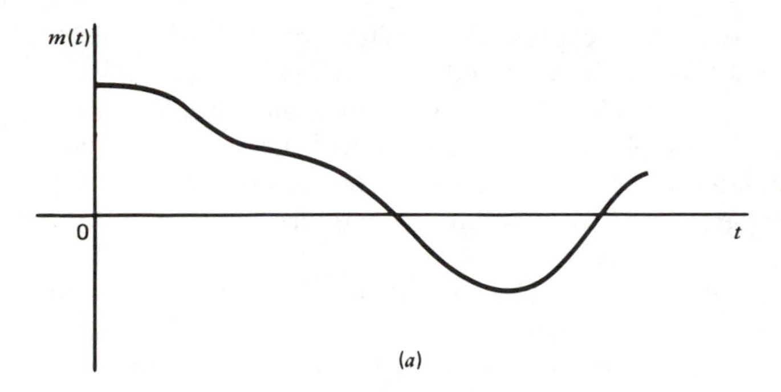

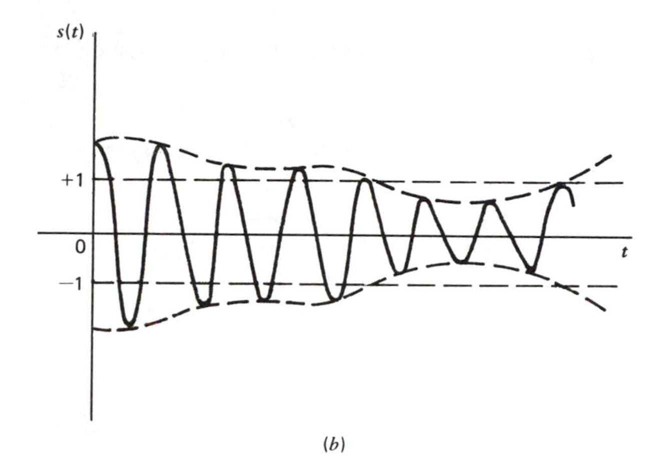

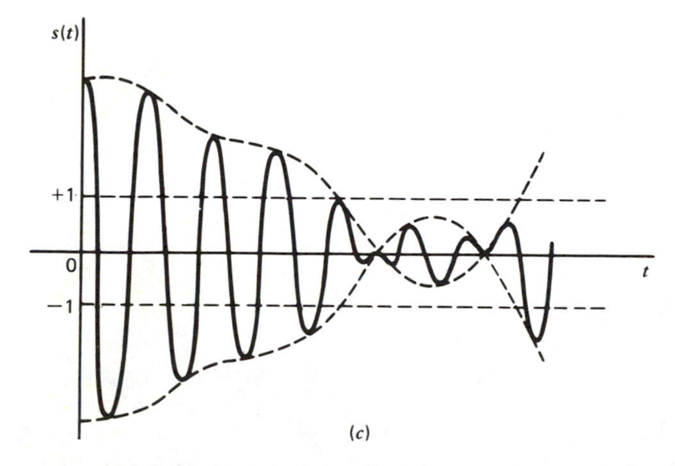

Figure 3.1 Illustrating the amplitude modulation process. *(a)* Baseband signal $m(t)$. *(b)* AM wave for $|k_a m(t)| < 1$ for all t. *(c)* AM wave for $|k_a m(t)| > 1$ for some t.

is always positive, and so we may express the envelope of the AM wave $s(t)$ of Eq. (3.2) as $A_c[1+k_a m(t)]$. When the amplitude sensitivity k_a of the modulator is large enough to make $|k_a m(t)|>1$ for any t, the carrier wave becomes *overmodulated*, resulting in carrier phase reversals whenever the factor $1+k_a m(t)$ crosses zero. The modulated wave then exhibits *envelope distortion*, as in Fig. 3.1(*c*). It is therefore apparent that by avoiding overmodulation, a one-to-one relationship is maintained between the envelope of the AM wave and the modulating wave for all values of time. The absolute maximum value of $k_a m(t)$ multiplied by 100 is referred to as the *percentage modulation*.

2. The carrier frequency f_c is much greater than the highest frequency component W of the message signal $m(t)$, that is,

$$f_c \gg W \tag{3.4}$$

We call W the *message bandwidth*. If the condition of Eq. (3.4) is not satisfied, an envelope cannot be visualized satisfactorily.

From Eq. (3.2), we find that the Fourier transform of the AM wave $s(t)$ is given by

$$S(f)=\frac{A_c}{2}[\delta(f-f_c)+\delta(f+f_c)]+\frac{k_a A_c}{2}[M(f-f_c)+M(f+f_c)] \tag{3.5}$$

Suppose that the baseband signal $m(t)$ is band-limited to the interval $-W \leqslant f \leqslant W$,

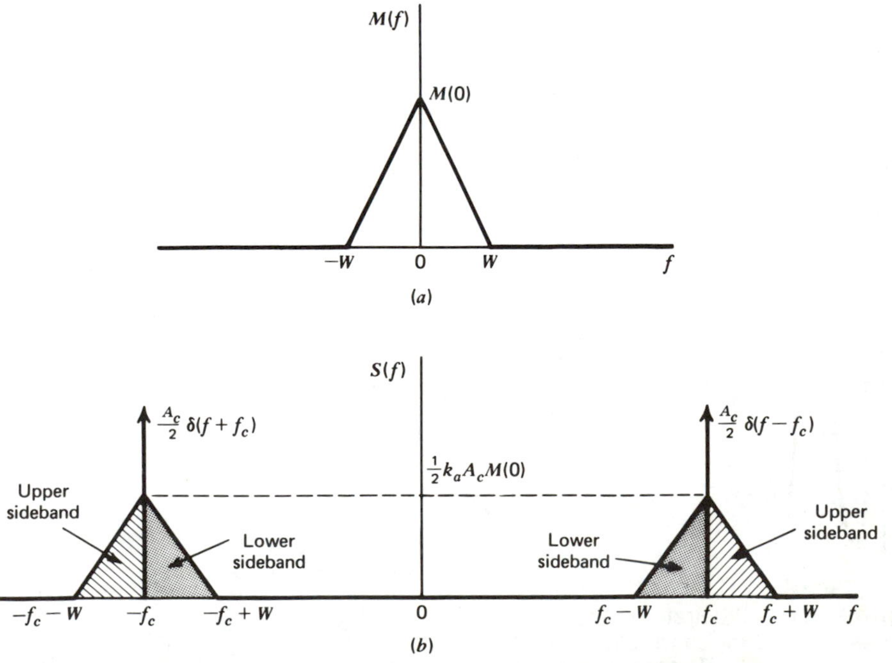

Figure 3.2 *(a)* Spectrum of baseband signal. *(b)* Spectrum of AM wave.

as in Fig. 3.2(*a*). The shape of the spectrum shown in this figure is intended for the purpose of illustration only. We find from Eq. (3.5) that the spectrum $S(f)$ of the AM wave is as shown in Fig. 3.2(*b*) for the case when $f_c > W$. This spectrum consists of two delta functions weighted by the factor $A_c/2$ and occurring at $\pm f_c$, and two versions of the baseband spectrum translated in frequency by $\pm f_c$ and scaled in amplitude by $k_a A_c/2$. From the spectrum of Fig. 3.2(*b*), we note the following:

1. For positive frequencies, the portion of the spectrum of an AM wave lying above the carrier frequency f_c is referred to as the *upper sideband*, whereas the symmetrical portion below f_c is referred to as the *lower sideband*. For negative frequencies, the upper sideband is represented by the portion of the spectrum below $-f_c$ and the lower sideband by the portion above $-f_c$. The condition $f_c > W$ ensures that the sidebands do not overlap.
2. For positive frequencies, the highest frequency component of the AM wave equals $f_c + W$, and the lowest frequency component equals $f_c - W$. The difference between these two frequencies defines the *transmission bandwidth* B_T for an AM wave, which is exactly twice the message bandwidth W, that is,

$$B_T = 2W \tag{3.6}$$

Example 1 Single-tone modulation

Consider a modulating wave $m(t)$ that consists of a single tone or frequency component, that is,

$$m(t) = A_m \cos(2\pi f_m t) \tag{3.7}$$

where A_m is the amplitude of the modulating wave and f_m is its frequency [see Fig. 3.3(*a*)]. The sinusoidal carrier wave has amplitude A_c and frequency f_c [see Fig. 3.3(*b*)]. The corresponding AM wave is therefore given by

$$s(t) = A_c[1 + \mu \cos(2\pi f_m t)]\cos(2\pi f_c t) \tag{3.8}$$

where

$$\mu = k_a A_m \tag{3.9}$$

The dimensionless constant μ is the *modulation factor*, or the percentage modulation when it is expressed numerically as a percentage. To avoid envelope distortion due to overmodulation, the modulation factor μ must be kept below unity.

Figure 3.3(*c*) shows a sketch of $s(t)$ for μ less than unity. Let A_{max} and A_{min} denote the maximum and minimum values of the envelope of the modulated wave. Then, from Eq. (3.8) we get

$$\frac{A_{max}}{A_{min}} = \frac{A_c(1 + \mu)}{A_c(1 - \mu)}$$

That is,

$$\mu = \frac{A_{max} - A_{min}}{A_{max} + A_{min}} \tag{3.10}$$

Expressing the product of the two cosines in Eq. (3.8) as the sum of two sinusoidal waves, one having frequency $f_c + f_m$ and the other having frequency $f_c - f_m$, we get

$$s(t) = A_c \cos(2\pi f_c t) + \tfrac{1}{2}\mu A_c \cos[2\pi(f_c + f_m)t] + \tfrac{1}{2}\mu A_c \cos[2\pi(f_c - f_m)t] \tag{3.11}$$

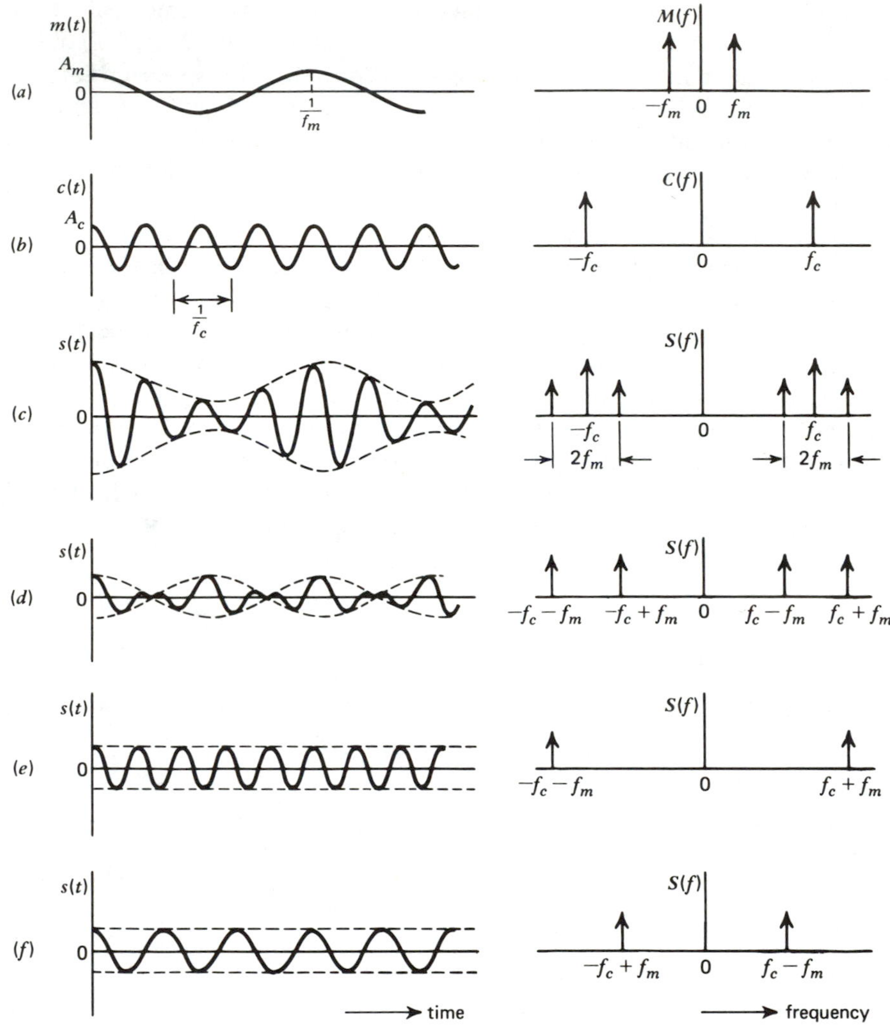

Figure 3.3 Illustrating the time-domain (on the left) and frequency-domain (on the right) characteristics of different modulated waves produced by a single tone. *(a)* Modulating wave. *(b)* Carrier wave. *(c)* AM wave. *(d)* DSBSC wave. *(e)* SSB wave with the upper-side frequency transmitted. *(f)* SSB wave with the lower-side frequency transmitted.

The Fourier transform of $s(t)$ is therefore

$$
\begin{aligned}
S(f) = {} & \tfrac{1}{2}A_c[\delta(f-f_c)+\delta(f+f_c)] \\
& +\tfrac{1}{4}\mu A_c[\delta(f-f_c-f_m)+\delta(f+f_c+f_m)] \\
& +\tfrac{1}{4}\mu A_c[\delta(f-f_c+f_m)+\delta(f+f_c-f_m)]
\end{aligned}
\tag{3.12}
$$

Thus the spectrum of an AM wave, for the special case of sinusoidal modulation, consists of delta functions at $\pm f_c$, $f_c \pm f_m$, and $-f_c \pm f_m$, as in Fig 3.3(*c*).

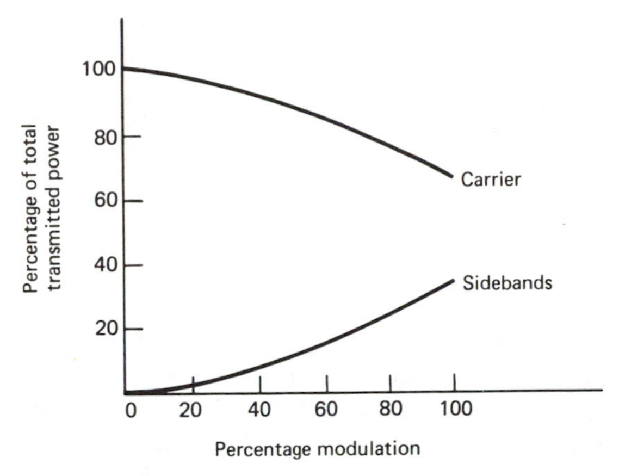

Figure 3.4 Variations of carrier power and total sideband power with percentage modulation.

In practice, the AM wave $s(t)$ is a voltage or current wave. In either case, the average power delivered to a 1-ohm resistor by $s(t)$ is comprised of three components:

Carrier power $=\frac{1}{2}A_c^2$
Upper side-frequency power $=\frac{1}{8}\mu^2 A_c^2$
Lower side-frequency power $=\frac{1}{8}\mu^2 A_c^2$

The ratio of the total sideband power to the total power in the modulated wave is therefore equal to $\mu^2/(2+\mu^2)$, which depends only on the modulation factor μ. If $\mu=1$, that is, 100 percent modulation is used, the total power in the two side-frequencies of the resulting AM wave is only one-third of the total power in the modulated wave.

Figure 3.4 shows the percentage of total power in both side-frequencies and in the carrier plotted versus the percentage modulation. Note that when the percentage modulation is less than 20 percent, the power in one side-frequency is less than 1 percent of the total power in the AM wave.

3.2 GENERATION OF AM WAVES

We next describe two devices for the generation of AM waves, namely, the *square-law modulator* and the *switching modulator*, both of which require the use of a nonlinear element for their implementation. These two devices are well-suited for low-power modulation purposes.

Square-Law Modulator

A square-law modulator requires three features: a means of summing the carrier and modulating waves, a nonlinear element, and a band-pass filter for extracting the desired modulation products. These features of the modulator are illustrated in Fig. 3.5. Semiconductor diodes and transistors are the most common nonlinear devices used for implementing square-law modulators. The filtering requirement is usually satisfied by using a single- or double-tuned filter.

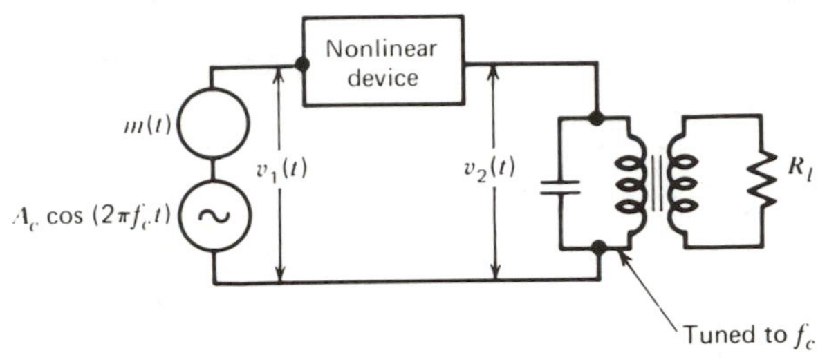

Figure 3.5 Square-law modulator.

When a nonlinear element such as a diode is suitably biased and operated in a restricted portion of its characteristic curve, that is, the signal applied to the diode is relatively weak, we find that the transfer characteristic of the diode-load resistor combination can be represented closely by a *square law*:

$$v_2(t)=a_1v_1(t)+a_2v_1^2(t) \tag{3.13}$$

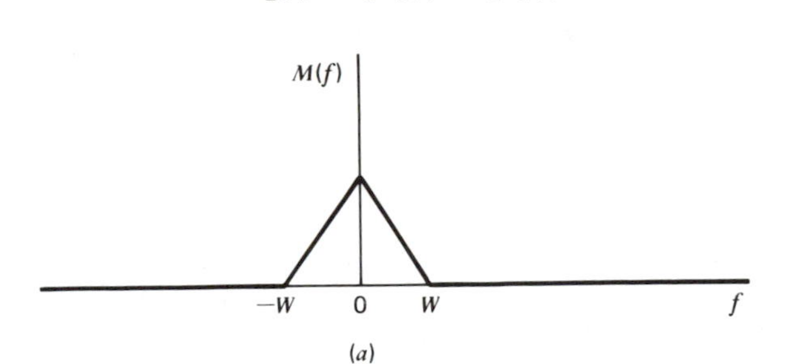

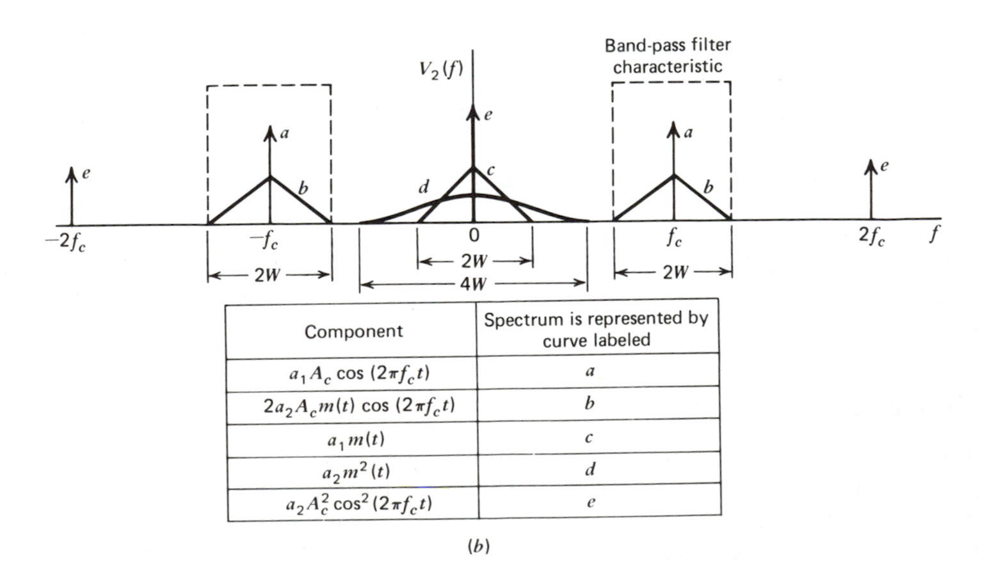

Component	Spectrum is represented by curve labeled
$a_1A_c \cos(2\pi f_c t)$	a
$2a_2A_c m(t) \cos(2\pi f_c t)$	b
$a_1 m(t)$	c
$a_2 m^2(t)$	d
$a_2 A_c^2 \cos^2(2\pi f_c t)$	e

Figure 3.6 Illustrating the spectral relationships in a square-law modulator. *(a)* Spectrum of baseband signal. *(b)* Spectrum of $v_2(t)$.

where a_1 and a_2 are constants. The input voltage $v_1(t)$ consists of the carrier wave plus the modulating wave, that is,

$$v_1(t) = A_c \cos(2\pi f_c t) + m(t) \tag{3.14}$$

Therefore, substituting Eq. (3.14) in (3.13), the resulting voltage developed across the primary winding of the output transformer is given by

$$v_2(t) = a_1 A_c \left[1 + \frac{2a_2}{a_1} m(t)\right] \cos(2\pi f_c t) + a_1 m(t) + a_2 m^2(t) + a_2 A_c^2 \cos^2(2\pi f_c t) \tag{3.15}$$

The first term in Eq. (3.15) is the desired AM wave with amplitude sensitivity $k_a = 2a_2/a_1$. The remaining three terms are unwanted terms that are removed by filtering.

We assume that the modulating wave $m(t)$ is band-limited to the interval $-W \leqslant f \leqslant W$, as in Fig. 3.6(*a*). Then, from Eq. (3.15) we find that the spectrum $V_2(f)$ of the voltage $v_2(t)$ is as shown in Fig. 3.6(*b*). It follows therefore that the unwanted terms may be removed from $v_2(t)$ by designing the tuned filter at the modulator output of Fig. 3.5 to have a mid-band frequency f_c and bandwidth $2W$, which satisfy the requirement $f_c > 3W$.

Switching Modulator

Consider next the arrangement shown in Fig. 3.7(*a*), where it is assumed that the carrier wave $c(t)$ applied to the diode is large in amplitude, so that it swings right across the characteristic curve of the diode. We assume that the diode acts as an *ideal switch*, that is, it presents zero impedance when it is forward-biased [corresponding to $c(t) > 0$] and infinite impedance when it is reverse-biased [corresponding to $c(t) < 0$]. We may thus approximate the transfer characteristic of the diode-load resistor combination by a *piecewise-linear* characteristic, as shown in Fig. 3.7(*b*). Accordingly, for an input voltage $v_1(t)$ given by

$$v_1(t) = A_c \cos(2\pi f_c t) + m(t) \tag{3.16}$$

where $|m(t)| \ll A_c$, the resulting load voltage $v_2(t)$ is

$$v_2(t) \simeq \begin{cases} v_1(t), & c(t) > 0 \\ 0, & c(t) < 0 \end{cases} \tag{3.17}$$

That is, the load voltage $v_2(t)$ varies periodically between the values $v_1(t)$ and zero at a rate equal to the carrier frequency f_c. In this way, by assuming a modulating wave that is weak compared with the carrier wave, we have effectively replaced the nonlinear behavior of the diode by an approximately equivalent linear time-varying operation.

We may express Eq. (3.17) mathematically as follows

$$v_2(t) \simeq [A_c \cos(2\pi f_c t) + m(t)] g_p(t) \tag{3.18}$$

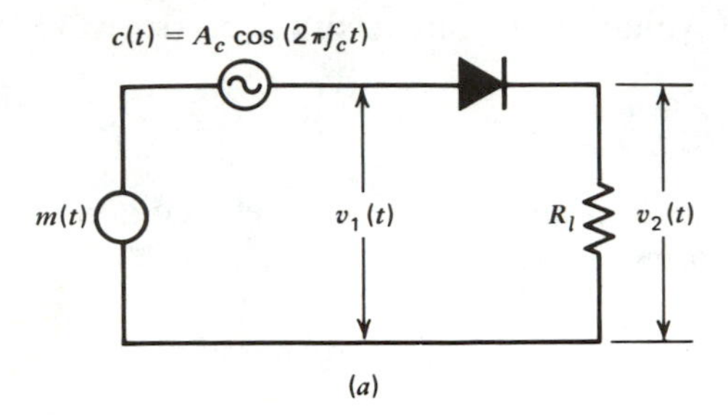

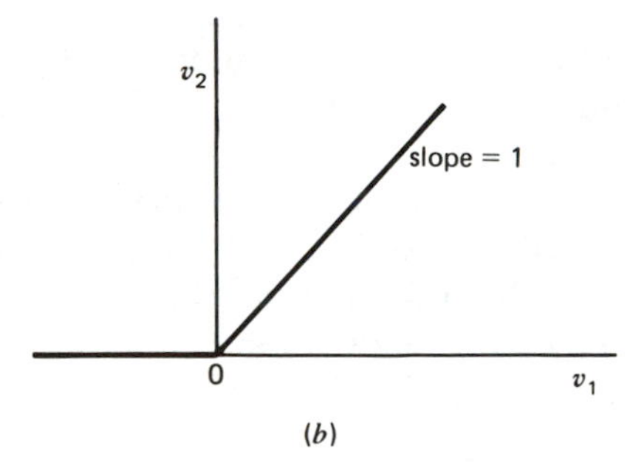

Figure 3.7 Switching modulator. *(a)* Circuit diagram. *(b)* Idealized input-output relation.

where $g_p(t)$ is a periodic pulse train of duty cycle equal to one-half, and period $T_0 = 1/f_c$, as in Fig. 3.8. Representing this $g_p(t)$ by its Fourier series, we have

$$g_p(t) = \frac{1}{2} + \frac{2}{\pi} \sum_{n=1}^{\infty} \frac{(-1)^{n-1}}{2n-1} \cos[2\pi f_c t(2n-1)] \tag{3.19}$$

Therefore, substituting Eq. (3.19) in (3.18), we find that the load voltage $v_2(t)$ consists of the sum of two components:

1. The component

$$\frac{A_c}{2}\left[1 + \frac{4}{\pi A_c} m(t)\right] \cos(2\pi f_c t)$$

which is the desired AM wave with amplitude sensitivity $k_a = 4/\pi A_c$.

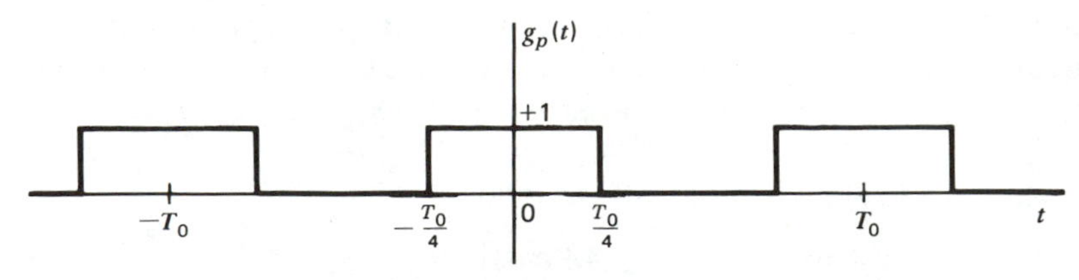

Figure 3.8 Periodic pulse train.

2. An unwanted component, the spectrum of which contains delta functions at 0, $\pm 2f_c$, $\pm 4f_c$, and so on, and which occupies frequency intervals of width $2W$ centered at 0, $\pm 3f_c$, $\pm 5f_c$, and so on, where W is the message bandwidth.

Here again, the unwanted terms are removed from the load voltage $v_2(t)$ by means of a band-pass filter with mid-band frequency f_c and bandwidth $2W$, provided $f_c > 2W$.

3.3 DEMODULATION OF AM WAVES

The process of *demodulation* provides a means of recovering from an incoming modulated wave an output that is proportional to the original modulating wave. In effect, demodulation is the reverse of modulation. We will describe two devices for the demodulation of AM waves, namely, the *square-law detector* and the *envelope detector*.

Square-Law Detector

A square-law detector is essentially obtained by using a square-law modulator for the purpose of demodulation. Consider Eq. (3.13) defining the transfer characteristic of a nonlinear device, which is reproduced here for convenience:

$$v_2(t) = a_1 v_1(t) + a_2 v_1^2(t) \tag{3.20}$$

where $v_1(t)$ and $v_2(t)$ are the input and output voltages, respectively, and a_1 and a_2 are constants. When such a device is used for the demodulation of an AM wave, we have for the input

$$v_1(t) = A_c[1 + k_a m(t)]\cos(2\pi f_c t) \tag{3.21}$$

Therefore, substituting Eq. (3.21) in (3.20), we get

$$\begin{aligned} v_2(t) = {} & a_1 A_c[1 + k_a m(t)]\cos(2\pi f_c t) \\ & + \tfrac{1}{2} a_2 A_c^2[1 + 2k_a m(t) + k_a^2 m^2(t)][1 + \cos(4\pi f_c t)] \end{aligned} \tag{3.22}$$

The desired signal, namely, $a_2 A_c^2 k_a m(t)$, is due to the $a_2 v_1^2(t)$ term—hence, the description "square-law detector." This component can be extracted by means of a low-pass filter. This is not the only contribution within the baseband spectrum, however, because the term $\frac{1}{2} a_2 A_c^2 k_a^2 m^2(t)$ will give rise to a plurality of similar frequency components. The ratio of wanted signal to distortion is equal to $2/k_a m(t)$. To make this ratio large we limit the percentage modulation, that is, we choose $|k_a m(t)|$ small compared with unity for all t. We conclude therefore that distortionless recovery of the baseband signal $m(t)$ is possible only if the applied AM wave is weak [so as to justify the use of a square-law input-output relation as in Eq. (3.20)] and if the percentage modulation is very small.

Envelope Detector

Consider the case of a narrow-band AM wave, that is, one in which the carrier frequency is large compared with the message bandwidth, and for which the percentage modulation is less than 100 percent. Then the desired demodulation can be accomplished by using a simple, yet highly effective device known as

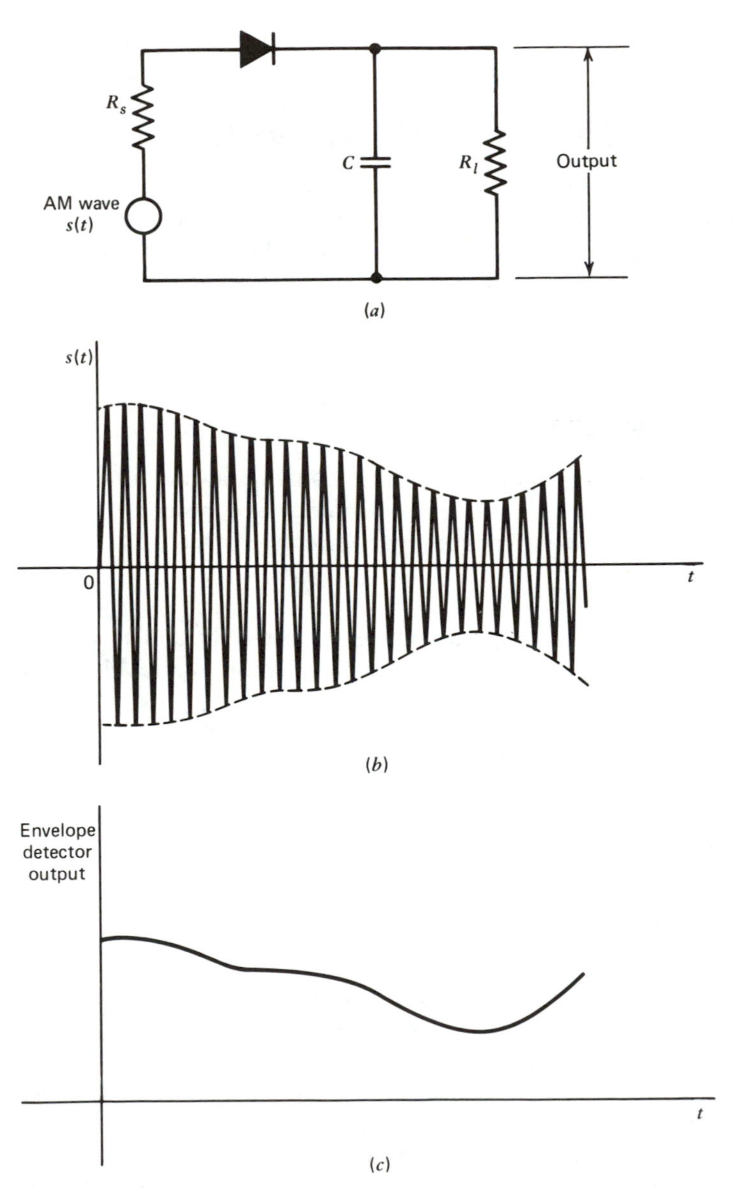

Figure 3.9 Envelope detector. *(a)* Circuit diagram. *(b)* AM wave input. *(c)* Envelope detector output.

envelope detector. Ideally, an envelope detector produces an output signal that follows the envelope of the input signal waveform exactly. Some version of this circuit is used in almost all comercial AM radio receivers.

An envelope detector of the series type is shown in Fig. 3.9(*a*), which consists of a diode and a resistor-capacitor filter. The operation of this envelope detector is as follows. On the positive half-cycle of the input signal, the diode is forward-biased and the capacitor C charges up rapidly to the peak value of the input signal. When the input signal falls below this value, the diode becomes reverse-biased and the capacitor C discharges slowly through the load resistor R_l. The discharging process continues until the next positive half-cycle. When the input signal becomes greater than the voltage across the capacitor, the diode conducts again and the process is repeated. We assume that the diode is ideal, presenting zero impedance to current flow in the forward-biased region, and infinite impedance in the reverse-biased region. We further assume that the AM wave applied to the envelope detector is supplied by a voltage source of internal impedance R_s. The charging time constant R_sC must be short compared with the carrier period $1/f_c$, that is,

$$R_sC \ll \frac{1}{f_c}, \tag{3.23}$$

so that the capacitor C charges rapidly and thereby follows the applied voltage up to the positive peak when the diode is conducting. On the other hand, the discharging time constant R_lC must be long enough to ensure that the capacitor discharges slowly through the load resistor R_l between positive peaks of the carrier wave, but not so long that the capacitor voltage will not discharge at the maximum rate of change of the modulating wave, that is,

$$\frac{1}{f_c} \ll R_lC \ll \frac{1}{W} \tag{3.24}$$

where W is the message bandwidth. The result is that the capacitor voltage or detector output is very nearly the same as the envelope of the AM wave, as illustrated in Fig. 3.9(*c*). The detector output usually has a small ripple [not shown in Fig. 3.9(*c*)] at the carrier frequency; this ripple is easily removed by low-pass filtering.

3.4 DOUBLE-SIDEBAND SUPPRESSED-CARRIER MODULATION (DSBSC)

The carrier wave $c(t)$ is completely independent of the information-carrying signal or baseband signal $m(t)$, which means that the transmission of the carrier wave represents a waste of power. This points to a shortcoming of amplitude modulation, namely, that only a fraction of the total transmitted power is affected by $m(t)$. To overcome this shortcoming, we may suppress the carrier component from the modulated wave, resulting in *double-sideband suppressed carrier modulation*

(DSBSC). Thus, by suppressing the carrier, we obtain a modulated wave that is proportional to the product of the carrier wave and the baseband signal.

To describe a DSBSC wave as a function of time, we may write

$$\begin{aligned} s(t) &= c(t)m(t) \\ &= A_c \cos(2\pi f_c t)m(t) \end{aligned} \tag{3.25}$$

This modulated wave undergoes a phase reversal whenever the baseband signal $m(t)$ crosses zero, as illustrated in Fig. 3.10. Accordingly, unlike amplitude modulation, the envelope of a DSBSC wave is different from the baseband signal.

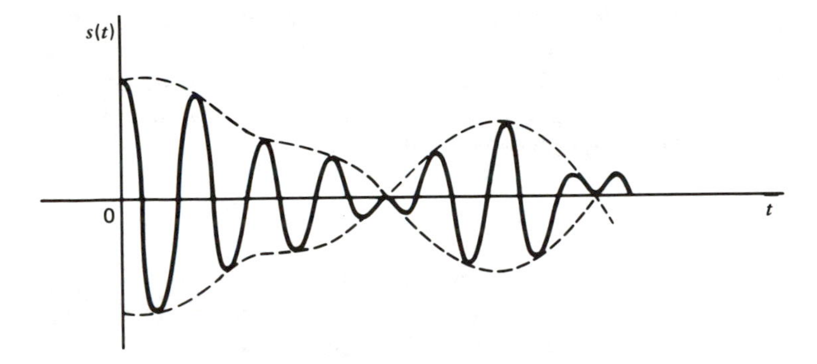

Figure 3.10 DSBSC wave.

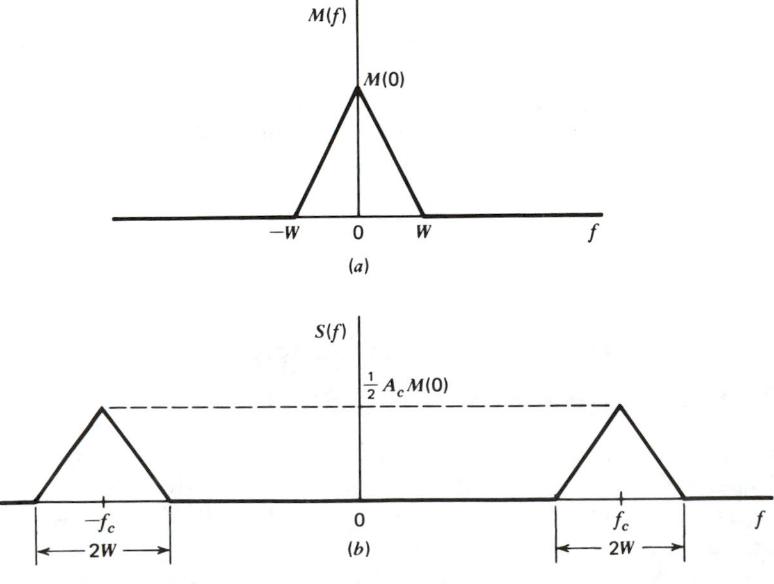

Figure 3.11 *(a)* Spectrum of baseband signal. *(b)* Spectrum of DSBSC wave.

From Eq. (3.25), the Fourier transform of $s(t)$ is obtained as

$$S(f)=\tfrac{1}{2}A_c[M(f-f_c)+M(f+f_c)] \tag{3.26}$$

For the case when the baseband signal $m(t)$ is limited to the interval $-W \leqslant f \leqslant W$, as in Fig. 3.11(*a*), we thus find that the spectrum $S(f)$ of the DSBSC wave $s(t)$ is as illustrated in part (*b*) of the figure. Except for a change in scale factor, the modulation process simply translates the spectrum of the baseband signal by $\pm f_c$. Of course, the transmission bandwidth required by DSBSC modulation is the same as that for amplitude modulation, namely, $2W$.

3.5 GENERATION OF DSBSC WAVES

A double-sideband suppressed-carrier modulated wave consists simply of the product of the baseband signal and the carrier wave, as shown by Eq. (3.25). A device for achieving this requirement is called a *product modulator*. In this section, we describe two forms of a product modulator—the *balanced modulator* and the *ring modulator*.

Balanced Modulator

One possible scheme for generating a DSBSC wave is to use two AM modulators arranged in a balanced configuration so as to suppress the carrier wave, as shown in the block diagram of Fig. 3.12. We assume that the two AM modulators are identical, except for the sign reversal of the modulating wave applied to the input of one of the modulators. Thus the outputs of the two AM modulators may be expressed as follows:

$$s_1(t)=A_c[1+k_a m(t)]\cos(2\pi f_c t) \tag{3.27}$$

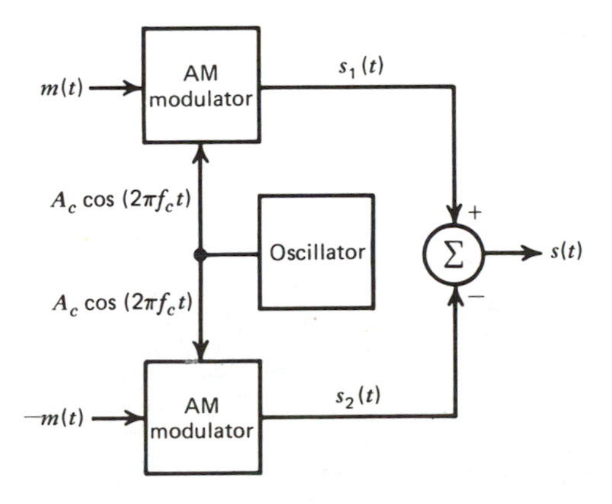

Figure 3.12 Balanced modulator.

and

$$s_2(t) = A_c[1 - k_a m(t)]\cos(2\pi f_c t) \tag{3.28}$$

Substracting $s_2(t)$ from $s_1(t)$, we obtain

$$\begin{aligned} s(t) &= s_1(t) - s_2(t) \\ &= 2k_a A_c \cos(2\pi f_c t) m(t) \end{aligned} \tag{3.29}$$

Hence, except for the scaling factor $2k_a$, the balanced modulator output is equal to the product of the modulating wave and the carrier, as required.

Ring Modulator

One of the most useful product modulators that is well-suited for generating a DSBSC wave is the ring modulator shown in Fig. 3.13(*a*). The four diodes form a ring in which they all point in the same way—hence, the name. The diodes are controlled by a square-wave carrier $c(t)$ of frequency f_c, which is applied by means of two center-tapped transformers. We assume that the diodes are ideal and the transformers are perfectly balanced. When the carrier supply is positive, the outer diodes are switched on, presenting zero impedance, whereas the inner diodes are switched off, presenting infinite impedance, as in Fig. 3.13(*b*), so that the modulator

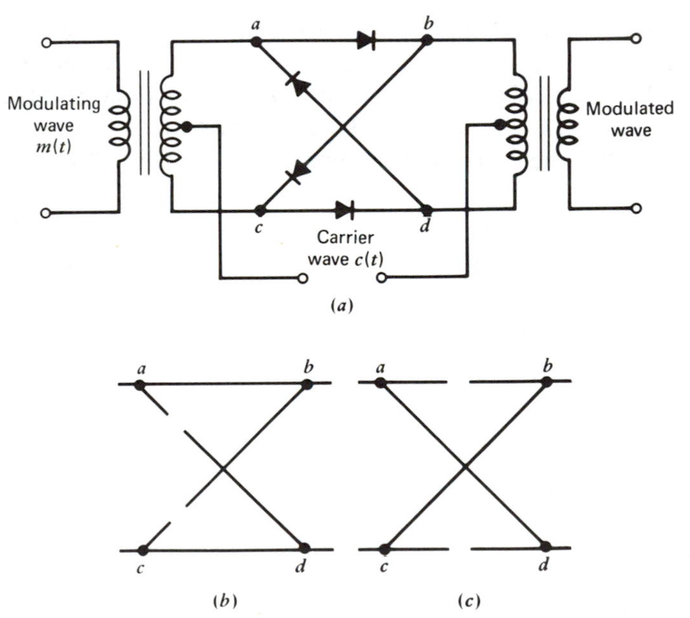

Figure 3.13 Ring modulator. *(a)* Circuit diagram. *(b)* Illustrating the condition when the outer diodes are switched on and the inner diodes are switched off. *(c)* Illustrating the condition when the outer diodes are switched off and the inner diodes are switched on.

multiplies the baseband signal $m(t)$ by $+1$. When the carrier supply is negative, the situation becomes reversed as in Fig. 3.13(c), and the modulator multiplies the baseband signal by -1. Thus the ring modulator, in its ideal form, is a product modulator for a square-wave carrier and the baseband signal, as illustrated in Fig. 3.14 for the case of a sinusoidal modulating wave.

The square-wave carrier $c(t)$ can be represented by a Fourier series as follows

$$c(t)=\frac{4}{\pi}\sum_{n=1}^{\infty}\frac{(-1)^{n-1}}{2n-1}\cos[2\pi f_c t(2n-1)] \tag{3.30}$$

The ring modulator output is therefore

$$\begin{aligned} s(t) &= c(t)m(t) \\ &= \frac{4}{\pi}\sum_{n=1}^{\infty}\frac{(-1)^{n-1}}{2n-1}\cos[2\pi f_c t(2n-1)]m(t) \end{aligned} \tag{3.31}$$

We see that there is no output from the modulator at the carrier frequency; that is, the modulator output consists entirely of modulation products. Thus the ring modulator is referred to as a *double-balanced modulator*, because it is balanced with respect to the baseband signal as well as the square-wave carrier.

Assuming that $m(t)$ is limited to the frequency band $-W \leq f \leq W$, the spectrum of the modulator output consists of sidebands around each of the odd harmonics of the square-wave carrier $c(t)$, as illustrated in Fig. 3.15. The desired pair of

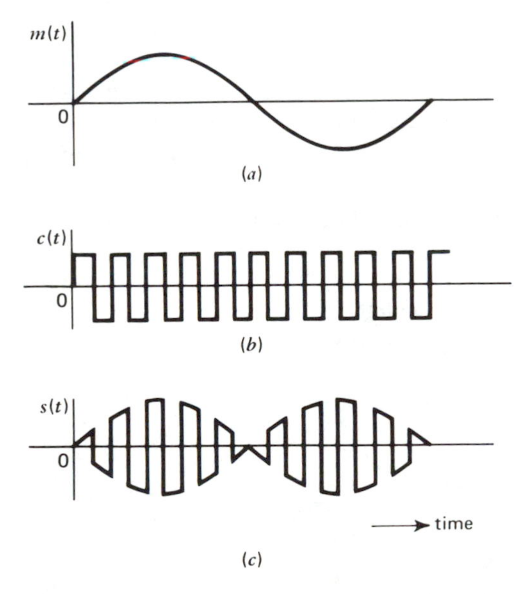

Figure 3.14 Waveforms illustrating the operation of the ring modulator for a sinusoidal modulating wave. *(a)* Modulating wave. *(b)* Square wave carrier. *(c)* Modulated wave.

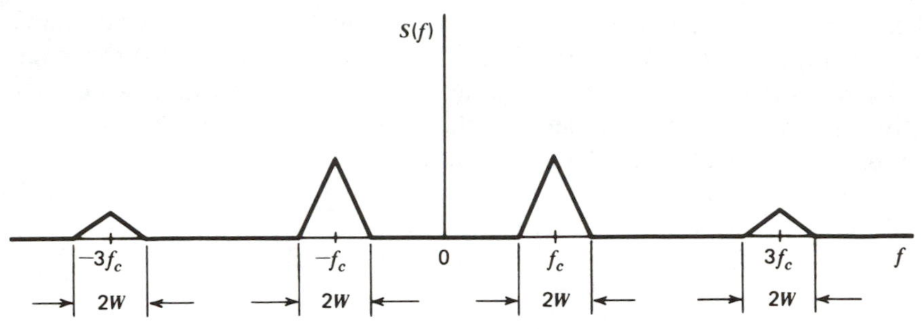

Figure 3.15 Illustrating the spectrum of ring modulator output.

sidebands around the carrier frequency f_c may be selected by using a band-pass filter of mid-band frequency f_c and bandwidth $2W$. It is apparent from Fig. 3.15 that to prevent *sideband overlap*, we must choose $f_c > W$.

3.6 COHERENT DETECTION OF DSBSC WAVES

The baseband signal $m(t)$ can be uniquely recovered from a DSBSC wave $s(t)$ by first multiplying $s(t)$ with a locally generated sine-wave and then low-pass filtering the product, as in Fig. 3.16. It is assumed that the local oscillator signal is exactly coherent or synchronized, in both frequency and phase, with the carrier wave $c(t)$ used in the product modulator to generate $s(t)$. This method of demodulation is known as *coherent detection* or *synchronous detection*.

It is instructive to derive coherent detection as a special case of the more general demodulation process using a local oscillator signal of the same frequency but arbitrary phase difference ϕ, measured with respect to the carrier wave $c(t)$. Thus, denoting the local oscillator signal by $A_c' \cos(2\pi f_c t + \phi)$, and using Eq. (3.25) for the DSBSC wave $s(t)$, we find that the product modulator output in Fig. 3.16 is

$$\begin{aligned} v(t) &= A_c' \cos(2\pi f_c t + \phi) s(t) \\ &= A_c A_c' \cos(2\pi f_c t) \cos(2\pi f_c t + \phi) m(t) \\ &= \tfrac{1}{2} A_c A_c' \cos(4\pi f_c t + \phi) m(t) + \tfrac{1}{2} A_c A_c' \cos \phi \, m(t) \end{aligned} \tag{3.32}$$

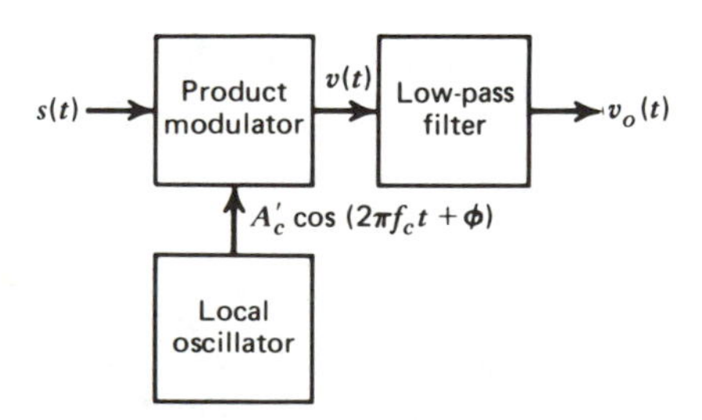

Figure 3.16 Coherent detection of DSBSC waves.

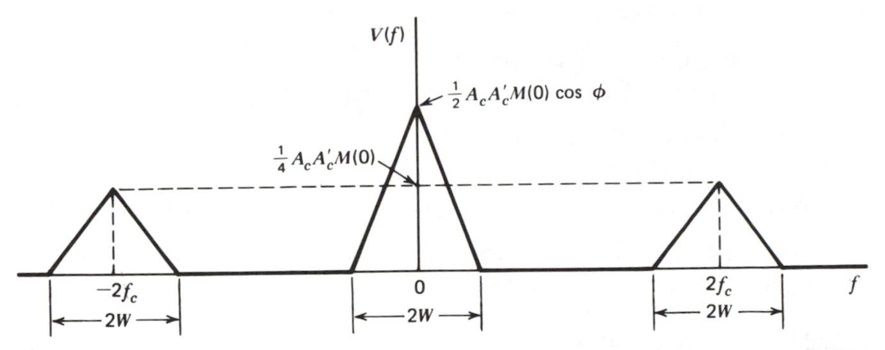

Figure 3.17 Illustrating the spectrum of a product modulator output with a DSBSC wave as input.

The first term in Eq. (3.32) represents a DSBSC wave with a carrier frequency $2f_c$, whereas the second term is proportional to the baseband signal $m(t)$. This is further illustrated by the spectrum $V(f)$ shown in Fig. 3.17, where it is assumed that the baseband signal $m(t)$ is limited to the interval $-W \leqslant f \leqslant W$. It is therefore apparent that the first term in Eq. (3.32) is removed by the low-pass filter in Fig. 3.16, provided that the cutoff frequency of this filter is greater than W but less than $2f_c - W$, which is satisfied by choosing $f_c > W$. At the filter output we then obtain a signal given by

$$v_o(t) = \tfrac{1}{2} A_c A'_c \cos \phi m(t) \tag{3.33}$$

The demodulated signal $v_o(t)$ is therefore proportional to $m(t)$ when the phase error ϕ is a constant. The amplitude of this demodulated signal is maximum when $\phi = 0$, and it is minimum (zero) when $\phi = \pm\pi/2$. The zero demodulated signal, which occurs for $\phi = \pm\pi/2$, represents the *quadrature null effect* of the coherent detector. Thus the phase error ϕ in the local oscillator causes the detector output to be attenuated by a factor equal to $\cos \phi$. As long as the phase error ϕ is constant, the detector output provides an undistorted version of the original baseband signal $m(t)$. In practice, however, we usually find that the phase error ϕ varies randomly with time, due to random variations in the communication channel. The result is that at the detector output, the multiplying factor $\cos \phi$ also varies randomly with time, which is obviously undesirable. Therefore, circuitry must be provided in the receiver to maintain the local oscillator in perfect synchronism, in both frequency and phase, with the carrier wave used to generate the DSBSC wave in the transmitter. The resulting receiver complexity is the price that must be paid for suppressing the carrier wave to save transmitter power.

Example 2 Single-tone modulation (continued)

Consider again the sinusoidal modulating signal

$$m(t) = A_m \cos(2\pi f_m t) \tag{3.34}$$

The corresponding DSBSC wave is given by

$$\begin{aligned} s(t) &= A_c A_m \cos(2\pi f_c t)\cos(2\pi f_m t) \\ &= \tfrac{1}{2} A_c A_m \cos[2\pi(f_c + f_m)t] + \tfrac{1}{2} A_c A_m \cos[2\pi(f_c - f_m)t] \end{aligned} \quad (3.35)$$

Figure 3.3(*d*) shows a sketch of this modulated wave.

The Fourier transform of $s(t)$ is therefore

$$S(f) = \tfrac{1}{4} A_c A_m [\delta(f - f_c - f_m) + \delta(f + f_c + f_m) + \delta(f - f_c + f_m) + \delta(f + f_c - f_m)] \quad (3.36)$$

Thus the spectrum of the DSBSC wave, for the case of a sinusoidal modulating wave, consists of delta functions at $f_c \pm f_m$ and $-f_c \pm f_m$, as in Fig. 3.3(*d*).

Assuming perfect synchronism between the local oscillator in Fig. 3.16 and the carrier wave, we find that the product modulator output is

$$\begin{aligned} v(t) &= A_c' \cos(2\pi f_c t)\{\tfrac{1}{2} A_c A_m \cos[2\pi(f_c - f_m)t] + \tfrac{1}{2} A_c A_m \cos[2\pi(f_c + f_m)t]\} \\ &= \tfrac{1}{4} A_c' A_c A_m \cos[2\pi(2f_c - f_m)t] + \tfrac{1}{4} A_c' A_c A_m \cos(2\pi f_m t) \\ &\quad + \tfrac{1}{4} A_c' A_c A_m \cos[2\pi(2f_c + f_m)t] + \tfrac{1}{4} A_c' A_c A_m \cos(2\pi f_m t) \end{aligned} \quad (3.37)$$

where the first two terms are produced by the lower side-frequency, whereas the last two terms are produced by the upper side-frequency. The first and third terms, of frequencies $2f_c - f_m$ and $2f_c + f_m$, respectively, are removed by the low-pass filter in Fig. 3.16. The coherent detector output thus reproduces the original modulating wave. Note, however, that this detector output appears as two equal terms, one derived from the upper side-frequency and the other from the lower side-frequency. We conclude, therefore, that for the transmission of information, only one side-frequency is necessary. We shall have more to say about this issue in Section 3.8.

Costas Receiver

One method of obtaining a practical synchronous receiving system, suitable for demodulating DSBSC waves, is to use the *Costas receiver* shown in Fig. 3.18.* This system consists of two coherent detectors supplied with the same input signal, namely, the incoming DSBSC wave $A_c \cos(2\pi f_c t) m(t)$, but with individual local oscillator signals that are in phase quadrature to each other. The frequency of the local oscillator is adjusted to be the same as the carrier frequency f_c, which is assumed known a priori. The detector in the upper path is referred to as the *in-phase coherent detector* or *I-channel*, and that in the lower path is referred to as the *quadrature-phase coherent detector* or *Q-channel*. These two detectors are coupled together to form a negative feedback system designed in such a way as to maintain the local oscillator synchronous with the carrier wave. To understand the operation of this receiver, suppose that the local oscillator signal is of the same phase as the carrier wave $A_c \cos(2\pi f_c t)$ used to generate the incoming DSBSC wave. Under these conditions, we find that the *I*-channel output contains the desired demodulated signal $m(t)$, whereas the *Q*-channel output is zero due to the quadrature null effect of the *Q*-channel. Suppose next the local oscillator phase drifts from its proper value by a small amount ϕ radians. The *I*-channel output will

* J. P. Costas, "Synchronous communications," *Proc. IRE*, vol. 44, pp. 1713–1718, Dec. 1956.

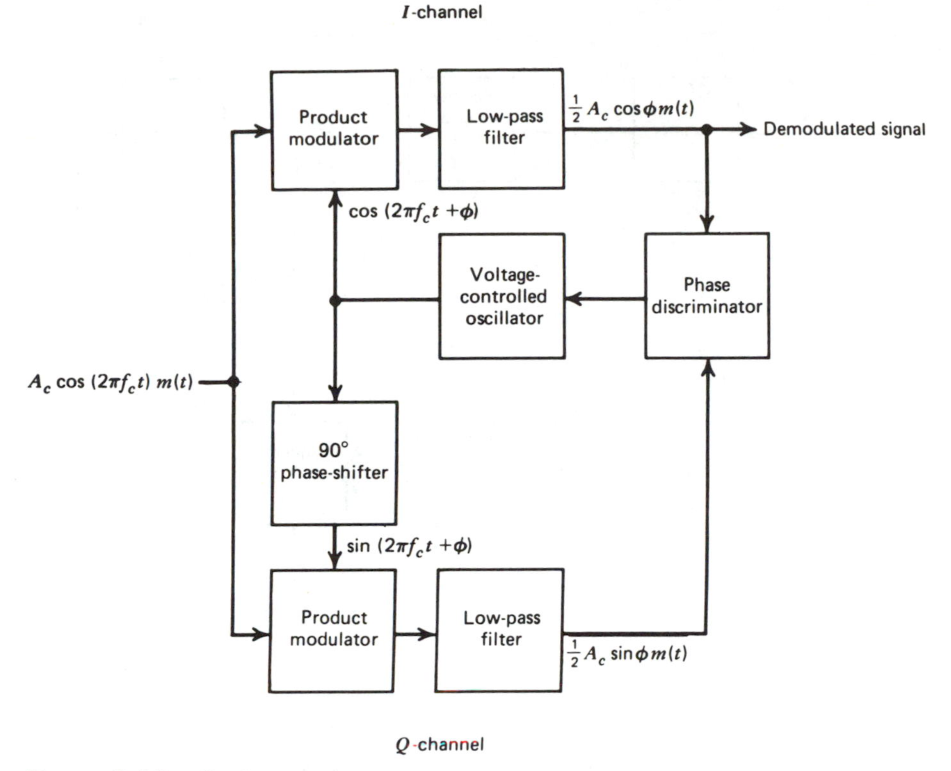

Figure 3.18 Costas receiver.

remain essentially unchanged, but there will now be some signal appearing at the Q-channel output, which is proportional to $\sin\phi \simeq \phi$. This Q-channel output will have the same polarity as the I-channel output for one direction of local oscillator phase drift and opposite polarity for the opposite direction of local oscillator phase drift. Thus, by combining the I- and Q-channel outputs in a *phase discriminator* (which consists of a multiplier followed by a low-pass filter), a dc control signal is obtained which automatically corrects for local oscillator phase errors.

It is apparent that phase control in the Costas receiver ceases with modulation, and that phase-lock has to be re-established with the reappearance of modulation. This is not a serious problem when receiving voice transmission, because the lock-up process normally occurs so rapidly that no perceptible distortion is observed.

Squaring Loop

Another method for generating a reference carrier from a DSBSC wave is to use a *squaring loop*, the mechanization of which is illustrated in Fig. 3.19(a). At the input end of this carrier-recovery circuit, we have a square-law device characterized by the relation

$$y(t) = s^2(t) \tag{3.38}$$

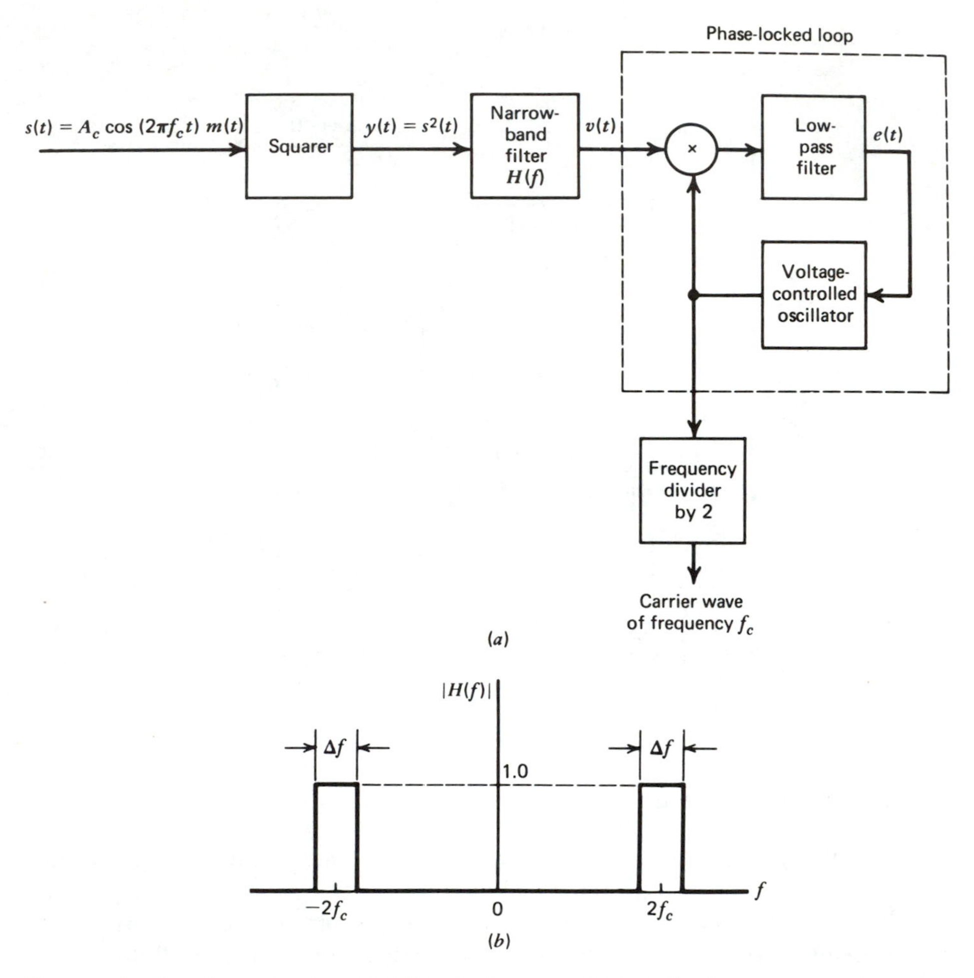

Figure 3.19 Squaring loop. *(a)* Block diagram. *(b)* Amplitude response of narrow-band filter.

Therefore, with the DSBSC wave

$$s(t) = A_c \cos(2\pi f_c t) m(t)$$

applied to the input of this square-law device, we obtain

$$\begin{aligned} y(t) &= A_c^2 \cos^2(2\pi f_c t) m^2(t) \\ &= \frac{A_c^2}{2} m^2(t)[1 + \cos(4\pi f_c t)] \end{aligned} \tag{3.39}$$

The signal $y(t)$ is next applied to a narrow-band filter centered about $2f_c$. Assuming that the amplitude response of this filter is idealized as in Fig. 3.19(*b*), and that the bandwidth Δf of the filter is small enough for the spectrum of $y(t)$ to be essentially constant inside the passband of the filter, we find the output is approximately sinusoidal as shown by (see Problem 3.16):

$$v(t) \simeq \frac{A_c^2}{2} E \, \Delta f \cos(4\pi f_c t) \tag{3.40}$$

where E is the energy of the message signal $m(t)$.

The resultant sinusoidal wave with twice the carrier frequency at the narrow-band filter output is tracked by means of a *phase-locked loop* (PLL). This loop consists of a multiplier, low-pass filter, and voltage-controlled oscillator (VCO) connected in the form of a negative feedback system. At the multiplier output we obtain two terms, one depending on the difference between the frequency and phase of the VCO output and those of the signal $v(t)$ at the PLL input, and the other depending on their sum. This latter term is removed by the loop filter. The resulting *error signal*, denoted by $e(t)$ in Fig. 3.19(a) is applied to the VCO input, causing the frequency of the VCO to coincide with that of the PLL input $v(t)$, and its phase to be in quadrature with that of $v(t)$, which in turn tends to reduce $e(t)$ to zero. In this way, the VCO is enabled to track changes in the frequency and phase of the PLL input $v(t)$.

Finally, the VCO output is frequency-divided by a factor of two to produce the desired carrier wave which is available for coherent detection of the DSBSC wave. As a result of this frequency division, however, we have a *phase ambiguity* of π radians. This phase ambiguity arises because a phase change of 2π radians at the frequency divider input produces a phase change of π radians at the output. Hence, the frequency divider output may equal $\cos(2\pi f_c t)$ or $\cos(2\pi f_c t + \pi)$. If the incorrect phase is used, the polarity of the demodulated wave will be reversed. This difficulty is not significant in the case of voice communications.

3.7 QUADRATURE-CARRIER MULTIPLEXING

The *quadrature-carrier multiplexing* or *quadrature-amplitude modulation* (QAM) scheme enables two DSBSC modulated waves (resulting from the application of two *independent* message signals) to occupy the same transmission bandwidth, and yet it allows for the separation of the two message signals at the receiver output. It is therefore a *bandwidth-conservation scheme*.

A block diagram of the quadrature-carrier multiplexing system is shown in Fig. 3.20. The transmitter part of the system, shown in part (a) of the figure, involves the use of two separate product modulators that are supplied with two carrier waves of the same frequency but differing in phase by 90 degrees. The transmitted signal $s(t)$ consists of the sum of these two product modulator outputs, as shown by

$$s(t) = A_c m_1(t) \cos(2\pi f_c t) + A_c m_2(t) \sin(2\pi f_c t) \tag{3.41}$$

where $m_1(t)$ and $m_2(t)$ denote the two different message signals applied to the product modulators. Thus $s(t)$ occupies a transmission bandwidth of $2W$, centered at the carrier frequency f_c, where W is the message bandwidth of $m_1(t)$ or $m_2(t)$.

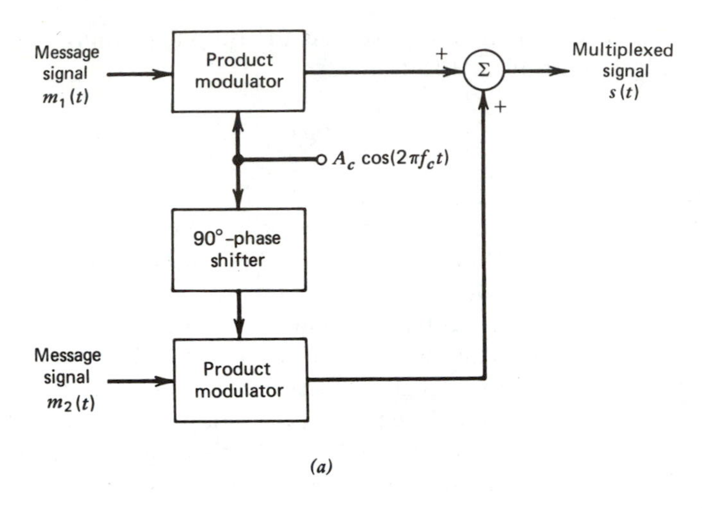

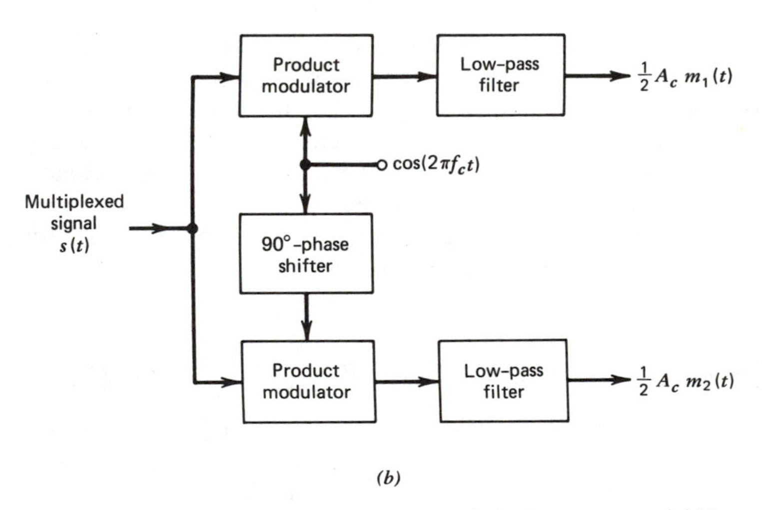

Figure 3.20 Quadrature-carrier multiplexing system. *(a)* Transmitter. *(b)* Receiver.

According to Eq. (3.41), we may view $A_c m_1(t)$ as the in-phase component of the multiplexed (band-pass) signal $s(t)$, and $-A_c m_2(t)$ as its quadrature component.

The receiver part of the system is shown in Fig. 3.20(*b*). The multiplexed signal $s(t)$ is applied simultaneously to two separate coherent detectors that are supplied with two local carriers of the same frequency, but differing in phase by 90 degrees. The output of one detector is $\frac{1}{2}A_c m_1(t)$, whereas the output of the second detector is $\frac{1}{2}A_c m_2(t)$. For the system to operate satisfactorily, it is important to maintain the correct phase and frequency relationships between the local oscillators used in the transmitter and receiver parts of the system.

Quadrature-carrier multiplexing is used in color television* (see Problem 3.20). In this system, synchronizing pulses are transmitted in order to maintain the local oscillator in the receiver at the correct frequency and phase with respect to the carrier used in the transmitter.

3.8 SINGLE-SIDEBAND MODULATION (SSB)

Amplitude modulation and double-sideband suppressed-carrier modulation are wasteful of bandwidth because they both require a transmission bandwidth equal to twice the message bandwidth. In either case, one-half of the transmission bandwidth is occupied by the upper sideband of the modulated wave, whereas the other half is occupied by the lower sideband. However, the upper and lower sidebands are uniquely related to each other by virtue of their symmetry about the carrier frequency; that is, given the amplitude and phase spectra of either sideband, we can uniquely determine the other. This means that insofar as the transmission of information is concerned, only one sideband is necessary, and if the carrier and the other sidebands are suppressed at the transmitter, no information is lost. Thus the communication channel needs to provide only the same bandwidth as the baseband signal, a conclusion which is intuitively satisfying. When only one sideband is transmitted, the modulation system is referred to as a *single-sideband* (SSB) system.

The precise frequency-domain description of an SSB wave depends on which sideband is transmitted. Consider a baseband signal $m(t)$ with a spectrum $M(f)$ that is limited to the band $-W \leqslant f \leqslant W$, as in Fig. 3.21(*a*). The spectrum of the DSBSC wave, obtained by multiplying $m(t)$ by the carrier wave $A_c \cos(2\pi f_c t)$, is as shown in Fig. 3.21(*b*). The upper sideband is represented in duplicate by the frequencies above f_c and those below $-f_c$; and when only the upper sideband is transmitted, the resulting SSB wave $s(t)$ has the spectrum shown in Fig. 3.21(*c*). Likewise, the lower sideband is represented in duplicate by the frequencies below f_c (for positive frequencies) and those above $-f_c$ (for negative frequencies); and when only the lower sideband is transmitted, the spectrum of the corresponding SSB wave is as shown in Fig. 3.21(*d*). Thus the essential function of an SSB modulation system is to translate the spectrum of the modulating wave, either with or without inversion, to a new location in the frequency domain; and the transmission bandwidth requirement of the system is one-half that of an AM or DSBSC modulation system. The benefit of using SSB modulation is therefore derived principally from the reduced bandwidth requirement and the elimination of the high-power carrier wave. The principal disadvantage of the SSB modulation system is its cost and complexity.

To describe an SSB wave in the time domain, we may use the representation developed in Section 2.14 for a band-pass signal. According to this representation,

* H. Stark and F. B. Tuteur, *Modern Electrical Communications: Theory and Systems*, pp. 285–297 (Prentice-Hall, 1979). F. E. Terman, *Electronic and Radio Engineering*, Fourth Edition, pp. 999–1003 (McGraw-Hill, 1955).

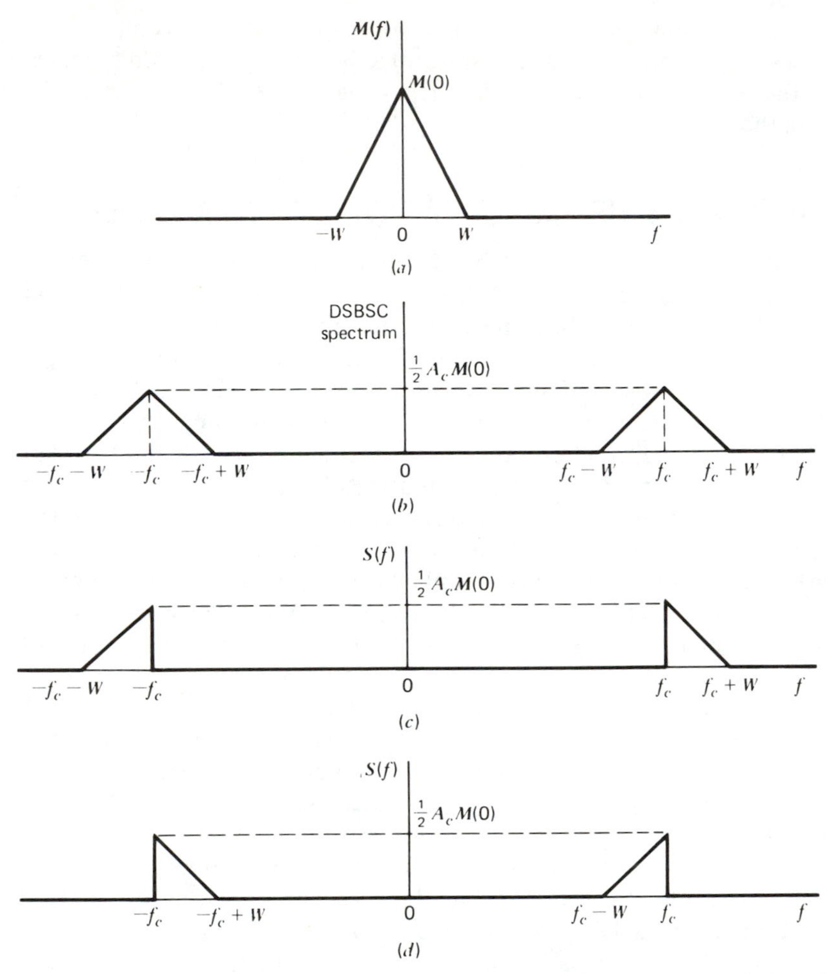

Figure 3.21 *(a)* Spectrum of baseband signal. *(b)* Spectrum of DSBSC wave. *(c)* Spectrum of SSB wave with the upper sideband transmitted. *(d)* Spectrum of SSB wave with the lower sideband transmitted.

we may express the time-domain description of an SSB wave $s(t)$ in the canonical form

$$s(t) = s_c(t)\cos(2\pi f_c t) - s_s(t)\sin(2\pi f_c t) \tag{3.42}$$

where $s_c(t)$ is the in-phase component of the SSB wave, and $s_s(t)$ is its quadrature component. The in-phase component $s_c(t)$, except for a scaling factor, may be derived from $s(t)$ by first multiplying $s(t)$ by $\cos(2\pi f_c t)$ and then passing the product through a low-pass filter. Similarly, the quadrature component $s_s(t)$, except for a scaling factor, may be derived from $s(t)$ by first multiplying $s(t)$ by $\sin(2\pi f_c t)$ and then passing the product through an identical low-pass identical filter. We thus find that the Fourier transforms of $s_c(t)$ and $s_s(t)$ are related to that of the SSB wave

$s(t)$ as follows, respectively (see Problem 2.62):

$$S_c(f) = \begin{cases} S(f-f_c) + S(f+f_c), & -W \leqslant f \leqslant W \\ 0, & \text{elsewhere} \end{cases} \tag{3.43}$$

and

$$S_s(f) = \begin{cases} j[S(f-f_c) - S(f+f_c)], & -W \leqslant f \leqslant W \\ 0, & \text{elsewhere} \end{cases} \tag{3.44}$$

where $-W \leqslant f \leqslant W$ defines the frequency band occupied by the message signal $m(t)$. Consider now the case of an SSB wave that is obtained by transmitting only the upper sideband. The spectrum of such a modulated wave is shown in Fig. 3.21(*c*), which is reproduced for convenience in Fig. 3.22(*a*). In parts (*b*) and (*c*) of this figure we have depicted two frequency-shifted spectra pertaining to $S(f-f_c)$ and $S(f+f_c)$, respectively. Therefore, from Eqs. (3.43) and (3.44) it follows that the corresponding spectra of the in-phase component $s_c(t)$ and the quadrature component $s_s(t)$ are as shown in parts (*d*) and (*e*) of Fig. 3.22, respectively. On the basis of Fig. 3.22(*d*), we may thus write

$$S_c(f) = \tfrac{1}{2} A_c M(f) \tag{3.45}$$

where $M(f)$ is the Fourier transform of the message signal $m(t)$. Accordingly, the in-phase component $s_c(t)$ is defined by

$$s_c(t) = \tfrac{1}{2} A_c m(t) \tag{3.46}$$

Next, on the basis of Fig. 3.22(*e*), we may write

$$S_s(f) = \begin{cases} -\dfrac{j}{2} A_c M(f), & f > 0 \\ 0, & f = 0 \\ \dfrac{j}{2} A_c M(f), & f < 0 \end{cases}$$

$$= -\frac{j}{2} A_c \operatorname{sgn}(f) M(f) \tag{3.47}$$

where $\operatorname{sgn}(f)$ is the signum function, equal to $+1$ for positive frequencies, zero for $f=0$, and -1 for negative frequencies. However, we note that (see Section 2.12)

$$-j \operatorname{sgn}(f) M(f) = \hat{M}(f) \tag{3.48}$$

where $\hat{M}(f)$ is the Fourier transform of $\hat{m}(t)$, the Hilbert transform of $m(t)$. Hence, substituting Eq. (3.48) in (3.47), we get

$$S_s(f) = \tfrac{1}{2} A_c \hat{M}(f) \tag{3.49}$$

which shows that the quadrature component $s_s(t)$ is defined by

$$s_s(t) = \tfrac{1}{2} A_c \hat{m}(t) \tag{3.50}$$

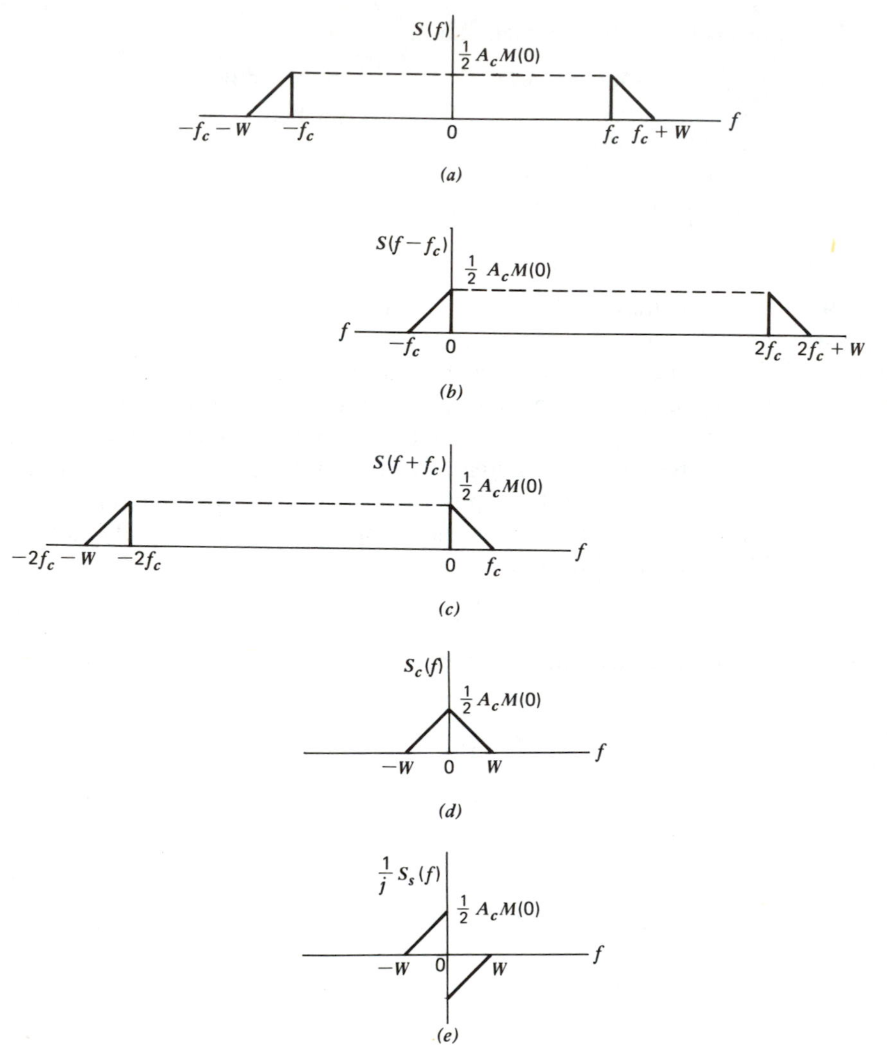

Figure 3.22 *(a)* Spectrum of SSB wave. *(b)* Spectrum of SSB wave shifted to the right by f_c. *(c)* Spectrum of SSB wave shifted to the left by f_c. *(d)* Spectrum of in-phase component. *(e)* Spectrum of quadrature component.

Thus substituting Eqs. (3.46) and (3.50) in (3.42), we find that the canonical representation of an SSB wave $s(t)$ obtained by transmitting only the upper sideband is as follows

$$s(t) = \frac{1}{2}A_c m(t)\cos(2\pi f_c t) - \frac{1}{2}A_c \hat{m}(t)\sin(2\pi f_c t) \tag{3.51}$$

Following a procedure similar to that described above, we find that the canonical representation for an SSB wave $s(t)$ obtained by transmitting only the lower sideband, as illustrated in Fig. 3.21(*d*), is of the same form as in Eq. (3.51), except that the minus sign on the right-hand side of the equation is replaced by a plus sign.

Example 3 Single-tone modulation (continued)

Consider again the sinusoidal modulating wave

$$m(t) = A_m \cos(2\pi f_m t) \tag{3.52}$$

The Hilbert transform of this signal is (sée Example 23 of Chapter 2)

$$\hat{m}(t) = A_m \sin(2\pi f_m t) \tag{3.53}$$

Therefore, substituting Eqs. (3.52) and (3.53) in (3.51), we find that the SSB wave, obtained by transmitting only the upper side-frequency, is defined by

$$\begin{aligned} s(t) &= \tfrac{1}{2} A_c A_m [\cos(2\pi f_m t)\cos(2\pi f_c t) - \sin(2\pi f_m t)\sin(2\pi f_c t)] \\ &= \tfrac{1}{2} A_c A_m \cos[2\pi(f_c + f_m)t] \end{aligned} \tag{3.54}$$

This is exactly the same as the result obtained by suppressing the lower side-frequency $f_c - f_m$ of the corresponding DSBSC wave of Eq. (3.35). The SSB wave of Eq. (3.54) and its spectrum are illustrated in Fig. 3.3(*e*).

Next, using Eq. (3.51) with the minus sign replaced by a plus sign, we find that the SSB wave, obtained by transmitting only the lower side-frequency, is defined by

$$\begin{aligned} s(t) &= \tfrac{1}{2} A_c A_m [\cos(2\pi f_m t)\cos(2\pi f_c t) + \sin(2\pi f_c t)\sin(2\pi f_m t)] \\ &= \tfrac{1}{2} A_c A_m \cos[2\pi(f_c - f_m)t] \end{aligned} \tag{3.55}$$

which is exactly the same as the result obtained by suppressing the upper side-frequency $f_c + f_m$ of the DSBSC wave of Eq. (3.35). The SSB wave of Eq. (3.55) and its spectrum are illustrated in Fig. 3.3(*f*).

3.9 GENERATION OF SSB WAVES

We next describe two methods of generating the SSB wave that are in common use, namely, the *frequency discrimination method* and the *phase discrimination method*. These two methods are based on the frequency-domain and time-domain descriptions of the SSB wave, respectively.

Frequency Discrimination Method

The frequency discrimination method may be used to generate an SSB wave when the baseband (i.e., the band of significant frequencies in the modulating wave) is restricted and appropriately related to the carrier frequency. Under these conditions the desired sideband will appear in a nonoverlapping interval in the spectrum in such a way that it may be selected by an appropriate filter. Thus an SSB modulator based on frequency discrimination consists basically of a product modulator (e.g., ring modulator) and a filter which is designed to pass the desired sideband of the DSBSC wave at the modulator output and reject the other sideband. A block diagram of this modulator is shown in Fig. 3.23(*a*). The most severe requirement of this method of SSB generation usually arises from the unwanted sideband, the nearest frequency component of which is separated from the desired sideband by twice the lowest frequency component of the modulating wave.

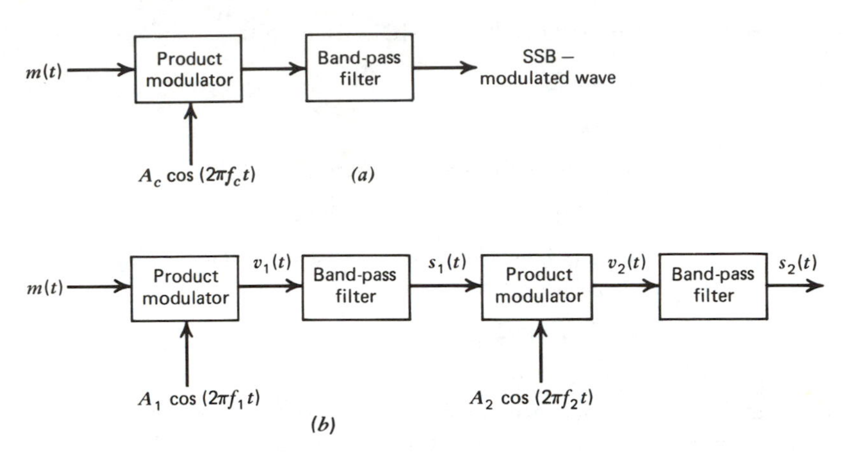

Figure 3.23 *(a)* Block diagram of the frequency discrimination method (single stage) for generating SSB waves. *(b)* Block diagram of a two-stage SSB modulator.

In designing the band-pass filter in the SSB modulation scheme of Fig. 3.23(*a*), we must therefore satisfy two basic requirements: (1) the passband of the filter occupies the same frequency range as the spectrum of the desired SSB wave, and (2) the width of the transition band of the filter, separating the passband from the stopband where the unwanted sideband of the filter input lies, is twice the lowest frequency component of the modulating wave. We usually find that this kind of frequency discrimination can be satisfied only by using highly selective filters, which can be realized using crystal resonators with a Q factor per resonator in the range of 1000 to 2000.*

When it is necessary to generate an SSB wave occupying a frequency band that is much higher than that of the baseband signal (e.g., translating a voice signal to the high-frequency region of the radio spectrum), it becomes very difficult to design an appropriate filter that will pass the desired sideband and reject the other, using the simple arrangement of Fig. 3.23(*a*). In such a situation it is necessary to resort to a multiple-modulation process so as to ease the filtering requirement. This approach is illustrated in Fig. 3.23(*b*) involving two stages of modulation. The SSB wave at the first filter output is used as the modulating wave for the second product modulator, which produces a DSBSC wave with a spectrum that is symmetrically spaced about the second carrier frequency f_2. The frequency separation between the sidebands of this DSBSC wave is effectively twice the first carrier frequency f_1, thereby enabling the removal of the unwanted sideband by the second filter.

* For a discussion of the filtering requirements in generating SSB waves, see C. F. Kurth, "Generation of single-sideband signals in multiplex communication systems," *IEEE Trans. Circuits and Systems*, vol. CAS-23, pp. 1–17, Jan. 1976.

Example 4

Consider the use of the two-stage modulation scheme of Fig. 3.23(*b*) to generate an SSB wave assuming that the input signal $m(t)$ consists of a voice signal occupying the frequency band 0.3 to 3.4 kHz, and the oscillator frequencies $f_1 = 100$ kHz and $f_2 = 10$ MHz. We wish to specify the sidebands of the various modulated waves appearing in this system, and also specify the filtering requirements of the system.

The spectrum of the DSBSC wave $v_1(t)$ appearing at the output of the first product modulator has a lower sideband occupying the frequency band 96.6 to 99.7 kHz and an upper sideband occupying the frequency band 100.3 to 103.4 kHz. We assume that only the upper sideband is transmitted, resulting in the SSB wave $s_1(t)$ the spectrum of which occupies the frequency band 100.3 to 103.4 kHz. To achieve this, the first band-pass filter must have a lower transition band equal to 99.7 to 100.3 kHz so as to suppress the lower sideband.

The spectrum of the DSBSC wave $v_2(t)$ appearing at the output of the second product modulator has a lower sideband occupying the frequency band 9.8966 to 9.8997 MHz and an upper sideband occupying the frequency band 10.1003 to 10.1034 MHz. Here again, assuming that only the upper sideband is transmitted, we find that the spectrum of the resulting SSB wave $s_2(t)$ occupies the frequency band 10.1003 to 10.1034 MHz. To achieve this, the second band-pass filter must have a lower transition band equal to 9.8997 to 10.1003 MHz so as to suppress the lower sideband. The signal $s_2(t)$ is the desired modulated wave.

The positive frequency bands occupied by the sidebands of the modulated waves $v_1(t)$, $s_1(t)$, $v_2(t)$, and $s_2(t)$ appearing at the various points in the two-stage modulation system of Fig. 3.23(*b*) are summarized in Table 3.1.

Table 3.1 Summary of the Positive Frequency Bands Occupied by the Sidebands of the Modulated Waves in Fig. 3.23(b).

Modulated Wave	Frequency Band Occupied by the Lower Sideband	Frequency Band Occupied by the Upper Sideband
$v_1(t)$	96.6–99.7 kHz	100.3–103.4 kHz
$s_1(t)$		100.3–103.4 kHz
$v_2(t)$	9.8966–9.8997 MHz	10.1003–10.1034 MHz
$s_2(t)$		10.1003–10.1034 MHz

Phase Discrimination Method

The phase discrimination method of generating an SSB wave involves two separate simultaneous modulation processes and subsequent combination of the resulting modulation products, as shown in Fig. 3.24. The derivation of this system follows directly from Eq. (3.51), which defines the canonical representation of SSB waves in the time domain for the case when only the upper sideband is transmitted. The system uses two product modulators *A* and *B* supplied with carrier waves in phase quadrature to each other. The incoming baseband signal $m(t)$ is applied to product modulator *A*, producing a DSBSC wave that contains *reference phase* sidebands symmetrically spaced about carrier frequency f_c. The Hilbert transform $\hat{m}(t)$ of $m(t)$ is applied to product modulator *B*, producing a DSBSC wave that contains side-

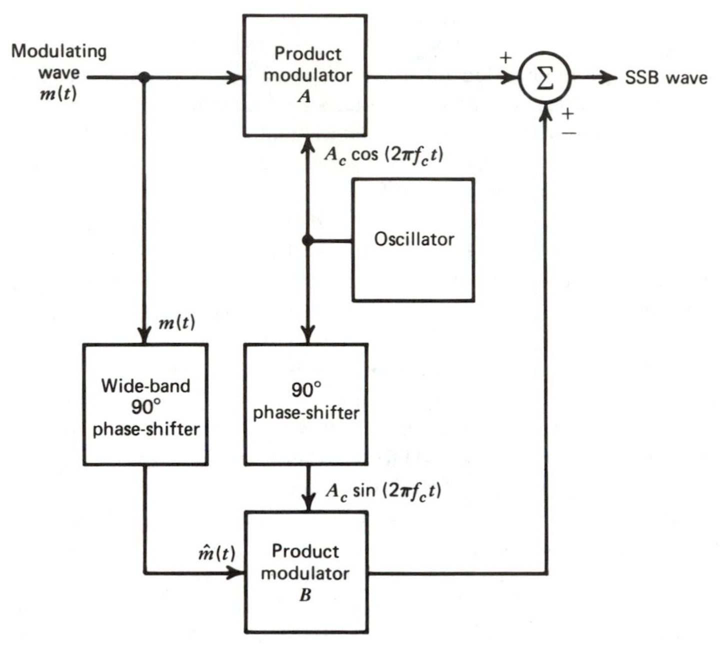

Figure 3.24 Block diagram of the phase discrimination method for generating SSB waves.

bands having identical amplitude spectra as those of modulator A but of such relative phase that vector addition or subtraction of the two modulator outputs in the summing device results in cancellation of one set of sidebands and reinforcement of the other set. The use of the product modulator B output with a plus sign at the summing junction yields an SSB wave with only the lower sideband, whereas the use of this output with a minus sign yields an SSB wave with the only the upper sideband. In this way the desired SSB wave is produced. The SSB modulator of Fig. 3.24 is also known as the *Hartley modulator*.*

To generate the Hilbert transform $\hat{m}(t)$ of the input baseband signal $m(t)$ in Fig. 3.24, we require a network that shifts the phase angle of every frequency component of $m(t)$ by 90 degrees but leaves the amplitude unchanged. In practice, it is difficult to design such a network over a wide enough frequency range of the modulating wave $m(t)$ to be useful for most applications. If, however, a phase-shifting network is included in each modulation path, the required constant phase-difference, in general, can be maintained with any desired tolerance over any prescribed frequency range.† Such an arrangement is shown in Fig. 3.25 where the phase-shifting networks are identified by α and β. The phase-shifts α and β are

* R. V. L. Hartley, U.S. Patent 1666206, April 17, 1928.

† S. Darlington, "Realization of a constant phase difference," *Bell Syst. Tech. J.*, vol. 29, pp. 94–104, Jan. 1950.

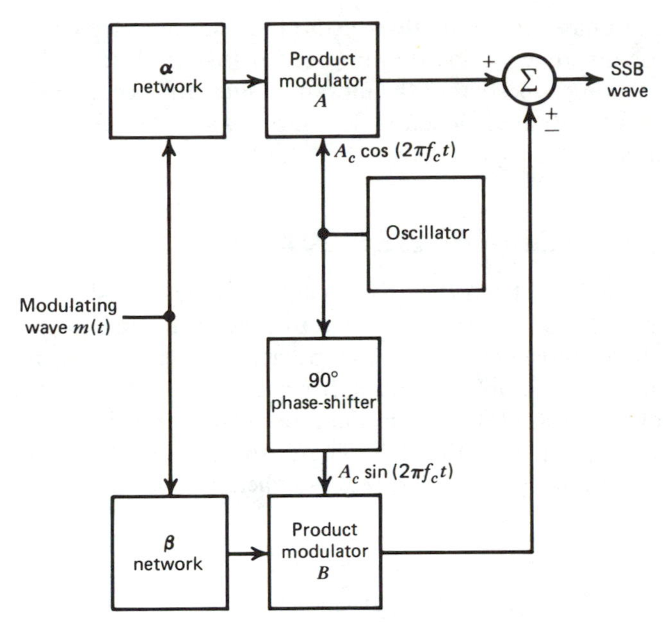

Figure 3.25 Block diagram of the phase discrimination method for generating SSB waves by using a pair of phase-shifting networks α and β to realize a constant 90-degree phase difference.

related by

$$\beta - \alpha = \frac{\pi}{2} \tag{3.56}$$

Since the SSB modulator of Fig. 3.25 does not require any sharp cutoff filters, it is possible to generate the desired sideband in a single frequency translational step regardless of how high the final frequency band of the modulated wave may be. However, the degree to which the unwanted sideband may be suppressed depends upon the following factors: (1) balancing accuracy of the product modulators, (2) control accuracy of the quadrature phase relationship of the two carrier waves, and (3) errors in the approximation of the constant 90 degree-phase difference between $m(t)$ and $\hat{m}(t)$. As a practical matter, it is quite easy to realize 20 dB suppression, reasonable to expect 30 dB suppression, and quite difficult to go beyond 40 dB, using the phase discrimination method.

The frequency discrimination and phase discrimination methods described above provide two well-known procedures for the generation of SSB waves. There is indeed a third method known as *Weaver's method*,* which may also be used to generate an SSB wave. Weaver's method is basically different from either the

* D. K. Weaver, Jr., "A third method of generation and detection of single-sideband signals," *Proc. IRE*, vol. 44, pp. 1703–1705, Dec. 1956.

frequency discrimination or phase discrimination method in that no sharp cutoff filters or wide-band phase-difference networks are needed (see Problem 3.25). This method has the interesting property that the unwanted sideband occupies the same band of frequencies as the desired sideband, and the unwanted sideband in the usual sense is not present.

3.10 DEMODULATION OF SSB WAVES

To recover the baseband signal $m(t)$ from the SSB wave $s(t)$, we have to shift the spectrum in Fig. 3.21(c) or (d) by the amounts $\pm f_c$ so as to convert the transmitted sideband back into the baseband signal. This can be accomplished by using coherent detection, which involves applying the SSB wave $s(t)$, together with a locally generated sine wave $A_c' \cos(2\pi f_c t)$, to a product modulator and then low-pass filtering the modulator output, as in Fig. 3.26. Thus, using Eq. (3.51), we find that the product modulator output is given by (for the case when the upper sideband only is transmitted)

$$\begin{aligned} v(t) &= \tfrac{1}{2}A_c A_c' \cos(2\pi f_c t)[m(t)\cos(2\pi f_c t) - \hat{m}(t)\sin(2\pi f_c t)] \\ &= \tfrac{1}{4}A_c A_c' m(t) + \tfrac{1}{4}A_c A_c'[m(t)\cos(4\pi f_c t) - \hat{m}(t)\sin(4\pi f_c t)] \end{aligned} \tag{3.57}$$

The first term in Eq. (3.57) is the desired demodulated signal, whereas the second term represents an SSB wave corresponding to a carrier frequency $2f_c$. The frequency translations involved in Eq. (3.57) are illustrated in Fig. 3.27. The high-

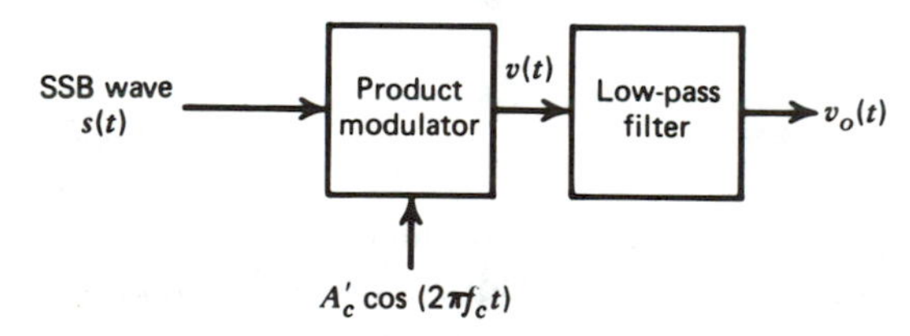

Figure 3.26 Coherent detection of an SSB wave.

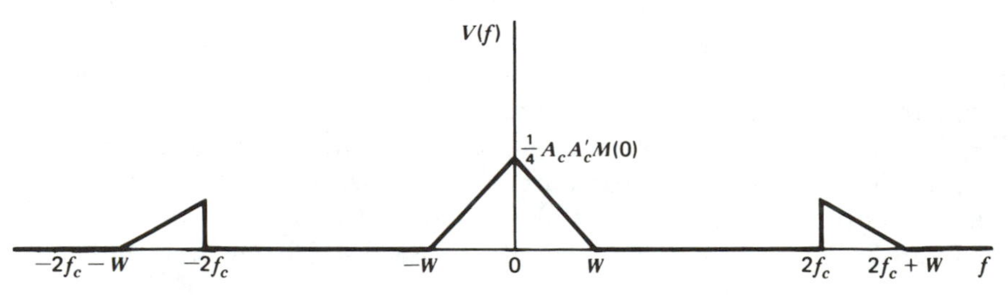

Figure 3.27 Spectrum of the product modulator output $v(t)$ in the coherent detector of Fig. 3.26.

frequency components in Eq. (3.57) are removed by the low-pass filter in Fig. 3.26, thereby yielding the desired baseband signal.

For this method of demodulation to work, there must of course be available at the receiver a sine-wave of the correct carrier frequency f_c and of the correct phase relationship with the carrier wave used to generate the incoming SSB wave. This can be provided by either transmitting a *pilot carrier* in addition to the selected sideband or by using a highly stable oscillator.

Any error in the frequency or the phase of the local oscillator signal in the receiver, with respect to the carrier wave, gives rise to distortion in the demodulated signal. To evaluate the nature of this distortion, assume a frequency error Δf so that the local oscillator signal is $A'_c \cos[2\pi(f_c+\Delta f)t]$. The resulting demodulated signal is given by (for the case when the upper sideband only is transmitted)

$$v_o(t)=\tfrac{1}{4}A_cA'_c[m(t)\cos(2\pi\,\Delta ft)+\hat{m}(t)\sin(2\pi\,\Delta ft)] \tag{3.58}$$

The demodulated signal in Eq. (3.58) represents an SSB wave corresponding to a carrier frequency Δf. The effect of frequency error Δf in the local oscillator may be interpreted as follows:

1. If the incoming SSB wave $s(t)$ contains the lower sideband and the frequency error Δf is positive, or equivalently if $s(t)$ contains the upper sideband and Δf is negative, then the frequency components of the demodulated signal $v_o(t)$ are shifted outward by the amount Δf, compared with the baseband signal $m(t)$. This is illustrated in Fig. 3.28(*b*) for the case of a baseband signal (e.g., voice) with an energy gap occupying the interval $-f_a \leqslant f \leqslant f_a$, as in part (*a*) of the figure.
2. If the incoming SSB wave $s(t)$ contains the upper sideband and the frequency error Δf is positive, or equivalently if $s(t)$ contains the lower sideband and Δf is negative, then the frequency components of the demodulated signal $v_o(t)$ are shifted inward by the amount Δf, compared with the baseband signal $m(t)$. This is illustrated in Fig. 3.28(*c*).

This type of distortion caused by frequency error in the demodulation process is unique to SSB modulation systems. In order to reduce the effect of frequency error distortion in telephone systems, we have to limit the frequency error to 2–5 Hz. This requires very precise and stable carrier frequencies at both the transmitting and receiving ends of the system.

Consider next the effect of a phase error ϕ in the local oscillator signal with reference to the carrier wave used to generate the incoming SSB wave $s(t)$. Thus, denoting the local oscillator signal by $A'_c \cos(2\pi f_c t+\phi)$, we find that the resulting demodulated signal is given by (for the case when the upper sideband only is transmitted)

$$v_o(t)=\tfrac{1}{4}A_cA'_c[m(t)\cos\phi+\hat{m}(t)\sin\phi] \tag{3.59}$$

It is evident that the demodulated signal $v_o(t)$ contains an unwanted component proportional to $\hat{m}(t)\sin\phi$, which cannot be removed by filtering. This distortion

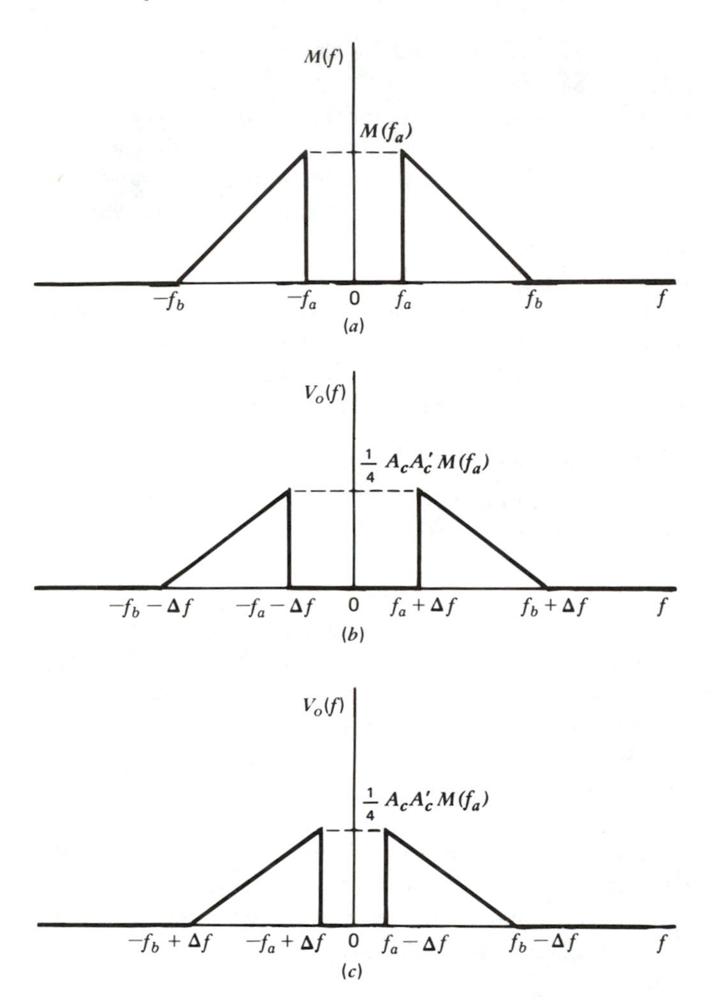

Figure 3.28 Illustrating the effect of a frequency error f on the output of the coherent detector with SSB wave $s(t)$ as input. *(a)* Spectrum of baseband signal with energy gap in the interval $-f_a < f < f_a$. *(b)* Spectrum of coherent detector output with $s(t)$ containing lower sideband and $\Delta f > 0$ or with $s(t)$ containing upper sideband and $\Delta f < 0$. *(c)* Spectrum of coherent detector output with $s(t)$ containing upper sideband and $\Delta f > 0$ or with $s(t)$ containing lower sideband and $\Delta f < 0$.

appears as a *phase distortion*. To show this, we take the Fourier transform of $v_o(t)$ in Eq. (3.59) to obtain

$$V_o(f) = \tfrac{1}{4} A_c A_c' [M(f)\cos\phi + \hat{M}(f)\sin\phi] \tag{3.60}$$

But from the definition of the Hilbert transform $\hat{m}(t)$, we know that

$$\hat{M}(f) = -j\,\text{sgn}(f) M(f) \tag{3.61}$$

Therefore, substituting Eq. (3.61) in (3.60), we get

$$V_o(f)=\begin{cases}\frac{1}{4}A_cA_c'M(f)\exp(-j\phi) & f>0\\ \frac{1}{4}A_cA_c'M(f)\exp(+j\phi) & f<0\end{cases} \tag{3.62}$$

Thus the error in the phase of the local oscillator signal results in phase distortion, where each frequency component of $m(t)$ undergoes a constant phase shift at the demodulator output. This phase distortion is usually not serious with voice communications because the human ear is relatively insensitive to phase distortion; the presence of phase distortion gives rise to a Donald Duck voice effect. In the transmission of music and video signals, on the other hand, phase distortion in the form of a constant phase difference in all components can be intolerable.

3.11 VESTIGIAL SIDEBAND MODULATION (VSB)

Single-sideband modulation is rather well-suited for the transmission of voice because of the energy gap that exists in the spectrum of voice signals between zero and a few hundred hertz. When the baseband signal contains significant components at extremely low frequencies (as in the case of television and telegraph signals), however, the upper and lower sidebands meet at the carrier frequency. This means that the use of SSB modulation is inappropriate for the transmission of such baseband signals due to the difficulty of isolating one sideband. This difficulty suggests another scheme known as *vestigial sideband* modulation (VSB), which is a compromise between SSB and DSBSC modulation. In this modulation scheme, one sideband is passed almost completely whereas just a trace, or *vestige*, of the other sideband is retained, as illustrated in Fig. 3.29. Specifically, the trans-

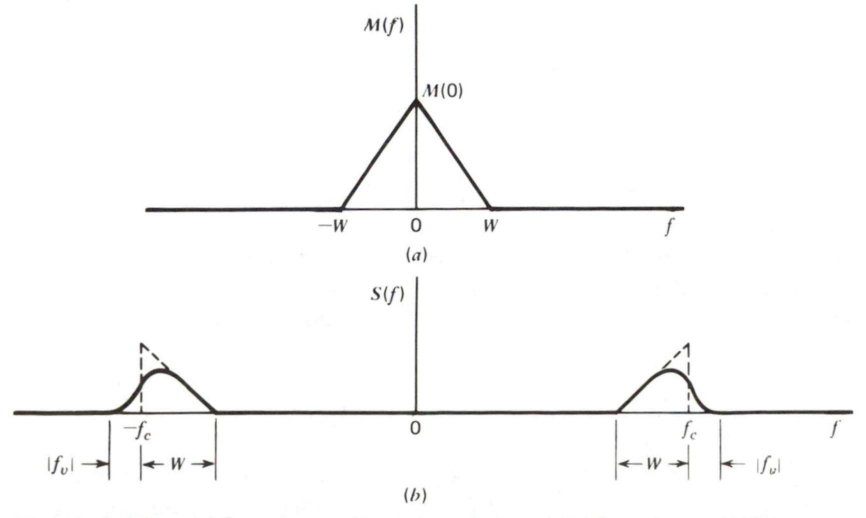

Figure 3.29 *(a)* Spectrum of baseband signal. *(b)* Spectrum of VSB wave.

mitted vestige of the unwanted sideband [upper sideband in Fig. 3.29(*b*)] compensates for the amount removed from the desired sideband [lower sideband in Fig. 3.29(*b*)]. The transmission bandwidth required by a VSB system is therefore

$$B_T = W + f_v \tag{3.63}$$

where W is the message bandwidth [see Fig. 3.29(*a*)] and f_v is the width of the vestigial sideband.

Vestigial sideband modulation can be generated by passing a DSBSC wave through an appropriate filter of transfer function $H(f)$, as in Fig. 3.30(*a*). The spectrum $S(f)$ of the resulting VSB wave $s(t)$ is therefore given by

$$S(f) = \frac{A_c}{2}[M(f - f_c) + M(f + f_c)]H(f) \tag{3.64}$$

where $M(f)$ is the Fourier transform of the baseband signal $m(t)$. We wish to determine the specification of the filter transfer function $H(f)$, so that $S(f)$ defines the spectrum of the desired VSB wave $s(t)$. This can be established by passing $s(t)$ through a coherent detector and then determining the necessary condition for the detector output to provide an undistorted version of the original baseband signal $m(t)$. Thus, multiplying $s(t)$ by a locally generated sine-wave $A'_c \cos(2\pi f_c t)$, which is synchronous with the carrier wave $A_c \cos(2\pi f_c t)$ in both frequency and phase, as in Fig. 3.30(*b*), we get

$$v(t) = A'_c \cos(2\pi f_c t) s(t) \tag{3.65}$$

Transforming this relation into the frequency domain gives the Fourier transform of $v(t)$ as

$$V(f) = \frac{A'_c}{2}[S(f - f_c) + S(f + f_c)] \tag{3.66}$$

Therefore, substitution of Eq. (3.64) in (3.66) yields

$$\begin{aligned} V(f) = {} & \frac{A_c A'_c}{4} M(f)[H(f - f_c) + H(f + f_c)] \\ & + \frac{A_c A'_c}{4}[M(f - 2f_c)H(f - f_c) + M(f + 2f_c)H(f + f_c)] \end{aligned} \tag{3.67}$$

The spectrum $V(f)$ is illustrated in Fig. 3.30(*c*). The second term in Eq. (3.67) represents a VSB wave corresponding to carrier frequency $2f_c$. This term is removed by the low-pass filter in Fig. 3.30(*b*) to produce an output $v_o(t)$, the spectrum $V_o(f)$ of which is given by

$$V_o(f) = \frac{A_c A'_c}{4} M(f)[H(f - f_c) + H(f + f_c)] \tag{3.68}$$

The spectrum $V_o(f)$ is illustrated in Fig. 3.30(*d*). For a distortionless reproduction of the original baseband signal $m(t)$ at the coherent detector output, we require $V_o(f)$ to be a scaled version of $M(f)$. This means, therefore, that the transfer

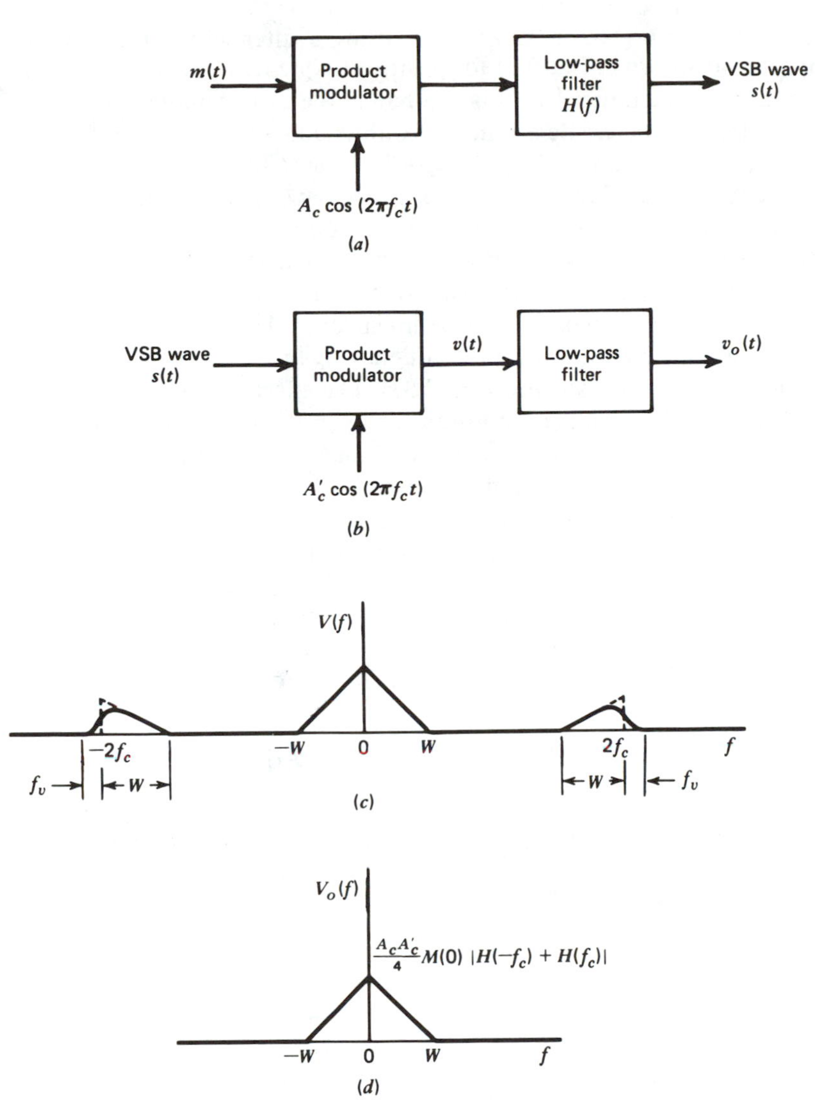

Figure 3.30 Scheme for the generation and demodulation of a VSB wave. *(a)* Block diagram of VSB modulator. *(b)* Block diagram of VSB demodulator. *(c)* Spectrum of the product modulator output *v(t)* in the demodulation scheme. *(d)* Spectrum of the demodulated signal $v_o(t)$.

function $H(f)$ must satisfy the condition

$$H(f-f_c)+H(f+f_c)=2H(f_c) \tag{3.69}$$

where $H(f_c)$ is a constant. When the baseband spectrum $M(f)$ is zero outside the interval $-W \leqslant f \leqslant W$, we need to satisfy Eq. (3.69) only for values of f in this interval.

The requirement of Eq. (3.69) is satisfied by using a filter with an amplitude response such as that shown in Fig. 3.31 for positive frequencies.* This response is normalized so that the magnitude $|H(f)|$ is one-half at the carrier frequency f_c. The cutoff portion of this characteristic around f_c exhibits odd symmetry in the sense that inside the transition interval $f_c - f_v \leq f \leq f_c + f_v$, the sum of the values of $|H(f)|$ at any two frequencies equally displaced above and below f_c is unity. Such a filter is much less elaborate than that required if one sideband is to be completely suppressed. Note, however, that in order to preserve the baseband spectrum, the phase response of the VSB filter in Fig. 3.30(a) must also exhibit odd symmetry about the carrier frequency f_c. In particular, it must be linear over the frequency intervals $f_c - W \leq |f| \leq f_c + W$, and its value at the frequency f_c has to equal zero or an integer multiple of 2π radians (see Problem 3.33). The effect of this linear phase characteristic is merely to introduce a constant delay in the recovery of the baseband signal $m(t)$ at the receiver output. Note also that the frequency response of Fig. 3.31 pertains to a VSB wave containing a vestige of the lower sideband.

Next, we wish to determine the time-domain description of a VSB wave. Here, again, the procedure we use for deriving this description is based on the canonical form for band-pass signals, which was presented in Section 2.14. To derive the in-phase component $s_c(t)$ of the VSB wave $s(t)$, we note that its Fourier transform is defined in terms of the Fourier transform of the modulated wave $s(t)$ by Eq. (3.43). Thus, substituting Eq. (3.64) into (3.43), we find that the Fourier transform of $s_c(t)$ is given by

$$S_c(f) = \tfrac{1}{2} A_c M(f)[H(f - f_c) + H(f + f_c)] \tag{3.70}$$

Assuming that the transfer function $H(f)$ of the VSB filter in Fig. 3.30(a) satisfies the requirement of Eq. (3.69) with $H(f_c) = \frac{1}{2}$, we find that Eq. (3.70) simplifies as

$$S_c(f) = \tfrac{1}{2} A_c M(f) \tag{3.71}$$

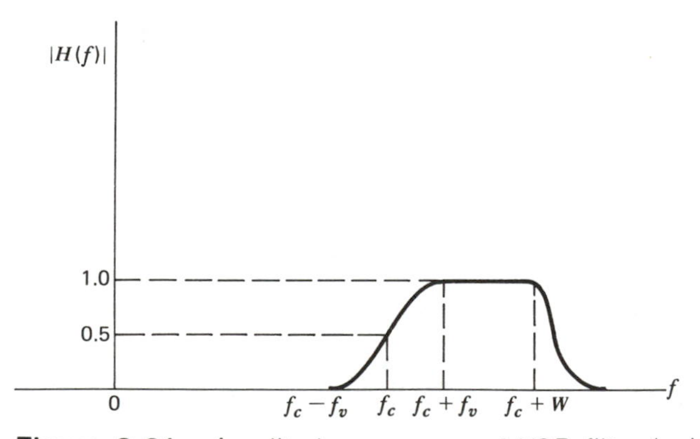

Figure 3.31 Amplitude response of VSB filter (only positive-frequency portion shown).

* The gradual transition in the frequency response of the VSB filter, as defined in Fig. 3.31 is of a form similar to that used to preserve freedom from a phenomenon known as intersymbol interference in data transmission systems, as described in Section 9.2.

This relation shows that the in-phase component of the VSB wave $s(t)$ is defined by

$$s_c(t)=\tfrac{1}{2}A_c m(t) \tag{3.72}$$

which, except for a scaling factor, is the same as the coherent detector output $v_o(t)$.

To determine the quadrature component $s_s(t)$ of the VSB wave $s(t)$, we first substitute Eq. (3.64) into Eq. (3.44), obtaining

$$S_s(f)=\frac{j}{2}\,A_c M(f)[H(f-f_c)-H(f+f_c)] \tag{3.73}$$

This equation suggests that we may generate $s_s(t)$, except for a scaling factor, by passing the message signal $m(t)$ through a filter whose transfer function is defined by

$$H_s(f)=j[H(f-f_c)-H(f+f_c)] \tag{3.74}$$

In Fig. 3.32, we have plotted the frequency response of this filter, assuming that $H(f)$ is as defined in Fig. 3.31. Using $m_s(t)$ to denote the output of this filter in response to the input $m(t)$, we may thus express the quadrature component of the VSB wave $s(t)$ as

$$s_s(t)=\tfrac{1}{2}A_c m_s(t) \tag{3.75}$$

Accordingly, substituting Eqs. (3.72) and (3.75) in (3.42), we find that a VSB wave containing a vestige of the lower sideband may be expressed in the form*

$$s(t)=\tfrac{1}{2}A_c m(t)\cos(2\pi f_c t)-\tfrac{1}{2}A_c m_s(t)\sin(2\pi f_c t) \tag{3.76}$$

The physical significance of this result is that the VSB modulator of Fig. 3.30(*a*) may be replaced by the equivalent network shown in Fig. 3.33. It is therefore apparent that, as with single-sideband modulation, the quadrature component is not independent of the in-phase component or message signal $m(t)$. The role of the

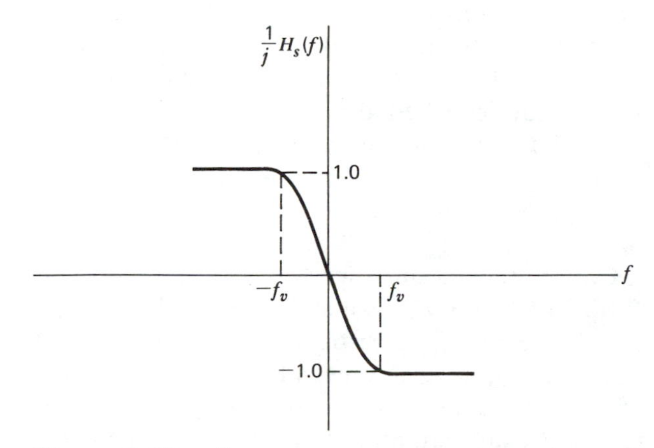

Figure 3.32 Frequency response of filter for producing the quadrature-component of the VSB wave.

* Another time-domain representation of a VSB signal consists of the product of a narrow-band "envelope" function and a SSB signal. For details of this representation, see F. S. Hill, Jr., "On time domain representations for vestigial sideband signals," *Proc. IEEE*, Vol. 62, pp. 1032–1033, July 1974.

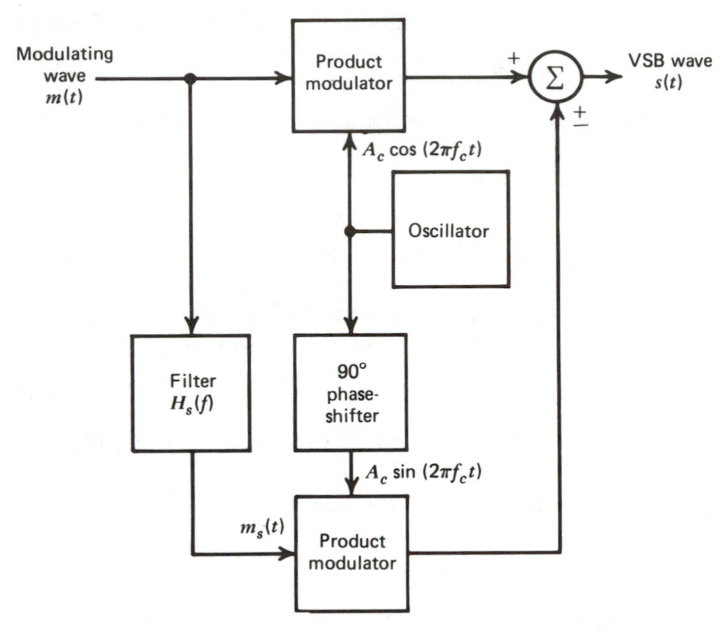

Figure 3.33 Block diagram of phase discrimination method for generating a VSB wave. When the minus sign (at the summing junction) is selected, a vestige of the lower sideband is transmitted; when the plus sign is selected, a vestige of the upper sideband is transmitted.

quadrature component is merely to interfere with the in-phase component, so as to eliminate power in one of the sidebands. Herein lies the reason for the fact that SSB- and VSB-modulated signals have favorable spectral properties. Note, however, that regardless of the nature of the quadrature component, the message signal $m(t)$ may be recovered from the modulated signal $s(t)$ by using coherent detection.

Following a procedure similar to that described above, we may show that the time-domain representation of a VSB wave containing a vestige of the upper sideband is of the same form as in Eq. (3.76), except that the minus sign on the right-hand side of this equation is replaced with a plus sign. Hence, we may also use the scheme shown in Fig. 3.33 for generating a VSB wave with a vestige of the upper sideband, except that the two product modulator outputs are now added at the summing junction. The DSBSC and SSB waves may be regarded as special cases of the VSB wave defined by Eq. (3.76). If the vestigial sideband is increased to the width of a full sideband, the resulting wave becomes a DSBSC wave with the result that $m_s(t)$ vanishes. If, on the other hand, the width of the vestigial sideband is reduced to zero, the resulting wave becomes an SSB wave containing the lower sideband, with the result that $m_s(t) = \hat{m}(t)$, where $\hat{m}(t)$ is the Hilbert transform of $m(t)$.

Vestigial sideband modulation has the virtue of conserving bandwidth almost as efficiently as single-sideband modulation, while retaining the excellent low-frequency baseband characteristics of double-sideband modulation. Thus VSB

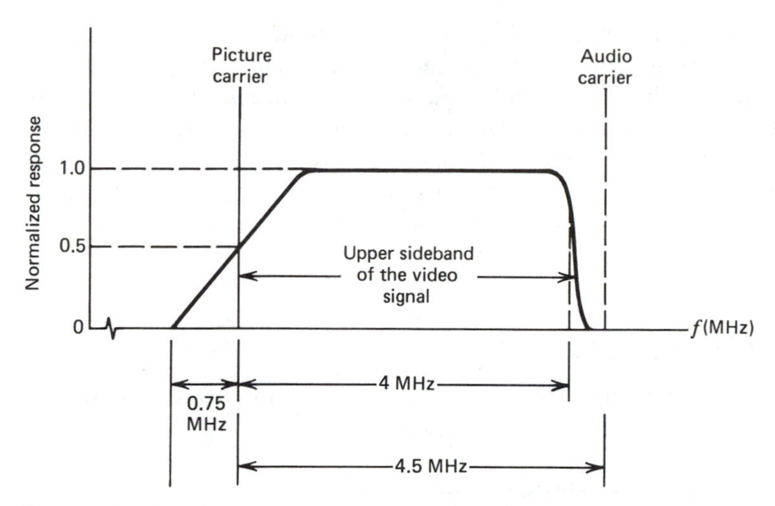

Figure 3.34 Frequency response of a VSB filter used in TV receivers.

modulation has become standard for the transmission of television and similar signals where good phase characteristics and transmission of low-frequency components are important, but the bandwidth required for double-sideband transmission is unavailable or uneconomical.

It is of interest, however, to note that in commercial television broadcasting the transmitted signal is not quite VSB-modulated, because the shape of the transition region is not rigidly controlled at the transmitter. Instead, a VSB filter is inserted in each receiver. The overall performance is the same as vestigial-sideband modulation except for some wasted power and bandwidth. Figure 3.34 shows the idealized frequency response for a VSB filter designed for television receivers.

Envelope Detection of a VSB Wave Plus Carrier

In commercial television broadcasting, a sizable carrier is transmitted together with the modulated wave. This makes it possible to demodulate the incoming modulated wave by an envelope detector in the receiver. It is, therefore, of interest to determine the distortion introduced by the envelope detector. Adding the carrier component $A_c \cos(2\pi f_c t)$ to Eq. (3.76), scaled by a factor k_a, modifies the modulated wave applied to the envelope detector input as

$$s(t) = A_c[1 + \tfrac{1}{2}k_a m(t)]\cos(2\pi f_c t) - \tfrac{1}{2}k_a A_c m_s(t)\sin(2\pi f_c t) \tag{3.77}$$

where the constant k_a determines the percentage modulation. The envelope detector output, denoted by $a(t)$, is therefore

$$\begin{aligned} a(t) &= A_c\{[1 + \tfrac{1}{2}k_a m(t)]^2 + [\tfrac{1}{2}k_a m_s(t)]^2\}^{1/2} \\ &= A_c[1 + \tfrac{1}{2}k_a m(t)]\left\{1 + \left[\frac{\tfrac{1}{2}k_a m_s(t)}{1 + \tfrac{1}{2}k_a m(t)}\right]^2\right\}^{1/2} \end{aligned} \tag{3.78}$$

Equation (3.78) indicates that the distortion is contributed by the quadrature component $m_s(t)$ of the incoming VSB wave. This distortion can be reduced by using two methods: (1) by reducing the percentage modulation to reduce k_a and (2) by increasing the width of the vestigial sideband to reduce $m_s(t)$. Both methods are used in practice. In commercial television broadcasting, the vestigial sideband occupies a width of about 0.75 MHz, or about one-sixth of a full sideband. This has been determined empirically as the width of vestigial sideband required to keep the distortion due to $m_s(t)$ within tolerable limits when the percentage modulation is nearly 100.

3.12 DISCUSSION

Having studied the characteristics of the different forms of amplitude modulation, we are now in a position to compare their practical merits:

1. In ordinary AM systems the sidebands are transmitted in full, accompanied by the carrier. Accordingly, demodulation is accomplished rather simply by using an envelope detector or square-law detector. On the other hand, in suppressed-carrier systems the receiver is more complex because additional circuitry must be provided for the purpose of carrier recovery. It is for this reason we find that in commercial AM radio *broadcast* systems, which involve one transmitter and numerous receivers, full AM is used in preference to DSBSC or SSB modulation.
2. Suppressed-carrier modulation systems have an advantage over full AM systems in that they require much less power to transmit the same amount of information, which makes the transmitters for such systems less expensive than those required for full AM. Suppressed-carrier systems are therefore well-suited for *point-to-point communication* involving one transmitter and one receiver, which would justify the use of increased receiver complexity.
3. Single-sideband modulation requires the minimum transmitter power and minimum transmission bandwidth possible for conveying a message signal from one point to another. We thus find that single-sideband modulation is the preferred method of modulation for long-distance transmission of voice signals over metallic circuits, because it permits longer spacing between the *repeaters*, which is a more important consideration here than simple terminal equipment. A repeater is simply a wide-band amplifier that is used at intermediate points along the transmission path so as to make up for the attenuation incurred during the course of transmission.
4. Vestigial-sideband modulation requires a transmission bandwidth that is intermediate between that required for SSB and DSBSC systems, and the saving can be significant if modulating waves with large bandwidths are being handled, as in the case of television signals and wide-band data.
5. Double-sideband suppressed-carrier modulation, single-sideband modulation, and vestigial-sideband modulation are all examples of *linear modulation*. The output of a linear modulator can be expressed in the canonical form

$$s(t) = s_c(t)\cos(2\pi f_c t) - s_s(t)\sin(2\pi f_c t)$$

The in-phase component $s_c(t)$ is a scaled version of the incoming message signal $m(t)$. The quadrature component $s_s(t)$ is derived from $m(t)$ by some linear filtering operation. Accordingly, the principle of superposition can be used to calculate the modulator output $s(t)$ as the sum of responses of the modulator to individual components of $m(t)$. In Table 3.2 we have summarized the definitions for $s_c(t)$ and $s_s(t)$ in terms of $m(t)$ for DSBSC, SSB, and VSB modulated waves, assuming a carrier of unit amplitude. In a strict sense, ordinary amplitude modulation fails to meet the definition of a linear modulator with respect to the message signal. If $s_1(t)$ is the AM wave produced by a message signal $m_1(t)$ and $s_2(t)$ is the AM wave produced by a second message signal $m_2(t)$, then the AM wave produced by $m_1(t)$ plus $m_2(t)$ is not equal to $s_1(t)$ plus $s_2(t)$. However, the departure from linearity in AM is of a rather mild sort, such that many of the mathematical procedures applicable to linear modulation may be retained. For example, the bandpass representation is still applicable to an AM wave, with the in-phase and quadrature components defined by, respectively,

$$s_c(t)=1+k_a m(t)$$

and

$$s_s(t)=0$$

where k_a is the amplitude sensitivity of the modulator.

6. The band-pass representation may also be used to describe quadrature amplitude modulation. In this case, we have (assuming a carrier of unit amplitude)

$$s_c(t)=m_1(t)$$

Table 3.2 Forms of Linear Modulation

Type of Modulation	In-phase component $s_c(t)$	Quadrature component $s_s(t)$	Comments
DSBSC	$m(t)$	0	$m(t)$ = message signal
SSB			
(a) Upper sideband transmitted	$\frac{1}{2}m(t)$	$\frac{1}{2}\hat{m}(t)$	$\hat{m}(t)$ = Hilbert transform of $m(t)$
(b) Lower sideband transmitted	$\frac{1}{2}m(t)$	$-\frac{1}{2}\hat{m}(t)$	
VSB			
(a) Vestige of lower sideband transmitted	$\frac{1}{2}m(t)$	$\frac{1}{2}m_s(t)$	$m_s(t)$ = Output of filter of transfer function $H_s(f)$, produced by $m(t)$.
(b) Vestige of upper sideband transmitted	$\frac{1}{2}m(t)$	$-\frac{1}{2}m_s(t)$	For the definition of $H_s(f)$, see Eq. (3.74)

and

$$s_s(t) = -m_2(t)$$

where $m_1(t)$ and $m_2(t)$ are the two independent message signals at the quadrature-modulator input [see Eq. (3.41)].

3.13 FREQUENCY TRANSLATION

In the processing of signals in communication systems, it is often convenient or necessary to translate the modulated wave upward or downward in frequency, so that it occupies a new frequency band. This frequency translation is accomplished by multiplication of the signal by a locally generated sine-wave, and subsequent filtering. For example, consider the DSBSC wave

$$s(t) = m(t)\cos(2\pi f_c t) \tag{3.79}$$

in which the modulating wave $m(t)$ is limited to the band $-W \leqslant f \leqslant W$. The spectrum of $s(t)$ therefore occupies the frequency bands $f_c - W \leqslant f \leqslant f_c + W$ and $-f_c - W \leqslant f \leqslant -f_c + W$, as in Fig. 3.35(*a*). Suppose that it is required to translate

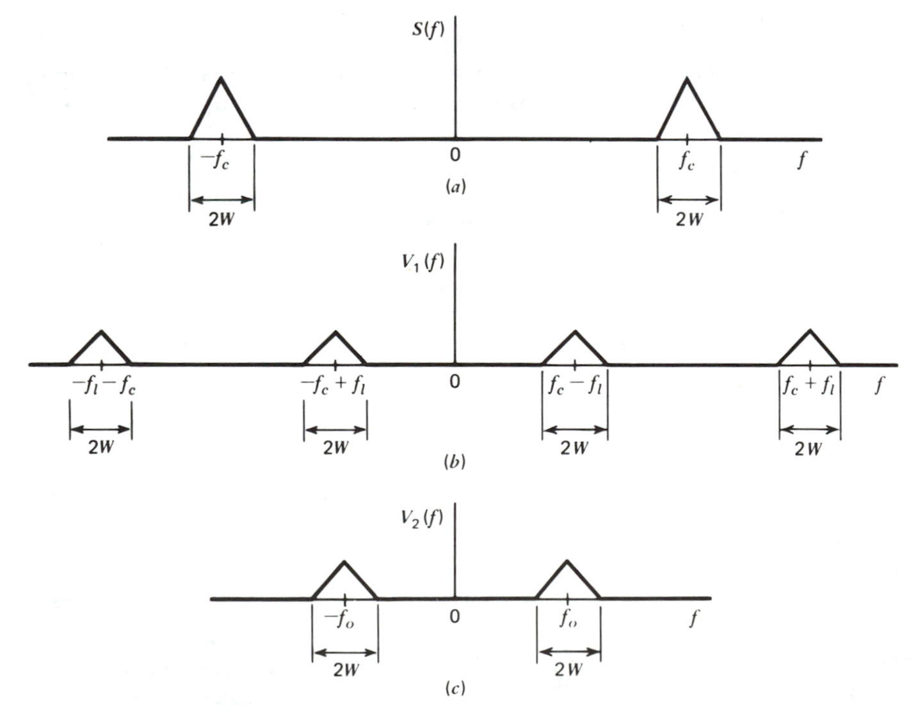

Figure 3.35 Illustrating the frequency translation process. *(a)* Spectrum of DSBSC wave. *(b)* Spectrum of signal obtained by multiplying DSBSC wave with a local carrier. *(c)* Spectrum of desired DSBSC wave, translated downward in frequency.

this modulated wave downward in frequency, so that its carrier frequency is changed from f_c to a new value f_o, where $f_o < f_c$. To accomplish this requirement, we first multiply the incoming modulated wave $s(t)$ by a sine-wave of frequency f_l supplied by a local oscillator to obtain

$$\begin{aligned} v_1(t) &= s(t)\cos(2\pi f_l t) \\ &= m(t)\cos(2\pi f_c t)\cos(2\pi f_l t) \\ &= \tfrac{1}{2}m(t)\cos[2\pi(f_c - f_l)t] + \tfrac{1}{2}m(t)\cos[2\pi(f_c + f_l)t] \end{aligned} \tag{3.80}$$

The multiplier output $v_1(t)$ consists of two DSBSC waves, one with a carrier frequency of $f_c - f_l$ and the other with a carrier frequency of $f_c + f_l$. The spectrum of $v_1(t)$ is therefore as shown in Fig. 3.35(*b*). Let the frequency f_l of the local oscillator be chosen so that

$$f_c - f_l = f_o \tag{3.81}$$

Then from Fig. 3.35(*b*) we see that the modulated wave with the desired carrier frequency f_o may be extracted by passing the multiplier output $v_1(t)$ through a band-pass filter of mid-band frequency f_o and bandwidth $2W$, provided

$$f_c + f_l - W > f_c - f_l + W$$

or

$$f_l > W \tag{3.82}$$

The filter output is therefore

$$\begin{aligned} v_2(t) &= \tfrac{1}{2}m(t)\cos[2\pi(f_c - f_l)t] \\ &= \tfrac{1}{2}m(t)\cos(2\pi f_o t) \end{aligned} \tag{3.83}$$

This output is the desired modulated wave, translated downward in frequency, as shown in Fig. 3.35(*c*).

A device that carries out the frequency translation of a modulated wave is called a *mixer*. The operation itself is called *mixing* or *heterodyning*. For the implementation of a mixer, we may use a multiplier and band-pass filter, as shown in Fig. 3.36. The multiplier is usually constructed by using nonlinear or switching devices, similar to modulators. Note that mixing is a linear operation in that it completely preserves the relation of the sidebands of the incoming modulated wave to the carrier.

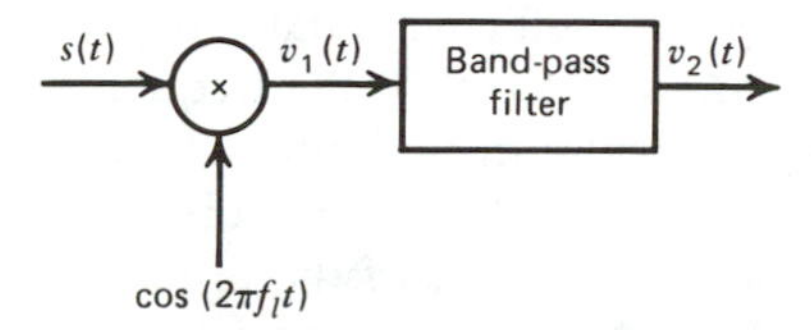

Figure 3.36 Block diagram of mixer.

Example 5

Consider an incoming narrow-band signal of bandwidth 10 kHz, and mid-band frequency which may lie in the range 0.535–1.605 MHz. It is required to translate this signal to a fixed frequency band centered at 0.455 MHz. The problem is to determine the range of tuning that must be provided in the local oscillator.

Let f_c denote the mid-band frequency of the incoming signal, and f_l denote the local oscillator frequency. Then we may write

$$0.535 < f_c < 1.605$$

and

$$f_c - f_l = 0.455$$

where both f_c and f_l are expressed in MHz. That is,

$$f_l = f_c - 0.455$$

When $f_c = 0.535$ MHz, we get $f_l = 0.08$ MHz; and when $f_c = 1.605$ MHz, we get $f_l = 1.15$ MHz. Thus the required range of tuning of the local oscillator is 0.08–1.15 MHz.

3.14 FREQUENCY-DIVISION MULTIPLEXING (FDM)

Multiplexing is a technique whereby a number of independent signals can be combined into a composite signal suitable for transmission over a common channel. Voice frequencies transmitted over telephone systems, for example, range from 300 to 3400 Hz. To transmit a number of these signals over the same channel, the signals must be kept apart so that they do not interfere with each other, and thus they can be separated at the receiving end. This is accomplished by separating the signals either in frequency or in time. The technique of separating the signals in frequency is referred to as *frequency-division multiplexing* (FDM), whereas the technique of separating the signals in time is called *time-division multiplexing* (TDM). In this section, we discuss FDM systems, whereas TDM systems are discussed in Chapter 7.

A block diagram of an FDM system is shown in Fig. 3.37. The incoming message signals are assumed to be of the low-pass type, but their spectra do not necessarily have nonzero values all the way down to zero frequency. Following each signal input, we have shown a low-pass filter which is designed to remove high-frequency components that do not contribute significantly to signal representation but are capable of disturbing other message signals that share the common channel. These low-pass filters may be omitted only if the input signals are sufficiently band-limited initially. The filtered signals are applied to modulators which shift the frequency ranges of the signals so as to occupy mutually exclusive frequency intervals. The necessary carrier frequencies, to perform these frequency translations, are obtained from a carrier supply. For the modulation, we may use any one of the processes described in previous sections of this chapter. However, the most widely used method of modulation in frequency-division multiplexing is single-sideband

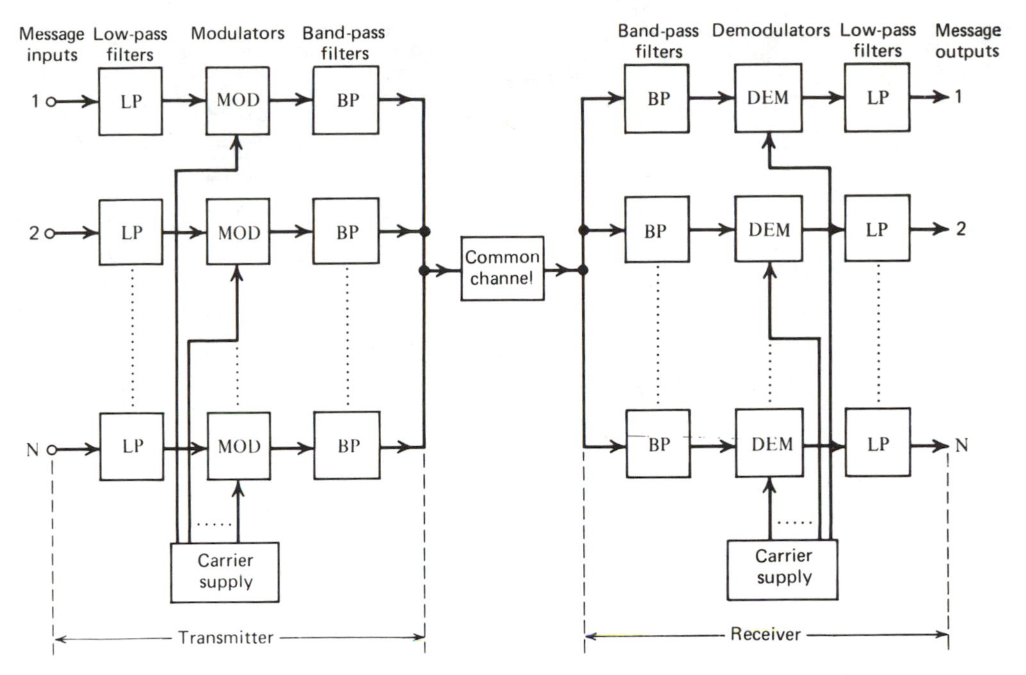

Figure 3.37 Block diagram of FDM system.

modulation which, in the case of voice signals, requires a bandwidth that is approximately equal to that of the original voice signal. In practice, each voice input is usually assigned a bandwidth of 4 kHz. The band-pass filters following the modulators are used to restrict the band of each modulated wave to its prescribed range. The resulting band-pass filter outputs are next combined in parallel to form the input to the common channel. At the receiving terminal, a bank of band-pass filters, with their inputs connected in parallel, is used to separate the message signals on a frequency-occupancy basis. Finally, the original message signals are recovered by individual demodulators. Note that the FDM system shown in Fig. 3.37 operates in only one direction. To provide for two-way transmission, as in telephony for example, we have to completely duplicate the multiplexing facilities, with the components connected in reverse order and with the signal waves proceeding from right to left.*

Example 6

The practical implementation of an FDM system usually involves many steps of modulation and demodulation, as illustrated in Fig. 3.38. The first multiplexing step combines 12 voice inputs into a *basic group*, which is formed by having the *n*th input modulate a carrier at fre-

* For a discussion of the performance of multiplex transmission, see W. R. Bennett, *Introduction to Signal Transmission*, pp. 213–218 (McGraw-Hill, 1970).

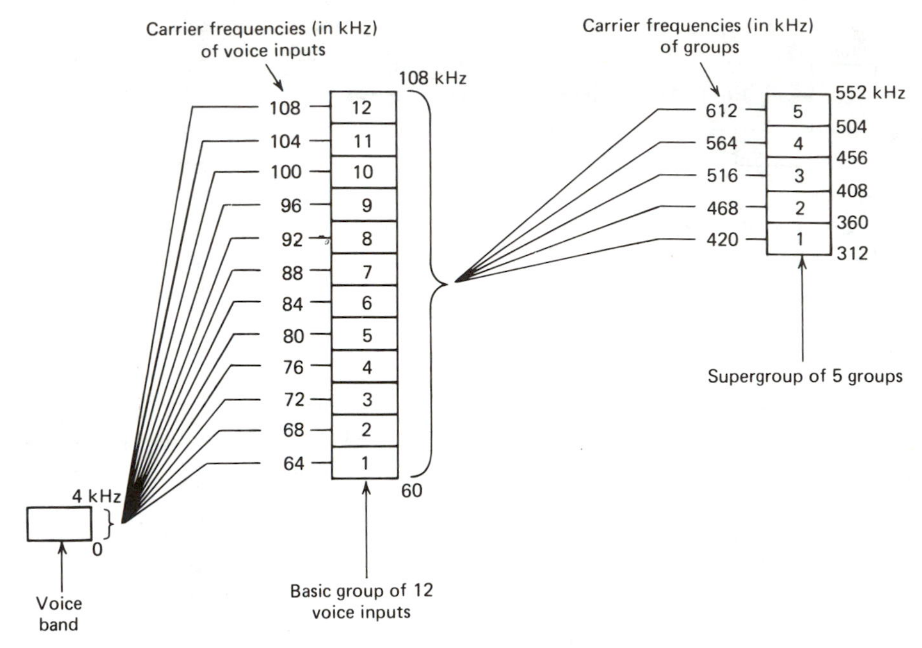

Figure 3.38 Illustrating the modulation steps in an FDM system.

quency $f_c = 112 - 4n$ kHz, where $n = 1, 2, \ldots, 12$. The lower sidebands are then selected by band-pass filtering and combined to form a group of 12 lower sidebands (one for each voice input). Thus the basic group occupies the frequency band 60–108 kHz. The next step in the FDM hierarchy involves the combination of 5 basic groups into a *supergroup*. This is accomplished by using the nth group to modulate a carrier at frequency $f_c = 372 + 48n$ kHz, where $n = 1, 2, \ldots, 5$. Here again the lower sidebands are selected by filtering and then combined to form a supergroup occupying the band 312–552 kHz. Thus a supergroup is designed to accommodate 60 independent voice inputs. The reason for forming the supergroup in this manner is that economical filters of the required characteristics are available only over a limited frequency range. In a similar manner, supergroups are combined into *mastergroups*, and mastergroups are combined into *very large groups*.*

Carrier Synchronization

Although SSB modulation permits economies in bandwidth and transmitted power, it complicates the transmission problem in that it requires synchronization between the carrier used in the transmitter for modulation and that locally generated in the receiver for demodulation. A method that is commonly used to achieve this synchronization requires a *pilot frequency* to be transmitted with the outgoing

* 1. "Transmission Systems for Communications," Bell Telephone Laboratories, pp. 128–137 (Western Electric, 1970).

2. "Reference Data for Radio Engineers," International Telephone and Telegraph Corporation, pp. 30–23 to 30–27 (H. W. Sams, 1968).

signal. This pilot frequency undergoes whatever frequency translations or phase shifts are imposed on the various voice-frequency signals. The transmitted pilot is used to modulate a local oscillator in the receiver to obtain the carrier frequencies used in demodulating the information-bearing signals. If the transmitted signals, including the pilot, undergo additional frequency translations during transmission, this method cancels these errors out, leaving only the possibility of a fixed error between the transmitter and receiver local oscillators. This error may be maintained within acceptable limits by properly controlling the frequency of the local oscillator.

Problems

Problem 3.1 A carrier wave of frequency 1 MHz is modulated 50 percent by a sine-wave of frequency 5 kHz. The resulting AM wave is transmitted through the resonant circuit of Fig. P3.1 which is tuned to the carrier frequency, and has a Q factor of 175. Determine the modulated wave after transmission through this circuit. What is the percentage modulation of this modulated wave?

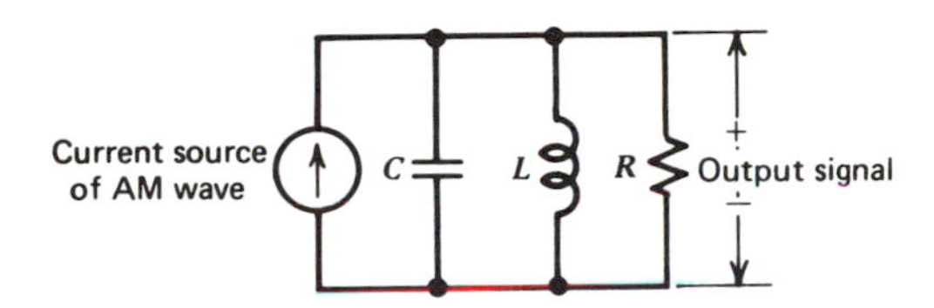

Figure P3.1

Problem 3.2 For a p-n junction diode, the current i through the diode and the voltage v across it are related by

$$i = I_0\left[\exp\left(-\frac{v}{V_T}\right) - 1\right]$$

where I_0 is the reverse saturation current and V_T is the volt-equivalent of temperature defined by

$$V_T = \frac{kT}{e}$$

where k is Boltzmann's constant in joules per degree Kelvin, T is the absolute temperature in degree Kelvin, and e is the charge of an electron. At room temperature, $V_T = 0.026$ volts.

(a) Expand i as a power series in v, retaining terms up to v^3.

(b) Let

$$v = 0.01\cos(2\pi f_m t) + 0.01\cos(2\pi f_c t) \text{ volts}$$

where $f_m = 1$ kHz and $f_c = 100$ kHz. Determine the spectrum of the resulting diode current i.

(c) Specify the band-pass filter required to extract from the diode current an AM wave with carrier frequency f_c.

(d) What is the percentage modulation of this AM wave?

Problem 3.3 Suppose nonlinear devices are available for which the output current i_o and input voltage v_i are related by

$$i_o = a_1 v_i + a_3 v_i^3$$

where a_1 and a_3 are constants. Explain how these devices may be used to provide: (1) a product modulator and (2) an amplitude modulator.

Problem 3.4 In *class C operation*, an amplifier is biased below cutoff so that if a sinusoidal voltage is applied to the amplifier input, the output current flows for less than 180 degrees of a cycle. An important advantage of class C operation is that it provides a high efficiency of power generation. Consider a class C amplifier in which the output current is defined by

$$i(t) = I_0[1 + k_a m(t)]g_p(t)$$

where I_0 and k_a are constants, $m(t)$ is the modulating wave applied to the amplifier input, and $g_p(t)$ is a periodic waveform of period $1/f_c$, defined by (see Fig. P3.2):

$$g_p(t) = \begin{cases} \cos(2\pi f_c t) - \cos(\theta_c), & -\dfrac{\theta_c}{2\pi f_c} \leqslant t \leqslant \dfrac{\theta_c}{2\pi f_c} \\ 0, & \text{for the remainder of the period} \end{cases}$$

(a) Assuming that $m(t)$ is limited to the band $-W \leqslant f \leqslant W$, determine the spectrum of $i(t)$.
(b) Specify a suitable filter for extracting an AM wave with carrier frequency f_c from $i(t)$.

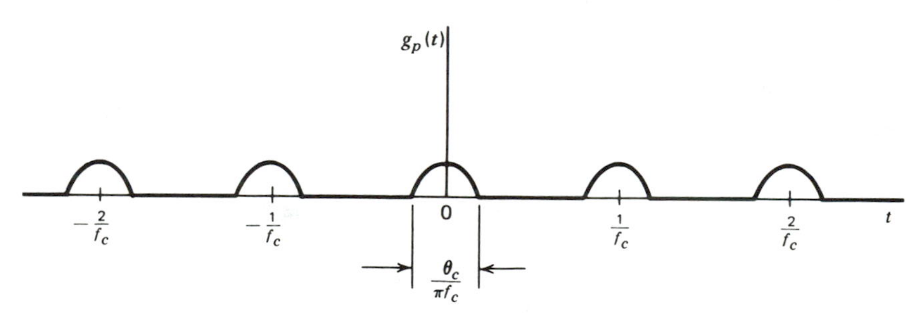

Figure P3.2

Problem 3.5 Consider the amplitude-modulated wave of Fig. P3.3 with a triangular envelope. This modulated wave is applied to an envelope detector with zero source resistance and a load resistance of 250 Ω. The carrier frequency $f_c = 40$ kHz. Suggest a suitable value for the capacitor C in order that the distortion (at the envelope detector output) is negligible for frequencies up to and including the eleventh harmonic of the modulating wave.

Problem 3.6 Consider a modulating wave $m(t)$ with $|m(t)| \leqslant 1/k_a$ so that $1 + k_a m(t) \geqslant 0$ for all t. Assume that the spectrum of $m(t)$ is zero for $|f| \geqslant W$. Let

$$s(t) = A_c[1 + k_a m(t)]\cos(2\pi f_c t),$$

where the carrier frequency $f_c > W$.

(a) The modulated wave $s(t)$ is applied to a full-wave rectifier, producing the output $v_1(t)$. Determine the spectrum of $v_1(t)$.

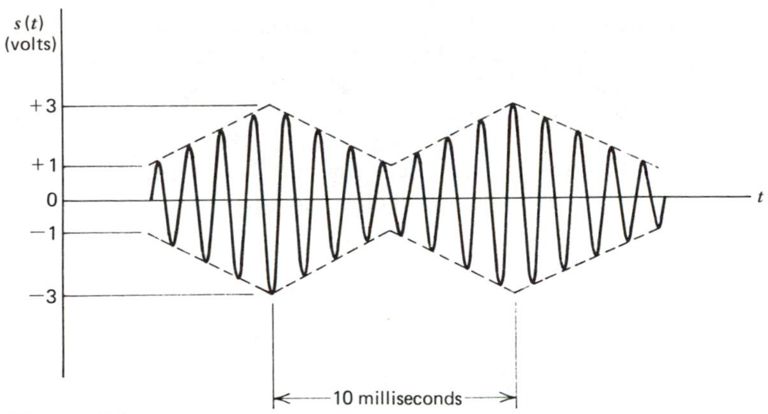

Figure P3.3.

(b) If the rectifier output $v_1(t)$ is passed through an ideal low-pass filter defined by the transfer function

$$H(f)=\begin{cases}1 & |f|<W \\ 0, & |f|>W\end{cases}$$

show that the filter output $v_2(t)$ is related to $m(t)$ by

$$v_2(t)=\frac{2A_c}{\pi}\,[1+k_a m(t)].$$

Problem 3.7 Consider the AM wave

$$s(t)=A_c[1+\mu\cos(2\pi f_m t)]\cos(2\pi f_c t)$$

produced by a sinusoidal modulating wave of frequency f_m. Assume that the modulation factor is $\mu=2$, and that the carrier frequency is $f_c \gg f_m$. The AM wave $s(t)$ is applied to an ideal envelope detector, producing the output $v(t)$.

(a) Determine the Fourier series representation of $v(t)$.
(b) What is the ratio of second-harmonic amplitude to fundamental amplitude in $v(t)$?

Problem 3.8 The AM wave

$$s(t)=A_c[1+k_a m(t)]\cos(2\pi f_c t)$$

is applied to the system shown in Fig. P3.4. Assuming that $|k_a m(t)|<1$ for all t and the baseband signal $m(t)$ is limited to the interval $-W \leqslant f \leqslant W$, and that the carrier frequency is $f_c > 2W$, show that $m(t)$ can be obtained from the square-rooter output.

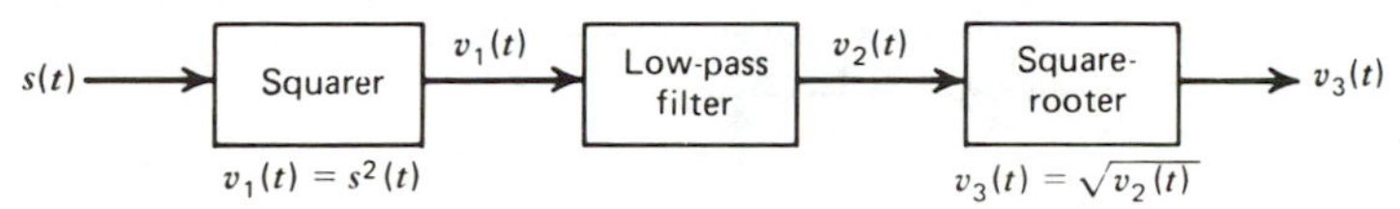

Figure P3.4

Problem 3.9 A sine-wave of frequency 5 kHz is applied to a product modulator, together with a carrier wave of frequency 1 MHz. The modulator output is next applied to the resonant circuit of Fig. P3.1. Determine the modulated wave after transmission through this circuit.

Problem 3.10 Consider a baseband signal $m(t)$ with the spectrum shown in Fig. P3.5. The message bandwidth $W = 1$ kHz. This signal is applied to a product modulator, together with a carrier wave $A_c \cos(2\pi f_c t)$, producing the DSBSC modulated wave $s(t)$. This modulated wave is next applied to a coherent detector. Assuming perfect synchronism between the carrier waves in the modulator and detector, determine the spectrum of the detector output when: (a) the carrier frequency $f_c = 1.25$ kHz and (b) the carrier frequency $f_c = 0.75$ kHz. What is the lowest carrier frequency in order that each component of the modulated wave $s(t)$ is uniquely determined by $m(t)$?

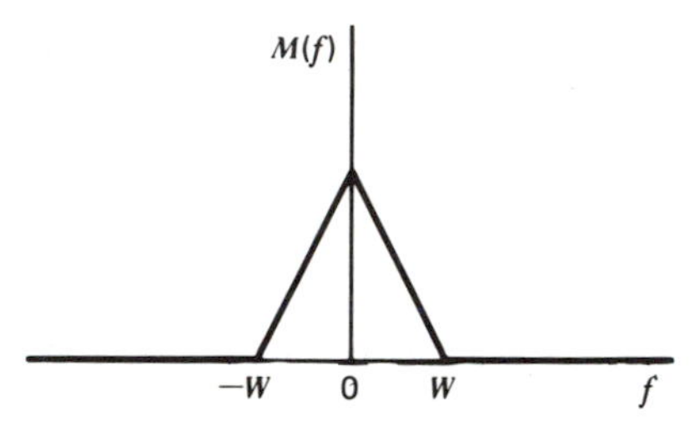

Figure P3.5

Problem 3.11 How would you recover the message signal from an AM wave that is over-modulated? Justify your answer.

Problem 3.12 A modulated wave with the spectrum

$$S(f) = M(f - f_c) + M(f + f_c)$$

is applied to a demodulator consisting of a product modulator and a filter. The product modulator is supplied with a periodic train of delta functions which alternate in sign with period $1/f_c$ and spacing $1/2f_c$. Assume that $f_c > W$, where W is the highest frequency component of $M(f)$.

(a) Determine the spectrum of the signal at the product modulator output.
(b) Discuss the choice of a suitable filter to produce an output whose spectrum is proportional to $M(f)$.

Problem 3.13 Figure P3.6 shows the block diagram of a *chopper-stabilized dc amplifier*. It uses a product modulator which shifts the spectrum of the input signal from the vicinity of zero frequency to the vicinity of the carrier frequency. The modulated signal can then be amplified in an ac amplifier and synchronously demodulated. The carrier consists of a square-wave of frequency f_c and unit amplitude.

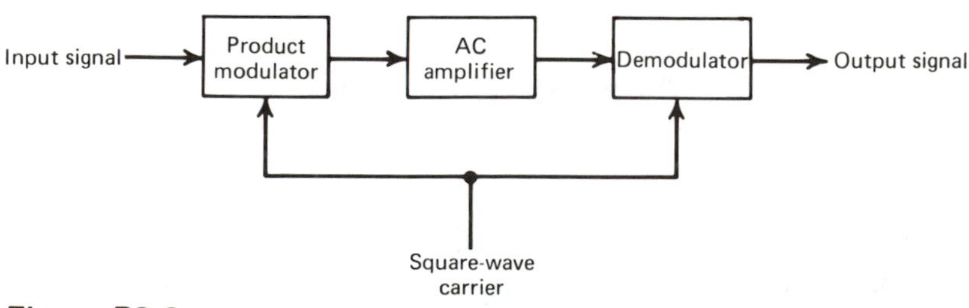

Figure P3.6

(a) Specify the frequency response of the ac amplifier in order that there be no distortion, assuming that the input signal is limited to the band $-W \leqslant f \leqslant W$.
(b) Determine the overall gain of the system, from input to output, assuming that the ac amplifier has an amplification equal to K.

Problem 3.14 Figure P3.7 shows the waveform of a chopper-modulated sinusoidal wave. This modulated wave is applied to a demodulator consisting of a product modulator and an ideal low-pass filter with a pass-band amplitude response of one and bandwidth B. The demodulator is supplied with a local carrier $10 \cos(2\pi f_c t + \phi)$ volts.

(a) Find the spectrum of the product modulator output, assuming $\phi = 0$.
(b) For $f_c = 100$ kHz, $\phi = 0$, and the modulation frequency f_m in the range 10 Hz to 1 kHz, specify the permissible range of values for the filter bandwidth B, which will allow only the modulating wave to appear at the filter output. What is the amplitude of this output?
(c) For $\phi = \pi/2$, and using the same filter as in part (b), what is the amplitude of the filter output?

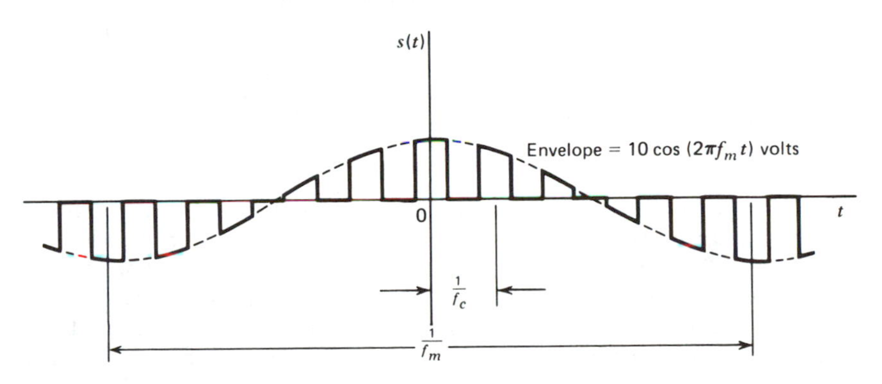

Figure P3.7

Problem 3.15 A DSBSC wave is demodulated by applying it to a coherent detector.

(a) Evaluate the effect of a frequency error Δf in the local carrier frequency of the detector, measured with respect to the carrier frequency of the incoming DSBSC wave.
(b) For the case of a sinusoidal modulating wave, show that because of this frequency error, the demodulated wave exhibits *beats* at the error frequency. Illustrate your answer with a sketch of this demodulated wave.

Problem 3.16 Consider the DSBSC wave

$$s(t) = A_c \cos(2\pi f_c t) m(t)$$

where $A_c \cos(2\pi f_c t)$ is the carrier wave and $m(t)$ is the message signal. This modulated wave is applied to a square-law device characterized by

$$y(t) = s^2(t).$$

The output $y(t)$ is next applied to a narrow-band filter with a passband amplitude response of one, mid-band frequency $2f_c$, and bandwidth Δf. Assume that Δf is small enough to treat the spectrum of $y(t)$ as essentially constant inside the passband of the filter.

(a) Determine the spectrum of the square-law device output $y(t)$.
(b) Show that the filter output $v(t)$ is approximately sinusoidal, given by

$$v(t) \simeq \frac{A_c^2}{2} E \, \Delta f \cos(4\pi f_c t)$$

where E is the energy of the message signal $m(t)$.

Problem 3.17 Consider a composite wave obtained by adding a noncoherent carrier $A_c \cos[2\pi f_c t + \phi]$ to a DSBSC wave $\cos(2\pi f_c t) m(t)$. This composite wave is applied to an ideal envelope detector. Find the resulting detector output. Evaluate this output for:

(a) $\phi = 0$.
(b) $\phi \neq 0$ and $|m(t)| \ll A_c/2$.

Problem 3.18 Consider the quadrature-carrier multiplex system of Fig. 3.20. The multiplexed signal $s(t)$ produced at the transmitter output in part (*a*) of this figure is applied to a communication channel of transfer function $H(f)$. The output of this channel is in turn applied to the receiver input in part (*b*) of Fig. 3.20. Prove that the condition

$$H(f_c + f) = H^*(f_c - f), \qquad 0 \leqslant f \leqslant W$$

is necessary for recovery of the message signals $m_1(t)$ and $m_2(t)$ at the receiver outputs; f_c is the carrier frequency, and W is the message bandwidth.

Hint: Evaluate the spectra of the two receiver outputs.

Problem 3.19 Suppose that in the receiver of the quadrature-carrier multiplex system of Fig. 3.20, the local carrier available for demodulation has a phase error ϕ with respect to the carrier source used in the transmitter. Assuming a distortionless communication channel between the transmitter and receiver, show that this phase error will cause *crosstalk* to arise between the two demodulated signals at the receiver outputs. By crosstalk we mean that a portion of one message signal appears at the receiver output pertaining to the other message signal, and vice versa.

Problem 3.20 The transmission of *color* in commercial television broadcasting is based on the premise that all colors found in nature can be approximated by mixing three additive *primary colors*: *red*, *green*, and *blue*. These three primary colors are represented by the video signals $m_R(t)$, $m_G(t)$, and $m_B(t)$, respectively. In order to conserve bandwidth and also produce a picture that can be viewed on a conventional black-and-white (monochrome) television receiver, the transmission of these three primary colors is accomplished by observing that they can be uniquely represented by any three signals that are independent linear combinations of $m_R(t)$, $m_G(t)$, and $m_B(t)$. In the standard color-television system, the three signals that are transmitted have the form

$$m_L(t) = 0.30 m_R(t) + 0.59 m_G(t) + 0.11 m_B(t)$$
$$m_I(t) = 0.60 m_R(t) - 0.28 m_G(t) - 0.32 m_B(t)$$
$$m_Q(t) = 0.21 m_R(t) - 0.52 m_G(t) + 0.31 m_B(t)$$

The signal $m_L(t)$ is called the *luminance signal*; when received on a conventional monochrome television receiver, it produces a black-and-white version of the color picture. The signals $m_I(t)$ and $m_Q(t)$ are called the *chrominance signals*; they indicate the way the color of the picture departs from shades of gray. With $m_L(t)$, $m_I(t)$, and $m_Q(t)$ defined as above, we have by simultaneous solution:

$$m_R(t) = m_L(t) - 0.96m_I(t) + 0.62m_Q(t)$$
$$m_G(t) = m_L(t) - 0.28m_I(t) - 0.64m_Q(t)$$
$$m_B(t) = m_L(t) - 1.10m_I(t) + 1.70m_Q(t)$$

The luminance signal $m_L(t)$ is assigned the entire 4.2 MHz bandwidth. Owing to certain properties of human vision, tests show that if the nominal bandwidths of the chrominance signals $m_I(t)$ and $m_Q(t)$ are 1.6 MHz and 0.6 MHz, respectively, then satisfactory color reproduction is possible.

Figure P3.8 shows a simplified block diagram of the color-television transmitter. The chrominance signals $m_I(t)$ and $m_Q(t)$ are combined using a variation of quadrature-multiplexing with a subcarrier having a frequency of 455/2 times the *line-scanning frequency*, or approximately 3.6 MHz. The standard television picture takes 1/60 second to go from the top to the bottom of the picture and return to the top; during this time, 262.5 lines are transmitted. The output resulting from the quadrature-multiplexing operation is next superimposed on the luminance signal $m_L(t)$ to give a combined video signal $m(t)$ covering the frequency range 6 Hz to 4.2 MHz, and to which standard blanking and synchronizing pulses are added. In addition, a "burst" of 8 cycles of the subcarrier is superimposed on the "back porch" of the blanking pulses for color subcarrier synchronization at the receiver. The resulting combined video signal is then amplified and amplitude-modulated in the usual manner, ready for transmission.

(a) Develop an expression for the combined video signal $m(t)$.
(b) Sketch the amplitude spectra for the three components that make up $m(t)$.
(c) Explain the fact that, by limiting the frequency range of the I-channel modulation output $x_I(t)$ to 2.0–4.2 MHz, the result is that the 0–0.6 MHz portion of the chrominance signal $m_I(t)$ is sent DSB, whereas the 0.6–1.6 MHz portion is sent SSB (lower sideband).

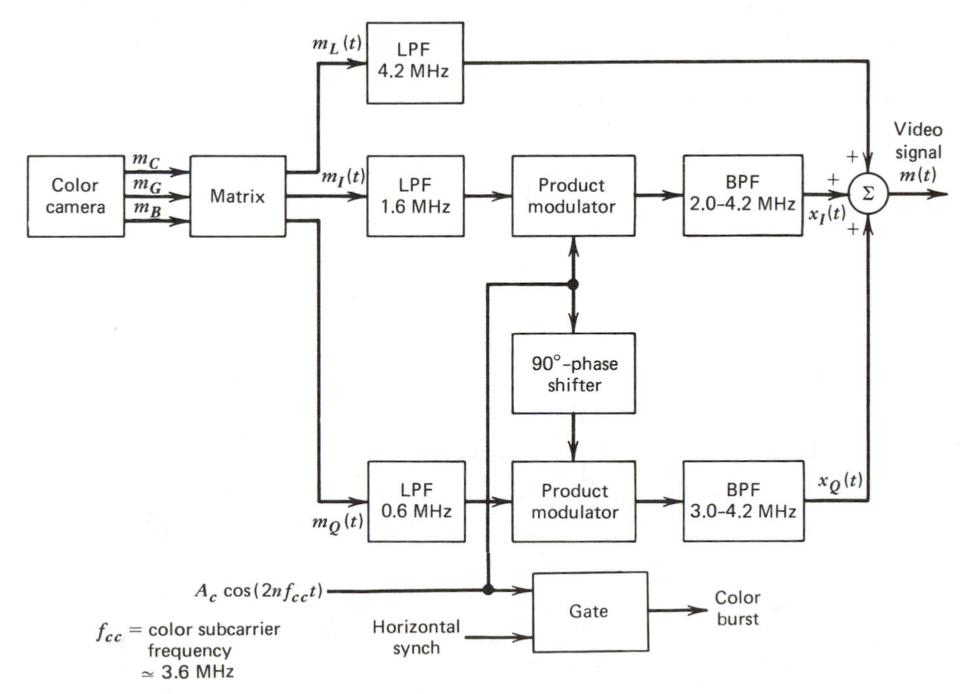

Figure P3.8

(d) Construct a block diagram for the color-television receiver so as to recover the primary color signals from the incoming modulated wave.

Problem 3.21

(a) Consider a pulse of amplitude A and duration T. This pulse is applied to an SSB modulator, producing the modulated wave $s(t)$. Determine the envelope of $s(t)$, and show that this envelope exhibits peaks at the beginning and end of the pulse.
(b) Suppose that the input consists of an RF pulse of amplitude A, duration T, and frequency f_c. Sketch the resulting modulator output.

Problem 3.22 Using the message signal

$$m(t)=\frac{t}{1+t^2}$$

determine and sketch the modulated waves for the following methods of modulation:

(a) Amplitude modulation with 50 percent modulation.
(b) Double-sideband suppressed-carrier modulation.
(c) Single-sideband modulation with only the upper sideband transmitted.
(d) Single-sideband modulation with only the lower sideband transmitted.

Problem 3.23 A voice signal occupying the frequency band 0.3–3.4 kHz is to be modulated onto a carrier wave of frequency 11.6 MHz using SSB modulation with only the upper sideband transmitted. Assume the availability of band-pass filters which provide an attenuation of 50 dB in a transition band that is 1 percent of the mid-band frequency, as illustrated in Fig. P3.9. Design a system to generate this SSB wave using the two-stage modulator of Fig. 3.23(*b*), so that the unwanted sideband is attenuated by no less than 50 dB.

Hint: Align the transition band of each filter with the band of frequencies separating the two sidebands of the modulated wave at the pertinent filter input.

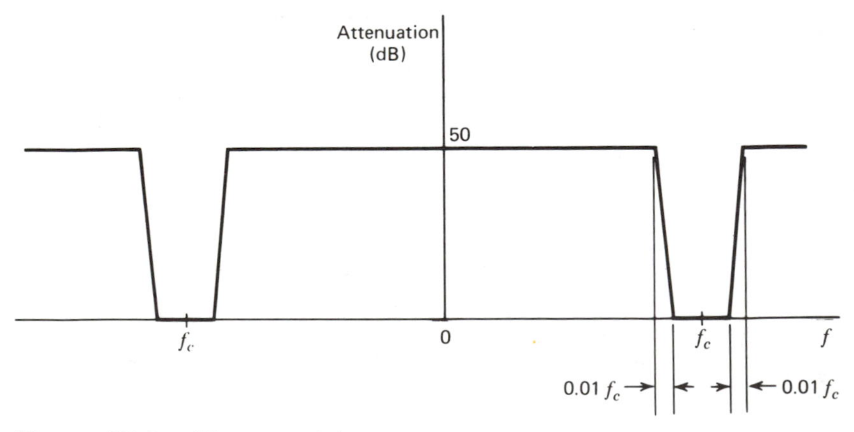

Figure P3.9 (Not to scale).

Problem 3.24 Consider the phase discrimination method of generating an SSB wave, shown in Fig. 3.25. Assume a modulating wave

$$m(t)=A_m\cos(2\pi f_m t)$$

Determine the ratio of the amplitude of the undesired side-frequency component to that of the desired side-frequency component when the system of Fig. 3.25 deviates from the ideal condition due to the following factors, considered one at a time:

(a) The phase difference $\beta-\alpha$ deviates from $\pi/2$ radians by δ, as shown by

$$\beta-\alpha=\frac{\pi}{2}+\delta$$

(b) The amplitudes of the modulating waves applied to the product modulators are unequal, that is

$$m_\alpha(t)=aA_m\cos(2\pi f_m t+\alpha)$$

and

$$m_\beta(t)=bA_m\cos(2\pi f_m t+\beta)$$

where a and b are constants.

(c) The carrier waves applied to the two product modulators are not exactly in phase quadrature; that is

$$c_1(t)=A_c\cos(2\pi f_c t)$$

and

$$c_2(t)=A_c\sin(2\pi f_c t+\Delta)$$

Problem 3.25 Figure P3.10 shows the block diagram of *Weaver's method* for generating SSB modulated waves. The message (modulating) wave $m(t)$ is limited to the frequency band $f_a \leq |f| \leq f_b$. The auxiliary carrier applied to the first pair of product modulators has a frequency f_0, which lies at the center of this band, as shown by

$$f_0=\frac{f_a+f_b}{2}$$

The low-pass filters in the in-phase and quadrature channels are identical, each with a cutoff frequency equal to $(f_b-f_a)/2$. The carrier applied to the second pair of product modulators has a frequency f_c that is greater than $(f_b-f_a)/2$.

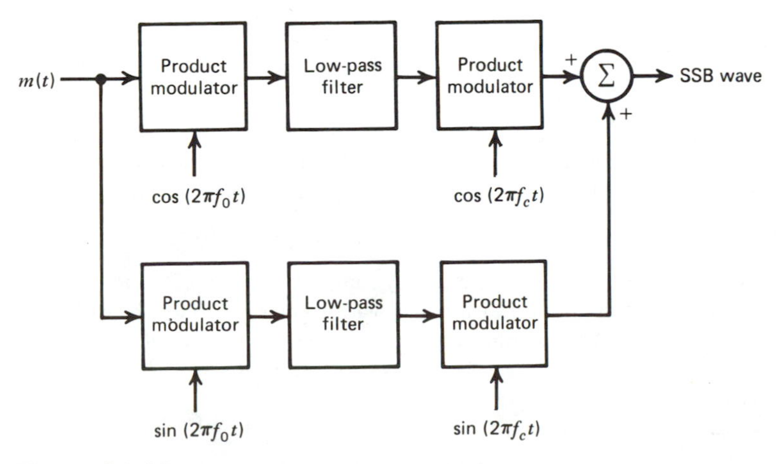

Figure P3.10

Sketch the spectra at the various points in the modulator of Fig. P3.10, and hence show that:

(a) For the lower sideband, the contributions of the in-phase and quadrature channels are of opposite polarity, and by adding them at the modulator output, the lower sideband is suppressed.
(b) For the upper sideband, the contributions of the in-phase and quadrature channels are of the same polarity, and by adding them, the upper sideband is transmitted.
(c) How would you modify the modulator of Fig. P3.10, so that only the lower sideband is transmitted?

Problem 3.26

(a) Let $s_u(t)$ denote the SSB wave obtained by transmitting only the upper sideband, and $\hat{s}_u(t)$ its Hilbert transform. Show that

$$m(t)=\frac{2}{A_c}\left[s_u(t)\cos(2\pi f_c t)+\hat{s}_u(t)\sin(2\pi f_c t)\right]$$

and

$$\hat{m}(t)=\frac{2}{A_c}\left[\hat{s}_u(t)\cos(2\pi f_c t)-s_u(t)\sin(2\pi f_c t)\right]$$

where $m(t)$ is the baseband signal, $\hat{m}(t)$ its Hilbert transform, f_c the carrier frequency, and A_c the carrier amplitude.

(b) Show that the corresponding equations in terms of the SSB wave $s_l(t)$ obtained by transmitting only the lower sideband are

$$m(t)=\frac{2}{A_c}\left[s_l(t)\cos(2\pi f_c t)+\hat{s}_l(t)\sin(2\pi f_c t)\right]$$

and

$$\hat{m}(t)=\frac{2}{A_c}\left[s_l(t)\sin(2\pi f_c t)-\hat{s}_l(t)\cos(2\pi f_c t)\right]$$

(c) Using the results of (a) and (b) above, set up the block diagram of a receiver for demodulating an SSB wave.

Problem 3.27 Consider the modulated wave

$$s(t)=A_c\cos(2\pi f_c t)+m(t)\cos(2\pi f_c t)-\hat{m}(t)\sin(2\pi f_c t)$$

which represents a carrier plus an SSB wave, with $m(t)$ denoting the message signal and $\hat{m}(t)$ its Hilbert transform. Determine the conditions for which an ideal envelope detector, with $s(t)$ as input, would produce a good approximation to the message signal $m(t)$.

Problem 3.28

(a) Consider a baseband signal $m(t)$ containing frequency components at 100, 200, and 400 Hz. This signal is applied to an SSB modulator together with a carrier at 100 kHz, with only the upper sideband retained. In the coherent detector used to recover $m(t)$, the local oscillator supplies a sine-wave of frequency 100.02 kHz. Determine the frequency components of the detector output.
(b) Repeat your analysis, assuming that only the lower sideband is transmitted.

Problem 3.29 The spectrum of a voice signal $m(t)$ is zero outside the interval $f_a \leq |f| \leq f_b$.

In order to ensure communication privacy, this signal is applied to a *scrambler* that consists of the following cascade of components: a product modulator, a high-pass filter, a second product modulator, and a low-pass filter. The carrier wave applied to the first product modulator has a frequency equal to f_c, whereas that applied to the second product modulator has a frequency equal to f_b+f_c; both of them have a unit amplitude. The high-pass and low-pass filters have the same cutoff frequency at f_c. Assume that $f_c>f_b$.

(a) Derive an expression for the scrambler output $s(t)$, and sketch its spectrum.
(b) Show that the original voice signal $m(t)$ may be recovered from $s(t)$ by using an *unscrambler* that is identical to the unit described above.

Problem 3.30 Consider the SSB wave

$$s(t)=m(t)\cos(2\pi f_c t)-\hat{m}(t)\sin(2\pi f_c t)$$

where f_c is the carrier frequency, $m(t)$ is the message signal, and $\hat{m}(t)$ is its Hilbert transform. This modulated wave is applied to a square-law device characterized by

$$y(t)=s^2(t)$$

Show that the output $y(t)$ contains a frequency component at twice the carrier frequency but that it has a time-varying phase, which makes it impractical to recover the carrier by squaring.

Problem 3.31 A method that is used for carrier recovery in SSB modulation systems involves transmitting two pilot frequencies that are appropriately positioned with respect to the transmitted sideband. This is illustrated in Fig. P3.11(*a*) for the case when only the lower sideband is transmitted. In this case, the two pilot frequencies f_1 and f_2 are defined by

$$f_1=f_c-W-\Delta f$$

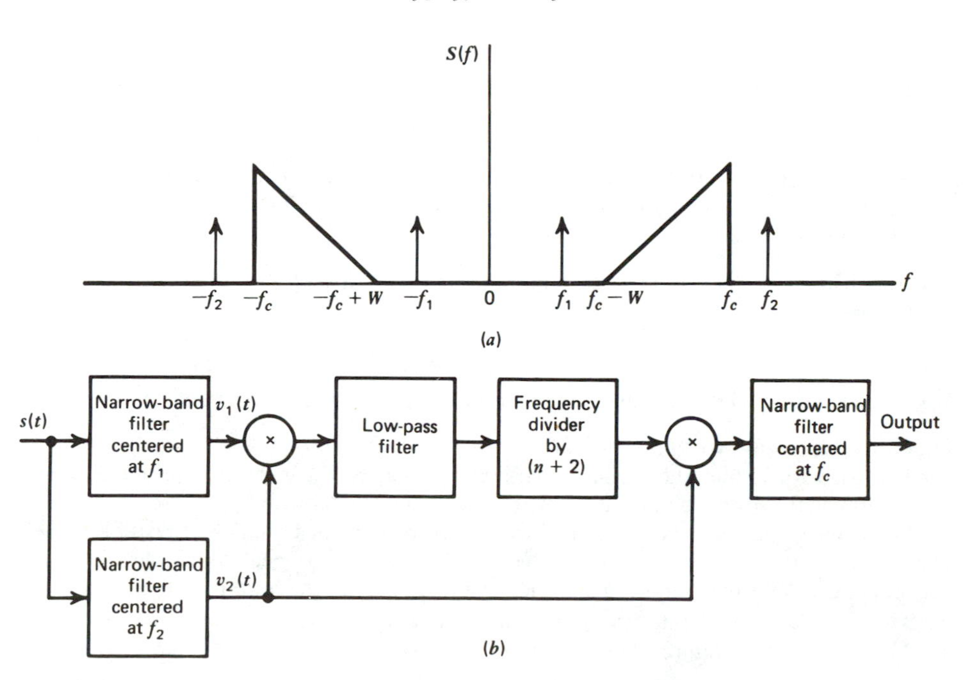

Figure P3.11

and

$$f_2 = f_c + \Delta f$$

where f_c is the carrier frequency and W is the message bandwidth. The Δf is chosen so as to satisfy the relation

$$n = \frac{W}{\Delta f}$$

where n is an integer. Carrier recovery is accomplished by using the scheme shown in Fig. P3.11(*b*). The outputs of the two narrow-band filters centered at f_1 and f_2 are defined by, respectively,

$$v_1(t) = A_1 \cos(2\pi f_1 t + \phi_1)$$

and

$$v_2(t) = A_2 \cos(2\pi f_2 t + \phi_2)$$

The low-pass filter is designed to select the difference frequency component of the first multiplier output due to $v_1(t)$ and $v_2(t)$.

(a) Show that the output signal of the circuit in Fig. P3.11(*b*) is proportional to the carrier wave $A_c \cos(2\pi f_c t)$ if the phase angles ϕ_1 and ϕ_2 satisfy the relation

$$\phi_2 = -\frac{\phi_1}{1+n}$$

(b) For the case when only the upper sideband is transmitted, the two pilot frequencies are defined by

$$f_1 = f_c - \Delta f$$

and

$$f_2 = f_c + W + \Delta f$$

How would you modify the carrier recovery circuit of Fig. P3.11(*b*) in order to deal with this case? What is the corresponding relation between ϕ_1 and ϕ_2 for the circuit output to be proportional to the carrier wave?

Problem 3.32 The single-tone modulating wave $m(t) = A_m \cos(2\pi f_m t)$ is used to generate the VSB wave

$$s(t) = \tfrac{1}{2} a A_m A_c \cos[2\pi(f_c + f_m)t] + \tfrac{1}{2} A_m A_c (1-a) \cos[2\pi(f_c - f_m)t]$$

where a is a constant, less than unity, representing the attenuation of the upper side-frequency.

(a) Find the quadrature component of the VSB wave $s(t)$.
(b) This VSB wave, plus the carrier $A_c \cos(2\pi f_c t)$, is passed through an envelope detector. Determine the distortion produced by the quadrature component.
(c) What is the value of constant a for which this distortion reaches its worst possible value?

Problem 3.33 For a distortionless reproduction of the baseband signal at the coherent detector output of a VSB receiver, show that the phase response of the filter used to generate the VSB wave in Fig. 3.30(*a*) has to satisfy two conditions (assuming that the amplitude response of the filter satisfies the requirements shown in Fig. 3.31):

(a) The phase is linear with frequency for $f_c - W \leqslant |f| \leqslant f_c + W$.
(b) The value of the phase at the carrier frequency is zero or an integer multiple of 2π radians.

Problem 3.34 Show that the quadrature component of a VSB signal may be derived by passing the Hilbert transform of the message signal through a high-pass filter. Sketch the frequency response of this filter.

Problem 3.35 Consider the receiver of Fig. P3.12. Both band-pass filters are assumed ideal, with the passband of filter 1 equal to 0.5–1.5 MHz and that of filter 2 having a width of 10 kHz. The input signal consists of a series of amplitude-modulated carriers spaced 10 kHz apart, and occupying the frequency band 0.5–1.5 MHz.

(a) Determine the range of permissible values of the mid-band frequency of filter 2 in order that the receiver may select any one of the amplitude-modulated carriers by adjusting the variable-frequency oscillator.
(b) For any setting of the mid-band frequency of filter 2, which lies inside the permissible range determined in part (a), find the required range of frequencies of the variable-frequency oscillator that would enable any one of the amplitude-modulated carriers in the input wave to be selected individually.

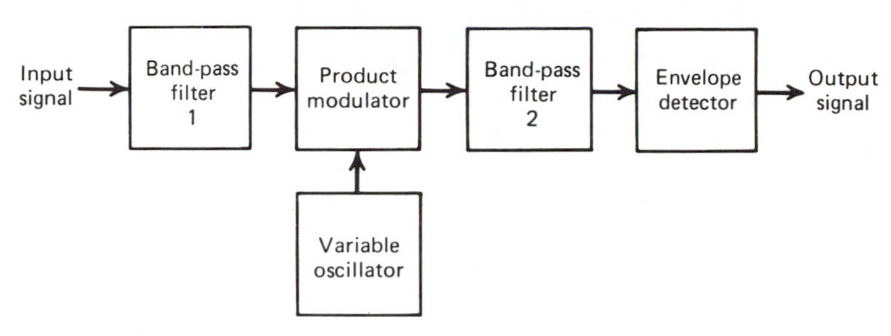

Figure P3.12

Problem 3.36 Figure P3.13 shows the block diagram of a *heterodyne spectrum analyzer*. It consists of a variable-frequency oscillator, multiplier, band-pass filter, and root mean-square (rms) meter. The oscillator has an amplitude A and operates over the range f_0 to f_0+W, where f_0 is the mid-band frequency of the filter and W is the signal bandwidth. Assume that $f_0=2W$, the filter bandwidth Δf is small compared with f_0, and the passband amplitude response of the filter is one. Determine the value of the rms meter output for a low-pass input signal $g(t)$.

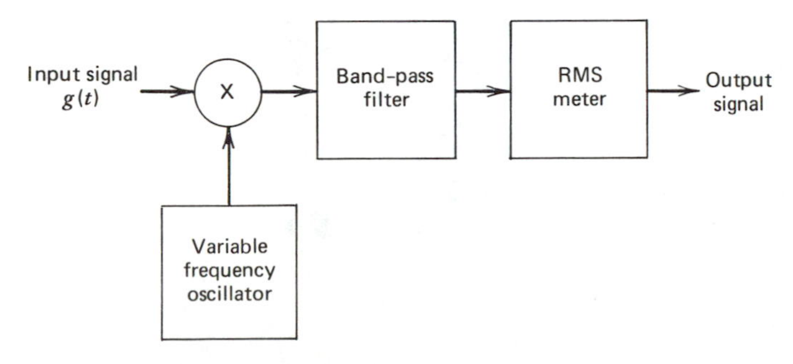

Figure P3.13

Problem 3.37 Consider a multiplex system in which four input signals $m_1(t)$, $m_2(t)$, $m_3(t)$, and $m_4(t)$ are respectively multiplied by the carrier waves

$$[\cos(2\pi f_a t)+\cos(2\pi f_b t)],$$
$$[\cos(2\pi f_a t+\alpha_1)+\cos(2\pi f_b t+\beta_1)],$$
$$[\cos(2\pi f_a t+\alpha_2)+\cos(2\pi f_b t+\beta_2)],$$

and

$$[\cos(2\pi f_a t+\alpha_3)+\cos(2\pi f_b t+\beta_3)],$$

and the resulting DSBSC waves are summed and then transmitted over a common channel. In the receiver, demodulation is achieved by multiplying the sum of the DSBSC waves by the four carrier waves separately, and then using filtering to remove the unwanted components.

(a) Determine the conditions which the phase angles α_1, α_2, α_3, and β_1, β_2, β_3 must satisfy in order that the output of the kth demodulator is $m_k(t)$, where $k=1, 2, 3, 4$.
(b) Determine the minimum separation of carrier frequencies f_a and f_b in relation to the bandwidth of the input signals so as to ensure a satisfactory operation of the system.

Problem 3.38 Figure P3.14 shows the block diagram of a *frequency synthesizer*, which enables the generation of many frequencies, each with the same high accuracy as the *master oscillator*. The master oscillator of frequency 1 MHz feeds two *spectrum generators*, one directly and the other through a *frequency divider*. Spectrum generator 1 produces a signal rich in the following harmonics: 1, 2, 3, 4, 5, 6, 7, 8, and 9 MHz. The frequency divider provides a 100 kHz output, in response to which spectrum generator 2 produces a second signal rich in the following harmonics: 100, 200, 300, 400, 500, 600, 700, 800, and 900 kHz. The harmonic selectors are designed to feed two signals into the mixer, one from spectrum generator 1 and the other from spectrum generator 2. Find the range of possible frequency outputs of this synthesizer and its resolution.

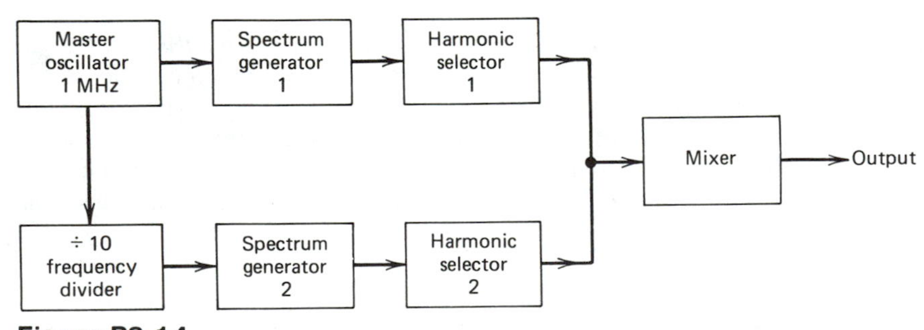

Figure P3.14

chapter 4 ANGLE MODULATION

In the previous chapter we investigated the effect of slowly varying the amplitude of a sinusoidal carrier wave in accordance with the baseband (information-carrying) signal. There is another way of modulating a sinusoidal carrier wave, namely, *angle modulation* in which the angle of the carrier wave is varied according to the baseband signal. In this method of modulation the amplitude of the carrier wave is maintained constant. An important feature of angle modulation is that it can provide better discrimination against noise and interference than amplitude modulation. As will be shown in Chapter 6, however, this improvement in performance is achieved at the expense of increased transmission bandwidth; that is, angle modulation provides us with a practical means of exchanging transmission bandwidth for improved noise performance. Such a tradeoff is not possible with amplitude modulation.

There are two forms of angle modulation that may be distinguished—*phase modulation* and *frequency modulation*. These two methods of modulation are closely related in that the properties of the one can be derived from those of the other. In this chapter we concentrate on frequency modulation.

4.1 BASIC DEFINITIONS: PHASE MODULATION (PM) AND FREQUENCY MODULATION (FM)

Let $\theta_i(t)$ denote the *angle* of a modulated sinusoidal carrier, which is a function of the message. We express the resulting *angle-modulated wave* as

$$s(t)=A_c \cos[\theta_i(t)] \tag{4.1}$$

where A_c is the carrier amplitude. A complete oscillation occurs whenever $\theta_i(t)$ changes by 2π radians. If $\theta_i(t)$ increases monotonically with time, the average frequency in hertz, over an inverval from t to $t+\Delta t$, is given by

$$f_{\Delta t}(t)=\frac{\theta_i(t+\Delta t)-\theta_i(t)}{2\pi\Delta t} \tag{4.2}$$

We may thus define the *instantaneous frequency* of the angle-modulated wave $s(t)$ by

$$\begin{aligned} f_i(t) &= \lim_{\Delta t \to 0} f_{\Delta t}(t) \\ &= \lim_{\Delta t \to 0} \left[\frac{\theta_i(t+\Delta t)-\theta_i(t)}{2\pi\Delta t}\right] \\ &= \frac{1}{2\pi}\frac{d\theta_i(t)}{dt} \end{aligned} \tag{4.3}$$

Thus, according to Eq. (4.1), we may interpret the angle-modulated wave $s(t)$ as a rotating phasor of length A_c and angle $\theta_i(t)$. The angular velocity of such a phasor is $d\theta_i(t)/dt$, in accordance with Eq. (4.3). In the simple case of an unmodulated carrier, the angle $\theta_i(t)$ is

$$\theta_i(t)=2\pi f_c t+\phi_c$$

and the corresponding phasor rotates with a constant angular velocity equal to $2\pi f_c$. The constant ϕ_c is the value of $\theta_i(t)$ at $t=0$.

There are an infinite number of ways in which the angle $\theta_i(t)$ may be varied in some manner with the baseband signal. However, we shall consider only two commonly used methods, phase modulation and frequency modulation, as defined below:

1. *Phase modulation (PM) is that form of angle modulation in which the angle* $\theta_i(t)$ *is varied linearly with the baseband signal* $m(t)$*, as shown by*

$$\theta_i(t) = 2\pi f_c t + k_p m(t) \tag{4.4}$$

The term $2\pi f_c t$ represents the angle of the *unmodulated* carrier, and the constant k_p represents the *phase sensitivity* of the modulator, expressed in radians per volt. This assumes that $m(t)$ is a voltage waveform. For convenience, we have assumed in Eq. (4.4) that the angle of the unmodulated carrier is zero at $t=0$. The phase-modulated wave $s(t)$ is thus described in the time domain by

$$s(t) = A_c \cos[2\pi f_c t + k_p m(t)] \tag{4.5}$$

2. *Frequency modulation (FM) is that form of angle modulation in which the instantaneous frequency* $f_i(t)$ *is varied linearly with the baseband signal* $m(t)$*, as shown by*

$$f_i(t) = f_c + k_f m(t) \tag{4.6}$$

The term f_c represents the frequency of the unmodulated carrier, and the constant k_f represents the *frequency sensitivity* of the modulator, expressed in hertz per volt. This assumes that $m(t)$ is a voltage waveform. Integrating Eq. (4.6) with respect to time and multiplying the result by 2π, we get

$$\theta_i(t) = 2\pi f_c t + 2\pi k_f \int_0^t m(t)\,dt \tag{4.7}$$

where, for convenience, we have assumed that the angle of the unmodulated carrier wave is zero at $t=0$. The frequency-modulated wave is therefore described in the time domain by

$$s(t) = A_c \cos\left[2\pi f_c t + 2\pi k_f \int_0^t m(t)\,dt\right] \tag{4.8}$$

A consequence of allowing the angle $\theta_i(t)$ to become dependent on the message signal $m(t)$ as in Eq. (4.4) or on its integral as in Eq. (4.7) is that the *zero crossings* of a PM wave or FM wave no longer have a perfect regularity in their spacing; zero crossings refer to the instants of time at which a waveform changes from a negative to a positive value or vice versa. This is one important feature which distinguishes both PM and FM waves from an AM wave. Another important difference is that the envelope of a PM or FM wave is constant (equal to the carrier amplitude), whereas the envelope of an AM wave is dependent on the message signal.

These differences between amplitude-modulated and angle-modulated waves are illustrated in Fig. 4.1 for the case of sinusoidal modulation. Parts (a) and (b) of this figure refer to the sinusoidal carrier and modulating waves, respectively.

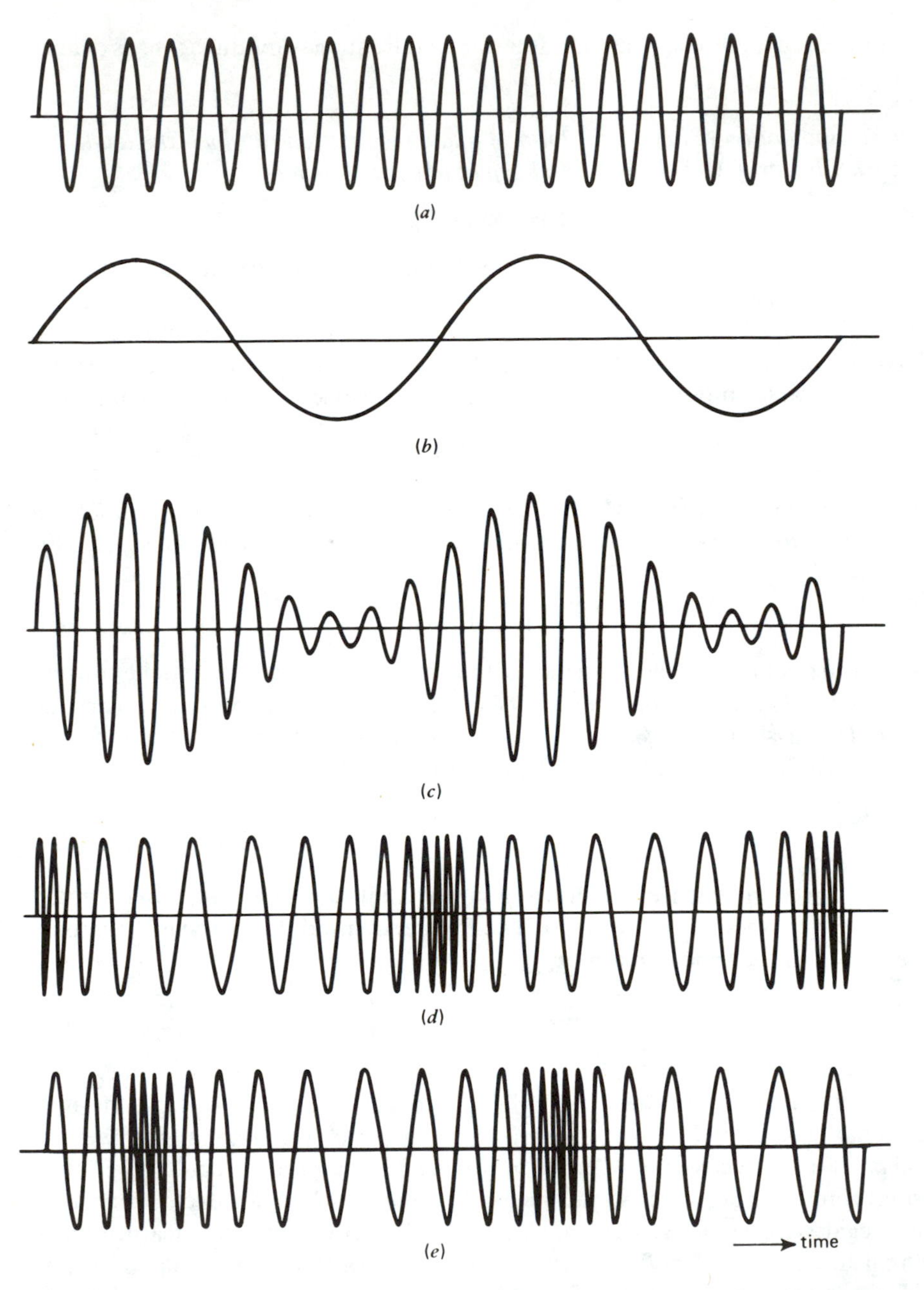

Figure 4.1 Illustrating AM, PM and FM waves produced by a single tone. *(a)* Carrier wave. *(b)* Sinusoidal modulating wave. *(c)* Amplitude-modulated wave. *(d)* Phase-modulated wave. *(e)* Frequency-modulated wave.

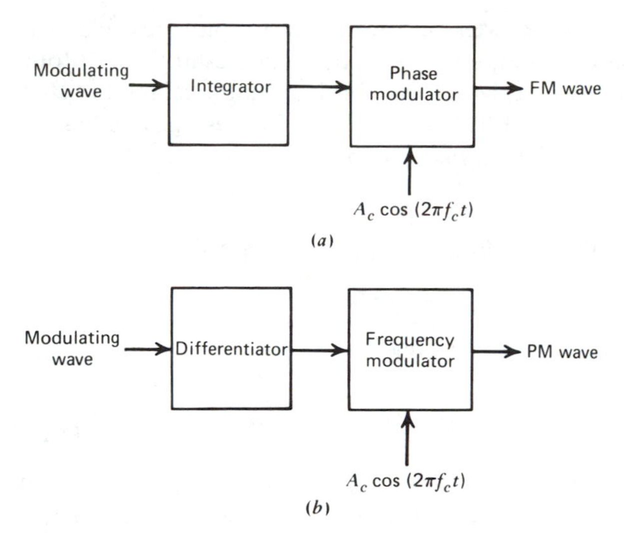

Figure 4.2 Illustrating the relationship between frequency modulation and phase modulation. *(a)* Scheme for generating an FM wave by using a phase modulator. *(b)* Scheme for generating a PM wave by using a frequency modulator.

Parts (*c*), (*d*) and (*e*) show the corresponding AM, PM, and FM waves, respectively. These waveforms indicate that a distinction can be made between PM and FM waves only when compared with the original modulating wave, which shows that there exists a close relationship between PM and FM waves.

Furthermore, comparing Eq. (4.5) with (4.8) reveals that an FM wave may be regarded as a PM wave in which the modulating wave is $\int_0^t m(t)dt$ in place of $m(t)$. This means that an FM wave can be generated by first integrating $m(t)$ and then using the result as the input to a phase modulator, as in Fig. 4.2(*a*). Conversely, a PM wave can be generated by first differentiating $m(t)$ and then using the result as the input to a frequency modulator, as in Fig. 4.2(*b*). We may thus deduce all the properties of PM waves from those of FM waves, and vice versa. Henceforth, we concentrate our attention on FM waves.

4.2 SINGLE-TONE FREQUENCY MODULATION

The FM wave $s(t)$ defined by Eq. (4.8) is a nonlinear function of the modulating wave $m(t)$. Hence, frequency modulation is a *nonlinear modulation process*. Consequently, unlike amplitude modulation, the spectrum of an FM wave is not related in a simple manner to that of the modulating wave. In fact, as we see later, the transmission bandwidth required by an FM wave can be much greater than that of a corresponding AM wave.

In the spectral analysis of FM waves we first consider the simplest possible case, namely, that of single-tone modulation, and then extend the result to multitone modulation. Our objective in this analysis is to establish an empirical relationship between the transmission bandwidth of an FM wave and the message bandwidth.

Consider then a sinusoidal modulating wave defined by

$$m(t)=A_m \cos(2\pi f_m t) \tag{4.9}$$

The instantaneous frequency of the resulting FM wave equals

$$\begin{aligned} f_i(t) &= f_c + k_f A_m \cos(2\pi f_m t) \\ &= f_c + \Delta f \cos(2\pi f_m t) \end{aligned} \tag{4.10}$$

where

$$\Delta f = k_f A_m \tag{4.11}$$

The quantity Δf is called the *frequency deviation*, representing the maximum departure of the instantaneous frequency of the FM wave from the carrier frequency f_c. A fundamental characteristic of an FM wave is that the frequency deviation Δf is proportional to the amplitude of the modulating wave, and is independent of the modulation frequency.

Using Eq. (4.10), the angle $\theta_i(t)$ of the FM wave is obtained as

$$\begin{aligned} \theta_i(t) &= 2\pi \int_0^t f_i(t)dt \\ &= 2\pi f_c t + \frac{\Delta f}{f_m} \sin(2\pi f_m t) \end{aligned} \tag{4.12}$$

The ratio of the frequency deviation Δf to the modulation frequency f_m is commonly called the *modulation index* of the FM wave. We denote it by β, so that we may write

$$\beta = \frac{\Delta f}{f_m} \tag{4.13}$$

and

$$\theta_i(t) = 2\pi f_c t + \beta \sin(2\pi f_m t) \tag{4.14}$$

From Eq. (4.14) we see that, in a physical sense, the parameter β represents the phase deviation of the FM wave, that is, the maximum departure of the angle $\theta_i(t)$ from the angle $2\pi f_c t$ of the unmodulated carrier.

The FM wave itself is given by

$$s(t) = A_c \cos[2\pi f_c t + \beta \sin(2\pi f_m t)] \tag{4.15}$$

Depending on the value of the modulation index β, we may distinguish two cases of frequency modulation: (1) *narrow-band FM* for which β is small, and (2) *wide-band FM* for which β is large, both compared to one radian. The reason for this distinction is that, as will be shown later, the transmission bandwidth of a narrow-band FM wave is closely equal to twice the message bandwidth, whereas in the case of a wide-band FM wave it is well in excess of this value.

4.3 NARROW-BAND FREQUENCY MODULATION

Consider Eq. (4.15), which defines an FM wave resulting from the use of a sinusoidal modulating wave. Expanding this relation, we get

$$s(t) = A_c \cos(2\pi f_c t)\cos[\beta \sin(2\pi f_m t)] - A_c \sin(2\pi f_c t)\sin[\beta \sin(2\pi f_m t)] \quad (4.16)$$

Assuming that the modulation index β is small compared to one radian, we may use the following approximations:

$$\cos[\beta \sin(2\pi f_m t)] \simeq 1 \quad (4.17)$$

and

$$\sin[\beta \sin(2\pi f_m t)] \simeq \beta \sin(2\pi f_m t) \quad (4.18)$$

Then Eq. (4.16) simplifies as follows

$$s(t) \simeq A_c \cos(2\pi f_c t) - \beta A_c \sin(2\pi f_c t)\sin(2\pi f_m t) \quad (4.19)$$

Equation (4.19) defines the approximate form of a narrow-band FM wave produced by a sinusoidal modulating wave $A_m \cos(2\pi f_m t)$. From this representation we obtain the modulator shown in block diagramic form in Fig. 4.3. This modulator involves splitting the carrier wave $A_c \cos(2\pi f_c t)$ into two paths. One path is direct, and the other contains a 90-degree phase-shifting network and a product modulator, which generates a DSBSC wave. A combination of these two signals produces a narrow-band FM wave with some distortion.

Ideally, an FM wave has a constant envelope and, for the case of a sinusoidal modulating wave of frequency f_m, the angle $\theta_i(t)$ is also sinusoidal with the same

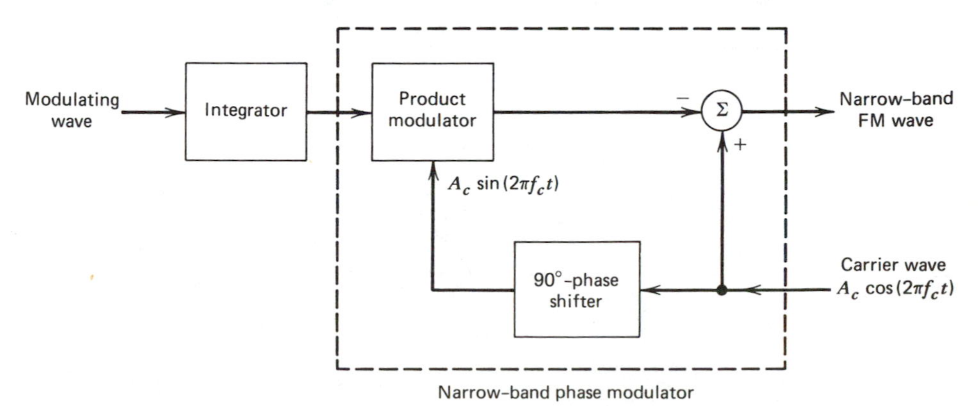

Figure 4.3 Block diagram of a method for generating a narrowband FM signal.

frequency. But the modulated wave produced by the narrow-band modulator of Fig. 4.3 differs from this ideal condition in two respects (see Problem 4.5):

1. The envelope contains a *residual* amplitude modulation and, therefore, varies with time.
2. For a sinusoidal modulating wave, the angle $\theta_i(t)$ contains *harmonic distortion* in the form of third- and higher-order harmonics of the modulation frequency f_m.

However, by restricting the modulation index to $\beta \leqslant 0.3$ radians, say, the effects of residual AM and harmonic PM are limited to negligible levels.

Returning to Eq. (4.19), we may expand it as follows

$$s(t) \simeq A_c \cos(2\pi f_c t) + \tfrac{1}{2}\beta A_c\{\cos[2\pi(f_c+f_m)t] - \cos[2\pi(f_c-f_m)t]\} \qquad (4.20)$$

This expression is somewhat similar to the corresponding one defining an AM wave, that is, Eq. (3.11) which is reproduced here for convenience

$$s_{AM}(t) = A_c \cos(2\pi f_c t) + \tfrac{1}{2}\mu A_c\{\cos[2\pi(f_c+f_m)t] + \cos[2\pi(f_c-f_m)t]\} \qquad (4.21)$$

where μ is the modulation factor of the AM wave. Comparing Eqs. (4.20) and (4.21) we see that, in the case of a sinusoidal modulating wave, the basic difference between an AM wave and a narrow-band FM wave is that the sign of the lower side-frequency in the narrow-band FM is reversed. Thus a narrow-band FM wave requires essentially the same transmission bandwidth (i.e., $2f_m$) as the AM wave.

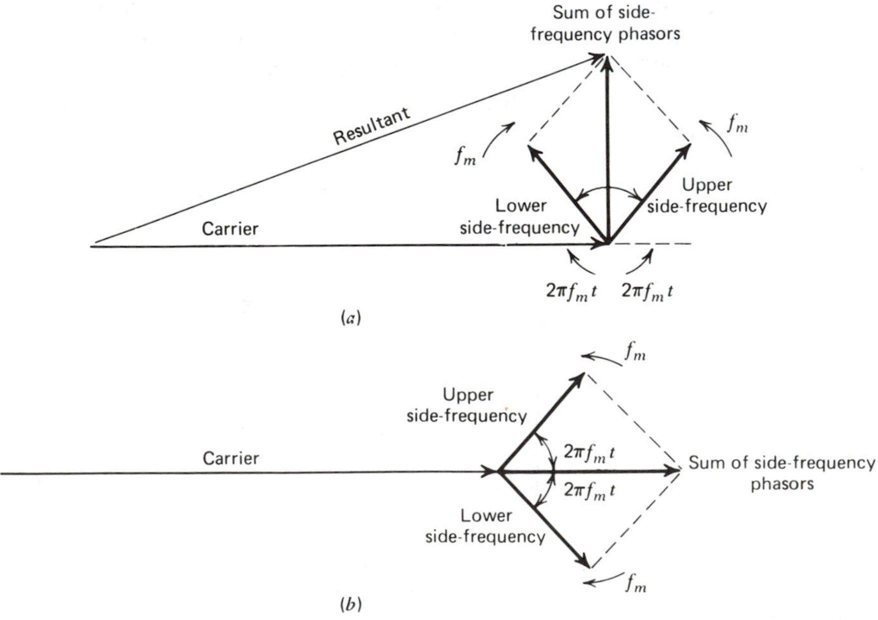

Figure 4.4 A phasor comparison of narrow-band FM and AM waves. *(a)* Narrow-band FM wave. *(b)* AM wave.

We may represent the narrow-band FM wave with a phasor diagram as shown in Fig. 4.4.(a) where we have used the carrier phasor as reference. We see that the resultant of the two side-frequency phasors is always at right angles to the carrier phasor. The effect of this is to produce a resultant phasor representing the narrow-band FM wave which is approximately of the same amplitude as the carrier phasor, but out of phase with respect to it. This phasor diagram should be contrasted with that of Fig. 4.4(b) which represents an AM wave. In this case we see that the resultant phasor representing the AM wave has an amplitude different from that of the carrier phasor, but always in phase with it.

4.4 WIDE-BAND FREQUENCY MODULATION

We next wish to determine the spectrum of the single-tone FM wave of Eq. (4.15) for an arbitrary value of the modulation index β. In general, an FM wave produced by a sinusoidal modulating wave, as in Eq. (4.15), is by itself nonperiodic, unless the carrier frequency f_c is an integral multiple of the modulation frequency f_m. However, by rewriting this equation in the form

$$\begin{aligned} s(t) &= \mathrm{Re}[A_c \exp(j2\pi f_c t + j\beta \sin(2\pi f_m t))] \\ &= \mathrm{Re}[\tilde{s}(t)\exp(j2\pi f_c t)] \end{aligned} \tag{4.22}$$

where $\tilde{s}(t)$ is the complex envelope of the FM wave $s(t)$, defined by

$$\tilde{s}(t) = A_c \exp[j\beta \sin(2\pi f_m t)], \tag{4.23}$$

we find that $\tilde{s}(t)$ is a periodic function of time, with a fundamental frequency equal to the modulation frequency f_m. We may therefore expand $\tilde{s}(t)$ in the form of a complex Fourier series as follows

$$\tilde{s}(t) = \sum_{n=-\infty}^{\infty} c_n \exp(j2\pi n f_m t) \tag{4.24}$$

where the complex Fourier coefficient c_n equals

$$\begin{aligned} c_n &= f_m \int_{-1/2f_m}^{1/2f_m} \tilde{s}(t)\exp(-j2\pi n f_m t)\,dt \\ &= f_m A_c \int_{-1/2f_m}^{1/2f_m} \exp[j\beta \sin(2\pi f_m t) - j2\pi n f_m t]\,dt \end{aligned} \tag{4.25}$$

For convenience, we define the variable

$$x = 2\pi f_m t, \tag{4.26}$$

in terms of which we may rewrite Eq. (4.25) as

$$c_n = \frac{A_c}{2\pi} \int_{-\pi}^{\pi} \exp[j(\beta \sin x - nx)]\,dx \tag{4.27}$$

The integral on the right-hand side of Eq. (4.27) is recognized as the nth *order Bessel function of the first kind* and argument β (see Appendix 4). This function is commonly denoted by the symbol $J_n(\beta)$, that is,

$$J_n(\beta)=\frac{1}{2\pi}\int_{-\pi}^{\pi} \exp[j(\beta \sin x-nx)]dx \tag{4.28}$$

Hence, we may rewrite Eq. (4.27) as

$$c_n=A_c J_n(\beta) \tag{4.29}$$

Substituting Eq. (4.29) in (4.24), we get, in terms of the Bessel function $J_n(\beta)$, the following expansion for the complex envelope of the FM wave:

$$\tilde{s}(t)=A_c \sum_{n=-\infty}^{\infty} J_n(\beta)\exp(j2\pi nf_m t) \tag{4.30}$$

Next, substituting Eq. (4.30) in (4.22), we get

$$s(t)=A_c\cdot \text{Re}\left[\sum_{n=-\infty}^{\infty} J_n(\beta)\exp[j2\pi(f_c+nf_m)t]\right] \tag{4.31}$$

Interchanging the order of summation and evaluation of the real part in the right-hand side of Eq. (4.31), we get

$$s(t)=A_c \sum_{n=-\infty}^{\infty} J_n(\beta)\cos[2\pi(f_c+nf_m)t] \tag{4.32}$$

This is the desired form for the Fourier series representation of the single-tone FM wave $s(t)$ for an arbitrary value of β. The discrete spectrum of $s(t)$ is obtained by taking the Fourier transforms of both sides of Eq. (4.32); thus

$$S(f)=\frac{A_c}{2}\sum_{n=-\infty}^{\infty} J_n(\beta)[\delta(f-f_c-nf_m)+\delta(f+f_c+nf_m)] \tag{4.33}$$

In Fig. 4.5 we have plotted the Bessel function $J_n(\beta)$ versus the modulation index β for different positive integer values of n. We can develop further insight into the behavior of the Bessel function $J_n(\beta)$ by making use of the following properties (see Appendix 4):

1. For n even, we have $J_n(\beta)=J_{-n}(\beta)$, whereas for n odd, we have $J_n(\beta)=-J_{-n}(\beta)$. That is,

$$J_n(\beta)=(-1)^n J_{-n}(\beta) \tag{4.34}$$

2. For small values of β, we have

$$\left.\begin{aligned} J_0(\beta)&\simeq 1\\ J_1(\beta)&\simeq\frac{\beta}{2}\\ J_n(\beta)&\simeq 0, \qquad n>1 \end{aligned}\right\} \tag{4.35}$$

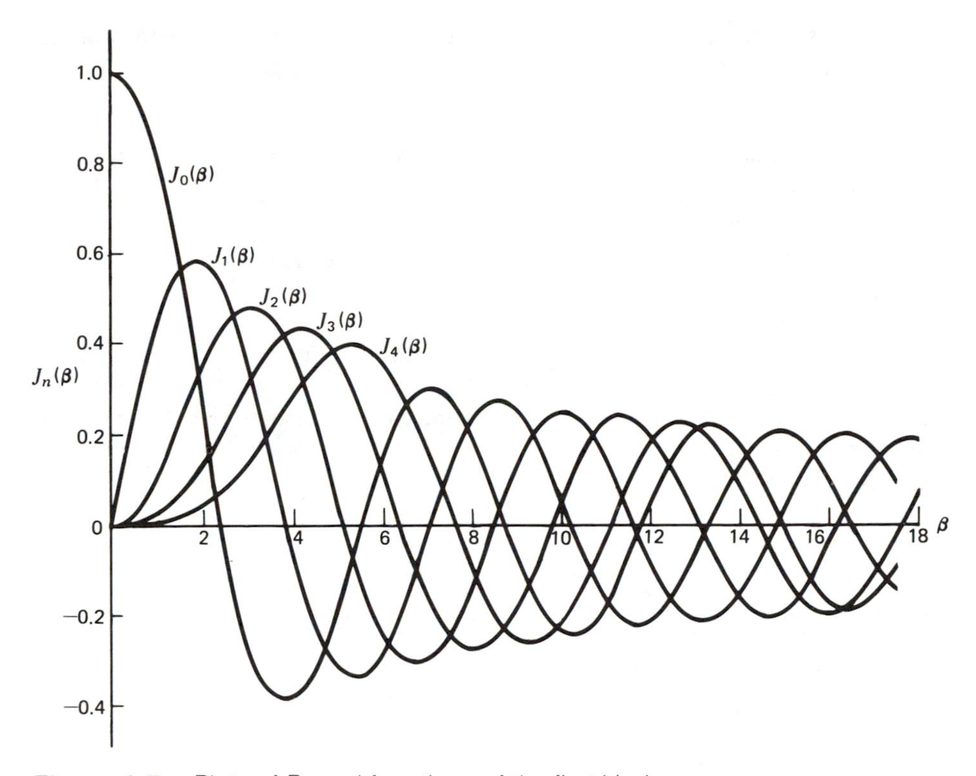

Figure 4.5 Plots of Bessel functions of the first kind.

3.
$$\sum_{n=-\infty}^{\infty} J_n^2(\beta) = 1 \tag{4.36}$$

Thus, using Eqs. (4.33) through (4.36), and the curves of Fig. 4.5, we may make the following observations:

1. The spectrum of an FM wave contains a carrier component and an infinite set of side-frequencies located symmetrically on either side of the carrier at frequency separations of f_m, $2f_m$, $3f_m$, ... In this respect, the result is unlike that which prevails in an AM system, since in an AM system a sinusoidal modulating signal gives rise to only one pair of side-frequencies.
2. For the special case of β small compared with unity, only the Bessel coefficients $J_0(\beta)$ and $J_1(\beta)$ have significant values, so that the FM wave is effectively composed of a carrier and a single pair of side-frequencies at $f_c \pm f_m$. This situation corresponds to the special case of narrow-band FM.
3. The amplitude of the carrier component varies with β according to $J_0(\beta)$. That is, unlike an AM wave, the amplitude of the carrier component of an FM wave is dependent on the modulation index β. The physical explanation for this property is that the envelope of an FM wave is constant, so that the average

power of such a wave developed across a 1-ohm resistor is also constant, as shown by

$$P=\frac{1}{2}A_c^2 \tag{4.37}$$

When the carrier is modulated to generate the FM wave, the power in the side-frequencies may appear only at the expense of the power originally in the carrier, thereby making the amplitude of the carrier component dependent on β. Note that the average power of an FM wave may also be determined from Eq. (4.32), obtaining

$$P=\frac{1}{2}A_c^2 \sum_{n=-\infty}^{\infty} J_n^2(\beta) \tag{4.38}$$

Substituting Eq. (4.36) in (4.38), the expression for P reduces to the same form as in Eq. (4.37).

Example 1

We wish to investigate the ways in which variations in the amplitude and frequency of a sinusoidal modulating wave affect the spectrum of the FM wave. Consider first the case when the frequency of the modulating wave is fixed, but its amplitude is varied, producing a corresponding variation in the frequency deviation Δf. Thus, keeping the modulation frequency f_m fixed, we find that the amplitude spectrum of the resulting FM wave is as shown plotted in Fig. 4.6 for $\beta=1$, 2, and 5. In this diagram we have normalized the spectrum with respect to the unmodulated carrier amplitude.

Consider next the case when the amplitude of the modulating wave is fixed; that is, the frequency deviation Δf is maintained constant, and the modulation frequency f_m is varied. In this case we find that the amplitude spectrum of the resulting FM wave is as shown plotted in Fig. 4.7 for $\beta=1$, 2, and 5. We see that when Δf is fixed and β is increased, we have an increasing number of spectral lines crowding into the fixed frequency interval $f_c-\Delta f<f<f_c+\Delta f$. That is, when β approaches infinity, the bandwidth of the FM wave approaches the limiting value of $2\,\Delta f$.

4.5 MULTITONE FM WAVES

In practice, the modulating wave $m(t)$ is usually multitone in nature in that it consists of a group of sine-waves of different frequencies, which may be completely unrelated or harmonically related. Consider a carrier wave $A_c \cos(2\pi f_c t)$ that is frequency modulated by two tones of frequencies f_1 and f_2, producing frequency deviations Δf_1 and Δf_2, respectively. The resulting FM wave may be written in the form

$$s(t)=A_c \cos[2\pi f_c t+\beta_1 \sin(2\pi f_1 t)+\beta_2 \sin(2\pi f_2 t)] \tag{4.39}$$

where $\beta_1=\Delta f_1/f_1$ denotes the modulation index of the first tone and $\beta_2=\Delta f_2/f_2$ denotes that of the second tone. Following a procedure similar to that used for analyzing the FM wave produced by a single modulating tone, we may expand the

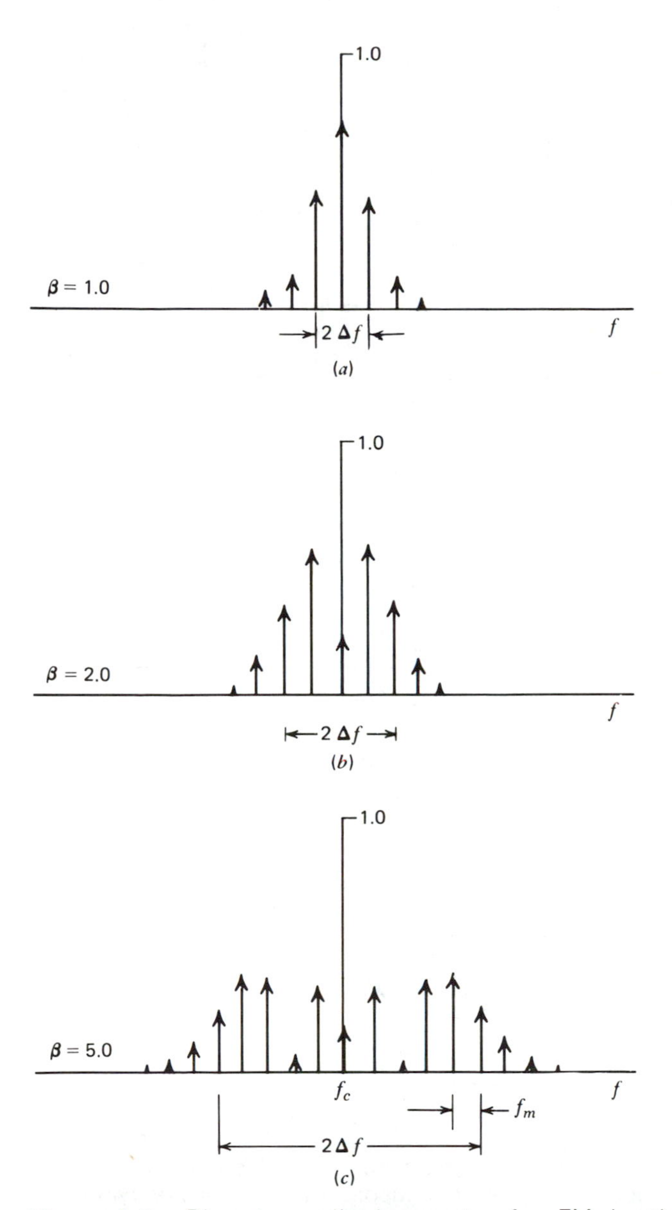

Figure 4.6 Discrete amplitude spectra of an FM signal, normalized with respect fo the carrier amplitude, for the case of sinusoidal modulation of fixed frequency and varying amplitude. Only the spectra for positive frequencies are shown.

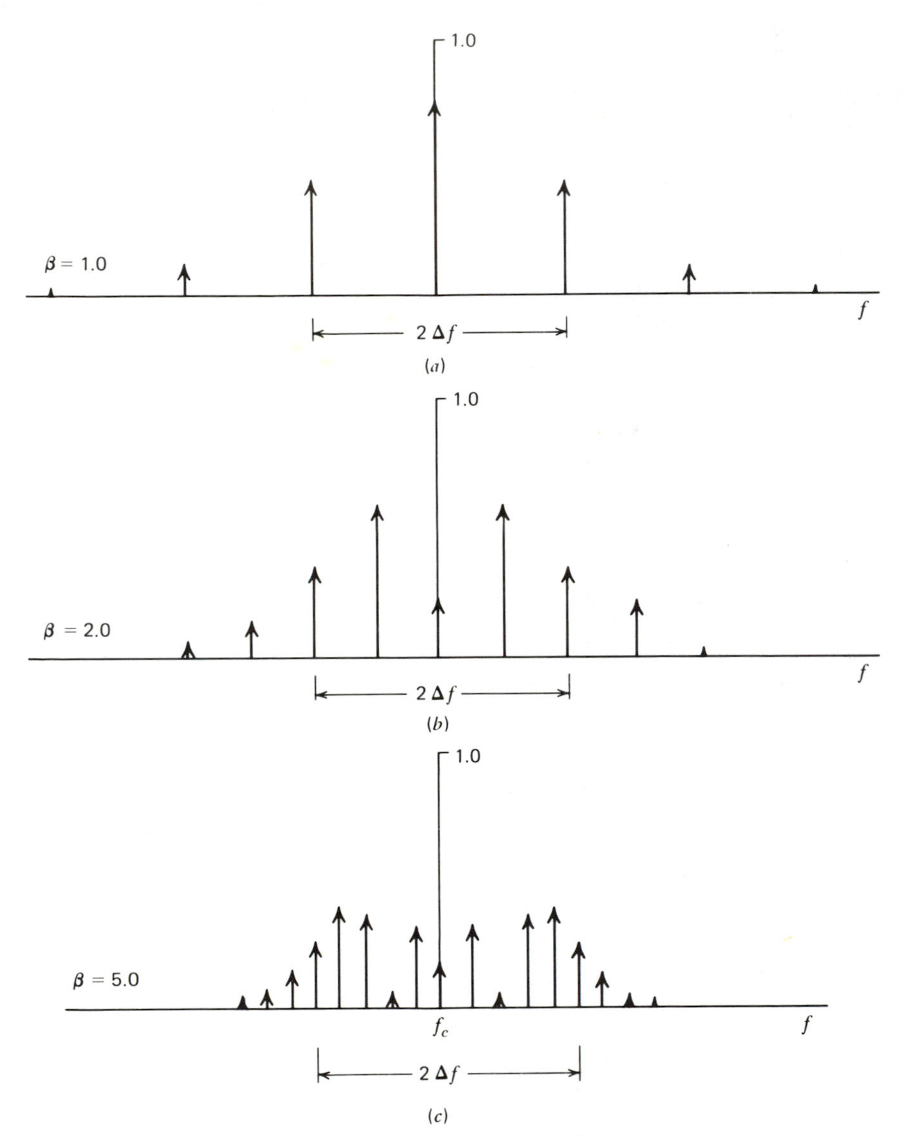

Figure 4.7 Discrete amplitude spectra of an FM signal, normalized with respect to the carrier amplitude, for the case of sinusoidal modulation of varying frequency and fixed amplitude. Only the spectra for positive frequencies are shown.

modulated wave of Eq. (4.39) as follows

$$s(t)=A_c \sum_{m=-\infty}^{\infty} \sum_{n=-\infty}^{\infty} J_m(\beta_1)J_n(\beta_2)\cos[2\pi(f_c+mf_1+nf_2)t] \tag{4.40}$$

Equation (4.40) shows that the spectrum of an FM wave, produced by a modulating wave with two tones of frequencies f_1 and f_2, consists of four types of terms as described below:

1. A carrier component of amplitude $J_0(\beta_1)J_0(\beta_2)$ and frequency f_c.
2. A set of side-frequencies corresponding to the modulation frequency f_1, with amplitudes $J_m(\beta_1)J_0(\beta_2)$ and frequencies $(f_c \pm mf_1)$, where $m=1, 2, 3, \ldots$.
3. A set of side-frequencies corresponding to the modulation frequency f_2, with amplitudes $J_0(\beta_1)J_n(\beta_2)$ and frequencies $(f_c \pm nf_2)$, where $n=1, 2, 3, \ldots$.
4. A set of cross-modulation terms with amplitudes $J_m(\beta_1)J_n(\beta_2)$ and frequencies $(f_c \pm mf_1 \pm nf_2)$, where $m=1, 2, 3, \ldots$ and $n=1, 2, 3, \ldots$.

It is interesting to compare this situation with the corresponding one involving amplitude modulation. When a carrier wave of frequency f_c is amplitude-modulated by a signal consisting of two tones of frequencies f_1 and f_2, the spectrum of the resulting AM wave consists of a carrier, a pair of upper side-frequencies at f_c+f_1

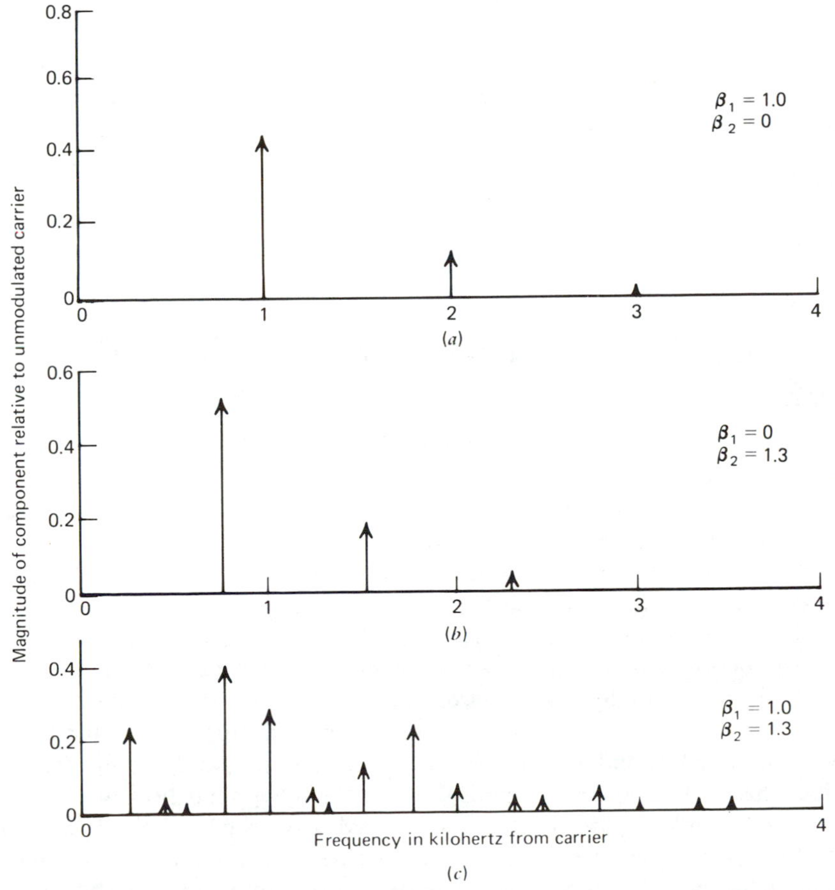

Figure 4.8 The spectra of upper side-frequencies of a two-tone FM wave for varying β_1 and β_2.

and f_c+f_2, and a pair of lower side-frequencies at f_c-f_1 and f_c-f_2. That is, superposition holds in amplitude modulation. On the other hand, the host of new side-frequencies in the FM wave of Eq. (4.40) emphasizes that superposition does not apply to frequency modulation. This is to be expected, because as pointed out earlier, frequency modulation is a nonlinear modulation process.

It is a straightforward matter to extend Eq. (4.40) to modulating waves with more than two tones; however, the analysis becomes more cumbersome.

Example 2

Consider a modulating wave that consists of two tones defined by

$$\beta_1=1.0$$
$$\beta_2=1.3$$
$$f_1=1 \text{ kHz}$$
$$f_2=0.77 \text{ kHz}$$

When a sinusoidal carrier wave is modulated by these two tones individually, the respective spectra of upper side-frequencies of the resulting FM waves are as shown plotted in parts (*a*) and (*b*) of Fig. 4.8. When the same carrier wave is frequency modulated by the two tones simultaneously, the resulting spectrum of upper side-frequencies is as shown plotted in Fig. 4.8(*c*). We see that new side-frequencies appear when both tones are applied. It is also apparent that the amplitudes of the side-frequencies which appeared when the modulating waves were applied individually are now changed. This is because each amplitude is determined by the product of two Bessel functions.

4.6 TRANSMISSION BANDWIDTH OF FM WAVES

In theory, an FM wave contains an infinite number of side-frequencies so that the bandwidth required to transmit such a signal is similarly infinite in extent. In practice, however, we find that the FM wave is effectively limited to a finite number of significant side-frequencies compatible with a specified amount of distortion. We may therefore specify an effective bandwidth required for the transmission of an FM wave. Consider first the case of an FM wave generated by a single-tone modulating wave of frequency f_m. In such an FM wave, the side-frequencies that are separated from the carrier frequency f_c by an amount greater than the frequency deviation Δf decrease rapidly toward zero, so that the bandwidth always exceeds the total frequency excursion, but nevertheless is limited. Specifically, for large values of the modulation index β, the bandwidth approaches, and is only slightly greater than, the total frequency excursion $2\,\Delta f$. On the other hand, for small values of the modulation index β, the spectrum of the FM wave is effectively limited to the carrier frequency f_c and one pair of side-frequencies at $f_c \pm f_m$, so that the bandwidth approaches $2f_m$. We may thus define an approximate rule for the transmission bandwidth of an FM wave generated by a single-tone modulating wave of frequency f_m as follows

$$B_T \simeq 2\Delta f + 2f_m = 2\,\Delta f\left(1+\frac{1}{\beta}\right) \tag{4.41}$$

This relation is known as *Carson's rule*.

For a more accurate assessment of the bandwidth requirement of an FM wave, we may use a definition based on retaining the maximum number of significant side-frequencies whose amplitudes are all greater than some selected value. A convenient choice for this value is 1 percent of the unmodulated carrier amplitude. *We may thus define the transmission bandwidth of an FM wave as the separation between the two frequencies beyond which none of the side-frequencies is greater than 1 percent of the carrier amplitude obtained when the modulation is removed.* That is, we define the transmission bandwidth as $2n_{\max}f_m$, where f_m is the modulation frequency and $n_{\max}$ is the maximum value of the integer n which satisfies the requirement $|J_n(\beta)|>0.01$. The value of $n_{\max}$ varies with the modulation index β and can be determined readily from tabulated values of the Bessel function $J_n(\beta)$. Table 4.1 shows the total number of significant side-frequencies (including both the upper and lower side-frequencies) for different values of β, calculated on the 1 percent basis explained above. The transmission bandwidth B_T calculated using this procedure can be presented in the form of a universal curve by normalizing it with respect to the frequency deviation Δf, and then plotting it versus β. This curve is shown in Fig. 4.9, which is drawn as a best fit through the set of points obtained by using Table 4.1. In Fig. 4.9 we note that as the modulation index β is increased, the bandwidth occupied by the significant side-frequencies drops toward that over which the carrier frequency actually deviates. This means that small values of the modulation index β are relatively more extravagant in transmission bandwidth than are the larger values of β.

Consider next an arbitrary modulating wave $m(t)$ with its highest frequency component denoted by W. The bandwidth required to transmit an FM wave generated by this modulating wave is estimated by using a worst-case tone-

Table 4.1

Modulation index β	Number of significant side-frequencies $2n_{\max}$
0.1	2
0.3	4
0.5	4
1.0	6
2.0	8
5.0	16
10.0	28
20.0	50
30.0	70

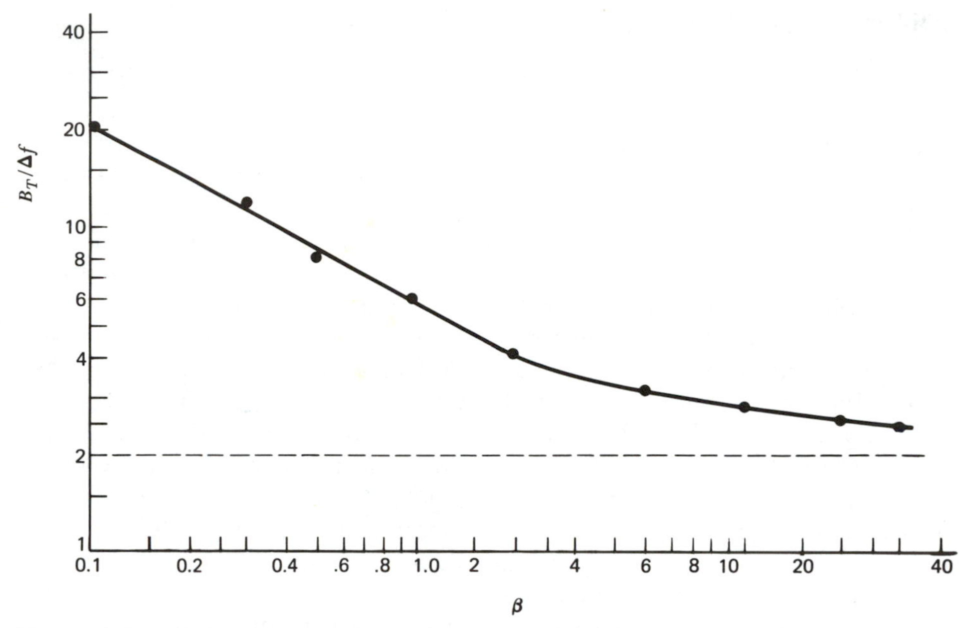

Figure 4.9 Universal curve for evaluating the 1 percent bandwidth of an FM wave.

modulation analysis. Specifically, we first determine the so-called *deviation ratio* D, defined as the ratio of the frequency deviation Δf, which corresponds to the maximum possible amplitude of the modulating wave $m(t)$, to the highest modulation frequency W; these conditions represent the extreme cases possible. *The deviation ratio D plays the same role for nonsinusoidal modulation that the modulation index β plays for the case of sinusoidal modulation.* Then, replacing β by D and replacing f_m by W, we use Carson's rule given by Eq. (4.41) or the universal curve of Fig. 4.9 to obtain a value for the transmission bandwidth of the FM wave. From a practical viewpoint, Carson's rule somewhat underestimates the bandwidth requirement of an FM system, whereas using the universal curve of Fig. 4.9 yields a somewhat conservative result. Thus the choice of a transmission bandwidth that lies between the bounds provided by these two rules of thumb is acceptable for most practical purposes.

Example 3

In North America, the maximum value of frequency deviation Δf is fixed at 75 kHz for commercial FM broadcasting by radio. If we take the modulation frequency $W = 15$ kHz, which is typically the maximum audio frequency of interest in FM transmission, we find that the corresponding value of the deviation ratio is

$$D = \frac{75}{15} = 5$$

Using Carson's rule of Eq. (4.41), replacing β by D and replacing f_m by W, the approximate value of the transmission bandwidth of the FM wave is obtained as

$$B_T = 2(75 + 15) = 180 \text{ kHz}$$

On the other hand, use of the curve of Fig. 4.9 gives the transmission bandwidth of the FM wave to be

$$B_T = 3.2\ \Delta f = 3.2 \times 75 = 240 \text{ kHz}$$

Thus Carson's rule underestimates the transmission bandwidth by 25 percent compared with the result of using the curve of Fig. 4.9.

4.7 GENERATION OF FM WAVES

There are essentially two basic methods of generating frequency-modulated waves, namely, the *indirect FM* and *direct FM*. In the indirect method of producing frequency modulation, the modulating wave is first used to produce a narrow-band FM wave, and *frequency multiplication* is next used to increase the frequency deviation to the desired level. On the other hand, in the direct method of producing frequency modulation the carrier frequency is directly varied in accordance with the input baseband signal. In this section, we describe the important features of both methods.

Indirect FM

A simple block diagram of an indirect FM system is shown in Fig. 4.10.* The baseband signal $m(t)$ is first integrated and then used to phase-modulate a crystal-controlled oscillator. In order to minimize the distortion inherent in the phase modulator, the maximum phase deviation or modulation index β is kept small,

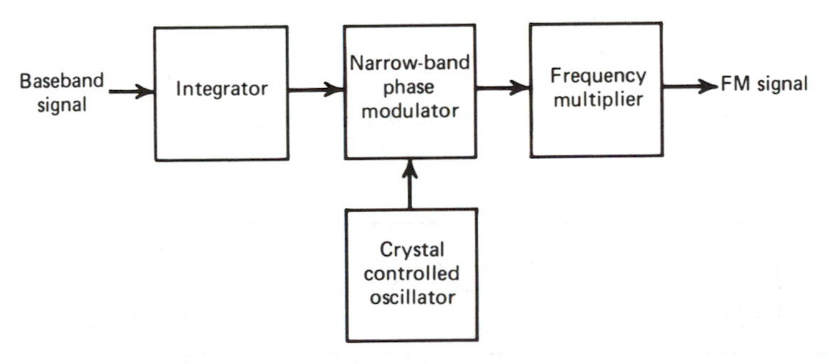

Figure 4.10 Block diagram of the indirect method of generating a wide-band FM signal.

* The indirect method of generating a wide-band FM wave was first proposed by Armstrong: E. H. Armstrong, "A method of reducing disturbances in radio signaling by a system of frequency modulation," *Proc. IRE*, vol. 24, pp. 689–740, May 1936. Armstrong was also the first to recognize the noise-cleaning properties of frequency modulation.

thereby resulting in a narrow-band FM wave. This signal is next multiplied in frequency by means of a frequency multiplier so as to produce the desired wide-band FM wave. For the implementation of the narrow-band phase modulator, we may use the arrangement described in Section 4.3.

Let $s_1(t)$ denote the output of the phase modulator,

$$s_1(t) = A_1 \cos\left[2\pi f_1 t + 2\pi k_f \int_0^t m(t)dt\right] \tag{4.42}$$

where f_1 is the frequency of the crystal-controlled oscillator, and k_f is a constant. For a sinusoidal modulating wave, the output $s_1(t)$ is given by

$$s_1(t) = A_1 \cos[2\pi f_1 t + \beta_1 \sin(2\pi f_m t)] \tag{4.43}$$

where β_1 is the modulation index; β_1 is kept small (less than 0.3 radians) to keep the distortion to a minimum.

The phase modulator output is next multiplied n times in frequency by the *frequency multiplier*, producing the desired wide-band FM wave:

$$s(t) = A_c \cos\left[2\pi f_c t + 2n\pi k_f \int_0^t m(t)dt\right] \tag{4.44}$$

where $f_c = nf_1$. In the case of a sinusoidal modulating wave, Eq. (4.44) reduces to

$$s(t) = A_c \cos[2\pi f_c t + \beta \sin(2\pi f_m t)] \tag{4.45}$$

where $\beta = n\beta_1$. Thus, by choosing n properly, we may set the final value of the modulation index β at any desired value.

Example 4

Figure 4.11 shows the simplified block diagram of a typical FM transmitter (based on the indirect method) used to transmit audio signals containing frequencies in the range 100 Hz to 15 kHz. The narrow-band phase modulator is supplied with a carrier wave of frequency $f_1 = 0.1$ MHz by a crystal-controlled oscillator. The desired FM wave at the transmitter output has a carrier frequency $f_c = 100$ MHz and the frequency deviation $\Delta f = 75$ kHz.

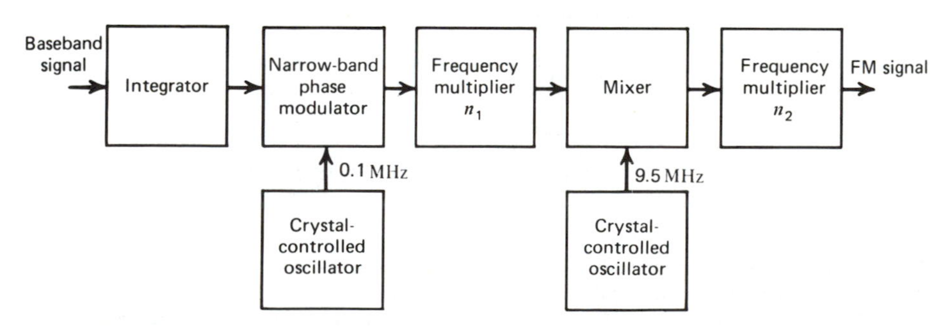

Figure 4.11 Block diagram of the wide-band frequency modulator for Example 4.

In order to limit the harmonic distortion produced by the narrow-band phase modulator, we restrict the modulation index β_1 to a maximum value of 0.3 radians. Consider then the value $\beta_1 = 0.2$ radians, which certainly satisfies this requirement. The lowest modulation frequencies of 100 Hz produce a frequency deviation $\Delta f_1 = 20$ Hz at the narrow-band phase modulator output, whereas the largest modulation frequencies of 15 kHz produce a frequency deviation $\Delta f_1 = 3$ kHz. The lowest modulation frequencies are therefore of immediate concern.

To produce a frequency deviation $\Delta f = 75$ kHz at the FM transmitter output, a frequency multiplication is required. Specifically, with $\Delta f_1 = 20$ Hz and $\Delta f = 75$ kHz, we require a total frequency multiplication ratio of 3750. However, using a straight frequency multiplication equal to this value would produce a much higher carrier frequency at the transmitter output than the desired value of 100 MHz. To generate an FM wave having both the desired frequency deviation and carrier frequency, we therefore need to use a *two-stage frequency multiplier* with an intermediate stage of frequency translation, as illustrated in Fig. 4.11. Let n_1 and n_2 denote the respective frequency multiplication ratios, so that

$$n_1 n_2 = \frac{\Delta f}{\Delta f_1} = \frac{75\,000}{20} = 3750 \tag{4.46}$$

The carrier frequency at the first frequency multiplier output is translated downward in frequency to $(f_2 - n_1 f_1)$ by mixing it with a sine-wave of frequency $f_2 = 9.5$ MHz, which is supplied by a second crystal-controlled oscillator. However, the carrier frequency at the input of the second frequency multiplier is equal to f_c/n_2. Equating these two frequencies, we get

$$f_2 - n_1 f_1 = \frac{f_c}{n_2}$$

Hence, with $f_1 = 0.1$ MHz, $f_2 = 9.5$ MHz, and $f_c = 100$ MHz, we have

$$9.5 - 0.1 n_1 = \frac{100}{n_2} \tag{4.47}$$

Table 4.2 Values of Carrier Frequency and Frequency Deviation at the Various Points in the Frequency Modulator of Fig. 4.11

	At the phase modulator output	At the first frequency multiplier output	At the mixer output	At the second frequency multiplier output
Carrier frequency	0.1 MHz	7.5 MHz	2.0 MHz	100 MHz
Frequency deviation	20 Hz	1.5 kHz	1.5 kHz	75 kHz

Solving Eqs. (4.46) and (4.47) for n_1 and n_2, we obtain

$$n_1 = 75$$
$$n_2 = 50$$

Using these frequency multiplication ratios, we get the set of values indicated in Table 4.2.

Direct FM

In a direct FM system, the instantaneous frequency of the carrier wave is varied directly in accordance with the baseband signal by means of a device known as a *voltage-controlled oscillator*. One way of implementing such a device is to use a sinusoidal oscillator having a relatively high-Q frequency-determining network and to control the oscillator by symmetrical incremental variation of the reactive components. An example of this scheme is shown in Fig. 4.12, representing a *Hartley oscillator*. We assume that the capacitive component of the frequency-determining network consists of a fixed capacitor shunted by a voltage-variable capacitor. The resultant capacitance is represented by $C(t)$ in Fig. 4.12. A voltage-variable capacitor, commonly called a *varactor* or *varicap*, is one whose capacitance varies with the voltage applied across its electrodes. The variable-voltage capacitance may be obtained, for example, by using a *p-n* junction diode that is biased in the reverse direction; the larger the reverse voltage applied to such a diode, the smaller the transition capacitance of the diode. The frequency of oscillation of the Hartley oscillator of Fig. 4.12 is given by

$$f_i(t) = \frac{1}{2\pi\sqrt{(L_1 + L_2)C(t)}} \tag{4.48}$$

where $C(t)$ is the total capacitance of the fixed capacitor and the variable-voltage capacitor, and L_1 and L_2 are the two inductances in the frequency-determining network. Assume that for a sinusoidal modulating wave of frequency f_m, the capacitance $C(t)$ is expressed as follows

$$C(t) = C_0 + \Delta C \cos(2\pi f_m t) \tag{4.49}$$

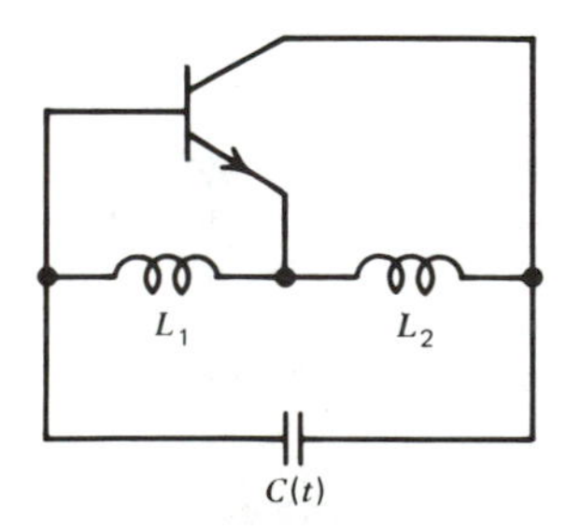

Figure 4.12 Hartley oscillator.

where C_0 is the total capacitance in the absence of modulation and ΔC is the maximum change. Substituting Eq. (4.49) in (4.48), we get

$$f_i(t) = f_0 \left[1 + \frac{\Delta C}{C_0} \cos(2\pi f_m t) \right]^{-1/2} \tag{4.50}$$

where f_0 is the unmodulated frequency of oscillation, that is,

$$f_0 = \frac{1}{2\pi\sqrt{C_0(L_1 + L_2)}} \tag{4.51}$$

Provided that the maximum change in capacitance ΔC is small compared with the unmodulated capacitance C_0, we may approximate Eq. (4.50) as follows

$$f_i(t) \simeq f_0 \left[1 - \frac{\Delta C}{2C_0} \cos(2\pi f_m t) \right] \tag{4.52}$$

Then, by defining

$$\frac{\Delta C}{2C_0} = -\frac{\Delta f}{f_0} \tag{4.53}$$

we obtain, for the instantaneous frequency of the oscillator, which is being frequency modulated by varying the capacitance of its frequency-determining resonant network, the following relation

$$f_i(t) \simeq f_0 + \Delta f \cos(2\pi f_m t) \tag{4.54}$$

Equation (4.54) is the desired relation for the instantaneous frequency of an FM wave.

In order to generate a wide-band FM wave with the required frequency deviation, we may use the configuration shown in Fig. 4.13 consisting of the voltage-controlled oscillator described above, followed by a series of frequency multipliers and mixers. This configuration permits the attainment of good oscillator stability, constant proportionality between output frequency change and input voltage change, and the necessary modulator bandwidth to achieve wide-band FM.

An FM transmitter using the direct method as described above, however, has the disadvantage that the carrier frequency is not obtained from a highly stable oscillator. It is therefore necessary, in practice, to provide some auxiliary means

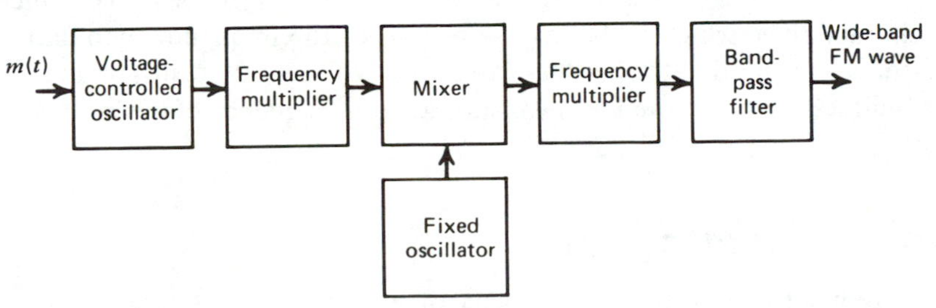

Figure 4.13 Block diagram of wide-band frequency modulator using a voltage-controlled oscillator.

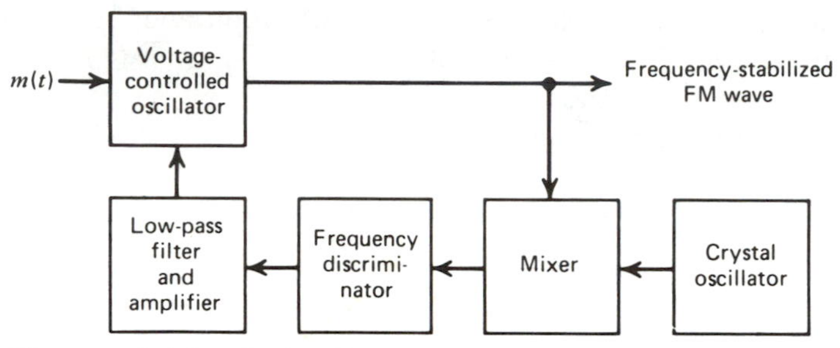

Figure 4.14 A feedback scheme for the frequency stabilization of a frequency modulator.

by which a very stable frequency generated by a crystal will be able to control the carrier frequency. One method of effecting this control is illustrated in Fig. 4.14. The output of the FM generator is applied to a mixer together with the output of a crystal-controlled oscillator, and the difference frequency term is extracted. The mixer output is next applied to a frequency discriminator and then low-pass filtered. A frequency discriminator is a device whose output voltage has an instantaneous amplitude that is proportional to the instantaneous frequency of the FM wave applied to its input; this device is described in detail in Section 4.8. When the FM transmitter has exactly the correct carrier frequency, the low-pass filter output is zero. However, deviations of the transmitter carrier frequency from its assigned value will cause the frequency discriminator-filter combination to develop a dc output voltage with a polarity determined by the sense of the transmitter frequency drift. This dc voltage, after suitable amplification, is applied to the voltage-controlled oscillator of the FM transmitter in such a way as to modify the frequency of the oscillator in a direction that tends to restore the carrier frequency to its required value.

4.8 DEMODULATION OF FM WAVES

Frequency demodulation is the process that enables us to recover the original modulating wave from the frequency-modulated wave. We describe two basic devices for performing frequency demodulation, namely, the *frequency discriminator* and the *phase-locked loop demodulator*. In both cases, the objective is to produce a transfer characteristic which is the inverse of that of the frequency modulator. That is, a frequency demodulator produces an output voltage whose instantaneous amplitude is directly proportional to the instantaneous frequency of the input FM wave.

Frequency Discriminator

Basically, a frequency discriminator consists of a *slope circuit* followed by an *envelope detector*. An ideal slope circuit is characterized by a transfer function that

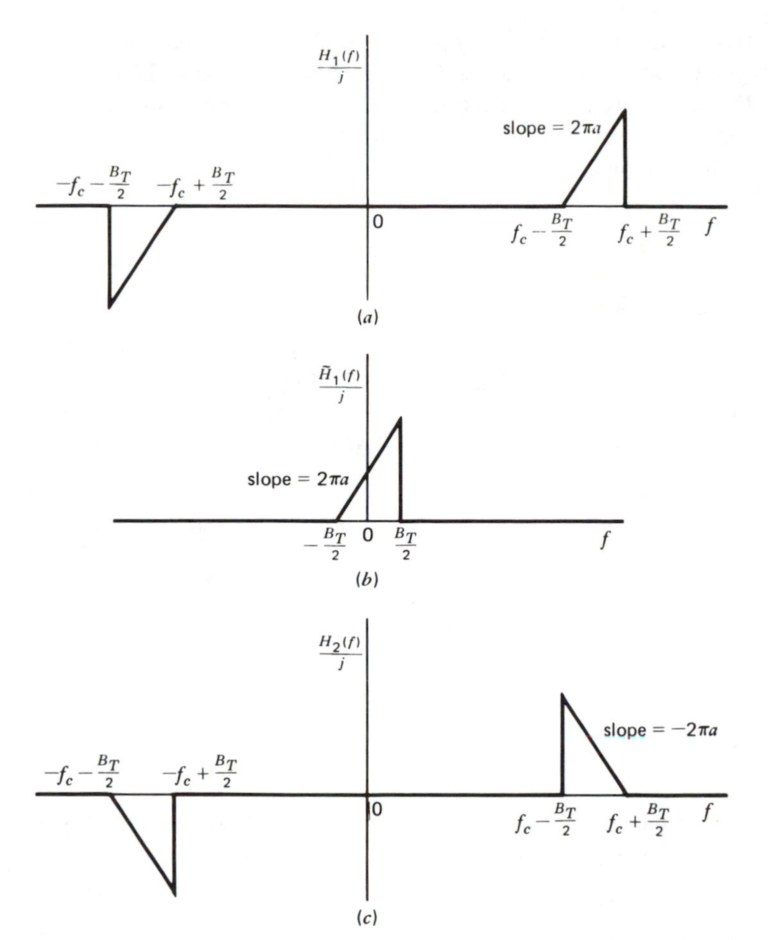

Figure 4.15 *(a)* Frequency response of ideal slope circuit. *(b)* Frequency response of complex low-pass filter equivalent to the slope circuit response of part *(a)*. *(c)* Frequency response of ideal slope circuit complementary to that of part *(a)*.

is purely imaginary, varying linearly with frequency inside a prescribed frequency interval. Consider the transfer function depicted in Fig. 4.15(*a*), which is defined by

$$H_1(f) = \begin{cases} j2\pi a\left(f - f_c + \dfrac{B_T}{2}\right), & f_c - \dfrac{B_T}{2} \leqslant f \leqslant f_c + \dfrac{B_T}{2} \\ j2\pi a\left(f + f_c - \dfrac{B_T}{2}\right), & -f_c - \dfrac{B_T}{2} \leqslant f \leqslant -f_c + \dfrac{B_T}{2} \\ 0, & \text{elsewhere} \end{cases} \tag{4.55}$$

where a is a constant. We wish to evaluate the response of this slope circuit, denoted by $s_1(t)$, which is produced by an FM wave $s(t)$ of carrier frequency f_c and transmission bandwidth B_T. It is assumed that the spectrum of $s(t)$ is essentially zero outside the frequency interval $f_c - B_T/2 \leqslant |f| \leqslant f_c + B_T/2$. For evaluation of the response $s_1(t)$, it is convenient to use the procedure described in Section 2.15, which involves replacing the slope circuit with an equivalent low-pass filter and driving this filter with the complex envelope of the input FM wave $s(t)$.

Let $\tilde{H}_1(f)$ denote the complex transfer function of the slope circuit defined by Fig. 4.15(*a*). This complex transfer function is related to $H_1(f)$ by

$$\tilde{H}_1(f-f_c) = H_1(f), \qquad f > 0 \tag{4.56}$$

Hence, using Eqs. (4.55) and (4.56), we get

$$\tilde{H}_1(f) = \begin{cases} j2\pi a\left(f + \dfrac{B_T}{2}\right), & -\dfrac{B_T}{2} \leqslant f \leqslant \dfrac{B_T}{2} \\ 0, & \text{elsewhere} \end{cases} \tag{4.57}$$

which is shown in Fig. 4.15(*b*).

The incoming FM wave $s(t)$ is defined by Eq. (4.8), which is reproduced here for convenience:

$$s(t) = A_c \cos\left[2\pi f_c t + 2\pi k_f \int_0^t m(t)dt\right] \tag{4.58}$$

The complex envelope of this FM wave is

$$\tilde{s}(t) = A_c \exp\left[j2\pi k_f \int_0^t m(t)dt\right] \tag{4.59}$$

Let $\tilde{s}_1(t)$ denote the complex envelope of the response of the slope circuit defined by Fig. 4.15(*a*). Then we may express the Fourier transform of $\tilde{s}_1(t)$ as follows:

$$\begin{aligned} \tilde{S}_1(f) &= \tilde{H}_1(f)\tilde{S}(f) \\ &= \begin{cases} j2\pi a\left(f + \dfrac{B_T}{2}\right)\tilde{S}(f), & -\dfrac{B_T}{2} \leqslant f \leqslant \dfrac{B_T}{2} \\ 0, & \text{elsewhere} \end{cases} \end{aligned} \tag{4.60}$$

where $\tilde{S}(f)$ is the Fourier transform of $\tilde{s}(t)$. Now, from Section 2.3 we recall that the multiplication of the Fourier transform of a signal by the factor $j2\pi f$ is equivalent to differentiating the signal in the time domain. We thus deduce from Eq. (4.60) that

$$\tilde{s}_1(t) = a\left[\frac{d\tilde{s}(t)}{dt} + j\pi B_T \tilde{s}(t)\right] \tag{4.61}$$

Substituting Eq. (4.59) in (4.61), we get

$$\tilde{s}_1(t) = j\pi B_T a A_c\left[1 + \frac{2k_f}{B_T} m(t)\right]\exp\left[j2\pi k_f \int_0^t m(t)dt\right] \tag{4.62}$$

The desired response of the slope circuit is therefore

$$s_1(t) = \text{Re}[\tilde{s}_1(t)\exp(j2\pi f_c t)]$$
$$= \pi B_T a A_c \left[1 + \frac{2k_f}{B_T} m(t)\right] \cos\left[2\pi f_c t + 2\pi k_f \int_0^t m(t)dt + \frac{\pi}{2}\right] \quad (4.63)$$

The signal $s_1(t)$ is a hybrid-modulated wave in which both the amplitude and frequency of the carrier wave vary with the baseband signal $m(t)$. However, provided that we choose

$$\left|\frac{2k_f}{B_T} m(t)\right| < 1$$

for all t, then we may use an envelope detector to recover the amplitude variations and thus, except for a bias term, obtain the original baseband signal. The resulting envelope detector output is therefore

$$|\tilde{s}_1(t)| = \pi B_T a A_c \left[1 + \frac{2k_f}{B_T} m(t)\right] \quad (4.64)$$

The bias term $\pi B_T a A_c$ in the right-hand side of Eq. (4.64) is proportional to the slope a of the transfer function of the slope circuit. This suggests that the bias may be removed by subtracting from the envelope detector output $|\tilde{s}_1(t)|$ the output of a second envelope detector preceded by the *complementary slope circuit* with a transfer function $H_2(f)$ as described in Fig. 4.15(*c*). That is, the respective complex transfer functions of the two slope circuits are related by

$$\tilde{H}_2(f) = \tilde{H}_1(-f) \quad (4.65)$$

Let $s_2(t)$ denote the response of the complementary slope circuit produced by the incoming FM wave $s(t)$. Then, following a procedure similar to that described above, we find that the envelope of $s_2(t)$ is

$$|\tilde{s}_2(t)| = \pi B_T a A_c \left[1 - \frac{2k_f}{B_T} m(t)\right] \quad (4.66)$$

where $\tilde{s}_2(t)$ is the complex envelope of the signal $s_2(t)$. The difference between the two envelopes in Eqs. (4.64) and (4.66) is

$$s_o(t) = |\tilde{s}_1(t)| - |\tilde{s}_2(t)|$$
$$= 4\pi k_f a A_c m(t) \quad (4.67)$$

which is free from bias.

We may thus model the *ideal frequency discriminator* as a pair of slope circuits with their complex transfer functions related by Eq. (4.65), followed by envelope detectors and a summer, as in Fig. 4.16. This scheme is called a *balanced frequency discriminator*.

The idealized scheme of Fig. 4.16 can be closely realized using the circuit shown in Fig. 4.17(*a*). The upper and lower resonant filter sections of this circuit are tuned to frequencies above and below the unmodulated carrier frequency f_c. In Fig. 4.17(*b*) we have plotted the amplitude responses of these two tuned filters, together with their total response, assuming that both filters have a high-Q factor. The

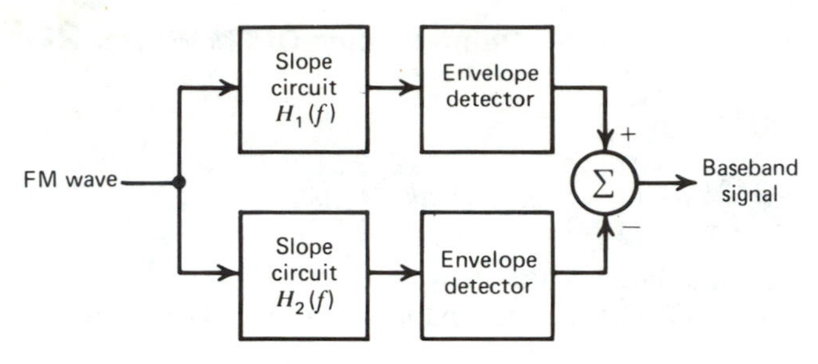

Figure 4.16 Idealized model of balanced frequency discriminator.

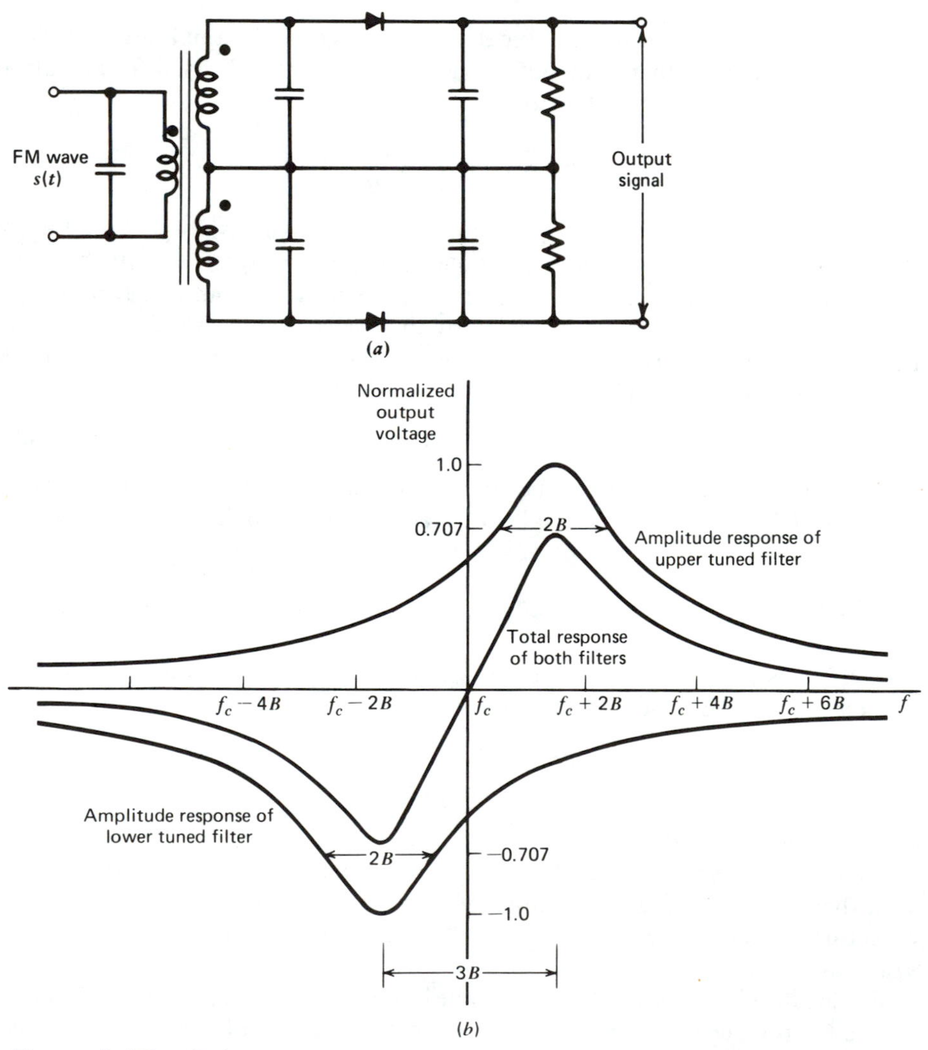

Figure 4.17 Balanced frequency discriminator. *(a)* Circuit diagram. *(b)* Frequency response.

linearity of the useful portion of this total response, centered at f_c, is determined by the separation of the two resonant frequencies. As illustrated in Fig. 4.17(*b*), a frequency separation of $3B$ gives satisfactory results, where $2B$ is the 3-dB bandwidth of either filter. However, there will be distortion in the output of this frequency discriminator due to the following factors:

1. The spectrum of the input FM wave $s(t)$ is not exactly zero for frequencies outside the range $f_c - B_T/2 \leq f \leq f_c + B_T/2$.
2. The tuned filter outputs are not strictly band-limited, and so some distortion is introduced by the low-pass RC filters following the diodes in the envelope detectors.
3. The tuned filter characteristics are not linear over the whole frequency band of the input FM wave $s(t)$.

Nevertheless, by proper design, it is possible to maintain the FM distortion produced by these factors within tolerable limits.

Phase-Locked Looped Demodulator

The *phased-locked loop* (PLL) is a negative feedback system which consists of three major components: a multiplier, a loop filter, and a voltage-controlled oscillator (VCO) connected together in the form of a feedback loop, as in Fig. 4.18. The VCO is a sine-wave generator whose frequency is determined by a voltage applied to it from an external source. In effect, any frequency modulator may serve as a VCO.

We assume that initially we have adjusted the VCO so that when the control voltage is zero, two conditions are satisfied: (1) the frequency of the VCO is precisely set at the unmodulated carrier frequency f_c, and (2) the VCO output has a 90-degree phase-shift with respect to the unmodulated carrier wave. Suppose that the input signal applied to the phase-locked loop is an FM wave defined by

$$s(t) = A_c \sin[2\pi f_c t + \phi_1(t)] \tag{4.68}$$

where A_c is the carrier amplitude. With a modulating wave $m(t)$, we have

$$\phi_1(t) = 2\pi k_f \int_0^t m(t)dt \tag{4.69}$$

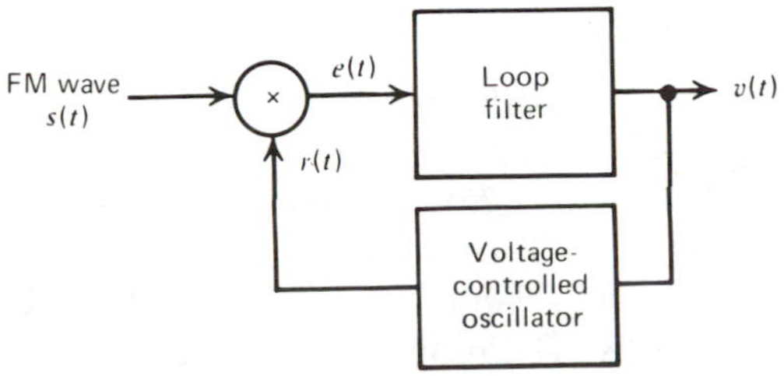

Figure 4.18 Phase-locked loop.

where k_f is the frequency sensitivity of the frequency modulator. Let the VCO output be defined by

$$r(t)=A_v \cos[2\pi f_c t+\phi_2(t)] \tag{4.70}$$

where A_v is the amplitude. With a control voltage $v(t)$ applied to the VCO input, we have

$$\phi_2(t)=2\pi k_v \int_0^t v(t)dt \tag{4.71}$$

where k_v is the frequency sensitivity of the VCO, measured in hertz per volt. The incoming FM wave $s(t)$ and the VCO output $r(t)$ are applied to the multiplier, producing two components: (1) a high-frequency component represented by

$$k_m A_c A_v \sin[4\pi f_c t+\phi_1(t)+\phi_2(t)],$$

and (2) a low-frequency component represented by $k_m A_c A_v \sin[\phi_1(t)-\phi_2(t)]$, where k_m is the *multiplier gain*, measured in volt^{-1}. The high-frequency component is eliminated by the combination of the filter and the VCO. Therefore, discarding the high-frequency component, the input to the loop filter is given by

$$e(t)=k_m A_c A_v \sin[\phi_e(t)] \tag{4.72}$$

where $\phi_e(t)$ is the *phase error* defined by

$$\begin{aligned}\phi_e(t)&=\phi_1(t)-\phi_2(t)\\ &=\phi_1(t)-2\pi k_v \int_0^t v(t)dt\end{aligned} \tag{4.73}$$

The loop filter operates on its input $e(t)$ to produce the output

$$v(t)=\int_{-\infty}^{\infty} e(\tau)h(t-\tau)d\tau \tag{4.74}$$

where $h(t)$ is the impulse response of the filter.

Using Eqs. (4.72) to (4.74) to relate $\phi_e(t)$ and $\phi_1(t)$, we obtain

$$\frac{d\phi_e(t)}{dt}=\frac{d\phi_1(t)}{dt}-2\pi K_0 \int_{-\infty}^{\infty} \sin[\phi_e(\tau)]h(t-\tau)d\tau \tag{4.75}$$

where K_0 is defined by

$$K_0=k_m k_v A_c A_v \tag{4.76}$$

The amplitudes A_c and A_v are both measured in volts, the multiplier gain in volt^{-1}, and the frequency sensitivity k_v in hertz per volt. Hence, it follows from Eq. (4.76) that K_0 has the dimensions of frequency. Equation (4.75) suggests the representation or model of Fig. 4.19. In this model we have also included the relationship between $v(t)$ and $e(t)$ as represented by Eqs. (4.72) and (4.74). We see that the block diagram of the model resembles Fig. 4.18. The multiplier is replaced by a subtractor and a sinusoidal nonlinearity, and the VCO by an integrator.

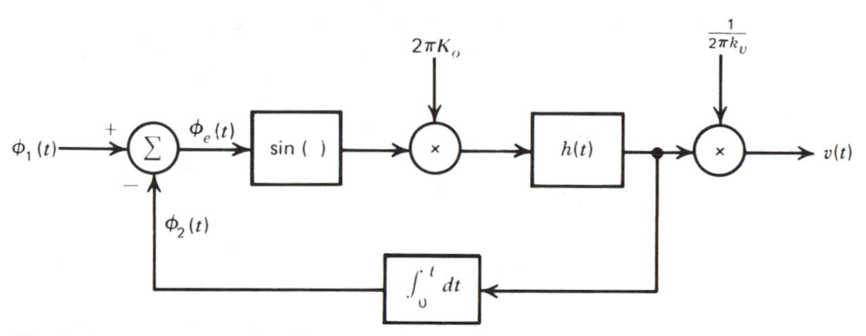

Figure 4.19 Nonlinear model of a phase-locked loop.

When the phase error $\phi_e(t)$ is zero, the phase-locked loop is said to be in *phase-lock*. When $\phi_e(t)$ is at all times small compared with one radian, we may use the approximation

$$\sin[\phi_e(t)] \simeq \phi_e(t) \tag{4.77}$$

which is accurate to within 4 percent for $\phi_e(t)$ less than 0.5 radians. In this case the loop is said to be near phase-lock and the sinusoidal nonlinearity of Fig. 4.19 may be disregarded. Thus we may represent the phase-locked loop by the linearized model shown in Fig. 4.20(*a*). According to this model, the phase error $\phi_e(t)$ is related to the input phase angle $\phi_1(t)$ by the integro-differential equation:

$$\frac{d\phi_e(t)}{dt} + 2\pi K_0 \int_{-\infty}^{\infty} \phi_e(\tau)h(t-\tau)d\tau = \frac{d\phi_1(t)}{dt} \tag{4.78}$$

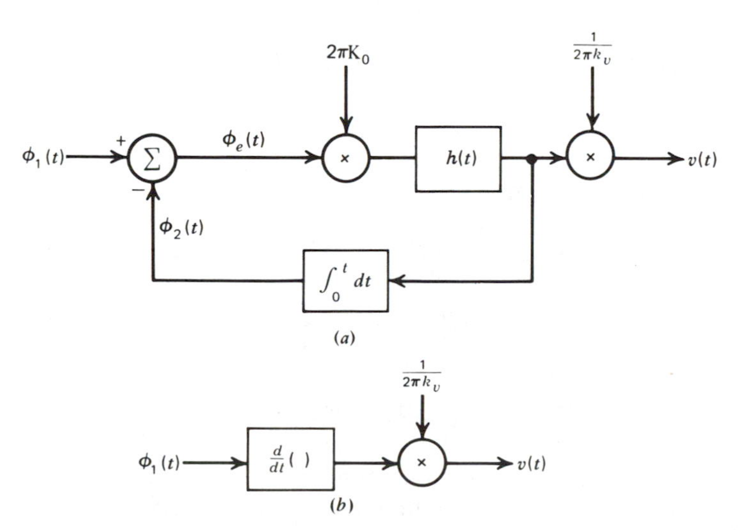

Figure 4.20 Models of a phase-locked loop. *(a)* Linearized model. *(b)* Simplified model when the loop gain is very large compared to unity.

Transforming Eq. (4.78) into the frequency domain and solving for $\Phi_e(f)$, the Fourier transform of $\phi_e(t)$, in terms of $\Phi_1(f)$, the Fourier transform of $\phi_1(t)$, we get

$$\Phi_e(f)=\frac{1}{1+L(f)}\,\Phi_1(f) \tag{4.79}$$

The function $L(f)$ in Eq. (4.79) is defined by

$$L(f)=K_0\,\frac{H(f)}{jf} \tag{4.80}$$

where $H(f)$ is the transfer function of the loop filter. The quantity $L(f)$ is called the *open-loop transfer function* of the phase-locked loop. Suppose that for all values of f inside the baseband, we make the magnitude of $L(f)$ very large compared with unity. Then from Eq. (4.79) we find that $\Phi_e(f)$ approaches zero. That is, the phase of the VCO becomes asymptotically equal to the phase of the incoming wave, and phase-lock is thereby established.

From Fig. 4.20(a) we see that $V(f)$, the Fourier transform of the phase-locked loop output $v(t)$, is related to $\Phi_e(f)$ by

$$V(f)=\frac{K_0}{k_v}\,H(f)\Phi_e(f) \tag{4.81}$$

or, equivalently,

$$V(f)=\frac{jf}{k_v}\,L(f)\Phi_e(f) \tag{4.82}$$

Therefore, substituting Eq. (4.79) in (4.82), we may write

$$V(f)=\frac{(jf/k_v)L(f)}{1+L(f)}\,\Phi_1(f) \tag{4.83}$$

Again, when we make $|L(f)|\gg 1$, we may approximate Eq. (4.83) as follows

$$V(f)\simeq\frac{jf}{k_v}\,\Phi_1(f) \tag{4.84}$$

The corresponding time-domain relation is

$$v(t)\simeq\frac{1}{2\pi k_v}\,\frac{d\phi_1(t)}{dt} \tag{4.85}$$

Thus provided the magnitude of $L(f)$ is very large for all frequencies of interest, the phase-locked loop may be modeled as a differentiator with its output scaled by the factor $1/2\pi k_v$, as in Fig. 4.20(b).

The simplified model of Fig. 4.20(b) provides the basis of using the phase-locked loop as a frequency demodulator. When the input signal is an FM wave as in Eq. (4.68), the phase $\phi_1(t)$ is related to the modulating wave $m(t)$ as in Eq. (4.69). Therefore, substituting Eq. (4.69) in (4.85), we find that the resulting output signal of the phase-locked loop is

$$v(t)\simeq\frac{k_f}{k_v}\,m(t) \tag{4.86}$$

That is, the output $v(t)$ of the phase-locked loop is approximately the same, except for the scale factor k_f/k_v, as the original baseband signal $m(t)$, and frequency demodulation is accomplished.

A significant feature of the phase-locked loop demodulator is that the bandwidth of the incoming FM wave can be much wider than that of the loop filter characterized by $H(f)$. The transfer function $H(f)$ can and should be restricted to the baseband. Then the control signal of the VCO has the bandwidth of the baseband signal $m(t)$, whereas the VCO output is a wide-band frequency-modulated wave whose instantaneous frequency tracks that of the incoming FM wave.

The complexity of the phase-locked loop is determined by the transfer function $H(f)$ of the loop filter. The simplest form of a phase-locked loop is obtained when $H(f)=1$; that is, there is no loop filter, and the resulting phase-locked loop is referred to as a *first-order phase-locked loop.* For higher-order loops, the transfer function $H(f)$ assumes a more complex form. The order of the phase-locked loop is determined by the order of the denominator polynomial of the *closed-loop transfer function,* which defines the output transform $V(f)$ in terms of the input transform $\Phi_1(f)$, as shown in Eq. (4.83). In the following two examples, we study the properties of first-order and second-order phase-locked loop demodulators using the linear model of Fig. 4.20(*a*).*

Example 5 First-Order Phase-Locked Loop

If the phase-locked loop has no loop filter, $H(f)=1$, the linearized model of the loop simplifies as in Fig. 4.21, and Eq. (4.79) becomes

$$\Phi_e(f)=\frac{1}{1+K_0/jf}\,\Phi_1(f) \tag{4.87}$$

We wish to investigate the loop behavior in the presence of a frequency-modulated input. In

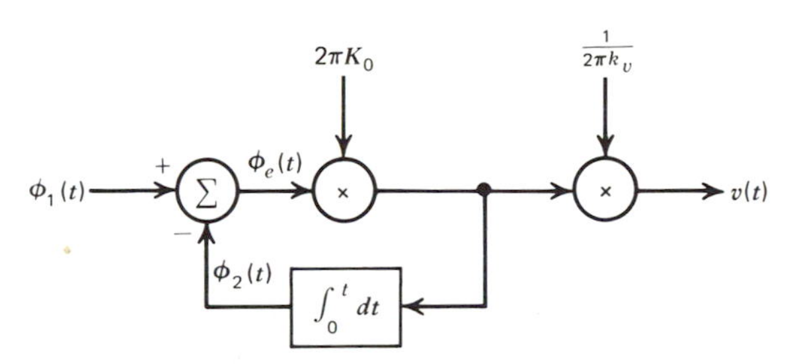

Figure 4.21 Linearized model of first-order phase-locked loop.

* When a phase-locked loop is used to demodulate an FM wave, the loop must first lock onto the incoming FM wave and then follow the variations in its phase. During the lock-up operation, the phase error $\phi_e(t)$ between the incoming FM wave and the VCO output will be large, which therefore requires the use of the nonlinear model of Fig. 4.19. For the nonlinear analysis of a phase-locked loop, see:

1. W. C. Lindsey, *Synchronization Systems in Communication and Control* (Prentice-Hall, 1972).
2. A. J. Viterbi, *Principles of Coherent Communication* (McGraw-Hill, 1966).

particular, we assume a single-tone modulating wave

$$m(t) = A_m \cos(2\pi f_m t) \tag{4.88}$$

with the corresponding FM wave given by

$$s(t) = A_c \sin[2\pi f_c t + \beta \sin(2\pi f_m t)] \tag{4.89}$$

where β is the modulation index. Thus,

$$\phi_1(t) = \beta \sin(2\pi f_m t) \tag{4.90}$$

Therefore, using Eq. (4.87), we find that the phase error $\phi_e(t)$ of the loop produced by the phase input $\phi_1(t)$ of Eq. (4.90) varies sinusoidally with time, as shown by

$$\phi_e(t) = \phi_{e0} \cos(2\pi f_m t + \psi) \tag{4.91}$$

The amplitude ϕ_{e0} and phase ψ of the phase error $\phi_e(t)$ are defined by

$$\phi_{e0} = \frac{\Delta f / K_0}{[1 + (f_m/K_0)^2]^{1/2}} \tag{4.92}$$

and

$$\psi = -\tan^{-1}(f_m/K_0) \tag{4.93}$$

where Δf is the frequency deviation; that is, $\Delta f = \beta f_m$.

In Fig. 4.22 we have plotted the phase error amplitude ϕ_{e0}, normalized with respect to $\Delta f / K_0$, versus the dimensionless parameter f_m/K_0. It is apparent that for a fixed frequency deviation Δf, the phase error amplitude has its largest value of $\Delta f / K_0$ at $f_m = 0$, and it decreases with increasing f_m.

For the loop to track the frequency modulation sufficiently closely, the phase error $\phi_e(t)$ should remain within the linear region of operation of the loop for all t. This means that the largest phase error amplitude should not exceed 0.5 radians, so that $\phi_e(t)$ satisfies the requirement of Eq. (4.77) for all t. That is,

$$\Delta f \leqslant 0.5 K_0 \tag{4.94}$$

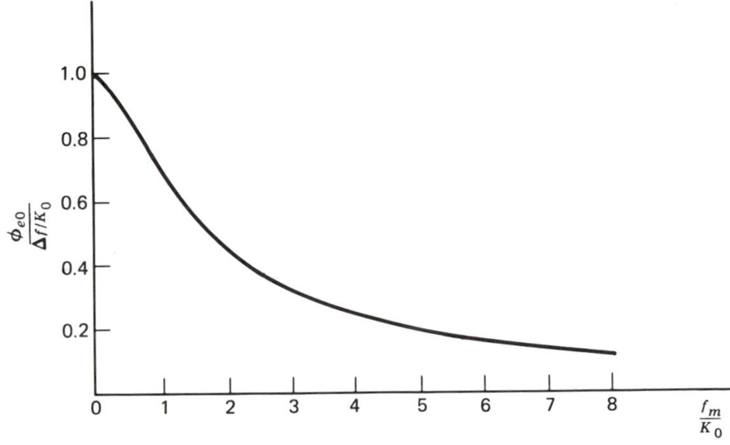

Figure 4.22 Phase-error amplitude characteristic of first-order phase-locked loop.

The output signal $v(t)$ of the loop is related to the phase error $\phi_e(t)$ by (see Fig. 4.21)

$$v(t)=\frac{K_0}{k_v}\phi_e(t) \tag{4.95}$$

Therefore, substituting Eq. (4.91) in (4.95), we get

$$v(t)=A_0\cos(2\pi f_m t+\psi) \tag{4.96}$$

where the amplitude A_0 is defined by

$$A_0=\frac{\Delta f/k_v}{[1+(f_m/K_0)^2]^{1/2}} \tag{4.97}$$

and the phase ψ is given by Eq. (4.93). From Eq. (4.97) we see that at $f_m=K_0$, the amplitude of the loop output $v(t)$ will have fallen by 3 dB below its value at $f_m=0$. The loop bandwidth of a first-order phase-locked loop is therefore K_0. We also see from Eq. (4.97) that a first-order phase-locked loop demodulator introduces distortion between the original modulating wave $m(t)$ and the signal $v(t)$ obtained at the loop output. This distortion is the same as the frequency distortion produced by passing the modulating wave $m(t)$ through a low-pass RC filter of time constant $1/2\pi K_0$.

We have thus far assumed that the phase error is sufficiently small to allow the loop to be considered linear in its operation. We next wish to evaluate the input frequency range over which the loop will hold lock. For this evaluation, consider the nonlinear differential equation (4.75), assuming a step change of frequency of magnitude f, so that

$$\phi_1(t)=2\pi ft \tag{4.98}$$

With this input applied to a first-order phase-locked loop, Eq. (4.75) becomes

$$\frac{d\phi_e(t)}{dt}+2\pi K_0\sin[\phi_e(t)]=2\pi f \tag{4.99}$$

The phase error $\phi_e(t)$ will have reached its steady-state value when the derivative $d\phi_e/dt$ is zero. Therefore, putting $d\phi_e/dt=0$ in Eq. (4.99) we obtain

$$\sin\phi_e=\frac{f}{K_0} \tag{4.100}$$

The sine of an angle cannot exceed unity in magnitude. Therefore, if $f>K_0$, Eq. (4.100) has no solution, Instead, the loop falls out of lock and the phase error becomes a beat-note rather than a dc level. The *hold-in frequency range* of a first-order phase-locked loop is therefore equal to $\pm K_0$.

From the above analysis, we conclude that the parameter K_0, defined by Eq. (4.76), uniquely determines the loop bandwidth as well as the hold-in frequency range of a first-order phase-locked loop. This is a limitation of the first-order phase-locked loop.

Example 6 Second-order phase-locked loop

Consider next a *second-order phase-locked loop* using a filter with the transfer function

$$H(f)=1+\frac{a}{jf} \tag{4.101}$$

where a is a constant. This filter consists of an integrator and a direct connection as shown in

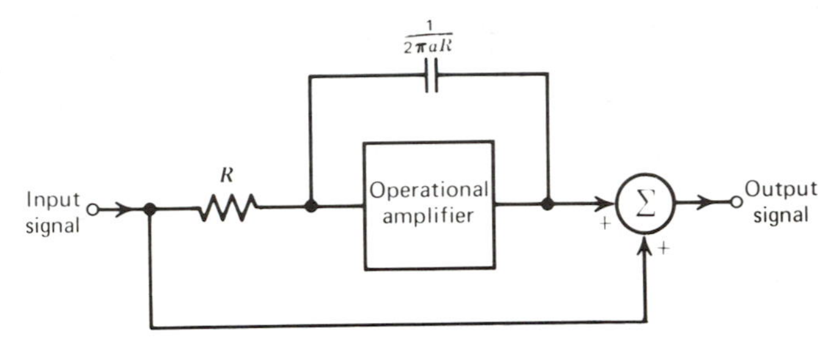

Figure 4.23 Loop filter for second-order phase-locked loop.

Fig. 4.23. For this phase-locked loop, Eqs. (4.79) and (4.101) give

$$\Phi_e(f)=\frac{(jf)^2/aK_0}{1+[(jf)/a]+[(jf)^2/aK_0]}\,\Phi_1(f) \tag{4.102}$$

Define:

$$f_n=\sqrt{aK_0} \tag{4.103}$$

and

$$\zeta=\sqrt{\frac{K_0}{4a}} \tag{4.104}$$

The f_n denotes the *natural frequency* of the loop and ζ denotes the *damping factor*. Then we may rewrite Eq. (4.102) in terms of f_n and ζ as follows

$$\Phi_e(f)=\frac{(jf/f_n)^2}{1+2\zeta(jf/f_n)+(jf/f_n)^2}\,\Phi_1(f) \tag{4.105}$$

As in Example 5, we assume that the incoming FM wave is produced by a single-tone modulating wave [see Eqs. (4.89) and (4.90)]. The phase input is

$$\phi_1(t)=\beta\sin(2\pi f_m t) \tag{4.106}$$

Hence, from Eq. (4.105) we find that the corresponding phase error is

$$\phi_e(t)=\phi_{e0}\cos(2\pi f_m t+\psi) \tag{4.107}$$

where the amplitude ϕ_{e0} and phase ψ are defined by

$$\phi_{e0}=\frac{(\Delta f/f_n)(f_m/f_n)}{\{[1-(f_m/f_n)^2]^2+4\zeta^2(f_m/f_n)^2\}^{1/2}} \tag{4.108}$$

and

$$\psi=\frac{\pi}{2}-\tan^{-1}\left[\frac{2\zeta(f_m/f_n)}{1-(f_m/f_n)^2}\right] \tag{4.109}$$

In Fig. 4.24 we have plotted the phase error amplitude ϕ_{e0}, normalized with respect to $\Delta f/f_n$, versus f_m/f_n for different values of ζ. It is apparent that for all values of the damping factor ζ, and assuming a fixed frequency deviation Δf, the phase error is small at low modulation frequencies, rises to a maximum at $f_m=f_n$, and then falls off at higher modulation frequencies. Note also that the maximum value of phase error amplitude decreases with increasing ζ.

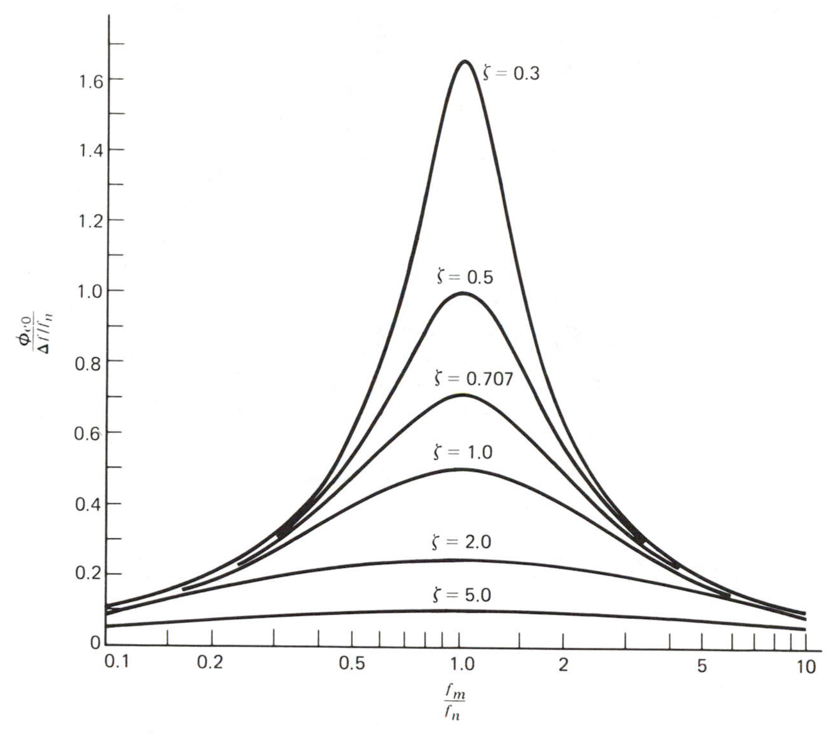

Figure 4.24 Phase-error amplitude characteristic of second-order phase-locked loop.

The Fourier transform of the loop output is related to $\Phi_e(f)$ by Eq. (4.81); that is

$$V(f)=\frac{K_0}{k_v}\left(1+\frac{a}{jf}\right)\Phi_e(f) \tag{4.110}$$

or

$$V(f)=\left(\frac{f_n^2}{jfk_v}\right)\left[1+2\zeta\left(\frac{jf}{f_n}\right)\right]\Phi_e(f) \tag{4.111}$$

Substituting Eq. (4.105) in (4.111), we get

$$V(f)=\frac{(jf/k_v)[1+2\zeta(jf/f_n)]}{1+2\zeta(jf/f_n)+(jf/f_n)^2}\,\Phi_1(f) \tag{4.112}$$

Therefore, for the phase input $\phi_1(t)$ of Eq. (4.106), we find that the corresponding loop output is

$$v(t)=A_0\cos(2\pi f_m t+\alpha) \tag{4.113}$$

where the amplitude A_0 and phase α are defined by

$$A_0=\frac{(\Delta f/k_v)[1+4\zeta^2(f_m/f_n)^2]^{1/2}}{\{[1-(f_m/f_n)^2]^2+4\zeta^2(f_m/f_n)^2\}^{1/2}} \tag{4.114}$$

and

$$\alpha = \tan^{-1}\left[2\zeta\left(\frac{f_m}{f_n}\right)\right] - \tan^{-1}\left[\frac{2\zeta(f_m/f_n)}{1-(f_m/f_n)^2}\right] \tag{4.115}$$

From Eq. (4.114), we see that the amplitude A_0 attains its maximum value of $\Delta f/k_v$ at $(f_m/f_n)=0$; it decreases with increasing f_m/f_n, dropping to zero at $(f_m/f_n)=\infty$.

The important feature of the second-order phase-locked loop is that with an incoming FM wave produced by a modulating sine wave of fixed amplitude (corresponding to fixed frequency deviation) and varying frequency, the frequency response that defines the phase error $\phi_e(t)$ is representative of a band-pass filter [see Eq. (4.108)], but the frequency response that defines the loop output $v(t)$ is representative of a low-pass filter [see Eq. (4.114)]. Therefore, by appropriately choosing the parameters ζ and f_n which determine the frequency response of the loop, it is possible to restrain the phase error to always remain small and thereby lie within the linear range of the loop, whereas at the same time the modulating wave is reproduced at the loop output with minimum distortion. This restraint is, however, conservative with respect to the hold-in capabilities of the loop. As a reasonable rule of thumb, the loop should remain locked if the maximum value of the phase error ϕ_{e0} (which occurs when the modulation frequency f_m is equal to the loop natural frequency f_n) is always less than 90 degrees.*

4.9 RESPONSE OF LINEAR FILTERS TO FM WAVES

When an FM wave is applied to a linear time-invariant filter that is characterized by amplitude and phase distortion, the resulting output will in general exhibit some residual amplitude modulation, in addition to distortion components in the phase angle of the FM wave. Whereas the unwanted amplitude modulation may be removed by means of limiters (see Problem 4.25), the distortion components in the phase angle of the FM wave will appear in the output of the frequency demodulator. This distortion arises from the fact that the sidebands of the FM wave are attenuated unequally by transmission through the filter. The calculation of the distortion produced by transmitting an FM wave through a linear filter has been one of the most vexatious problems in nonlinear modulation theory.† This calculation is made difficult by the nonlinear nature of the FM wave.

* F. M. Gardner, *Phaselock Techniques*, p. 40 (Wiley, 1966).

† The problem of distortion introduced by transmitting an FM wave through a linear filter was first analysed by Carson and Fry: J. R. Carson and T. C. Fry, "Variable frequency electric circuit theory with application to the theory of frequency modulation," *Bell Syst. Tech. J.*, vol. 16, pp. 513–540, Oct. 1937. Since then, numerous workers in the field have extended results and added new approaches. Many of these are cited by:

1. E. J. Baghdady, *Lectures on Communication Systems Theory*, pp. 459–483 (McGraw,Hill, 1961).
2. J. J. Downing, *Modulation Systems and Noise*, pp. 110–112 (Prentice-Hall, 1964).
3. P. F. Panter, *Modulation, Noise, and Spectral Analysis*, pp. 273–349 (McGraw-Hill, 1965).
4. H. E. Rowe, *Signals and Noise in Communication Systems*, pp. 202–203 (Van Nostrand, 1965).
5. J. H. Roberts, "Angle Modulation: The Theory of System Assessment," pp. 90–160, IEE Telecommunication Series 5 (Institution of Electrical Engineers, England, 1977).

Some exact results pertaining to FM distortion produced by linear filtering are presented by:

1. A. Mircea, "FM distortion theory," *Proc. IEEE*, vol. 54, pp. 705–706, April 1966.
2. A. Mircea, "FM intermodulation noise theory," *Proc. IEEE*, vol. 54, pp. 1463–1465, Oct. 1966.
3. E. Bedrosian and S. O. Rice, "Distortion and crosstalk of linearly filtered, angle-modulated signals," *Proc. IEEE*, vol. 56, pp. 2–13, Jan. 1968.

Because large-scale digital computers are widely available today, it should now be possible to analyze the response of a linear filter produced by an FM signal, relegating many of the tedious analytical operations to the computer. All we need to do is formulate the computations such that the digital computer can perform efficiently. This exercise is ultimately restricted in accuracy only by capacity of the computer and the cost of each run on the computer.

In order to develop a mathematical model of the problem suitable for simulation on a digital computer, we may use the procedure described in Section 2.15. According to this procedure: (1) we replace the FM wave by its in-phase and quadrature components, (2) we replace the band-pass filter by its complex low-pass equivalent, (3) evaluate the response of this equivalent filter produced by the combination of the in-phase and quadrature components of the FM wave, and (4) use this result to evaluate the desired response of the original band-pass filter produced by the FM input.

4.10 NONLINEAR EFFECTS IN FM SYSTEMS

Nonlinearities, in one form or another, are present in all electrical networks. There are two basic forms of nonlinearity to consider:

1. The nonlinearity is said to be *strong* when it is introduced intentionally and in a controlled manner for some specific application. Examples of strong nonlinearity include square-law modulators and limiters.
2. The nonlinearity is said to be *weak* when a linear performance is desired, and any nonlinearities are viewed as parasitic in nature.* The effect of such weak nonlinearities is to limit the useful signal levels in a system and thereby become an important design consideration.

In this section we examine the effects of weak nonlinearities on frequency modulation.

Consider a communications channel the transfer characteristic of which is defined by the nonlinear relation

$$v_o(t) = a_1 v_i(t) + a_2 v_i^2(t) + a_3 v_i^3(t) \tag{4.116}$$

where $v_i(t)$ and $v_o(t)$ are the input and output signals, respectively, and a_1, a_2, and a_3 are constants. It is assumed that the channel is *memoryless*; that is, the output signal $v_o(t)$ is an instantaneous function of the input signal $v_i(t)$. We wish to determine the effect of transmitting a frequency-modulated wave through such a channel. The FM wave is defined by

$$v_i(t) = A_c \cos[2\pi f_c t + \phi(t)] \tag{4.117}$$

* For a detailed discussion of the characterization and system effects of weak nonlinearities, see "Transmission Systems for Communication," Bell Telephone Laboratories, pp. 237–278 (Western Electric, 1970).

where

$$\phi(t) = 2\pi k_f \int_0^t m(t)dt$$

Substituting Eq. (4.117) in (4.116), we get

$$v_o(t) = a_1 A_c \cos[2\pi f_c t + \phi(t)] + a_2 A_c^2 \cos^2[2\pi f_c t + \phi(t)] + a_3 A_c^3 \cos^3[2\pi f_c t + \phi(t)] \quad (4.118)$$

Expanding and then collecting the terms in Eq. (4.118), we get

$$v_o(t) = \tfrac{1}{2} a_2 A_c^2 + (a_1 A_c + \tfrac{3}{4} a_3 A_c^3)\cos[2\pi f_c t + \phi(t)] + \tfrac{1}{2} a_2 A_c^2 \cos[4\pi f_c t + 2\phi(t)] + \tfrac{1}{4} a_3 A_c^3 \cos[6\pi f_c t + 3\phi(t)] \quad (4.119)$$

Thus the channel output consists of a dc component and three frequency-modulated waves with carrier frequencies of f_c, $2f_c$, and $3f_c$, respectively.

To extract the desired FM wave from the channel output $v_o(t)$, it is necessary to separate the FM with carrier frequency f_c from the one with carrier frequency $2f_c$. Let Δf denote the frequency deviation of the incoming FM wave $v_i(t)$, and W denote the highest frequency component of the modulating wave $m(t)$. Then, applying Carson's rule and noting that the frequency deviation about the second harmonic of the carrier frequency is doubled, we find that the necessary condition for separating the desired FM wave with the carrier frequency f_c from that with the carrier frequency $2f_c$ is

$$2f_c - (2\,\Delta f + W) > f_c + \Delta f + W$$

or

$$f_c > 3\,\Delta f + 2W \quad (4.120)$$

Thus by using a band-pass filter of mid-band frequency f_c and bandwidth $2\,\Delta f + 2W$, the channel output becomes

$$v_o(t) = (a_1 A_c + \tfrac{3}{4} a_3 A_c^3)\cos[2\pi f_c t + \phi(t)] \quad (4.121)$$

We see therefore that the only effect of passing an FM wave through a channel with amplitude nonlinearities, followed by appropriate filtering, is simply to modify its amplitude. That is, unlike amplitude modulation, frequency modulation is not affected by distortion produced by transmission through a channel with amplitude nonlinearities. It is for this reason that we find frequency modulation is widely used in microwave radio and satellite communication systems, because it permits the use of highly nonlinear amplifiers and power transmitters, which is particularly important from the standpoint of maximum power output at radio frequencies.

An FM system is extremely sensitive to phase nonlinearities, however. A common type of phase nonlinearity that is encountered in microwave radio systems is known as *AM-to-PM conversion*. This is the result of the phase characteristic of repeaters or amplifiers used in the system being dependent upon the instantaneous amplitude of the input signal. In practice, AM-to-PM conversion is characterized by a constant, K, which is measured in degrees per dB and may be interpreted as the peak phase change at the output for a 1-dB change in envelope at the input. When an FM wave is transmitted through a microwave radio link, it picks up spurious amplitude variations due to noise and interference during the course of transmission, and when such an FM wave is passed through a repeater with AM-to-PM conversion, the output will contain unwanted phase modulation and resultant distortion. It is therefore important to keep AM-to-PM conversion at a low level. For example, for a good microwave repeater, the AM-to-PM conversion constant K is less than 2 degrees per dB.

Problems

Problem 4.1 Sketch the PM and FM waves produced by the sawtooth wave shown in Fig. P4.1.

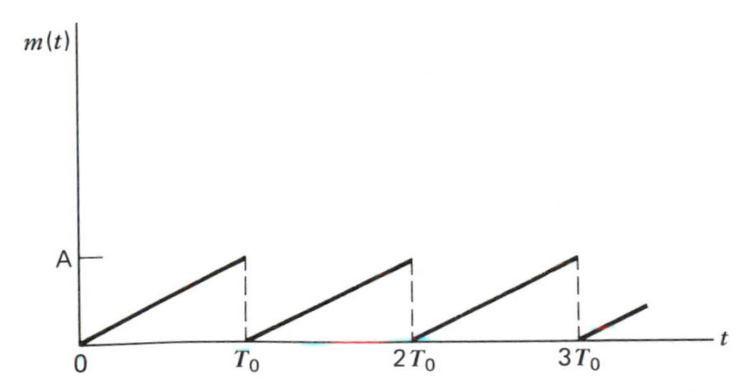

Figure P4.1

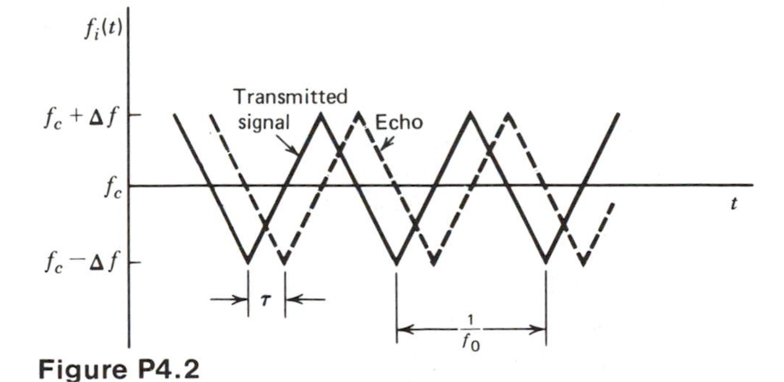

Figure P4.2

Problem 4.2 In a *frequency-modulated radar*, the instantaneous frequency of the transmitted carrier is varied as in Fig. P4.2, which is obtained by using a triangular modulating wave. The instantaneous frequency of the received echo signal is shown dotted in Fig. P4.2, where τ is the round-trip delay time. The transmitted and received echo signals are applied to a mixer and the difference frequency component is retained. Assuming that $f_0\tau \ll 1$, determine the number of beat cycles at the mixer output, averaged over one second, in terms of the peak deviation Δf of the carrier frequency, the delay τ, and the repetition frequency f_0 of the transmitted signal.

Problem 4.3 The instantaneous frequency of a sine-wave is equal to $f_c+\Delta f$ for $|t|\leqslant T/2$, and f_c for $|t|>T/2$. Determine the spectrum of this frequency-modulated wave.

Hint: Divide up the time interval of interest into three regions: $-\infty<t<-T/2$, $-T/2\leqslant t \leqslant T/2$, and $T/2<t<\infty$.

Problem 4.4 Single-sideband modulation may be viewed as a hybrid form of amplitude modulation and frequency modulation. Evaluate the envelope and instantaneous frequency of an SSB wave for the following two cases:

(a) When only the upper sideband is transmitted.
(b) When only the lower sideband is transmitted.

Problem 4.5 Consider a narrow-band FM wave approximately defined by

$$s(t)\simeq A_c\cos(2\pi f_c t)-\beta A_c\sin(2\pi f_c t)\sin(2\pi f_m t)$$

(a) Determine the envelope of this modulated wave. What is the ratio of the maximum to the minimum value of this envelope? Plot this ratio versus β, assuming that β is restricted to the interval $0\leqslant\beta\leqslant 0.3$.
(b) Determine the average power of the narrow-band FM wave, expressed as a percentage of the average power of the unmodulated carrier wave. Plot this result versus β, assuming that β is restricted to the interval $0\leqslant\beta\leqslant 0.3$.
(c) By expanding the angle $\theta_i(t)$ of the narrow-band FM wave $s(t)$ in the form of a power series, and restricting the modulation index β to a maximum value of 0.3 radians, show that

$$\theta_i(t)\simeq 2\pi f_c t+\beta\sin(2\pi f_m t)-\frac{\beta^3}{3}\sin^3(2\pi f_m t)$$

What is the value of the harmonic distortion for $\beta=0.3$?

Problem 4.6 The sinusoidal modulating wave

$$m(t)=A_m\cos(2\pi f_m t)$$

is applied to a phase modulator with phase sensitivity k_p. The unmodulated carrier wave has frequency f_c and amplitude A_c.

(a) Determine the spectrum of the resulting phase-modulated wave, assuming that the maximum phase deviation $\beta_p=k_pA_m$ does not exceed 0.3 radians.
(b) Construct a phasor diagram for this modulated wave, and compare it with that of the corresponding narrow-band FM wave.

Problem 4.7 Suppose that the phase-modulated wave of Problem 4.6 has an arbitrary value for the maximum phase deviation β_p. This modulated wave is applied to an ideal band-pass filter with mid-band frequency f_c and a passband extending from $f_c-1.5f_m$ to $f_c+1.5f_m$. Determine the envelope, phase, and instantaneous frequency of the modulated wave at the filter output as functions of time.

Problem 4.8 A carrier wave is frequency-modulated using a sinusoidal signal of frequency f_m and amplitude A_m.

(a) Determine the values of the modulation index β for which the carrier component of the FM wave is reduced to zero. For this calculation you may use the values of $J_0(\beta)$ given in Table A4.1 of Appendix 4.
(b) In a certain experiment conducted with $f_m=1$ kHz and increasing A_m (starting from 0

volts), it is found that the carrier component of the FM wave is reduced to zero for the first time when $A_m = 2$ volts. What is the frequency sensitivity of the modulator? What is the value of A_m for which the carrier component is reduced to zero for the second time?

Problem 4.9 An FM wave with modulation index $\beta = 1$ is transmitted through an ideal band-pass filter with mid-band frequency f_c and bandwidth $5f_m$, where f_c is the carrier frequency and f_m is the frequency of the sinusoidal modulating wave. Determine the amplitude spectrum of the filter output.

Problem 4.10 Consider the wide-band FM wave

$$s(t) = A_c \cos(2\pi f_c t + \beta \sin 2\pi f_m t)$$

produced by a sinusoidal modulating wave of frequency f_m, with β denoting the modulation index and A_c denoting the unmodulated carrier amplitude. Construct a phasor diagram for $s(t)$ for arbitrary β.

Problem 4.11 Consider the square-wave shown in Fig. P4.3, which is used to frequency-modulate the carrier wave $A_c \cos(2\pi f_c t)$. Assume a frequency sensitivity of k_f hertz per volt.

(a) Determine the waveform defining the instantaneous frequency of the FM wave.
(b) Determine the waveform defining the phase of the FM wave.
(c) Evaluate the complex envelope of the FM wave. Hence, show that this FM wave may be expanded as follows:

$$s(t) = A_c \sum_{n=-\infty}^{\infty} \alpha_n \cos\left(2\pi f_c t + \frac{2n\pi t}{T_0}\right)$$

where

$$\alpha_n = \frac{1}{2}\left[\operatorname{sinc}\left(\frac{\beta - n}{2}\right) + (-1)^n \operatorname{sinc}\left(\frac{\beta + n}{2}\right)\right]$$

$$\beta = k_f T_0$$

and T_0 is the period of the square wave.

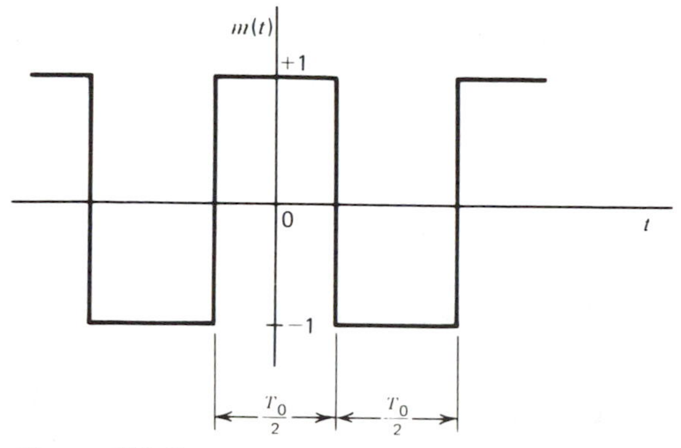

Figure P4.3

Problem 4.12 Consider the FM wave

$$s(t) = \cos[2\pi f_c t + \beta_1 \sin(2\pi f_1 t) + \beta_2 \sin(2\pi f_2 t)]$$

where

$$f_c = 45 \text{ MHz}$$
$$f_1 = 5 \text{ kHz}$$
$$f_2 = 3 \text{ kHz}$$
$$\beta_1 = \beta_2 = 2$$

(a) Find the amplitude spectrum of this FM wave, retaining only those components whose amplitudes exceed 1 percent of the unmodulated carrier amplitude.
(b) Calculate the average power of the components represented in this amplitude spectrum, expressing your answer as a percentage of the average power of the FM wave.

For the pertinent values of the Bessel functions, use Table A4.1, Appendix 4.

Problem 4.13 A carrier wave of frequency 100 MHz is frequency-modulated by a sine-wave of amplitude 20 volts and frequency 100 kHz. The frequency sensitivity of the modulator is 25 kHz per volt.

(a) Determine the approximate bandwidth of the FM wave, using Carson's rule.
(b) Determine the bandwidth by transmitting only those side-frequencies whose amplitudes exceed 1 percent of the unmodulated carrier amplitude. Use the universal curve of Fig. 4.9 for this calculation.
(c) Repeat your calculations, assuming that the amplitude of the modulating wave is doubled.
(d) Repeat your calculations, assuming that the modulation frequency is doubled.

Problem 4.14 Consider a wide-band PM wave produced by a sinusoidal modulating wave $A_m \cos(2\pi f_m t)$, using a modulator with a phase sensitivity equal to k_p randians per volt.

(a) Show that if the maximum phase deviation of the PM wave is large compared with one radian, the bandwidth of the PM wave varies linearly with the modulation frequency f_m.
(b) Compare this characteristic of a wide-band PM wave with that of a wide-band FM wave.

Problem 4.15 Figure P4.4 shows the block diagram of a real-time *spectrum analyzer* working on the principle of modulation. The given signal $g(t)$ and a frequency-modulated signal $s(t)$ are applied to a multiplier and the output $g(t)s(t)$ is fed into a filter of impulse response $h(t)$. The $s(t)$ and $h(t)$ are *linear FM signals* whose instantaneous frequencies vary at opposite rates, as shown by

$$s(t) = \cos(2\pi f_c t - \pi k t^2),$$

and

$$h(t) = \cos(2\pi f_c t + \pi k t^2)$$

where k is a constant. Show that the envelope of the filter output is proportional to the amplitude spectrum of the input signal $g(t)$ with kt playing the role of frequency f.

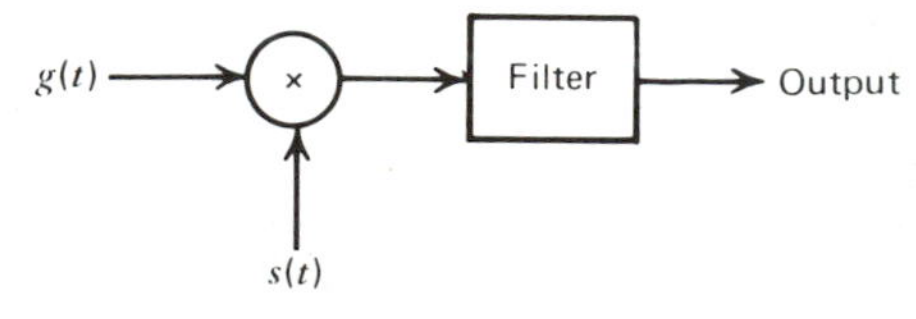

Figure P4.4

Hint: Use the complex notations described in Section 2.14 and 2.15 for band-pass signals and band-pass filters, respectively.

Problem 4.16 The block diagram of Fig. P4.5(a) shows the modulation system for transmitting *stereophonic FM waves*. The input signals $l(t)$ and $r(t)$ represent the left-hand and right-hand audio signals, respectively. These signals are added and subtracted to generate $l(t)+r(t)$ and $l(t)-r(t)$. The difference signal $l(t)-r(t)$ is used to generate a DSBSC wave with a carrier frequency of 38 kHz. This carrier is derived by applying a pilot carrier of frequency 19 kHz to a frequency doubler. The DSBSC wave, the sum signal $l(t)+r(t)$, and the pilot carrier are summed to produce a composite signal $m(t)$. The pilot carrier is transmitted for the purpose of receiver synchronization. The composite signal $m(t)$ is used to frequency modulate a carrier wave of frequency f_c, and the resulting FM wave is transmitted.

(a) Sketch the amplitude spectrum of the composite signal $m(t)$, assuming that the input signals $l(t)$ and $r(t)$ have the spectra shown in Fig. P4.5(b), and that $f_2 = 15$ kHz.
(b) Assuming a frequency deviation of 75 kHz, determine the transmission bandwidth of the FM wave.
(c) Develop the block diagram of a receiver for recovering the left-hand and right-hand audio signals from the incoming FM wave.

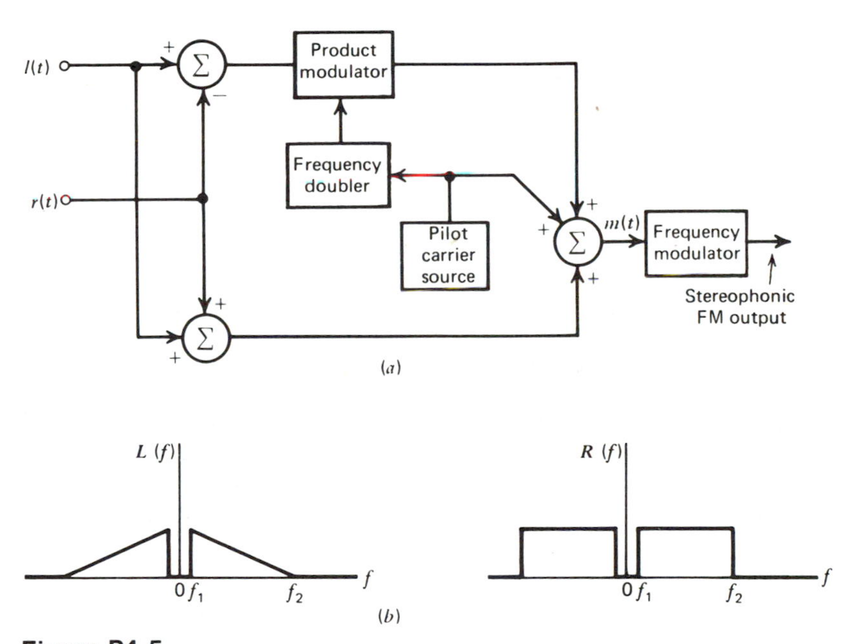

Figure P4.5

Problem 4.17 An FM wave with a frequency deviation of 10 kHz at a modulation frequency of 5 kHz is applied to two frequency multipliers connected in cascade. The first multiplier doubles the frequency and the second multiplier triples the frequency. Determine the frequency deviation and the modulation index of the FM wave obtained at the second multiplier output. What is the frequency separation of the adjacent side-frequencies of this FM wave?

Problem 4.18 An FM wave is applied to a square-law device with output voltage v_2 related to input voltage v_1 by

$$v_2 = av_1^2$$

where a is a constant. Explain how such a device can be used to obtain an FM wave with a greater frequency deviation than available at the input.

Problem 4.19 Figure P4.6 shows the frequency-determining network of a voltage-controlled oscillator. Frequency modulation is produced by applying the modulating wave $A_m \sin(2\pi f_m t)$ (plus a bias V_b) to a pair of varactor diodes connected across the parallel combination of a 200 μH inductor and 100 pF capacitor. The capacitance of each varactor diode is related to the voltage V (in volts) applied across its electrodes by

$$C = 100V^{-1/2} pF$$

The unmodulated frequency of oscillation is 1 MHz. The VCO output is applied to a frequency multiplier to produce an FM wave with a carrier frequency of 64 MHz and a modulation index of 5.

Determine: (a) the magnitude of the bias voltage V_b, and (b) the amplitude A_m of the modulating wave, given that $f_m = 10$ kHz.

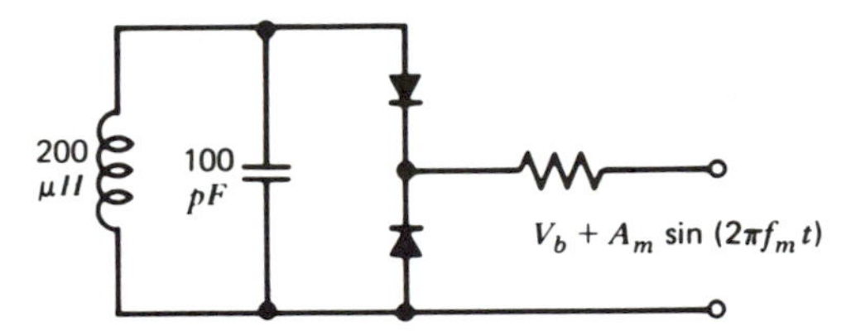

Figure P4.6

Problem 4.20 The FM wave

$$s(t) = A_c \cos\left[2\pi f_c t + 2\pi k_f \int_0^t m(t)dt\right]$$

is applied to the system shown in Fig. P4.7 consisting of a high-pass RC filter and an envelope detector. Assume that: (a) the resistance R is small compared with the reactance of the capacitor C for all significant frequency components of $s(t)$, and (b) the envelope detector does not load the filter. Determine the resulting signal at the envelope detector output, assuming that $k_f|m(t)| < f_c$ for all t.

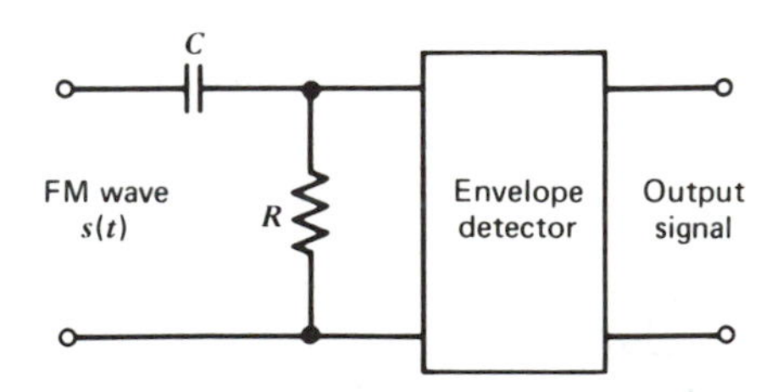

Figure P4.7

Problem 4.21 In the frequency discriminator of Fig. 4.17, let the frequency separation between the resonant frequencies of the two parallel tuned LC filters be denoted by $2kB$, where $2B$ is the 3-dB bandwidth of either filter and k is a scaling factor. Assume that both filters have a high-Q factor.

(a) Show that the total response of both filters has a slope equal to $2k/B(1+k^2)^{3/2}$ at the center frequency f_c.
(b) Let D denote the deviation of the total response measured with respect to a straight line passing through $f=f_c$ with this slope. Plot D versus δ for $k=1.5$ and $-kB \leqslant \delta \leqslant kB$, where $\delta = f - f_c$.

Problem 4.22 Consider the frequency demodulation scheme shown in Fig. P4.8 in which the incoming FM wave $s(t)$ is passed through a delay line that produces a phase shift of $\pi/2$ radians at the carrier frequency f_c. The delay-line output is subtracted from the incoming FM wave, and the resulting composite wave is then envelope-detected. This demodulator finds wide application in demodulating microwave FM waves. Assuming that

$$s(t) = A_c \cos[2\pi f_c t + \beta \sin(2\pi f_m t)]$$

analyze the operation of this demodulator when the modulation index β is less than unity and the delay T produced by the delay line is sufficiently small to justify making the approximations:

$$\cos(2\pi f_m T) \simeq 1$$

and

$$\sin(2\pi f_m T) \simeq 2\pi f_m T$$

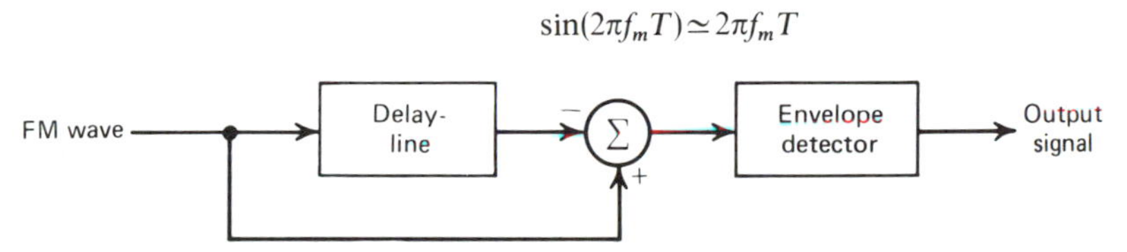

Figure P4.8

Problem 4.23 Figure P4.9 shows the block diagram of a *zero-crossing detector* for demodulating an FM wave. It consists of a limiter, a pulse generator for producing a short pulse at each zero-crossing of the input, and a low-pass filter for extracting the modulating wave.

(a) Show that the instantaneous frequency of the input FM wave is proportional to the number of zero crossings in the time interval $t-(T_1/2)$ to $t+(T_1/2)$, divided by T_1. Assume that the modulating wave is essentially constant during this time interval.
(b) Illustrate the operation of this demodulator, using the sawtooth wave of Fig. P4.1 as the modulating wave.

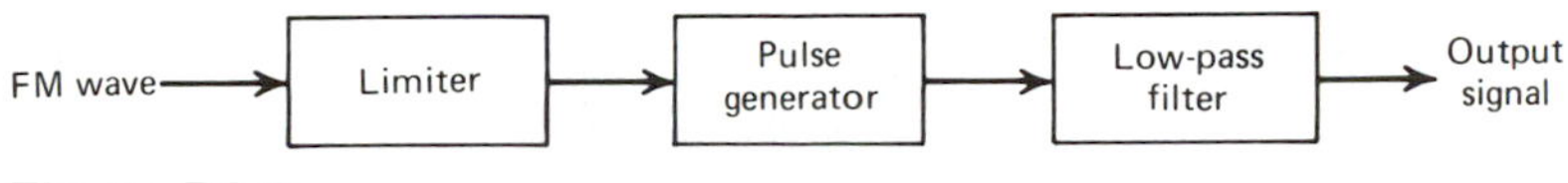

Figure P4.9

Problem 4.24 Suppose that the received signal in an FM system contains some residual amplitude modulation of positive amplitude $a(t)$, as shown by

$$s(t) = a(t)\cos[2\pi f_c t + \phi(t)]$$

where f_c is the carrier frequency. The phase $\phi(t)$ is related to the modulating wave $m(t)$ by

$$\phi(t)=2\pi k_f \int_0^t m(t)dt$$

where k_f is a constant. Assume that the signal $s(t)$ is restricted to a frequency band of width B_T, centered at f_c, where B_T is the transmission bandwidth of the FM wave in the absence of amplitude modulation, and that the amplitude modulation is slowly varying compared with $\phi(t)$. Show that the output of an ideal frequency discriminator produced by $s(t)$ is proportional to $a(t)m(t)$.

Hint: Use the complex notation described in Section 2.14 to represent the modulated wave $s(t)$.

Problem 4.25

(a) Let the modulated wave $s(t)$ in Problem 4.24 be applied to an ideal *limiter*, whose output $z(t)$ is defined by

$$\begin{aligned} z(t) &= \text{sgn}[s(t)] \\ &= \begin{cases} +1 & s(t)>0 \\ -1, & s(t)<0 \end{cases} \end{aligned}$$

Show that the limiter output may be expressed in the form of a Fourier series as follows:

$$z(t)=\frac{4}{\pi}\sum_{n=0}^{\infty}\frac{(-1)^n}{2n+1}\cos[2\pi f_c t(2n+1)+(2n+1)\phi(t)]$$

(b) Suppose that the limiter output is applied to a band-pass filter with a pass-band amplitude response of one and bandwidth B_T centered about the carrier frequency f_c, where B_T is the transmission bandwidth of the FM wave in the absence of amplitude modulation. Assuming that f_c is much greater than B_T, show that the resulting filter output equals

$$y(t)=\frac{4}{\pi}\cos[2\pi f_c t+\phi(t)]$$

By comparing this output with the original modulated wave $s(t)$, as defined in Problem 4.24, comment on the practical usefulness of the result.

Problem 4.26

(a) Consider an FM wave of carrier frequency f_c, which is produced by a modulating wave $m(t)$. Assume that f_c is large enough to justify treating this FM wave as a narrow-band signal. Find an approximate expression for its Hilbert transform.
(b) For the special case of a sinusoidal modulating wave $m(t)=A_m\cos(2\pi f_m t)$, find the exact expression for the Hilbert transform of the resulting FM wave. For this case, what is the error in the approximation used in part (a)?

Problem 4.27 The *single sideband version of angle modulation* is defined by

$$s(t)=\exp[-\hat{\phi}(t)]\cos[2\pi f_c t+\phi(t)]$$

where $\hat{\phi}(t)$ is the Hilbert transform of the phase function $\phi(t)$ and f_c is the carrier frequency.

(a) Show that the spectrum of the modulated wave $s(t)$ contains no frequency components in the interval $-f_c<f<f_c$, and is of infinite extent.

(b) Given that the phase function

$$\phi(t)=\beta \sin(2\pi f_m t)$$

where β is the modulation index and f_m is the modulation frequency, derive the corresponding expression for the modulated wave $s(t)$.

chapter 5 RANDOM PROCESSES

In Chapter 2 we dealt with the frequency analysis of *deterministic signals* and the transmission of such signals through linear filters. By deterministic signals we mean the class of signals that may be modeled as completely specified functions of time. There is, however, another important class of signals known as *random signals*, examples of which are encountered in every practical communication system. We say a signal is random if it is not possible to predict its precise value in advance. Consider, for example, a radio communication system. The received signal in such a system usually consists of an *information-bearing signal* component, a random *interference* component, and *receiver noise*. The information-bearing signal component may represent, for example, a voice signal that, typically, consists of randomly spaced bursts of energy of random duration. The interference component represents the extraneous electromagnetic waves produced by other communication systems and atmospheric electricity. A major source of noise is *thermal noise* which is caused by the random motion of the electrons in conductors and devices at the front end of the receiver. We thus find that the received signal is completely random in nature.

Although it is not possible to predict the precise value of a random signal in advance, nevertheless, it may be described in terms of its *statistical* properties such as the average power in the random signal, or the spectral distribution of this power on the average. The mathematical discipline that deals with the statistical characterization of random signals is *probability theory*. We begin this chapter by reviewing some basic definitions in probability theory.*

5.1 PROBABILITY

One way to approach the notion of probability is through the phenomenon of *statistical regularity*. There are many repeating situations in nature for which we can predict from previous experience roughly what will happen on the average, but not exactly what will happen. We say in such cases the occurrences are random. Whatever the reason for the randomness, in many situations leading to random occurrences, a definite average pattern of results may be observed when the situation is recreated a large number of times.

To use the idea of statistical regularity to explain the concept of probability, we proceed as follows:

1. *We prescribe a basic experiment*. For example, the experiment may be the observation of the result of the tossing of a fair coin, or the measurement of the amplitude of a noise voltage at a particular instant of time.
2. *We specify all possible outcomes of the basic experiment*. For example, in the case of a coin-tossing experiment, the possible outcomes of a trial are "heads" or "tails," whereas in a noise voltage measurement the instantaneous amplitude may have any value between plus and minus infinity.
3. *We repeat the basic experiment a large number of times, say n, under uniform conditions and observe the results of the experiment.*

* For a more detailed treatment of probability theory, see J. B. Thomas, *An Introduction to Applied Probability, Random Variables, and Stochastic Processes* (McGraw-Hill, 1965). W. B. Davenport and W. L. Root, *Random Signals and Noise* (McGraw-Hill, 1958). T. C. Fry, *Probability and its Engineering Uses* (Van Nostrand-Reinhold, 1965). W. Feller, *An Introduction to Probability Theory and Its Applications*, Third Edition, (Wiley, 1968). A. Papoulis, *Probability, Random Variables, and Stochastic Processes* (McGraw-Hill, 1965). J. M. Wozencraft and I. M. Jacobs, *Principles of Communication Engineering*, pp. 1–128 (Wiley, 1965). I. F. Blake, *An Introduction to Applied Probability* (Wiley, 1979).

Consider one of the possible outcomes of the experiment and denote it as *event A*. For example, event A may denote the observation that the amplitude of a noise voltage is less than or equal to some specified level. Suppose that in the n trials of the experiment, event A occurs n_A times. Then we may assign to the event A a nonnegative real number called the *probability of occurrence*, defined as follows:

$$P(A)=\lim_{n\to\infty}\left(\frac{n_A}{n}\right) \tag{5.1}$$

Of course, for a *certain event* we have $n_A=n$, so that the probability of a certain event is one. On the other hand, the probability of an *impossible event* is zero, because $n_A=0$. *The probability of an event is a nonnegative real number less than or equal to one.*

Consider next a basic experiment in which there are N possible outcomes denoted by $A_1, A_2, \ldots, A_N$, which are *mutually exclusive* in that the occurrence of any one event excludes the occurrence of all others. For example, in a coin-tossing experiment, the occurrence of "heads" excludes the occurrence of "tails" in a trial, and vice versa. Clearly, in general, if we include all the possible events A_1 to A_N, we have

$$\sum_{k=1}^{N} P(A_k)=1 \tag{5.2}$$

We have thus far been concerned with the outcomes of a single basic experiment. In many problems that are encountered in practice, however, we may be concerned with the outcomes of several basic experiments. It is therefore appropriate to extend our definition of probability to include the *joint probability* of the occurrence of two or more events. For example, we may be interested to know the joint probability that the noise voltage at one instant of time t_1 exceeds a certain value v_1 and that the noise voltage at another instant of time t_2 exceeds another value v_2. Assume that we perform an experiment and examine the occurrence of a pair of events A and B, in that order. Let n_{AB} denote the number of times the joint event (A, B) occurs in a total of n trials of the experiment. The joint probability of first A and then B occurring is

$$P(A, B)=\lim_{n\to\infty}\left(\frac{n_{AB}}{n}\right)$$

Suppose that, in the same n trials, the event A occurs n_A times and the event B occurs n_B times. Since we have assumed that the joint event (A, B) corresponds to A occurring first and then B, it follows that n_A must include n_{AB}. In other words,

$$\frac{n_{AB}}{n_A}\leqslant 1$$

The ratio n_{AB}/n_A represents the relative frequency of occurrence of event B given that event A has occurred. For large n, the ratio n_{AB}/n_A defines the probability of occurrence of event B, given that event A has occurred. This probability is referred to as the *conditional probability*, and is denoted as $P(B|A)$. That is,

$$P(B|A) = \lim_{n\to\infty}\left(\frac{n_{AB}}{n_A}\right)$$

or

$$P(B|A) = \lim_{n\to\infty}\left(\frac{n_{AB}/n}{n_A/n}\right)$$

Therefore, we may write

$$P(B|A) = \frac{P(A, B)}{P(A)} \tag{5.3}$$

or

$$P(A, B) = P(B|A)P(A) \tag{5.4}$$

It is apparent that we may also write

$$P(A, B) = P(A|B)P(B) \tag{5.5}$$

Equations (5.4) *and* (5.5) *state that the joint probability of two events may be expressed as the product of the conditional probability of one event, given the other, times the elementary probability of the other.* Note that the conditional probabilities $P(B|A)$ and $P(A|B)$ have essentially the same properties as the various probabilities previously defined.

Situations may exist where the conditional probability $P(A|B)$ and the probabilities $P(A)$ and $P(B)$ are easily determined directly, but the conditional probability $P(B|A)$ is desired. From Eqs. (5.4) and (5.5), it follows that, provided $P(A) \neq 0$, we may determine $P(B|A)$ by using the relation

$$P(B|A) = \frac{P(A|B)P(B)}{P(A)} \tag{5.6}$$

This relation is a special form of *Bayes rule.*

Suppose that the conditional probability $P(B|A)$ is simply equal to the elementary probability of occurrence of event B, that is,

$$P(B|A) = P(B) \tag{5.7}$$

It then follows that the probability of occurrence of the joint event (A, B) is equal to the product of the elementary probabilities of the events A and B:

$$P(A, B) = P(A)P(B), \tag{5.8}$$

so that

$$P(A|B) = P(A) \tag{5.9}$$

That is, the conditional probability of the event A, assuming the occurrence of the event B, is simply equal to the elementary probability of the event A. We thus see that in this case a knowledge of the occurrence of one event tells us no more about the probability of occurrence of the other event than we knew without that knowledge. Events A and B that satisfy such relations as Eq. (5.7), (5.8), or (5.9) are said to be *statistically independent*.

5.2 RANDOM VARIABLES

In Section 5.1 we discussed experiments, events (i.e., possible outcomes of experiments), and probabilities of events. In such discussions, it is convenient to think of an experiment and its possible outcomes as defining a space and its points. With each basic possible outcome of the experiment we may associate a point called the *sample point*, denoted by s. The totality of sample points $\{s\}$, corresponding to the aggregate of all possible outcomes of the experiment, is called the *sample space*, denoted by S. An event may correspond to a single sample point or a set of sample points. For example, in an experiment involving the throw of a die, there are six possible outcomes: the showing of one, two, three, four, five, or six dots on the upper face of the die. By assigning a sample point to each of these possible outcomes, we have a sample space that consists of six sample points. The basic event "a six shows" corresponds to a single sample point. On the other hand, the compound event "an even number of dots shows" corresponds to the set of three sample points, namely, the showing of two, four, and six dots.

It is customary, particularly when using the language of sample space, to think of the outcome of an experiment as a variable that can wander over the set of sample points, and whose value is determined by the experiment. *A function whose domain is a sample space and whose range is some set of real numbers is called a random variable of the experiment.** Thus when the outcome of the experiment is s, the random variable is denoted as $X(s)$ or simply X. For example, the sample space representing the outcomes of the throw of a die is a set of six sample points which may be taken to be the integers $1, 2, \ldots, 6$. Then if we identify the sample point k with the event that k dots show when the die is thrown, the function $X(k) = k$ is a random variable such that $X(k)$ equals the number of dots that show when the die is thrown. In this example, the random variable takes on only a discrete set of values. In such a case we say that we are dealing with a *discrete random variable*. *More precisely, the random variable X is a discrete random variable if X can take on only a finite number of values in any finite observation interval. If, however, the random variable X can take on any value in a whole observation interval, X is called a continuous random variable*. For example, the random variable that represents the amplitude of a noise voltage at a particular instant of time is a continuous random variable because it may take on any value between plus and minus infinity.

To proceed further, we need a probabilistic description of random variables that

* The term "random variable" is somewhat confusing: first, because the word "random" is not used in the sense of equal probability of occurrence, for which it should be reserved. Second, the word "variable" does not imply dependence (upon the experimental outcome), which is an essential part of the meaning. Nevertheless, the term is so deeply imbedded in the literature of probability that its usage has persisted.

works equally well for discrete as well as continuous random variables. Let us consider the random variable X and the probability of the event $X \leqslant x$. We denote this probability by $P(X \leqslant x)$. It is apparent that this probability is a function of the *dummy variable* x. To simplify our notation, we write

$$F_X(x) = P(X \leqslant x) \tag{5.10}$$

The function $F_X(x)$ is called the *cumulative distribution function* or simply the *distribution function* of the random variable X. Note that $F_X(x)$ is a function of x, not of the random variable X. However, it depends on the assignment of the random variable X, which accounts for the use of X as subscript. For any point x, the distribution function $F_X(x)$ expresses a probability.

The distribution function $F_X(x)$ has the following properties, which follow directly from Eq. (5.10):

1. The distribution function $F_X(x)$ is bounded between zero and one.
2. The distribution function $F_X(x)$ is a monotone-nondecreasing function of x; that is,

$$F_X(x_1) \leqslant F_X(x_2), \qquad \text{if } x_1 < x_2 \tag{5.11}$$

An alternative description of the probability distribution of the random variable X is often useful. This is the derivative of the distribution function, as shown by

$$f_X(x) = \frac{d}{dx} F_X(x) \tag{5.12}$$

which is called the *probability density function*. Note that the differentiation in Eq. (5.12) is with respect to the dummy variable x. The name, density function, arises from the fact that the probability of the event $x_1 < X \leqslant x_2$ equals

$$\begin{aligned} P(x_1 < X \leqslant x_2) &= P(X \leqslant x_2) - P(X \leqslant x_1) \\ &= F_X(x_2) - F_X(x_1) \\ &= \int_{x_1}^{x_2} f_X(x)\, dx \end{aligned} \tag{5.13}$$

Since $F_X(\infty) = 1$, corresponding to the probability of a certain event, and $F_X(-\infty) = 0$, corresponding to the probability of an impossible event, it follows immediately from Eq. (5.13) that

$$\int_{-\infty}^{\infty} f_X(x)\, dx = 1 \tag{5.14}$$

Also, as mentioned earlier, a distribution function must always be monotone nondecreasing. Hence, its derivative or the probability density function must always be nonnegative. *A probability density function must always be a nonnegative function, and with a total area of one.*

Several Random Variables

Thus far we have focused attention on situations involving a single random variable. However, we find frequently that the outcome of an experiment requires several random variables for its description. We now consider situations involving two random variables. The probabilistic description developed in this way may be readily extended to any number of random variables.

Consider two random variables X and Y. *We define the joint distribution function $F_{X,Y}(x, y)$ as the probability that the random variable X is less than or equal to a specified value x and that the random variable Y is less than or equal to a specified value y.* The variables X and Y may be two separate one-dimensional random variables or the components of a single two-dimensional random variable. In either case, the joint sample space is the xy-plane. The joint distribution function $F_{X,Y}(x, y)$ is the probability that the outcome of an experiment will result in a sample point lying inside the quadrant $(-\infty < X \leqslant x,\ -\infty < Y \leqslant y)$ of the joint-sample space. That is,

$$F_{X,Y}(x, y) = P(X \leqslant x, Y \leqslant y) \tag{5.15}$$

Suppose that the joint distribution function $F_{X,Y}(x, y)$ is continuous everywhere, and that the partial derivative

$$f_{X,Y}(x, y) = \frac{\partial^2 F_{X,Y}(x, y)}{\partial x \partial y} \tag{5.16}$$

exists and is continuous everywhere. We call the function $f_{X,Y}(x, y)$ the *joint probability density function* of the random variables X and Y. The joint distribution function $F_{X,Y}(x, y)$ is a monotone-nondecreasing function of both x and y. Therefore, from Eq. (5.16) it follows that the joint probability density function $f_{X,Y}(x, y)$ is always nonnegative. Also the total volume under the graph of a joint probability density function must be unity, as shown by

$$\int_{-\infty}^{\infty} \int_{-\infty}^{\infty} f_{X,Y}(\xi, \eta) d\xi d\eta = 1 \tag{5.17}$$

The probability density function for a single random variable (X, say) can be obtained from its joint probability density function with a second random variable (Y, say) in the following way. We first note that

$$F_X(x) = \int_{-\infty}^{\infty} \int_{-\infty}^{x} f_{X,Y}(\xi, \eta) d\xi \, d\eta \tag{5.18}$$

Therefore, differentiating both sides of Eq. (5.18) with respect to x, we get the desired relation:

$$f_X(x) = \int_{-\infty}^{\infty} f_{X,Y}(x, \eta) d\eta \tag{5.19}$$

Thus the probability density function $f_X(x)$ may be obtained from the joint probability density function $f_{X,Y}(x, y)$ by simply integrating it over all possible values of the undesired random variable, Y. The use of similar arguments in the other dimension yields $f_Y(y)$. The probability density functions $f_X(x)$ and $f_Y(y)$ are called *marginal densities*. Hence, the joint probability density function $f_{X,Y}(x, y)$ contains all the possible information about the joint random variables X and Y.

Suppose that X and Y are two continuous random variables with joint probability density function $f_{X,Y}(x, y)$. The *conditional probability density function* of Y given that $X = x$ is defined by

$$f_{Y|X}(y|x) = \frac{f_{X,Y}(x, y)}{f_X(x)} \tag{5.20}$$

provided that $f_X(x) > 0$, where $f_X(x)$ is the marginal density of X. The function $f_{Y|X}(y|x)$ may be thought of as a function of the variable y, with the variable x arbitrary, but fixed. Accordingly, it satisfies all the requirements of an ordinary probability density function, as shown by

$$f_{Y|X}(y|x) \geq 0$$

and

$$\int_{-\infty}^{\infty} f_{Y|X}(y|x)\, dy = 1$$

If the random variables X and Y are *statistically independent*, then knowledge of the outcome of X can in no way affect the distribution of Y. The result is that the condition probability density function $f_{Y|X}(y|x)$ reduces to the marginal density $f_Y(y)$, as shown by

$$f_{Y|X}(y|x) = f_Y(y)$$

In such a case, we may express the joint probability density function of the random variables X and Y as the product of their respective marginal densities, as shown by

$$f_{X,Y}(x, y) = f_X(x) f_Y(y)$$

Statistical Averages

Having discussed probability and some of its ramifications, we now seek ways for determining the *average* behavior of the outcomes arising in random experiments.

The *mean value* or *expected value* of a random variable X is commonly defined by

$$m_X = E[X] = \int_{-\infty}^{\infty} x f_X(x) dx \tag{5.21}$$

where E denotes the *expectation operator*. That is, the mean value m_X locates the center of gravity of the area under the probability density curve of the random variable X. Similarly, the expected value of a function of X, namely, $g(X)$, is defined by

$$E[g(X)] = \int_{-\infty}^{\infty} g(x) f_X(x) dx \tag{5.22}$$

For the special case of $g(X)=X^n$ we obtain the nth *moment* of the probability distribution of the random variable X; that is,

$$E[X^n]=\int_{-\infty}^{\infty} x^n f_X(x)dx \tag{5.23}$$

By far the most important moments of X are the first two moments. Thus putting $n=1$ in Eq. (5.23) gives the mean value of the random variable as discussed above, whereas putting $n=2$ gives the *mean-square value* of X:

$$E[X^2]=\int_{-\infty}^{\infty} x^2 f_X(x)dx \tag{5.24}$$

We may also define *central moments*, which are simply the moments of the difference between a random variable X and its mean value m_X. Thus the nth central moment is

$$E[(X-m_X)^n]=\int_{-\infty}^{\infty} (x-m_X)^n f_X(x)dx \tag{5.25}$$

For $n=1$, the central moment is, of course, zero, whereas for $n=2$ the second central moment is referred to as the *variance* of the random variable:

$$\text{Var}[X]=E[(X-m_X)^2]=\int_{-\infty}^{\infty} (x-m_X)^2 f_X(x)dx \tag{5.26}$$

The variance of a random variable X is commonly denoted as σ_X^2. The square root of the variance, namely, σ_X, is called the *standard deviation* of the random variable X.

The variance σ_X^2 of a random variable X in some sense is a measure of the variable's "randomness." By specifying the variance σ_X^2, we essentially constrain the effective width of the probability density function $f_X(x)$ of the random variable X about the mean m_X. A precise statement of this constraint is due to Chebyshev. The *Chebyshev inequality* states that for any positive number ε, we have

$$P(|X-m_X|\geqslant\varepsilon)\leqslant\frac{\sigma_X^2}{\varepsilon^2} \tag{5.27}$$

From this inequality we see that the mean and variance of a random variable give a partial description of its probability distribution.

We note from Eqs. (5.24) and (5.26) that the variance σ_X^2 and mean-square value $E[X^2]$ are related by

$$\begin{aligned}\sigma_X^2&=E[X^2-2m_X X+m_X^2]\\&=E[X^2]-2m_X E[X]+m_X^2\\&=E[X^2]-m_X^2\end{aligned} \tag{5.28}$$

Therefore, if the mean m_X is zero, then the variance σ_X^2 and the mean-square value $E[X^2]$ of the random variable X are equal.

Another important statistical average is the *characteristic function* $\phi_X(v)$ of the probability distribution of the random variable X, which is defined as the expectation of $\exp(jvX)$:

$$\phi_X(v) = E[\exp(jvX)]$$
$$= \int_{-\infty}^{\infty} f_X(x)\exp(jvx)dx \tag{5.29}$$

where v is real. In other words, the characteristic function $\phi_X(v)$ is (except for a sign change in the exponent) the Fourier transform of the probability density function $f_X(x)$. In this relation we have used $\exp(jvx)$ rather than $\exp(-jvx)$, so as to conform with the convention adopted in probability theory. Recognizing that v and x play analogous roles to the variables $2\pi f$ and t of Fourier transforms, respectively, we deduce the following inverse relation from analogy with the inverse Fourier transform:

$$f_X(x) = \frac{1}{2\pi} \int_{-\infty}^{\infty} \phi_X(v)\exp(-jvx)dv \tag{5.30}$$

This relation may be used to evaluate the probability density function $f_X(x)$ of the random variable X from its characteristic function $\phi_X(v)$.

Joint Moments

Consider next a pair of random variables X and Y. A set of statistical averages of importance in this case are the *joint moments*, namely, the expected value of $X^j Y^k$, where j and k may assume any positive integer values. We may thus write

$$E[X^j Y^k] = \int_{-\infty}^{\infty} \int_{-\infty}^{\infty} x^j y^k f_{X,Y}(x, y) dx\, dy \tag{5.31}$$

A joint moment of particular importance is the *correlation* defined by $E[XY]$, which corresponds to $j = k = 1$ in Eq. (5.31).

The correlation of the centered random variables $X - E[X]$ and $Y - E[Y]$, that is, the joint moment

$$\text{Cov}[XY] = E[(X - E[X])(Y - E[Y])] \tag{5.32}$$

is called the *covariance* of X and Y. Letting $m_X = E[X]$ and $m_Y = E[Y]$, we may expand Eq. (5.32) to obtain

$$\text{Cov}[XY] = E[XY] - m_X m_Y \tag{5.33}$$

Let σ_X^2 and σ_Y^2 denote the variances of X and Y, respectively. Then the covariance of X and Y normalized with respect to $\sigma_X \sigma_Y$ is called the *correlation coefficient* of X and Y:

$$\rho = \frac{\text{Cov}[XY]}{\sigma_X \sigma_Y} \tag{5.34}$$

We say that the two random variables X and Y are uncorrelated if and only if their covariance is zero, that is, if and only if

$$\mathrm{Cov}[XY]=0$$

We say that they are orthogonal if and only if their correlation is zero, that is, if and only if

$$E[XY]=0$$

From Eq. (5.33) we observe that if one of the random variables X and Y or both have zero means, and if they are orthogonal random variables, then they are uncorrelated, and vice versa. Note also that if X and Y are statistically independent, then they are uncorrelated; however, the converse of this statement is not necessarily true.

Transformation of Random Variables

Consider the problem of determining the probability density function of a random variable Y, which is obtained by a one-to-one transformation of a given random variable X. The simplest possible case is when the new random variable Y is a monotone-increasing differentiable function g of the random variable X (see Fig. 5.1):

$$Y=g(X) \tag{5.35}$$

In this case we have

$$\begin{aligned} F_Y(y) &= P(Y \leqslant y) \\ &= P(X \leqslant h(y)) \end{aligned}$$

where h is the inverse transformation

$$h(y)=g^{-1}(y) \tag{5.36}$$

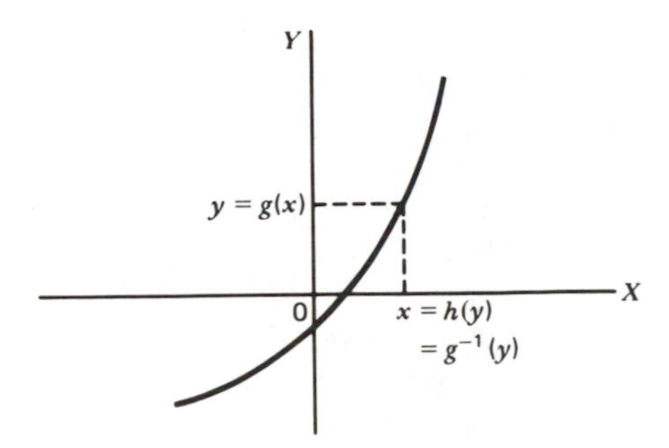

Figure 5.1 A one-to-one transformation of a random variable X.

This inverse transformation exists for all y, because x and y are related one to one. Assuming that the given random variable X has a probability density function $f_X(x)$, we may write

$$F_Y(y) = \int_{-\infty}^{h(y)} f_X(x)\,dx$$

Differentiating both sides of this relation, we get

$$f_Y(y) = f_X(h(y)) \frac{dh}{dy} \tag{5.37}$$

Consider next the case when g is a differentiable monotone-decreasing function with an inverse h. We may then write

$$F_Y(y) = \int_{h(y)}^{\infty} f_X(x)\,dx$$

which, on differentiation, yields

$$f_Y(y) = -f_X(h(y)) \frac{dh}{dy} \tag{5.38}$$

Since the derivative dh/dy is negative in this case, whereas it is positive in Eq. (5.37), we may express both results by the single formula

$$f_Y(y) = f_X(h(y)) \left|\frac{dh}{dy}\right| \tag{5.39}$$

This is the desired formula for finding the probability density function of a one-to-one differentiable function of a given random variable.

5.3 RANDOM PROCESSES

A basic concern in the statistical analysis of communication systems is the characterization of random signals such as voice signals, television signals, telegraph signals, digital computer data, and electrical noise. These random signals have two properties: first, the signals are functions of time, defined on some observation interval; second, the signals are random in the sense that before conducting an experiment, it is not possible to describe exactly the waveforms that will be observed. Accordingly, in describing random signals we find that each sample point in our sample space is a function of time. For example, in studying the fluctuations in the output of a transistor, we may assume the simultaneous testing of an indefinitely large number of identical transistors as a conceptual model of our problem. The output (measured as a function of time) of a particular transistor in the collection is then one sample point in our sample space. The sample space or ensemble comprised of functions of time is called a *random* or *stochastic process*. As an integral part of this notion, we assume the existence of a probability distribution defined over an appropriate class of sets in the sample space, so that we may speak with confidence of the probability of various events. *We may thus define a random process as an ensemble of time functions together with a probability rule which assigns a probability to any meaningful event associated with an observation of one of these functions.*

Consider a random process $X(t)$ represented by the set of *sample functions* $\{x_j(t)\}$, $j=1, 2, \ldots, n$, as illustrated in Fig. 5.2. Sample function or waveform $x_1(t)$, with probability of occurrence $P(s_1)$, corresponds to *sample point* s_1 of the *sample space* S, and so on for the other sample functions $x_2(t), \ldots, x_n(t)$. Now suppose we observe the set of waveforms $\{x_j(t)\}$, $j=1, 2, \ldots, n$, simultaneously at some time instant, $t=t_1$, as shown in the figure. Since each sample point s_j of the sample space S has associated with it a number $x_j(t_1)$ and a probability $P(s_j)$, we find that the resulting collection of numbers $\{x_j(t_1)\}$, $j=1, 2, \ldots, n$, forms a *random variable*. We denote this random variable by $X(t_1)$. By observing the given set of waveforms simultaneously at a second time instant, say t_2, we obtain a different collection of numbers, hence a different random variable $X(t_2)$. Indeed, the set of waveforms $\{x_j(t)\}$ defines a different random variable for each choice of observation instant. The difference between a random variable and a random process is that for a random variable the outcome of an experiment is mapped into a number, whereas for a random process the outcome is mapped into a waveform that is a function of time.

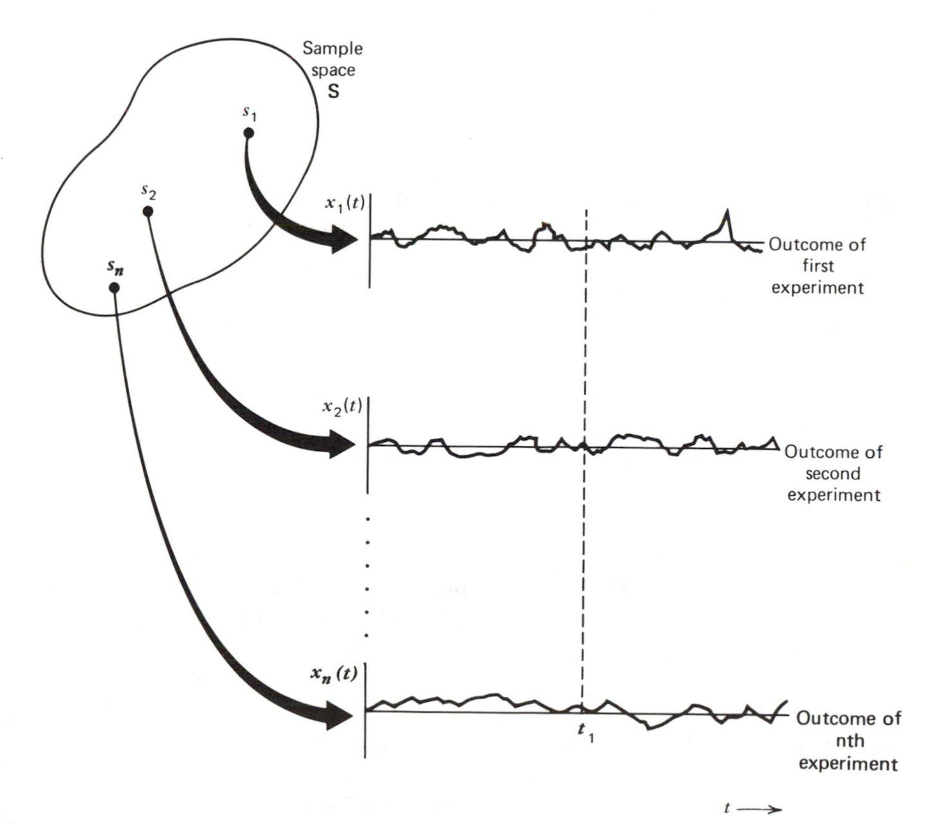

Figure 5.2 An ensemble of sample functions.

Random Vectors Obtained From Random Processes

By definition, a random process $X(t)$ implies the existence of an infinite number of random variables, one for each value of time t in the range $-\infty < t < \infty$. Thus we may speak of the distribution function $F_{X(t_1)}(x_1)$ of the random variable $X(t_1)$ obtained by observing the random process $X(t)$ at time t_1. More generally, for k time instants $t_1, t_2, \ldots, t_k$ we define the k random variables $X(t_1), X(t_2), \ldots, X(t_k)$, respectively, and express their *joint distribution function* as the probability of the joint event $X(t_1) \leqslant x_1, X(t_2) \leqslant x_2, \ldots, X(t_k) \leqslant x_k$ as shown by

$$F_{X(t_1),X(t_2),\ldots,X(t_k)}(x_1, x_2, \ldots, x_k) = P(X(t_1) \leqslant x_1, X(t_2) \leqslant x_2, \ldots, X(t_k) \leqslant x_k) \quad (5.40)$$

For convenience of notation, we write this joint distribution function simply as $F_{\mathbf{X(t)}}(\mathbf{x})$ where the *random vector* $\mathbf{X(t)}$ equals

$$\mathbf{X(t)} = \begin{bmatrix} X(t_1) \\ X(t_2) \\ \vdots \\ X(t_k) \end{bmatrix} \quad (5.41)$$

and the *dummy vector* $\mathbf{x}$ equals

$$\mathbf{x} = \begin{bmatrix} x_1 \\ x_2 \\ \vdots \\ x_k \end{bmatrix} \quad (5.42)$$

For a particular sample point s_j, the components of the random vector $\mathbf{X(t)}$ represent the values of the sample functions $x_j(t)$ observed at times $t_1, t_2 \ldots, t_k$. Note also that the joint distribution function $F_{\mathbf{X(t)}}(\mathbf{x})$ depends on the random process $X(t)$ and the set of times $\{t_j\}, j = 1, 2, \ldots, k$.

The joint probability density function of the random vector $\mathbf{X(t)}$ equals

$$f_{\mathbf{X(t)}}(\mathbf{x}) = \frac{\partial^k}{\partial x_1 \, \partial x_2 \ldots \partial x_k} F_{\mathbf{X(t)}}(\mathbf{x}) \quad (5.43)$$

This function is always nonnegative, with a total volume of one.

Example 1

Consider the probability of obtaining a sample function or waveform $x(t)$ of the random process $X(t)$ that passes through a set of k "windows," as illustrated in Fig. 5.3 for the case of $k=3$. That is, we wish to find the probability of the joint event

$$A = \{a_i < X(t_i) \leqslant b_i\}, \qquad i = 1, 2, \ldots, k$$

Given the joint probability density function $f_{\mathbf{X(t)}}(\mathbf{x})$, this probability equals

$$P(A) = \int_{a_1}^{b_1} \int_{a_2}^{b_2} \cdots \int_{a_k}^{b_k} f_{\mathbf{X(t)}}(\mathbf{x}) dx_1 \, dx_2 \ldots dx_k$$

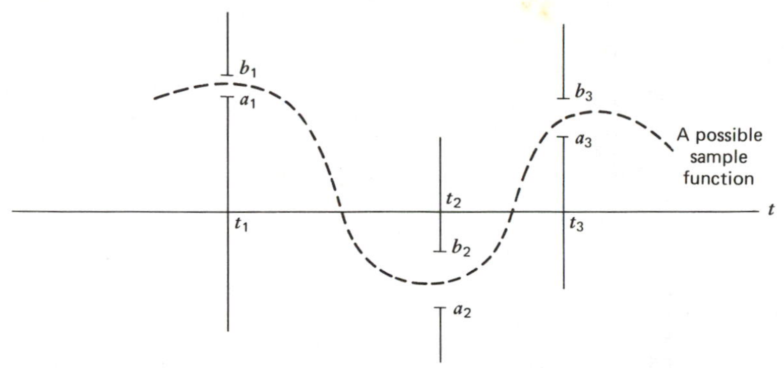

Figure 5.3 Illustrating the probability of a joint event.

5.4 STATIONARITY

Consider a set of times $t_1, t_2, \ldots, t_k$ in the interval in which a random process $X(t)$ is defined. A complete characterization of the random process $X(t)$ enables us to specify the joint probability density function $f_{\mathbf{X}(\mathbf{t})}(\mathbf{x})$. The random process $X(t)$ is said to be *strictly stationary* or *stationary in the strict sense* if this joint probability density function is invariant under shifts of the time origin, that is, if the equality

$$f_{\mathbf{X}(\mathbf{t})}(\mathbf{x}) = f_{\mathbf{X}(\mathbf{t}+\mathbf{T})}(\mathbf{x}) \tag{5.44}$$

holds for every finite set of time instants $\{t_j\}, j = 1, 2, \ldots, k$, and for every time shift T and dummy vector $\mathbf{x}$. The components of the random vector $\mathbf{X}(\mathbf{t})$ are obtained by observing the random process $X(t)$ at times $t_1, t_2, \ldots, t_k$, as in Eq (5.41). Correspondingly, the components of the random vector $\mathbf{X}(\mathbf{t}+\mathbf{T})$ are obtained by observing the random process $X(t)$ at times $t_1 + T, t_2 + T, \ldots, t_k + T$, where T is the time shift.

Stationary processes are of great importance for at least two reasons:

1. They are frequently encountered in practice or approximated to a high degree of accuracy. In actual fact, from a practical point of view, it is not necessary that a random process be stationary for all time, but only for some observation interval that is long enough for the particular situation.
2. Many of the important properties of stationary processes commonly encountered are described by first and second moments. Consequently, it is relatively easy to develop a simple but useful theory to describe these processes.

Random processes that are not stationary are called *nonstationary*.

Example 2

Suppose we have a random process $X(t)$ that is known to be strictly stationary. An implication of stationarity is that the probability of the set of sample functions of this process which pass

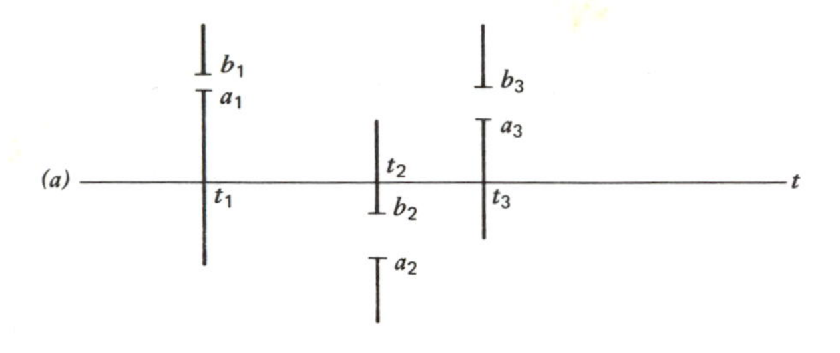

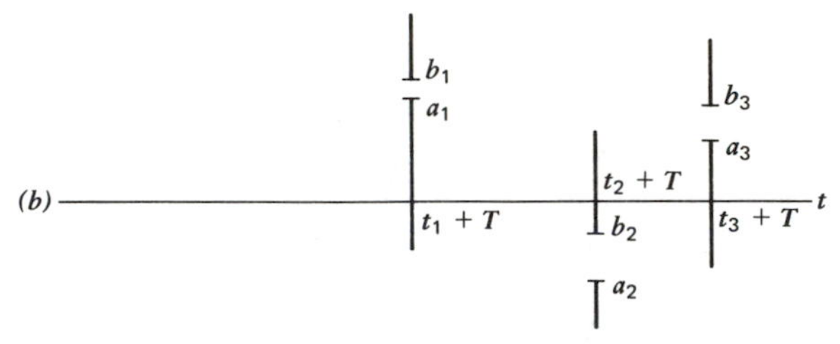

Figure 5.4 Illustrating the concept of stationarity.

through the windows of Fig. 5.4(*a*) is equal to the probability of the set of sample functions which pass through the corresponding time-shifted windows of Fig. 5.4(*b*). Note, however, that it is not necessary that these two sets consist of the same sample functions.

5.5 MEAN, CORRELATION, AND COVARIANCE FUNCTIONS

In many practical situations we find that it is not possible to determine (by means of suitable measurements, say) the probability distribution of a random process. Then we must content ourselves with a *partial description* of the distribution of the process. Ordinarily, the mean, autocorrelation function, and autocovariance function of the random process are taken to give a crude but, nevertheless, useful description of the distribution.

Consider a real-valued random process $X(t)$. We define the *mean* of the process $X(t)$ as

$$m_X(t_k) = E[X(t_k)] \tag{5.45}$$

where E denotes the expectation operator, and $X(t_k)$ is the random variable obtained by observing the random process $X(t)$ at time t_k. Denoting the probability density function of this random variable by $f_{X(t_k)}(x)$, we may rewrite Eq. (5.45) as

$$m_X(t_k) = \int_{-\infty}^{\infty} x f_{X(t_k)}(x)\,dx \tag{5.46}$$

We define the *autocorrelation function* of the random process $X(t)$ to be a function of two time variables t_k and t_i, as shown by

$$R_X(t_k, t_i) = E[X(t_k)X(t_i)] \tag{5.47}$$

where $X(t_k)$ and $X(t_i)$ are random variables obtained by observing the random process $X(t)$ at times t_k and t_i, respectively. Denoting the joint probability density function of these two random variables by $f_{X(t_k),X(t_i)}(x, y)$, we may rewrite Eq. (5.47) as

$$R_X(t_k, t_i) = \int_{-\infty}^{\infty} \int_{-\infty}^{\infty} xy f_{X(t_k),X(t_i)}(x, y) dx\, dy \tag{5.48}$$

The *autocovariance function* of the random process $X(t)$ is defined as

$$K_X(t_k, t_i) = E[(X(t_k) - m_X(t_k))(X(t_i) - m_X(t_i))] \tag{5.49}$$

This may be expanded to yield a useful relation between the mean, autocorrelation and autocovariance functions, namely,

$$K_X(t_k, t_i) = R_X(t_k, t_i) - m_X(t_k) m_X(t_i) \tag{5.50}$$

If the random process $X(t)$ has zero mean, then $K_x(t_k, t_i) = R_x(t_k, t_i)$; otherwise, they are unequal.

For a strictly stationary process, all three quantities described above take on simpler forms. In particular, we find that the mean of the random process is a constant m_X (say), so that we may write

$$m_X(t_k) = m_X, \qquad \text{for all } t_k \tag{5.51}$$

Also we find that both the autocorrelation and autocovariance functions depend only on the *time difference* $t_k - t_i$, as shown by:

$$R_X(t_k, t_i) = R_X(t_k - t_i) \tag{5.52}$$

and

$$K_X(t_k, t_i) = K_X(t_k - t_i) \tag{5.53}$$

The conditions of Eq. (5.51), (5.52) and (5.53) are *not* sufficient to guarantee that the random process $X(t)$ is strictly stationary. However, a random process $X(t)$ which is not strictly stationary but for which these conditions hold is said to be *wide-sense stationary* or *stationary in the wide sense.* Thus wide-sense stationarity represents a *weak* kind of stationarity in that all strictly stationary processes are also wide-sense stationary, but the converse is not necessarily true.

For convenience of notation, we define the autocorrelation function of a stationary process $X(t)$ as

$$R_X(\tau) = E[X(t+\tau)X(t)] \tag{5.54}$$

This autocorrelation function has several important properties:

1. The mean-square value of the process may be obtained from $R_X(\tau)$ simply by putting $\tau = 0$ in Eq. (5.54) as shown by

$$R_X(0) = E[X^2(t)] \tag{5.55}$$

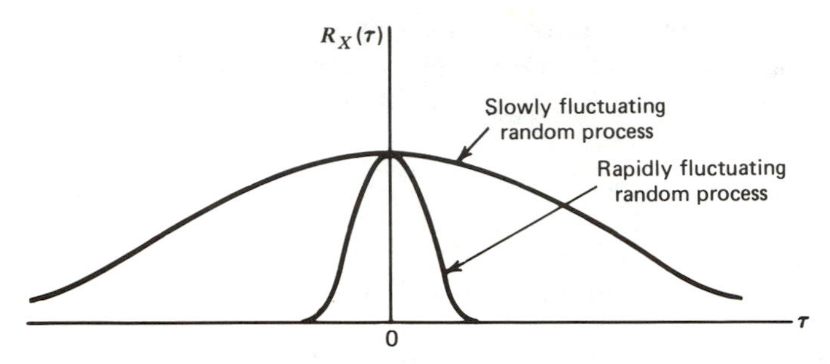

Figure 5.5 Illustrating the autocorrelation functions of slowly and rapidly fluctuating random processes.

2. The autocorrelation function $R_X(\tau)$ is an even function of τ, that is,

$$R_X(\tau) = R_X(-\tau) \tag{5.56}$$

3. The autocorrelation function $R_X(\tau)$ has its maximum magnitude at $\tau = 0$, that is,

$$|R_X(\tau)| \leqslant R_X(0) \tag{5.57}$$

The physical significance of the autocorrelation function $R_X(\tau)$ is that it provides a means of describing the interdependence of two random variables obtained by observing a random process $X(t)$ at times τ seconds apart. It is therefore apparent that the more rapidly the random process $X(t)$ changes with time, the more rapidly will the autocorrelation function $R_X(\tau)$ decrease from its maximum $R_X(0)$ as τ increases, as illustrated in Fig. 5.5. This decrease may be characterized by a *decorrelation time* τ_0, such that for $\tau > \tau_0$, the magnitude of the autocorrelation function $R_X(\tau)$ remains below some prescribed value. We may thus define the decorrelation time τ_0 of a wide-sense stationary random process $X(t)$ of zero mean as the time taken for the magnitude of the autocorrelation function $R_X(\tau)$ to decrease to 1 percent of its maximum value $R_X(0)$.

Example 3 Sine wave with random phase

Consider a sinusoidal signal with random phase, defined by

$$X(t) = A \cos(2\pi f_c t + \Theta) \tag{5.58}$$

where A and f_c are constants, and Θ is a random variable that is *uniformly distributed* over a range of 0 to 2π, that is,

$$f_\Theta(\theta) = \begin{cases} \dfrac{1}{2\pi}, & 0 \leqslant \theta \leqslant 2\pi \\ 0, & \text{elsewhere} \end{cases} \tag{5.59}$$

This means that the random variable Θ is equally likely to have any value in the range 0 to 2π. The autocorrelation function of $X(t)$ is

$$\begin{aligned}R_X(\tau) &= E[X(t+\tau)X(t)] \\ &= E[A^2 \cos(2\pi f_c t + 2\pi f_c \tau + \Theta)\cos(2\pi f_c t + \Theta)] \\ &= \frac{A^2}{2} E[\cos(4\pi f_c t + 2\pi f_c \tau + 2\Theta)] + \frac{A^2}{2} E[\cos(2\pi f_c \tau)] \\ &= \frac{A^2}{2} \int_0^{2\pi} \frac{1}{2\pi} \cos(4\pi f_c t + 2\pi f_c \tau + 2\theta) d\theta + \frac{A^2}{2} \cos(2\pi f_c \tau)\end{aligned}$$

The first term integrates to zero, and so we get

$$R_X(\tau) = \frac{A^2}{2} \cos(2\pi f_c \tau) \tag{5.60}$$

which is shown plotted in Fig. 5.6. We see, therefore, that the autocorrelation function of a sinusoidal wave with random phase is another sinusoid at the same frequency in the "τ-domain" rather than the time domain.

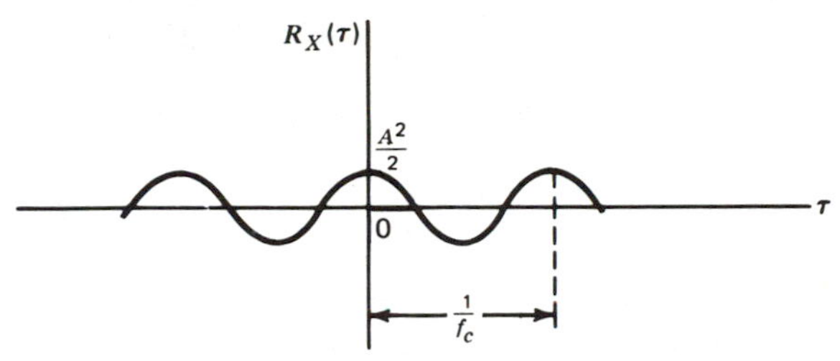

Figure 5.6 Autocorrelation function of a sine wave with random phase.

Example 4 Random binary wave

Figure 5.7 shows the sample function $x(t)$ of a process $X(t)$ consisting of a random sequence of *binary symbols* 1 and 0. It is assumed that:

1. The symbols 1 and 0 are represented by pulses of amplitude $+A$ and $-A$ volts, respectively, and duration T seconds.

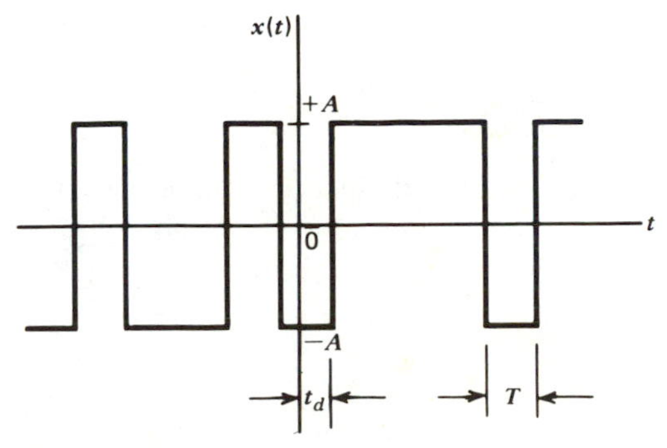

Figure 5.7 Sample function of random binary wave.

2. The pulses are not synchronized, so that the starting time of the first pulse, t_d, is equally likely to lie anywhere between zero and T seconds. That is, t_d is the sample value of a uniformly distributed random variable T_d, with its probability density function defined by

$$f_{T_d}(t_d) = \begin{cases} \dfrac{1}{T}, & 0 \leq t_d \leq T \\ 0, & \text{elsewhere} \end{cases} \tag{5.61}$$

3. During any time interval $(n-1)T < t - t_d < nT$, where n is an integer, the presence of a 1 or a 0 is determined by tossing a fair coin; specifically, if the outcome is "heads," we have a 1 and if the outcome is "tails," we have a 0. These two symbols are thus equally likely, and the presence of a 1 or 0 in any one interval is independent of all other intervals.

Since the amplitude levels $-A$ and $+A$ occur with equal probability, it follows immediately that $E[X(t)] = 0$, for all t, and the mean of the process is therefore zero.

To find the autocorrelation function $R_X(t_k, t_i)$, we have to evaluate $E[X(t_k)X(t_i)]$, where $X(t_k)$ and $X(t_i)$ are random variables obtained by observing the random process $X(t)$ at times t_k and t_i, respectively.

Consider first the case when $|t_k - t_i| > T$. Then the random variables $X(t_k)$ and $X(t_i)$ occur in different pulse intervals and are therefore independent. We thus have

$$E[X(t_k)X(t_i)] = E[X(t_k)]E[X(t_i)] = 0, \qquad |t_k - t_i| > T.$$

Consider next the case when $|t_k - t_i| < T$, with $t_k = 0$ and $t_i < t_k$. In such a situation we observe from Fig. 5.7 that the random variables $X(t_k)$ and $X(t_i)$ occur in the same pulse interval if and only if the delay t_d satisfies the condition $t_d < T - |t_k - t_i|$. We thus obtain the *conditional expectation*:

$$E[X(t_k)X(t_i)|t_d] = \begin{cases} A^2, & t_d < T - |t_k - t_i| \\ 0, & \text{elsewhere} \end{cases}$$

Averaging this result over all possible values of t_d, we get

$$\begin{aligned} E[X(t_k)X(t_i)] &= \int_0^{T-|t_k - t_i|} A^2 f_{T_d}(t_d)\,dt_d \\ &= \int_0^{T-|t_k - t_i|} \frac{A^2}{T}\,dt_d \\ &= A^2\left(1 - \frac{|t_k - t_i|}{T}\right), \qquad |t_k - t_i| < T \end{aligned}$$

By similar reasoning for any other value of t_k, we conclude that the autocorrelation function of a random binary wave, represented by the sample function shown in Fig. 5.7, is only a function of the time difference $\tau = t_k - t_i$, as shown by

$$R_X(\tau) = \begin{cases} A^2\left(1 - \dfrac{|\tau|}{T}\right), & |\tau| < T \\ 0, & |\tau| \geq T \end{cases} \tag{5.62}$$

This result is shown plotted in Fig. 5.8.

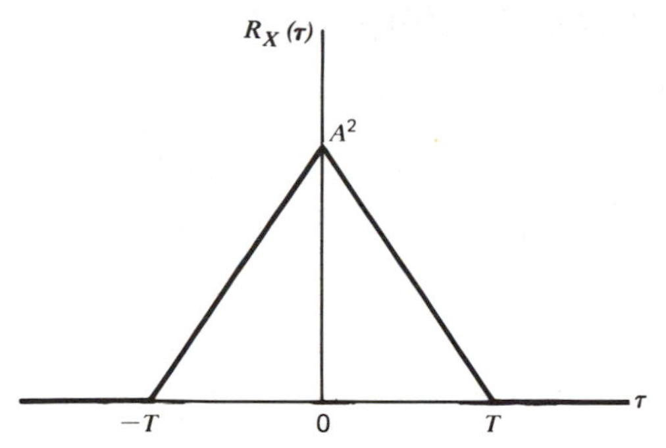

Figure 5.8 Autocorrelation function of random binary wave.

Cross-correlation Functions

Consider two random processes $X(t)$ and $Y(t)$ with autocorrelation functions $R_X(t, u)$ and $R_Y(t, u)$, respectively. The two *cross-correlation functions* of $X(t)$ and $Y(t)$ may be defined by

$$R_{XY}(t, u) = E[X(t)Y(u)]$$

and (5.63)

$$R_{YX}(t, u) = E[Y(t)X(u)]$$

where t and u denote two values of time at which the processes are observed. In this case, the correlation properties of the two random process $X(t)$ and $Y(t)$ may be displayed conveniently in matrix form as follows:

$$\mathbf{R}(t, u) = \begin{bmatrix} R_X(t, u) & R_{XY}(t, u) \\ R_{YX}(t, u) & R_Y(t, u) \end{bmatrix} \tag{5.64}$$

which is called the *correlation matrix* of the random processes $X(t)$ and $Y(t)$. If the correlation matrix can be written as

$$\mathbf{R}(t-u) = \begin{bmatrix} R_X(t-u) & R_{XY}(t-u) \\ R_{YX}(t-u) & R_Y(t-u) \end{bmatrix}, \tag{5.65}$$

then the random processes $X(t)$ and $Y(t)$ are each wide-sense stationary and, in addition, they are said to be jointly wide-sense stationary.

The cross-correlation function is not generally an even function of τ as was true for the autocorrelation function, nor does it have a maximum at the origin. However, it does obey a certain symmetry relationship as follows (see Problem 5.11):

$$R_{XY}(\tau) = R_{YX}(-\tau) \tag{5.66}$$

5.6 TIME AVERAGES AND ERGODICITY

If the theory of random processes is to be useful as a method for describing communication systems, we have to be able to estimate from observations of a random process $X(t)$ such probabilistic quantities as the mean and autocorrelation function of the process. For a stationary process, the mean is defined by

$$m_X = E[X(t)] = \int_{-\infty}^{\infty} x f_{X(t)}(x) dx$$

and the autocorrelation function is defined by

$$R_X(\tau) = E[X(t+\tau)X(t)] = \int_{-\infty}^{\infty} xy f_{X(t+\tau),X(t)}(x, y) dx\, dy$$

To compute m_X and $R_X(\tau)$ by *ensemble averaging*, as defined above, we have to average across all the sample functions of the process. In particular, this computation requires complete knowledge of the first-order and second-order joint probability density functions of the process. In many practical situations, however, these probability density functions are not available. Indeed, the only thing that we may usually find available is the recording of one (or at best, a small number) of sample functions of the random process. It seems natural then to consider also *time averages* of individual sample functions of the process.

We define the *time-averaged mean value* of the sample function $x(t)$ of a random process $X(t)$ as

$$\langle x(t) \rangle = \lim_{T \to \infty} \frac{1}{2T} \int_{-T}^{T} x(t) dt \tag{5.67}$$

where the symbol $\langle \ \rangle$ denotes *time-averaging*. In a similar way, we may define the *time-averaged autocorrelation function* of the sample function $x(t)$ as

$$\begin{aligned}\langle x(t+\tau)x(t) \rangle &= \lim_{T \to \infty} \frac{1}{2T} \int_{-T}^{T} x(t+\tau)x(t) dt \\ &= \lim_{T \to \infty} \frac{1}{2T} \int_{-T}^{T} x(t)x(t-\tau) dt \end{aligned} \tag{5.68}$$

The definitions given in Eqs. (2.5) and (2.152) for the mean value and autocorrelation function of periodic signals (when they are real-valued) may be viewed as special cases of Eqs. (5.67) and (5.68), respectively.

It is important to observe that the time averages $\langle x(t) \rangle$ and $\langle x(t+\tau)x(t) \rangle$ are random variables in that their values depend on which sample function of the random process $X(t)$ is used in the time-averaging evaluations. On the other hand, m_X is a constant, and $R_X(\tau)$ is an ordinary function of the variable τ.

In general, ensemble averages and time averages are not equal except for a very special class of random processes known as *ergodic processes.** *A random process $X(t)$ is said to be ergodic in the most general form if all of its statistical properties can be determined (with probability one) from a sample function representing one possible realization of the process.* We note here that it is necessary for a random process to be stationary in the strict sense for it to be ergodic. However, the converse is not always true; that is, not all stationary processes are ergodic.

Usually, we are not interested in estimating all the ensemble averages of a random process but rather only certain averages such as the mean and the autocorrelation function of the process. Accordingly, we may define ergodicity in a more limited sense, as described below:

(1) Ergodicity of the Mean

The time average $(1/2T) \int_{-T}^{T} x(t)dt$ is a random variable with a mean and variance of its own. For a stationary process, we find that its mean value is equal to

$$\begin{aligned} E\left[\frac{1}{2T}\int_{-T}^{T} x(t)dt\right] &= \frac{1}{2T}\int_{-T}^{T} E[x(t)]dt \\ &= \frac{1}{2T}\int_{-T}^{T} m_X dt \\ &= m_X \end{aligned} \tag{5.69}$$

Therefore, this time average provides an *unbiased estimate* of m_X. An estimator is said to be unbiased if the expected value of the estimate is exactly the same as the true value of the pertinent parameter. We say that the random process $X(t)$ is *ergodic in the mean* if

$$\lim_{T\to\infty} \frac{1}{2T}\int_{-T}^{T} x(t)dt = m_X \tag{5.70}$$

with probability one. That is, for a random process to be ergodic in the mean, its time-averaged and ensemble-averaged mean values must be equal with probability one. The necessary and sufficient condition for the ergodicity of the mean

* The problem of determining conditions under which time averages computed from a sample function of a random process can be ultimately identified with corresponding ensemble averages first arose in statistical mechanics. Physical systems possessing properties of this kind were called *ergodic* by L. Boltzmann in 1887. The word "ergodic" is of Greek origin; see D. ter Haar, *Elements of Statistical Mechanics*, p. 356 (Holt, Rinehart and Winston, 1954.)

is that the variance of the estimator $(1/2T)\int_{-T}^{T} x(t)dt$ approach zero as T approaches infinity.*

(2) Ergodicity of the Autocorrelation Function

Consider next the time average $(1/2T)\int_{-T}^{T} x(t+\tau)x(t)dt$, which is also a random variable with a mean and variance. Its mean value is equal to

$$
\begin{aligned}
E\left[\frac{1}{2T}\int_{-T}^{T} x(t+\tau)x(t)dt\right] &= \frac{1}{2T}\int_{-T}^{T} E[x(t+\tau)x(t)]dt \\
&= \frac{1}{2T}\int_{-T}^{T} R_X(\tau)dt \\
&= R_X(\tau)
\end{aligned}
\tag{5.71}
$$

Accordingly, this time average provides an unbiased estimate of the ensemble-averaged autocorrelation function $R_X(\tau)$ of the random process $X(t)$. We say that the random process $X(t)$ is *ergodic in the autocorrelation function* if

$$
\lim_{T\to\infty} \frac{1}{2T}\int_{-T}^{T} x(t+\tau)x(t)dt = R_X(\tau) \tag{5.72}
$$

with probability one. The necessary and sufficient condition for the ergodicity of the autocorrelation function is that the variance of the estimator $(1/2T)\int_{-T}^{T} x(t)x(t+\tau)dt$ approach zero as T approaches infinity.

In order to test for the ergodicity of the mean, it suffices to know the mean m_X and autocorrelation function $R_X(\tau)$ of the process. However, to test for the ergodicity of the autocorrelation function, we have to know fourth-order moments of the process.† Therefore, except for certain simple cases, it is usually very difficult to establish if a random process meets the conditions for the ergodicity of the mean

* If we have a random variable X such that its mean-square value is zero, then this random variable equals zero *with probability one*. In other words, if $E[X^2]=0$, then the probability $P(X\neq 0)=0$. Suppose that this is not true. We can then find an $\varepsilon>0$ such that

$$
P(|X|>\varepsilon) = \int_{|x|>\varepsilon} f_X(x)dx \neq 0
$$

where $f_X((x)$ is the probability density function of the random variable X. However, using the definition of the mean-square of a random variable, we may write

$$
E[X^2] = \int_{-\infty}^{\infty} x^2 f_X(x)dx \geq \int_{|X|>\varepsilon} x^2 f_X(x)dx \geq \varepsilon^2 \int_{|X|>\varepsilon} f_X(x)dx > 0
$$

Clearly, this is wrong. Hence, the statement $P[X\neq 0]=0$ is true.

Similarly, if we have a random variable X such that $E[(X-m_X)^2]=0$, then $X=m_X$ with probability one. That is, if the variance of a random variable X is zero, then X is a constant equal to the mean value m_X for almost all outcomes of the pertinent experiment.

† For a derivation of the conditions for the ergodicity of the mean and autocorrelation function, see:

1. A. Papoulis, pp. 327–332, op. cit.
2. A. Papoulis, *Signal Analysis*, pp. 352–363, (McGraw-Hill, 1977).

and autocorrelation function. Thus, in practice, we are usually forced to consider the physical origin of the random process, and thereby make a somewhat intuitive judgment as to whether it is reasonable to interchange time and ensemble averages.

Example 5 Sine wave with random phase (continued)

Consider again the random process $X(t)$, defined by

$$X(t) = A\cos(2\pi f_c t + \Theta)$$

where A and f_c are constants and Θ is a uniformly distributed random variable:

$$f_\Theta(\theta) = \begin{cases} \dfrac{1}{2\pi}, & 0 \leqslant \theta \leqslant 2\pi \\ 0, & \text{elsewhere} \end{cases}$$

The mean of this random process is

$$\begin{aligned} m_X &= \int_{-\infty}^{\infty} A\cos(2\pi f_c t + \theta) f_\Theta(\theta)\,d\theta \\ &= \int_0^{2\pi} \frac{A}{2\pi}\cos(2\pi f_c t + \theta)\,d\theta \\ &= 0 \end{aligned}$$

The autocorrelation function of the process was determined in Example 3; the result is reproduced here for convenience

$$R_X(\tau) = \frac{A^2}{2}\cos(2\pi f_c \tau)$$

Let $x(t)$ denote a sample function of the process; thus

$$x(t) = A\cos(2\pi f_c t + \theta)$$

The time-averaged mean and time-averaged autocorrelation function of the process are as follows, respectively,

$$\begin{aligned} \langle x(t) \rangle &= \lim_{T\to\infty} \frac{1}{2T} \int_{-T}^{T} A\cos(2\pi f_c t + \theta)\,dt \\ &= 0 \end{aligned}$$

$$\begin{aligned} \langle x(t+\tau)x(t) \rangle &= \lim_{T\to\infty} \frac{1}{2T} \int_{-T}^{T} A^2 \cos(2\pi f_c t + 2\pi f_c \tau + \theta)\cos(2\pi f_c t + \theta)\,dt \\ &= \lim_{T\to\infty} \frac{A^2}{4T} \int_{-T}^{T} [\cos(2\pi f_c \tau) + \cos(4\pi f_c t + 2\pi f_c \tau + 2\theta)]\,dt \\ &= \frac{A^2}{2}\cos(2\pi f_c \tau) \end{aligned}$$

which are exactly the same as the corresponding ensemble averages. This random process is therefore ergodic in both the mean and autocorrelation function.

Measurement of the Mean and Autocorrelation Function

When we are satisfied that it is reasonable to interchange time and ensemble averages, we may use Eqs. (5.67) and (5.68) to devise practical procedures for measuring the mean and autocorrelation function of a random process, respectively. Specifically, to measure the mean, we simply pass a *finite record* of the sample function of the process through an integrator, giving an estimate of the mean value of the process at its output.

However, the estimation of the autocorrelation function $R_X(\tau)$ by using a time average of a finite record presents certain problems.* Suppose that the sample function $x(t)$ of the process is available only for $|t|<T$. If $\tau>0$, then $t-\tau<-T$ for values of $t<-T+\tau$. In this situation, the product $x(t)x(t-\tau)$ can be integrated only in the interval from $-T+\tau$ to T. Thus in place of the time average defined in Eq. (5.68) for all values of τ, we have at our disposal the following time average

$$\frac{1}{2T-\tau}\int_{-T+\tau}^{T} x(t)x(t-\tau)dt, \qquad \tau>0 \tag{5.73}$$

whose expected value equals the autocorrelation function $R_X(\tau)$. Based on this time average, we may use the scheme shown in Fig. 5.9 to measure the autocorrelation function of the process. A finite record of the sample function $x(t)$ is split into two parts, one of which is passed through an adjustable delay network to produce a time delay τ. The two parts are then fed to a multiplier, followed by an integrator. An estimate of the autocorrelation function $R_X(\tau)$ is thereby obtained at the integrator output for the specific value of τ determined by the delay network.

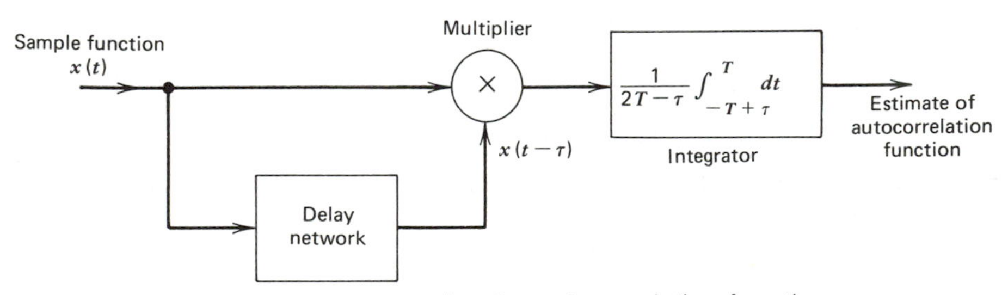

Figure 5.9 A scheme for measuring the autocorrelation function.

5.7 TRANSMISSION OF A RANDOM PROCESS THROUGH A LINEAR FILTER

Suppose that a random process $X(t)$ is applied as input to a linear time-invariant filter of impulse response $h(t)$, producing a random process $Y(t)$ at the filter output,

* 1. R. B. Blackman and J. W. Tukey, *The Measurement of Power Spectra*, p. 11 (Dover Publications, 1958).

2. G. M. Jenkins and D. G. Watts, *Spectral Analysis and Its Applications*, p. 174 (Holden-Day, 1969).

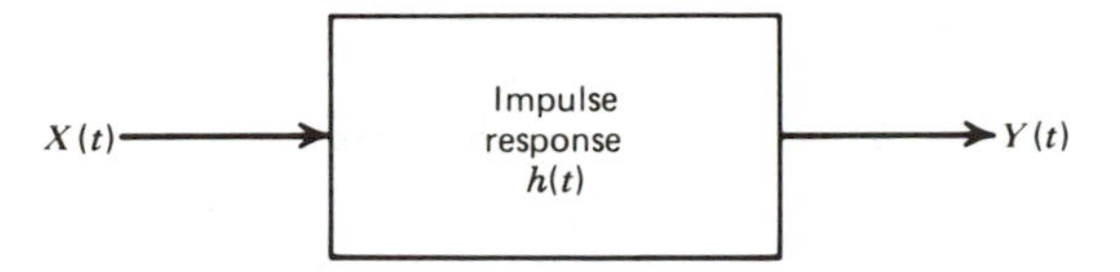

Figure 5.10 Transmission of a random process through a linear filter.

as in Fig. 5.10. In general, it is difficult to describe the probability distribution of the output random process $Y(t)$, even when the probability distribution of the input random process $X(t)$ is completely specified for $-\infty < t < \infty$.

In this section, we wish to determine the time-domain form of the input–output relations of the filter for defining the mean and autocorrelation functions of the output random process $Y(t)$ in terms of those of the input $X(t)$, assuming that $X(t)$ is a wide-sense stationary random process.

Consider first the mean of the output random process $Y(t)$. By definition, we have

$$m_Y(t) = E[Y(t)] = E\left[\int_{-\infty}^{\infty} h(\tau)X(t-\tau)d\tau\right] \tag{5.74}$$

Provided that the expectation $E[X(t)]$ is finite for all t, and the system is stable, we may interchange the order of the expectation and the integration with respect to τ in Eq. (5.74), and so write

$$\begin{aligned} m_Y(t) &= \int_{-\infty}^{\infty} h(\tau)E[X(t-\tau)]d\tau \\ &= \int_{-\infty}^{\infty} h(\tau)m_X(t-\tau)d\tau \end{aligned} \tag{5.75}$$

When the input random process $X(t)$ is wide-sense stationary, the mean $m_X(t)$ is a constant m_X, so that we may simplify Eq. (5.75) as follows

$$\begin{aligned} m_Y &= m_X \int_{-\infty}^{\infty} h(\tau)d\tau \\ &= m_X H(0) \end{aligned} \tag{5.76}$$

where $H(0)$ is the zero-frequency response of the system.

Consider next the autocorrelation function of the output random process $Y(t)$. By definition, we have

$$R_Y(t, u) = E[Y(t)Y(u)]$$

where t and u denote two values of the time at which the output process is observed. We may therefore use the convolution integral to write

$$R_Y(t, u) = E\left[\int_{-\infty}^{\infty} h(\tau_1)X(t-\tau_1)d\tau_1 \int_{-\infty}^{\infty} h(\tau_2)X(u-\tau_2)d\tau_2\right] \tag{5.77}$$

Here again, provided that $E[X^2(t)]$ is finite for all t and the system is stable, we may interchange the order of the expectation and the integrations with respect to τ_1

and τ_2 in Eq. (5.77), obtaining

$$R_Y(t, u) = \int_{-\infty}^{\infty} d\tau_1 h(\tau_1) \int_{-\infty}^{\infty} d\tau_2 h(\tau_2) E[X(t-\tau_1)X(u-\tau_2)]$$

$$= \int_{-\infty}^{\infty} d\tau_1 h(\tau_1) \int_{-\infty}^{\infty} d\tau_2 h(\tau_2) R_X(t-\tau_1, u-\tau_2) \tag{5.78}$$

When the input $X(t)$ is a wide-sense stationary random process, the autocorrelation function of $X(t)$ is only a function of the difference between the observation times $t-\tau_1$ and $u-\tau_2$. Thus, putting $\tau = t-u$ in Eq. (5.78), we may write

$$R_Y(\tau) = \int_{-\infty}^{\infty} \int_{-\infty}^{\infty} h(\tau_1)h(\tau_2)R_X(\tau-\tau_1+\tau_2)d\tau_1\, d\tau_2 \tag{5.79}$$

On combining this result with that involving the mean m_Y, we see that if the input to a stable linear time-invariant filter is a wide-sense stationary random process, then the output of the filter is also a wide-sense stationary random process.

Since $R_Y(0) = E[Y^2(t)]$, it follows that the mean-square value of the output random process $Y(t)$ is obtained by putting $\tau = 0$ in Eq. (5.79). We thus get the result:

$$E[Y^2(t)] = \int_{-\infty}^{\infty} \int_{-\infty}^{\infty} h(\tau_1)h(\tau_2)R_X(\tau_2-\tau_1)d\tau_1\, d\tau_2 \tag{5.80}$$

which is a constant.

5.8 POWER SPECTRAL DENSITY

Thus far we have considered the characterization of wide-sense stationary random processes in linear systems in the time domain. We turn next to the characterization of random processes in linear systems by using frequency-domain ideas. In particular, we wish to derive the frequency-domain equivalent to the result of Eq. (5.80) defining the mean-square value of the filter output.

By definition, the impulse response of a linear time-invariant filter is equal to the inverse Fourier transform of the transfer function of the system. We may thus write

$$h(\tau_1) = \int_{-\infty}^{\infty} H(f)\exp(j2\pi f\tau_1)df \tag{5.81}$$

Substituting this expression for $h(\tau_1)$ in Eq. (5.80), we get

$$E[Y^2(t)] = \int_{-\infty}^{\infty} \int_{-\infty}^{\infty} \left[\int_{-\infty}^{\infty} H(f)\exp(j2\pi f\tau_1)df\right] h(\tau_2)R_X(\tau_2-\tau_1)d\tau_1\, d\tau_2$$

$$= \int_{-\infty}^{\infty} df\, H(f) \int_{-\infty}^{\infty} d\tau_2 h(\tau_2) \int_{-\infty}^{\infty} R_X(\tau_2-\tau_1)\exp(j2\pi f\tau_1)d\tau_1 \tag{5.82}$$

In the last integral on the right-hand side of Eq. (5.82), define a new variable

$$\tau = \tau_2 - \tau_1$$

Then we may rewrite Eq. (5.82) in the form

$$E[Y^2(t)] = \int_{-\infty}^{\infty} df\, H(f) \int_{-\infty}^{\infty} d\tau_2 h(\tau_2) \exp(j2\pi f \tau_2) \int_{-\infty}^{\infty} R_X(\tau) \exp(-j2\pi f \tau) d\tau \quad (5.83)$$

However, the middle integral on the right-hand side in Eq. (5.83) is simply the complex conjugate $H^*(f)$ of the transfer function $H(f)$ of the filter, and so we may rewrite this equation in the form

$$E[Y^2(t)] = \int_{-\infty}^{\infty} df\, |H(f)|^2 \int_{-\infty}^{\infty} R_X(\tau) \exp(-j2\pi f \tau) d\tau \quad (5.84)$$

We may further simplify Eq. (5.84) by recognizing that the last integral is simply the Fourier transform of the autocorrelation function $R_X(\tau)$ of the input random process $X(t)$. Let this transform be denoted by $S_X(f)$:

$$S_X(f) = \int_{-\infty}^{\infty} R_X(\tau) \exp(-j2\pi f \tau) d\tau \quad (5.85)$$

The function $S_X(f)$ is called the *power spectral density* or *power spectrum* of the wide-sense stationary random process $X(t)$. Thus substituting Eq. (5.85) in (5.84), we obtain the desired relation

$$E[Y^2(t)] = \int_{-\infty}^{\infty} |H(f)|^2 S_X(f) df \quad (5.86)$$

Equation (5.86) *states that the mean-square value of the output of a stable linear time-invariant filter in response to a wide-sense stationary random process is equal to the integral over all frequencies of the power spectral density of the input random process multiplied by the squared magnitude of the transfer function of the filter.* This is the desired frequency-domain equivalent to the time-domain relation of Eq. (5.80).

To investigate the physical significance of the power spectral density, suppose that the random process $X(t)$ is passed through an ideal narrow-band filter with an amplitude response defined by (see Fig. 5.11)

$$|H(f)| = \begin{cases} 1, & |f \pm f_c| < \frac{1}{2}\Delta f \\ 0, & |f \pm f_c| > \frac{1}{2}\Delta f \end{cases} \quad (5.87)$$

Then from Eq. (5.86) we find that if the filter bandwidth Δf is sufficiently small and $S_X(f)$ is a continuous function, the mean-square value of the filter output is approximately

$$E[Y^2(t)] \simeq (2\Delta f) S_X(f_c) \quad (5.88)$$

The filter, however, passes only those frequency components of the input random process $X(t)$ that lie inside a narrow frequency band of width Δf centered about the frequency $\pm f_c$. Thus $S_X(f_c)$ represents the frequency density of the average power

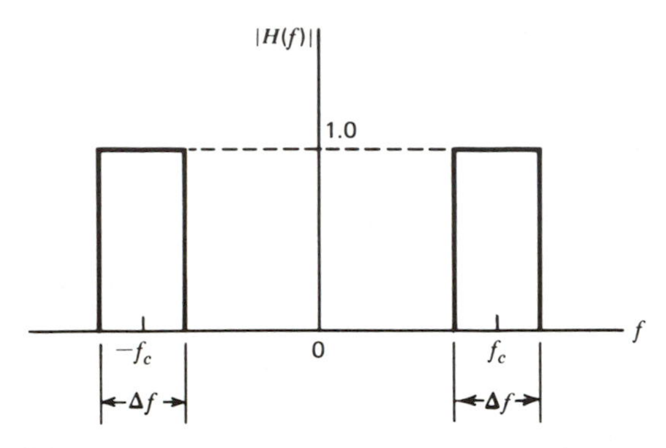

Figure 5.11 Amplitude response of ideal narrow-band filter.

in the random process $X(t)$, evaluated at the frequency $f=f_c$. The dimensions of the power spectral density are therefore in watts per hertz.

Properties of the Power Spectral Density

The power spectral density $S_X(f)$ and the autocorrelation function $R_X(\tau)$ of a wide-sense stationary random process $X(t)$ form a Fourier transform pair, as shown by the pair of relations:

$$S_X(f)=\int_{-\infty}^{\infty} R_X(\tau)\exp(-j2\pi f\tau)d\tau \tag{5.89}$$

$$R_X(\tau)=\int_{-\infty}^{\infty} S_X(f)\exp(j2\pi f\tau)df \tag{5.90}$$

Equations (5.89) and (5.90) are basic relations in the theory of spectral analysis of random processes, and together they constitute what are usually called the *Wiener–Khintchine relations.** As mentioned in Chapter 2, this pair of relations also apply to periodic signals [see Eqs. (2.159) and (2.160)].

The Wiener–Khintchine relations show that if either the autocorrelation function or power spectral density of a random process is known, the other can be found exactly. However, these functions display different aspects of the correlation information about the process. It is commonly accepted that for practical purposes, however, the power spectral density is the more useful "parameter."

We now wish to use this pair of relations to derive some general properties of the power spectral density of a wide-sense stationary process:

* Equations (5.89) and (5.90) are called the Wiener–Khintchine relations in commemoration of the pioneering work of Wiener and Khintchine: N. Wiener, "Generalized harmonic analysis," *Acta. Math.*, vol. 55, p. 117, 1930. A. I. Khintchine, "Korrelationstheorie der stationören stochastischen prozesse," *Math. Ann.*, vol. 109, pp. 415–458, 1934.

Property 1

The zero-frequency value of the power spectral density of a wide-sense stationary random process equals the total area under the graph of the autocorrelation function; that is,

$$S_X(0)=\int_{-\infty}^{\infty} R_X(\tau)d\tau \tag{5.91}$$

This property follows directly from Eq. (5.89) by putting $f=0$.

Property 2

The mean-square value of a wide-sense stationary random process equals the total area under the graph of the power spectral density; that is,

$$E[X^2(t)]=\int_{-\infty}^{\infty} S_X(f)df \tag{5.92}$$

This property follows directly from Eq. (5.90) by putting $\tau=0$, and noting that $R_X(0)=E[X^2(t)]$.

Property 3

The power spectral density of a wide-sense stationary random process is always nonnegative; that is,

$$S_X(f)\geqslant 0, \qquad \text{for all } f \tag{5.93}$$

This property is an immediate consequence of the fact that, in Eq. (5.88), the mean-square value $E[Y^2(t)]$ must always be nonnegative.

Property 4

The power spectral density of a real-valued random process is an even function of frequency; that is,

$$S_X(-f)=S_X(f) \tag{5.94}$$

This property is readily obtained by substituting $-f$ for f in Eq. (5.89):

$$S_X(-f)=\int_{-\infty}^{\infty} R_X(\tau)\exp(j2\pi f\tau)d\tau$$

Next, substituting $-\tau$ for τ, and recognizing that $R_X(-\tau)=R_X(\tau)$, we get

$$S_X(-f)=\int_{-\infty}^{\infty} R_X(\tau)\exp(-j2\pi f\tau)d\tau=S_X(f)$$

which is the desired result.

Example 6 Sine wave with random phase (continued)

Consider the random process $X(t)=A\cos(2\pi f_c t+\Theta)$, where the phase Θ is a uniformly distributed random variable over the range 0 to 2π. The autocorrelation function of this random process is given by Eq. (5.60) which is reproduced here for convenience:

$$R_X(\tau)=\frac{A^2}{2}\cos(2\pi f_c\tau)$$

Taking the Fourier transform of both sides of this relation, we find that the power spectral density of the random process $X(t)$ is

$$S_X(f)=\frac{A^2}{4}[\delta(f-f_c)+\delta(f+f_c)] \qquad (5.95)$$

which consists of a pair of delta functions weighted by the factor $A^2/4$ and located at $\pm f_c$, as in Fig. 5.12. We note that the total area under a delta function is one. Hence, the total area under the $S_X(f)$ of Eq. (5.95) is equal to $A^2/2$, as expected.

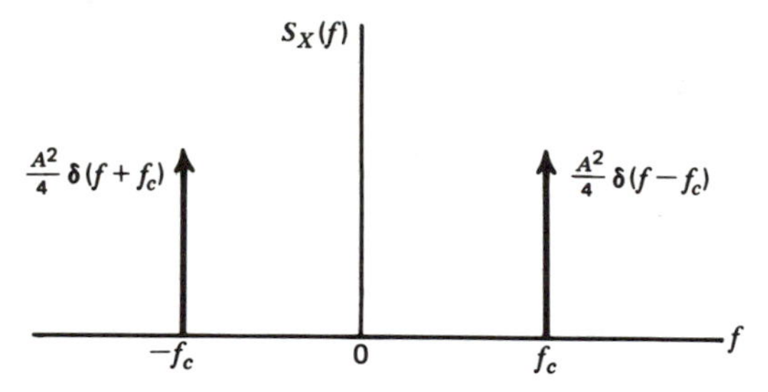

Figure 5.12 Power spectral density of sine wave with random phase.

Example 7 Random binary wave (continued)

Consider again a random binary wave consisting of a sequence of 1's and 0's represented by the values $+A$ and $-A$, respectively. In Example 4 we showed that the autocorrelation function of this random process equals

$$R_X(\tau)=\begin{cases} A^2\left(1-\dfrac{|\tau|}{T}\right), & |\tau|<T \\ 0, & |\tau|\geq T \end{cases}$$

The power spectral density of the process is therefore

$$S_X(f)=\int_{-T}^{T} A^2\left(1-\frac{|\tau|}{T}\right)\exp(-j2\pi f\tau)d\tau$$

Using the Fourier transform of a triangular function evaluated in Example 14 of Chapter 2, we obtain

$$S_X(f)=A^2T\,\mathrm{sinc}^2(fT) \qquad (5.96)$$

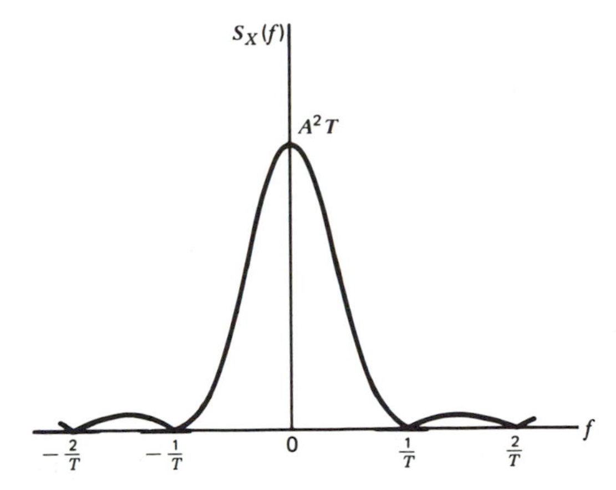

Figure 5.13 Power spectral density of random binary wave.

which is shown plotted in Fig. 5.13. Here again, we see that the power spectral density is non-negative for all f and that it is an even function of f. We note from Example 20, Chapter 2, that

$$\int_{-\infty}^{\infty} \text{sinc}^2(fT)df = \frac{1}{T}$$

Therefore, the total area under $S_X(f)$, or the average power of the random binary wave described above, is A^2.

The result of Eq. (5.96) may be generalized as follows. We note that the energy spectral density of a rectangular pulse $g(t)$ of amplitude A and duration T is given by

$$\Psi_g(f) = A^2T^2 \text{ sinc}^2(fT)$$

We may therefore rewrite Eq. (5.96) in terms of $\Psi_g(f)$ as follows

$$S_X(f) = \frac{\Psi_g(f)}{T} \tag{5.97}$$

Equation (5.97) states that, for a random binary wave in which binary symbols 1 and 0 are represented by pulses $g(t)$ and $-g(t)$, respectively, the power spectral density $S_X(f)$ is equal to the energy spectral density $\Psi_g(f)$ of the *symbol shaping pulse* $g(t)$ divided by the *symbol duration* T.

Example 8 Linear maximal sequences

There exists a class of deterministic sequences known as *maximum length sequences* with many of the properties of a random binary sequence and yet requiring simple instrumentation. A maximum-length sequence is a cyclic binary sequence generated by a *feedback shift register*, and has the longest possible period for this method of generation. A shift register of degree m is a device consisting of m consecutive 2-state memory stages regulated by a single timing clock. At each clock pulse, the state (represented by binary symbol 1 or 0) of each memory stage is shifted to the next stage down the line. In order to prevent the shift register from emptying by the end of m clock pulses, we use a logical (i.e., Boolean) function of the states of the m memory stages to compute a *feedback term*, and apply it to the first memory stage of the shift register.

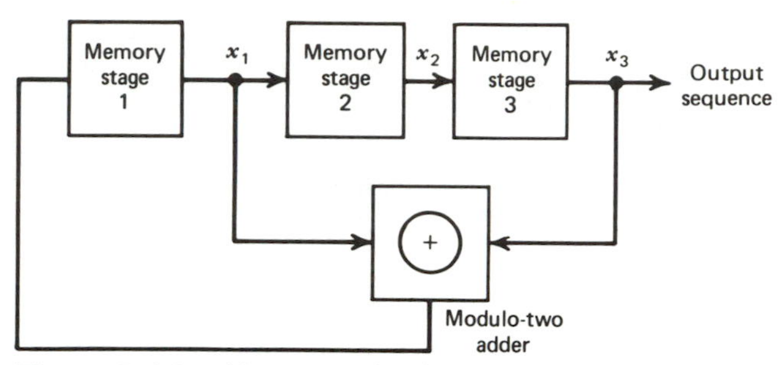

Figure 5.14 Linear-maximal-sequence generator.

The most important special form of this feedback shift register is the *linear* case in which the feedback function is obtained by using *modulo-two adders* to combine the outputs of the various memory stages. This operation is illustrated in Fig. 5.14 for the case of $m=3$. Representing the states of the three memory stages as x_1, x_2, and x_3, we see that in Fig. 5.14 the feedback function is equal to the modulo-two sum of x_1 and x_3.* A maximum length sequence generated by a feedback shift register using a linear feedback function is called a *linear maximal sequence*. This sequence is always periodic with a period defined by

$$N=2^m-1 \tag{5.98}$$

where m is the degree of the shift register. Assuming, for example, that the three memory stages of the shift register shown in Fig. 5.14 are in the initial states 0, 0, and 1, respectively, we find that the resulting output sequence is 1001110, repeating with period 7.

Representing the symbols 1 and 0 by the values $+A$ and $-A$, respectively, we find that the autocorrelation function of a linear maximal sequence is periodic with period NT, and that for the interval $-NT/2 \leqslant \tau \leqslant NT/2$ it is defined by

$$R_X(\tau)=\begin{cases} A^2\left(1-\dfrac{N+1}{NT}|\tau|\right), & |\tau|\leqslant T \\ -\dfrac{A^2}{N}, & \text{for the remainder of the period} \end{cases} \tag{5.99}$$

where T is the duration for which the symbol 1 or 0 is defined. This result is shown plotted in Fig. 5.15(a) for the case of $m=3$ or $N=7$. We thus see that the autocorrelation function of a linear maximal sequence is somewhat similar to that of a random binary sequence.

Linear maximal sequences are also referred to as *pseudo-random* or *pseudo-noise* (*PN*) *sequences*. The "random" comes from the fact that they have many of the randomness properties of a random binary sequence, specifically, the following:†

* In modulo-two addition, the sum of x_1 and x_3 takes the value 1 only when x_1 or x_3, but not both, takes the value 1. In other words, the carry is ignored. This operation is equivalent to the logical EXCLUSIVE OR.

† For further details of linear maximal sequences, see G. W. Golomb (Ed.), *Digital Communications with Space Applications*, pp. 1–32 (Prentice-Hall, 1964). M. P. Ristenbatt, "Pseudo-random binary coded waveforms" in the book entitled *Modern Radar*, edited by R. S. Berkowitz, pp. 274–313 (Wiley, 1965). See also the review paper: D. V. Sarwate and M. B. Pursley, "Crosscorrelation properties of pseudorandom and related sequences," Proc. IEEE, vol. 68, pp. 593–619, May 1980.

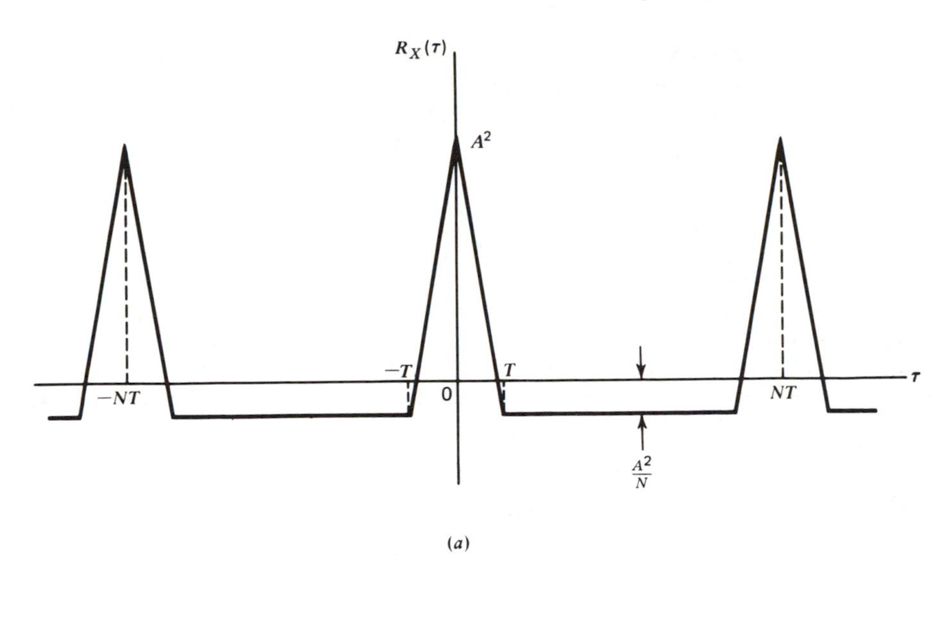

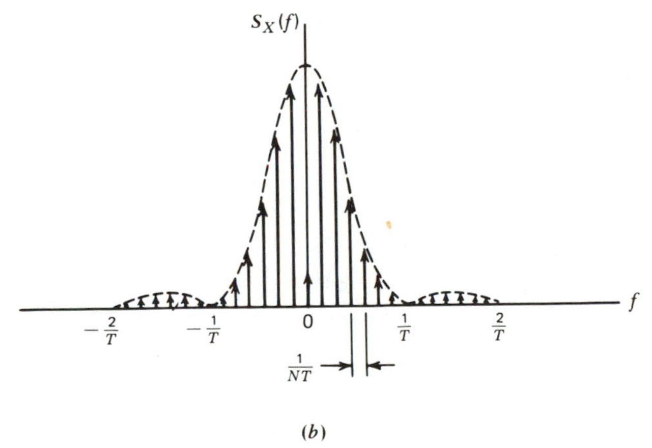

Figure 5.15 Characteristics of linear maximal sequence. *(a)* Autocorrelation function. *(b)* Power spectral density.

1. The number of 1's per period is always one more than the number of 0's.
2. In every period, half the *runs* (consecutive outputs of the same kind) are of length one, one-fourth of length two, one-eighth of length three, and so on, as long as the number of runs so indicated exceeds one.
3. The autocorrelation function is two-valued.

From Fig. 5.15(a), we note that the autocorrelation function of the sequence consists of a constant term equal to $-A^2/N$ plus a periodic train of triangular pulses of amplitude $A^2 + A^2/N$, pulse width $2T$ and period NT in the τ-domain. Therefore, using the results of Problem 2.25, we find that the power spectral density of a linear maximal sequence is given by

$$S_X(f) = -\frac{A^2}{N}\delta(f) + \frac{A^2}{N}\left(1+\frac{1}{N}\right)\sum_{n=-\infty}^{\infty}\text{sinc}^2\left(\frac{n}{N}\right)\delta\left(f-\frac{n}{NT}\right)$$

$$= \frac{A^2}{N^2}\delta(f) + A^2\left(\frac{1+N}{N^2}\right)\sum_{\substack{n=-\infty \\ n\neq 0}}^{\infty}\text{sinc}^2\left(\frac{n}{N}\right)\delta\left(f-\frac{n}{NT}\right) \tag{5.100}$$

which is shown plotted in Fig. 5.15(*b*) for the case of $m=3$ or $N=7$. Comparing this power spectral density characteristic with that of Fig. 5.13 for a random binary sequence, we see that they both have an envelope of the same form, namely, $\text{sinc}^2(fT)$, which depends only on the duration T. The fundamental difference, of course, is that whereas the random binary sequence has a continuous spectral density characteristic, the corresponding characteristic of a linear maximal sequence consists of delta functions spaced $1/NT$ hertz apart.

Example 9 Mixing of a random process with a sine wave

A situation that often arises in practice is that of *mixing* (i.e., multiplication) of a wide-sense stationary random process $X(t)$ with a sinusoidal wave $\cos(2\pi f_c t + \Theta)$, where the phase Θ is a random variable that is uniformly distributed over the interval 0 to 2π. The addition of the random phase Θ in this manner merely recognizes the fact that the time origin is arbitrarily chosen when $X(t)$ and $\cos(2\pi f_c t + \Theta)$ come from physically independent sources, as is usually the case. We are interested in determining the power spectral density of the random process $Y(t)$ defined by

$$Y(t) = X(t)\cos(2\pi f_c t + \Theta) \tag{5.101}$$

We note that the autocorrelation of $Y(t)$ is given by

$$\begin{aligned}
R_Y(\tau) &= E[Y(t+\tau)Y(t)] \\
&= E[X(t+\tau)\cos(2\pi f_c t + 2\pi f_c \tau + \Theta)X(t)\cos(2\pi f_c t + \Theta)] \\
&= E[X(t+\tau)X(t)]E[\cos(2\pi f_c t + 2\pi f_c \tau + \Theta)\cos(2\pi f_c t + \Theta)] \\
&= \tfrac{1}{2}R_X(\tau)E[\cos(2\pi f_c \tau) + \cos(4\pi f_c t + 2\pi f_c \tau + 2\Theta)] \\
&= \tfrac{1}{2}R_X(\tau)\cos(2\pi f_c \tau)
\end{aligned}$$

Because the power spectral density is the Fourier transform of the autocorrelation function, we find that the power spectral densities of the random process $X(t)$ and $Y(t)$ are related as follows:

$$S_Y(f) = \tfrac{1}{4}[S_X(f-f_c) + S_X(f+f_c)] \tag{5.102}$$

That is, to obtain the power spectral density of the random process $Y(t)$, we shift the given power spectral density $S_X(f)$ of random process $X(t)$ to the right by f_c, shift it to the left by f_c, add the two shifted power spectra, and divide the result by 4.

Relation Among the Power Spectral Densities of the Input and Output Random Processes

Let $S_Y(f)$ denote the power spectral density of the output random process $Y(t)$ obtained by passing the random process $X(t)$ through a linear filter of transfer function $H(f)$. Then, recognizing by definition that the power spectral density of a random process is equal to the Fourier transform of its autocorrelation function and using Eq. (5.79), we obtain

$$S_Y(f)=\int_{-\infty}^{\infty} R_Y(\tau)\exp(-j2\pi f\tau)d\tau$$

$$=\int_{-\infty}^{\infty}\int_{-\infty}^{\infty}\int_{-\infty}^{\infty} h(\tau_1)h(\tau_2)R_X(\tau-\tau_1+\tau_2)\exp(-j2\pi f\tau)d\tau_1\, d\tau_2\, d\tau \qquad (5.103)$$

Let $\tau-\tau_1+\tau_2=\tau_0$, or, equivalently, $\tau=\tau_0+\tau_1-\tau_2$. Then, by making this substitution in Eq. (5.103), we find that $S_Y(f)$ may be expressed as the product of three terms: the transfer function $H(f)$ of the filter, the complex conjugate of $H(f)$, and the power spectral density $S_X(f)$ of the input random process $X(t)$, as shown by

$$S_Y(f)=H(f)H^*(f)S_X(f) \qquad (5.104)$$

However, $|H(f)|^2=H(f)H^*(f)$. We thus find that the relationship among the power spectral densities of the input and output random processes is simply expressed in the frequency domain by writing

$$S_Y(f)=|H(f)|^2 S_X(f) \qquad (5.105)$$

That is, the output power spectral density equals the input power spectral density multiplied by the squared magnitude of the transfer function of the filter. By using this relation, we can therefore determine the effect of passing a random process through a linear filter.

Example 10 Comb filter

Consider the filter of Fig. 5.16(*a*) consisting of a delay line and a summing device. We wish to evaluate the power spectral density of the filter output $Y(t)$, given that the power spectral density of the filter input $X(t)$ is $S_X(f)$.

The transfer function of this filter is

$$\begin{aligned}H(f)&=1-\exp(-j2\pi fT)\\&=1-\cos(2\pi fT)+j\sin(2\pi fT)\end{aligned} \qquad (5.106)$$

The squared magnitude of $H(f)$ is

$$\begin{aligned}|H(f)|^2&=[1-\cos(2\pi fT)]^2+\sin^2(2\pi fT)\\&=2[1-\cos(2\pi fT)]\\&=4\sin^2(\pi fT)\end{aligned}$$

which is shown plotted in Fig. 5.16(*b*). Because of the periodic form of this frequency response, the filter of Fig. 5.16(*a*) is sometimes referred to as a *comb filter*.

The power spectral density of the filter output is therefore

$$S_Y(f)=4\sin^2(\pi fT)S_X(f) \qquad (5.107)$$

For values of frequency which are small compared to $1/T$, we have $\sin(\pi fT)\simeq\pi fT$, and so we may approximate Eq. (5.107) as follows:

$$S_Y(f)\simeq 4\pi^2 f^2 T^2 S_X(f) \qquad (5.108)$$

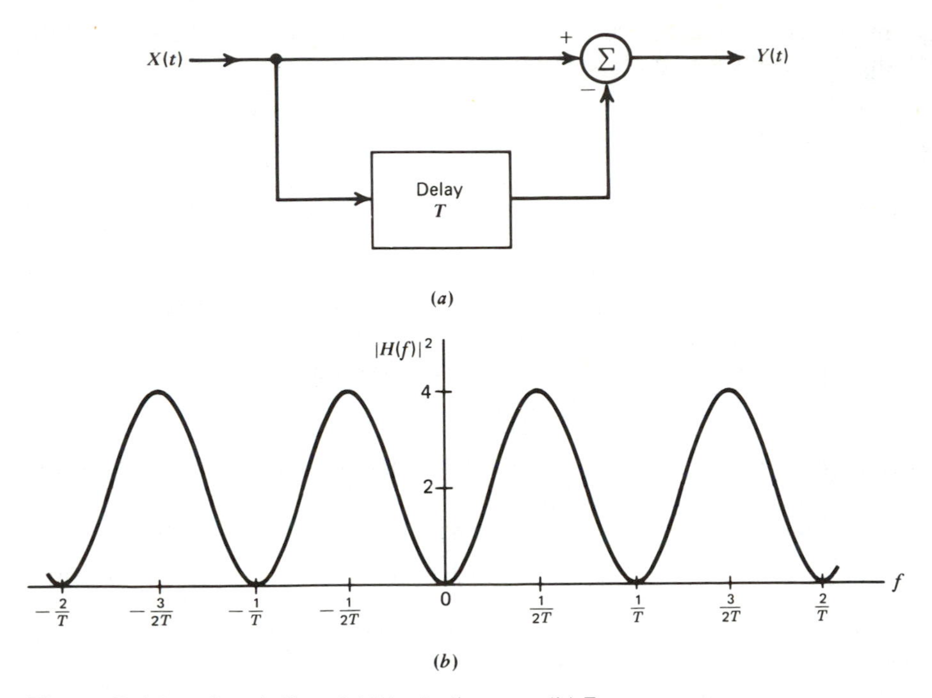

Figure 5.16 Comb filter. *(a)* Block diagram. *(b)* Frequency response.

Because differentiation in the time domain corresponds to multiplication by $j2\pi f$ in the frequency domain, we see from Eq. (5.108) that the comb filter of Fig. 5.16(*a*) acts as a differentiator for low-frequency inputs.

Relation Among the Power Spectral Density and the Amplitude Spectrum of a Sample Function

We now wish to relate the power spectral density $S_X(f)$ directly to the spectral properties of a sample function $x(t)$ of a wide-sense stationary process $X(t)$ that is ergodic. For the sample function $x(t)$ to be Fourier transformable, however, it must be absolutely integrable; that is,

$$\int_{-\infty}^{\infty} |x(t)| dt < \infty \tag{5.109}$$

This condition can never be satisfied by any sample function from a wide-sense stationary random process. Therefore, in order to use the Fourier transform technique, we have to modify the sample function $x(t)$ in such a way that its Fourier transform exists. There are several ways to achieve this, but the simplest way is to define a new sample function $x_T(t)$ of finite duration, as shown by

$$x_T(t) = \begin{cases} x(t), & -T \leq t \leq T \\ 0, & \text{otherwise} \end{cases} \tag{5.110}$$

The *truncated sample function* $x_T(t)$ will satisfy the condition of Eq. (5.109) and therefore be Fourier transformable, provided that T remains finite and that the wide-sense stationary random process to which $x(t)$ belongs has a finite mean-square value. Thus, using $X_T(f)$ to denote the Fourier transform of the truncated sample function $x_T(t)$, we may write

$$\begin{aligned} X_T(f) &= \int_{-\infty}^{\infty} x_T(t)\exp(-j2\pi ft)dt \\ &= \int_{-T}^{T} x(t)\exp(-j2\pi ft)dt \end{aligned} \tag{5.111}$$

Assuming that the wide-sense stationary random process is also ergodic, we may evaluate the autocorrelation function $R_X(\tau)$ of the random process using the time average formula

$$R_X(\tau) = \lim_{T\to\infty} \frac{1}{2T} \int_{-\infty}^{\infty} x_T(t+\tau)x_T(t)dt \tag{5.112}$$

which is based on the truncated sample function $x_T(t)$. From Section 2.8, we recall that the deterministic autocorrelation function $\int_{-\infty}^{\infty} x_T(t+\tau)x_T(t)dt$ and the energy spectral density $|X_T(f)|^2$ form a Fourier transform pair. We may therefore express Eq. (5.112) in the equivalent form

$$R_X(\tau) = \lim_{T\to\infty} \int_{-\infty}^{\infty} \frac{1}{2T} |X_T(f)|^2 \exp(j2\pi f\tau)df \tag{5.113}$$

The quantity $|X_T(f)|^2/2T$ is defined for a continuous range of frequencies $-\infty \leqslant f \leqslant \infty$ and has the same dimensions as the power spectral density. It is called the *periodogram*. This is a misnomer since the quantity is a function of frequency not period. Nevertheless, it has wide usage. The quantity was first used by statisticians to look for periodicities such as seasonal trends in data.

For a fixed value of the frequency f, the periodogram is a random variable in that its value varies in a random manner from one sample function of the random process to another. Thus, for a given sample function $x(t)$, the corresponding quantity $|X_T(f)|^2/2T$ does not converge in any statistical sense to a limiting value as T tends to infinity. As such, therefore, it would be incorrect to interchange the order of the integration and limiting operations in Eq. (5.113). If, however, we take the expectation of both sides of Eq. (5.113) over the ensemble of all sample functions of the random process, and recognize that for an ergodic process the autocorrelation function $R_X(\tau)$ is unchanged by such an operation, because each sample function of an ergodic process eventually takes on nearly all the modes of behavior of each other sample function, we get

$$R_X(\tau) = \lim_{T\to\infty} \int_{-\infty}^{\infty} \frac{1}{2T} E[|X_T(f)|^2]\exp(j2\pi f\tau)df \tag{5.114}$$

Now we may interchange the order of the integration and limiting operations, and so obtain

$$R_X(\tau)=\int_{-\infty}^{\infty}\left\{\lim_{T\to\infty}\frac{1}{2T}E[|X_T(f)|^2]\right\}\exp(j2\pi f\tau)df \tag{5.115}$$

Hence, comparing Eqs. (5.115) and (5.90), we obtain the desired relation between the power spectral density $S_X(f)$ of an ergodic process and the amplitude spectrum $|X_T(f)|$ of a truncated sample function $x_T(t)$ of the process:

$$\begin{aligned} S_X(f) &= \lim_{T\to\infty}\frac{1}{2T}E[|X_T(f)|^2] \\ &= \lim_{T\to\infty}\frac{1}{2T}E\left[\left|\int_{-T}^{T}x(t)\exp(-j2\pi ft)dt\right|^2\right] \end{aligned} \tag{5.116}$$

It is important to note that in Eq. (5.116) it is not possible to let $T\to\infty$ before taking the expectation.*

Cross-spectral Densities

Just as the power spectral density provides a measure of the frequency distribution of a single random process, cross-spectral densities provide a measure of the frequency interrelationship between two random processes. In particular, let $X(t)$ and $Y(t)$ be two jointly wide-sense stationary random processes with their cross-correlation functions denoted by $R_{XY}(\tau)$ and $R_{YX}(\tau)$. We then define the *cross-spectral densities* $S_{XY}(f)$ and $S_{YX}(f)$ of this pair of random processes to be the Fourier transforms of the respective cross-correlation functions, as shown by

$$S_{XY}(f)=\int_{-\infty}^{\infty}R_{XY}(\tau)\exp(-j2\pi f\tau)d\tau \tag{5.117}$$

and

$$S_{YX}(f)=\int_{-\infty}^{\infty}R_{YX}(\tau)\exp(-j2\pi f\tau)d\tau \tag{5.118}$$

The cross-correlation functions and cross-spectral densities thus form Fourier transform pairs. Accordingly, we may write

$$R_{XY}(\tau)=\int_{-\infty}^{\infty}S_{XY}(f)\exp(j2\pi f\tau)df \tag{5.119}$$

and

$$R_{YX}(\tau)=\int_{-\infty}^{\infty}S_{YX}(f)\exp(j2\pi f\tau)df \tag{5.120}$$

* Equation (5.116) provides the basis of a method for estimating the power spectral density of a random process, assuming that the process is ergodic. For further details, see G. M. Jenkins and D. G. Watts, *Spectral Analysis and Its Applications*, pp. 230–257 (Holden-Day, 1969). A. V. Oppenheim and R. W. Schafer, *Digital Signal Processing*, pp. 541–570 (Prentice-Hall, 1975).

The cross-spectral densities $S_{XY}(f)$ and $S_{YX}(f)$ are not necessarily real functions of the frequency f. However, substituting the relationship $R_{XY}(\tau)=R_{YX}(-\tau)$ in Eq. (5.117), we find that $S_{XY}(f)$ and $S_{YX}(f)$ are related by

$$S_{XY}(f)=S_{YX}(-f)=S^*_{YX}(f) \tag{5.121}$$

Example 11

Suppose that the random processes $X(t)$ and $Y(t)$ have zero mean, and they are individually stationary in the wide sense. Consider the sum random process

$$Z(t)=X(t)+Y(t) \tag{5.122}$$

The problem is to determine the power spectral density of $Z(t)$.

The autocorrelation function of $Z(t)$ is given by

$$\begin{aligned}R_Z(t, u)&=E[Z(t)Z(u)]\\&=E[(X(t)+Y(t))(X(u)+Y(u))]\\&=E[X(t)X(u)]+E[X(t)Y(u)]+E[Y(t)X(u)]+E[Y(t)Y(u)]\\&=R_X(t, u)+R_{XY}(t, u)+R_{YX}(t, u)+R_Y(t, u)\end{aligned} \tag{5.123}$$

Defining $\tau=t-u$, we may therefore write

$$R_Z(\tau)=R_X(\tau)+R_{XY}(\tau)+R_{YX}(\tau)+R_Y(\tau), \tag{5.124}$$

when the random processes $X(t)$ and $Y(t)$ are also jointly stationary in the wide sense. Accordingly, taking the Fourier transform of both sides of Eq. (5.124), we get

$$S_Z(f)=S_X(f)+S_{XY}(f)+S_{YX}(f)+S_Y(f) \tag{5.125}$$

We thus see that the cross-spectral densities $S_{XY}(f)$ and $S_{YX}(f)$ represent the spectral components that must be added to the individual power spectral densities of a pair of correlated random processes in order to obtain the power spectral density of their sum.

When the wide-sense stationary random processes $X(t)$ and $Y(t)$ are uncorrelated, the cross-spectral densities $S_{XY}(f)$ and $S_{YX}(f)$ are zero, and so Eq. (5.125) reduces as follows

$$S_Z(f)=S_X(f)+S_Y(f) \tag{5.126}$$

We may generalize this result by stating that when there is a multiplicity of zero-mean wide-sense stationary random processes that are uncorrelated with each other, the power spectral density of their sum is equal to the sum of their individual power spectral densities.

Example 12

Consider next the problem of passing two jointly wide-sense stationary random processes through a pair of separate, stable, linear, time-invariant filters, as shown in Fig. 5.17. In particular, suppose that the random process $X(t)$ is the input to the filter of impulse response $h_1(t)$ and that the random process $Y(t)$ is the input to the filter of impulse response $h_2(t)$. Let $V(t)$ and $Z(t)$ denote the random processes at the respective filter outputs. The cross-

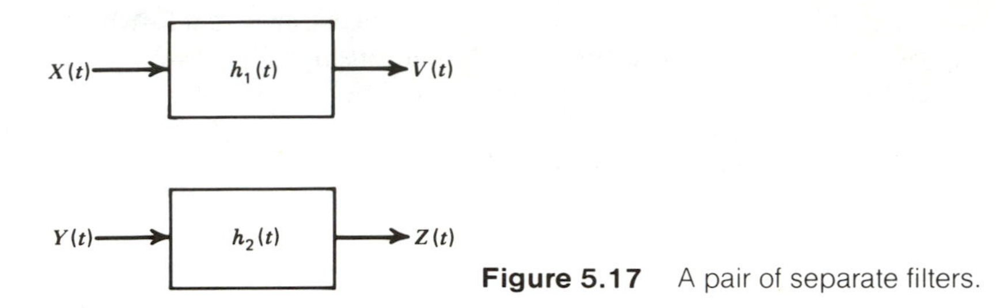

Figure 5.17 A pair of separate filters.

correlation function of $V(t)$ and $Z(t)$ is therefore,

$$\begin{aligned} R_{VZ}(t, u) &= E[V(t)Z(u)] \\ &= E\left[\int_{-\infty}^{\infty} h_1(\tau_1)X(t-\tau_1)d\tau_1 \int_{-\infty}^{\infty} h_2(\tau)Y(u-\tau_2)d\tau_2\right] \\ &= \int_{-\infty}^{\infty}\int_{-\infty}^{\infty} h_1(\tau_1)h_2(\tau_2)E[X(t-\tau_1)Y(u-\tau_2)]d\tau_1\, d\tau_2 \\ &= \int_{-\infty}^{\infty}\int_{-\infty}^{\infty} h_1(\tau_1)h_2(\tau_2)R_{XY}(t-\tau_1, u-\tau_2)d\tau_1\, d\tau_2 \end{aligned} \tag{5.127}$$

where $R_{XY}(t, u)$ is the cross-correlation function of $X(t)$ and $Y(t)$. Because the input random processes are jointly wide-sense stationary (by hypothesis), we may define $\tau = t - u$ and so rewrite Eq. (5.127) as follows:

$$R_{VZ}(\tau) = \int_{-\infty}^{\infty}\int_{-\infty}^{\infty} h_1(\tau_1)h_2(\tau_2)R_{XY}(\tau-\tau_1+\tau_2)d\tau_1\, d\tau_2 \tag{5.128}$$

Taking the Fourier transform of both sides of Eq. (5.128) and using a procedure similar to that which led to the development of Eq. (5.104), we finally get

$$S_{VZ}(f) = H_1(f)H_2^*(f)S_{XY}(f) \tag{5.129}$$

where $H_1(f)$ and $H_2(f)$ are the transfer functions of the respective filters in Fig. 5.17, and $H_2^*(f)$ is the complex conjugate of $H_2(f)$. This is the desired relationship between the cross-spectral density of the output processes and that of the input processes.

5.9 GAUSSIAN PROCESS

Let us suppose that we observe a random process $X(t)$ for an interval that starts at time $t=0$ and lasts until $t=T$. Suppose also that we weight the random process $X(t)$ by some function $g(t)$ and then integrate the product $g(t)X(t)$ over this observation interval, thereby obtaining a random variable Y defined by

$$Y = \int_0^T g(t)X(t)dt \tag{5.130}$$

We refer to Y as a *linear functional* of $X(t)$. The distinction between a function and a functional should be carefully noted. For example, the sum $Y = \sum_{i=1}^{N} a_i X_i$, where the a_i are constants and the X_i are random variables, is a linear *function* of the X_i; for each observed set of values for the random variables X_i, we have a corresponding value for the random variable Y. On the other hand, in Eq. (5.130) the value of the random variable Y depends on the course of the *argument function* $g(t)X(t)$ over the observation interval 0 to T. Thus a functional is a quantity that depends on the entire course of one or more functions rather than on a number of discrete variables. In other words, the domain of a functional is a set or space of admissible functions rather than a region of a coordinate space.

If in Eq. (5.130) the weighting function $g(t)$ is such that the mean-square value of the random variable Y is finite, and if the random variable Y is a *Gaussian-distributed* random variable for every $g(t)$ in this class of functions, then the process $X(t)$ is said to be a *Gaussian process*.* In other words, the process $X(t)$ is a Gaussian process if every linear functional of $X(t)$ is a Gaussian random variable.

We say that the random variable Y has a Gaussian distribution if its probability density function has the form

$$f_Y(y)=\frac{1}{\sqrt{2\pi}\sigma_Y}\exp\left[-\frac{(y-m_Y)^2}{2\sigma_Y^2}\right] \tag{5.131}$$

where m_Y is the mean and σ_Y^2 is the variance of the random variable Y. We thus see that a Gaussian-distributed random variable is completely characterized by specifying its mean and variance. A plot of this probability density function is given in Fig. 5.18 for the case when the mean m_Y is zero and the variance σ_Y^2 equals one.

A Gaussian process has two main virtues. First, the Gaussian process has many properties that make analytic results possible. Second, the random processes

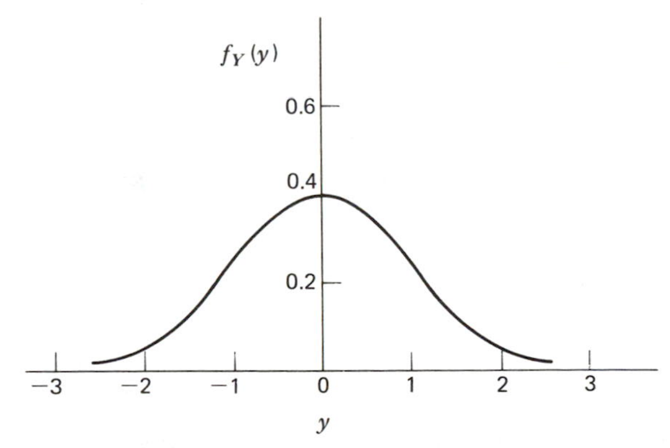

Figure 5.18 Normalized Gaussian distribution.

* The Gaussian distribution and Gaussian process are named after the great mathematician C. F. Gauss. At age 18, Gauss invented the *method of least squares* for finding the best value of a sequence of measurements of the same quantity. Gauss later used the method of least squares in fitting orbits of planets to data measurements, a procedure which was published in 1809 in his book entitled: *Theory of Motion of the Heavenly Bodies*. In connection with the error of observation, he developed the *Gaussian distribution*. This distribution is also known as the *normal distribution*. Partly for historical reasons, mathematicians commonly use normal, while engineers and physicists commonly use Gaussian.

produced by physical phenomena are often such that a Gaussian model is appropriate. The *central limit theorem** provides the mathematical justification for using a Gaussian process as a model of a large number of different physical phenomena in which the observed random variable, at a particular instant of time, is the result of a large number of individual random events. Furthermore, the use of a Gaussian model to describe such physical phenomena is usually confirmed by experiments. Thus the widespread occurrence of physical phenomena for which a Gaussian model is appropriate, together with the ease with which a Gaussian process is handled mathematically, make the Gaussian process very important in the study of communication systems.

Some of the important properties of a Gaussian process are as follows:

Property 1

If a Gaussian process $X(t)$ is applied to a stable linear filter, then the random process $Y(t)$ developed at the output of the filter is also Gaussian.

* A rather important result in probability theory, which is closely related to the Gaussian distribution, is the *central limit theorem*. Let X_i, $i=1, 2, \ldots, N$, be a sequence of random variables that satisfies the following requirements:

1. The X_i are statistically independent.
2. The X_i have the same probability distribution with mean m and variance σ^2.

Define a set of normalized random variables

$$Y_i=\frac{1}{\sigma}(X_i-m), \qquad i=1, 2, \ldots, N$$

so that we have

$$E[Y_i]=0$$

and

$$\text{Var}[Y_i]=1$$

Define the random variables

$$U_N=\sum_{i=1}^{N} X_i$$

and

$$V_N=\frac{1}{N}\sum_{i=1}^{N} Y_i$$

The central limit theorem states that the probability distribution of V_N approaches a normalized Gaussian distribution with zero mean and unit variance as N approaches infinity. Note that since we have

$$U_N=\sigma N V_N+Nm$$

the probability distribution of U_N approaches that of a Gaussian random variable with mean Nm and variance $N\sigma^2$. For a proof of the central limit theorem, see the references listed in the footnote on page.

It is important to realize, however, that the central limit theorem gives only the "limiting" form of the probability distribution of the normalized random variable V_N as N approaches infinity. When N is finite, it is sometimes found that the Gaussian limit gives a relatively poor approximation for the actual probability distribution of V_N even though N may be quite large.

This property is readily derived by using the definition of a Gaussian process based on Eq. (5.130). Consider the situation depicted in Fig. 5.10 where we have a linear time-invariant filter of impulse response $h(t)$, with the random process $X(t)$ as input and the random process $Y(t)$ as output. We assume that $X(t)$ is a Gaussian process. The random processes $Y(t)$ and $X(t)$ are related by the convolution integral

$$Y(t)=\int_0^T h(t-\tau)X(\tau)d\tau, \qquad 0\leqslant t<\infty \tag{5.132}$$

We assume that the impulse response $h(t)$ is such that the mean-square value of the output random process $Y(t)$ is finite for all t in the range $0\leqslant t<\infty$ for which $Y(t)$ is defined. To demonstrate that the output process $Y(t)$ is Gaussian, we must show that any linear functional of it is a Gaussian random variable. That is, if we define the random variable

$$Z=\int_0^\infty g_Y(t)Y(t)dt \tag{5.133}$$

or, equivalently,

$$Z=\int_0^\infty g_Y(t)\int_0^T h(t-\tau)X(\tau)d\tau\, dt, \tag{5.134}$$

then Z must be a Gaussian random variable for every function $g_Y(t)$, such that the mean-square value of Z is finite. Interchanging the order of integration in Eq. (5.134), we get

$$Z=\int_0^T g(\tau)X(\tau)d\tau \tag{5.135}$$

where

$$g(\tau)=\int_0^\infty g_Y(t)h(t-\tau)dt \tag{5.136}$$

Since $X(t)$ is a Gaussian process by hypothesis, it follows from Eq. (5.135) that Z must be a Gaussian random variable. We have thus shown that if the input $X(t)$ to a linear filter is a Gaussian process, then the output $Y(t)$ is also a Gaussian process. Note, however, that although our proof was carried out assuming a time-invariant linear filter, this property is true for any arbitrary stable linear system.

Property 2

Consider the set of random variables or samples $X(t_1)$, $X(t_2)$, . . . , $X(t_n)$, obtained by observing a random process $X(t)$ at times t_1, t_2, . . . , t_n. If the process $X(t)$ is Gaussian, then this set of random variables are jointly Gaussian for any n, with their *n-fold joint probability density function* being completely determined by specifying the set of means*

$$m_{X(t_i)} = E[X(t_i)], \qquad i = 1, 2, \ldots, n$$

and the set of autocovariance functions

$$K_X(t_k, t_i) = E[(X(t_k) - m_{X(t_k)})(X(t_i) - m_{X(t_i)})], \qquad k, i = 1, 2, \ldots, n$$

Property 2 is frequently used as the definition of a Gaussian process. However, this definition is more difficult to use than that based on Eq. (5.130) for evaluating the effects of filtering on a Gaussian process.

We may extend Property 2 to two (or more) random processes as follows. Consider the composite set of random variables or samples $X(t_1), X(t_2), \ldots, X(t_n)$, $Y(u_1), Y(u_2), \ldots, Y(u_m)$ obtained by observing a random process $X(t)$ at times $\{t_i\}$, $i = 1, 2, \ldots, n$, and a second random process $Y(t)$ at times $\{u_k\}$, $k = 1, 2, \ldots, m$. We say that the processes $X(t)$ and $Y(t)$ are *jointly Gaussian* if this composite set of random variables are jointly Gaussian for any n and m. Note that in addition to the mean and correlation functions of the random processes $X(t)$ and $Y(t)$ individually, we must also know the cross-covariance

$$E[X(t_i)Y(u_k)] - m_{X(t_i)}m_{Y(u_k)} = R_{XY}(t_i, u_k) - m_{X(t_i)}m_{Y(u_k)}$$

for any pair of observation instants (t_i, u_k). This additional knowledge is embodied in the cross-correlation function, $R_{XY}(t_i, u_k)$, of the two processes $X(t)$ and $Y(t)$.

* The joint probability density function of the Gaussian vector

$$\mathbf{X} = \begin{bmatrix} X(t_1) \\ X(t_2) \\ \vdots \\ X(t_n) \end{bmatrix}$$

is defined by

$$f_{\mathbf{X}}(\mathbf{x}) = \frac{1}{(2\pi)^{n/2}|\mathbf{K_X}|^{1/2}} \exp[-\tfrac{1}{2}(\mathbf{x} - \mathbf{m_X})\mathbf{K_X}^{-1}(\mathbf{x} - \mathbf{m_X})^T]$$

where the superscript T denotes transposition, and

$\mathbf{m_X}$ = mean vector
$= E[\mathbf{X}]$
$\mathbf{K_X}$ = covariance matrix
$= E[(\mathbf{X} - \mathbf{m_X})^T(\mathbf{X} - \mathbf{m_X})]$
$\mathbf{K_X}^{-1}$ = inverse of the covariance matrix
$|\mathbf{K_X}|$ = determinant of the covariance matrix

For a derivation of this function, see:

1. J. B. Thomas, pp. 128–144, *op. cit.*
2. W. B. Davenport and W. L. Root, pp. 147–154, *op. cit.*
3. D. J. Sakrison, *Communication Theory: Transmission of Waveforms and Digital Information*, pp. 87–97 (Wiley, 1968).

Property 3

If a Gaussian process is wide-sense stationary, then the process is also stationary in the strict sense.

This follows directly from Property 2.

Property 4

If the set of random variables $X(t_1), X(t_2), \ldots, X(t_n)$, obtained by sampling a Gaussian process $X(t)$ at times $t_1, t_2, \ldots, t_n$, are uncorrelated, that is,

$$E[(X(t_k)-m_{X(t_k)})(X(t_i)-m_{X(t_i)})]=0, \qquad i \neq k$$

then this set of random variables are statistically independent.

The implication of this property is that the joint probability density function of the set of random variables $X(t_1), X(t_2), \ldots, X(t_n)$ can be expressed as the product of the probability density functions of the individual random variables in the set.

5.10 NOISE

The term *noise* is used customarily to designate unwanted waves that tend to disturb the transmission and processing of signals in communication systems, and over which we have incomplete control. In practice, we find that there are many potential sources of noise in a communication system. The sources of noise may be external to the system (e.g., atmospheric noise, galactic noise, man-made noise), or internal to the system. The second category includes an important type of noise that arises due to *spontaneous fluctuations* of current or voltage in electrical circuits. This type of noise, in one way or another, is present in every communication system and represents a basic limitation on the transmission or detection of signals. The two most common examples of spontaneous fluctuations in electrical circuits are *shot noise* and *thermal noise*.*

Shot Noise

Shot noise arises in electronic devices because of the discrete nature of current flow in the device. Consider for example, a *temperature-limited diode*, as shown in Fig. 5.19. It consists of two electrodes enclosed in a vacuum: a *cathode* which is heated so that it emits electrons, and an *anode* or *plate* which is maintained at a positive potential with respect to the cathode so that it gathers the electrons. We assume that the cathode-plate potential difference is large enough to cause the

* For a detailed treatment of electrical noise, see A. Van der Ziel, *Noise* (Prentice-Hall, 1954), W. R. Bennett, *Electrical Noise* (McGraw-Hill, 1960). D. A. Bell, *Electrical Noise* (Van Nostrand, 1960) E. M. Cherry and D. E. Hooper, *Amplifying Devices and Low-Pass Amplifier Design* (Wiley, 1968). A. Van Der Ziel, *Noise: Sources, Characterization, Measurement* (Prentice-Hall, 1970).

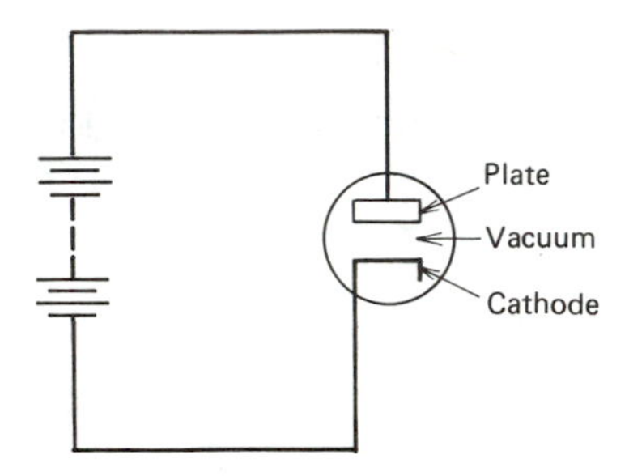

Figure 5.19 Diode.

electrons emitted thermionically by the heated cathode to be pulled to the plate with such high velocities that space-charge effects are negligible. The plate current is then determined effectively by the rate at which electrons are emitted from the cathode. By considering the plate current as the sum of a succession of current pulses, with each pulse caused by the transit of one electron through the cathode-plate space, we find that the mean-square value of the randomly fluctuating component of the current is given by (see Problem 5.29)

$$E[I_{SN}^2] = 2eI_0 \, \Delta f \text{ amps}^2 \tag{5.137}$$

where e is the *electron charge* equal to 1.59×10^{-19} coulombs, I_0 is the mean value of the current in amperes, and Δf is the bandwidth of the measuring instrument in hertz. Equation (5.137) is called the *Schottky formula*. The typical transit time of an electron from cathode to plate is of the order of 10^{-9} seconds. The Schottky formula holds provided that the operating frequency is small compared with the reciprocal of the transit time, so that we may neglect transit time effects.

Another important characteristic of shot noise is that it is Gaussian-distributed with zero mean. This follows from the fact that the number of electrons contributing to the shot noise current is very large, and their random emissions from the cathode are, for all practical purposes, statistically independent of each other. Hence, the central limit theorem predicts a Gaussian distribution for shot noise.

Thermal Noise

Thermal noise* is the name given to the electrical noise arising from the random motion of electrons in a conductor. The mean-square value of the thermal noise voltage appearing across the terminals of a resistor, measured in a bandwidth of Δf hertz, is, for all practical purposes, given by (see Problem 5.30)

$$E[V_{TN}^2] = 4kTR \, \Delta f \text{ volts}^2 \tag{5.138}$$

* Thermal noise was first studied experimentally by Johnson in 1928, and for this reason it is sometimes referred to as the "Johnson noise". See J. B. Johnson, "Thermal agitation of electricity in conductors," *Physical Review*, vol. 32, pp. 97–109, July 1928.
Johnson's experiments were confirmed theoretically by Nyquist. See H. Nyquist, "Thermal agitation of electric charge in conductors," *Physical Review*, vol. 32, pp. 110–113, July 1928.

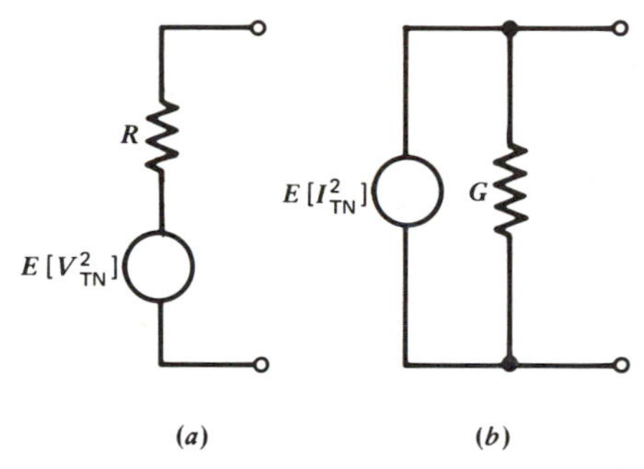

Figure 5.20 Models of a noisy resistor. *(a)* Thévenin equivalent circuit. *(b)* Norton equivalent circuit.

where k is *Boltzmann's constant* equal to 1.38×10^{-23} joules per degree Kelvin, T is the *absolute temperature* in degrees Kelvin, and R is the resistance in ohms. We may thus model a noisy resistor by the *Thévenin equivalent circuit* consisting of a noise voltage generator of mean-square value $E[V^2_{TN}]$ in series with a noiseless resistor, as in Fig. 5.20(*a*). Alternatively, we may use the *Norton equivalent circuit* consisting of a noise current generator in parallel with a noiseless conductance, as in Fig. 5.20(*b*). The mean-square value of the noise current generator is

$$\begin{aligned} E[I^2_{TN}] &= \frac{1}{R^2} E[V^2_{TN}] \\ &= 4kTG\,\Delta f \text{ amps}^2 \end{aligned} \tag{5.139}$$

where $G = 1/R$ is the conductance. It is also of interest to note that because the number of electrons in a resistor is very large and their random motions inside the resistor are statistically independent of each other, the central limit theorem indicates that thermal noise is Gaussian-distributed with zero mean.

Noise calculations involve the transfer of power, and so we find that the use of the *maximum-power transfer theorem* is applicable. This theorem states that the maximum possible power is transferred from a source of internal resistance R to a load of resistance R_l when $R_l = R$. Under this *matched condition*, the power produced by the source is divided equally between the internal resistance of the source and the load resistance, and the power delivered to the load is referred to as the *available power*. Applying the maximum-power transfer theorem to the Thévenin equivalent circuit of Fig. 5.20(*a*) or the Norton equivalent circuit of Fig. 5.20(*b*), we find that a noisy resistor produces an *available noise power* equal to $kT\,\Delta f$ watts.

White Noise

The noise analysis of communication systems is customarily based on an idealized form of noise called *white noise*, the power spectral density of which is independent of the operating frequency. The adjective *white* is used in the sense that white light

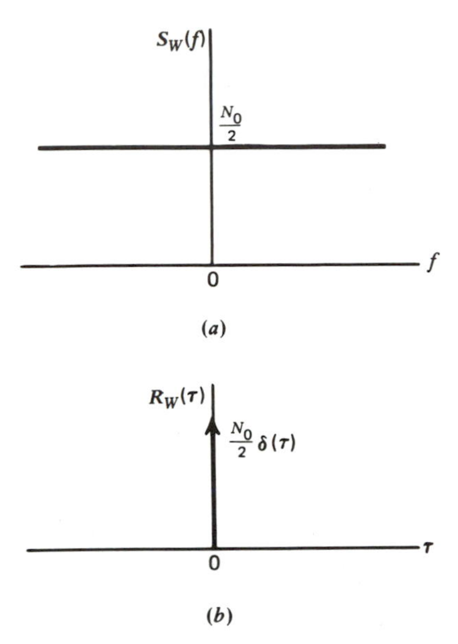

Figure 5.21 Characteristics of white noise. *(a)* Power spectral density. *(b)* Autocorrelation function.

contains equal amounts of all frequencies within the visible band of electromagnetic radiation. We denote the power spectral density of white noise $w(t)$ as

$$S_W(f)=\frac{N_0}{2} \tag{5.140}$$

where the factor 1/2 has been included to indicate that half the power is associated with positive frequency and half with negative frequency, as illustrated in Fig. 5.21(*a*). The dimensions of N_0 are in watts per hertz. The parameter N_0 is usually referenced to the input stage of the receiver of a communication system. It may be expressed as

$$N_0=kT_e \tag{5.141}$$

where k is Boltzmann's constant and T_e is the *equivalent noise temperature* of the receiver. *The equivalent noise temperature of a system is defined as the temperature at which a noisy resistor has to be maintained such that, by connecting the resistor to the input of a noiseless version of the system, it produces the same available noise power at the output of the system as that produced by all the sources of noise in the actual system.** The important feature of the equivalent noise temperature is that it depends only on the parameters of the system.

Since the autocorrelation function is the inverse Fourier transform of the power

* The noisiness of a system may also be measured in terms of the so-called *noise figure*. The relationship between the equivalent noise temperature and noise figure is developed in Appendix 3.

spectral density, it follows that for white noise,

$$R_W(\tau)=\frac{N_0}{2}\,\delta(\tau) \tag{5.142}$$

That is, the autocorrelation function of white noise consists of a delta function weighted by the factor $N_0/2$ and occurring at $\tau=0$, as in Fig. 5.21(*b*). We note that $R_W(\tau)$ is zero for $\tau\neq 0$. Accordingly, any two different samples of white noise, no matter how closely together in time they are taken, are uncorrelated. If the white noise $w(t)$ is also Gaussian, then the two samples are statistically independent. In a sense, white Gaussian noise represents the ultimate in "randomness."

Strictly speaking, white noise has infinite average power and, as such, it is not physically realizable. Nevertheless, white noise has convenient mathematical properties and therefore is useful in system analysis.

The utility of a white noise process is parallel to that of an impulse function or delta function in the analysis of linear systems. Just as we may observe the effect of an impulse only after it has been passed through a system with a finite bandwidth, so it is with white noise whose effect is observed only after passing through a similar system. We may state, therefore, that as long as the bandwidth of a noise process at the input of a system is appreciably larger than that of the system itself, then we may model the noise process as white noise.

Example 13 Ideal low-pass filtered white noise

Suppose that a white Gaussian noise $w(t)$ of zero mean and power spectral density $N_0/2$ is applied to an ideal low-pass filter of bandwidth B and a passband amplitude response of one. The power spectral density of the noise $n(t)$ appearing at the filter output is therefore [see Fig. 5.22(*a*)].

$$S_N(f)=\begin{cases}\dfrac{N_0}{2}, & -B<f<B\\[2mm] 0, & |f|>B\end{cases}$$

The autocorrelation function of $n(t)$ is the inverse Fourier transform of the power spectral density shown in Fig. 5.22(*a*):

$$\begin{aligned}R_N(\tau)&=\int_{-B}^{B}\frac{N_0}{2}\exp(j2\pi f\tau)\,df\\&=N_0B\,\mathrm{sinc}(2B\tau)\end{aligned} \tag{5.143}$$

This autocorrelation function is shown plotted in Fig. 5.22(*b*). We see that $R_N(\tau)$ has its maximum value of N_0B at the origin, and it passes through zero at $\tau=\pm n/2B$, where $n=1, 2, 3, \ldots$

Since the input noise $w(t)$ is Gaussian (by hypothesis), it follows that the band-limited noise $n(t)$ at the filter output is also Gaussian. Suppose now that $n(t)$ is sampled at the rate of $2B$ times per second. From 5.22(*b*), we see that the resulting noise samples are uncorrelated and, being Gaussian, they are statistically independent. Accordingly, the joint probability density function of a set of noise samples obtained in this way is equal to the product of the individual probability density functions. Note that each such noise sample has a mean of zero and variance of N_0B.

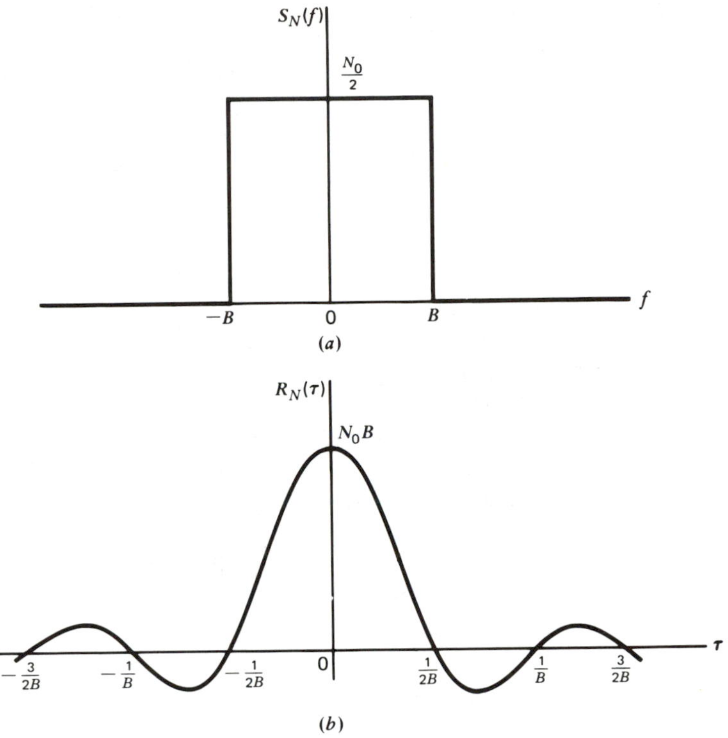

Figure 5.22 Characteristics of low-pass filtered white noise. *(a)* Power spectral density. *(b)* Autocorrelation function.

Example 14 RC low-pass filtered white noise

Consider next a white Gaussian noise $w(t)$ of zero mean and power spectral density $N_0/2$ applied to a low-pass RC filter, as in Fig. 5.23(*a*). The transfer function of the filter is

$$H(f)=\frac{1}{1+j2\pi f RC} \tag{5.144}$$

The power spectral density of the noise $n(t)$ appearing at the low-pass RC filter output is therefore [see Fig. 5.23(*b*)]

$$S_N(f)=\frac{N_0/2}{1+(2\pi f RC)^2} \tag{5.145}$$

From Example 4 of Chapter 2, we have

$$\exp(-|\tau|) \rightleftharpoons \frac{2}{1+(2\pi f)^2} \tag{5.146}$$

Therefore, using the time-scaling property of the Fourier transform, we find that the autocorrelation function of the filtered noise $n(t)$ equals

$$R_N(\tau)=\frac{N_0}{4RC}\exp\left(-\frac{|\tau|}{RC}\right) \tag{5.147}$$

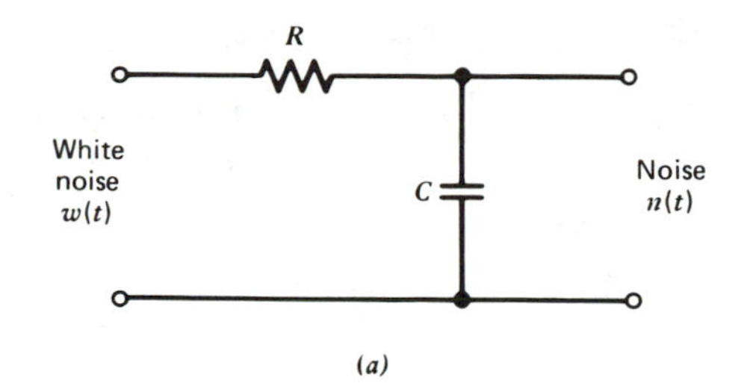

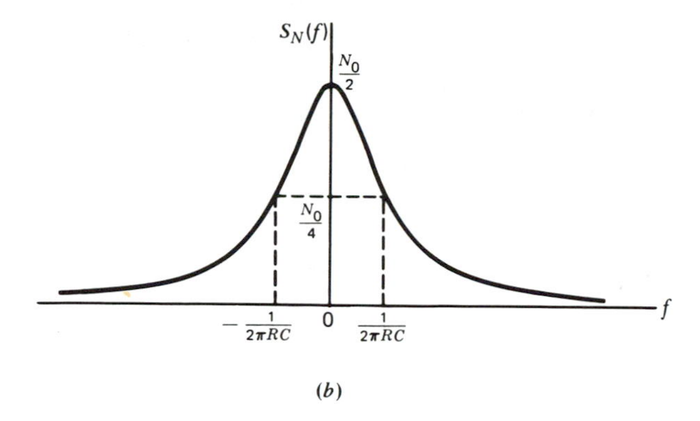

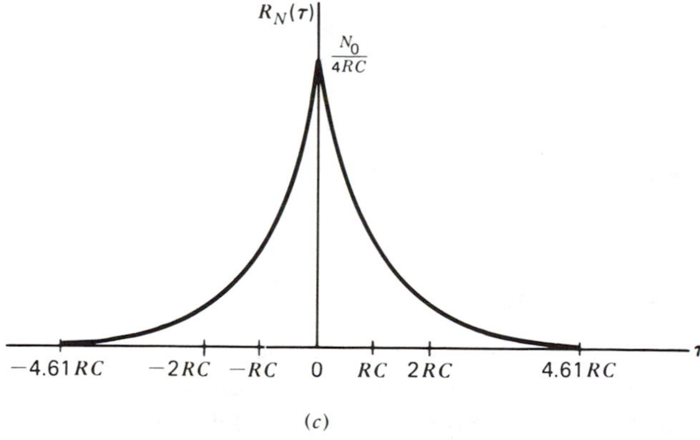

Figure 5.23 Characteristics of RC-filtered white noise. *(a)* Low-pass RC filter. *(b)* Power spectral density of filter output *n(t)*. *(c)* Autocorrelation function of *n(t)*.

which is shown plotted in Fig. 5.23(*c*). The decorrelation time τ_0 for which $R_N(\tau)$ drops to 1 percent of its maximum value of $N_0/4RC$ is equal to $4.61RC$. Thus, if the noise appearing at the filter output is sampled at a rate equal to or less than $0.217/RC$ samples per second, the resulting samples are effectively uncorrelated and, being Gaussian, they are statistically independent.

Example 15 Autocorrelation of a sine wave plus white noise

Consider a random process $X(t)$ consisting of a sinusoidal wave component $A\cos(2\pi f_c t+\Theta)$ and a white noise process $W(t)$ of zero mean and power spectral density $N_0/2$, as shown by

$$X(t)=A\cos(2\pi f_c t+\Theta)+W(t) \tag{5.148}$$

where Θ is a uniformly distributed random variable. The problem is to determine the autocorrelation function of the random process $X(t)$.

It is clear that the two components of $X(t)$ are independent. Therefore, the autocorrelation function of $X(t)$ is the sum of the individual autocorrelation functions of the sinusoidal wave component $A\cos(2\pi f_c t+\Theta)$ and the white noise component. In Example 3, we showed that the autocorrelation function of the sinusoidal component is equal to $(A^2/2)\cos(2\pi f_c\tau)$. The autocorrelation function of the white noise component is equal to $(N_0/2)\delta(\tau)$. Therefore,

$$R_X(\tau)=\frac{A^2}{2}\cos(2\pi f_c\tau)+\frac{N_0}{2}\delta(\tau) \tag{5.149}$$

which is shown plotted in Fig. 5.24. We thus see that for $|\tau|>0$, the autocorrelation function of the random process $X(t)$ is the same as that of the sinusoidal wave component. This shows that by determining the autocorrelation function of $X(t)$, we can detect the presence of a periodic signal component which is corrupted by additive white noise.

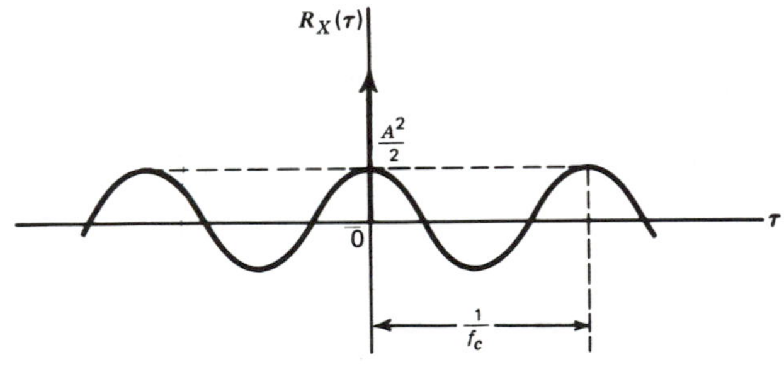

Figure 5.24 Autocorrelation function of sinusoidal wave plus white noise.

Noise Equivalent Bandwidth

In Example 13 we observe that when a source of white noise of zero mean and power spectral density $N_0/2$ is connected across the input of an ideal low-pass filter of bandwidth B and passband amplitude response of one, the average output noise power [or equivalently $R_N(0)$] is equal to N_0B. In Example 14 we observe that when such a noise source is connected to the input of the simple RC low-pass filter of Fig. 5.23(*a*), the corresponding value of the average output noise power is equal to $N_0/(4RC)$. For this filter, the half-power or 3-dB bandwidth is equal to $1/(2\pi RC)$. Here again we find that the average output noise power of the filter is proportional to the bandwidth.

We may generalize this statement to include all kinds of low-pass filters by defining a noise equivalent bandwidth as follows. Suppose that we have a source of white noise of zero mean and power spectral density $N_0/2$ connected to the input of an arbitrary low-pass filter of transfer function $H(f)$. The resulting average output noise power is therefore

$$N = \frac{N_0}{2} \int_{-\infty}^{\infty} |H(f)|^2 \, df$$

$$= N_0 \int_{0}^{\infty} |H(f)|^2 \, df \tag{5.150}$$

where, in the last line, we have made use of the fact that the amplitude response $|H(f)|$ is an even function of frequency.

Consider next the same source of white noise connected to the input of an ideal low-pass filter of zero-frequency response $H(0)$ and bandwidth B. In this case, the average output noise power is

$$N = N_0 B H^2(0) \tag{5.151}$$

Therefore, equating this average output noise power to that in Eq. (5.150), we may define the *noise equivalent bandwidth* as

$$B = \frac{\int_0^{\infty} |H(f)|^2 \, df}{H^2(0)} \tag{5.152}$$

Thus the procedure for calculating the noise equivalent bandwidth consists of replacing the arbitrary low-pass filter of transfer function $H(f)$ by an equivalent ideal low-pass filter of zero-frequency response $H(0)$ and bandwidth B, as illustrated in Fig. 5.25.

In a similar way, we may define a noise equivalent bandwidth for a band-pass filter.

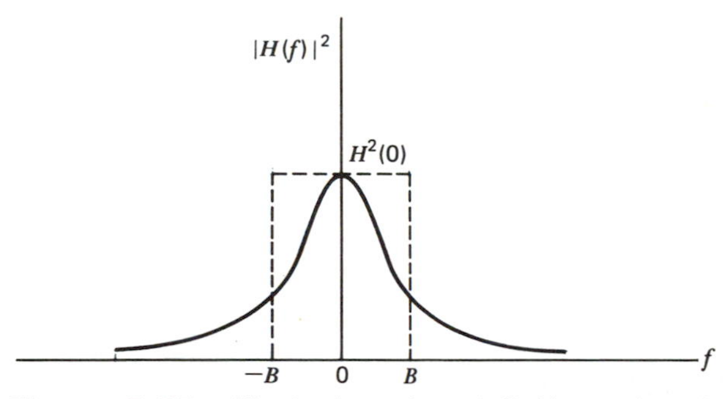

Figure 5.25 Illustrating the definition of noise equivalent bandwidth.

5.11 NARROW-BAND NOISE

At the receiver front-end of practical communication systems using carrier modulation, we find that the signals of interest and the noise are usually processed by frequency-selective filters designed with large enough bandwidth to pass the signals of interest essentially undistorted but not so large as to admit excessive noise through the receiver. In a typical receiver, we usually find that the filter used for such a purpose is narrow-band in that its mid-band frequency is large compared to its bandwidth. The noise process appearing at the output of such a filter is called *narrow-band noise*. With the spectral components of narrow-band noise concentrated about some mid-band frequency $\pm f_c$, as in Fig. 5.26(*a*), we find that the sample function $n(t)$ of such a process appears somewhat similar to a sine-wave of frequency f_c, which undulates slowly in both amplitude and phase, as in Fig. 5.26(*b*).

Consider then the noise $n(t)$ produced at the output of a narrow-band filter in response to the sample function $w(t)$ of a white Gaussian noise process of zero mean and unit power spectral density applied to the filter input. Let $H(f)$ denote the transfer function of this filter. Accordingly, we may express the power spectral density $S_N(f)$ of the noise $n(t)$ in terms of $H(f)$ as follows

$$S_N(f) = |H(f)|^2 \tag{5.153}$$

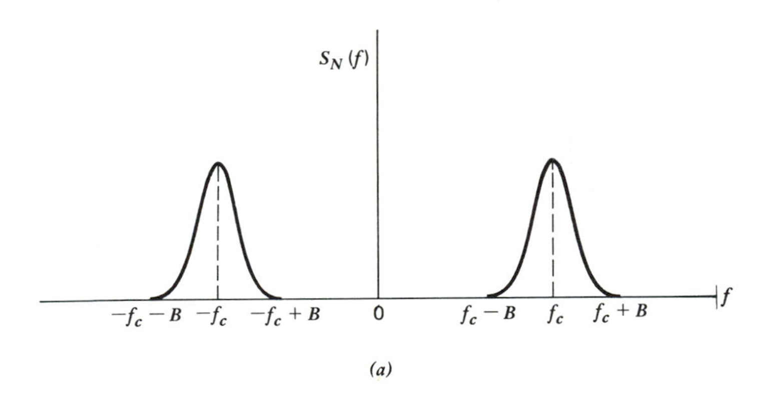

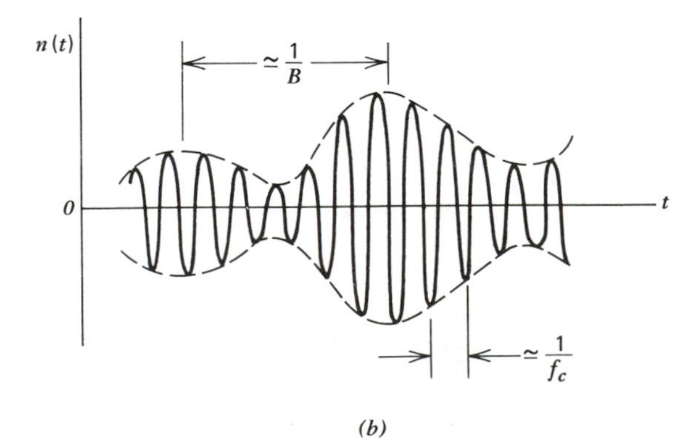

Figure 5.26 *(a)* Power spectral density of narrow-band noise. *(b)* Sample function of narrow-band noise.

Indeed, any narrow-band noise encountered in practice may be modeled by applying a white noise to a suitable filter in the manner described above (see Problem 5.36).

In this section we wish to represent the narrow-band noise $n(t)$ in terms of its in-phase and quadrature components in a manner similar to that described for a narrow-band signal, as in Section 2.14. The derivation presented here is based on the idea of a pre-envelope and related concepts, which were discussed in Chapter 2.

Let $n_+(t)$ and $\tilde{n}(t)$ denote the pre-envelope and complex envelope of the narrow-band noise $n(t)$, respectively. We assume that the power spectrum of $n(t)$ is centered around the frequency f_c. Then we may write

$$n_+(t) = n(t) + j\hat{n}(t) \tag{5.154}$$

and

$$\tilde{n}(t) = n_+(t)\exp(-j2\pi f_c t) \tag{5.155}$$

where $\hat{n}(t)$ is the Hilbert transform of $n(t)$. The complex envelope $\tilde{n}(t)$ may itself be expressed as

$$\tilde{n}(t) = n_c(t) + jn_s(t) \tag{5.156}$$

Hence, combining Eqs. (5.154) through (5.156), we find that the in-phase and quadrature components of the narrow-band noise $n(t)$ are as follows, respectively,

$$n_c(t) = n(t)\cos(2\pi f_c t) + \hat{n}(t)\sin(2\pi f_c t) \tag{5.157}$$

and

$$n_s(t) = \hat{n}(t)\cos(2\pi f_c t) - n(t)\sin(2\pi f_c t) \tag{5.158}$$

Eliminating $\hat{n}(t)$ between Eqs. (5.157) and (5.158), we get the desired *canonical form* for representing the narrow-band noise $n(t)$, as shown by

$$n(t) = n_c(t)\cos(2\pi f_c t) - n_s(t)\sin(2\pi f_c t) \tag{5.159}$$

Using Eqs. (5.157) and (5.158), we may derive the following properties of the quadrature components of a narrow-band noise:

Property 1

If the narrow-band noise $n(t)$ has zero mean, then its in-phase component $n_c(t)$ and quadrature component $n_s(t)$ have zero mean too.

To prove this property, we first observe that the noise $\hat{n}(t)$ is obtained by passing $n(t)$ through a linear filter (i.e., Hilbert transformer). Accordingly, $\hat{n}(t)$ will have zero mean if $n(t)$ has zero mean. Furthermore, from Eqs. (5.157) and (5.158) we see that $n_c(t)$ and $n_s(t)$ are weighted sums of $n(t)$ and $\hat{n}(t)$. It follows therefore that if the original narrow-band noise $n(t)$ has zero mean, so will its quadrature components $n_c(t)$ and $n_s(t)$.

Property 2

If the narrow-band noise $n(t)$ is Gaussian, then its in-phase component $n_c(t)$ and quadrature component $n_s(t)$ are jointly Gaussian.

To prove this property, we observe that $\hat{n}(t)$ is derived from $n(t)$ by a linear filtering operation. Then, if $n(t)$ is Gaussian, $\hat{n}(t)$ is Gaussian, and $n(t)$ and $\hat{n}(t)$ are jointly Gaussian. Therefore, the quadrature components $n_c(t)$ and $n_s(t)$ are jointly Gaussian, since they are weighted sums of jointly Gaussian processes.

Property 3

If the narrow-band noise $n(t)$ is wide-sense stationary, then its in-phase component $n_c(t)$ and quadrature component $n_s(t)$ are jointly wide-sense stationary.

If $n(t)$ is wide-sense stationary, so is its Hilbert transform $\hat{n}(t)$. However, since the quadrature components $n_c(t)$ and $n_s(t)$ are weighted sums of $n(t)$ and $\hat{n}(t)$, and the weighting functions, namely, $\cos(2\pi f_c t)$, and $\sin(2\pi f_c t)$, vary with time, we cannot directly assert that $n_c(t)$ and $n_s(t)$ are wide-sense stationary. To prove this, we have to evaluate their correlation functions.

Using Eqs. (5.157) and (5.158), we find that the quadrature components $n_c(t)$ and $n_s(t)$ of a narrow-band noise $n(t)$ have the same autocorrelation function, as shown by (see Problem 5.38)

$$R_{N_c}(\tau)=R_{N_s}(\tau)=R_N(\tau)\cos(2\pi f_c\tau)+\hat{R}_N(\tau)\sin(2\pi f_c\tau) \tag{5.160}$$

and their cross-correlation functions are given by

$$R_{N_cN_s}(\tau)=-R_{N_sN_c}(\tau)=R_N\sin(2\pi f_c\tau)-\hat{R}_N(\tau)\cos(2\pi f_c\tau) \tag{5.161}$$

where $R_N(\tau)$ is the autocorrelation function of $n(t)$, and $\hat{R}_N(\tau)$ is the Hilbert transform of $R_N(\tau)$. From Eqs. (5.160) and (5.161), we see that the correlation functions $R_{N_c}(\tau)$, $R_{N_s}(\tau)$, and $R_{N_cN_s}(\tau)$ of the quadrature components $n_c(t)$ and $n_s(t)$ depend only on the time shift τ. This proves that $n_c(t)$ and $n_s(t)$ are wide-sense stationary if the original narrow-band noise $n(t)$ is wide-sense stationary.

Property 4

Both the in-phase noise $n_c(t)$ and quadrature noise $n_s(t)$ have the same power spectral density, which is related to the power spectral density $S_N(f)$ of the original narrow-band noise $n(t)$ as follows

$$S_{N_c}(f)=S_{N_s}(f)=\begin{cases}S_N(f-f_c)+S_N(f+f_c), & -B\leq f\leq B\\ 0, & \text{elsewhere}\end{cases} \tag{5.162}$$

where it is assumed that $S_N(f)$ occupies the frequency interval $f_c-B\leq|f|\leq f_c+B$ and $f_c>B$.

To prove this property, we take the Fourier transforms of both sides of Eq. (5.160), and use the fact that

$$F[\hat{R}_N(\tau)] = -j\,\mathrm{sgn}(f)F[R_N(\tau)]$$
$$= -j\,\mathrm{sgn}(f)S_N(f) \tag{5.163}$$

We thus obtain the result

$$S_{N_c}(f) = S_{N_s}(f)$$
$$= \tfrac{1}{2}[S_N(f-f_c)+S_N(f+f_c)] - \tfrac{1}{2}[S_N(f-f_c)\mathrm{sgn}(f-f_c) - S_N(f+f_c)\mathrm{sgn}(f+f_c)]$$
$$= \tfrac{1}{2}S_N(f-f_c)[1-\mathrm{sgn}(f-f_c)] + \tfrac{1}{2}S_N(f+f_c)[1+\mathrm{sgn}(f+f_c)] \tag{5.164}$$

Now, with the power spectral density $S_N(f)$ of the original narrow-band noise $n(t)$ occupying the frequency interval $f_c - B \leqslant |f| \leqslant f_c + B$, where $f_c > B$, as illustrated in part (*a*) of Fig. 5.27, we find that the corresponding shapes of $S_N(f-f_c)$ and

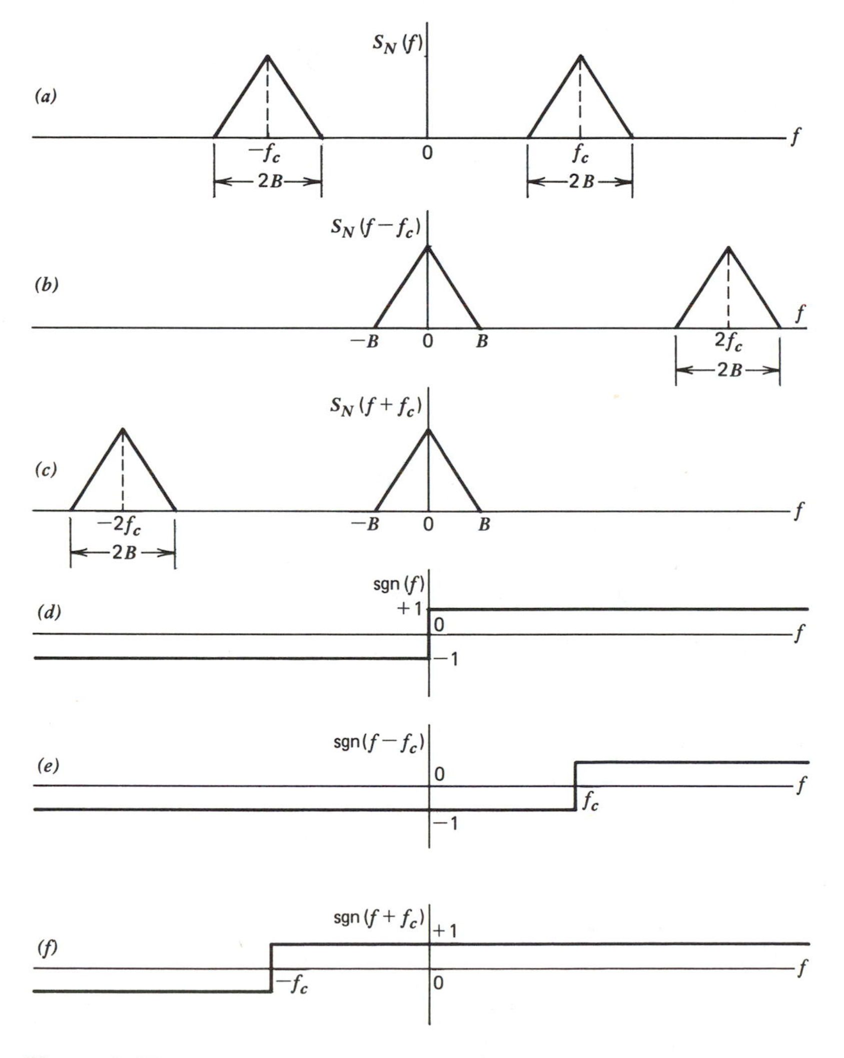

Figure 5.27

$S_N(f+f_c)$ are as in parts (*b*) and (*c*) of Fig. 5.27, respectively. Parts (*d*), (*e*), and (*f*) of this figure show the shapes of $\text{sgn}(f)$, $\text{sgn}(f-f_c)$ and $\text{sgn}(f+f_c)$, respectively. Accordingly, we may make the following observations:

1. For frequencies defined by $-B \leqslant f \leqslant B$, we have

$$\text{sgn}(f-f_c) = -1$$

   and

$$\text{sgn}(f+f_c) = +1$$

   Hence, substituting these results in Eq. (5.164), we obtain

$$\begin{aligned} S_{N_c}(f) &= S_{N_s}(f) \\ &= S_N(f-f_c) + S_N(f+f_c), \qquad -B \leqslant f \leqslant B \end{aligned}$$

2. For $2f_c - B \leqslant f \leqslant 2f_c + B$, we have

$$\text{sgn}(f-f_c) = 1$$

   and

$$S_N(f+f_c) = 0$$

   with the result that $S_{N_c}(f)$ and $S_{N_s}(f)$ are both zero.
3. For $-2f_c - B \leqslant f \leqslant -2f_c + B$, we have

$$S_N(f-f_c) = 0$$

   and

$$\text{sgn}(f+f_c) = -1$$

   with the result that, here too, $S_{N_c}(f)$ and $S_{N_s}(f)$ are both zero.
4. Outside the frequency intervals defined in 1, 2, and 3, both $S_N(f-f_c)$ and $S_N(f+f_c)$ are zero, and in a corresponding way, $S_{N_c}(f)$ and $S_{N_s}(f)$ are also zero.

Combining these results, we obtain the relationship defined in Eq. (5.162).

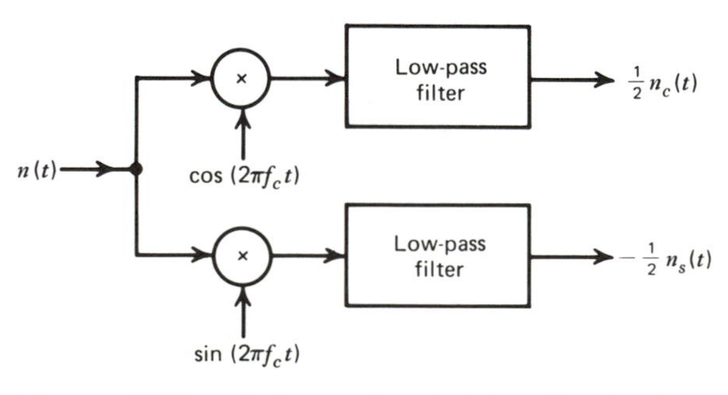

Figure 5.28 Scheme for deriving the in-phase component $n_c(t)$ and quadrature component $n_s(t)$ of narrow-band noise $n(t)$.

As a consequence of this property, we may extract the in-phase component $n_c(t)$ and quadrature component $n_s(t)$, except for scaling factors, from the narrow-band noise $n(t)$ by using the scheme shown in Fig. 5.28, where both low-pass filters have a cutoff frequency at B (see Problem 5.42).

Property 5

If a narrow-band noise $n(t)$ has zero mean, then the quadrature components $n_c(t)$ and $n_s(t)$ have the same variance as $n(t)$ itself.

This property follows directly from Eq. (5.162), according to which the total area under the power spectral density curve of $n_c(t)$ or $n_s(t)$ is the same as the total area under the power spectral density curve of $n(t)$. Hence, $n_c(t)$ and $n_s(t)$ have the same mean-square value as $n(t)$. Earlier we showed that if $n(t)$ has zero mean, then $n_c(t)$ and $n_s(t)$ have zero mean too. It follows therefore if $n(t)$ has zero mean, then $n_c(t)$ and $n_s(t)$ have the same variance as $n(t)$ itself.

Property 6

The cross-spectral densities of the quadrature components of a narrow-band noise are purely imaginary, as shown by

$$S_{N_cN_s}(f) = -S_{N_cN_s}(f) = \begin{cases} j[S_N(f+f_c) - S_N(f-f_c)], & -B \leqslant f \leqslant B \\ 0, & \text{elsewhere} \end{cases} \tag{5.165}$$

To prove this property, we take the Fourier transforms of both sides of Eq. (5.161), and use the relation of Eq. (5.163), obtaining

$$\begin{aligned} S_{N_cN_s}(f) &= -S_{N_sN_c}(f) \\ &= -\frac{j}{2}[S_N(f-f_c) - S_N(f+f_c)] + \frac{j}{2}[S_N(f-f_c)\text{sgn}(f-f_c) + S_N(f+f_c)\text{sgn}(f+f_c)] \\ &= \frac{j}{2}S_N(f+f_c)[1+\text{sgn}(f+f_c)] - \frac{j}{2}S_N(f-f_c)[1-\text{sgn}(f-f_c)] \end{aligned} \tag{5.166}$$

Following a procedure similar to that described for proving Property 4, we may show that Eq. (5.166) reduces to the form shown in Eq. (5.165).

Property 7

If a narrow band noise $n(t)$ is Gaussian with zero mean, and its power spectral density $S_N(f)$ is locally symmetric about the mid-band frequency $\pm f_c$, then the in-phase noise $n_c(t)$ and the quadrature noise $n_s(t)$ are statistically independent.

To prove this property, we observe that if $S_N(f)$ is locally symmetric about $\pm f_c$, then

$$S_N(f-f_c) = S_N(f+f_c) \tag{5.167}$$

Consequently, we find from Eq. (5.165) that the cross-spectral densities of the quadrature components $n_c(t)$ and $n_s(t)$ are zero for all frequencies. This, in turn, means that the cross-correlation functions $R_{N_cN_s}(\tau)$ and $R_{N_sN_c}(\tau)$ are zero for all τ. That is,

$$E[N_c(t_k+\tau)N_s(t_k)]=0, \tag{5.168}$$

which implies that the random variables $N_c(t_k+\tau)$ and $N_s(t_k)$ (obtained by observing the in-phase noise at time $t_k+\tau$ and the quadrature noise at time t_k, respectively) are orthogonal for all τ.

The narrow-band noise $n(t)$ is assumed to be Gaussian with zero mean; hence, from Properties 1 and 2 it follows that both $N_c(t_k+\tau)$ and $N_s(t_k)$ are also Gaussian with zero mean. We thus conclude that because $N_c(t_k+\tau)$ and $N_s(t_k)$ are orthogonal and have zero mean, they are uncorrelated, and being Gaussian, they are statistically independent for all τ. In other words, the in-phase noise $n_c(t)$ and the quadrature noise $n_s(t)$ are statistically independent.

In accordance with Property 7, we may therefore express the joint-probability density function of the random variables $N_c(t_k+\tau)$ and $N_s(t_k)$ (for any delay τ) as the product of their individual probability density functions, as shown by

$$\begin{aligned} f_{N_c(t_k+\tau),N_s(t_k)}(n_c, n_s) &= f_{N_c(t_k+\tau)}(n_c)f_{N_s(t_k)}(n_s) \\ &= \frac{1}{\sqrt{2\pi}\sigma}\exp\left(-\frac{n_c^2}{2\sigma^2}\right)\frac{1}{\sqrt{2\pi}\sigma}\exp\left(-\frac{n_s^2}{2\sigma^2}\right) \\ &= \frac{1}{2\pi\sigma^2}\exp\left(-\frac{n_c^2+n_s^2}{2\sigma^2}\right) \end{aligned} \tag{5.169}$$

where σ^2 is the variance of the original narrow-band noise $n(t)$. Equation (5.169) holds only if the spectral density $S_N(f)$ of $n(t)$ is locally symmetrical about $\pm f_c$. Otherwise, this relation holds only for $\tau=0$ or those values of τ for which $n_c(t)$ and $n_s(t)$ are uncorrelated.

To sum up, if the narrow-band noise $n(t)$ is zero-mean, stationary, and Gaussian, then its quadrature components $n_c(t)$ and $n_s(t)$ are both zero-mean, jointly stationary, and jointly Gaussian. To evaluate the power spectral density of $n_c(t)$ or $n_s(t)$, we may proceed as follows:

1. We shift the positive frequency portion of the power spectral density $S_N(f)$ of the original narrow-band noise $n(t)$ to the left by f_c.
2. We shift the negative frequency portion of $S_N(f)$ to the right by f_c.
3. We add these two shifted spectra to obtain the desired $S_{N_c}(f)$ or $S_{N_s}(f)$.

Example 16 Ideal band-pass filtered white noise

Consider a white Gaussian noise of zero mean and power spectral density $N_0/2$ which is passed through an ideal band-pass filter of passband amplitude response equal to one, mid-band frequency f_c, and bandwidth $2B$. The power spectral density characteristic of the filtered noise

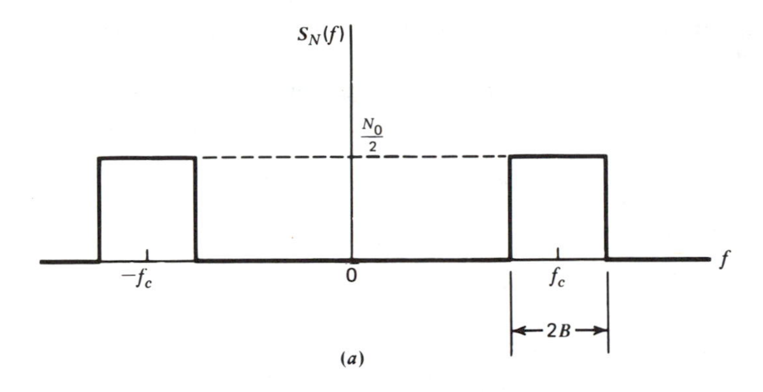

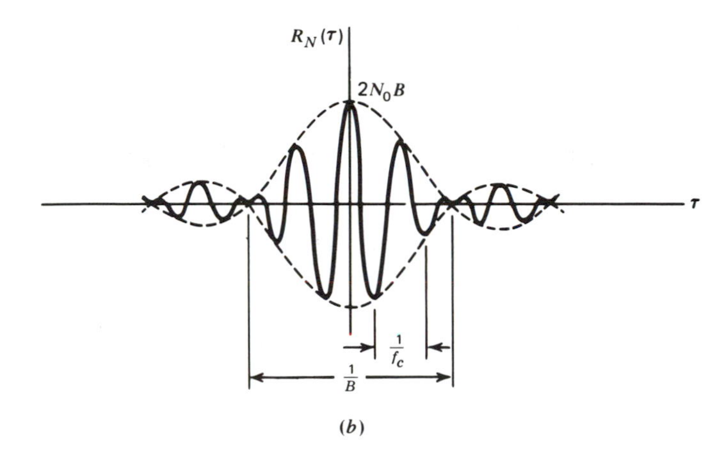

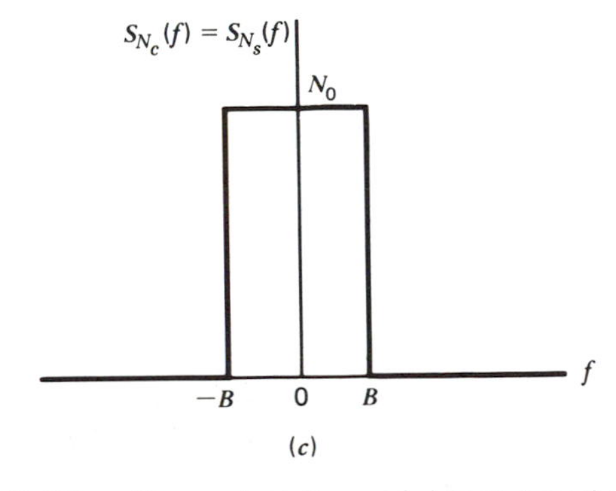

Figure 5.29 Characteristics of ideal band-pass filtered white noise. *(a)* Power spectral density. *(b)* Autocorrelation function. *(c)* Power spectral density of in-phase and quadrature components.

$n(t)$ will therefore be as shown in Fig. 5.29(a). The problem is to determine the autocorrelation functions of $n(t)$ and its in-phase and quadrature components.

The autocorrelation function of $n(t)$ is the inverse Fourier transform of the power spectral density characteristic shown in Fig. 5.29(a):

$$\begin{aligned} R_N(\tau) &= \int_{-f_c-B}^{-f_c+B} \frac{N_0}{2} \exp(j2\pi f\tau)df + \int_{f_c-B}^{f_c+B} \frac{N_0}{2} \exp(j2\pi f\tau)df \\ &= N_0 B \operatorname{sinc}(2B\tau)[\exp(-j2\pi f_c\tau) + \exp(j2\pi f_c\tau)] \\ &= 2N_0 B \operatorname{sinc}(2B\tau)\cos(2\pi f_c\tau) \end{aligned} \tag{5.170}$$

which is shown plotted in Fig. 5.29(b).

The spectral density characteristic of Fig. 5.29(a), is symmetrical about $\pm f_c$. Therefore, we find that the corresponding spectral density characteristic of the in-phase noise component $n_c(t)$ or the quadrature noise component $n_s(t)$ is as shown in Fig. 5.29(c). The autocorrelation function of $n_c(t)$ or $n_s(t)$ is therefore (see Example 13):

$$R_{N_c}(\tau) = R_{N_s}(\tau) = 2N_0 B \operatorname{sinc}(2B\tau) \tag{5.171}$$

Example 17 Transmission of white noise through a high-Q tuned filter

Consider next the band-pass LCR filter of Fig. 5.30(a). The transfer function of this filter, relating the output voltage to the input voltage, is given by

$$H(f) = \frac{R}{R + j2\pi f L + (1/j2\pi f C)} \tag{5.172}$$

Defining

$$f_c = \frac{1}{2\pi\sqrt{LC}}, \tag{5.173}$$

and

$$Q = \frac{1}{R}\sqrt{\frac{L}{C}}, \tag{5.174}$$

we may rewrite Eq. (5.172) as

$$H(f) = \frac{1}{1 + jQ[(f/f_c) - (f_c/f)]} \tag{5.175}$$

If the Q-factor of the filter is high compared with unity, we may approximate Eq. (5.175) as follows

$$H(f) \simeq \begin{cases} \dfrac{1}{1 + j2Q(f - f_c)/f_c}, & f > 0 \\ \dfrac{1}{1 + j2Q(f + f_c)/f_c} & f < 0 \end{cases} \tag{5.176}$$

Suppose that we connect, across the input terminals of this filter, a voltage source generating white Gaussian noise of zero mean and power spectral density $N_0/2$. The power spectral density of the resulting noise $n(t)$ at the filter output is therefore

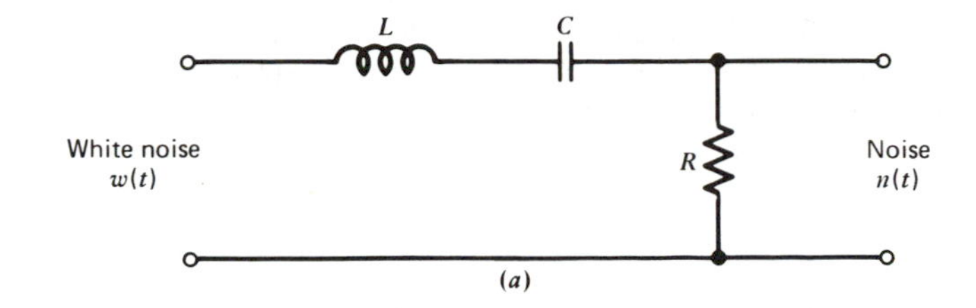

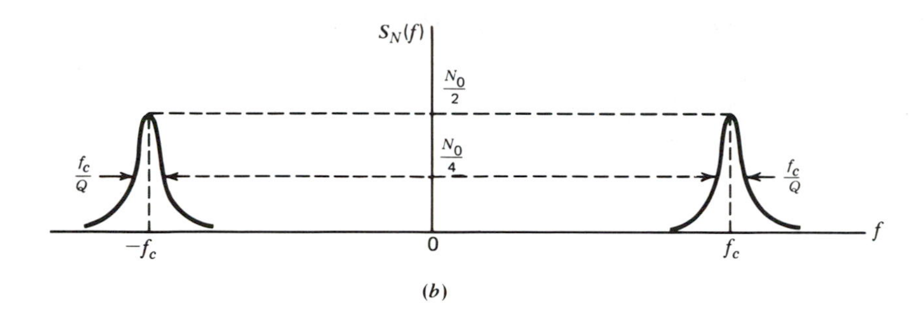

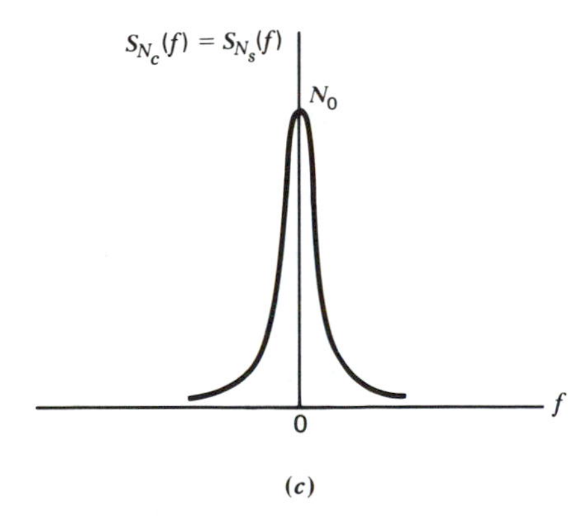

Figure 5.30 *(a)* LCR filter. *(b)* Power spectral density of filter output *n(t)* produced by white noise input. *(c)* Power spectral density of in-phase and quadrature components of *n(t)*.

$$S_N(f) \simeq \begin{cases} \dfrac{N_0/2}{1+4Q^2(f-f_c)^2/f_c^2}, & f<0 \\[2ex] \dfrac{N_0/2}{1+4Q^2(f+f_c)^2/f_c^2}, & f>0 \end{cases} \tag{5.177}$$

which is shown plotted in Fig. 5.30(*b*). Thus the corresponding power spectral density of the in-phase noise component $n_c(t)$ or quadrature noise component $n_c(t)$ is

$$S_{N_c}(f) = S_{N_s}(f) \simeq \frac{N_0}{1+(2Qf/f_c)^2} \tag{5.178}$$

which is shown plotted in Fig. 5.30(*c*). Comparing this last relation with Eq. (5.145), we see that they are both basically of a similar form. This means that the in-phase or quadrature component of the noise $n(t)$ at the output of the narrow-band filter of Fig. 5.30(*a*) has effectively the same characteristics as a noise process produced by passing a white noise through a corresponding low-pass *RC* filter.

Representation of Narrow-Band Noise in Terms of Envelope and Phase Components

We considered above the representation of a narrow-band noise $n(t)$ in terms of its in-phase and quadrature components. We may also represent the noise $n(t)$ in terms of its envelope and phase components as follows:

$$n(t)=r(t)\cos[2\pi f_c t+\psi(t)] \tag{5.179}$$

where

$$r(t)=[n_c^2(t)+n_s^2(t)]^{1/2} \tag{5.180}$$

and

$$\psi(t)=\tan^{-1}\left[\frac{n_s(t)}{n_c(t)}\right] \tag{5.181}$$

The function $r(t)$ is called the *envelope* of $n(t)$, and the function $\psi(t)$ is called the *phase* of $n(t)$.

The probability distributions of $r(t)$ and $\psi(t)$ may be obtained from those of $n_c(t)$ and $n_s(t)$ as follows. Let N_c and N_s represent the random variables obtained by observing (at some fixed time) the random processes represented by the sample functions $n_c(t)$ and $n_s(t)$, respectively. We note that N_c and N_s are independent Gaussian random variables of zero mean and variance σ^2, and so we may express their joint probability density function by

$$f_{N_c,N_s}(n_c, n_s)=\frac{1}{2\pi\sigma^2}\exp\left(-\frac{n_c^2+n_s^2}{2\sigma^2}\right) \tag{5.182}$$

Accordingly, the probability of the joint event that N_c lies between n_c+dn_c, and that N_s lies between n_s and n_s+dn_s [i.e., the pair of random variables N_c and N_s lies jointly inside the shaded area of Fig. 5.31(*a*)] is given by

$$f_{N_c,N_s}(n_c, n_s)dn_c dn_s=\frac{1}{2\pi\sigma^2}\exp\left(-\frac{n_c^2+n_s^2}{2\sigma^2}\right)dn_c\, dn_s \tag{5.183}$$

Define the transformation (see Fig. 5.31)

$$n_c=r\cos\psi \tag{5.184}$$

and

$$n_s=r\sin\psi \tag{5.185}$$

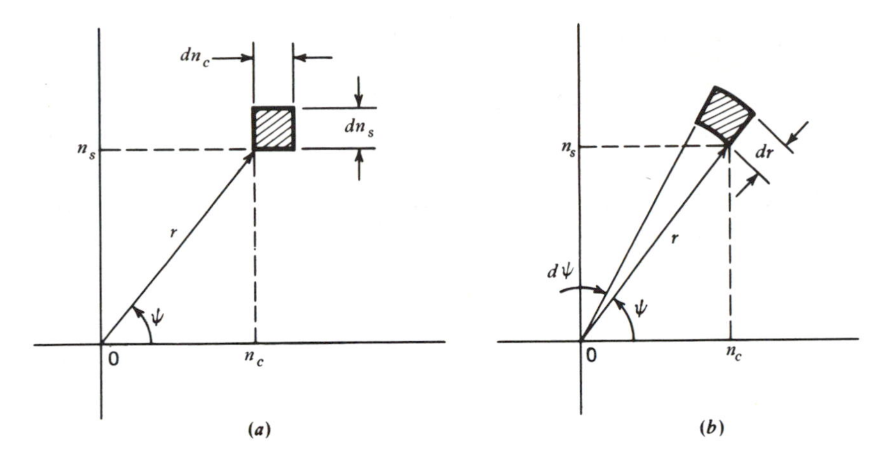

Figure 5.31 Illustrating the coordinate system for representation of narrow-band noise: *(a)* in terms of in-phase and quadrature components, and *(b)* in terms of envelope and phase.

In a limiting sense, we may equate the two incremental areas shown shaded in parts (*a*) and (*b*) of Fig. 5.31, and thus write

$$dn_c\, dn_s = r\, dr\, d\psi \tag{5.186}$$

Now, let R and Ψ denote the random variables obtained by observing (at some time) the random processes represented by the envelope $r(t)$ and phase $\psi(t)$, respectively. Then, substituting Eqs. (5.182), (5.185), and (5.186) in (5.183), we find the probability of the random variables R and Ψ lying jointly inside the shaded area of Fig. 5.31(*b*) is equal to

$$\frac{r}{2\pi\sigma^2}\exp\left(-\frac{r^2}{2\sigma^2}\right)dr\, d\psi$$

That is, the joint probability density function of R and Ψ is

$$f_{R,\Psi}(r, \psi) = \frac{r}{2\pi\sigma^2}\exp\left(-\frac{r^2}{2\sigma^2}\right) \tag{5.187}$$

This probability density function is independent of the angle ψ, which means that the random variables R and Ψ are statistically independent. We may thus express $f_{R,\Psi}(r, \psi)$ as the product of $f_R(r)$ and $f_\Psi(\psi)$. In particular, the random variable Ψ is uniformly distributed inside the range 0 to 2π, as shown by

$$f_\Psi(\psi) = \begin{cases} \dfrac{1}{2\pi}, & 0 \leq \psi \leq 2\pi \\ 0, & \text{elsewhere} \end{cases} \tag{5.188}$$

This leaves the probability density function of R as

$$f_R(r)=\begin{cases}\dfrac{r}{\sigma^2}\exp\left(-\dfrac{r^2}{2\sigma^2}\right), & r\geq 0\\ 0, & \text{elsewhere}\end{cases} \tag{5.189}$$

where σ^2 is the variance of the original narrow-band noise $n(t)$. A random variable having the probability density function of Eq. (5.189) is said to be *Rayleigh-distributed.**

For convenience of graphical presentation, let

$$v=\frac{r}{\sigma} \tag{5.190}$$

and

$$f_V(v)=\sigma f_R(r) \tag{5.191}$$

Then we may rewrite the Rayleigh distribution of Eq. (5.189) in the normalized form

$$f_V(v)=\begin{cases}v\exp\left(-\dfrac{v^2}{2}\right), & v\geq 0\\ 0, & \text{elsewhere}\end{cases} \tag{5.192}$$

Equation (5.192) is shown plotted in Fig. 5.32. The peak value of the distribution $f_V(v)$ occurs at $v=1$ and is equal to 0.607. Note also that, unlike the Gaussian distribution, the Rayleigh distribution is zero for negative values of v. This is because the envelope $r(t)$ can only assume positive values.

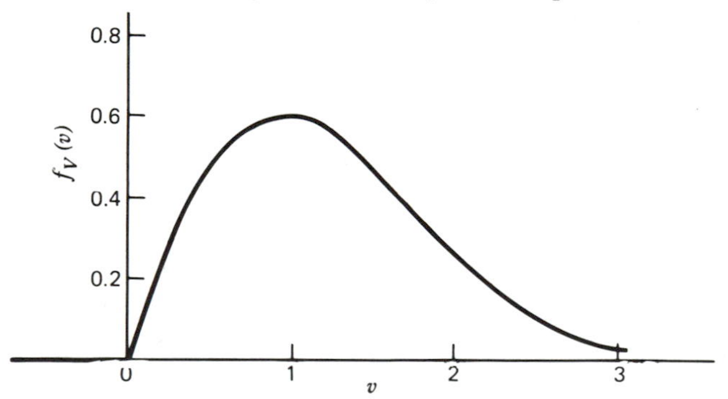

Figure 5.32 Rayleigh distribution.

5.12 ENVELOPE OF SINE WAVE PLUS NARROW-BAND NOISE

Suppose next that we add the sinusoidal wave $A\cos(2\pi f_c t)$ to the narrow-band noise $n(t)$, where A and f_c are both constants. We assume that the frequency of the

* The Rayleigh distribution is named after the English physicist J. W. Strutt, Lord Rayleigh.

sinusoidal wave is the same as the nominal mid-band frequency for the noise. A sample function of the sine wave-plus-noise is then expressed by

$$x(t)=A\cos(2\pi f_c t)+n(t) \tag{5.193}$$

Representing the narrow-band noise $n(t)$ in terms of its in-phase and quadrature components, we may write

$$x(t)=n_c'(t)\cos(2\pi f_c t)-n_s(t)\sin(2\pi f_c t) \tag{5.194}$$

where

$$n_c'(t)=A+n_c(t) \tag{5.195}$$

We assume that $n(t)$ is Gaussian with zero mean and variance σ^2. Accordingly, we may state that:

1. Both $n_c'(t)$ and $n_s(t)$ are Gaussian and statistically independent.
2. The mean of $n_c'(t)$ is A and that of $n_s(t)$ is zero.
3. The variance of both $n_c'(t)$ and $n_s(t)$ is σ^2.

We may therefore express the joint probability density function of the random variables N_c' and N_s corresponding to $n_c'(t)$ and $n_s(t)$, as follows:

$$f_{N_c',N_s}(n_c', n_s)=\frac{1}{2\pi\sigma^2}\exp\left[-\frac{(n_c'-A)^2+n_s^2}{2\sigma^2}\right] \tag{5.196}$$

Let $r(t)$ denote the envelope of $x(t)$ and $\psi(t)$ denote its phase. From Eq. (5.194), we thus find that

$$r(t)=\{[n_c'(t)]^2+n_s^2(t)\}^{1/2} \tag{5.197}$$

and

$$\psi(t)=\tan^{-1}\left[\frac{n_s(t)}{n_c'(t)}\right] \tag{5.198}$$

Following a procedure similar to that described in Section 5.11 for the derivation of the Rayleigh distribution, we find that the joint probability density function of the random variables R and Ψ, corresponding to $r(t)$ and $\psi(t)$, is given by

$$f_{R,\Psi}(r, \psi)=\frac{r}{2\pi\sigma^2}\exp\left(-\frac{r^2+A^2-2Ar\cos\psi}{2\sigma^2}\right) \tag{5.199}$$

We see that in this case, however, we cannot express the joint probability density function $f_{R,\Psi}(r, \psi)$ as a product $f_R(r)f_\Psi(\psi)$. This is because we now have a term involving the values of both random variables multiplied together as $r\cos\psi$. Hence, R and Ψ are dependent random variables for nonzero values of the amplitude A of the sinusoidal wave component.

We are interested, in particular, in the probability density function of R. To determine this probability density function, we integrate Eq. (5.199) over all possible values of ψ, obtaining the marginal density

$$f_R(r) = \int_0^{2\pi} f_{R,\Psi}(r, \psi)\, d\psi$$

$$= \frac{r}{2\pi\sigma^2} \exp\left(-\frac{r^2 + A^2}{2\sigma^2}\right) \int_0^{2\pi} \exp\left(\frac{Ar}{\sigma^2} \cos\psi\right) d\psi \tag{5.200}$$

The integral in the right-hand side of Eq. (5.200) can be identified in terms of the defining integral for the *modified Bessel function of the first-kind of zero order* (see Appendix 4); that is,

$$I_0(x) = \frac{1}{2\pi} \int_0^{2\pi} \exp(x \cos\psi)\, d\psi \tag{5.201}$$

Thus, letting $x = Ar/\sigma^2$, we may rewrite Eq. (5.200) in the compact form:

$$f_R(r) = \frac{r}{\sigma^2} \exp\left(-\frac{r^2 + A^2}{2\sigma^2}\right) I_0\left(\frac{Ar}{\sigma^2}\right) \tag{5.202}$$

This relation is called the *Rician distribution.**

As with the Rayleigh distribution, the graphical presentation of the Rician distribution is simplified by putting

$$v = \frac{r}{\sigma}, \tag{5.203}$$

$$a = \frac{A}{\sigma}, \tag{5.204}$$

and

$$f_V(v) = \sigma f_R(r) \tag{5.205}$$

Then we may express the Rician distribution of Eq. (5.202) in the *normalized* form

$$f_V(v) = v \exp\left(-\frac{v^2 + a^2}{2}\right) I_0(av) \tag{5.206}$$

which is shown plotted in Fig. 5.33 for the values 0, 1, 2, 3, 5 of the parameter a. Based on these curves, we observe that:

1. When a is zero, we get the Rayleigh distribution.
2. The envelope distribution is approximately Gaussian in the vicinity of $v = a$ when a is large, that is, when the sine-wave amplitude A is large compared with σ, the square root of the average power of the noise $n(t)$.

* S. O. Rice, "Mathematical analysis of random noise," *Bell System Tech J.*, vol. 23, pp. 282–333, July 1944; vol. 24, pp. 96–157, Jan. 1945.

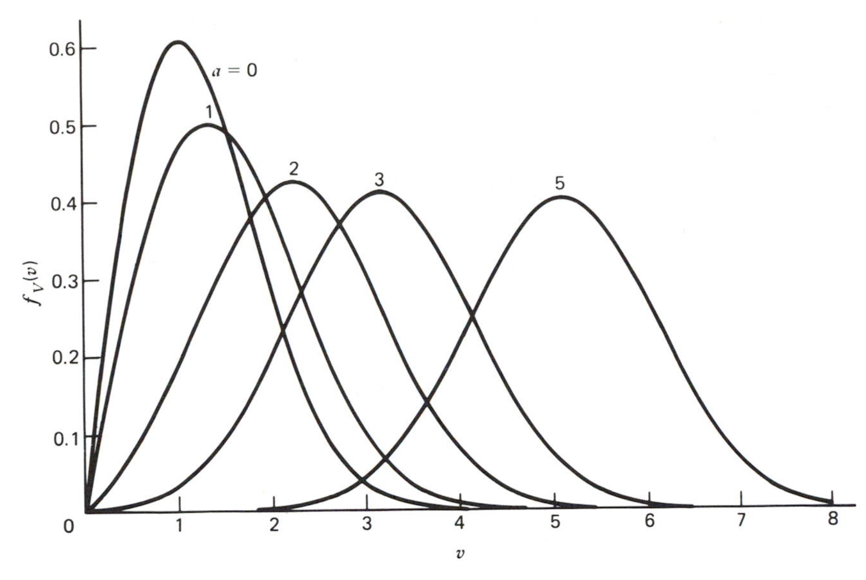

Figure 5.33 Rician distribution.

Problems

Problem 5.1 A Gaussian-distributed random variable X, of zero mean and variance σ_X^2, is transformed by a piecewise-linear rectifier characterized by the input–output relation (see Fig. P5.1)

$$Y = \begin{cases} X, & X \geq 0 \\ 0, & X < 0. \end{cases}$$

The probability density function of the new random variable Y is described by

$$f_Y(y) = \begin{cases} 0, & y < 0 \\ k\delta(f), & y = 0 \\ \dfrac{1}{\sqrt{2\pi}\sigma_X} \exp\left(-\dfrac{y^2}{2\sigma_X^2}\right), & y > 0 \end{cases}$$

(a) Explain the reasons for this result.
(b) Determine the value of the constant k by which the delta function $\delta(f)$ is weighted.

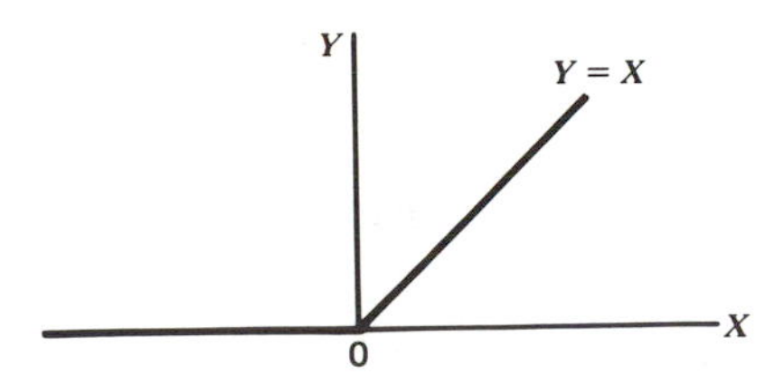

Figure P5.1

Problem 5.2 A Gaussian-distributed random variable X, of zero mean and variance σ_X^2, is transformed by a square-law device defined by (see Fig. P5.2)

$$Y = X^2$$

Show that the probability density function of the new random variable Y equals

$$f_Y(y) = \begin{cases} \dfrac{1}{\sqrt{2\pi y}\sigma_X} \exp\left(-\dfrac{y}{2\sigma_X^2}\right), & y \geq 0 \\ 0, & y < 0 \end{cases}$$

Hint: Evaluate the probability $P(Y \leq y)$ for the two intervals: (a) $y < 0$, and (b) $y \geq 0$.

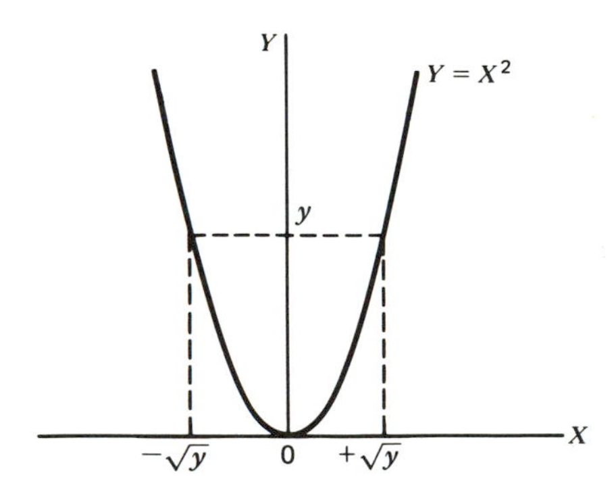

Figure P5.2

Problem 5.3 Consider a random process $X(t)$ defined by

$$X(t) = \sin(2\pi F t)$$

in which the frequency F is a random variable with the probability density function

$$f_F(f) = \begin{cases} \dfrac{1}{W}, & 0 \leq f \leq W \\ 0, & \text{otherwise} \end{cases}$$

Show that $X(t)$ is nonstationary.

Hint: Examine specific sample functions of the random process $X(t)$ for the frequency $f = W/4$, $W/2$, and W, say.

Problem 5.4 Consider the sine-wave process

$$X(t) = A\cos(2\pi f_c t)$$

where the frequency f_c is constant and the amplitude A is uniformly distributed:

$$f_A(a) = \begin{cases} 1, & 0 \leq a \leq 1 \\ 0, & \text{otherwise} \end{cases}$$

Determine whether or not this process is stationary in the strict sense.

Problem 5.5 A random process $X(t)$ is defined by

$$X(t) = A\cos(2\pi f_c t)$$

where A is a Gaussian-distributed random variable of zero mean and variance σ_A^2. This random process is applied to an ideal integrator, producing an output $Y(t)$ defined by

$$Y(t) = \int_0^t X(\tau)d\tau$$

(a) Determine the probability density function of the output $Y(t)$ at a particular time t_k.
(b) Determine whether or not $Y(t)$ is stationary.
(c) Determine whether or not $Y(t)$ is ergodic.

Problem 5.6 Let X and Y be statistically independent Gaussian-distributed random variables, each with zero mean and unit variance. Define the Gaussian process

$$Z(t) = X\cos(2\pi t) + Y\sin(2\pi t)$$

(a) Determine the joint probability density function of the random variables $Z(t_1)$ and $Z(t_2)$ obtained by observing $Z(t)$ at times t_1 and t_2.
(b) Is the process $Z(t)$ stationary? Why?

Problem 5.7 Prove the following two properties of the autocorrelation function $R_X(\tau)$ of a random process $X(t)$:

(a) If $X(t)$ contains a dc component equal to A, then $R_X(\tau)$ will contain a constant component equal to A^2.
(b) If $X(t)$ contains a sinusoidal component, then $R_X(\tau)$ will also contain a sinusoidal component of the same frequency.

Problem 5.8 The square wave $x(t)$ of Fig. P5.3 of constant amplitude A, period T_0, and delay t_d, represents the sample function of a random process $X(t)$. The delay is random, described by the probability density function

$$f_{T_d}(t_d) = \begin{cases} \dfrac{1}{T_0}, & -\frac{1}{2}T_0 \le t_d \le \frac{1}{2}T_0 \\ 0, & \text{otherwise} \end{cases}$$

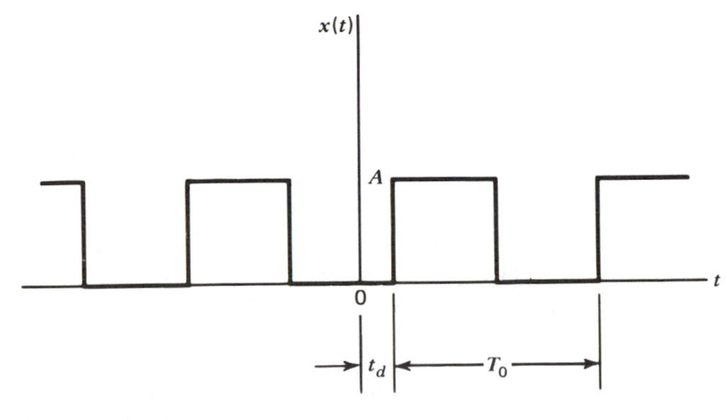

Figure P5.3

(a) Determine the probability density function of the random variable $X(t_k)$ obtained by observing the random process $X(t)$ at time t_k.
(b) Determine the mean and autocorrelation function of $X(t)$ using ensemble-averaging.
(c) Determine the mean and autocorrelation function of $X(t)$ using time-averaging.
(d) Establish whether or not $X(t)$ is wide-sense stationary. In what sense is it ergodic?

Problem 5.9 A binary wave consists of a random sequence of symbols 1 and 0, similar to that described in Example 4, with one basic difference: symbol 1 is now represented by a pulse of amplitude A volts and symbol 0 is represented by zero volts. All other parameters are the same as before. Show that for this new random binary wave $X(t)$:

(a) The autocorrelation function equals

$$R_X(\tau)=\begin{cases} \dfrac{A^2}{4}+\dfrac{A^2}{4}\left(1-\dfrac{|\tau|}{T}\right), & |\tau|<T \\ \dfrac{A^2}{4}, & |\tau|\geq T \end{cases}$$

(b) The power spectral density equals

$$S_X(f)=\frac{A^2}{4}\,\delta(f)+\frac{A^2T}{4}\,\mathrm{sinc}^2(fT)$$

What is the percentage power contained in the dc component of the binary wave?

Problem 5.10 A random process $Y(t)$ consists of a dc component of $\sqrt{3/2}$ volts, a periodic component $g_p(t)$, and a random component $X(t)$. The autocorrelation function of $Y(t)$ is shown in Fig. P5.4.

(a) What is the average power of the periodic component $g_p(t)$?
(b) What is the average power of the random component $X(t)$?

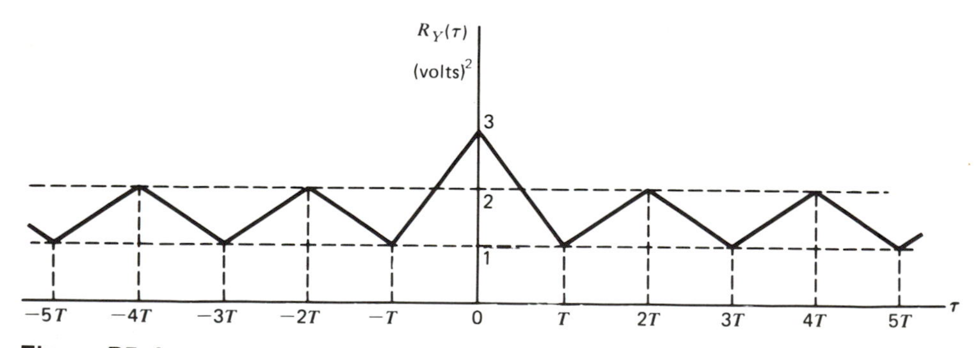

Figure P5.4

Problem 5.11 Consider a pair of wide-sense stationary random processes $X(t)$ and $Y(t)$. Show that the cross-correlations $R_{XY}(\tau)$ and $R_{YX}(\tau)$ of these processes have the following properties:

(a) $R_{XY}(-\tau)=R_{YX}(\tau)$
(b) $|R_{XY}(\tau)|\leq\frac{1}{2}[R_X(0)+R_Y(0)]$
where $R_X(\tau)$ and $R_Y(\tau)$ are the autocorrelation functions of $X(t)$ and $Y(t)$, respectively.

Problem 5.12 Consider two linear filters connected in cascade as in Fig. P5.5. Let $X(t)$ be a

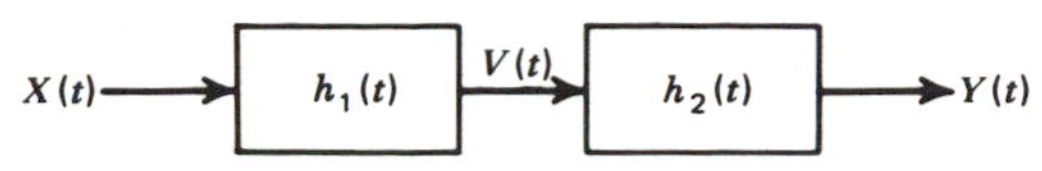

Figure P5.5

wide-sense stationary process with autocorrelation function $R_X(\tau)$. The random process appearing at the first filter output is $V(t)$ and that at the second filter output is $Y(t)$.

(a) Find the autocorrelation function of $Y(t)$.
(b) Find the cross-correlation function $R_{VY}(\tau)$ of $V(t)$ and $Y(t)$.

Problem 5.13 A wide-sense stationary random process $X(t)$ is applied to a linear time-invariant filter of impulse response $h(t)$, producing an output $Y(t)$.

(a) Show that the cross-correlation function $R_{YX}(\tau)$ of the output $Y(t)$ and input $X(t)$ is equal to the impulse response $h(\tau)$ convolved with the autocorrelation function $R_X(\tau)$ of the input, as shown by

$$R_{YX}(\tau)=\int_{-\infty}^{\infty} h(u)R_X(\tau-u)du$$

(b) Show that the second cross-correlation function $R_{XY}(\tau)$ equals

$$R_{XY}(\tau)=\int_{-\infty}^{\infty} h(-u)R_X(\tau-u)du$$

(c) Assuming that $X(t)$ is a white noise process with zero mean and power spectral density $N_0/2$, show that

$$R_{YX}(\tau)=\frac{N_0}{2}h(\tau)$$

Comment on the practical significance of this result.

Problem 5.14 Given that a stationary random process $X(t)$ has the autocorrelation function $R_X(\tau)$ and power spectral density $S_X(f)$, show that:

(a) The autocorrelation function of $dX(t)/dt$, the first derivative of $X(t)$, is equal to minus the second derivative of $R_X(\tau)$.
(b) The power spectral density of $dX(t)/dt$ is equal to $4\pi^2 f^2 S_X(f)$.

Problem 5.15 Consider the function $\sigma(f)$ defined by

$$\sigma(f)=\frac{S(f)}{R(0)}$$

where $S(f)$ is the power spectral density of a random process and $R(0)$ is the value of its autocorrelation function for a lag of zero (i.e., $\tau=0$). Explain why $\sigma(f)$ has the properties usually associated with a probability density function.

Problem 5.16 Consider a wide-sense stationary random process $X(t)$ having the power spectral density $S_X(f)$ shown in Fig. P5.6. Find the autocorrelation function $R_X(\tau)$ of the process $X(t)$.

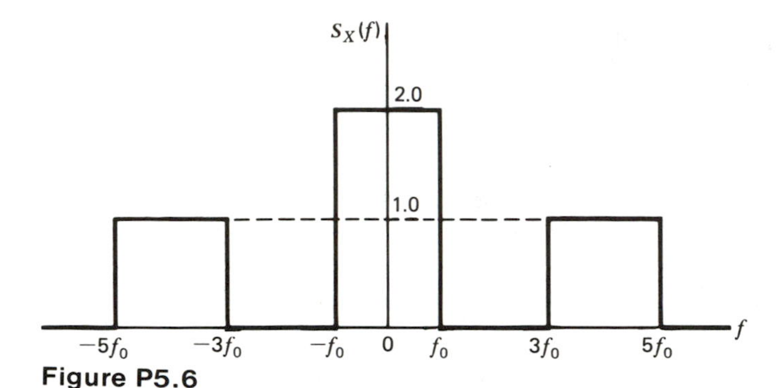

Figure P5.6

Problem 5.17 The power spectral density of a random process $X(t)$ is shown in Fig. P5.7.

(a) Determine and sketch the autocorrelation function $R_X(\tau)$ of $X(t)$.
(b) What is the dc power contained in $X(t)$?
(c) What is the ac power contained in $X(t)$?
(d) What sampling rates will give uncorrelated samples of $X(t)$? Are the samples statistically independent?

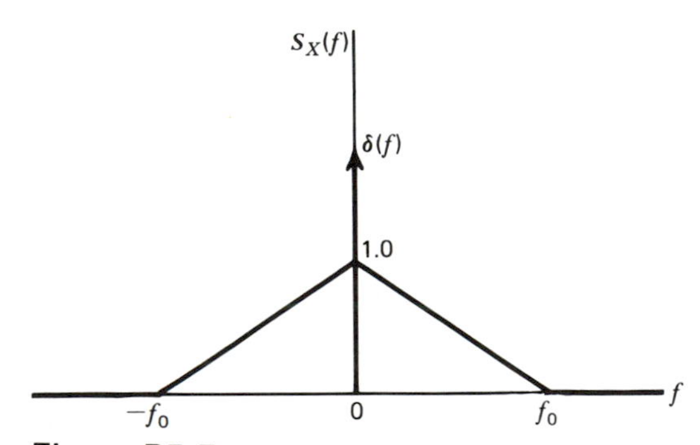

Figure P5.7

Problem 5.18 A pair of noise processes $n_1(t)$ and $n_2(t)$ are related by

$$n_2(t) = n_1(t)\cos(2\pi f_c t + \theta) - n_1(t)\sin(2\pi f_c t + \theta)$$

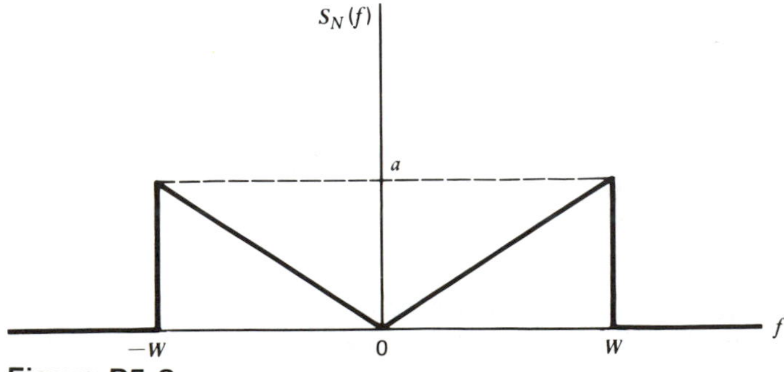

Figure P5.8

where f_c is a constant and θ is the value of a random variable defined by

$$f_\Theta(\theta)=\begin{cases}\dfrac{1}{2\pi}, & 0\leqslant\theta\leqslant 2\pi\\ 0, & \text{otherwise}\end{cases}$$

The noise process $n_1(t)$ is stationary and its power spectral density is as shown in Fig. P5.8. Find and plot the corresponding power spectral density of $n_2(t)$.

Problem 5.19 A *random telegraph signal* $X(t)$, characterized by the autocorrelation function

$$R_X(\tau)=\exp(-2v|\tau|)$$

where v is a constant, is applied to the low-pass RC filter of Fig. P5.9. Determine the power spectral density and autocorrelation function of the random process at the filter output.

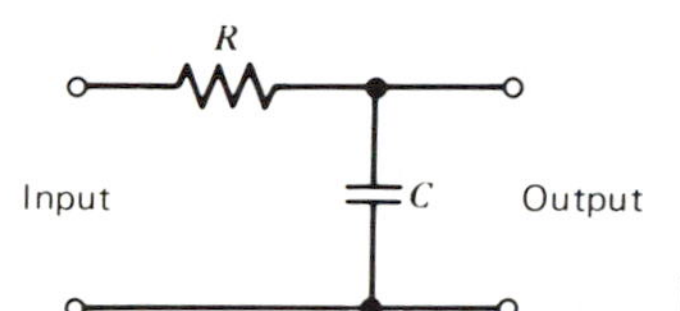

Figure P5.9

Problem 5.20 The output of an oscillator is described by

$$X(t)=A\cos(2\pi Ft+\Theta),$$

where A is a constant, and F and Θ are independent random variables. The probability density function of F is denoted by $f_F(f)$, and that of Θ is defined by

$$f_\Theta(\theta)=\begin{cases}\dfrac{1}{2\pi}, & 0\leqslant\theta\leqslant 2\pi\\ 0, & \text{otherwise}\end{cases}$$

Show that the power spectral density of $X(t)$ is given by

$$S_X(f)=\frac{A^2}{2}f_F(f)$$

What happens to this power spectrum when the frequency f assumes a constant value?

Problem 5.21 Consider a sample function $x(t)$ of a random process $X(t)$, which is observed for the interval $0\leqslant t\leqslant T$. An *estimate* of the autocorrelation function of the process is defined by

$$\hat{R}_X(\tau)=\begin{cases}\dfrac{1}{T}\displaystyle\int_0^{T-|\tau|}x(t+\tau)x(t)dt, & |\tau|\leqslant T\\ 0, & \text{otherwise}\end{cases}$$

(a) Determine the expected value of $\hat{R}_X(\tau)$ and show that it approaches the autocorrelation function $R_X(\tau)$ of the process as the observation interval T approaches infinity.
(b) Show that $\hat{R}_X(\tau)$ equals $1/T$ times the (deterministic) autocorrelation function of the truncated sample function $x_T(t)$, where

$$x_T(t)=\begin{cases}x(t), & 0\leqslant t\leqslant T\\ 0, & \text{otherwise}\end{cases}$$

Hence, show that the Fourier transform of $\hat{R}_X(\tau)$ equals the periodogram $|X_T(f)|^2/T$, where $x_T(t) \rightleftharpoons X_T(f)$.

Problem 5.22 A stationary, Gaussian process $X(t)$ with mean m_X and variance σ_X^2 is passed through two linear filters with impulse responses $h_1(t)$ and $h_2(t)$, yielding processes $Y(t)$ and $Z(t)$, as shown in Fig. P5.10.

(a) Determine the joint probability density function of the random variables $Y(t_1)$ and $Z(t_2)$.
(b) What conditions are necessary and sufficient to ensure that $Y(t_1)$ and $Z(t_2)$ are statistically independent?

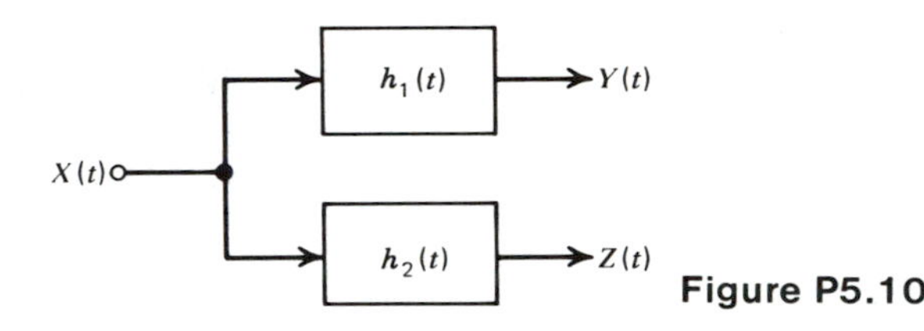

Figure P5.10

Problem 5.23 Let $X(t)$ be a stationary, Gaussian process with zero mean, variance σ^2, and autocorrelation function $R_X(\tau)$. This process is applied to an ideal linear rectifier characterized by the input–output relation shown in Fig. P5.1. The random process at the rectifier output is denoted by $Y(t)$.

(a) Using the result of Problem 5.1, show that the mean of $Y(t)$ equals

$$E[Y(t)] = \frac{\sigma}{\sqrt{2\pi}}$$

(b) Show that the autocorrelation function of $Y(t)$ is

$$R_Y(\tau) = \frac{\sigma^2}{2\pi}\{[1-\rho^2(\tau)]^{1/2} + \rho(\tau)\cos^{-1}[-\rho(\tau)]\}$$

where

$$\rho(\tau) = \frac{R_X(\tau)}{\sigma^2}$$

Hint: You may use the following result

$$\int_0^\infty \int_0^\infty uv \exp(-u^2 - v^2 - 2uv\cos\theta)\,du\,dv = \frac{1-\theta\cot\theta}{4\sin^2\theta}$$

Problem 5.24 A Gaussian process $X(t)$ of zero mean and variance σ^2 is passed through a full-wave rectifier, described by the input–output relation of Fig. P5.11.

(a) Show that the probability density function of the random variable $Y(t_k)$, obtained by observing the random process $Y(t)$ at the rectifier output at time t_k, is as follows

$$f_{Y(t_k)}(y) = \begin{cases} \sqrt{\dfrac{2}{\pi}}\dfrac{1}{\sigma}\exp\left(-\dfrac{y^2}{2\sigma^2}\right), & y \geq 0 \\ 0, & y < 0 \end{cases}$$

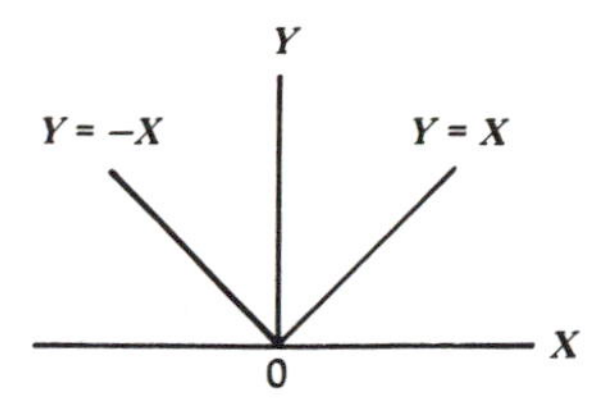

Figure P5.11

(b) Determine the mean and variance of the random variable $Y(t_k)$.

Problem 5.25 Let $X(t)$ be a zero-mean, stationary, Gaussian process with autocorrelation function $R_X(\tau)$. This process is applied to a square-law device defined by the input–output relation (see Fig. P5.2).

$$Y(t) = X^2(t)$$

where $Y(t)$ is the output.

(a) Using the result of Problem 5.2, show that the mean of $Y(t)$ is

$$E[Y(t)] = R_X(0)$$

(b) Show that the autocovariance function of $Y(t)$ is

$$K_Y(\tau) = 2R_X^2(\tau)$$

Problem 5.26 A stationary, Gaussian process $X(t)$ with zero mean and power spectral density $S_X(f)$ is applied to a linear filter whose impulse response $h(t)$ is shown in Fig. P5.12. A sample Y is taken of the random process at the filter output at time T.

(a) Determine the mean and variance of Y.
(b) What is the probability density function of Y?

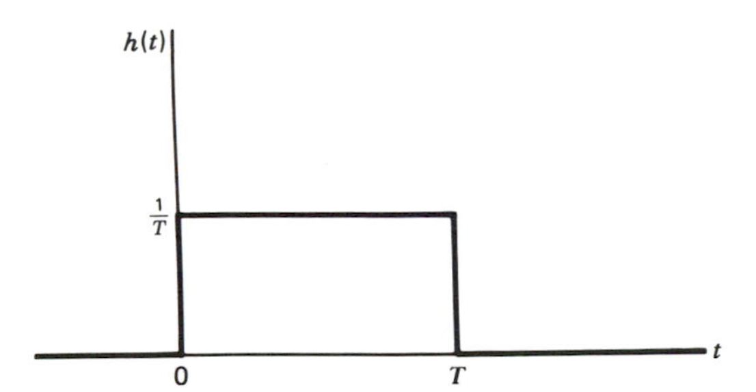

Figure P5.12

Problem 5.27 Consider the modulated random process

$$X(t) = A_c \cos[2\pi f_c t + \Theta + \Phi(t)]$$

where A_c and f_c are constants. The carrier phase Θ is a random variable uniformly distributed over the range 0–2π. Assume that the spectrum of $\exp[j\Phi(t)]$ is zero for frequencies greater than f_c.

(a) Show that the autocorrelation function of $X(t)$ is

$$R_X(\tau)=\frac{A_c^2}{2}\,\mathrm{Re}\{\exp(j2\pi f_c\tau)E[\exp(jQ_{t,\tau})]\}$$

where $Q_{t,\tau}=\Phi_{t+\tau}-\Phi_t$

(b) Assuming that $\Phi(t)$ is a zero-mean Gaussian process, show that

$$R_X(\tau)=\frac{A_c^2}{2}\cos(2\pi f_c\tau)\exp[-\tfrac{1}{2}\sigma_Q^2(t+\tau,\, t)]$$

where $\sigma_Q^2(t+\tau,\, t)=\mathrm{Var}[\Phi_{t+\tau}-\Phi_t]$

Hint: Express $X(t)$ as the real part of $A_c\exp(j2\pi f_c t)\exp(j\Theta)\exp[j\Phi(t)]$, and use the characteristic function of a Gaussian random variable.

Problem 5.28

(a) Suppose that the function $\Phi(t)$ in Problem 5.27 corresponds to frequency modulation, as shown by

$$\Phi(t)=2\pi k_f\int_0^t M(t)dt$$

where k_f is a constant and $M(t)$ is a zero-mean Gaussian process representing the modulating signal. Show that $\sigma_Q^2(t+\tau,\, t)$ is a function only of τ and may be expressed as

$$\sigma_Q^2(\tau)=4\pi^2k_f^2\int_0^\tau\int_0^\tau R_M(\tau_1-\tau_2)d\tau_1 d\tau_2$$

where $R_M(\tau)$ is the autocorrelation function of $M(t)$.

(b) Let

$$g(\tau)=\exp[-\tfrac{1}{2}\sigma_Q^2(\tau)]$$

and denote the Fourier transform of this function by $G(f)$. Show that the power spectral density of $X(t)$ is

$$S_X(f)=\frac{A_c^2}{4}[G(f-f_c)+G(f+f_c)]$$

(c) The root mean-square (rms) bandwidth of $G(f)$ is defined by

$$B_{\mathrm{rms}}=\left[\frac{\int_{-\infty}^{\infty}f^2G(f)df}{\int_{-\infty}^{\infty}G(f)df}\right]^{1/2}$$

Show that

$$B_{\mathrm{rms}}=k_f\sqrt{R_M(0)}$$

Hints: (1) Use properties of the Fourier transform, pertaining to differentiation in the time domain and the area under the Fourier transform, to express B_{rms} in terms of operations involving $g(\tau)$.

(2) For the differentiation of an integral, use *Leibnitz's rule*:

$$\frac{d}{dx}\int_{a(x)}^{b(x)}z(x,\, y)dy=\frac{db(x)}{dx}z[x,\, b(x)]-\frac{da(x)}{dx}z[x,\, a(x)]+\int_{a(x)}^{b(x)}\left[\frac{d}{dx}z(x,\, y)\right]dy$$

Problem 5.29 The current pulse induced in the plate circuit of a *temperature-limited diode* by one electron is given by

$$i_e(t)=\begin{cases}\dfrac{2e}{\tau_a^2}\,t, & 0\leqslant t\leqslant \tau_a\\ 0, & \text{otherwise}\end{cases}$$

where e is the electron charge, and τ_a is the time required for an electron to travel from the cathode to the plate. The total plate current is

$$i_T(t)=I_0+i(t)$$

where I_0 is the mean value of the plate current and $i(t)$ is the shot-noise current. The power spectral density of $i(t)$ is defined by

$$S_I(f)=\bar{n}|I_e(f)|^2$$

where $\bar{n}$ is the average number of electrons emitted per second, and $I_e(f)$ is the Fourier transform of $i_e(t)$.

(a) Show that the power spectral density of the shot-noise current is

$$S_I(f)=\frac{4eI_0}{(f/f_0)^4}\left\{\left(\frac{f}{f_0}\right)^2+2\left[1-\cos\left(\frac{f}{f_0}\right)-\frac{f}{f_0}\sin\left(\frac{f}{f_0}\right)\right]\right\}$$

where

$$f_0=\frac{1}{2\pi\tau_a}$$

(b) What is the value of $S_I(f)$ for frequencies small compared with f_0? What is the corresponding mean-square value of the shot-noise current $i(t)$ measured in a bandwidth Δf?

Problem 5.30 The *equipartition theorem* of statistical mechanics states that, for any physical system considered as a whole, the mean value of the kinetic energy for every independent direction of motion is $\frac{1}{2}kT$, where k is Boltzmann's constant and T is the absolute temperature. The independent directions of motion represent *degrees of freedom* of the system. The equipartition theorem is applicable to any system in which the total energy is expressible as the sum of a large number of individual terms, each one of which is proportional to the square of a different independent generalized coordinate describing the configuration of the system. Consider a conductor of resistance R connected in parallel with a lossless inductor L, as in Fig. P5.13. By applying the equipartition theorem to this circuit, and assuming that the power spectral density of the thermal noise voltage produced by the resistor is a constant, show that this constant is equal to $2kTR$. What is the mean-square value of the thermal noise voltage measured in a bandwidth Δf?

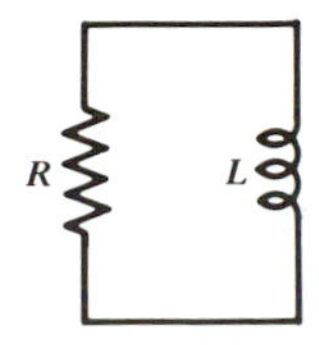

Figure P5.13

Problem 5.31 Consider a white Gaussian noise process of zero mean and power spectral density $N_0/2$ which is applied to the input of the system shown in Fig. P5.14.

(a) Find the power spectral density of the random process at the output of the system.
(b) What are the mean and variance of this output?

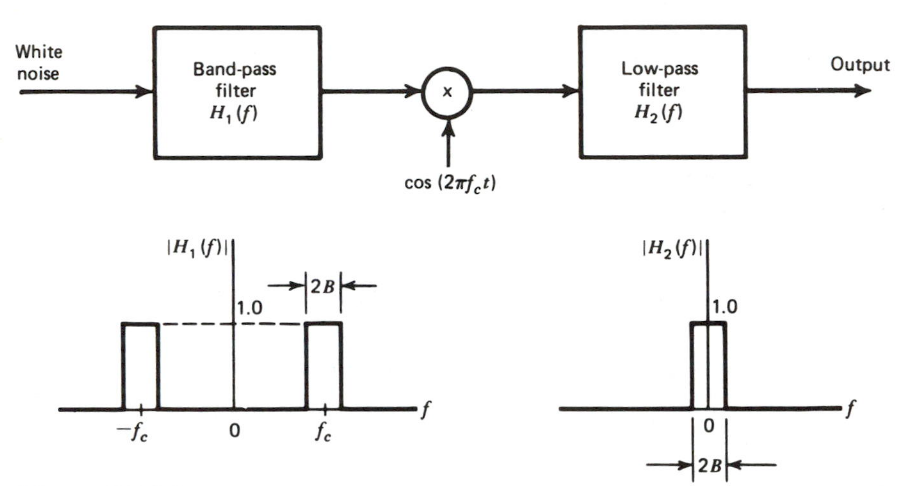

Figure P5.14

Problem 5.32 A white Gaussian noise $w(t)$ of zero mean and power spectral density $N_0/2$ is applied to the low-pass RC filter of Fig. P5.9 producing the noise $n(t)$. Determine the probability density function of the random variable obtained by observing the filter output at time t_k.

Problem 5.33 A white noise $w(t)$ of power spectral density $N_0/2$ is applied to a *Butterworth* low-pass filter of order n, whose amplitude response is defined by

$$|H(f)| = \frac{1}{[1+(f/f_0)^{2n}]^{1/2}}$$

(a) Determine the noise equivalent bandwidth for this low-pass filter.
(b) What is the limiting value of the noise equivalent bandwidth as n approaches infinity?

Problem 5.34 A white Gaussian noise $w(t)$ of zero mean and power spectral density $N_0/2$ is applied to the high-pass RL filter shown in Fig. P5.15. Determine the autocorrelation function of the output.

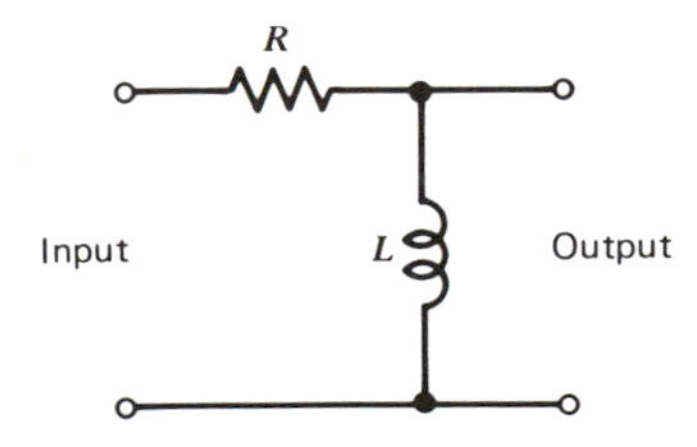

Figure P5.15

Problem 5.35 A voltage source generating white Gaussian noise of zero mean and power spectral density $N_0/2$ is connected to the input of the low-pass RL filter shown in Fig. P5.16. The noise at the filter output is denoted by $n(t)$.

(a) Find the autocorrelation function of $n(t)$.
(b) Find the variance of $n(t)$.
(c) What is the maximum rate at which $n(t)$ can be sampled so that the resulting samples are essentially uncorrelated?

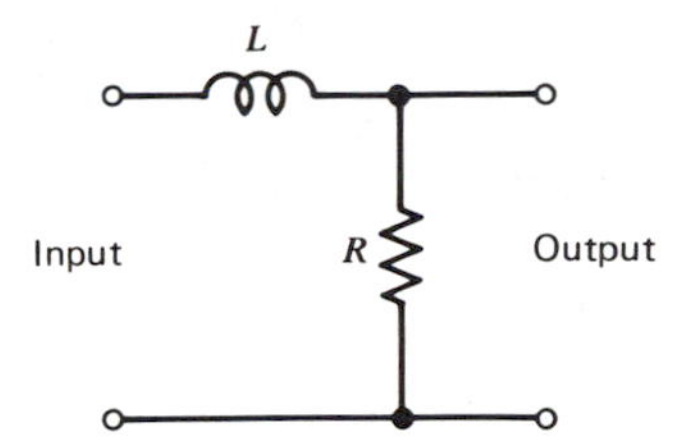

Figure P5.16

Problem 5.36 Let $X(t)$ be a stationary process with zero mean, autocorrelation function $R_X(\tau)$, and power spectral density $S_X(f)$. We are required to find a linear filter with impulse response $h(t)$, such that the filter output is $X(t)$ when the input is white noise of power spectral density $N_0/2$.

(a) Determine the condition that the impulse response $h(t)$ must satisfy in order to achieve this requirement.
(b) What is the corresponding condition on the transfer function $H(f)$ of the filter?
(c) Using the Paley–Wiener criterion (see Section 2.10), find the requirement on $S_X(f)$ for the filter to be causal.

Problem 5.37 Consider a narrow-band noise $n(t)$ with its Hilbert transform denoted by $\hat{n}(t)$.

(a) Show that the cross-correlation functions of $n(t)$ and $\hat{n}(t)$ are given by

$$R_{N\hat{N}}(\tau) = -\hat{R}_N(\tau)$$

and

$$R_{\hat{N}N}(\tau) = \hat{R}_N(\tau)$$

where $\hat{R}_N(\tau)$ is the Hilbert transform of the autocorrelation function $R_N(\tau)$ of $n(t)$.

Hint: Use the formula

$$\hat{n}(t) = \frac{1}{\pi}\int_{-\infty}^{\infty} \frac{n(\lambda)}{t-\lambda}\, d\lambda$$

(b) Show that $R_{N\hat{N}}(0) = 0$.

Problem 5.38 A narrow-band noise $n(t)$ has zero mean and autocorrelation function $R_N(\tau)$. Its power spectral density $S_N(f)$ is centered about $\pm f_c$. The quadrature components $n_c(t)$ and $n_s(t)$ of $n(t)$ are defined by the weighted sums

$$n_c(t) = n(t)\cos(2\pi f_c t) + \hat{n}(t)\sin(2\pi f_c t)$$

and

$$n_s(t) = \hat{n}(t)\cos(2\pi f_c t) - n(t)\sin(2\pi f_c t)$$

Using the result of part (a) of Problem 5.37, show that $n_c(t)$ and $n_s(t)$ have the correlation functions:

$$R_{N_c}(\tau) = R_{N_s}(\tau) = R_N(\tau)\cos(2\pi f_c \tau) + \hat{R}_N(\tau)\sin(2\pi f_c \tau)$$

and

$$R_{N_c N_s}(\tau) = -R_{N_s N_c}(\tau) = R_N(\tau)\sin(2\pi f_c \tau) - \hat{R}_N(\tau)\cos(2\pi f_c \tau)$$

Problem 5.39 The power spectral density of a narrow-band noise $n(t)$ is as shown in Fig. P5.17.

(a) Find the power spectral densities of the in-phase and quadrature components of $n(t)$.
(b) Find their cross-spectral densities.

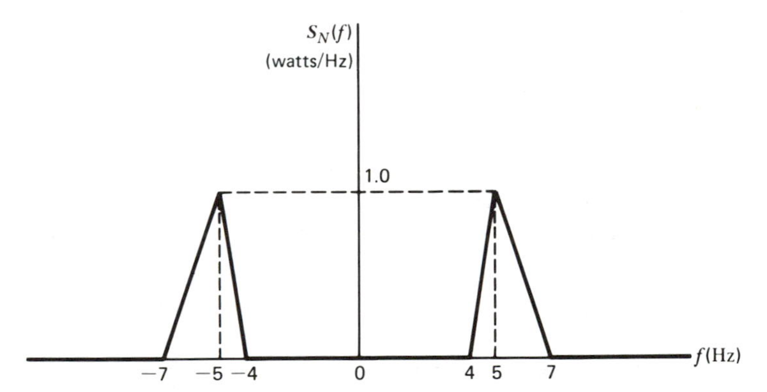

Figure P5.17

Problem 5.40 Consider a Gaussian noise $n(t)$ with zero mean and power spectral density $S_N(f)$ shown in Fig. P5.18.

(a) Find the probability density function of the envelope of $n(t)$.
(b) What are the mean and variance of this envelope?

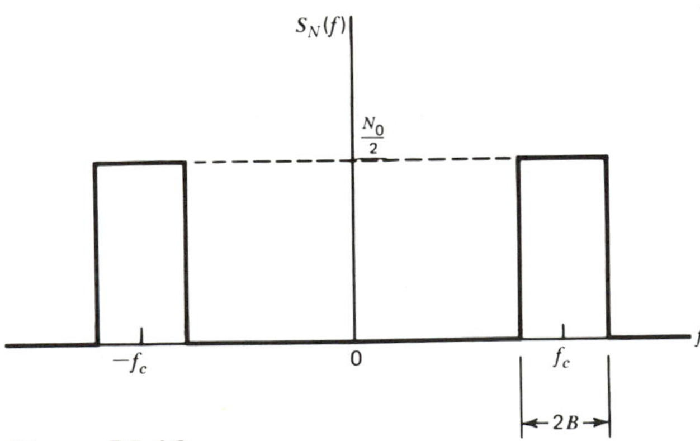

Figure P5.18

Problem 5.41 Consider a device whose output is equal to the square of the envelope of the input. Let a Gaussian noise of zero mean and the power spectral density shown in Fig. P5.18 be applied to the input of this device, producing the random process $Z(t)$. Determine the probability density function of the random variable $Z(t_k)$.

Problem 5.42 Consider a narrow-band noise $n(t)$ whose power spectral density is limited to $f_c - B \leqslant |f| \leqslant f_c + B$. Show that, except for scaling factors, the in-phase component $n_c(t)$ and quadrature component of $n(t)$ can be derived from $n(t)$ by using the scheme shown in Fig. 5.28.

Problem 5.43 Consider a random process consisting of the sum of a narrow-band Gaussian noise (of zero mean and variance σ^2) and a sinusoidal wave $A\cos(2\pi f_c t)$. The frequency f_c is the same as the mid-band frequency of the noise. In Section 5.12 we determined the probability density function of the envelope of this sum random process. The purpose of this problem is to determine the probability density function of the phase, namely, $f_\Psi(\psi)$. Show that

$$f_\Psi(\psi) = \frac{1}{2\pi}\exp\left(-\frac{a^2}{2}\right) + \frac{a\cos\psi}{2\sqrt{2\pi}}\exp\left(-\frac{a^2}{2}\sin^2\psi\right)\mathrm{erfc}\left(-\frac{a}{\sqrt{2}}\cos\psi\right)$$

where

$$a = A/\sigma$$

and $\mathrm{erfc}(-a\cos\psi\sqrt{2})$ is the complementary error function (see Appendix 2). What is the value of ψ for which $f_\Psi(\psi)$ is a maximum? What is this maximum value? What happens to $f_\Psi(\psi)$ when the parameter a goes to zero? What happens to it when a goes to infinity?

Problem 5.44 Consider the problem of propagating signals through so-called *random* or *fading communication channels*. Examples of such channels include the *ionosphere* from which short-wave (high-frequency) signals are reflected back to the earth producing long-range radio transmission, *troposcatter, underwater communications*, and so forth. A simple model of such a channel is shown in Fig. P5.19, which consists of a large collection of *random scatterers*, with the result that a single incident beam is converted into a correspondingly large number of scattered beams at the receiver. The transmitted signal is equal to $A\exp(j2\pi f_c t)$. Assume that all scattered beams travel at the same mean velocity. However, each beam differs in amplitude and phase from the incident beam, so that the kth scattered beam is given by $A_k\exp(j2\pi f_c t$

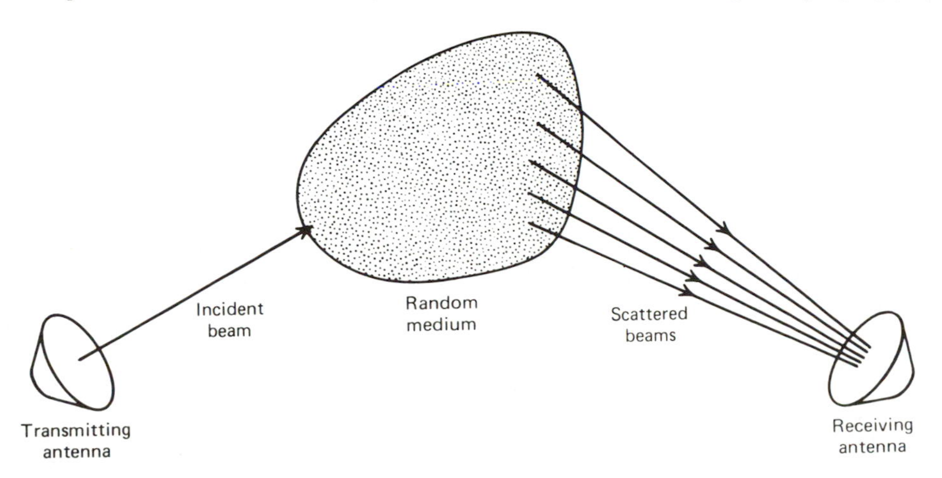

Figure P5.19

$+j\Theta_k)$ where the amplitude A_k and phase Θ_k vary slowly and randomly with time. In particular, assume that the Θ_k's are all independent of one another and uniformly distributed random variables.

(a) With the received signal denoted by

$$x(t)=r(t)\exp[j2\pi f_c t+\psi(t)],$$

show that the random variable R, obtained by observing the envelope of the received signal at time t, is Rayleigh-distributed, and that the random variable Ψ, obtained by observing the phase at some fixed time, is uniformly distributed.

(b) Assuming that the channel includes a line-of-sight path, so that the received signal contains a sinusoidal component of frequency f_c, show that in this case the envelope of the received signal is Rician-distributed.

chapter 6 NOISE IN CW MODULATION

One of the basic issues in the study of modulation systems is the analysis of the effects of noise on the performance of the receiver, and the use of the results of this analysis in system design. Another matter of interest is the comparison of the noise performances of different modulation-demodulation schemes. In order to carry out this analysis, we obviously need a criterion that describes in a meaningful manner the noise performance of the particular modulation system under study. In the case of continuous-wave (CW) modulation systems, the customary practice is to use the *output signal-to-noise ratio* as an intuitive measure for describing the fidelity with which the demodulation process in the receiver recovers the original message from the received modulated signal in the presence of noise. *Output signal-to-noise ratio is defined as the ratio of the average power of the message signal to the average power of the noise, both measured at the receiver output.* Such a criterion is perfectly well-defined as long as the message signal and noise appear additively at the receiver output. In this chapter we use this criterion to study the noise performance of different CW modulation systems. In our discussion, however, we assume that the noise is stationary, white, and Gaussian. These simplifying assumptions enable us to obtain a basic understanding of the way in which noise affects the performance of different receivers.

We consider both amplitude modulation (AM) and frequency modulation (FM) receivers. In the AM case, we consider double-sideband suppressed carrier (DSBSC) and single-sideband (SSB) receivers using coherent detection, and full AM receivers using envelope detection.

6.1 AM RECEIVERS

The usual AM radio receiver is of the so-called *superheterodyne* type, which is represented schematically in Fig. 6.1. Basically the receiver consists of a radio-frequency (RF) section, a mixer and local oscillator, an intermediate frequency (IF) section, and demodulator. Typical frequency parameters of commercial AM radio are as follows:

RF carrier range = 0.535–1.605 MHz
Mid-band frequency of IF section = 0.455 MHz
If bandwidth = 10 kHz

The incoming amplitude-modulated wave is picked up by the receiving antenna and amplified in the RF section which is tuned to the carrier frequency of the in-

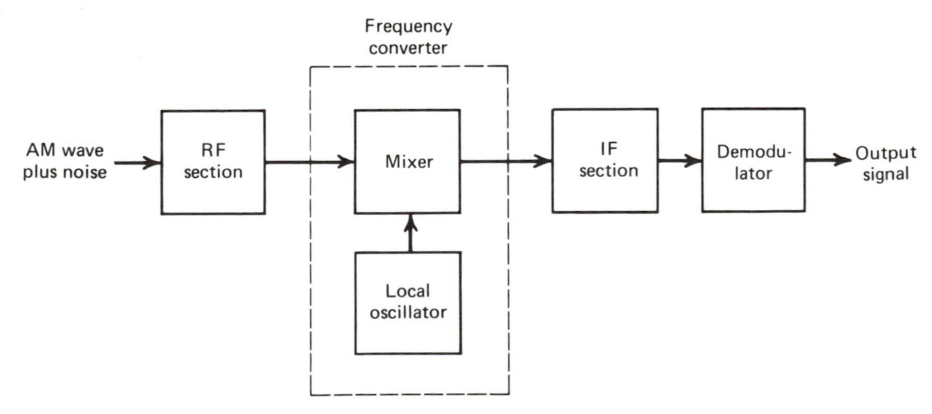

Figure 6.1 Basic elements of an AM receiver of the superheterodyne type.

coming wave. The combination of mixer and local oscillator (of adjustable frequency) provides a *frequency conversion* or *heterodyning* function, whereby the incoming signal is converted to a predetermined fixed *intermediate frequency*, usually lower than the signal frequency. This frequency conversion is achieved without disturbing the relation of the sidebands to the carrier. The result of this conversion is to produce an intermediate-frequency carrier defined by

$$f_{IF} = f_{RF} - f_{LO} \tag{6.1}$$

where f_{LO} is the frequency of the local oscillator and f_{RF} is the carrier frequency of the incoming RF signal. We refer to f_{IF} as the intermediate frequency (IF), because the signal is neither at the original input frequency nor at the final baseband frequency. The mixer-local oscillator combination is sometimes referred to as the *first detector*, in which case the demodulator is called the *second detector*.

The IF section consists of one or more stages of tuned amplification, with a bandwidth corresponding to that required for the particular type of signal that the receiver is intended to handle. This section provides most of the amplification and selectivity in the receiver. The output of the IF section is applied to a demodulator, the purpose of which is to recover the baseband signal. If coherent detection is used, then a coherent signal source must be provided in the receiver. The final operation in the receiver is the power amplification of the recovered message.

In a superheterodyne receiver the mixer will develop an intermediate frequency output when the input signal frequency is greater or less than the local oscillator frequency by an amount equal to the intermediate frequency. That is, there are two input frequencies, namely, $|f_{LO} \pm f_{IF}|$, which will result in f_{IF} at the mixer output. This introduces the possibility of simultaneous reception of two signals differing in frequency by twice the intermediate frequency. Accordingly, it is necessary to employ selective stages in the RF section (i.e., between the antenna and the mixer) in order to favor the desired signal and discriminate against the undesired or *image signal*. The effectiveness of suppressing unwanted image signals increases as the number of selective stages in the radio-frequency section increases, and as the ratio of intermediate to signal frequency increases.

AM Receiver Model

For the purpose of evaluating the effects of noise on the performance of a superheterodyne receiver, we may consider the model shown in Fig. 6.2(*a*), which consists of an equivalent IF filter and demodulator. The equivalent IF filter represents the cascade filtering characteristics of the RF and IF sections of the receiver. At the filter input, we have a signal that consists of the received modulated signal $s(t)$, translated in frequency and amplified, and the additive *front-end receiver noise* $w(t)$. The noise $w(t)$ is modeled as white Gaussian noise of zero mean and power spectral density $N_0/2$, defined for both positive and negative frequencies; *that is, N_0 is the average noise power per unit bandwidth measured at the front end of the receiver.*

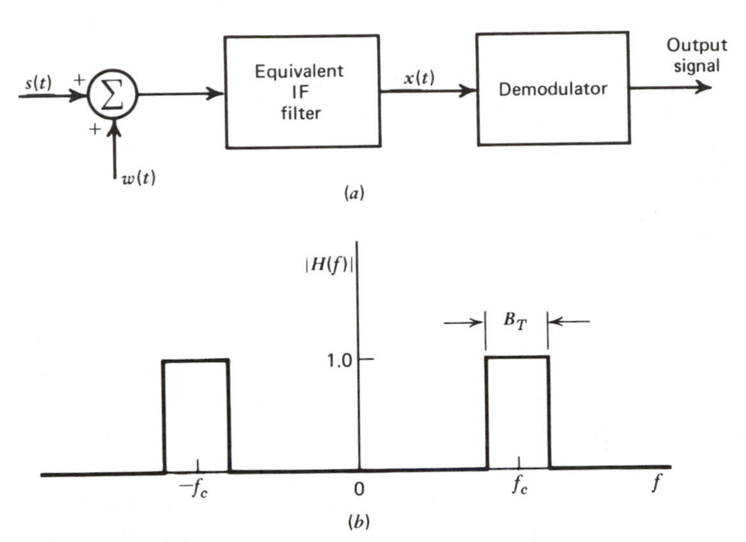

Figure 6.2 Modeling of an AM receiver. *(a)* Model. *(b)* Idealized characteristic of equivalent IF filter.

The IF filter has a bandwidth that is just wide enough to accommodate the bandwidth of the modulated signal $s(t)$. The IF filter is usually tuned so that its mid-band frequency is the same as the carrier frequency of the modulated signal $s(t)$. An exception to this is the single-sideband modulated wave, as will be explained later. For convenience in signal-to-noise analysis, we assume that the equivalent IF filter has an ideal band-pass characteristic, as shown in Fig. 6.2(*b*), where f_c refers to the carrier frequency measured at the mixer output; that is, $f_c = f_{\text{IF}}$, and B_T refers to the transmission bandwidth of the modulated signal $s(t)$.

The composite signal $x(t)$, at the IF filter output, is defined by

$$x(t) = s(t) + n(t) \tag{6.2}$$

where $n(t)$ is a band-limited white noise with the following power spectral density:

$$S_N(f) = \begin{cases} \dfrac{N_0}{2}, & f_c - \dfrac{B_T}{2} \leq |f| \leq f_c + \dfrac{B_T}{2} \\ 0, & \text{otherwise} \end{cases} \tag{6.3}$$

The filtered noise $n(t)$ may be regarded narrow-band, because the IF filter usually has a bandwidth that is small compared with its mid-band frequency.

6.2 SIGNAL-TO-NOISE RATIOS (SNR)

A common and useful measure of the fidelity of the received message is the *output signal-to-noise ratio* defined as

$$(\text{SNR})_O = \frac{\text{average power of message signal at the receiver output}}{\text{average power of noise at the receiver output}} \tag{6.4}$$

The output signal-to-noise ratio is unambiguous as long as the recovered message and noise at the demodulator output are additive. This requirement is satisfied exactly in the case of linear receivers using coherent detection, and approximately in the case of nonlinear receivers (e.g., using envelope detection) provided that the average noise power is small compared with the average carrier power.

The output signal-to-noise ratio depends, among other factors, on the type of modulation used in the transmitter and the type of demodulation used in the receiver. Thus it may be informative to compare the output signal-to-noise ratios of different modulation-demodulation systems. We will make this comparison on the basis that each system produces the same transmitted or modulated message signal power and must cope with the same noise power in the bandwidth of the message. As a frame of reference, we define the *channel signal-to-noise ratio:*

$$(\mathrm{SNR})_C = \frac{\text{average power of modulated message signal at the receiver input}}{\text{average power of noise in message bandwidth at the receiver input}} \tag{6.5}$$

This ratio may be viewed as the signal-to-noise ratio which results from *baseband* or *direct transmission* of the message without modulation, as in Fig. 6.3. Here, it is assumed that: (1) the message power at the low-pass filter input is the same as the modulated message signal power, and (2) the low-pass filter passes the message signal and rejects out-of-band noise.

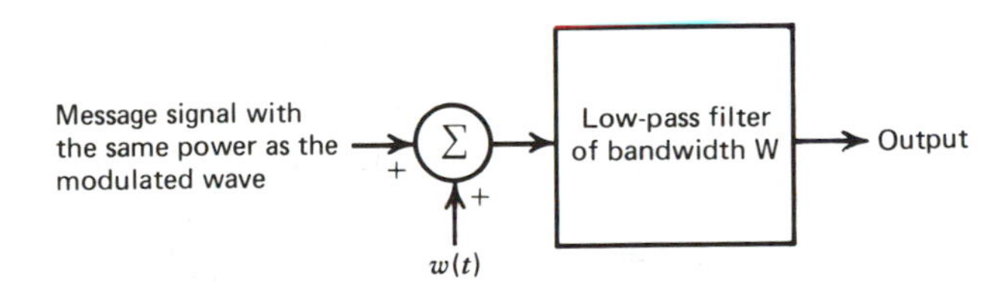

Figure 6.3 Illustrating the baseband transmission of message signal of bandwidth W, for calculating channel signal-to-noise ratio.

For the purpose of comparing different modulation systems, we normalize the receiver performance by dividing the output signal-to-noise ratio by the channel signal-to-noise ratio. We may thus define a *figure of merit* for the receiver as follows:

$$\text{Figure of merit} = \frac{(\mathrm{SNR})_O}{(\mathrm{SNR})_C} \tag{6.6}$$

Clearly, the higher the value that the figure of merit has, the better the noise performance of the receiver.

6.3 SIGNAL-TO-NOISE RATIOS FOR COHERENT RECEPTION WITH DSBSC MODULATION

We begin the noise analysis by evaluating the output and channel signal-to-noise ratios for an AM receiver using coherent detection, with an incoming double-sideband suppressed-carrier (DSBSC) modulated wave. The use of coherent detection requires multiplication of the IF filter output $x(t)$ by a locally generated sinusoidal wave $\cos(2\pi f_c t)$ and then low-pass filtering the product, as in Fig. 6.4. For convenience, we assume that the amplitude of the locally generated sine wave is unity. For this demodulation scheme to operate satisfactorily, however, it is necessary that the local oscillator be synchronized both in phase and frequency with the oscillator generating the carrier wave in the transmitter. We assume that this synchronization has been achieved.

We show presently that coherent detection has the unique feature that for any input signal-to-noise ratio, an output strictly proportional to the original message signal is always present. It is this property of coherent detection, namely, that the output message component is unmutilated and the noise component always appears additively with the message irrespective of the input signal-to-noise ratio, which distinguishes coherent detection from all other demodulation techniques.

Consider a DSBSC wave defined by

$$s(t) = A_c \cos(2\pi f_c t) m(t) \tag{6.7}$$

where $A_c \cos(2\pi f_c t)$ is the sinusoidal carrier wave and $m(t)$ is the message signal. We assume that $m(t)$ is the sample function of a stationary process of zero mean and power spectral density $S_M(f)$ limited to a maximum frequency W, as in Fig. 6.5(a). That is, W is the message bandwidth. For a stationary process the area under the curve of power spectral density equals the average power of the process. Hence, the average power of the message signal is

$$P = \int_{-W}^{W} S_M(f)\,df \tag{6.8}$$

The carrier wave is usually independent of the message signal. To emphasize this

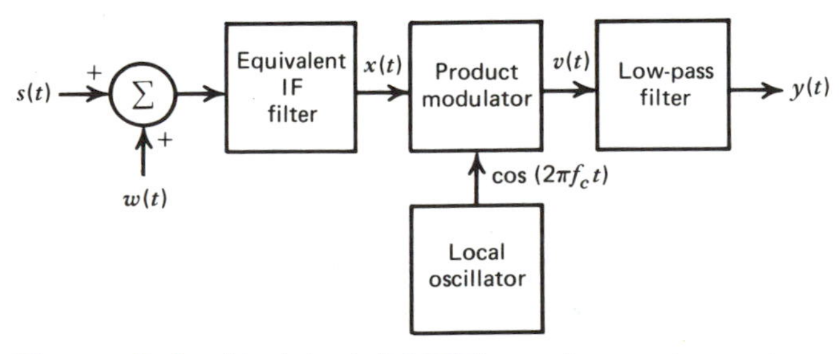

Figure 6.4 Model of DSBSC receiver using coherent detection.

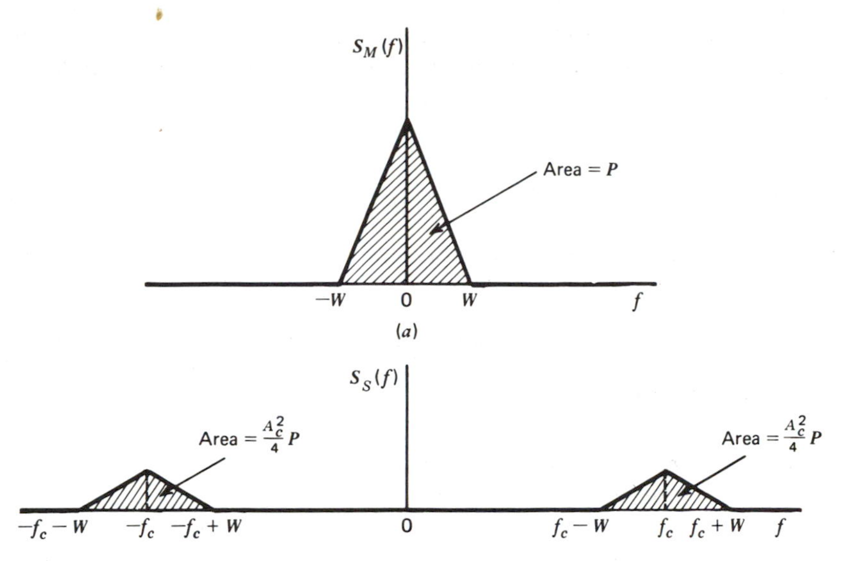

Figure 6.5 *(a)* Power spectral density of message signal *m(t)*. *(b)* Power spectral density of DSBSC modulated signal.

independence, the carrier should include a random phase that is uniformly distributed over 2π radians. In Eq. (6.7) this random phase angle has been omitted for convenience. Using the result of Example 9, Chapter 5, we find that the power spectral density $S_S(f)$ of the modulated wave $s(t)$ is related to that of $m(t)$ as follows:

$$S_S(f) = \frac{A_c^2}{4}\left[S_M(f-f_c) + S_M(f+f_c)\right] \tag{6.9}$$

Accordingly, the power spectral density $S_S(f)$ of the modulated signal $s(t)$ is as shown in Fig. 6.5(*b*), where we see that:

1. The transmission bandwidth B_T is equal to $2W$.
2. The average power of the modulated signal $s(t)$ is equal to $A_c^2 P/2$.

With a noise spectral density of $N_0/2$, the average noise power in the message bandwidth W is equal to WN_0. The channel signal-to-noise ratio of the system is therefore

$$(\text{SNR})_{\text{C,DSB}} = \frac{\frac{1}{2}A_c^2 P}{2WN_0} \tag{6.10}$$

Next, we wish to determine the output signal-to-noise ratio of the system. Using the narrow-band representation of the filtered noise $n(t)$, the total signal at the coherent detector input may be expressed as:

$$\begin{aligned} x(t) &= s(t) + n(t) \\ &= A_c \cos(2\pi f_c t)m(t) + n_c(t)\cos(2\pi f_c t) - n_s(t)\sin(2\pi f_c t) \end{aligned} \tag{6.11}$$

where $n_c(t)$ and $n_s(t)$ are the in-phase and quadrature components of $n(t)$, with respect to the carrier $\cos(2\pi f_c t)$, respectively. The output of the product-modulator component of the coherent detector is therefore

$$\begin{aligned} v(t) &= x(t)\cos(2\pi f_c t) \\ &= \tfrac{1}{2}A_c m(t) + \tfrac{1}{2}n_c(t) + \tfrac{1}{2}[A_c m(t) + n_c(t)]\cos(4\pi f_c t) - \tfrac{1}{2}A_c n_s(t)\sin(4\pi f_c t) \end{aligned} \tag{6.12}$$

The low-pass filter in the coherent detector removes the high-frequency components of $v(t)$, yielding a receiver output

$$y(t) = \tfrac{1}{2}A_c m(t) + \tfrac{1}{2}n_c(t) \tag{6.13}$$

Equation (6.13) indicates that:

1. The message $m(t)$ and in-phase noise component $n_c(t)$ of the narrow-band noise $n(t)$ appear additively at the receiver output.
2. The quadrature component $n_s(t)$ of the noise $n(t)$ is completely rejected by the coherent detector.

The message signal component at the receiver output equals $A_c m(t)/2$. Hence, we have

$$\text{average power of message signal at the receiver output} = \frac{A_c^2 P}{4} \tag{6.14}$$

where P is the average power of the original message signal $m(t)$.

The noise component at the receiver output equals $n_c(t)/2$. Hence, the power spectral density of the output noise equals $(1/2)^2$ times that of $n_c(t)$. To calculate the average power of the noise at the receiver output, we first determine the power spectral density of the in-phase noise component $n_c(t)$. The IF filter has a bandwidth B_T equal to $2W$, twice the message bandwidth, in order to accommodate the upper and lower sidebands of the modulated wave $s(t)$. The power spectral density $S_N(f)$ of the narrow-band noise $n(t)$ thus takes on the ideal form shown in Fig. 6.6(*a*). Hence, the power spectral density of $n_c(t)$ is as shown in Fig. 6.6(*b*) (see Example 16 of Chapter 5). Evaluating the area under the curve of power spectral density of Fig. 6.6(*b*), and multiplying the result by $(1/2)^2$, we get

$$\begin{aligned} \text{average power of noise at the receiver output} &= \left(\frac{1}{2}\right)^2 2WN_0 \\ &= \frac{1}{2}WN_0 \end{aligned} \tag{6.15}$$

Dividing Eq. (6.14) by (6.15) gives the output signal-to-noise ratio

$$(\text{SNR})_{O,\text{DSB}} = \frac{A_c^2 P}{2WN_0} \tag{6.16}$$

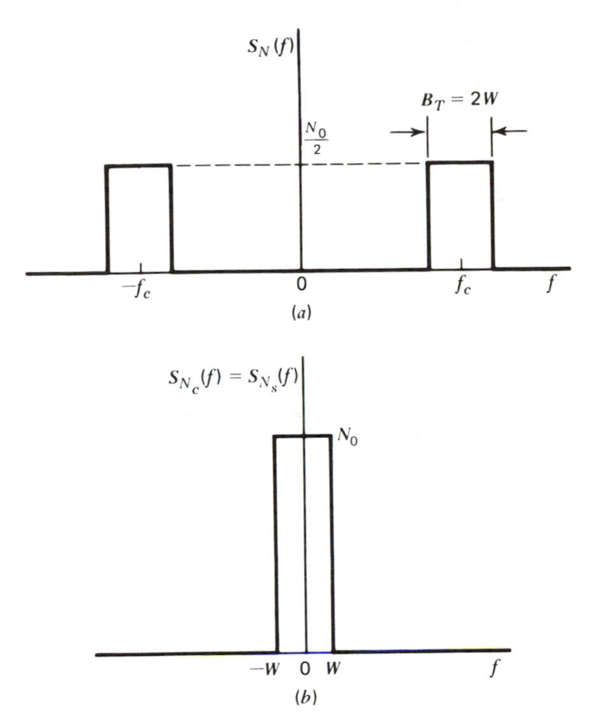

Figure 6.6 Noise analysis of DSBSC modulation system using coherent detection. *(a)* Power spectral density of narrow-band noise $n(t)$ at IF filter output. *(b)* Power spectral density of in-phase component $n_c(t)$ and quadrature component $n_s(t)$ of noise $n(t)$.

Using Eqs. (6.10) and (6.16), we obtain the figure of merit

$$\left.\frac{(\mathrm{SNR})_O}{(\mathrm{SNR})_C}\right|_{\mathrm{DSB}} = 1 \tag{6.17}$$

Note that at the coherent detector output in the receiver of Fig. 6.4 using DSBSC modulation, the translated signal sidebands sum coherently, whereas the translated noise sidebands sum incoherently. This means that the output signal-to-noise ratio in this receiver is twice the signal-to-noise ratio at the coherent detector input.

6.4 SIGNAL-TO-NOISE RATIOS FOR COHERENT RECEPTION WITH SSB MODULATION

Consider next the case of an AM receiver using coherent detection, with an incoming single-sideband (SSB) modulated wave. We assume that only the lower sideband is transmitted, so that we may express the modulated wave as

$$s(t)=\frac{A_c}{2}\cos(2\pi f_c t)m(t)+\frac{A_c}{2}\sin(2\pi f_c t)\hat{m}(t) \tag{6.18}$$

where $\hat{m}(t)$ is the Hilbert transform of the message signal $m(t)$. We may make the following observations concerning the in-phase and quadrature components of $s(t)$ in Eq. (6.18):

1. The two components $m(t)$ and $\hat{m}(t)$ are orthogonal to each other. Therefore, with the message signal $m(t)$ assumed to have zero mean, it follows that $m(t)$ and $\hat{m}(t)$ are uncorrelated, so that their power spectral densities are additive.
2. The Hilbert transform $\hat{m}(t)$ is obtained by passing $m(t)$ through a linear filter with a transfer function $-j\ \text{sgn}(f)$. The squared magnitude of this transfer function is equal to one for all f. Accordingly, we find that both $m(t)$ and $\hat{m}(t)$ have the same power spectral density.

Thus, using a procedure similar to that in Section 6.3, we find that the in-phase and quadrature components of the modulated signal $s(t)$ contribute an average power of $A_c^2P/8$ each. The average power of $s(t)$ is therefore $A_c^2P/4$. This result is half that in the DSBSC case, which is intuitively satisfying.

The average noise power in the message bandwidth W is WN_0, as in the DSBSC case. Thus the channel signal-to-noise ratio of a coherent-receiver with SSB modulation is

$$(\text{SNR})_{\text{C,SSB}}=\frac{A_c^2P}{4WN_0} \tag{6.19}$$

As illustrated in Fig. 6.7(a), the transmission bandwidth $B_T=W$ and the mid-band frequency of the power spectral density $S_N(f)$ of the narrow-band noise $n(t)$ differs from the carrier frequency f_c by $W/2$. Therefore, we may express $n(t)$ as:

$$n(t)=n_c(t)\cos\left[2\pi\left(f_c-\frac{W}{2}\right)t\right]-n_s(t)\sin\left[2\pi\left(f_c-\frac{W}{2}\right)t\right] \tag{6.20}$$

The output of the coherent detector, due to the combined influence of the modulated signal $s(t)$ and noise $n(t)$, is thus given by

$$y(t)=\frac{A_c}{4}m(t)+\frac{1}{2}n_c(t)\cos(\pi Wt)+\frac{1}{2}n_s(t)\sin(\pi Wt) \tag{6.21}$$

As expected, we see that the quadrature component $\hat{m}(t)$ of the modulated message signal $s(t)$ has been eliminated from the detector output, but unlike the case of DSBSC modulation, the quadrature component of the narrow-band noise $n(t)$ now appears at the output.

The message component in the receiver output is $A_c m(t)/4$ so that the average power of the recovered message is $A_c^2P/16$. The noise component in the receiver output is $[n_c(t)\cos(\pi Wt)+n_s(t)\sin(\pi Wt)]/2$. To determine the average power of the output noise, we note the following:

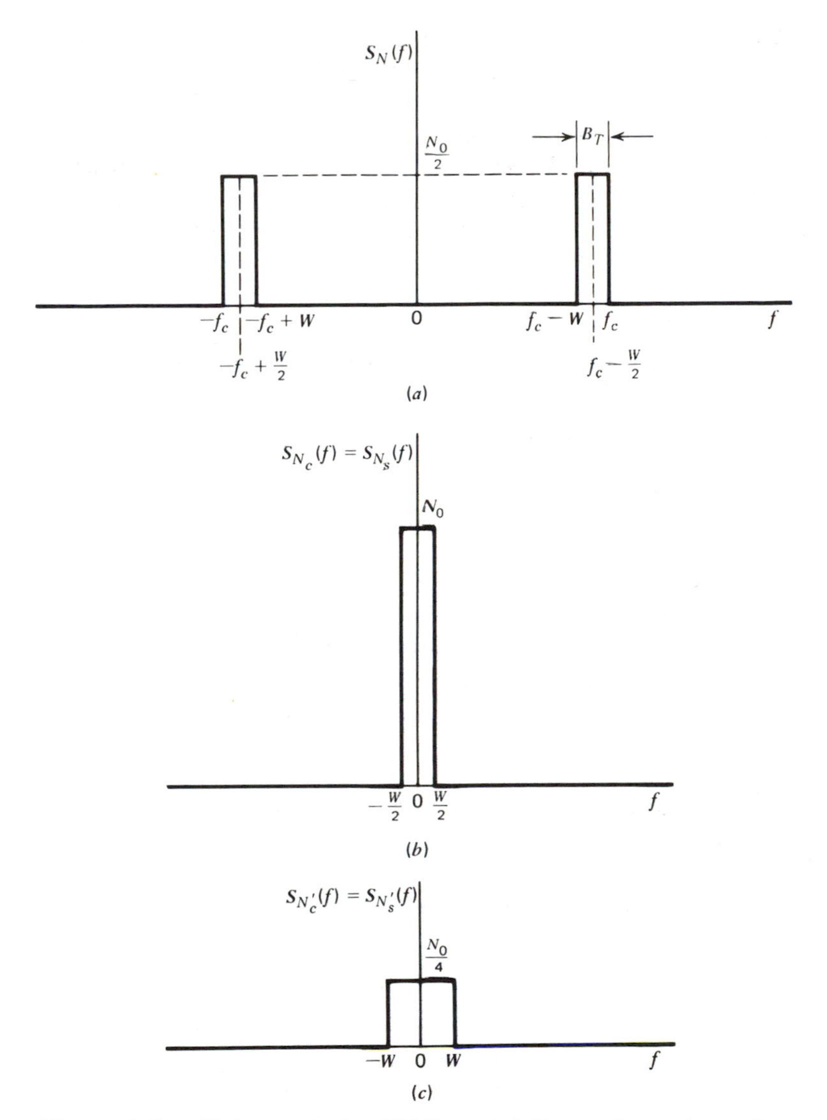

Figure 6.7 Noise analysis of SSB modulation system using coherent detection: *(a)* Power spectral density of narrow-band noise $n(t)$ at IF filter output. *(b)* Power spectral density of in-phase and quadrature components of $n(t)$ with respect to $f_c - W/2$. *(c)* Power spectral density of $n_c'(t) = n_c(t)\cos(\pi W t)$ and $n_s'(t) = n_s(t)\sin(\pi W t)$.

1. The power spectral density of both $n_c(t)$ and $n_s(t)$ is as shown in Fig. 6.7(*b*).
2. The sinusoidal wave $\cos(\pi W t)$ is independent of both $n_c(t)$ and $n_s(t)$. Hence, the power spectral density of $n_c'(t) = n_c(t)\cos(\pi W t)$ is obtained by shifting $S_{N_c}(f)$ to the left by $W/2$, shifting it to the right by $W/2$, adding the shifted spectra, and dividing the result by 4. The power spectral density of $n_s'(t) = n_s(t)\sin(\pi W t)$ is

obtained in a similar way. The power spectral density of both $n_c'(t)$ and $n_s'(t)$, obtained in this manner, is shown sketched in Fig. 6.7(c).

From Fig. 6.7(c) we see that the average power of the noise component $n_c'(t)$ or $n_s'(t)$ is $WN_0/2$. Therefore, from Eq. (6.21), the average output noise power is $WN_0/4$. We thus find that the output signal-to-noise ratio of a system, using SSB modulation in the transmitter and coherent detection in the receiver, is given by

$$(\text{SNR})_{\text{O,SSB}} = \frac{A_c^2 P}{4WN_0} \tag{6.22}$$

Hence, from Eqs. (6.19) and (6.22), the figure of merit of such a system equals

$$\left.\frac{(\text{SNR})_{\text{O}}}{(\text{SNR})_{\text{C}}}\right|_{\text{SSB}} = 1 \tag{6.23}$$

Comparing Eqs. (6.17) *and* (6.23), *we conclude that for the same average transmitted (or modulated message) signal power and the same average noise power in the message bandwidth, an SSB receiver will have exactly the same output signal-to-noise ratio as a DSBSC receiver, when both receivers use coherent detection for the recovery of the message signal.* Furthermore, in both cases, the noise performance of the receiver is the same as that obtained by simply transmitting the message signal itself in the presence of the same noise. The only effect of the modulation process is to translate the message signal to a different frequency band.

6.5 NOISE IN AM RECEIVERS USING ENVELOPE DETECTION

In a full amplitude-modulated (AM) wave both sidebands and the carrier are transmitted, as shown by

$$s(t) = A_c[1 + k_a m(t)]\cos(2\pi f_c t) \tag{6.24}$$

where $A_c \cos(2\pi f_c t)$ is the carrier wave, $m(t)$ is the message signal, and k_a is a constant that determines the percentage modulation. In this section, we evaluate the noise performance of an AM receiver using an envelope detector. As explained in Section 3.3, an envelope detector consists simply of a nonlinear device (usually a diode) followed by a low-pass RC filter.

From Eq. (6.24), the average power in the modulated message signal $s(t)$ is equal to $A_c^2(1 + k_a^2 P)/2$, where P is the average power of the message signal. With an average noise power of WN_0 in the message bandwidth, the channel signal-to-noise ratio is therefore

$$(\text{SNR})_{\text{C,AM}} = \frac{A_c^2(1 + k_a^2 P)}{2WN_0} \tag{6.25}$$

The received signal $x(t)$ at the envelope detector input consists of the modulated message signal $s(t)$ and narrow-band noise $n(t)$. Representing $n(t)$ in terms of its

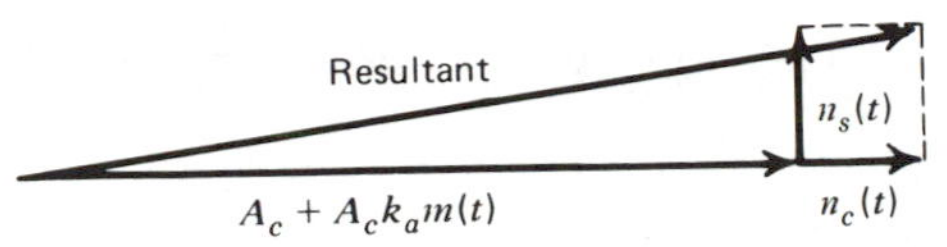

Figure 6.8 Phasor diagram for AM wave plus narrow-band noise for the case of high carrier-to-noise ratio.

in-phase and quadrature components, namely, $n_c(t)$ and $n_s(t)$, we may express $x(t)$ as

$$\begin{aligned} x(t) &= s(t) + n(t) \\ &= [A_c + A_c k_a m(t) + n_c(t)]\cos(2\pi f_c t) - n_s(t)\sin(2\pi f_c t) \end{aligned} \tag{6.26}$$

It is informative to represent the components that comprise the signal $x(t)$ by means of phasors, as in Fig. 6.8. From this phasor diagram, the receiver output is obtained as

$$\begin{aligned} y(t) &= \text{envelope of } x(t) \\ &= \{[A_c + A_c k_a m(t) + n_c(t)]^2 + n_s^2(t)\}^{1/2} \end{aligned} \tag{6.27}$$

The signal $y(t)$ defines the output of an ideal envelope detector. The phase of $x(t)$ is of no interest to us, because an ideal envelope detector is totally insensitive to variations in the phase of $x(t)$.

The expression defining $y(t)$ is somewhat complex and needs to be simplified in some manner. Specifically, we would like to approximate the output $y(t)$ as the sum of a message term plus a term due to noise and distortion. In general, this is quite difficult to achieve. However, when the average carrier power is large compared with the average noise power, so that the receiver is operating satisfactorily, then the signal term $A_c[1 + k_a m(t)]$ will be large compared with the noise terms $n_c(t)$ and $n_s(t)$, at least most of the time. Then we may approximate the output $y(t)$ as (see Problem 6.12):

$$y(t) \simeq A_c + A_c k_a m(t) + n_c(t) \tag{6.28}$$

The presence of the dc or constant term A_c in the envelope detector output $y(t)$ of Eq. (6.28) is due to demodulation of the transmitted carrier wave. We may ignore this term, however, because it bears no relation whatsoever to the message signal $m(t)$. In any case, it may be removed simply by means of a blocking capacitor. Thus if we neglect the term A_c in Eq. (6.28), we find that the remainder has, except for scaling factors, the same form as the output of a DSBSC receiver using coherent detection. Accordingly, the output signal-to-noise ratio of an AM receiver using an envelope detector is approximately

$$(\text{SNR})_{\text{O,AM}} \simeq \frac{A_c^2 k_a^2 P}{2W N_0} \tag{6.29}$$

This expression is, however, valid only if:

1. The noise, at the receiver input, is small compared to the signal.

2. The amplitude sensitivity k_a is adjusted for a percentage modulation less than or equal to 100 percent.

Using Eqs. (6.25) and 6.29), we obtain the figure of merit:

$$\left.\frac{(\text{SNR})_O}{(\text{SNR})_C}\right|_{\text{AM}} \simeq \frac{k_a^2 P}{1 + k_a^2 P} \tag{6.30}$$

Thus, whereas the figure of merit of a DSBSC or SSB receiver using coherent detection is always unity, the corresponding figure of merit of an AM receiver using envelope detection is always less than unity. *In other words, the noise performance of an AM receiver is always inferior to that of a DSBSC or SSB receiver.* This is due to the wasteage of transmitter power which results from transmitting the carrier as a component of the AM wave.

Example 1 Single-tone modulation

Consider the special case of a sine-wave of frequency f_m and amplitude A_m as the modulating wave, as shown by

$$m(t) = A_m \cos(2\pi f_m t)$$

The corresponding AM wave is

$$s(t) = A_c[1 + \mu \cos(2\pi f_m t)]\cos(2\pi f_c t)$$

where $\mu = k_a A_m$ is the modulation factor. The average power of the modulating wave $m(t)$ is

$$P = \tfrac{1}{2} A_m^2$$

Therefore, using Eq. (6.30), we get

$$\left.\frac{(\text{SNR})_O}{(\text{SNR})_C}\right|_{\text{AM}} = \frac{\frac{1}{2}k_a^2 A_m^2}{1 + \frac{1}{2}k_a^2 A_m^2}$$

$$= \frac{\mu^2}{2 + \mu^2} \tag{6.31}$$

When $\mu = 1$, which corresponds to 100 percent modulation, we get a figure of merit equal to 1/3. This means that, other factors being equal, this AM system must transmit three times as much average power as a suppressed-carrier system in order to achieve the same quality of noise performance.

Threshold Effect

When the carrier-to-noise ratio is small compared with unity, the noise term dominates and the performance of the envelope detector changes completely from that described above. In this case it is more convenient to represent the narrowband noise $n(t)$ in terms of its envelope $r(t)$ and phase $\psi(t)$, as shown by

$$n(t) = r(t)\cos[2\pi f_c t + \psi(t)] \tag{6.32}$$

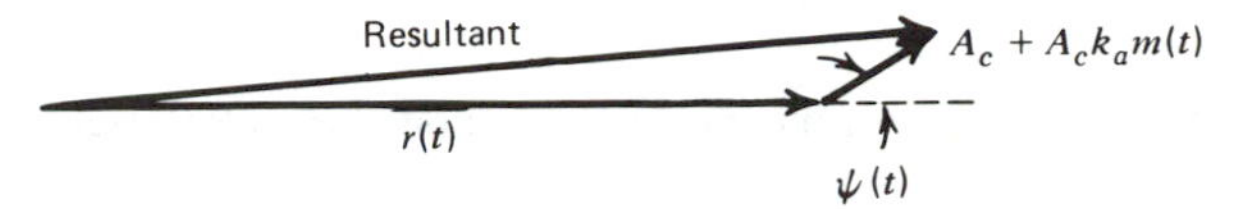

Figure 6.9 Phasor diagram for AM wave plus narrow-band noise for the case of low carrier-to-noise ratio.

The corresponding phasor diagram for the detector input $x(t) = s(t) + n(t)$ is shown in Fig. 6.9 where we have used the noise as reference, because it is now the dominant term. To the noise phasor $r(t)$ we have added a phasor representing the signal term $A_c[1 + k_a m(t)]$, with the angle between them equal to the phase $\psi(t)$ of the noise $n(t)$. In Fig. 6.9 it is assumed that the carrier-to-noise ratio is so low that the carrier amplitude A_c is small compared with the noise envelope $r(t)$, at least most of the time. Then we may neglect the quadrature component of the signal with respect to the noise, and thus find directly from Fig. 6.9 that the envelope detector output is approximately

$$y(t) \simeq r(t) + A_c \cos[\psi(t)] + A_c k_a m(t) \cos[\psi(t)] \tag{6.33}$$

This relation reveals that when the carrier-to-noise ratio is low, the detector output has no component strictly proportional to the message signal $m(t)$. The last term of the expression defining $y(t)$ contains the message signal $m(t)$ multiplied by noise in the form of $\cos[\psi(t)]$. From Section 5.11 we recall that the phase $\psi(t)$ of the narrow-band noise $n(t)$ is uniformly distributed over 2π radians. It follows therefore that we have a complete loss of information in that the detector output does not contain the message signal $m(t)$ at all. The loss of a message in an envelope detector that operates at a low carrier-to-noise ratio is referred to as the *threshold effect. By threshold we mean a value of the carrier-to-noise ratio below which the noise performance of a detector deteriorates much more rapidly than proportionately to the carrier-to-noise ratio.* It is important to recognize that every nonlinear detector (e.g., envelope detector) exhibits a threshold effect. On the other hand, such an effect does not occur in a coherent detector.

A detailed analysis of the threshold effect in envelope detectors is complicated.* We may develop some insight into the threshold effect, however, by using the following qualitative approach.† Let R denote the random variable obtained by observing the process, with sample function $r(t)$, at some fixed time. Intuitively, an envelope detector is expected to be operating well into the threshold region if the probability that the random variable R exceeds the carrier amplitude A_c is, say, 0.5. On the other hand, if this same probability is only 0.01, the envelope detector is expected to be relatively free of loss of message and the threshold effects. The evaluation of the carrier-to-noise ratios, corresponding to these probabilities, is best illustrated by way of an example.

* D. Middleton, *An Introduction to Statistical Communication Theory*, pp. 563–574 (McGraw-Hill, 1960).

† J. J. Downing, *Modulation Systems and Noise*, p. 71 (Prentice-Hall, 1964).

Example 2

From Section 5.11, we recall that the envelope $r(t)$ of the narrow-band noise $n(t)$ is Rayleigh-distributed; that is,

$$f_R(r)=\frac{r}{\sigma_N^2}\exp\left(-\frac{r^2}{2\sigma_N^2}\right) \tag{6.34}$$

where σ_N^2 is the variance of the noise $n(t)$. For an AM system, we have $\sigma_N^2=2WN_0$. Therefore, probability of the event $R\geqslant A_c$ is defined by

$$\begin{aligned}P(R\geqslant A_c)&=\int_{A_c}^{\infty} f_R(r)dr\\&=\int_{A_c}^{\infty}\frac{r}{2WN_0}\exp\left(-\frac{r^2}{4WN_0}\right)dr\\&=\exp\left(-\frac{A_c^2}{4WN_0}\right)\end{aligned} \tag{6.35}$$

Define a *carrier-to-noise ratio* as

$$\rho=\frac{\text{average carrier power}}{\text{average noise power in bandwidth of the modulated message signal}} \tag{6.36}$$

Since the bandwidth of an AM wave is $2W$, and the average carrier power is $A_c^2/2$, we have

$$\rho=\frac{A_c^2}{4WN_0} \tag{6.37}$$

Then we may rewrite Eq. (6.35) in the form

$$P(R\geqslant A_c)=\exp(-\rho) \tag{6.38}$$

Solving for $P(R\geqslant A_c)=0.5$, we get

$$\rho=\ln 2=0.69$$

Similarly, for $P(R\geqslant A_c)=0.01$, we get

$$\rho=\ln 100=4.6$$

Thus with a carrier-to-noise ratio of -1.6 dB the envelope detector is expected to be well into the threshold region, whereas with a carrier-to-noise ratio of 6.6 dB the detector is expected to be operating satisfactorily. We ordinarily need a signal-to-noise ratio considerably greater than 6.6 dB for satisfactory intelligibility, which means therefore that threshold effects are seldom of great importance in AM receivers using envelope detection.

6.6 FM RECEIVERS

We turn next to study the effects of noise on the performance of frequency modulation (FM). We begin this study by briefly describing the important elements of an FM receiver. Like AM radio receivers, most FM ratio receivers are also of the superheterodyne type. The basic difference between AM and FM receivers is that the AM demodulator is replaced by an FM demodulator such as the limiter-

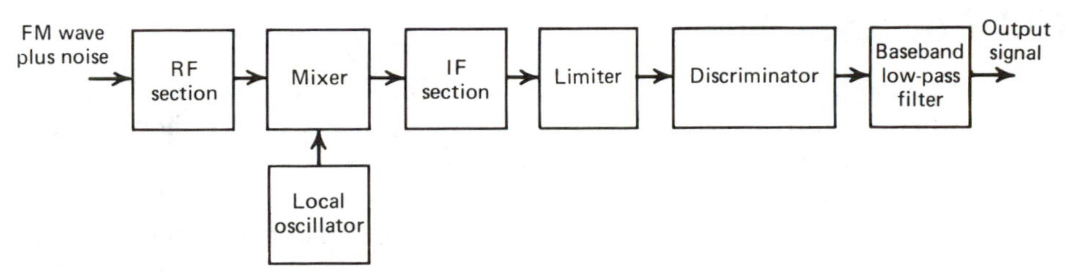

Figure 6.10 Basic elements of FM receiver of the superheterodyne type.

discriminator, as shown in Fig. 6.10. Typical frequency parameters for commercial FM radio are:

RF carrier range = 88–108 MHz
Mid-band frequency of IF section = 10.7 MHz
IF bandwidth = 0.2 MHz

In an FM system, the message information is transmitted by variations of the instantaneous frequency of a sinusoidal carrier wave, and its amplitude is maintained constant. Therefore, any variations of the carrier amplitude at the receiver input must result from noise or interference. The *amplitude limiter*, following the IF section, is used to remove amplitude variations by clipping the modulated wave at the IF section output almost to the zero axis. The resulting rectangular wave is rounded off by a band-pass filter which suppresses harmonics of the carrier frequency. Thus the filter output is again sinusoidal, with an amplitude that is practically independent of the carrier amplitude at the receiver input (see Problem 4.25). The amplitude limiter and filter usually form an integral unit.

The discriminator consists of two components:

1. A *slope network* or *differentiator* with a purely imaginary transfer function that varies linearly with frequency. It produces a hybrid-modulated wave in which both amplitude and frequency vary in accordance with the message signal.
2. An envelope detector that recovers the amplitude variation and thus reproduces the message signal.

The slope network and envelope detector are usually implemented as integral parts of a single physical unit.

The *post-detection filter*, labeled "baseband low-pass filter" in Fig. 6.10, has a bandwidth that is just large enough to accommodate the highest frequency component of the message signal. This filter removes the out-of-band components of the noise at the discriminator output and thereby keeps the output noise as small as possible.

FM Receiver Model

For the purpose of evaluating the performance of an FM receiver in the presence of additive front-end receiver noise $w(t)$, we may use the model shown in Fig. 6.11(a). As before, the noise $w(t)$ is modeled as white Gaussian noise of zero mean and power spectral density $N_0/2$. The received FM signal $s(t)$, translated in frequency and amplitude, has a carrier frequency f_c and transmission bandwidth B_T, so that only a negligible amount of power lies outside the frequency band $f_c \pm B_T/2$.

As in the AM case, the equivalent IF filter represents the combined filtering effects of the RF and IF sections of the receiver. This filter has a mid-band frequency f_c and bandwidth B_T, and therefore passes the FM signal essentially without distortion. We assume that the equivalent IF filter has an ideal bandpass characteristic, with the bandwidth B_T small compared with the mid-band frequency f_c, as in Fig. 6.11(b). We may thus use the usual narrow-band representation for the filtered noise $n(t)$ in terms of its in-phase and quadrature components.

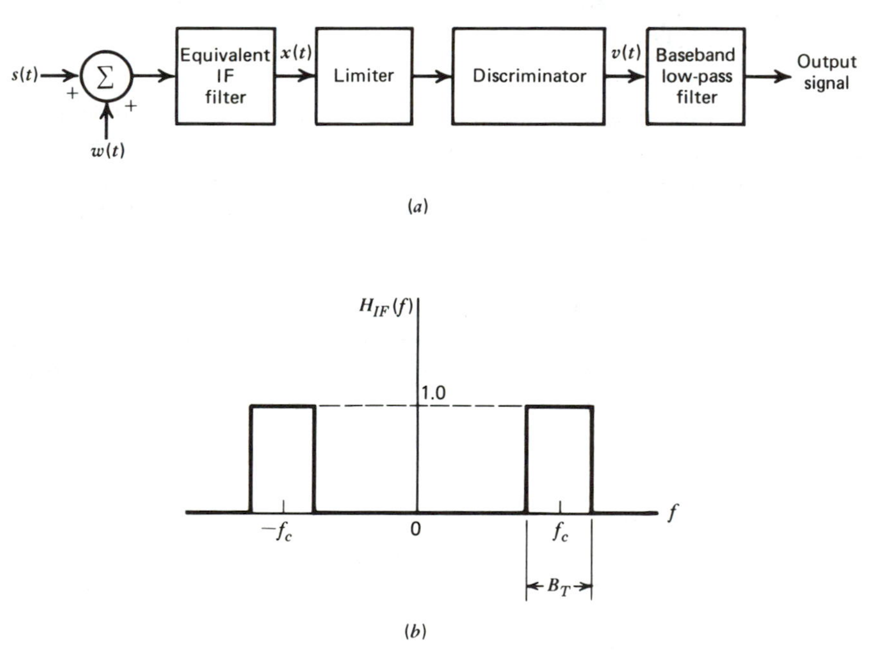

Figure 6.11 Modeling of an FM receiver. *(a)* Model. *(b)* Idealized IF filter characteristic.

The limiter removes any amplitude variations at the IF output. The discriminator is assumed to be ideal in the sense that its output is proportional to the deviation in the instantaneous frequency of the carrier away from f_c. Also, the post-detection filter is assumed to be an ideal low-pass filter with a bandwidth equal to the message bandwidth W.

6.7 NOISE IN FM RECEPTION

The narrow-band noise $n(t)$ at the IF filter output is defined in terms of its in-phase and quadrature components by

$$n(t) = n_c(t)\cos(2\pi f_c t) - n_s(t)\sin(2\pi f_c t) \tag{6.39}$$

Equivalently, we may express $n(t)$ in terms of its envelope and phase as

$$n(t) = r(t)\cos[2\pi f_c t + \psi(t)] \tag{6.40}$$

where

$$r(t) = [n_c^2(t) + n_s^2(t)]^{1/2} \tag{6.41}$$

and

$$\psi(t) = \tan^{-1}\left[\frac{n_s(t)}{n_c(t)}\right] \tag{6.42}$$

The envelope $r(t)$ is Rayleigh-distributed and the phase $\psi(t)$ is uniformly distributed over 2π radians (see Section 5.11). We assume that the FM signal at the IF filter output is

$$s(t) = A_c \cos\left[2\pi f_c t + 2\pi k_f \int_0^t m(t)dt\right] \tag{6.43}$$

where A_c is the carrier amplitude, f_c is the carrier frequency, k_f is the frequency sensitivity, and $m(t)$ is the message or modulating wave. For convenience, we define,

$$\phi(t) = 2\pi k_f \int_0^t m(t)dt, \tag{6.44}$$

so that we may express $s(t)$ in the simple form

$$s(t) = A_c \cos[2\pi f_c t + \phi(t)] \tag{6.45}$$

The total signal (i.e., signal plus noise) at the IF section output is therefore

$$\begin{aligned} x(t) &= s(t) + n(t) \\ &= A_c \cos[2\pi f_c t + \phi(t)] + r(t)\cos[2\pi f_c t + \psi(t)] \end{aligned} \tag{6.46}$$

It is informative to represent $x(t)$ by means of a phasor diagram, as in Fig. 6.12. In this diagram we have used the signal term as reference. The relative phase $\theta(t)$ of

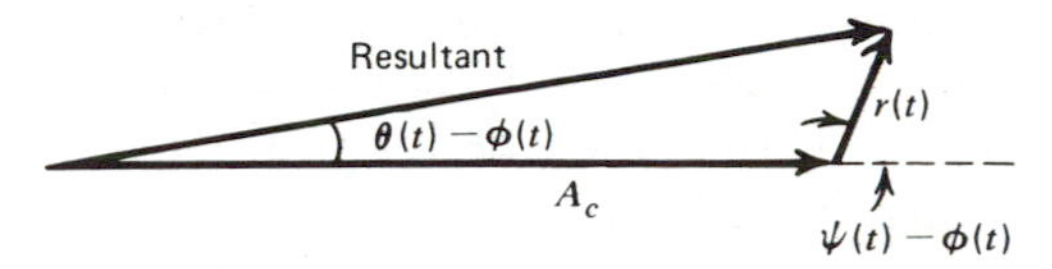

Figure 6.12 Phasor diagram for FM wave plus narrow-band noise for the case of high carrier-to-noise ratio.

the resultant phasor representing $x(t)$ is obtained directly from Fig. 6.12 as

$$\theta(t)=\phi(t)+\tan^{-1}\left\{\frac{r(t)\sin[\psi(t)-\phi(t)]}{A_c+r(t)\cos[\psi(t)-\phi(t)]}\right\} \tag{6.47}$$

The envelope of $x(t)$ is of no interest to us, because any envelope variations at the IF section output are removed by the limiter.

Our motivation is to determine the error in the instantaneous frequency of the carrier wave caused by the presence of the narrow-band noise $n(t)$. With the discriminator assumed ideal, its output is proportional to $\theta'(t)/2\pi$ where $\theta'(t)$ is the derivative of $\theta(t)$ with respect to time. In view of the complexity of the expression defining $\theta(t)$, however, we need to make certain simplifying approximations so that our analysis may yield useful results.

We assume that the carrier-to-noise ratio measured at the discriminator input is large compared with unity. Let R denote the random variable obtained by observing the process, with sample function $r(t)$, at some fixed time. Then, at least most of the time, the random variable R is small compared with the carrier amplitude A_c, and so the expression for the relative phase $\theta(t)$ simplifies considerably as follows:

$$\theta(t)\simeq\phi(t)+\frac{r(t)}{A_c}\sin[\psi(t)-\phi(t)] \tag{6.48}$$

or

$$\theta(t)\simeq 2\pi k_f\int_0^t m(t)dt+\frac{r(t)}{A_c}\sin[\psi(t)-\phi(t)] \tag{6.49}$$

The discriminator output is therefore

$$\begin{aligned} v(t) &= \frac{1}{2\pi}\frac{d\theta(t)}{dt} \\ &\simeq k_f m(t)+n_d(t) \end{aligned} \tag{6.50}$$

where the noise term $n_d(t)$ is defined by

$$n_d(t)=\frac{1}{2\pi A_c}\frac{d}{dt}\{r(t)\sin[\psi(t)-\phi(t)]\} \tag{6.51}$$

We thus see that provided the carrier-to-noise ratio is high, the discriminator output $v(t)$ consists of the original message or modulating wave $m(t)$, multiplied by the constant factor k_f, plus an additive noise component $n_d(t)$. Accordingly, we may use the output signal-to-noise ratio as previously defined to assess the quality of performance of the FM receiver. Before doing this, however, it is instructive to see if we can simplify the expression defining the noise $n_d(t)$.

From the phasor diagram of Fig. 6.12, we note that the effect of variations in the phase $\psi(t)$ of the narrow-band noise appear referred to the signal term $\phi(t)$. We know that the phase $\psi(t)$ is uniformly distributed over 2π radians. It is therefore tempting to assume that the phase difference $\psi(t)-\phi(t)$ is also uniformly distributed over 2π radians. If such an assumption were true, then the noise $n_d(t)$ at the discriminator output will be independent of the modulating signal, and will depend only on the characteristics of the carrier and narrow-band noise. It has been shown by Rice that this assumption is justified provided that the carrier-to-noise ratio is high.* Then we may simplify Eq. (6.51) as:

$$n_d(t) \simeq \frac{1}{2\pi A_c} \frac{d}{dt} \{r(t)\sin[\psi(t)]\} \tag{6.52}$$

However, from the definitions of $r(t)$ and $\psi(t)$ given by Eqs. (6.41) and (6.42), we note that the quadrature component $n_s(t)$ of the narrow-band noise $n(t)$ is

$$n_s(t) = r(t)\sin[\psi(t)] \tag{6.53}$$

Therefore, we may rewrite Eq. (6.52) as

$$n_d(t) \simeq \frac{1}{2\pi A_c} \frac{dn_s(t)}{dt} \tag{6.54}$$

This means that the additive noise $n_d(t)$ appearing at the discriminator output is determined effectively by the carrier amplitude A_c and the quadrature component $n_s(t)$ of the narrow-band noise $n(t)$.

The output signal-to-noise ratio is defined as the ratio of the average output signal power to the average output noise power. From Eq. (6.50), we see that the message component in the discriminator output, and therefore the low-pass filter output, is $k_f m(t)$. Hence, the average output signal power is equal to $k_f^2 P$, where P is the average power of the message signal $m(t)$.

To determine the average output noise power, we note that the noise $n_d(t)$ at the discriminator output is proportional to the time derivative of the quadrature noise component $n_s(t)$. From Section 2.3, we recall that differentiation of a function with respect to time corresponds to multiplication of its Fourier transform by $j2\pi f$. It follows therefore that we may obtain the noise process $n_d(t)$ by passing $n_s(t)$ through a linear filter with a transfer function equal to

$$\frac{j2\pi f}{2\pi A_c} = \frac{jf}{A_c}$$

This means that the power spectral density $S_{N_d}(f)$ of the noise $n_d(t)$ is related to the power spectral density $S_{N_s}(f)$ of the quadrature noise component $n_s(t)$ as follows:

$$S_{N_d}(f) = \frac{f^2}{A_c^2} S_{N_s}(f) \tag{6.55}$$

With the equivalent IF filter in Fig. 6.11 having an ideal band-pass characteristic of bandwidth B_T and mid-band frequency f_c, it follows that the narrow-band noise $n(t)$ will have a power spectral density characteristic that is similarly shaped. This means that the quadrature component $n_s(t)$ of the narrow-band noise $n(t)$ will have

* S. O. Rice, "Noise in FM Receivers," in the *Symposium Proceedings of Time Series Analysis*, edited by M. Rosenblatt, pp. 395–422 (Wiley, 1963).

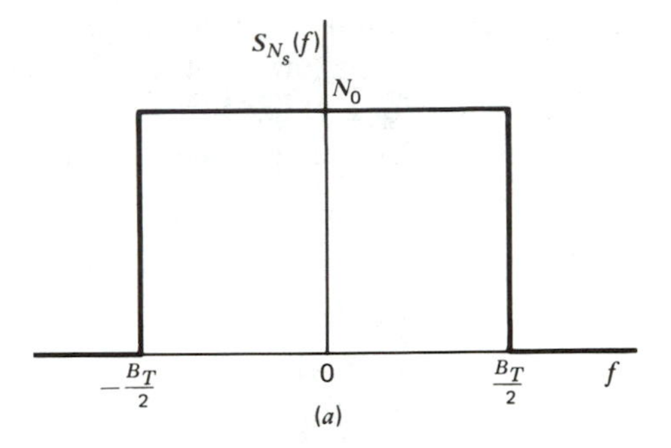

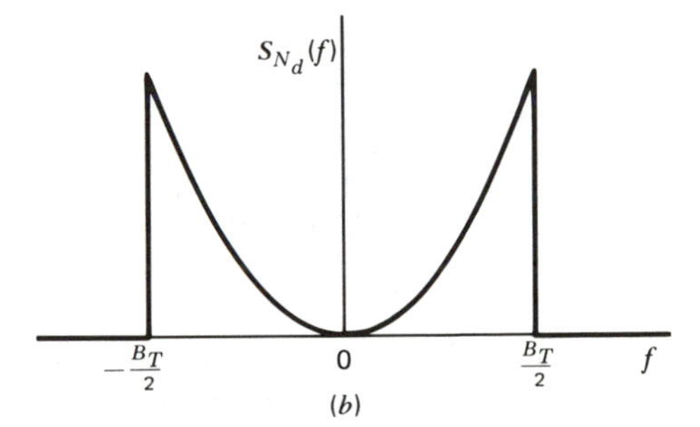

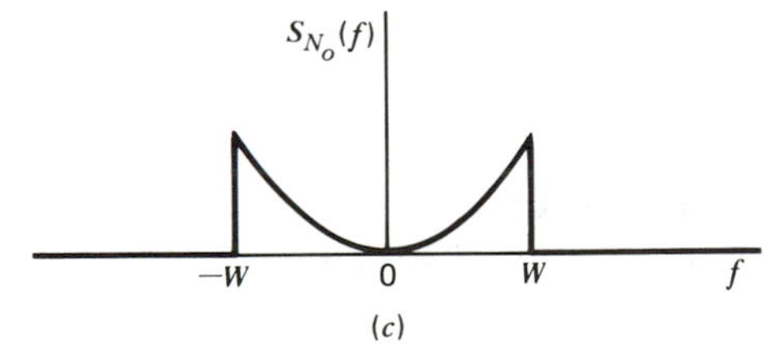

Figure 6.13 Noise analysis of FM receiver. *(a)* Power spectral density of quadrature component $n_s(t)$ of narrow-band noise $n(t)$. *(b)* Power spectral density of noise $n_d(t)$ at discriminator output. *(c)* Power spectral density of noise $n_o(t)$ at receiver output.

the ideal low-pass characteristic shown in Fig. 6.13(*a*). The corresponding power spectral density of the noise $n_d(t)$ is shown in Fig. 6.13(*b*). That is,

$$S_{N_d}(f)=\begin{cases}\dfrac{N_0 f^2}{A_c^2}, & |f|\leqslant\dfrac{B_T}{2}\\ 0, & \text{otherwise}\end{cases} \tag{6.56}$$

The discriminator output is followed by a low-pass filter with a bandwidth equal to the message bandwidth W. For wide-band FM, we usually find that W is smaller than $B_T/2$, where B_T is the transmission bandwidth of the FM signal. This means that the out-of-band components of noise $n_d(t)$ will be rejected. Therefore, the power spectral density $S_{N_o}(f)$ of the noise $n_o(t)$ appearing at the receiver output is defined by

$$S_{N_o}(f)=\begin{cases}\dfrac{N_0 f^2}{A_c^2}, & |f|\leq W \\ 0, & \text{otherwise}\end{cases} \tag{6.57}$$

as shown in Fig. 6.13(*c*). The average output noise power is determined by integrating the power spectral density $S_{N_o}(f)$ from $-W$ to W. We thus get

$$\begin{aligned}\text{average power of output noise} &= \frac{N_0}{A_c^2}\int_{-W}^{W} f^2\,df \\ &= \frac{2N_0 W^3}{3A_c^2}\end{aligned} \tag{6.58}$$

Note that the average output noise power is inversely proportional to the average carrier power $A_c^2/2$. Accordingly, in an FM system, increasing the carrier power has a *noise-quieting effect*.

Earlier we determined the average output signal power as $k_f^2 P$. Therefore, provided the carrier-to-noise ratio is high, we may divide this average output signal power by the average output noise power of Eq. (6.58) to obtain the output signal-to-noise ratio

$$(\text{SNR})_{\text{O,FM}}=\frac{3A_c^2 k_f^2 P}{2N_0 W^3} \tag{6.59}$$

The average power in the modulated signal $s(t)$ is $A_c^2/2$, and the average noise power in the message bandwidth is WN_0. Thus the channel signal-to-noise ratio is

$$(\text{SNR})_{\text{C,FM}}=\frac{A_c^2}{2WN_0} \tag{6.60}$$

Dividing the output signal-to-noise ratio by the channel signal-to-noise ratio, we get the figure of merit

$$\left.\frac{(\text{SNR})_\text{O}}{(\text{SNR})_\text{C}}\right|_{\text{FM}}=\frac{3k_f^2 P}{W^2} \tag{6.61}$$

The frequency deviation Δf is proportional to the frequency sensitivity k_f of the modulator. Also, by definition, the deviation ratio D is equal to the frequency deviation Δf divided by the message bandwidth W. Therefore, it follows from Eq. (6.61) that the figure of merit of a wideband FM system is a quadratic function of the deviation ratio. Now, in wideband FM, the transmission bandwidth B_T is approximately proportional to the deviation ratio D. *Accordingly, we may state that when the carrier-to-noise ratio is high, an increase in the transmission bandwidth* B_T *provides a corresponding quadratic increase in the output signal-to-noise ratio or figure of merit of the FM system.*

Example 3 Single-tone modulation

Consider the case of a sine-wave of frequency f_m as the modulating wave, and assume a peak frequency deviation Δf. The modulated wave is thus defined by

$$s(t) = A_c \cos\left[2\pi f_c t + \frac{\Delta f}{f_m}\sin(2\pi f_m t)\right]$$

Therefore, we may write

$$2\pi k_f \int_0^t m(t)dt = \frac{\Delta f}{f_m}\sin(2\pi f_m t)$$

Differentiating both sides with respect to time:

$$m(t) = \frac{\Delta f}{k_f}\cos(2\pi f_m t)$$

Hence, the average power of the message signal $m(t)$ is

$$P = \frac{(\Delta f)^2}{2k_f^2}$$

Substituting this result into the formula for the output signal-to-noise ratio given by Eq. (6.59), we get

$$(\text{SNR})_{\text{O,FM}} = \frac{3A_c^2(\Delta f)^2}{4N_0W^3}$$

$$= \frac{3A_c^2\beta^2}{4N_0W} \qquad (6.62)$$

where $\beta = \Delta f/W$ is the modulation index. Using Eq. (6.61) to evaluate the corresponding figure of merit, we get

$$\left.\frac{(\text{SNR})_\text{O}}{(\text{SNR})_\text{C}}\right|_{\text{FM}} = \frac{3}{2}\left(\frac{\Delta f}{W}\right)^2$$

$$= \frac{3}{2}\beta^2 \qquad (6.63)$$

It is important to note that the modulation index $\beta = \Delta f/W$ is determined by the bandwidth W of the post-detection low-pass filter, and is not related to the sinusoidal message frequency f_m, except insofar as this filter is usually chosen so as to pass the spectrum of the desired message; this is merely a matter of consistent design. For a specified bandwidth W the sinusoidal message frequency f_m may lie anywhere between 0 and W, and would yield the same output signal-to-noise ratio.

It is of particular interest to compare the performance of AM and FM systems. One way of making this comparison is to consider the figures of merit of the two systems based on a sinusoidal modulating signal. For an AM system operating with a sinusoidal modulating signal and 100 percent modulation, we have (from Example 1):

$$\left.\frac{(\text{SNR})_\text{O}}{(\text{SNR})_\text{C}}\right|_{\text{AM}} = \frac{1}{3}$$

Comparing this figure of merit with the corresponding result obtained above for an FM system, we see that the use of frequency modulation offers the possibility of improved signal-to-noise ratio over amplitude modulation when

$$\tfrac{3}{2}\beta^2 > \tfrac{1}{3}$$

that is,

$$\beta > 0.5$$

We may therefore consider $\beta = 0.5$ as defining roughly the transition between narrow-band FM and wideband FM. This statement, based on noise considerations, further confirms a similar observation that was made in Chapter 4 when considering the bandwidth of FM waves.

Capture Effect

The inherent ability of an FM system to minimize the effects of unwanted signals (e.g., noise, as discussed above) also applies to *interference* produced by another frequency-modulated signal whose frequency content is close to the carrier frequency of the desired FM wave. However, interference suppression in an FM receiver works well only when the interference is weaker than the desired FM input. When the interference is the stronger one of the two, the receiver locks on to the stronger signal and thereby suppresses the desired FM input. When they are of nearly equal strength, the receiver fluctuates back and forth between them. This phenomenon is known as the *capture effect*.

6.8 FM THRESHOLD EFFECT

The formula of Eq. (6.59), defining the output signal-to-noise ratio of an FM receiver, is valid only if the carrier-to-noise ratio, measured at the discriminator input, is high compared with unity. It is found experimentally that as the input noise is increased so that the carrier-to-noise ratio is decreased, the FM receiver *breaks*. At first, individual clicks are heard in the receiver output, and as the carrier-to-noise ratio decreases still further, the clicks rapidly merge into a *crackling* or *sputtering sound*. Near the breaking point, Eq. (6.59) begins to fail by predicting values of output signal-to-noise ratio larger than the actual ones. This phenomenon is known as the *threshold effect*.* *The threshold is defined as the minimum carrier-to-noise ratio yielding an FM improvement which is not significantly deteriorated from the value predicted by the usual signal-to-noise formula assuming small noise.*

For a qualitative discussion of the FM threshold effect, consider first the case when there is no signal present, so that the carrier wave is unmodulated. Then the composite signal at the frequency discriminator input is

$$x(t) = [A_c + n_c(t)]\cos(2\pi f_c t) - n_s(t)\sin(2\pi f_c t) \qquad (6.64)$$

where $n_c(t)$ and $n_s(t)$ are the in-phase and quadrature components of the narrow-

* 1. S. O. Rice, *op. cit.*

2. M. Schwartz, W. R. Bennett, and S. Stein, *Communication Systems and Techniques*, pp. 129–163 (McGraw-Hill, 1966).

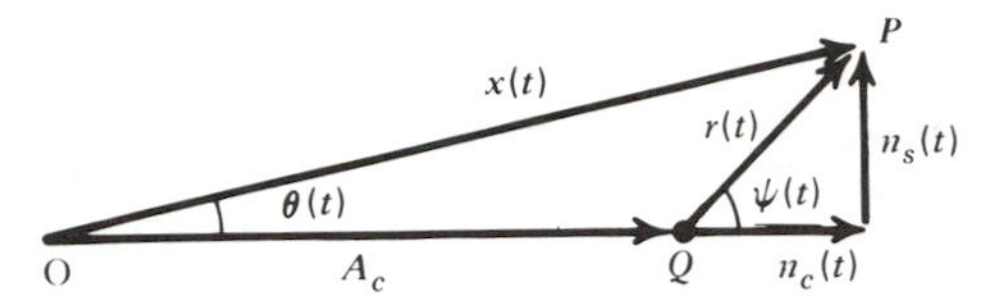

Figure 6.14 A phasor diagram interpretation of Eq. (6.64).

band noise $n(t)$ with respect to the carrier wave $\cos(2\pi f_c t)$. The phasor diagram of Fig. 6.14 shows the phase relations between the various components of $x(t)$ in Eq. (6.64). As the amplitudes and phases of $n_c(t)$ and $n_s(t)$ change with time in a random manner, the point P wanders around the point Q. When the carrier-to-noise ratio is large, $n_c(t)$ and $n_s(t)$ are usually much smaller than the carrier amplitude A_c, and so the wandering point P in Fig. 6.14 spends most of its time near point Q. Thus the angle $\theta(t)$ is approximately $n_s(t)/A_c$ to within a multiple of 2π. The wandering point P occasionally sweeps around the origin and $\theta(t)$ increases or decreases by 2π radians. Figure 6.15 illustrates how, in a rough way, these excursions in $\theta(t)$ produce impulse-like components in $\theta'(t) = d\theta/dt$. The discriminator

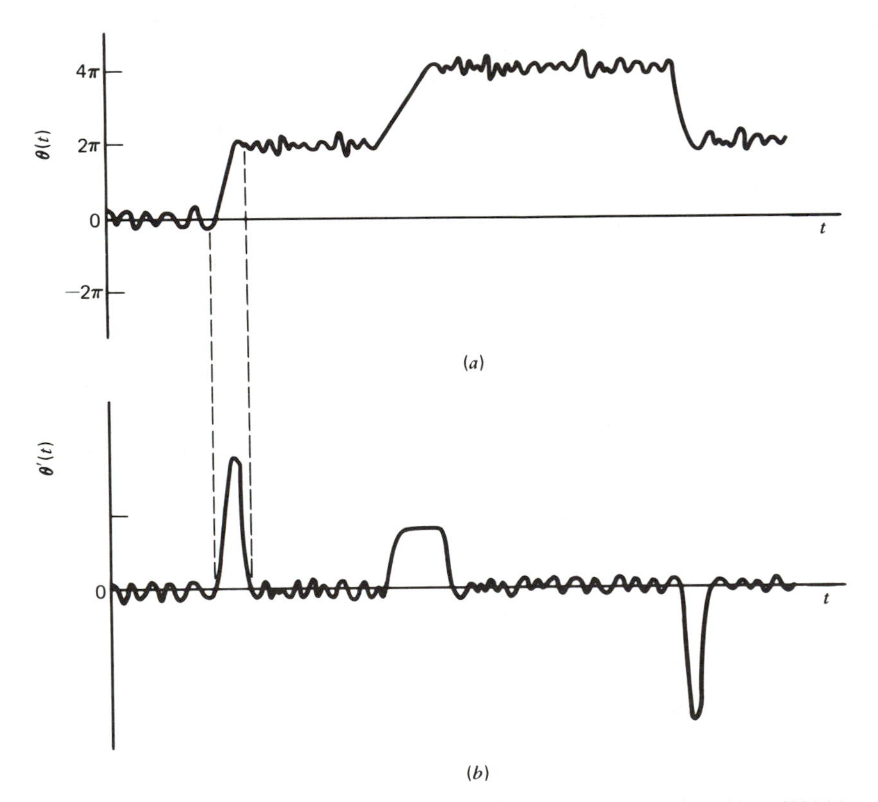

Figure 6.15 Sketch showing impulse-like components in $\theta'(t) = d\theta(t)/dt$ produced by changes of 2π in $\theta(t)$.

output $v(t)$ is equal to $\theta'(t)/2\pi$. These impulse-like components have different heights depending on how close the wandering point P comes to the origin O, but all have areas nearly equal to $\pm 2\pi$ radians. When the signal shown in Fig. 6.15(b) is passed through the post-detection low-pass filter, corresponding but wider impulse-like components are excited in the receiver output and are heard as clicks. The clicks are produced only when $\theta(t)$ changes by $\pm 2\pi$.

From the phasor diagram of Fig. 6.14, we may deduce the conditions required for clicks to occur. A positive-going click occurs when the envelope $r(t)$ and phase $\psi(t)$ of the narrow-band noise $n(t)$ satisfy the following conditions:

$$r(t) > A_c$$

$$\psi(t) < \pi < \psi(t) + d\psi(t)$$

$$\frac{d\psi(t)}{dt} > 0$$

These conditions ensure that the phase $\theta(t)$ of the resultant phasor $x(t)$ changes by 2π radians in the time increment dt, during which the phase of the narrow-band noise increases by the incremental amount $d\psi(t)$. Similarly, the conditions for a negative-going click to occur are as follows

$$r(t) > A_c$$

$$\psi(t) > -\pi > \psi(t) + d\psi(t)$$

$$\frac{d\psi(t)}{dt} < 0$$

These conditions ensure that $\theta(t)$ changes by -2π radians during the time increment dt.

As the carrier-to-noise ratio is decreased, the average number of clicks per unit time increases. When this number becomes appreciably large, the threshold is said to occur.

In the case of an unmodulated carrier, the average number of positive-going clicks per second, N_+ is the same as the average number of negative-going clicks per second, N_-. Furthermore, assuming an ideal IF filter of bandwidth B_T centered on f_c, as in Fig. 6.11(b), we have*

$$N_+ = \frac{B_T}{4\sqrt{3}} \operatorname{erfc}(\sqrt{\rho}) \tag{6.65}$$

where ρ is the *carrier-to-noise ratio*:

$$\rho = \frac{A_c^2}{2B_T N_0} \tag{6.66}$$

* S. O. Rice, *op. cit.*

and $\text{erfc}(\sqrt{\rho})$ is the *complementary error function*, (see Appendix 2):

$$\text{erfc}(\sqrt{\rho})=\frac{2}{\sqrt{\pi}}\int_{\sqrt{\rho}}^{\infty}\exp(-z^2)dz \tag{6.67}$$

Equation (6.65) shows that N_+ is proportional to the IF filter bandwidth.

The output signal-to-noise ratio is calculated as follows:

1. The output signal is taken as the receiver output measured in the absence of noise. The average output signal power is calculated assuming a sinusoidal modulation that produces a frequency deviation Δf equal to $B_T/2$, so that the carrier swings back and forth across the entire input frequency band.
2. The average output noise power is calculated when there is no signal present; that is, the carrier is unmodulated, with no restriction imposed on the value of the carrier-to-noise ratio ρ.

The resulting output signal-to-noise ratio is given by*

$$(\text{SNR})_{\text{O,FM}}=\frac{3\rho(B_T/2W)^3}{1+4\sqrt{3}\rho(B_T/2W)^2\,\text{erfc}(\sqrt{\rho})} \tag{6.68}$$

As the carrier-to-noise ratio ρ approaches infinity, the denominator of Eq. (6.68) tends to unity, and so the output signal-to-noise ratio approaches the limiting value of $3\rho(B_T/2W)^3$. This limiting value is in accordance with Eq. (6.62) calculated for the case when the frequency deviation $\Delta f=B_T/2$ (as described above), and the modulation index

$$\beta=\frac{\Delta f}{W}=\frac{B_T}{2W}$$

Curve I of Fig. 6.16 is a plot of Eq. (6.68) when the ratio $B_T/2W$ is equal to 5. This curve shows that the output signal-to-noise ratio deviates appreciably from a linear function of the carrier-to-noise ratio ρ when ρ becomes less than 10 dB.

Curve II of Fig. 6.16† shows the effect of modulation on the output signal-to-noise ratio when the modulating signal (assumed sinusoidal) and the noise are present at the same time. The presence of noise tends to reduce the average output signal power by a small amount; however, this reduction is usually small enough to be omitted, so that the average output signal power may be taken to be effectively the same as for curve I. The average output noise power, on the other hand, is strongly dependent on the presence of the modulating signal, which accounts for the noticeable deviation of curve II from curve I. In particular, we find that as ρ decreases from infinity, the output signal-to-noise ratio deviates appreciably from

* S. O. Rice, *op. cit*.

† Curve II is also predicted from theory; see S. O. Rice, *op. cit*. The validity of this curve has been confirmed experimentally by D. L. Schilling and J. Billig; see M. Schwartz, W. R. Bennett, and S. Stein, p. 153, *op. cit*. For some earlier experimental work on the threshold phenomenon in FM, see M. G. Crosby, "Frequency modulation noise characteristics," *Proc. IRE*, vol. 25, pp. 472–514, April 1937.

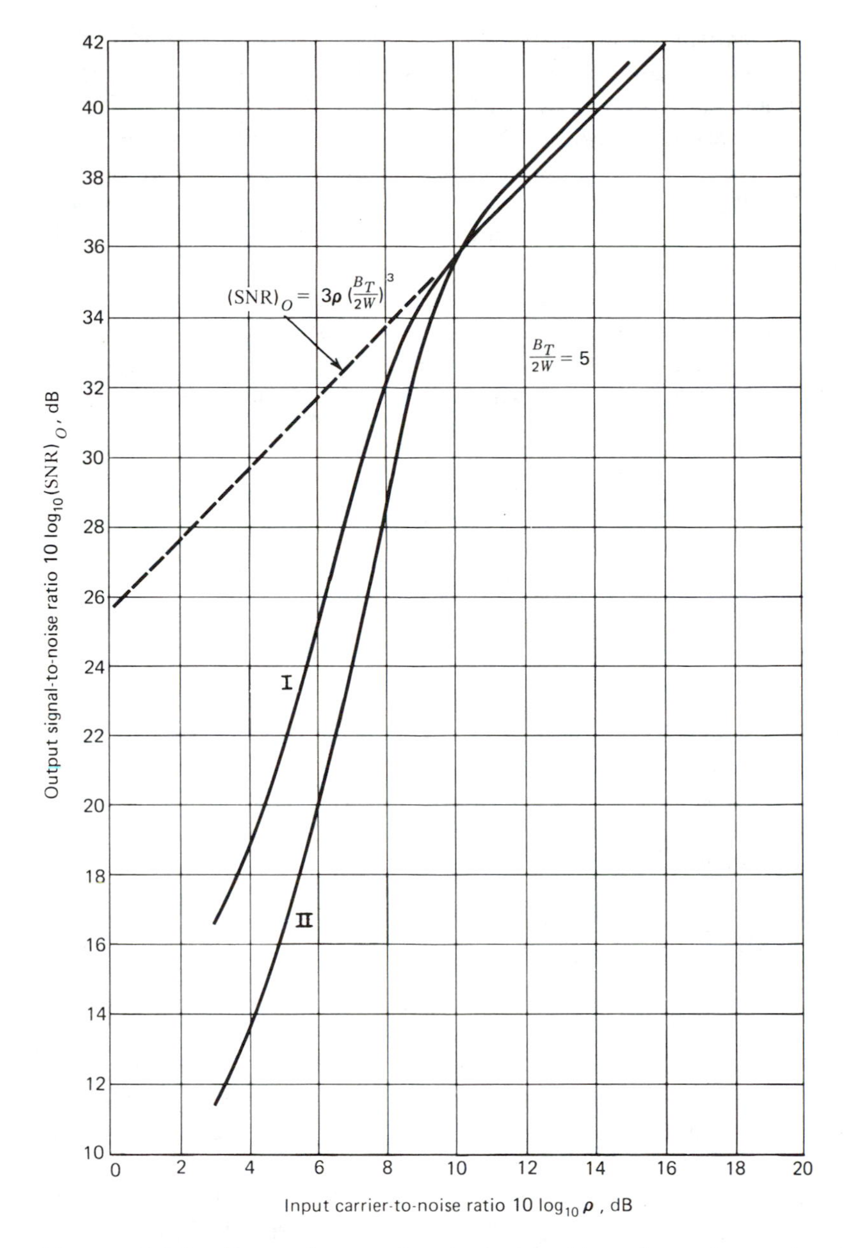

Figure 6.16 Dependence of output signal-to-noise ratio on input carrier-to-noise ratio. In curve I, the average output noise power is calculated assuming an unmodulated carrier. In curve II, the average output noise power is calculated assuming a sinusoidally modulated carrier.

a linear function of ρ when ρ is about 11 dB. Also when the signal is present, the resulting modulation of the carrier tends to increase the average number of clicks per second. Experimentally, it is found that occasional clicks are heard in the receiver output at a carrier-to-noise ratio of 13 dB, which appears to be only slightly higher than theory indicates. Also it is of interest to note that the enhanced increase in the average number of clicks per second tends to cause the output signal-to-noise ratio to fall off somewhat more sharply just below the threshold level in the presence of modulation.

From the above discussion we may conclude that threshold effects in FM receivers may be avoided in most practical cases of interest if the carrier-to-noise ratio ρ is equal to or greater than 20 or, equivalently, 13 dB. Thus, using Eq. (6.66), we find that the loss of message at the discriminator output is negligible if

$$\frac{A_c^2}{2B_T N_0} \geqslant 20 \tag{6.69}$$

or, equivalently, if the average transmitted power $A_c^2/2$ satisfies the condition

$$\frac{A_c^2}{2} \geqslant 20 B_T N_0 \tag{6.70}$$

To use this formula, we may proceed as follows:

1. For a specified modulation index β and message bandwidth W, we determine the transmission bandwidth of the FM wave, B_T, using the universal curve of Fig. 4.9 or Carson's rule.
2. For a specified average noise power per unit bandwidth, N_0, we use Eq. (6.70) to determine the minimum value of the average transmitted power $A_c^2/2$ that is necessary to operate above threshold.

FM Threshold Reduction

In certain applications such as space communications, there is a particular interest in reducing the noise threshold in an FM receiver so as to satisfactorily operate the receiver with the minimum signal power possible. Threshold reduction in FM receivers may be achieved by using an FM demodulator with negative feedback (commonly referred to as an *FMFB demodulator*),* or by using a phase-locked loop demodulator.

The block diagram of an FMFB demodulator is shown in Fig. 6.17. We see that the local oscillator of the conventional FM receiver has been replaced by a voltage-controlled oscillator (VCO) whose instantaneous output frequency is controlled by the demodulated signal. In order to understand the operation of this receiver, suppose for the moment that the VCO is removed from the circuit and the feedback

* The idea of using feedback around an FM demodulator was originally proposed by Chaffee, long before the advent of space communications: J. G. Chaffee, "The application of negative feedback to frequency-modulation systems," *Bell Syst. Tech. J.*, vol. 18, pp. 403–437, July 1939.

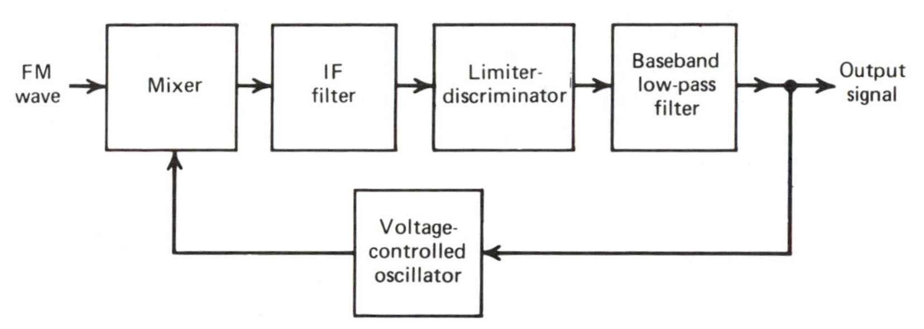

Figure 6.17 FMFB demodulator.

loop is left open.* Assume that a wide-band FM wave is applied to the receiver input, and a second FM wave, from the same source but whose modulation index is a fraction smaller, is applied to the VCO terminal of the mixer. The output of the mixer would consist of the difference frequency component, because the sum frequency component is removed by the IF filter. The frequency deviation of the mixer output would be small, although the frequency deviation of both input FM waves is large, since the difference between their instantaneous deviations is small. Hence, the modulation indices would subtract and the resulting FM wave at the mixer output would have a smaller modulation index. The FM wave with reduced modulation index may be passed through an IF filter, whose bandwidth need be only a fraction of that required for either wide-band FM wave, and then frequency demodulated. It is now apparent that the second wide-band FM wave applied to the mixer may be obtained by feeding the output of the frequency discriminator back to the VCO.

It will now be shown that the signal-to-noise ratio of an FMFB receiver is the same as that of a conventional FM receiver with the same input signal and noise power if the carrier-to-noise ratio is sufficiently large. Assume for the moment that there is no feedback around the demodulator. In the combined presence of an unmodulated carrier $A_c \cos(2\pi f_c t)$ and narrow-band noise $n(t) = n_c(t)\cos(2\pi f_c t) - n_s(t)\sin(2\pi f_c t)$, the phase of the composite signal $x(t)$ at the limiter-discriminator input is approximately equal to $n_s(t)/A_c$, assuming that the carrier-to-noise ratio is high. The envelope of $x(t)$ is of no interest to us, because the limiter removes all variations in the envelope. Thus the composite signal at the frequency discriminator input consists of a small index phase-modulated wave with the modulation derived from the component $n_s(t)$ of noise which is in phase quadrature with the carrier. When feedback is applied, the VCO generates a wave that reduces the phase-modulation index of the wave in the IF filter output, that is, the quadrature component $n_s(t)$ of noise. Thus we see that as long as the carrier-to-noise ratio is sufficiently large, the FMFB receiver does not respond to the in-phase noise component $n_c(t)$, but that it would demodulate the quadrature noise component

* Our treatment of the FMFB demodulator is based on: L. H. Enole, "Decreasing the threshold in FM by frequency feedback," *Proc. IRE*, vol. 50, pp. 18–30, Jan. 1962. See, also, J. H. Roberts, "Angle modulation: The theory of system assessment," *IEEE Telecommunication Series* 5, pp. 166–181. (Institution of Electrical Engineers, England, 1977).

$n_s(t)$ in exactly the same fashion as it would demodulate signal modulation. Signal and quadrature noise are reduced in the same proportion by the applied feedback, with the result that the baseband signal-to-noise ratio is independent of feedback. For large carrier-to-noise ratios the baseband signal-to-noise ratio of an FMFB receiver is then the same as that of a conventional FM receiver.

The reason why an FMFB receiver is able to extend the threshold is that, unlike a conventional FM receiver, it uses a very important piece of *a priori* information, namely, that even though the carrier frequency of the incoming FM wave will usually have large frequency deviations, its rate of change will be at the baseband rate. An FMFB demodulator is essentially a *tracking filter* which can track only the slowly varying frequency of wide-band FM waves, and consequently it responds only to a narrow band of noise centered about the instantaneous carrier frequency. The bandwidth of noise to which the FMFB receiver responds is precisely the band of noise which the VCO tracks. The net result is that an FMFB receiver is capable of realizing a threshold reduction of the order of 5–7 dB, which represents a significant improvement in the design of minimum power FM systems.

The phase-locked loop demodulator, which was described in Section 4.8, exhibits threshold reduction properties that are similar to those of the FMFB demodulator. Thus, like the FMFB demodulator, a phase-locked loop is a tracking filter and, as such, the bandwidth of noise to which it responds is precisely the band of noise which the VCO tracks. However, while the thresholds of the phase-locked loop and FMFB demodulators occur because of the same basic mechanism, the details by which they occur are, of course, different.* Practical experience with the phase-locked loop, however, confirms the conclusion that very comparable performance with the FMFB demodulator is obtained in many situations, so that the choice between these two types of threshold-extension devices is often made in favor of the phase-locked loop because of its simpler construction.

6.9 PRE-EMPHASIS AND DE-EMPHASIS IN FM

In Section 6.7 we showed that the power spectral density of the noise at the receiver output has a square-law dependence on the operating frequency; this is illustrated in Fig. 6.18(*a*). In part (*b*) of this figure we have included the power spectral density of a typical message source; audio and video signals typically have spectra of this form. We see that the power spectral density of the message usually falls off appreciably at higher frequencies. On the other hand, the power spectral density of the output noise increases rapidly with frequency. Thus, at $f = \pm W$, the relative spectral density of the message is quite low, whereas that of the output noise is quite high in comparison. Clearly, the message is not utilizing the frequency band allowed to it in an efficient manner. It may appear that one way of improving the noise per-

* H. Taub and D. L. Schilling, *Principles of Communication Systems*, pp. 354–357 (McGraw-Hill, 1971). See, also, J. H. Roberts, *op. cit.*, pp. 200–202.

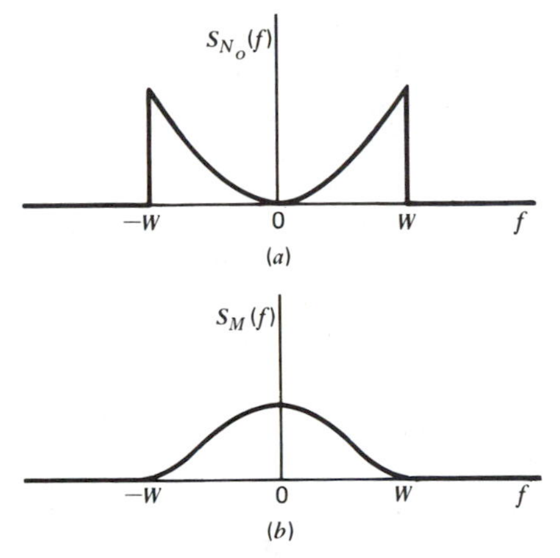

Figure 6.18 *(a)* Power spectral density of noise at FM receiver output. *(b)* Power spectral density of a typical message source.

formance of the system is to slightly reduce the bandwidth of the post-detection low-pass filter so as to reject a large amount of noise power while losing only a small amount of message power. Such an approach, however, is usually not satisfactory because the distortion of the message caused by the reduced filter bandwidth, even though slight, may not be tolerable. For example, in the case of music we find that although the high-frequency notes contribute only a very small fraction of the total power, nonetheless, they contribute a great deal from an aesthetic viewpoint.

A more satisfactory approach to the efficient utilization of the allowed frequency band is based on the use of *pre-emphasis* in the transmitter and *de-emphasis* in the receiver, as illustrated in Fig. 6.19. In this method, we artificially emphasize the high-frequency components of the message signal prior to modulation in the transmitter, and therefore before the noise is introduced in the receiver. In effect, the low-frequency and high-frequency portions of the power spectral density of the message are equalized in such a way that the message fully occupies the frequency band allotted to it. Then, at the discriminator output in the receiver, we perform the inverse operation by de-emphasizing the high-frequency components, so as to restore the original signal-power distribution of the message. In such a process the

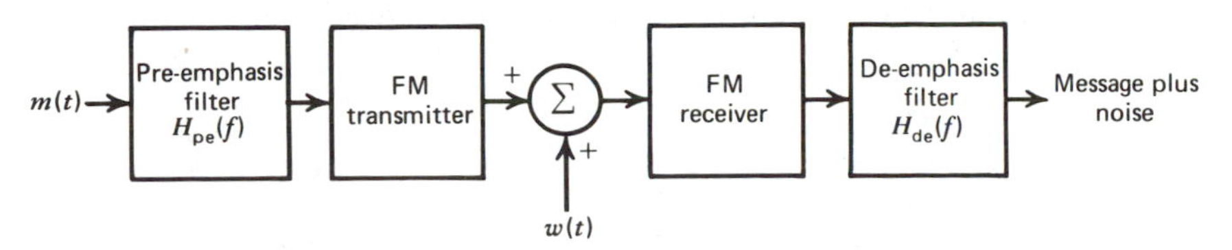

Figure 6.19 Use of pre-emphasis and de-emphasis in an FM system.

high-frequency components of the noise at the discriminator output are also reduced, thereby effectively increasing the output signal-to-noise ratio of the system. Such a pre-emphasis and de-emphasis process is widely used in FM transmission and reception.

In order to produce an undistorted version of the original message at the receiver output, the pre-emphasis filter in the transmitter and the de-emphasis filter in the receiver must ideally have transfer functions that are the inverse of each other. That is, if $H_{pe}(f)$ designates the transfer function of the pre-emphasis filter, then the transfer function $H_{de}(f)$ of the de-emphasis filter must ideally be

$$H_{de}(f)=\frac{1}{H_{pe}(f)}, \qquad -W\leqslant f\leqslant W \tag{6.71}$$

This choice of transfer functions makes the average message power at the receiver output independent of the pre-emphasis and de-emphasis procedure.

From our previous noise analysis in FM systems, assuming a high carrier-to-noise ratio, the power spectral density of the noise $n_d(t)$ at the discriminator output is

$$S_{N_d}(f)=\begin{cases}\dfrac{N_0 f^2}{A_c^2}, & |f|\leqslant\dfrac{B_T}{2}\\ 0, & \text{otherwise}\end{cases} \tag{6.72}$$

Therefore, the modified power spectral density of the noise at the de-emphasis filter output is equal to $|H_{de}(f)|^2 S_{N_d}(f)$. Recognizing, as before, that the post-detection low-pass filter has a bandwidth W which is, in general, less than $B_T/2$, we find that the average power of the modified noise at the receiver output is

$$\begin{pmatrix}\text{average output noise}\\ \text{power with de-emphasis}\end{pmatrix}=\frac{N_0}{A_c^2}\int_{-W}^{W} f^2|H_{de}(f)|^2\,df \tag{6.73}$$

Because the average message power at the receiver output is ideally unaffected by the pre-emphasis and de-emphasis procedure, it follows that the improvement in output signal-to-noise ratio produced by the use of pre-emphasis in the transmitter and de-emphasis in the receiver is defined by

$$D=\frac{\text{average output noise power without pre-emphasis and de-emphasis}}{\text{average output noise power with pre-emphasis and de-emphasis}} \tag{6.74}$$

Earlier we showed that the average output noise power without pre-emphasis and de-emphasis is equal to $(2N_0W^3/3A_c^2)$. Therefore, after cancellation of common terms, we may write

$$D=\frac{2W^3}{3\int_{-W}^{W} f^2|H_{de}(f)|^2\,df} \tag{6.75}$$

It must be emphasized that this improvement factor assumes the use of a high carrier-to-noise ratio at the discriminator input.

Example 4

A simple pre-emphasis filter that emphasizes high frequencies and which is commonly used in practice is defined by the transfer function

$$H_{pe}(f)=1+\frac{jf}{f_0} \tag{6.76}$$

This transfer function is closely realized by the RC-amplifier network shown in Fig. 6.20(*a*), provided that $R \ll r$ and $2\pi f CR \ll 1$ inside the frequency band of interest. The amplifier in Fig. 6.20(*a*) is intended to make up for the attenuation introduced by the RC network at low frequencies. The frequency parameter $f_0=1/(2\pi Cr)$.

The corresponding de-emphasis filter in the receiver is defined by the transfer function

$$H_{de}(f)=\frac{1}{1+jf/f_0} \tag{6.77}$$

which can be realized using the simple RC network of Fig. 6.20(*b*).

The improvement in output signal-to-noise ratio of the FM receiver, resulting from use of the pre-emphasis and de-emphasis filters of Fig. 6.20, is therefore

$$\begin{aligned} D &= \frac{2W^3}{3\displaystyle\int_{-W}^{W}\frac{f^2\,df}{1+(f/f_0)^2}} \\ &= \frac{(W/f_0)^3}{3[(W/f_0)-\tan^{-1}(W/f_0)]} \end{aligned} \tag{6.78}$$

In commercial FM broadcasting, we typically have $f_0=2.1$ kHz, and we may reasonably assume $W=15$ kHz. This set of values yields $D=22$, which corresponds to an improvement of 13 dB in the output signal-to-noise ratio of the receiver. The output signal-to-noise ratio of an FM receiver without pre-emphasis and de-emphasis is typically 40–50 dB. We see therefore that by using the simple pre-emphasis and de-emphasis filters shown in Fig. 6.20, we can obtain a significant improvement in the noise performance of the receiver.

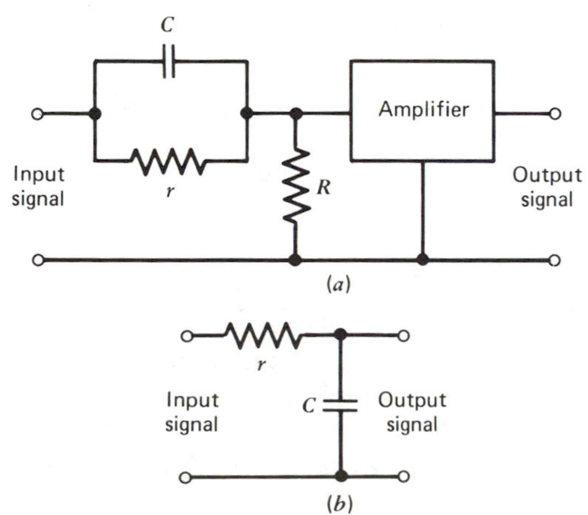

Figure 6.20 *(a)* Pre-emphasis filter. *(b)* De-emphasis filter.

The use of the simple *linear* pre-emphasis and de-emphasis filters described above is an example of how the performance of an FM system may be improved by utilizing the differences between characteristics of signals and noise in the system. These simple filters also find application in audio tape-recording. In recent years *nonlinear* pre-emphasis and de-emphasis techniques have been applied successfully to tape-recording. These techniques (known as *Dolby-A*, *Dolby-B*, and *DBX* systems) use a combination of filtering and dynamic range compression to reduce the effects of noise, particularly when the signal level is low.*

6.10 DISCUSSION

We conclude the noise analysis of CW modulation systems by presenting a comparison of the relative merits of the different modulation techniques. For the purpose of this comparison, we assume that the modulation is produced by a single sine-wave. For the comparison to be meaningful, we also assume that all the different modulation systems operate with exactly the same channel signal-to-noise ratio [see Eq. (6.5)]. In making the comparison, it is informative to keep in mind the transmission bandwidth requirement of the modulation system in question. In this regard, we use a *normalized transmission bandwidth* defined by

$$B_n = \frac{B_T}{W} \tag{6.79}$$

where B_T is the transmission bandwidth of the modulated wave and W is the message bandwidth. We may thus make the following observations:

1. In a full AM system using envelope detection, the output signal-to-noise ratio, assuming sinusoidal modulation, is given by [see Eq. (6.31)]

   $$(\mathrm{SNR})_O = \frac{\mu^2}{2+\mu^2}(\mathrm{SNR})_C$$

   This relation is shown plotted as curve I in Fig. 6.21, assuming $\mu = 1$. In this curve we have also included the AM threshold effect. Since in a full AM system both sidebands are transmitted, the normalized transmission bandwidth B_n equals 2.
2. In the case of a DSBSC or SSB modulation system using coherent detection, the output signal-to-noise ratio is given by [see Eqs. (6.17) and (6.23)]:

   $$(\mathrm{SNR})_O = (\mathrm{SNR})_C$$

   This relation is shown plotted as curve II in Fig. 6.21. We see, therefore, that the noise performance of a DSBSC or SSB system, using coherent detection, is superior to that of a full AM system using envelope detection by 4.8 dB. It should also be noted that neither the DSBSC nor the SSB system exhibits a threshold effect. With regard to transmission bandwidth requirement, we have

* For a detailed description of Dolby systems, see F. G. Stremler, *Introduction to communication systems*, pp. 671–673 (Addison-Wesley, 1982).

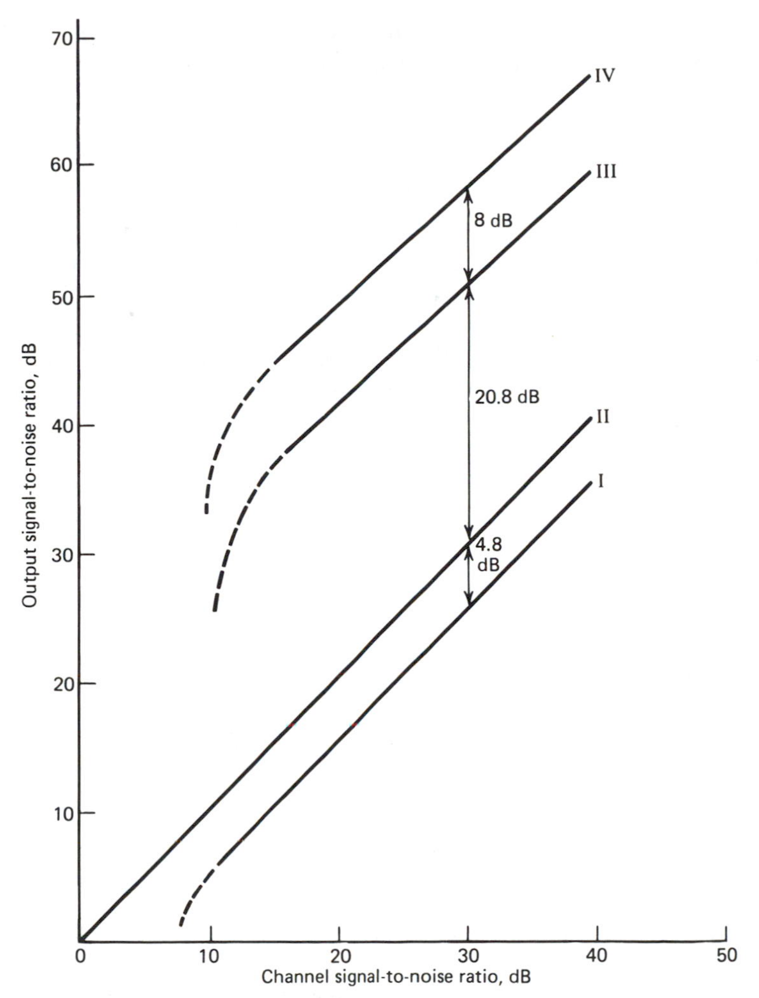

Figure 6.21 Comparison of the noise performance of various CW modulation systems. Curve I: Full AM. $\mu=1$.(Curve II: DSBSC. SSB. Curve III. FM. $\beta=2$. Curve IV FM. $\beta=5$.(Curves III and IV include 13-dB pre-emphasis, de-emphasis improvement).

$B_n=2$ for the DSBSC system and $B_n=1$ for the SSB system. Thus, among the family of AM systems, SSB modulation is optimum with regard to noise performance as well as bandwidth conservation.

3. In an FM system using a conventional discriminator, the output signal-to-noise ratio, assuming sinusoidal modulation, is given by [see Eq. (6.63)]

$$(\mathrm{SNR})_O=\tfrac{3}{2}\beta^2(\mathrm{SNR})_C$$

where β is the modulation index. This relation is shown plotted as curves III and IV in Fig. 6.21, corresponding to $\beta=2$ and $\beta=5$, respectively. In each case, we have included a 13-dB improvement which is typically obtained by using pre-emphasis in the transmitter and de-emphasis in the receiver. To determine the transmission bandwidth requirement, we use the universal curve of Fig. 4.9 and so find that:

$$B_n = 8 \qquad \text{for } \beta = 2$$
$$B_n = 16 \qquad \text{for } \beta = 5$$

Thus we see that, compared with the SSB system which is the optimum form of linear modulation, by using wide-band FM we obtain an improvement in output signal-to-noise ratio equal to 20.8 dB for a normalized bandwidth $B_n = 8$, and an improvement of 28.8 db for $B_n = 16$. This clearly illustrates the improvement in noise performance that is achievable by using wide-band FM. However, the price that we have to pay for this improvement is excessive transmission bandwidth. It is, of course, assumed that the FM system operates above threshold for the noise improvement to be realizable as described above.

Problems

Problem 6.1 Consider the sample function of a random process

$$x(t) = A + w(t)$$

where A is a constant and $w(t)$ is a white Gaussian noise of zero mean and power spectral density $N_0/2$. The sample function $x(t)$ is passed through the low-pass RC filter shown in Fig. P6.1. Find an expression for the output signal-to-noise ratio, with the dc component A regarded as the signal of interest.

Problem 6.2 The sample function

$$x(t) = A_c \cos(2\pi f_c t) + w(t)$$

is applied to the low-pass RC filter of Fig. P6.1. The amplitude A_c and frequency f_c of the sinusoidal components are constants, and $w(t)$ is a white Gaussian noise of zero mean and power spectral density $N_0/2$. Find an expression for the output signal-to-noise ratio with the sinusoidal component of $x(t)$ regarded as the signal of interest.

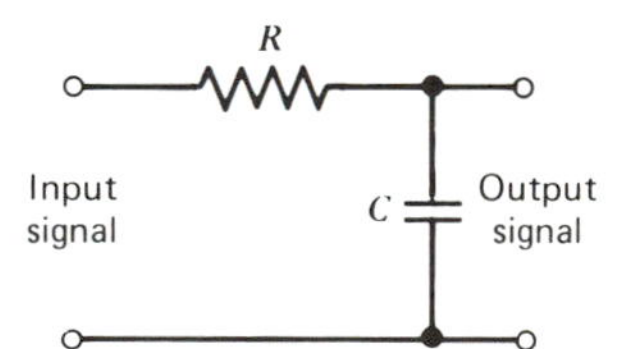

Figure P6.1

Problem 6.3 Suppose next the sample function $x(t)$ of Problem 6.2 is applied to the band-pass LCR filter of Fig. P6.2, which is tuned to the frequency f_c of the sinusoidal component. Assume that the Q factor of the filter is high compared with unity. Find an expression for the output signal-to-noise ratio, by treating the sinusoidal component of $x(t)$ as the signal of interest.

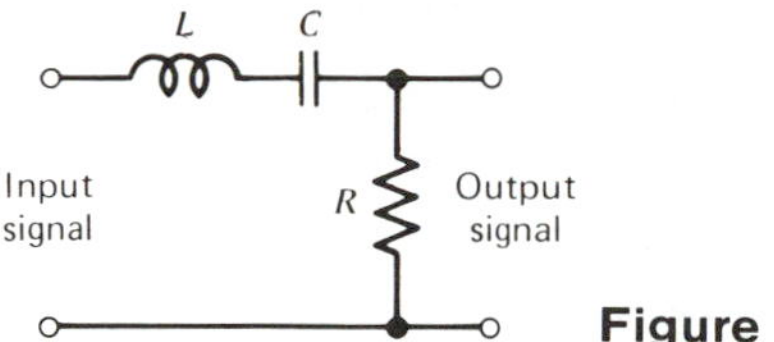

Figure P6.2

Problem 6.4 The input to the low-pass RC filter of Fig. P6.1 consists of a white Gaussian noise of zero mean and power spectral density $N_0/2$, plus a signal which is a sequence of constant-amplitude rectangular pulses. The pulse amplitude is A, the pulse duration is T, and the period of the sequence is T_0, where $T \ll T_0$. Derive an expression for the output signal-to-noise ratio of the filter, defined as the ratio of the square of the maximum amplitude of the output signal with no noise at the input to the average power of the output noise.

Problem 6.5 Evaluate the autocorrelation functions and cross-correlation functions of the in-phase and quadrature components of the narrow-band noise at the coherent detector input for: (a) the DSBSC system, (b) an SSB system using the lower sideband, and (c) an SSB system using the upper sideband.

Problem 6.6 In a receiver using coherent detection, the sine-wave generated by the local oscillator suffers from a phase error $\theta(t)$ with respect to the carrier wave $\cos(2\pi f_c t)$. Assuming that $\theta(t)$ is a sample function of a zero-mean Gaussian process of variance σ_Θ^2, and that most of the time the maximum value of $\theta(t)$ is small compared with unity, find the mean-square error of the receiver output for: (a) DSBSC modulation and (b) SSB modulation. The mean-square error is defined as the expected value of the difference between the receiver output and the message signal component of the receiver output.

Problem 6.7 The block diagram of Fig. P6.3 shows an SSB modulation system supplied with a pilot signal, which is harmonically related to the carrier. In the receiver, a band-pass filter of bandwidth Δf is used to extract the pilot signal from the noisy received wave, and the pilot signal is then frequency divided to provide the local carrier for demodulation. The predetection filter is included to limit the spectrum of the coherent detector input to the minimum possible frequency band. The additive white noise at the receiver input has zero mean and a power spectral density equal to $N_0/2$. Assuming a large signal-to-noise ratio, determine the phase error variance of the local carrier applied to the coherent detector.

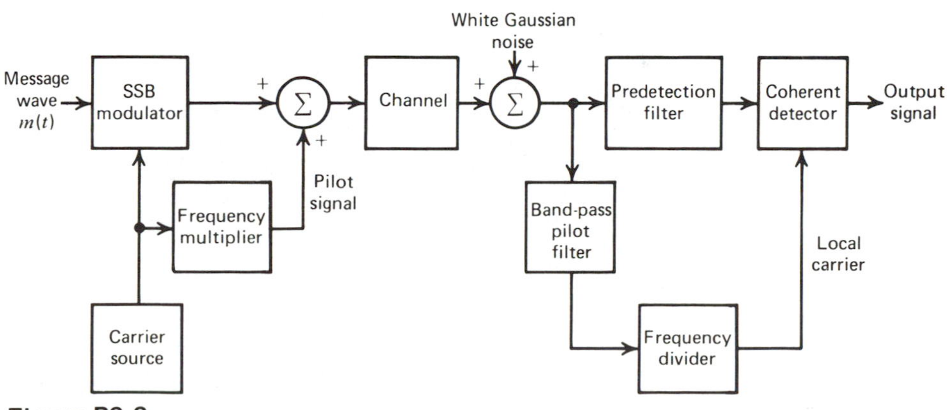

Figure P6.3

Problem 6.8 Let a message signal $m(t)$ be transmitted using single-sideband modulation. The power spectral density of $m(t)$ is

$$S_M(f) = \begin{cases} a\dfrac{|f|}{W}, & |f| \leq W \\ 0, & \text{otherwise} \end{cases}$$

where a and W are constants. White Gaussian noise of zero mean and power spectral density $N_0/2$ is added to the SSB-modulated wave at the receiver input. Find an expression for the output signal-to-noise ratio of the receiver.

Problem 6.9 An SSB-modulated wave is transmitted over a noisy channel, with the power spectral density of the noise being as shown in Fig. P6.4. The message bandwidth is 4 kHz and the carrier frequency is 200 kHz. Assuming that only the upper-sideband is transmitted, and that the average power of the modulated wave is 10 watts, determine the output signal-to-noise ratio of the receiver for the case when the predetection filter characteristic is ideal.

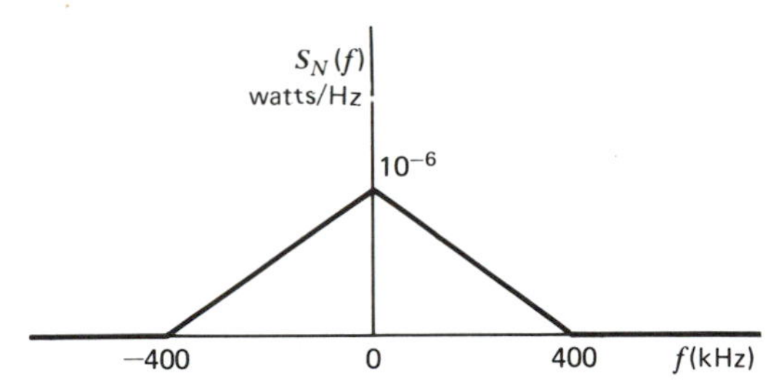

Figure P6.4

Problem 6.10 The average noise power per unit bandwidth measured at the front end of an AM receiver is 10^{-3} watts per hertz. The modulating wave is sinusoidal, with a carrier power of 80 kilo-watts, and a sideband power of 10 kilo-watts per sideband. The message bandwidth is 4 kHz. Assuming the use of an envelope detector in the receiver, determine the output signal-to-noise ratio of the system. By how many decibels is this system inferior to a DSBSC modulation system?

Problem 6.11 Consider an AM receiver using a square-law detector whose output is proportional to the square of the input, as indicated in Fig. P6.5. The AM wave is defined by

$$s(t) = A_c[1 + \mu\cos(2\pi f_m t)]\cos(2\pi f_c t)$$

Assume that the additive noise at the detector input is Gaussian with zero mean; it is defined by

$$n(t) = n_c(t)\cos(2\pi f_c t) - n_s(t)\sin(2\pi f_c),$$

where $E[n^2(t)] = E[n_c^2(t)] = E[n_s^2(t)] = \sigma_N^2$.

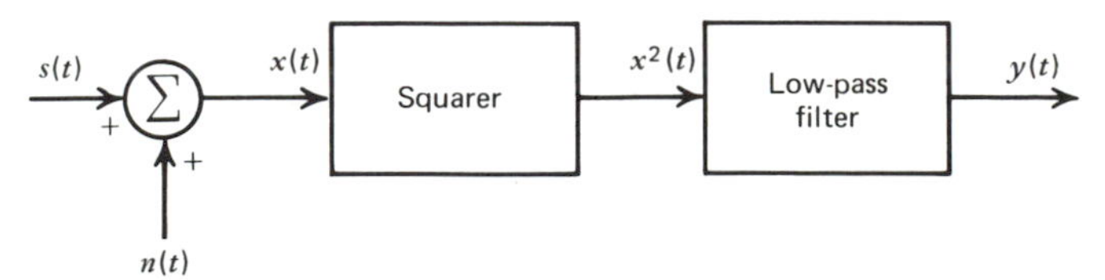

Figure P6.5

(a) Show that the output signal-to-noise ratio of the receiver is given by

$$(\mathrm{SNR})_O = \frac{2\mu^2\rho^2}{1+\rho(2+\mu^2)}$$

where ρ is the carrier-to-noise ratio.

(b) Evaluate the asymptotic behavior of $(\mathrm{SNR})_O$ with respect to ρ.

(c) Plot the dependence of $(\mathrm{SNR})_O$ on ρ for the case of 100 percent modulation.

Problem 6.12 Consider the output of an envelope detector defined by Eq. (6.27), which is reproduced here for convenience

$$y(t) = \{[A_c + A_c k_a m(t) + n_c(t)]^2 + n_s^2(t)\}^{1/2}$$

(a) Assume that the probability of the event

$$|n_s(t)| > \varepsilon A_c |1 + k_a m(t)|$$

is equal to or less than δ_1, where $\varepsilon \ll 1$. What is the probability that the effect of the quadrature component $n_s(t)$ is negligible?

(b) Suppose that k_a is adjusted relative to the message signal $m(t)$ such that the probability of the event

$$A_c[1 + k_a m(t)] + n_c(t) < 0$$

is equal to δ_2. What is the probability that the approximation

$$y(t) \simeq A_c[1 + k_a m(t)] + n_c(t)$$

is valid?

(c) Comment on the significance of the result in (b) for the case when δ_1 and δ_2 are both small compared with unity.

Problem 6.13 Consider an AM receiver with a received signal consisting of an unmodulated carrier $A_c \cos(2\pi f_c t)$ and an interfering signal $A_i \cos(2\pi f_i t + \theta_i)$.

(a) Construct a phasor diagram for the composite signal $x(t)$.

(b) Assuming the use of envelope detection, determine the receiver output when A_i is small compared with A_c.

Problem 6.14 An unmodulated carrier of amplitude A_c and frequency f_c and band-limited white noise are summed and then passed through an ideal envelope detector. Assume the noise spectral density to be of height $N_0/2$ and bandwidth $2W$, centered about the carrier frequency f_c. Determine the output signal-to-noise ratio for the case when the carrier-to-noise ratio is high.

Problem 6.15 Consider the AM receiver shown in Fig. P6.6 with the input signal $x(t)$ consisting of an AM wave plus white noise $w(t)$, as shown by

$$x(t) = [1 + m(t)]\cos(2\pi f_c t) + w(t)$$

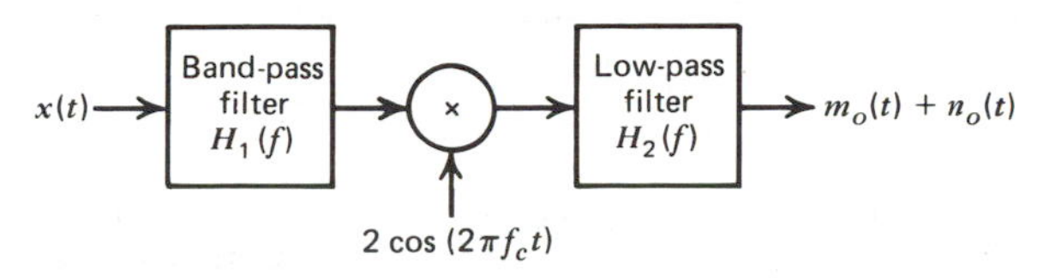

Figure P6.6

where $\cos(2\pi f_c t)$ is the carrier, $m(t)$ is the message signal, and $w(t)$ has zero mean and power spectral density $N_0/2$. The band-pass filter is centered at f_c with conjugate symmetry about this frequency. Let $m_o(t)$ and $n_o(t)$ denote the components of the receiver output due to the input AM wave and noise, respectively. Assume that the band-pass filter at the input and the low-pass filter at the output introduce distortion. As a measure of this distortion and that produced by the noise, define the mean-square error

$$\mathscr{E} = E[(m_o(t) + n_o(t) - m(t))^2]$$

(a) Show that

$$\mathscr{E} = \int_{-\infty}^{\infty} |1 - A_c H(f)|^2 S_M(f) df + N_0 \int_{-\infty}^{\infty} |H(f)|^2 \, df$$

where $S_M(f)$ is the power spectral density of the message signal $m(t)$. The transfer function $H(f)$ is defined by

$$H(f) = \tilde{H}_1(f) H_2(f)$$

where $H_2(f)$ is the transfer function of the low-pass filter and $\tilde{H}_1(f)$ is the complex transfer function of the band-pass filter, that is,

$$H_1(f) = \tilde{H}_1(f - f_c) + \tilde{H}_1^*(-f - f_c).$$

(b) Given that

$$S_M(f) = \frac{S_0}{1 + (f/f_0)^2}$$

and

$$H(f) = \begin{cases} 1, & |f| \leq B \\ 0, & \text{otherwise} \end{cases}$$

find the value of B that minimizes the mean-square error $\mathscr{E}$.

Problem 6.16 Consider the amplitude-modulated wave

$$s(t) = [1 + m(t)] \cos(2\pi f_c t)$$

where $m(t)$ is a sample function of a zero-mean, stationary, Gaussian random process with power spectral density

$$S_M(f) = \begin{cases} S_0, & -W \leq f \leq W \\ 0, & \text{elsewhere} \end{cases}$$

Assume that $2W < f_c$. The AM wave $s(t)$ is applied to an ideal square-law detector, yielding the output

$$v(t) = s^2(t)$$

The signal $v(t)$ is next applied to an ideal low-pass filter with transfer function

$$H(f) = \begin{cases} 1, & -W \leq f \leq W \\ 0, & \text{elsewhere} \end{cases}$$

When S_0 is sufficiently small, the filter output is approximately equal to $m(t)$. Determine the mean-square value of the error, defined as the difference between the filter output and the message signal $m(t)$.

Problem 6.17 An AM receiver, operating with a sinusoidal modulating wave and 80 percent modulation, has an output signal-to-noise ratio of 30 dB.

(a) What is the corresponding carrier-to-noise ratio?
(b) By how many decibels can we decrease the carrier-to-noise ratio so that the system is operating just above threshold?

Problem 6.18 Evaluate the output signal-to-noise ratio of a vestigial sideband system, the receiver of which uses coherent detection. The additive noise at the detector input is narrow-band.

Problem 6.19 Consider a phase modulation (PM) system, with the modulated wave defined by

$$s(t) = A_c \cos[2\pi f_c t + k_p m(t)]$$

where k_p is a constant and $m(t)$ is the message signal. The additive noise $n(t)$ at the phase detector input is

$$n(t) = n_c(t)\cos(2\pi f_c t) - n_s(t)\sin(2\pi f_c t)$$

Assuming that the carrier-to-noise ratio at the detector input is high compared with unity, determine: (a) the output signal-to-noise ratio, and (b) the figure of merit of the system. Compare your results with the FM system for the case of sinusoidal modulation.

Problem 6.20 The input signal of an FM receiver consists of an unmodulated carrier accompanied by an interfering sine wave. The interference is 20 dB below the carrier level, and the frequency separation between them is 15 kHz. Assuming that the receiver uses a frequency discriminator with a sensitivity of 0.2 volts per KHz, determine the receiver output voltage.

Problem 6.21 Suppose that the spectrum of a modulating signal occupies the frequency interval $f_1 \leqslant |f| \leqslant f_2$. To accommodate this signal, the receiver of an FM system (without pre-emphasis) uses an ideal band-pass filter connected to the output of the frequency discriminator; the filter passes frequencies in the interval $f_1 \leqslant |f| \leqslant f_2$. Determine the output signal-to-noise ratio and figure of merit of the system in the presence of additive white Gaussian noise at the receiver input.

Problem 6.22 An FDM system uses single-sideband modulation to combine 12 independent voice signals and then uses frequency modulation to transmit the composite baseband signal. Each voice signal has a power P and occupies the frequency band 0.3–3.4 kHz; the system allocates it a bandwidth of 4 kHz. For each voice signal, only the lower sideband is transmitted. The subcarrier waves used for the first stage of modulation are defined by

$$c_k(t) = A_k \cos(2\pi k f_0 t), \qquad 0 \leqslant k \leqslant 11$$

The received signal consists of the transmitted FM signal plus white Gaussian noise of zero mean and power spectral density $N_0/2$.

(a) Sketch the power spectral density of the signal produced at the frequency discriminator output, showing both the signal and noise components.
(b) Find the relationship between the subcarrier amplitudes A_k so that the modulated voice signals have equal signal-to-noise ratios.

Problem 6.23 By using the pre-emphasis filter shown in Fig. 6.20(a) and with a voice signal as the modulating wave, an FM transmitter produces a signal that is essentially frequency-modulated by the lower audio frequencies and phase-modulated by the higher audio frequencies. Explain the reasons for this phenomenon.

Problem 6.24 The insertion of a pre-emphasis filter in the transmitter of an FM system may cause an increase in the transmission bandwidth of the system. For this evaluation it is analytically convenient to use the root mean-square (rms) bandwidth of the FM signal (see Problem 5.28).

(a) Let B_{rms} denote the rms bandwidth of an FM signal without pre-emphasis, and B'_{rms} denote the rms bandwidth of the pre-emphasized signal. Show that the ratio $B'_{\mathrm{rms}}/B_{\mathrm{rms}}$ is equal to the square root of the ratio of the average power of the emphasized message signal to that of the original message signal $m(t)$.

(b) Assume that the power spectral density of the message signal $m(t)$ is

$$S_M(f)=\begin{cases}\dfrac{S_0}{1+(f/f_0)^2}, & -W \leqslant f \leqslant W \\ 0, & \text{elsewhere}\end{cases}$$

and the transfer function of the pre-emphasis filter is

$$H_{pe}(f)=1+\frac{jf}{f_0}$$

Evaluate the bandwidth ratio $B'_{\mathrm{rms}}/B_{\mathrm{rms}}$.

Problem 6.25 Suppose that the transfer functions of the pre-emphasis and de-emphasis filters of an FM system are scaled as follows

$$H_{pe}(f)=k\left(1+\frac{jf}{f_0}\right)$$

and

$$H_{de}(f)=\frac{1}{k}\left(\frac{1}{1+jf/f_0}\right)$$

The scaling factor k is to be chosen so that the average power of the emphasized message signal is the same as that of the original message signal $m(t)$.

(a) Find the value of k that satisfies this requirement for the case when the power spectral density of the message signal $m(t)$ is

$$S_M(f)=\begin{cases}\dfrac{S_0}{1+(f/f_0)^2}, & -W \leqslant f \leqslant W \\ 0, & \text{elsewhere}\end{cases}$$

(b) What is the corresponding value of the improvement ratio D produced by using this pair of pre-emphasis and de-emphasis filters? Compare this ratio with that obtained in Example 4. The improvement factor is defined by Eq. (6.74).

Problem 6.26 A phase modulation (PM) system uses a pair of pre-emphasis and de-emphasis filters defined by the transfer functions:

$$H_{pe}(f)=1+\frac{jf}{f_0}$$

and

$$H_{de}(f) = \frac{1}{1 + (jf/f_0)}$$

Show that the improvement in output signal-to-noise ratio produced by using this pair of filters is

$$D = \frac{W/f_0}{\tan^{-1}(W/f_0)}$$

Evaluate this improvement for the case when $W = 15$ kHz and $f_0 = 2.1$ kHz, and compare it with the corresponding value for an FM system.

chapter 7 PULSE-ANALOG MODULATION

In *CW modulation*, some parameter of a sinusoidal carrier wave is varied continuously in accordance with the message. This is in direct contrast to *pulse modulation* where some parameter of a pulse train is varied in accordance with the message. We may distinguish two basic types of pulse modulation, namely, *pulse-analog modulation* and *pulse-code modulation.* In pulse-analog modulation systems, a periodic pulse train is used as the carrier wave, and some characteristic feature of each pulse (e.g., amplitude, duration, or position) is varied in a continuous manner in accordance with the pertinent sample value of the message. On the other hand, in pulse-code modulation, a discrete-time, discrete-amplitude representation is used for the signal, and as such it has no CW counterpart. In this chapter we study different methods of pulse-analog modulation, while pulse-code modulation is examined in Chapter 8.

An operation that is basic to the design of all pulse modulation systems is the *sampling process* whereby an analog signal is converted into a corresponding sequence of numbers that are usually uniformly spaced in time. For such a procedure to have practical utility, it is necessary that we choose the sampling rate properly, so that this sequence of numbers uniquely defines the original analog signal. This is the essence of the sampling theorem, which is derived next.

7.1 SAMPLING THEOREM

Consider an arbitrary signal $g(t)$ of finite energy, which is specified for all time, as shown in Fig. 7.1(*a*). Suppose that we sample the signal $g(t)$ instantaneously and at

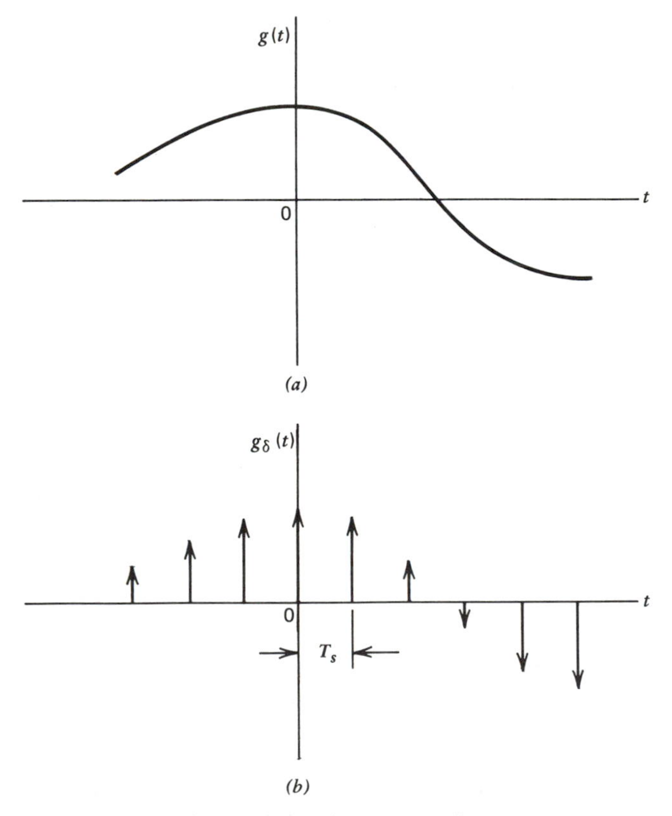

Figure 7.1 Illustration of the sampling process. *(a)* Analog signal. *(b)* Instantaneously sampled signal.

a uniform rate, once every T_s seconds. As a result of this sampling process, we obtain an infinite sequence of numbers spaced T_s seconds apart and denoted by $\{g(nT_s)\}$ where n takes on all possible integer values. We refer to T_s as the *sampling period*, and to its reciprocal $1/T_s$ as the *sampling rate*. This ideal form of sampling is called *instantaneous sampling*.

Let $g_\delta(t)$ denote the signal obtained by multiplying the sequence of numbers $\{g(nT_s)\}$ by a corresponding sequence of delta functions spaced T_s seconds apart and then summing the individual contributions, as shown by [see Fig. 7.1(*b*)]

$$g_\delta(t) = \sum_{n=-\infty}^{\infty} g(nT_s)\delta(t - nT_s) \tag{7.1}$$

We refer to $g_\delta(t)$ as the *ideal sampled signal*. Equivalently, we may express $g_\delta(t)$ as the product of the original analog signal $g(t)$ and an *ideal sampling function* or *Dirac comb*, $\delta_{T_s}(t)$ of period T_s. That is,

$$\begin{aligned} g_\delta(t) &= g(t)\delta_{T_s}(t) \\ &= g(t) \sum_{n=-\infty}^{\infty} \delta(t - nT_s) \end{aligned} \tag{7.2}$$

We may therefore determine the Fourier transform of the sampled signal $g_\delta(t)$ by convolving the Fourier transform of $g(t)$ with that of the ideal sampling function $\delta_{T_s}(t)$. From Example 19 of Chapter 2, we note that the Fourier transform of $\delta_{T_s}(t)$ consists of another series of delta functions spaced $1/T_s$ hertz apart. Hence, we may express the Fourier transform of the sampled signal $g_\delta(t)$ as

$$G_\delta(f) = G(f) \otimes \frac{1}{T_s} \sum_{n=-\infty}^{\infty} \delta\left(f - \frac{n}{T_s}\right) \tag{7.3}$$

where $\otimes$ denotes convolution. Interchanging the order of summation and convolution in Eq. (7.3), we get

$$G_\delta(f) = \frac{1}{T_s} \sum_{n=-\infty}^{\infty} G(f) \otimes \delta\left(f - \frac{n}{T_s}\right) \tag{7.4}$$

However, the convolution of the frequency function $G(f)$ with a delta function reproduces the function itself; that is,

$$G(f) \otimes \delta\left(f - \frac{n}{T_s}\right) = G\left(f - \frac{n}{T_s}\right)$$

We may therefore rewrite Eq. (7.4) in the form

$$G_\delta(f) = \frac{1}{T_s} \sum_{n=-\infty}^{\infty} G\left(f - \frac{n}{T_s}\right) \tag{7.5}$$

From Eq. (7.5), we see that $G_\delta(f)$ represents a continuous spectrum which is periodic with a period equal to $1/T_s$. *In other words, the process of uniformly sampling a signal in the time domain results in a periodic spectrum in the frequency domain with a period equal to the sampling rate.*

Another useful expression for the Fourier transform $G_\delta(f)$ may be obtained by taking the Fourier transform of both sides of Eq. (7.1) and noting that the Fourier transform of the delta function $\delta(t-nT_s)$ is equal to $\exp(-j2\pi nf T_s)$. We may thus write

$$G_\delta(f)=\sum_{n=-\infty}^{\infty} g(nT_s)\exp(-j2\pi nf T_s) \tag{7.6}$$

This relation may be viewed as a complex Fourier series representation of the periodic frequency function $G_\delta(f)$, with the sequence of samples $\{g(nT_s)\}$ defining the coefficients of the expansion. That is,

$$g(nT_s)=T_s\int_0^{1/T_s} G_\delta(f)\exp(j2\pi nf T_s)df \tag{7.7}$$

Note that in the Fourier series defined by Eqs. (7.6) and (7.7) the usual roles of time and frequency have been interchanged.

The relations, as derived above, apply to any continuous-time signal $g(t)$ of finite energy and infinite duration. Suppose, however, that the signal is strictly band-limited, with no frequency components higher than W hertz. That is, the Fourier transform $G(f)$ of the signal $g(t)$ has the property that $G(f)$ is zero for $|f|>W$, as illustrated in Fig. 7.2(*a*); the shape of the spectrum shown in this figure is intended for the purpose of illustration only. Suppose also that we choose the sampling period $T_s=1/2W$. Then the corresponding spectrum $G_\delta(f)$ of the sampled signal $g_\delta(t)$ is as shown in Fig. 7.2(*b*). Putting $T_s=1/2W$ in Eq. (7.6) yields

$$G_\delta(f)=\sum_{n=-\infty}^{\infty} g\left(\frac{n}{2W}\right)\exp\left(-\frac{j\pi nf}{W}\right) \tag{7.8}$$

Since from Eq. (7.5) we have

$$G(f)=\frac{1}{2W}G_\delta(f), \qquad -W\leqslant f\leqslant W \tag{7.9}$$

it follows from Eq. (7.8) that we may also write

$$G(f)=\frac{1}{2W}\sum_{n=-\infty}^{\infty} g\left(\frac{n}{2W}\right)\exp\left(-\frac{j\pi nf}{W}\right), \qquad -W\leqslant f\leqslant W \tag{7.10}$$

Therefore, if the sample values $g(n/2W)$ of the signal $g(t)$ are specified for all time, then the Fourier transform $G(f)$ of the signal is uniquely determined by using the Fourier series of Eq. (7.10). Because $g(t)$ is related to $G(f)$ by the inverse Fourier transform, it follows that the signal $g(t)$ is itself uniquely determined by the sample values $g(n/2W)$ for $-\infty\leqslant n\leqslant\infty$. In other words, the sequence $\{g(n/2W)\}$ contains all of the information of $g(t)$.

Consider next the problem of reconstructing the signal $g(t)$ from the sequence of sample values $\{g(n/2W)\}$. Substituting Eq. (7.10) in the formula for the inverse

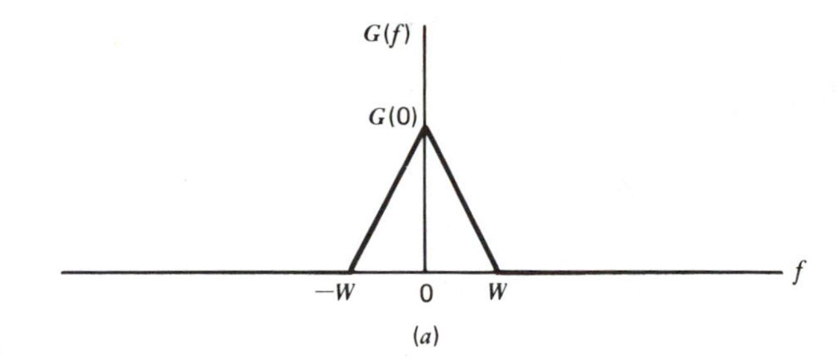

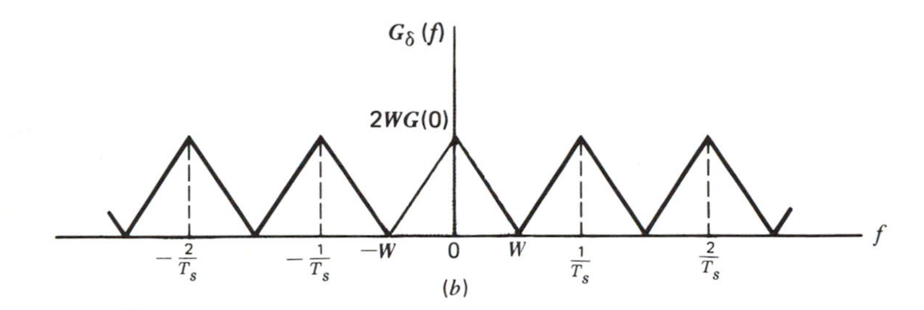

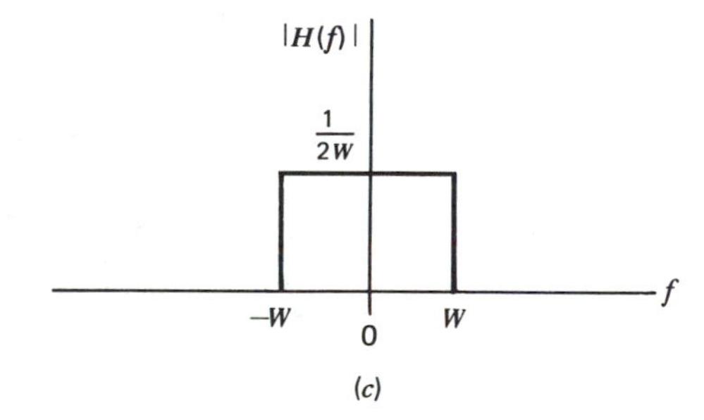

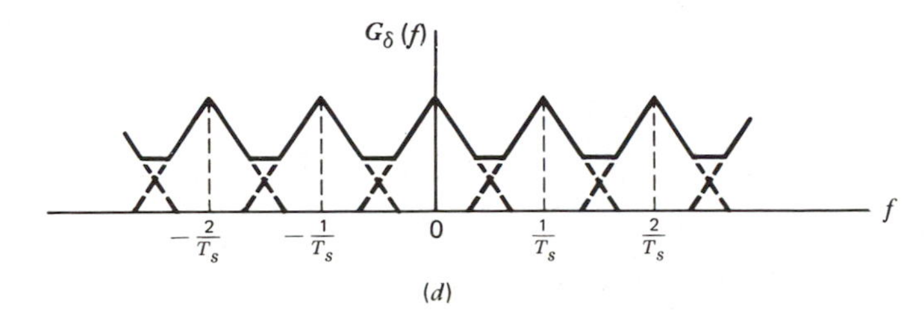

Figure 7.2 Illustration of the effect of varying the sampling rate $1/T_s$ on the spectrum of a sampled signal. *(a)* Spectrum of signal $g(t)$. *(b)* Spectrum of sampled signal $g_\delta(t)$ for $1/T_s = 2W$. *(c)* Amplitude response of ideal reconstruction filter. *(d)* Spectrum of $g_\delta(t)$ for $1/T_s < 2W$.

Fourier transform defining $g(t)$ in terms of $G(f)$, we get

$$g(t)=\int_{-\infty}^{\infty} G(f)\exp(j2\pi ft)df$$
$$=\int_{-W}^{W}\frac{1}{2W}\sum_{n=-\infty}^{\infty} g\left(\frac{n}{2W}\right)\exp\left(-\frac{j\pi nf}{W}\right)\exp(j2\pi ft)df$$

Interchanging the order of summation and integration:

$$g(t)=\sum_{n=-\infty}^{\infty} g\left(\frac{n}{2W}\right)\frac{1}{2W}\int_{-W}^{W}\exp\left[j2\pi f\left(t-\frac{n}{2W}\right)\right]df \tag{7.11}$$

The integral term in Eq. (7.11) may be readily evaluated, yielding

$$g(t)=\sum_{n=-\infty}^{\infty} g\left(\frac{n}{2W}\right)\frac{\sin(2\pi Wt-n\pi)}{(2\pi Wt-n\pi)}$$
$$=\sum_{n=-\infty}^{\infty} g\left(\frac{n}{2W}\right)\text{sinc}(2Wt-n), \qquad -\infty\leqslant t\leqslant\infty \tag{7.12}$$

Equation (7.12) provides an *interpolation formula* for reconstructing the original signal $g(t)$ from the sequence of sample values $\{g(n/2W)\}$, with the sinc function $\text{sinc}(2Wt)$ playing the role of an *interpolation function*. Each sample is multiplied by a delayed version of the interpolation function, and all the resulting waveforms are added to obtain $g(t)$. It is noteworthy that Eq. (7.12) also represents the response of an ideal low-pass filter of bandwidth W and zero transmission delay, which is produced by an input signal consisting of the sequence of samples $\{g(n/2W)\}$ for $-\infty\leqslant n\leqslant\infty$. This is intuitively satisfying, since, by inspection of the spectrum of Fig. 7.2(*b*), we see that the original signal $g(t)$ may be recovered exactly from the sequence of samples $\{g(n/2W)\}$ by passing it through an ideal low-pass filter of bandwidth W. This is illustrated in block diagrammatic form in Fig. 7.3. The amplitude response of the ideal reconstruction filter is shown in Fig. 7.2(*c*).

We may develop another important interpretation of Eq. (7.12) by using the property that the function $\text{sinc}(2Wt-n)$, where n is an integer, is one of a family of shifted sinc functions that are mutually orthogonal. To prove this property, we use the formula (see part f of Problem 2.12):

$$\int_{-\infty}^{\infty} g_1(t)g_2^*(t)dt=\int_{-\infty}^{\infty} G_1(f)G_2^*(f)df$$

where $g_1(t)\rightleftharpoons G_1(f)$ and $g_2(t)\rightleftharpoons G_2(f)$. This relation may be viewed as a generalization of Rayleigh's energy theorem. Put

$$g_1(t)=\text{sinc}(2Wt-n)=\text{sinc}\left[2W\left(t-\frac{n}{2W}\right)\right]$$

and

$$g_2(t)=\text{sinc}(2Wt-m)=\text{sinc}\left[2W\left(t-\frac{m}{2W}\right)\right]$$

Since $\text{sinc}(2Wt)\rightleftharpoons\frac{1}{2W}\text{rect}\left(\frac{f}{2W}\right)$, and using the time-shifting property of the

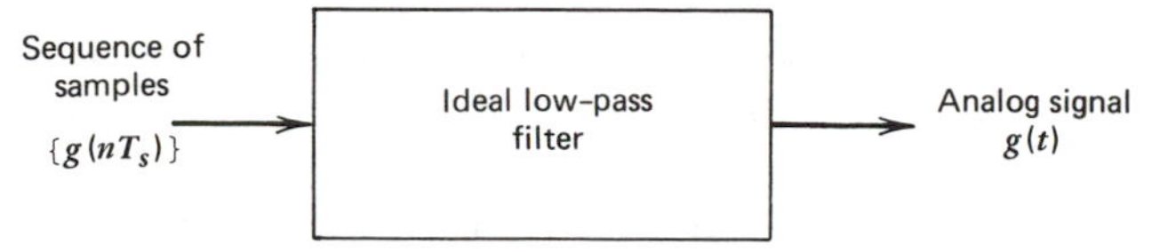

Figure 7.3 Reconstruction filter.

Fourier transform, we find that

$$G_1(f)=\frac{1}{2W}\,\mathrm{rect}\left(\frac{f}{2W}\right)\exp\left(-\frac{j\pi nf}{W}\right)$$

and

$$G_2(f)=\frac{1}{2W}\,\mathrm{rect}\left(\frac{f}{2W}\right)\exp\left(-\frac{j\pi mf}{W}\right)$$

Hence,

$$\begin{aligned}\int_{-\infty}^{\infty}\mathrm{sinc}(2Wt-n)\mathrm{sinc}(2Wt-m)dt&=\left(\frac{1}{2W}\right)^2\int_{-W}^{W}\exp\left[-\frac{j\pi f}{W}(n-m)\right]df\\&=\frac{\sin[\pi(n-m)]}{2W\pi(n-m)}\end{aligned}$$

This result equals $1/2W$ when $n=m$, and zero when $n\neq m$. We therefore have

$$\int_{-\infty}^{\infty}\mathrm{sinc}(2Wt-n)\mathrm{sinc}(2Wt-m)dt=\begin{cases}\dfrac{1}{2W}, & n=m\\ 0, & n\neq m\end{cases}\tag{7.13}$$

Accordingly, Eq. (7.12) represents the expansion of the signal $g(t)$ as an infinite sum of orthogonal functions with the coefficients of the expansion, $g(n/2W)$, defined by

$$g\left(\frac{n}{2W}\right)=2W\int_{-\infty}^{\infty}g(t)\mathrm{sinc}(2Wt-n)dt\tag{7.14}$$

We may thus view the coefficients $g(n/2W)$ of this expansion as coordinates in an infinite-dimensional *signal space*. In this space each signal corresponds to precisely one point and each point to one signal. We will have more to say on the idea of signal space in Chapter 10.

We may now state the *sampling theorem** for band-limited signals of finite energy in two equivalent ways:

* The sampling theorem was introduced to communication theory by Shannon in 1949: C. E. Shannon, "Communication in the presence of noise," *Proc. IRE*, vol. 37, pp. 10–21, January 1949.
It is for this reason that the theorem is sometimes referred to in the literature as the "Shannon sampling theorem." However, the interest of communication engineers in the sampling theorem may be traced back to Nyquist in 1928: H. Nyquist, "Certain topics in telegraph transmission theory," *AIEE Trans.*, vol. 47, pp. 617–644, 1928.
Indeed, the sampling theorem was known to mathematicians at least since 1915: E. T. Whittaker, "On the functions which are represented by the expansion of interpolation theory," *Proc. Royal Society*, Edinburgh, vol. 35, pp. 181–194, 1915.

1. *A band-limited signal of finite energy, which has no frequency components higher than W hertz, is completely described by specifying the values of the signal at instants of time separated by $1/2W$ seconds.*
2. *A band-limited signal of finite energy, which has no frequency components higher than W hertz, may be completely recovered from a knowledge of its samples taken at the rate of $2W$ per second.*

The sampling rate of $2W$ samples per second, for a signal bandwidth of W hertz, is often called the *Nyquist rate*. The sampling theorem serves as the basis for the interchangeability of analog signals and digital sequences, which is so valuable in digital communication systems.

The derivation of the sampling theorem, as described above, is based on the assumption that the signal $g(t)$ is strictly band-limited. Such a requirement, however, can be satisfied only if $g(t)$ has infinite duration. In other words, a strictly band-limited signal cannot be simultaneously time-limited, and vice versa. Nevertheless, we may consider a time-limited signal as *essentially* band-limited in the sense that its frequency components outside some band of interest have negligible effects.* We may then justify the practical application of the sampling theorem.

Aliasing

When the sampling rate $1/T_s$ exceeds the Nyquist rate $2W$, all replicas of $G(f)$ involved in the construction of $G_\delta(f)$ move farther apart, and there is no problem in recovering the original signal $g(t)$ from its sampled version $g_\delta(t)$. When, however, the sampling rate $1/T_s$ is less than $2W$, we find that in constructing the spectrum $G_\delta(f)$ of the sampled signal, the frequency-shifted replicas of the original spectrum $G(f)$ overlap, as illustrated in Fig. 7.2(*d*). In this case, the high frequencies in $G(f)$ get reflected into the low frequencies in $G_\delta(f)$. The phenomenon of a high-frequency component in the spectrum of the original signal $g(t)$ seemingly taking on the identity of a lower frequency in the spectrum of its sampled version $g_\delta(t)$ is called the *aliasing effect*. It is, therefore, evident that if the sampling rate $1/T_s$ is less than the Nyquist rate $2W$, the original signal $g(t)$ cannot be recovered exactly from its sampled version $g_\delta(t)$, and information is thereby lost in the sampling process.

Another factor that contributes to aliasing is the fact that a signal cannot be finite in both time and frequency, as mentioned previously. Since this violates the strict band-limited requirement of the sampling theorem, we find that whenever a time-limited signal is sampled, there always will be some aliasing produced by the sampling process. Accordingly, in order to combat the effects of aliasing in practice,

* For a discussion of the interplay between time-domain and frequency-domain descriptions of analog signals, see:

1. J. B. Thomas, *An Introduction to Statistical Communication Theory*, pp. 382–400, (Wiley, 1949).
2. D. Slepian, "On bandwidth," *Proc. IEEE*, vol. 64, pp. 292–300, March 1976.
3. A. Papoulis, *Signal Analysis*, pp. 183–220 (McGraw-Hill, 1977).

we use two corrective measures:

1. Prior to sampling, a low-pass *pre-alias filter* is used to attenuate those high-frequency components of the signal that lie outside the band of interest.
2. The filtered signal is sampled at a rate slightly higher than the Nyquist rate.

It is of interest to note that the use of a sampling rate $1/T_s$ higher than the Nyquist rate $2W$ has also the desirable effect of making it somewhat easy to design the low-pass reconstruction filter so as to recover the original analog signal from its sampled version. With such a sampling rate, we find that there are gaps, each of width $(1/T_s)-2W$ between the frequency-shifted replicas of $G(f)$. Accordingly, we may choose the bandwidth B of the ideal reconstruction filter to satisfy the condition

$$W < B < \frac{1}{T_s} - W$$

7.2 SAMPLING OF BAND-PASS SIGNALS

Up to this point in our discussion of the sampling process, we have focused attention on low-pass signals. However, many of the signals encountered in practice are intrinsically band-pass in nature. In the case of such signals, particularly when the bandwidth is small compared to the highest frequency component, it is indeed possible to use a sampling rate that is less than twice the highest frequency component present in the signal.

To explore this issue in detail,* consider a band-pass signal $g(t)$ with carrier frequency f_c and bandwidth $2W$ (or half-bandwidth W), so that its spectrum occupies frequency intervals defined by $f_c - W \leqslant |f| \leqslant f_c + W$, as shown in Fig. 7.4($a$). Expressing this signal in terms of its in-phase and quadrature components, we have

$$g(t) = g_c(t)\cos(2\pi f_c t) - g_s(t)\sin(2\pi f_c t) \tag{7.15}$$

where $g_c(t)$ is the in-phase component and $g_s(t)$ is the quadrature component. Initially, let us assume that the highest frequency component in the band-pass signal $g(t)$, that is, $f_c + W$, is an integer multiple of the bandwidth $2W$, so that we may write

$$f_c + W = 2kW$$

or

$$f_c = (2k-1)W \tag{7.16}$$

where k is a positive integer. At the sampling instant $t = nT_s$, where T_s is the sampling

* Our analysis is based on J. L. Brown, Jr., "First-order sampling of bandpass signals—A new approach," *IEEE Trans. Information Theory*, vol. IT-26, pp. 613–615, September 1980.

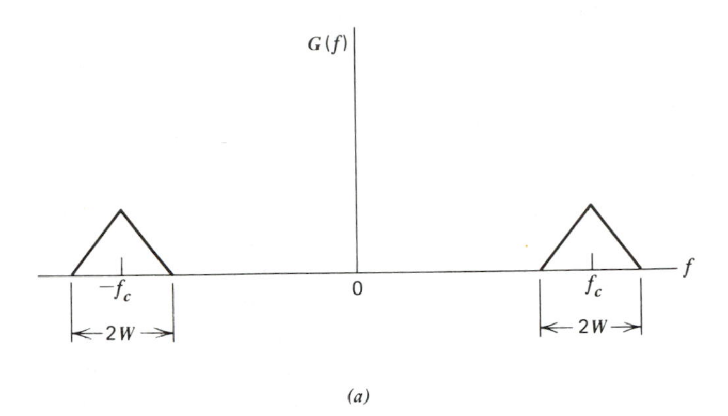

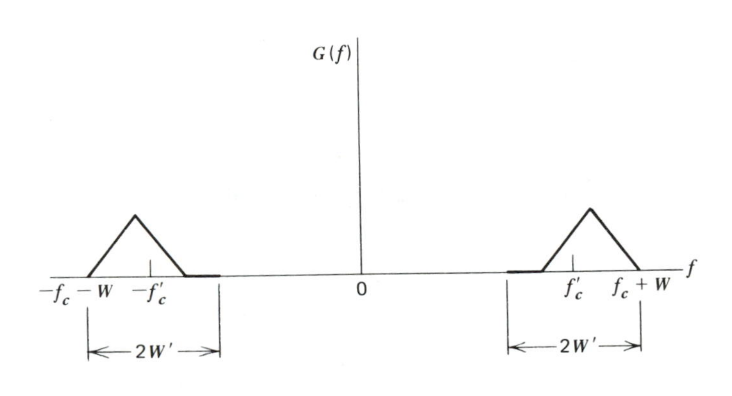

Figure 7.4 *(a)* Spectrum of band-pass signal *g(f)*. *(b)* Illustrating the definitions of the new mid-band frequency f_c' and bandwidth $2W'$.

period and $n=0, \pm 1, \pm 2, \ldots$, we find from Eq. (7.15) that

$$g(nT_s)=g_c(nT_s)\cos(2n\pi f_c T_s)-g_s(nT_s)\sin(2n\pi f_c T_s) \tag{7.17}$$

For a sampling rate equal to twice the bandwidth of the signal $g(t)$, that is,

$$\frac{1}{T_s}=4W$$

we have

$$g\left(\frac{n}{4W}\right)=g_c\left(\frac{n}{4W}\right)\cos\left(\frac{n\pi f_c}{2W}\right)-g_s\left(\frac{n}{4W}\right)\sin\left(\frac{n\pi f_c}{2W}\right) \tag{7.18}$$

Therefore, substituting Eq. (7.16) in (7.18), we get

$$g\left(\frac{n}{4W}\right)=g_c\left(\frac{n}{4W}\right)\cos\left[\frac{n\pi}{2}(2k-1)\right]-g_s\left(\frac{n}{4W}\right)\sin\left[\frac{n\pi}{2}(2k-1)\right] \tag{7.19}$$

For n even, say $n=2v$, we find that for all k,

$$\cos\left[\frac{n\pi}{2}(2k-1)\right]=\cos[v\pi(2k-1)]=(-1)^v$$

and

$$\sin\left[\frac{n\pi}{2}(2k-1)\right]=\sin[v\pi(2k-1)]=0,$$

for which case Eq. (7.19) simplifies as

$$g\left(\frac{v}{2W}\right)=(-1)^v g_c\left(\frac{v}{2W}\right) \tag{7.20}$$

Next, for n odd, say $n=2v-1$, we find that

$$\cos\left[\frac{n\pi}{2}(2k-1)\right]=\cos\left[\frac{\pi}{2}(2v-1)(2k-1)\right]=0$$

and

$$\sin\left[\frac{n\pi}{2}(2k-1)\right]=\sin\left[\frac{\pi}{2}(2v-1)(2k-1)\right]=(-1)^{v+k}$$

Therefore, Eq. (7.19) simplifies as follows

$$g\left(\frac{2v-1}{4W}\right)=(-1)^{v+k+1} g_s\left(\frac{2v-1}{4W}\right) \tag{7.21}$$

However, the components $g_c(t)$ and $g_s(t)$ are low-pass signals, both limited to $|f|\leqslant W$. Therefore, by the sampling theorem, it follows that both $g_c(t)$ and $g_s(t)$ are determined by their samples taken at a uniform rate of $2W$ samples per second. Accordingly, we may state that:

1. The samples $g_c(v/2W)$, $v=0, \pm 1, \pm 2, \ldots$, are sufficient to determine the in-phase component $g_c(t)$, as shown by

$$\begin{aligned} g_c(t) &= \sum_{v=-\infty}^{\infty} g_c\left(\frac{v}{2W}\right)\operatorname{sinc}(2Wt-v) \\ &= \sum_{v=-\infty}^{\infty} (-1)^v g\left(\frac{v}{2W}\right)\operatorname{sinc}(2Wt-v) \end{aligned} \tag{7.22}$$

where we have made use of Eq. (7.20).

2. The samples $g_s((2v-1)/4W)$, $v=0, \pm 1, \pm 2, \ldots$, are sufficient to determine the quadrature component $g_s(t)$, as shown by

$$\begin{aligned} g_s(t) &= \sum_{v=-\infty}^{\infty} g_s\left(\frac{2v-1}{4W}\right)\operatorname{sinc}(2Wt-v+\tfrac{1}{2}) \\ &= \sum_{v=-\infty}^{\infty} (-1)^{v+k+1} g\left(\frac{2v-1}{4W}\right)\operatorname{sinc}(2Wt-v+\tfrac{1}{2}) \end{aligned} \tag{7.23}$$

where we have made use of Eq. (7.21). Note that the samples of the signal $g(t)$ used to reconstruct the quadrature component $g_s(t)$, as in Eq. (7.23), are delayed by $1/4W$ seconds with respect to the corresponding samples of $g(t)$ used to reconstruct the in-phase component $g_c(t)$.

Next, substituting Eqs. (7.22) and (7.23) in (7.15), we get the following representation for the band-pass signal $g(t)$:

$$g(t)=\sum_{v=-\infty}^{\infty}(-1)^v g\left(\frac{v}{2W}\right)\operatorname{sinc}(2Wt-v)\cos(2\pi f_c t)$$
$$+\sum_{v=-\infty}^{\infty}(-1)^{v+k} g\left(\frac{2v-1}{4W}\right)\operatorname{sinc}(2Wt-v+\tfrac{1}{2})\sin(2\pi f_c t) \tag{7.24}$$

where k is a positive integer. But, with $(1/T_s)=4W$ and $(f_c/W)=2k-1$, we may write

$$(-1)^v\cos(2\pi f_c t)=\cos[2\pi f_c(t-2vT_s)]$$

and

$$(-1)^{v+k}\sin(2\pi f_c t)=\sin[2\pi f_c(t-(2v-1)T_s)]$$

Accordingly, we may reassemble the two summations in Eq. (7.24) to yield

$$g(t)=\sum_{n=-\infty}^{\infty} g(nT_s)\operatorname{sinc}\left(2Wt-\frac{n}{2}\right)\cos[2\pi f_c(t-nT_s)] \tag{7.25}$$

which is the desired form of the expansion for a band-pass signal $g(t)$ for the special case when the highest frequency component present in $g(t)$ is an integer multiple of the bandwidth $2W$. In such a case, Eq. (7.25) states that we may use a minimum sampling rate equal to twice the bandwidth of the signal $g(t)$, namely, $4W$ samples per second.

Consider next the general case where only $f_c>W\geq 0$ is assumed; that is, the band of significant frequencies contained in the band-pass signal $g(t)$ occupies an arbitrary position along the frequency axis. Let r denote the largest integer contained in the number $(f_c+W)/2W$, so that we may write

$$r\leq\frac{f_c+W}{2W}<r+1 \tag{7.26}$$

Holding the highest frequency f_c+W contained in the signal $g(t)$ constant, suppose we increase the bandwidth to a new value $2W'$ which makes

$$\frac{f_c+W}{2W'}=r \tag{7.27}$$

Also, define a new carrier frequency f'_c as

$$f'_c=\left(1-\frac{1}{2r}\right)(f_c+W) \tag{7.28}$$

The definitions of f'_c and W' are illustrated in Fig. 7.4(b). Clearly, this spectrum still corresponds to the original band-pass signal $g(t)$. However, the highest frequency component f_c+W is now a multiple of $2W'$. Accordingly, the representation of Eq. (7.25) applies with the original half-bandwidth W replaced by the new value W', the sampling period T_s replaced by $T_s'=1/4W'$, and the carrier frequency f_c replaced by the new value f_c'. We may thus write

$$g(t)=\sum_{n=-\infty}^{\infty} g(nT_s')\text{sinc}\left(2W't-\frac{n}{2}\right)\cos[2\pi f_c'(t-nT_s')] \tag{7.29}$$

which shows that a representation of the band-pass signal $g(t)$ is possible with uniform samples taken at a rate of $1/T_s'$ samples per second. Equation (7.29) may be put in a more standard form by convolving both sides of this equation with the function:

$$\phi_1(t)=4W\ \text{sinc}(2Wt)\cos(2\pi f_c t) \tag{7.30}$$

The Fourier transform of this function occupies the same frequency intervals as the band-pass signal $g(t)$, as shown by

$$\Phi_1(f)=\begin{cases}1, & f_c-W\leqslant|f|\leqslant f_c+W\\ 0, & \text{otherwise}\end{cases} \tag{7.31}$$

It is clear, therefore, that the convolution operation described above leaves the band-pass signal $g(t)$ unchanged. Hence, we may rewrite Eq. (7.29) as

$$g(t)=\sum_{n=-\infty}^{\infty} g(nT_s')\phi_1(t)\otimes\phi_{2,n}(t) \tag{7.32}$$

where $\otimes$ signifies convolution, and

$$\phi_{2,n}(t)=\text{sinc}[2W'(t-nT_s')]\cos[2\pi f_c'(t-nT_s')] \tag{7.33}$$

The convolution of the functions $\phi_1(t)$ and $\phi_2(t, n)$ is given by (see Problem 7.14)

$$\phi_1(t)\otimes\phi_{2,n}(t)=\frac{W}{W'}\text{sinc}[2W(t-nT_s')]\cos[2\pi f_c(t-nT_s')] \tag{7.34}$$

Hence, noting that $T_s'=1/4W'$, and substituting Eq. (7.34) in (7.32), we get the desired representation:

$$g(t)=4WT_s'\sum_{n=-\infty}^{\infty} g(nT_s')\text{sinc}[2W(t-nT_s')]\cos[2\pi f_c(t-nT_s')] \tag{7.35}$$

The minimum sampling rate is defined by

$$\frac{1}{T_s'}=\frac{2(f_c+W)}{r}\geqslant 4W \tag{7.36}$$

where the integer r is itself defined by Eq. (7.26). In Fig. 7.5 we have plotted the minimum sampling rate versus the highest frequency present in the band-pass signal $g(t)$. We see that, regardless of band location, the minimum sampling rate necessary for signal reconstruction always lies between $4W$ and $8W$ samples per second.

Here again it is usually desirable to sample a bandpass signal at a rate higher

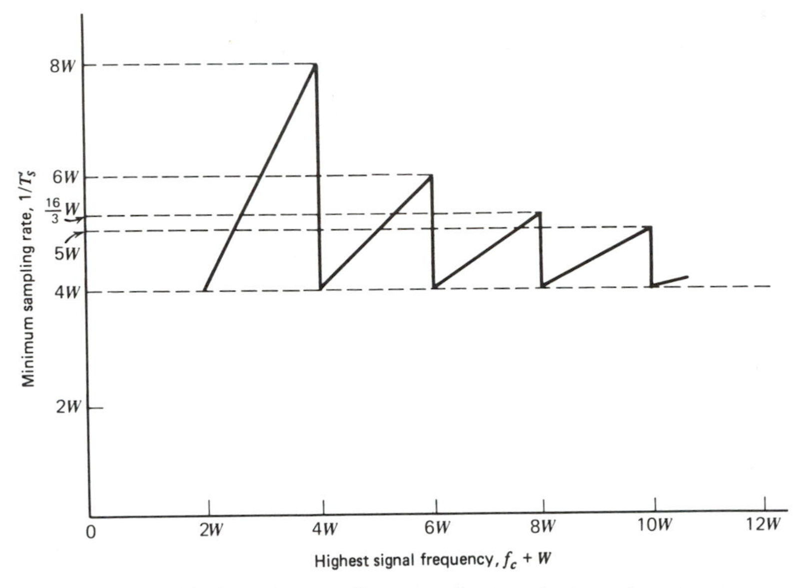

Figure 7.5 Minimum sampling rate for band-pass signals.

than that computed from Fig. 7.5 in order to account for the fact that the spectrum is not precisely band-limited in practice.

7.3 PRACTICAL ASPECTS OF SAMPLING

In practice, the sampling of an analog signal is accomplished by means of high-speed switching transistor circuits. Accordingly, we find that the resulting sampled waveform deviates from the ideal form of instantaneous sampling described in Section 7.1 because the operation of a physical switching circuit, however fast, still requires a nonzero interval of time. Furthermore, we often find that samples of an analog signal are lengthened intentionally for convenience in instrumentation or transmission. In this section we evaluate the effects of these practical deviations from the ideal condition,

Ordinary Samples of Finite Duration

Let an arbitrary analog signal $g(t)$ be applied to a switching circuit controlled by a sampling function $c(t)$ that consists of an infinite succession of rectangular pulses of amplitude A, duration T, and occurring with period T_s. The output of the switching circuit is denoted by $s(t)$. The waveforms $g(t)$, $c(t)$, and $s(t)$ are illustrated in parts (a), (b), and (c) of Fig. 7.6, respectively. We see that the switching operation merely extracts from the analog signal $g(t)$ successive portions of predetermined duration T, taken regularly at the rate $1/T_s$. Accordingly, the sampled signal $s(t)$

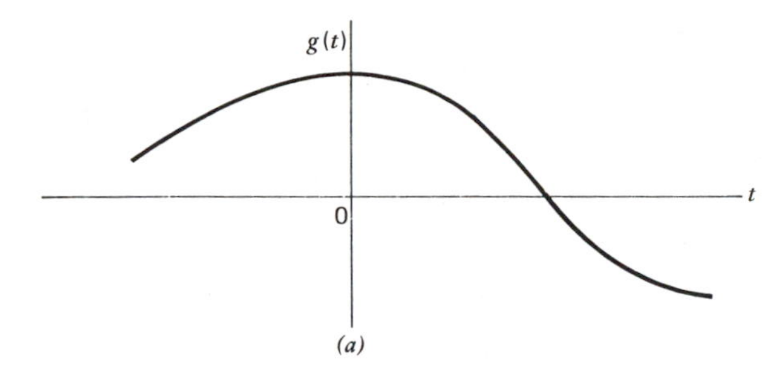

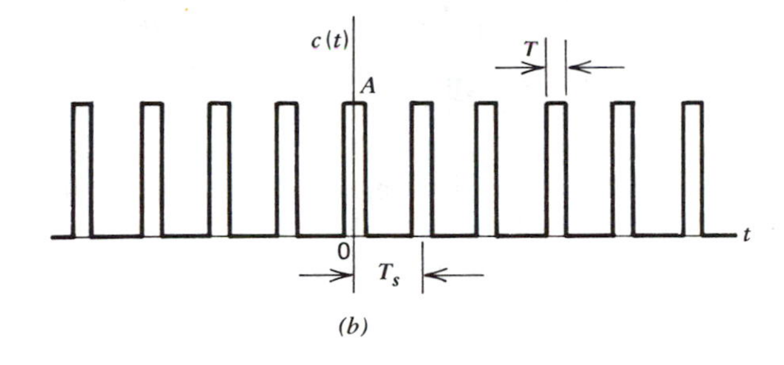

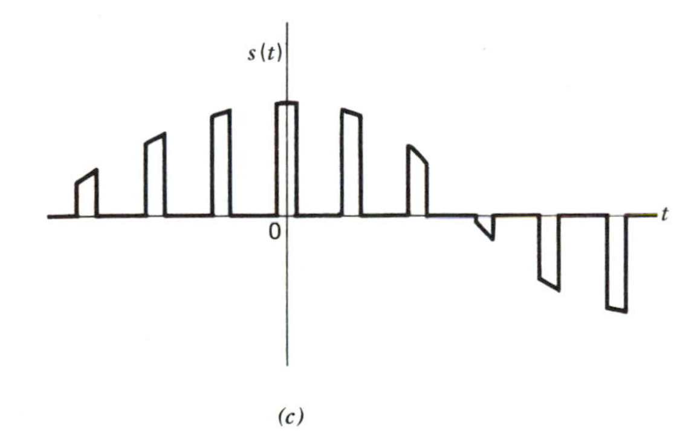

Figure 7.6 *(a)* Analog signal. *(b)* Sampling function. *(c)* Sampled signal.

consists of a sequence of positive and negative pulses, as in Fig. 7.6(c). Mathematically, we have

$$s(t) = c(t)g(t) \tag{7.37}$$

However, $c(t)$ may be expressed in the form of a complex Fourier series as follows (see Example 1 of Chapter 2);

$$c(t) = \frac{TA}{T_s} \sum_{n=-\infty}^{\infty} \operatorname{sinc}\left(\frac{nT}{T_s}\right) \exp\left(\frac{j2\pi nt}{T_s}\right) \tag{7.38}$$

Therefore, substituting Eq. (7.38) in (7.37), we get

$$s(t)=\frac{TA}{T_s}\sum_{n=-\infty}^{\infty}\operatorname{sinc}\left(\frac{nT}{T_s}\right)\exp\left(\frac{j2\pi nt}{T_s}\right)g(t) \tag{7.39}$$

Taking the Fourier transform of both sides of Eq. (7.39), and using the frequency-shifting property of the Fourier transform, we get

$$S(f)=\frac{TA}{T_s}\sum_{n=-\infty}^{\infty}\operatorname{sinc}\left(\frac{nT}{T_s}\right)G\left(f-\frac{n}{T_s}\right) \tag{7.40}$$

where $S(f)=F[s(t)]$ and $G(f)=F[g(t)]$.

The relationship between the spectra $G(f)$ and $S(f)$ is illustrated in Fig. 7.7, assuming that $g(t)$ contains no frequencies lying outside the band $-W$ to W, and that the sampling rate $1/T_s$ is greater than the Nyquist rate $2W$, so that there is no aliasing. We see that the effect of the finite duration of the sampling pulses is to multiply the nth lobe of the spectrum $S(f)$ by $(TA/T_s)\operatorname{sinc}(nT/T_s)$. The original signal $g(t)$ can be recovered from $s(t)$ with no distortion by passing $s(t)$ through an ideal low-pass filter whose bandwidth B satisfies the condition $W<B<(1/T_s)-W$, as before. We conclude, therefore, that the use of sampling pulses of finite duration has no important effects on the sampling process.

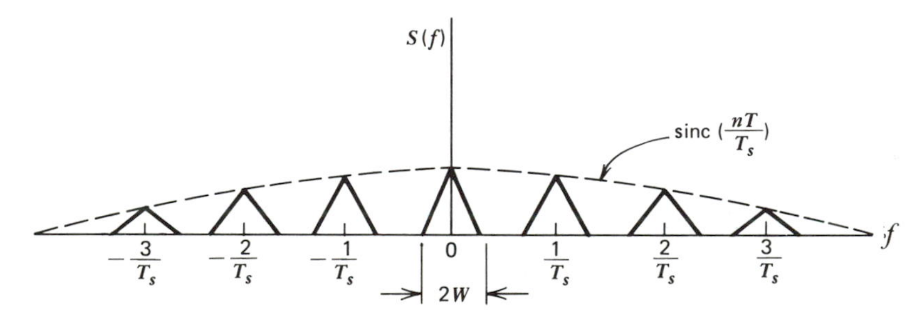

Figure 7.7 Illustrating the effect of using ordinary samples of finite duration on the spectrum of the sampled signal.

Consider the case when $TA=1$, so that each rectangular pulse of the sampling function $c(t)$ has unit area. Then, comparing Eq. (7.40) with (7.5), we see that as the pulse duration T approaches zero, $S(f)$ approaches $G_\delta(f)$. In other words, the ideal sampled signal $g_\delta(t)$ represents the limiting form of the sampled signal $s(t)$ as T approaches zero, with $TA=1$.

Flat-Top Samples

Consider next the situation where the analog signal $g(t)$ is sampled instantaneously at the rate $1/T_s$, and that the duration of each sample is lengthened to T, as illustrated in Fig. 7.8. One reason for intentionally lengthening the duration of each sample is to avoid the use of an excessive transmission bandwidth, since bandwidth is inversely proportional to pulse duration.

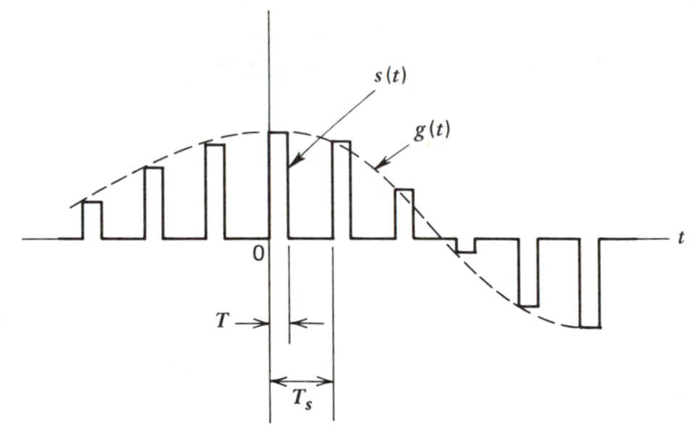

Figure 7.8 Flat-top samples.

Using $s(t)$ to denote the sequence of flat-top pulses generated in this way, we may write

$$s(t) = \sum_{n=-\infty}^{\infty} g(nT_s)h(t - nT_s) \tag{7.41}$$

The $h(t)$ is a rectangular pulse of unit amplitude and duration T, defined as follows [see Fig. 7.9(a)]

$$h(t) = \begin{cases} 1, & 0 < t < T \\ \frac{1}{2}, & t = 0, t = T \\ 0, & \text{otherwise} \end{cases} \tag{7.42}$$

From Eq. (7.1), the instantaneously sampled version of $g(t)$ is given by

$$g_\delta(t) = \sum_{n=-\infty}^{\infty} g(nT_s)\delta(t - nT_s) \tag{7.43}$$

Convolving $g_\delta(t)$ with the pulse $h(t)$, we get

$$\begin{aligned} g_\delta(t) \otimes h(t) &= \int_{-\infty}^{\infty} g_\delta(\tau)h(t-\tau)d\tau \\ &= \int_{-\infty}^{\infty} \sum_{n=-\infty}^{\infty} g(nT_s)\delta(\tau - nT_s)h(t-\tau)d\tau \\ &= \sum_{n=-\infty}^{\infty} g(nT_s) \int_{-\infty}^{\infty} \delta(\tau - nT_s)h(t-\tau)d\tau \end{aligned} \tag{7.44}$$

Using the sifting property of the delta function, we thus obtain

$$g_\delta(t) \otimes h(t) = \sum_{n=-\infty}^{\infty} g(nT_s)h(t - nT_s) \tag{7.45}$$

Therefore, from Eqs. (7.41) and (7.45) it follows that $s(t)$ is mathematically equivalent to the convolution of $g_\delta(t)$, the instantaneously sampled version of $g(t)$, and the

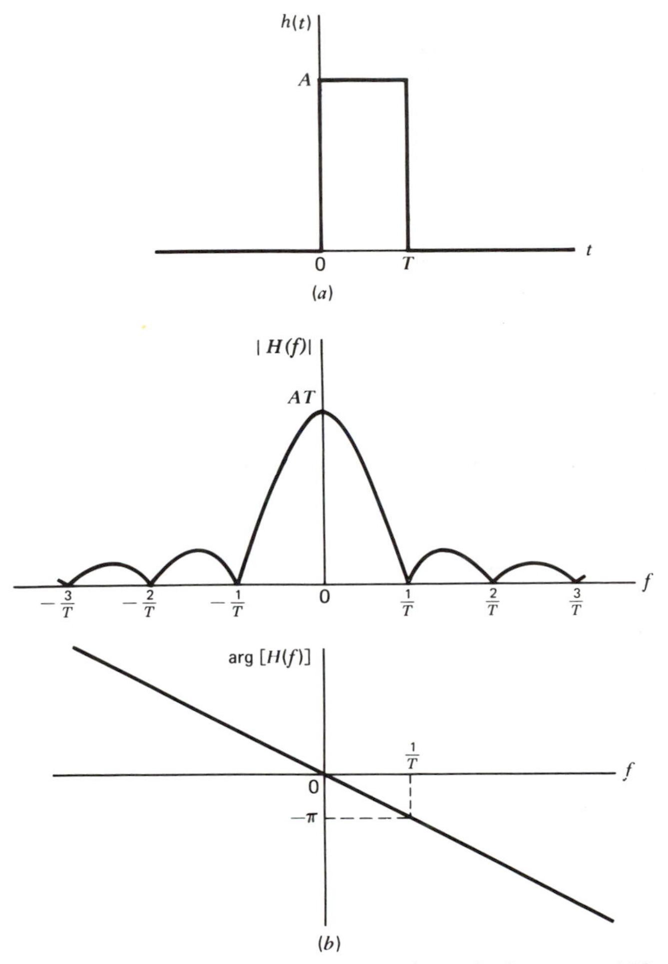

Figure 7.9 *(a)* Rectangular pulse *h(t)*. *(b)* Spectrum *H(f)*.

pulse $h(t)$, as shown by

$$s(t)=g_\delta(t)\otimes h(t) \tag{7.46}$$

Taking the Fourier transform of both sides of Eq. (7.46) and recognizing that the convolution of two time functions is transformed into the multiplication of their respective Fourier transforms, we get

$$S(f)=G_\delta(f)H(f) \tag{7.47}$$

where $S(f)=F[s(t)]$, $G_\delta(f)=F[g_\delta(t)]$, and $H(f)=F[h(t)]$. Therefore, substitution of Eq. (7.5) into (7.47) yields

$$S(f)=\frac{1}{T_s}\sum_{n=-\infty}^{\infty} G\left(f-\frac{n}{T_s}\right)H(f) \tag{7.48}$$

where $G(f)=F[g(t)]$.

Finally, suppose that $g(t)$ is strictly band-limited and that the sampling rate $1/T_s$ is greater than the Nyquist rate. Then, passing $s(t)$ through a low-pass reconstruction filter, we find that the spectrum of the resulting filter output is equal to $G(f)H(f)$. This is equivalent to passing the original analog signal $g(t)$ through a low-pass filter of transfer function $H(f)$.

From Eq. (7.42) we find that

$$H(f) = T\,\mathrm{sinc}(fT)\exp(-j\pi fT) \tag{7.49}$$

which is shown plotted in Fig. 7.9(*b*). Hence, we see that by using flat-top samples, we have introduced *amplitude distortion* as well as a *delay* of $T/2$. This effect is similar to the variation in transmission with frequency that is caused by the finite size of the scanning aperture in television and facsimile. Accordingly, the distortion caused by lengthening the samples, as in Fig. 7.8, is referred to as the *aperture effect*.

This distortion may be corrected by connecting an *equalizer* in cascade with the low-pass reconstruction filter. The equalizer has the effect of decreasing the in-band loss of the reconstruction filter as the frequency increases in such a manner as to compensate for the aperture effect. Ideally, the frequency response of the equalizer is given by

$$\frac{1}{|H(f)|} = \frac{1}{T\,\mathrm{sinc}(fT)} = \frac{1}{T}\frac{\pi fT}{\sin(\pi fT)}$$

The amount of equalization needed in practice is usually small.

Example 1

At $f = 1/2T_s$, which corresponds to the highest frequency component of the message signal for a sampling rate equal to the Nyquist rate, we find from Eq. (7.49) that the amplitude response

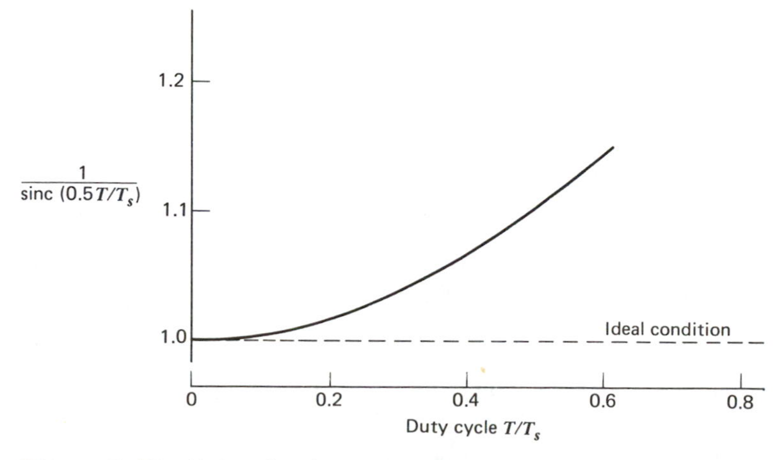

Figure 7.10 Normalized equalization (to compensate for aperture effect) plotted versus T/T_s.

of the equalizer normalized to that at zero frequency is equal to

$$\frac{1}{\text{sinc}(0.5T/T_s)} = \frac{(\pi/2)(T/T_s)}{\sin[(\pi/2)(T/T_s)]}$$

where the ratio T/T_s is equal to the duty cycle of the sampling pulses. In Fig. 7.10, this result is plotted as a function of T/T_s. Ideally, it should be equal to one for all values of T/T_s. For a duty cycle of 10 percent, it is equal to 1.0041. It follows therefore that for duty cycles of less than 10 percent, the aperture effect becomes negligible, and the need for equalization may be omitted altogether.

7.4 RECONSTRUCTION OF A MESSAGE PROCESS FROM ITS SAMPLES

A study of the sampling process would be incomplete without considering the reconstruction of a stationary message process from its samples. In this section we show that when a wide-sense stationary message process (whose power spectrum is band-limited) is reconstructed from the sequence of its samples taken at a rate equal to twice the highest frequency component, the reconstructed process equals the original process in the *mean-square sense* for all time.

Consider then a wide-sense stationary message process $X(t)$ with autocorrelation function $R_X(\tau)$ and power spectral density $S_X(f)$. We assume that

$$S_X(f)=0, \qquad \text{for } |f|>W$$

Suppose we have available an infinite sequence of samples of the process taken at a uniform rate equal to $2W$, that is, twice the highest frequency component of the process. Using $X'(t)$ to denote the reconstructed process, based on this infinite sequence of samples, we may write [see Eq. (7.12)]

$$X'(t) = \sum_{n=-\infty}^{\infty} X\left(\frac{n}{2W}\right) \text{sinc}(2Wt-n), \qquad -\infty \leqslant t \leqslant \infty \tag{7.50}$$

where $X(n/2W)$ is the random variable obtained by sampling or observing the message process $X(t)$ at time $t=n/2W$. The *mean-square value of the error* between the original message process $X(t)$ and the reconstructed message process $X'(t)$ equals

$$\begin{aligned}\mathscr{E} &= E[(X(t)-X'(t))^2] \\ &= E[X^2(t)] - 2E[X(t)X'(t)] + E[(X'(t))^2]\end{aligned} \tag{7.51}$$

We recognize the first expectation term on the right-hand side of Eq. (7.51) as the mean-square value of $X(t)$, which equals $R_X(0)$; thus

$$E[X^2(t)] = R_X(0) \tag{7.52}$$

For the second expectation term, we use Eq. (7.50) and so write

$$E[X(t)X'(t)] = E\left[X(t) \sum_{n=-\infty}^{\infty} X\left(\frac{n}{2W}\right) \text{sinc}(2Wt-n)\right]$$

Interchanging the order of summation and expectation:

$$E[X(t)X'(t)] = \sum_{n=-\infty}^{\infty} E\left[X(t)X\left(\frac{n}{2W}\right)\right]\text{sinc}(2Wt-n)$$

$$= \sum_{n=-\infty}^{\infty} R_X\left(t-\frac{n}{2W}\right)\text{sinc}(2Wt-n) \qquad (7.53)$$

For a stationary process the expectation $E[X(t)X'(t)]$ is independent of time t. Hence, putting $t=0$ in the right-hand side of Eq. (7.53) and recognizing that $R_X(-n/2W)=R_X(n/2W)$, we may write

$$E[X(t)X'(t)] = \sum_{n=-\infty}^{\infty} R_X\left(\frac{n}{2W}\right)\text{sinc}(-n) \qquad (7.54)$$

The term $R_X(n/2W)$ represents a sample of the autocorrelation function $R_X(\tau)$ taken at $\tau=n/2W$. Now, since the power spectral density $S_X(f)$ or equivalently the Fourier transform of $R_X(\tau)$ is zero for $|f|>W$, we may represent $R_X(\tau)$ in terms of its samples taken at $\tau=n/2W$ as follows [see Eq. (7.12)]

$$R_X(\tau) = \sum_{n=-\infty}^{\infty} R_X\left(\frac{n}{2W}\right)\text{sinc}(2W\tau-n), \qquad -\infty \leqslant \tau \leqslant \infty \qquad (7.55)$$

Accordingly, we deduce from Eqs. (7.54) and (7.55) that

$$E[X(t)X'(t)] = R_X(0) \qquad (7.56)$$

For the third and final expectation term on the right-hand side of Eq. (7.51), we again use Eq. (7.50) and so write

$$E[(X'(t))^2] = E\left[\sum_{n=-\infty}^{\infty} X\left(\frac{n}{2W}\right)\text{sinc}(2Wt-n) \sum_{k=-\infty}^{\infty} X\left(\frac{k}{2W}\right)\text{sinc}(2Wt-k)\right]$$

$$= E\left[\sum_{n=-\infty}^{\infty} \text{sinc}(2Wt-n) \sum_{k=-\infty}^{\infty} X\left(\frac{n}{2W}\right)X\left(\frac{k}{2W}\right)\text{sinc}(2Wt-k)\right]$$

Interchanging the order of expectation and inner summation;

$$E[(X'(t))^2] = \sum_{n=-\infty}^{\infty} \text{sinc}(2Wt-n) \sum_{k=-\infty}^{\infty} E\left[X\left(\frac{n}{2W}\right)X\left(\frac{k}{2W}\right)\right]\text{sinc}(2Wt-k)$$

$$= \sum_{n=-\infty}^{\infty} \text{sinc}(2Wt-n) \sum_{k=-\infty}^{\infty} R_X\left(\frac{n-k}{2W}\right)\text{sinc}(2Wt-k) \qquad (7.57)$$

However, in view of Eq. (7.55), we recognize that the inner summation on the right-hand side of Eq. (7.57) equals $R_X(t-n/2W)$. Hence, we may simplify Eq. (7.57) as follows

$$E[(X'(t))^2] = \sum_{n=-\infty}^{\infty} R_X\left(t-\frac{n}{2W}\right)\text{sinc}(2Wt-n)$$

$$= R_X(0) \qquad (7.58)$$

Finally, substituting Eqs. (7.52), (7.56), and (7.58) into Eq. (7.51), we get the result

$$\mathscr{E} = 0$$

In other words, the mean-square value of the difference between the original message process $X(t)$ and the reconstructed message process $X'(t)$ is zero, as prevously stated.

7.5 TIME-DIVISION MULTIPLEXING (TDM)

The sampling theorem enables us to transmit the complete information contained in a band-limited message signal $g(t)$ by using samples of $g(t)$ taken uniformly at a rate that is usually slightly higher than the Nyquist rate. An important feature of the sampling process is a conservation of time. That is, the transmission of the message samples engages the transmission channel for only a fraction of the sampling interval on a periodic basis, and in this way some of the time interval between adjacent samples is cleared for use by other independent message sources on a time-shared basis. We thereby obtain a *time-division multiplex system* (TDM) which enables the joint utilization of a common transmission channel by a plurality of independent message sources without mutual interference.

The concept of TDM is illustrated by the block diagram shown in Fig. 7.11. Each input message signal is first restricted in bandwidth by a low-pass filter to remove the frequencies that are nonessential to an adequate signal representation. The low-pass filter outputs are then applied to a *commutator* which is usually implemented using electronic switching circuitry. The function of the commutator is twofold: (1) to take a narrow sample of each of the N input messages at a rate $1/T_s$ that is slightly higher than $2W$, where W is the cutoff frequency of the low-pass input filter, and (2) to sequentially interleave these N samples inside a sampling interval T_s. Indeed, this latter function is the essence of the time-division multiplexing operation. Following the commutation process, the multiplexed signal is

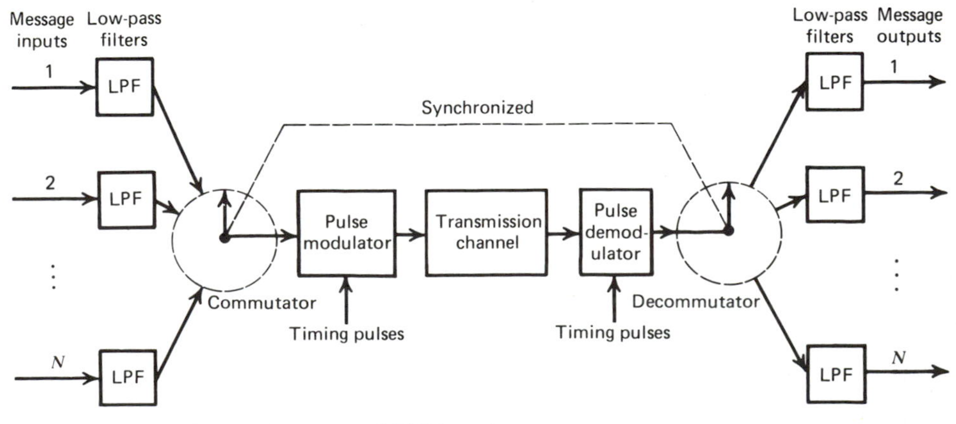

Figure 7.11 Block diagram of TDM system.

applied to a *pulse modulator*, the purpose of which is to transform the multiplexed signal into a form suitable for transmission over the common channel. It is clear that the use of time-division multiplexing introduces a bandwidth expansion factor N, because the scheme must squeeze N samples derived from N independent message sources into a time slot equal to one sampling interval. At the receiving end of the system, the received signal is applied to a *pulse demodulator* which performs the reverse operation of the pulse modulator. The narrow samples produced at the pulse demodulator output are distributed to the appropriate low-pass reconstruction filters by means of a *decommutator*, which operates in synchronism with the commutator in the transmitter. This synchronization is essential for a satisfactory operation of the system. The way this synchronization is implemented, however, depends on the method of pulse modulation used to transmit the multiplexed sequence of samples. We shall have more to say on this issue in later sections of the chapter, and in Chapter 8.

The TDM system is highly sensitive to linear distortion in the common transmission channel, that is, to variations of amplitude with frequency or lack of proportionality of phase with frequency. Accordingly, accurate equalization of both the amplitude and phase responses of the channel is necessary to ensure a satisfactory operation of the system. This issue is discussed in detail in Chapter 9. To a first approximation, however, TDM is immune to nonlinear distortion in the channel as a source of crosstalk because the different message signals are not simultaneously impressed on the channel.

Time-division multiplexing (described here) and frequency-division multiplexing (described in Section 3.14) constitute the two standard signal multiplexing formats used in telephony. With the coexistence of analog and digital facilities in the telephone network, which is expected to continue for the foreseeable future, provisions have to be made for an interface between analog and digital sections of the network. The so-called *transmultiplexer* is an interface facility specifically designed for this purpose. This device utilizes sophisticated digital signal processing techniques to convert between time-division multiplexed and frequency-division multiplexed signals.*

7.6 PULSE-AMPLITUDE MODULATION (PAM)

In pulse-amplitude modulation, the amplitudes of regularly spaced rectangular pulses vary with the instantaneous sample values of a continuous message signal in a one-to-one fashion. This method of modulation is illustrated in Fig. 7.12 where parts (*a*), (*b*), and (*c*) represent a message signal, the pulse carrier, and the corre-

* For a description of the various methods of designing transmultiplexers, see

1. S. L. Freeny, "TDM/FDM translation as an application of digital signal processing," *IEEE Communications Society Magazine*, vol. 18, pp. 5–15, January 1980.
2. H. Scheuermann and H. Göckler, "A comprehensive survey of digital transmultiplexing methods," *Proc IEEE*, vol. 69, pp. 1419–1450, November 1981.

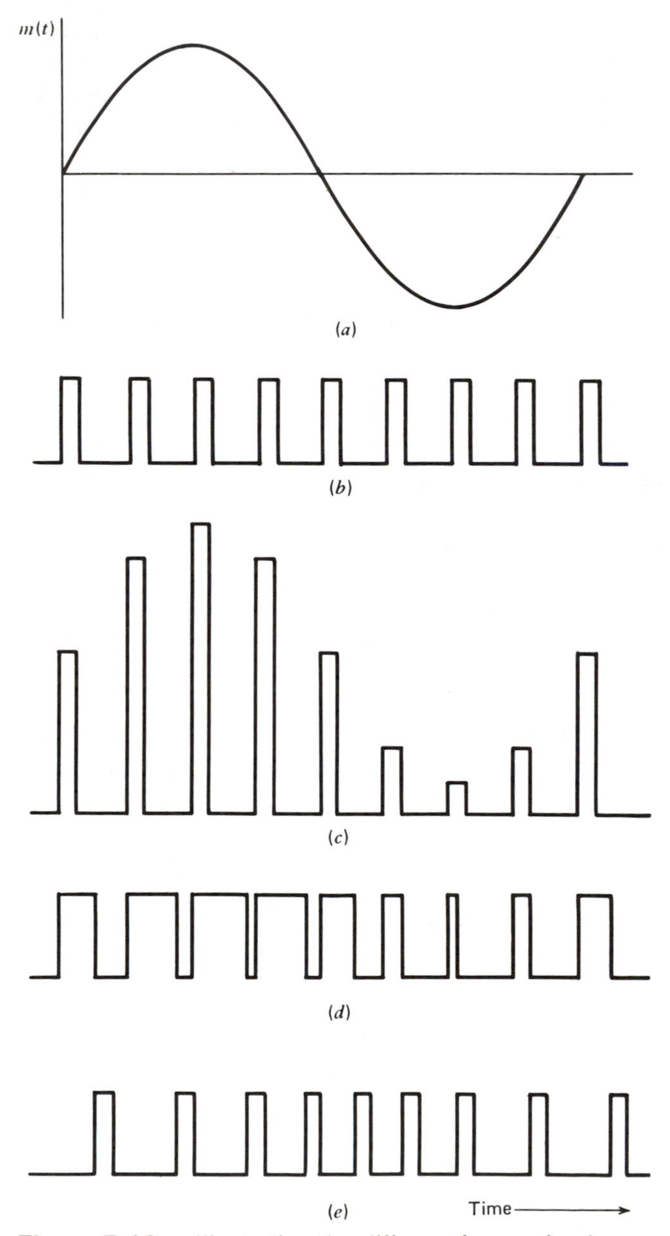

Figure 7.12 Illustrating the different forms of pulse-analog modulation for the case of a sinusoidal modulating wave. *(a)* Modulating wave. *(b)* Pulse carrier. *(c)* PAM wave. *(d)* PDM wave. *(e)* PPM wave.

sponding pulse-amplitude-modulated (PAM) wave, respectively. Thus a PAM wave is defined by:

$$s(t)=\sum_{n=-\infty}^{\infty}[1+k_a m(nT_s)]g(t-nT_s) \tag{7.59}$$

where $m(nT_s)$ represents the nth sample of the message signal $m(t)$, T_s is the sampling period, k_a is a constant called *amplitude sensitivity*, and $g(t)$ denotes the pulse. As with AM, the constant k_a is chosen so as to maintain a single polarity, that is, $1+k_a m(nT_s)>0$ for all n. The sampling rate $1/T_s$ must be equal to or greater than twice the highest frequency component of the message signal $m(t)$ in accordance with the sampling theorem. In practice, it is customary to choose $1/T_s$ greater than twice the highest frequency component of $m(t)$.

The PAM wave $s(t)$ is easily demodulated by a low-pass filter with a cutoff frequency just large enough to accommodate the highest frequency component of the message signal $m(t)$. However, the reconstructed signal exhibits a slight amplitude distortion caused by the aperture effect due to lengthening of the samples. The aperture effect and its equalization were described in Section 7.3.

The transmission of a PAM wave imposes rather stringent requirements on the amplitude- and phase-frequency responses of the channel because of the relatively short duration of the transmitted pulses. Furthermore, despite an excessive transmission bandwidth requirement, the noise performance of a PAM system can never be better than baseband-signal transmission (see Problem 7.23). Accordingly, we find that for transmission over long distances, PAM would be used only as a means of message processing for TDM, from which conversion to some other form of pulse modulation can subsequently be made.

Synchronization in PAM Systems

Most pulse systems require synchronization of the receiver to the transmitter. In the so-called *start-stop* method of synchronization, we maintain synchronization on a per frame basis. This method involves transmitting some information, in addition to the message-bearing pulses, to serve as a *time mark* within each frame interval so that certain gates in the receiver structure may be made to open and close at the appropriate instants of time. In some cases, the necessary time mark is established by transmitting a distinctive *marker* per frame, whereas in other cases it is established by omitting a pulse in the pertinent time slot. Clearly, when markers are used, they must differ from the message-bearing pulses in some recognizable fashion.

In a PAM system, the marker pulse may be identified by making its amplitude exceed that of all possible message pulses, as illustrated in Fig. 7.13(*b*) for a PAM system involving three independent message sources; part (*a*) of this figure shows the sampling times of the message sources and the times of synchronizing or marker pulses. Such a marker can be readily located at the receiver by applying the received pulses to a *slicer*, with a slicing level that is just in excess of the maximum

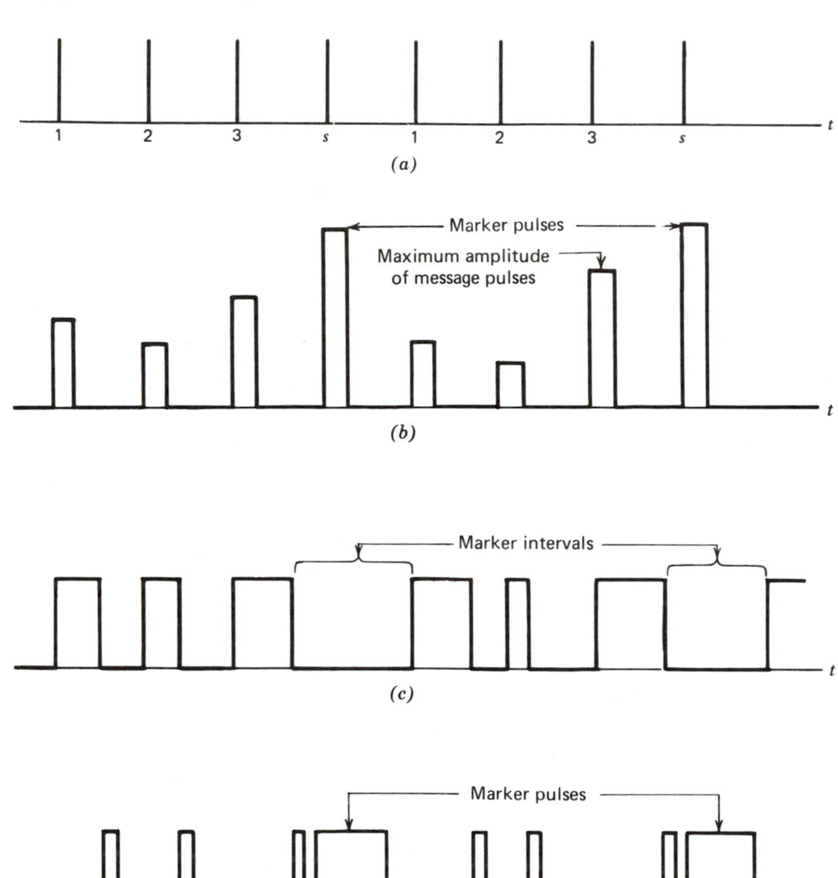

Figure 7.13 Illustrating synchronization in pulse-analog modulation systems. *(a)* Sampling times of message sources (numbered 1, 2 and 3) and times of synchronising pulses (identified by letter *s*). *(b)* Multiplexed PAM wave. *(c)* Multiplexed PDM wave. *(d)* Multiplexed PPM wave.

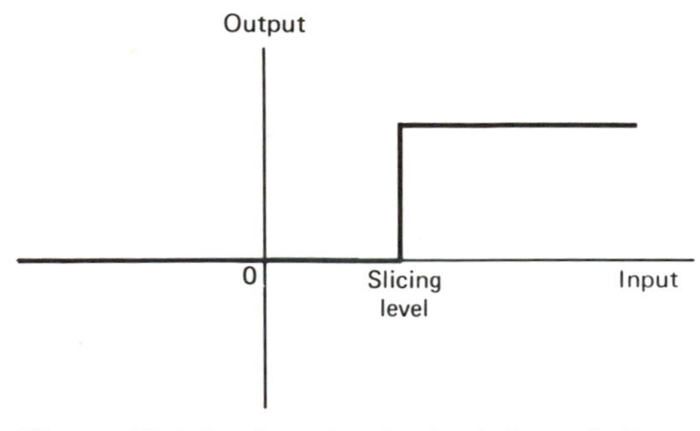

Figure 7.14 Input-output relation of slicer.

amplitude of the message pulses, so that these pulses produce zero output. An ideal slicer has the property that its output is zero whenever its input is below the slicing level, and is constant whenever the input exceeds this level, as illustrated in Fig. 7.14. The pulses observed at the slicer output will thus be due to the markers only.

7.7 PULSE-TIME MODULATION

In a pulse modulation system, we may use the increased bandwidth consumed by pulses to obtain an improvement in noise performance by representing the sample values of the message signal by some property of the pulse other than by amplitude. *In pulse-duration modulation (PDM), the samples of the message signal are used to vary the duration of the individual pulses.* This form of modulation is also referred to as *pulse-width modulation* or *pulse-length modulation.* The modulating wave may vary the time of occurrence of the leading edge, the trailing edge, or both edges of the pulse. In Fig. 7.12(*d*) the trailing edge of each pulse is varied in accordance with the message signal.

In PDM, long pulses expend considerable power during the pulse while bearing no additional information. If this unused power is subtracted from PDM, so that only time transitions are preserved, we obtain a more efficient type of pulse modulation known as *pulse-position modulation* (PPM). *In PPM the position of a pulse relative to its unmodulated time of occurrence is varied in accordance with the message signal, as shown in Fig.* 7.12(*e*).

Figure 7.15 depicts a simple procedure for generating a PDM wave, wherein the trailing edge alone is modulated. The message wave and a saw-tooth sweep wave are summed, and the combination is applied to a slicer. The message wave varies during the sweeping process, and the duration of a duration-modulated pulse is proportional to the magnitude of the message wave at the trailing edge of the pulse. Note that in Fig. 7.15 the duration of each pulse is dictated by the value of the message wave at the time of occurrence of the trailing edge, rather than the value of the message wave at the regularly recurring time of the leading edge. We refer to this method of sampling as *natural sampling* in contrast to the uniform sampling procedure described in Section 7.1. A consequence of the use of natural sampling, as in Fig. 7.15, is that it distorts the reconstructed signal by a certain amount. However, this distortion may be acceptable in some instances (e.g., voice transmission) if high fidelity is not essential.

We may generate a PPM wave from a PDM wave, for example, by simply using a *monostable multivibrator*, as in Fig. 7.16. This device has one absolutely stable state and one quasi-stable state into which it is triggered by an externally applied pulse. The monostable multivibrator is designed to trigger on the trailing edges of the duration-modulated pulses, assuming that the time of occurrence of the trailing edge of each duration-modulated pulse is varied in accordance with the message signal. The fixed pulse duration of the PPM wave at the monostable multivibrator output can be set by appropriately choosing the resistance-capacitance combination in the timing circuitry of the device.

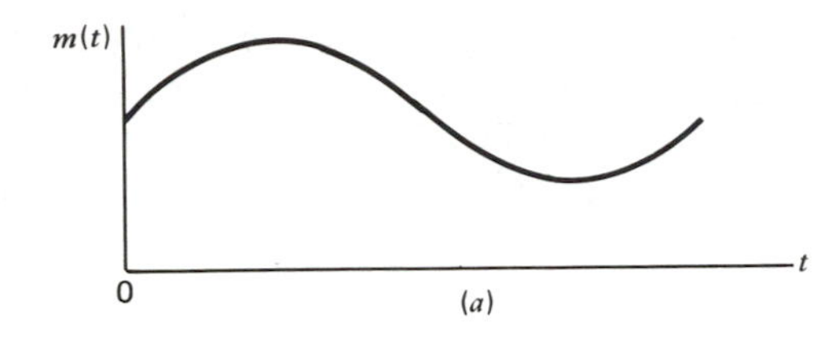

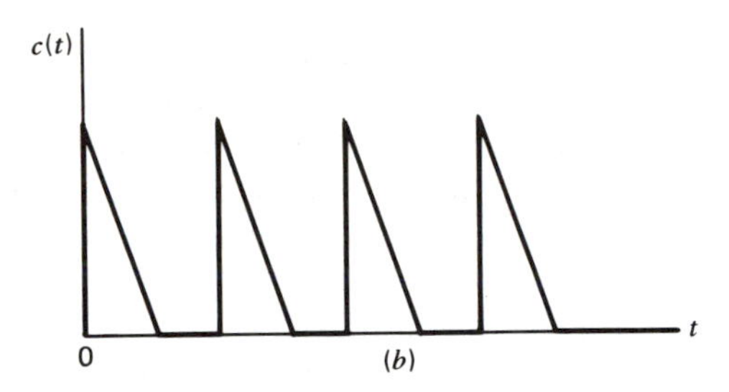

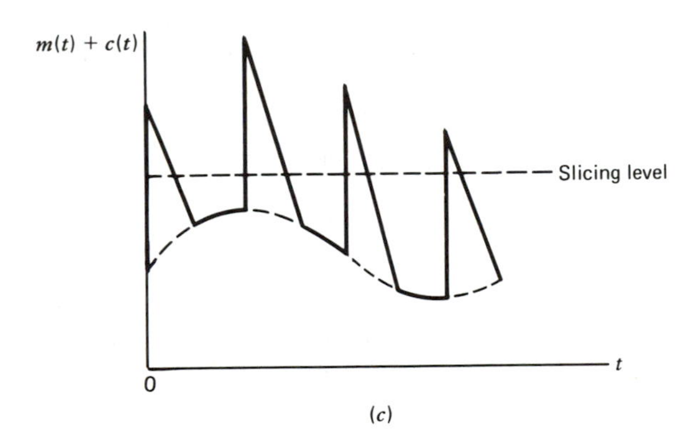

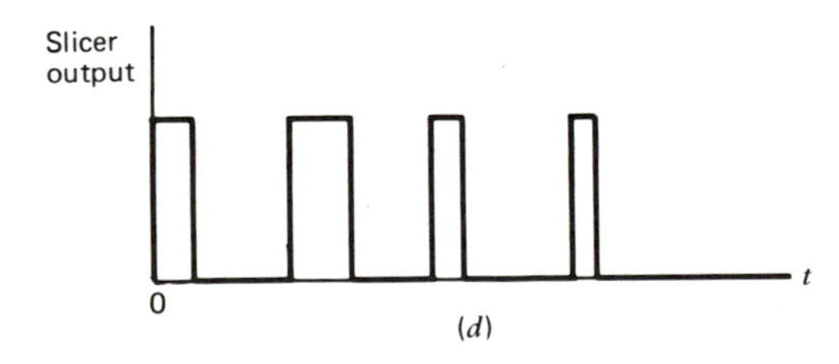

Figure 7.15 Illustration of a procedure for generating PDM waves. *(a)* Modulating wave. *(b)* Saw-tooth sweep wave. *(c)* Modulating wave plus saw-tooth wave. *(d)* Slicer output.

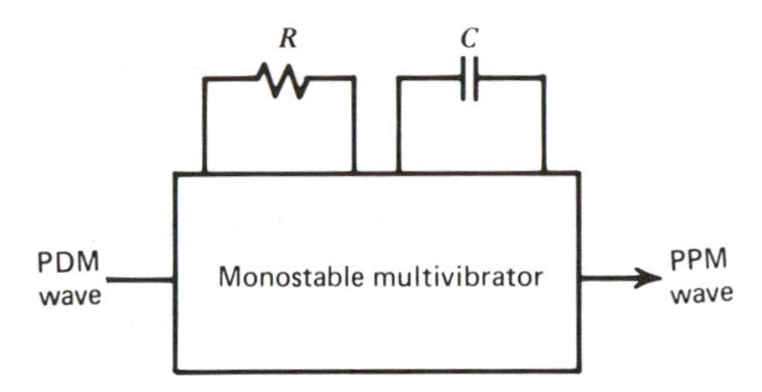

Figure 7.16 A device for converting a PDM wave into a PPM wave.

Spectra of PDM and PPM Waves

The spectral analysis of a PDM or PPM wave is complicated.* We present here only a qualitative description of the spectra of PDM and PPM waves. Let T_s denote the time separation between the leading edges of duration-modulated pulses obtained by natural sampling, with the modulation superimposed on the trailing edges. Then assuming a sinusoidal modulating wave of frequency f_m, we find that the spectrum of a naturally sampled PDM wave consists of the following components:

1. A dc component equal to the average value of the pulses.
2. Sinusoidal components of frequencies equal to integer multiples of $1/T_s$, corresponding to spectral lines at $\pm n/T_s$, where $n=1, 2, 3, \ldots$. These sinusoidal components as well as the dc component are contributed by the unmodulated pulse train which may be regarded as the carrier of the PDM wave.
3. A sinusoidal component of frequency f_m and in phase with the modulating wave, corresponding to spectral lines at $\pm f_m$
4. Sinusoidal components of frequencies equal to $(n/T_s) \pm l f_m$, where n, $l=1, 2, 3, \ldots$, corresponding to pairs of side-frequencies centered around each spectral line of the unmodulated pulse train, except the dc component. These components represent the cross-modulation products between the sinusoidal modulation and sampling frequencies.

The message signal may be recovered by passing the PDM wave through a low-pass filter. However, the reconstruction is accomplished with a certain amount of distortion caused by the cross-modulation products that fall in the signal band. The frequencies of the important in-band distortion components are $(1/T_s) - 2f_m$, $(1/T_s) - 3f_m$, and so on. To prevent undue distortion of the reconstructed message signal, it is necessary to restrict the maximum excursion of the trailing edge of a duration-modulated pulse. It is of interest to note that when uniform sampling is

* 1. M. Schwartz, W. R. Bennett, and S. Stein, *Communication Systems and Techniques*, pp. 248–255 (McGraw-Hill, 1966).
2. H. S. Black, *Modulation Theory*, pp. 266–276 and 282–287 (Van Nostrand, 1953).

used to generate a PDM wave, the output of the low-pass reconstruction filter contains not only the desired message wave, but also its harmonics. With natural sampling these harmonics are missing. As with natural sampling, the other in-band distortion products of the form $(1/T_s) - lf_m$ are present. Thus we may expect a net deterioration of quality in the reconstructed message signal when the sampling is uniform instead of natural. Detailed studies show this to be true.

In the case of a PPM wave, we may assume that each pulse has a very small duration compared with the sampling interval T_s, so that it may be approximated as an impulse. Then it turns out that the spectrum of a PPM wave obtained by natural sampling, and with a sinusoidal modulating wave, is similar in form to that of a PDM wave, except that it contains a component proportional to the derivative of the modulating wave rather than the modulating wave itself. Thus we may demodulate a PPM wave by passing it through a low-pass filter and then integrating it so as to restore the wanted signal component to its original waveform. Alternatively, we may first convert a PPM into a PDM wave, and then pass the resulting PDM wave through a low-pass filter. In this way, a greater signal amplitude with less distortion can be obtained at the receiver output.

Noise in Pulse-Time Systems

In a pulse-time modulation system, the transmitted information is contained in the relative positions of modulated pulse edges. The presence of noise affects the performance of such a system by falsifying the time at which the modulated pulses are judged to begin or end. Immunity to noise can be established by making the pulse build up so rapidly that the time interval during which noise can exert any perturbation is very short. Indeed, additive noise would have no effect on the pulse edges if the received pulses were perfectly rectangular, because the presence of noise introduces only vertical perturbations. However, the reception of perfectly rectangular pulses implies infinite transmission bandwidth. Thus, with a finite transmission bandwidth in practice, we find that the received pulses have a finite rise time, and so the performance of the receiver is affected by noise.

As with CW modulation systems, the noise performance of a PPM or PDM system may be described in terms of the output signal-to-noise ratio. Also, to compare the noise performances of PPM and PDM systems we may use the figure of merit, which is equal to the output signal-to-noise ratio divided by the channel signal-to-noise ratio. We illustrate these evaluations by first considering the example of a PPM system using a raised cosine pulse and sinusoidal modulation. We then consider the corresponding case of a PDM system.

Example 2 Signal-to-noise ratios of a PPM system using sinusoidal modulation

Consider a PPM system whose pulse train, in the absence of modulation, is as shown in Fig. 7.17(a). The standard pulse is assumed to be a *raised cosine pulse*, which is a convenient type

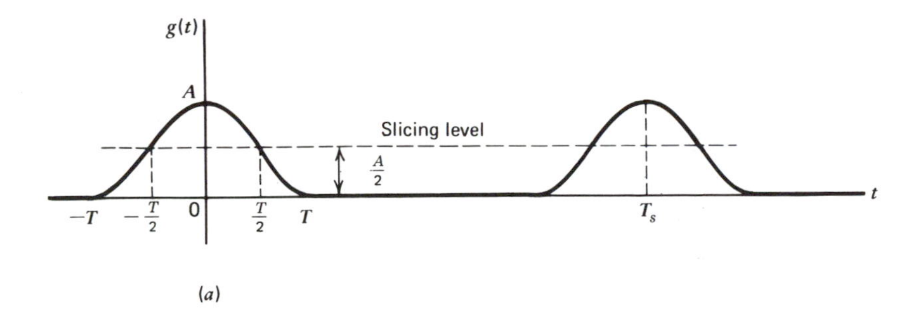

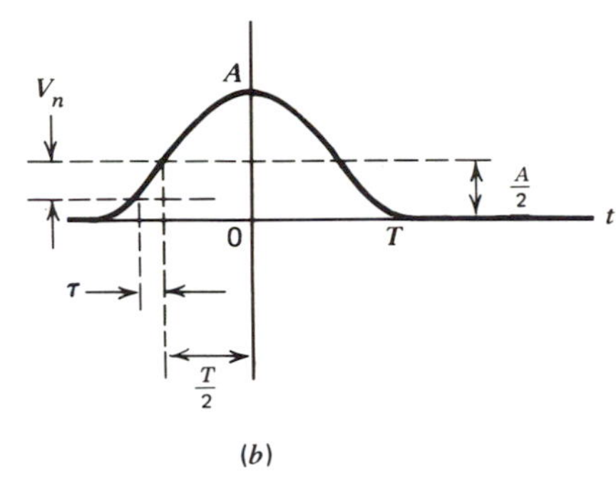

Figure 7.17 Noise analysis of PPM system. *(a)* Unmodulated pulse train. *(b)* Illustrating the effect of noise on pulse detection time.

of pulse for analysis. This pulse, denoted as $g(t)$, is defined by

$$g(t)=\frac{A}{2}\left[1+\cos(\pi B_T t)\right], \qquad -T \leqslant t \leqslant T \tag{7.60}$$

where $B_T = 1/T$. The time of occurrence of such a pulse may be determined by applying the pulse to an ideal slicer with the input–output amplitude characteristic shown in Fig. 7.14, and then observing the slicer output. We assume that the slicing level is set at half the peak pulse amplitude, namely, $A/2$, as in Fig. 7.17(*a*). For inputs below the slicing level the output is zero, and for inputs above the slicing level the output is constant.

The Fourier transform of the pulse $g(t)$ is defined by

$$G(f)=\frac{A\sin(2\pi f/B_T)}{2\pi f(1-4f^2/B_T^2)} \tag{7.61}$$

As indicated in Fig. 7.18, this transform has its first nulls at $f=\pm B_T$ and is small outside this interval, so that the transmission bandwidth required to pass such a pulse may be taken as essentially equal to B_T.

Let the peak-to-peak swing in the position of a pulse be denoted by T_s. Then, in response to a full-load sinusoidal modulating wave, the peak-to-peak amplitude of the receiver output will be KT_s, where K is a constant determined by the receiver circuitry. The corresponding average signal power at the receiver output is therefore $K^2T_s^2/8$.

In the presence of additive noise, both the amplitude and position of the pulse will be per-

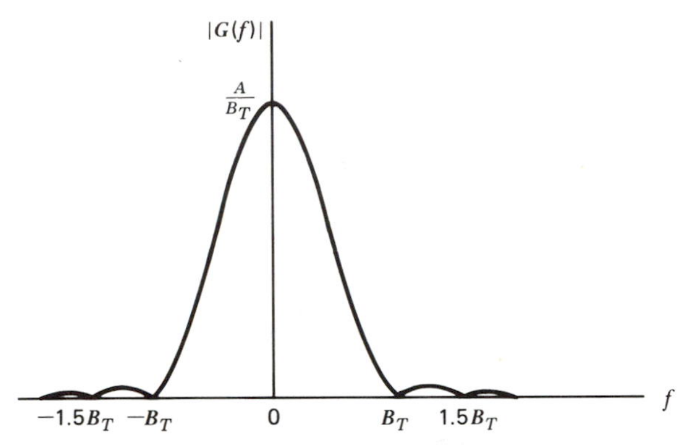

Figure 7.18 Amplitude spectrum of a raised-cosine pulse.

turbed. Random variations in the pulse amplitude are removed by the slicer. Random variations in the pulse position due to noise will remain, however, thereby contributing to noise at the receiver output. We assume that, at the receiver input, the noise power is small compared with the peak pulse power. Then, if at a particular instant of time the noise amplitude is V_n, the time of pulse detection will be displaced by a small amount τ, as depicted in Fig. 7.17(*b*). To a first order of approximation, V_n/τ is equal to the slope of the pulse $g(t)$ at time $t = -T/2$. Thus, using Eq. (7.60), we get

$$\frac{V_n}{\tau} = \left.\frac{dg(t)}{dt}\right|_{t=-T/2}$$

$$= \frac{\pi B_T A}{2}$$

That is,

$$\tau = \frac{2V_n}{\pi B_T A} \tag{7.62}$$

The error τ in the position of the pulse $g(t)$ will produce an average noise power at the receiver output equal to $K^2 E[\tau^2]$. Assuming that the noise at the front end of the receiver has a power spectral density $N_0/2$, we find that, with a transmission bandwidth B_T,

$$E[V_n^2] = B_T N_0 \tag{7.63}$$

Using Eqs. (7.62) and (7.63), we obtain:

$$\text{average power of output noise} = K^2 E[\tau^2]$$

$$= \frac{4K^2 N_0}{\pi^2 B_T A^2} \tag{7.64}$$

The output signal-to-noise ratio, assuming a full-load sinusoidal modulation, is therefore

$$(\text{SNR})_O = \frac{K^2 T_s^2/8}{4K^2 N_0/\pi^2 B_T A^2}$$

$$= \frac{\pi^2 B_T T_s^2 A^2}{32 N_0} \tag{7.65}$$

The average transmitted power P in a PPM system is independent of the applied modulation. Accordingly, we may determine P by averaging the power in a single pulse of the PPM wave over the sampling period T_s, as shown by,

$$P=\frac{1}{T_s}\int_{-T_s/2}^{T_s/2} g^2(t)dt$$

$$=\frac{3A^2}{4T_sB_T} \tag{7.66}$$

The average noise power in a message bandwidth W is equal to WN_0. The channel signal-to-noise ratio is therefore

$$(\text{SNR})_C=\frac{3A^2/4T_sB_T}{WN_0}$$

$$=\frac{3A^2}{4T_sB_TWN_0} \tag{7.67}$$

Thus the figure of merit of a PPM system using a raised cosine pulse is

$$\text{Figure of merit}=\frac{(\text{SNR})_O}{(\text{SNR})_C}$$

$$=\frac{\pi^2}{24}B_T^2T_s^3W \tag{7.68}$$

Assuming that the message signal is sampled at its Nyquist rate, we have $T_s=1/2W$. Then, we find from Eq. (7.68) that the corresponding value of the figure of merit is $(\pi^2/192)(B_T/W)^2$, which is greater than unity if $B_T>4.41W$. We also see that, as is the case with an FM system, the figure of merit of a PPM system is proportional to the square of the normalized transmission bandwidth B_T/W.

Example 3 Signal-to-noise ratios of a PDM system using sinusoidal modulation

Consider next a PDM system whose pulse train, in the absence of modulation, is as shown in Fig. 7.19. The PDM pulse consists of a duration D which is preceded and followed by leading and trailing segments that are identical to the corresponding halves of the PPM pulse. The slicer in the receiver is set at half the peak pulse amplitude, removing all noise effects except for

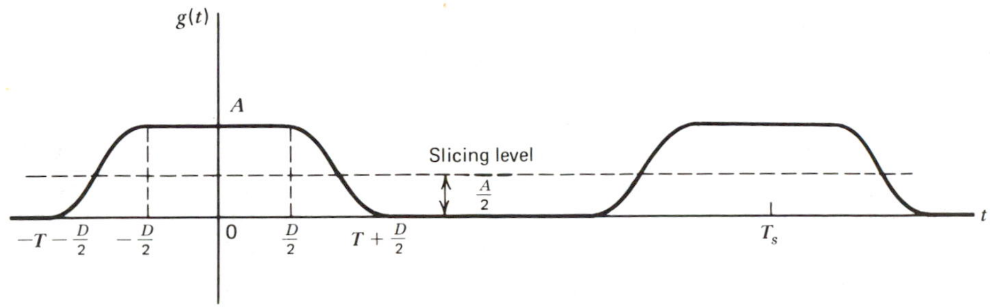

Figure 7.19 Unmodulated pulse train for PDM.

the displacement of edge detection time by a small amount τ, similar to that evaluated above for the PPM system. It is assumed that one edge of the duration-modulated pulse is fixed by means of a noise-free reference. We thus find that the output signal-to-noise ratio of this PDM system is the same as in Eq. (7.65).

From Fig. 7.19, we see that the average power of the PDM wave is equal to that of the PPM pulse, given by Eq. (7.66), plus the amount contributed by the rectangular portion of duration D. This latter contribution is equal to A^2D/T_s for a fixed value of D. However, the duration D varies in accordance with the applied modulation. Therefore, assuming a full-load sinusoidal modulating wave of frequency f_m, the modulation of duration D is defined by

$$D=\frac{T_s}{2}\left[1+\cos(2\pi f_m t+\Theta)\right] \tag{7.69}$$

where Θ is a random variable that is uniformly distributed over the range 0 to 2π. Accordingly, the average power of the transmitted PDM wave is given by the ensemble average

$$\begin{aligned} P &= E\left[\frac{1}{T_s}\int_{-T_s/2}^{T_s/2} g^2(t)dt\right] \\ &= E\left[\frac{3A^2}{4T_sB_T}+\frac{A^2D}{T_s}\right] \\ &= \frac{3A^2}{4T_sB_T}+\frac{A^2}{2} \end{aligned} \tag{7.70}$$

With an average noise power of WN_0 in the message bandwidth W, it follows therefore that the channel signal-to-noise ratio of the PDM system is

$$(\text{SNR})_C=\frac{A^2(3+2T_sB_T)}{4T_sB_TWN_0} \tag{7.71}$$

From Eqs. (7.65) and (7.71), the corresponding figure of merit of the PDM system is obtained as

$$\text{Figure of merit}=\frac{\pi^2B_T^2T_s^3W}{24(1+2T_sB_T/3)} \tag{7.72}$$

Comparing the figures of merit given by Eqs. (7.68) and (7.72), we see that for sinusoidal modulation the noise performance of a PPM system, using a raised cosine pulse, is better than that of a PDM system, using the pulse shown in Fig. 7.19, by a factor equal to $1+2T_sB_T/3$, where T_s is the sampling period and B_T is the transmission bandwidth. The physical reason for this is that a PPM system preserves all the signal information contained at the instants at which the pulses terminate, and yet it avoids the wastage of a considerable amount of power which a PDM system expends during the pulse. Thus PPM is more efficient than PDM in utilizing the transmitted power.

In the noise analysis presented above for both PPM and PDM systems, we have assumed that the average power of the additive noise at the front end of the receiver is small compared with the peak pulse power. In particular, it is assumed that there are two crossings of the slicing level for each pulse, one for the leading edge and one for the trailing edge. A Gaussian noise will have occasional peaks that produce additional crossings of the slicing level, however, and so the occasional noise peaks are mistaken for message pulses. The analysis neglects the *false pulses* produced by high noise peaks. It is apparent that these false pulses have a finite though small probability of occurrence when the noise is Gaussian, no matter how small its standard deviation is compared with the peak amplitude of the pulses. As the transmission bandwidth is increased indefinitely, the accompanying increase in average noise power eventually causes the false pulses to occur often enough, thereby causing loss of the wanted message signal at the receiver output. *We thus find, in practice, that both PPM and PDM systems suffer from a threshold effect similar to that experienced in FM systems.*

Synchronization in Pulse-Time Modulation Systems

As with PAM systems, synchronization in pulse-time modulation systems is established by transmitting a distinctive marker per frame. In a PDM system, the marker may be identified by omitting a pulse, as illustrated in Fig. 7.13(*c*) for a PDM system involving three independent message sources. One method of identifying such a marker in the receiver is to utilize the charging time of a simple resistor-capacitor circuit to measure the duration of the intervals between duration-modulated pulses. The time constant of the circuit is chosen so that, during a marker interval, the voltage across the capacitor rises to a value considerably higher than that during the normal charging interval. Thus, by applying the output of the circuit to a slicer with an appropriate slicing level, the presence of a marker is detected.

In a PPM system, the marker pulse may be identified by making its duration several times longer than that of the message pulses, as illustrated in Fig. 7.13(*d*). At the receiver, the marker pulses may be separated from the message pulses by using a procedure essentially similar to that described for the PDM system. In this case, however, the capacitor is charged during the time of occurrence of each pulse, and discharged during the intervening intervals. Accordingly, the voltage across the capacitor reaches its highest value during the presence of a marker pulse, and the marker pulses are thereby separated from the message pulses.

Problems

Problem 7.1 The signal

$$g(t) = 10\cos(20\pi t)\cos(200\pi t)$$

is sampled at the rate of 250 samples per second.

(a) Determine the spectrum of the resulting sampled signal.
(b) Specify the cutoff frequency of the ideal reconstruction filter so as to recover $g(t)$ from its sampled version.
(c) What is the Nyquist rate for $g(t)$?
(d) By treating $g(t)$ as a band-pass signal, determine the lowest permissible sampling rate for this signal.

***Problem* 7.2** The signals

$$g_1(t) = 10\cos(100\pi t)$$

and

$$g_2(t) = 10\cos(50\pi t)$$

are both sampled at the rate of 75 samples per second. Show that the two sequences of samples thus obtained are identical. What is the reason for this phenomenon?

***Problem* 7.3** The signal

$$g(t) = 10\cos(60\pi t)\cos^2(160\pi t)$$

is sampled at the rate of 400 samples per second. Determine the range of permissible cutoff frequencies for the ideal reconstruction filter that may be used to recover $g(t)$ from its sampled version.

***Problem* 7.4** A signal $g(t)$ consists of two frequency components $f_1 = 3.9$ kHz and $f_2 = 4.1$ kHz in such a relationship that they just cancel each other out when the signal $g(t)$ is sampled at the instants $t = 0, T, 2T, \ldots$, where $T = 125\ \mu s$. The signal $g(t)$ is defined by

$$g(t) = \cos\left(2\pi f_1 t + \frac{\pi}{2}\right) + A\cos(2\pi f_2 t + \phi)$$

Find the values of amplitude A and phase ϕ of the second frequency component.

***Problem* 7.5** Let E denote the energy of a strictly band-limited signal $g(t)$. Show that E may be expressed in terms of the sample values of $g(t)$, taken at the Nyquist rate, as follows

$$E = \frac{1}{2W} \sum_{n=-\infty}^{\infty} \left| g\left(\frac{n}{2W}\right) \right|^2$$

where W is the highest frequency component of $g(t)$.

***Problem* 7.6** The spectrum of a signal $g(t)$ is shown in Fig. P7.1. This signal is sampled at the Nyquist rate with a periodic train of rectangular pulses of duration 50/3 milliseconds. Plot the spectrum of the sampled signal for frequencies up to 50 hertz.

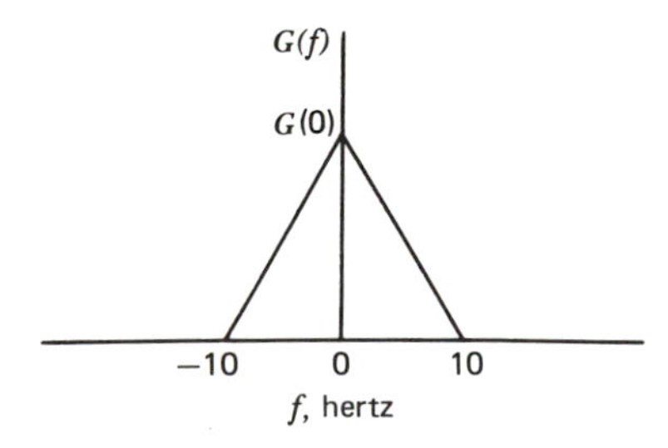

Figure P7.1

***Problem* 7.7** This problem is aimed at investigating the fact that practical electronic switching circuits will not produce a sampling function that consists of exactly rectangular pulses. Let $h(t)$ denote some arbitrary pulse shape, so that the sampling function $c(t)$ may be expressed as

$$c(t)=\sum_{n=-\infty}^{\infty} h(t-nT_s)$$

where T_s is the sampling period. The sampled version of an incoming analog signal $g(t)$ is defined by

$$s(t)=c(t)g(t)$$

(a) Show that the Fourier transform of $s(t)$ is given by

$$S(f)=\frac{1}{T_s}\sum_{n=-\infty}^{\infty} G\left(f-\frac{n}{T_s}\right)H\left(\frac{n}{T_s}\right)$$

where $G(f)=F[g(t)]$ and $H(f)=F[h(t)]$.

(b) What is the effect of using the arbitrary pulse shape $h(t)$?

Problem 7.8 Consider a continuous-time signal $g(t)$ of finite energy, with a continuous spectrum $G(f)$. Assume that $G(f)$ is sampled uniformly at the discrete frequencies $f=kF_s$, thereby obtaining the sequence of frequency samples $G(kF_s)$, where k is an integer in the entire range $-\infty<k<\infty$, and F_s is the frequency sampling interval. Show that if $g(t)$ is duration-limited, so that it is zero outside the interval $-T\leqslant t\leqslant T$, then the signal is completely defined by specifying $G(f)$ at frequencies spaced $1/2T$ hertz apart.

Problem 7.9

(a) Consider a stationary process $X(t)$ that is *not* strictly band-limited in the band W; that is,

$$S_X(f)\neq 0, \qquad |f|>W$$

where $S_X(f)$ is the power spectral density of the process. The process $X(t)$ is applied to an ideal low-pass filter defined by the transfer function

$$H(f)=\begin{cases}1, & |f|<W\\ 0, & |f|>W\end{cases}$$

producing the process $X_l(t)$. This process is next sampled at a rate equal to $2W$, producing the sequence of samples $X_l(n/2W)$. An approximate reconstruction of the original process is defined by

$$Y(t)=\sum_{n=-\infty}^{\infty} X_l\left(\frac{n}{2W}\right)\text{sinc}\left[2W\left(t-\frac{n}{2W}\right)\right]$$

Show that the mean-square value of the *sampling error* is

$$\begin{aligned}\mathscr{E}&=E[(X(t)-Y(t))^2]\\ &=2\int_W^{\infty} S_X(f)df\end{aligned}$$

(b) Given that

$$S_X(f)=\frac{f_0}{f^2+f_0^2}$$

determine the corresponding value of the mean-square error $\mathscr{E}$, and plot it as a function of W/f_0.

Problem 7.10 Consider a sequence of samples $x(nT_s)$ obtained by sampling a continuous-time signal $x(t)$ at the rate $1/T_s$. It is required to increase the sampling period T_s to a new value

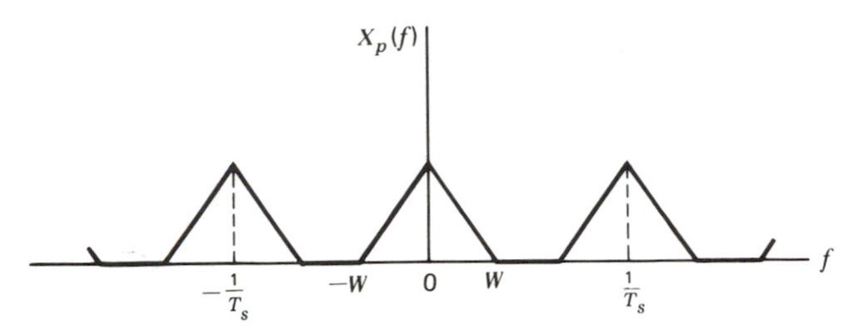

Figure P7.2

$T'_s = MT_s$, where M is an integer. Such a process is called *decimation*: Let $y(nT'_s)$ denote the new sequence with the sampling period T'_s.

(a) Find the relationship between the Fourier transforms of the new sequence $y(nT'_s)$ and the original sequence $x(nT_s)$.
(b) What is the condition required to avoid the occurrence of aliasing in the decimation process?
(c) For a sequence $x(nT_s)$ with the amplitude spectrum shown in Fig. P7.2, find the amplitude spectrum of $y(nT'_s)$, assuming that $1/4W < T_s < 1/2W$, and $M = 2$. How would you avoid the occurrence of aliasing in this example?

Problem 7.11 In the *interpolation* process, which is the dual of the decimation process, we increase the sampling rate by an integer ratio. Consider a sequence $x(nT_s)$ the sampling period of which is to be reduced to a new value $T'_s = T_s/L$, where L is an integer. This requirement is achieved by using the arrangement shown in Fig. P7.3(*a*), where we first insert $(L-1)$ zero-valued samples between each sample of $x(nT_s)$, and then pass the resulting sequence $v(nT'_s)$ through a low-pass filter with a periodic frequency response, as in Fig. P7.3(*b*).

(a) Determine the relation between the Fourier transform of $v(nT'_s)$ and that of $x(nT_s)$.
(b) Specify the zero-frequency response $H(0)$ and bandwidth B of the low-pass filter so that the output sequence $y(nT'_s)$ has a spectrum that is periodic with period $1/T'_s$, and a dc value equal to that of the input sequence $x(nT_s)$.

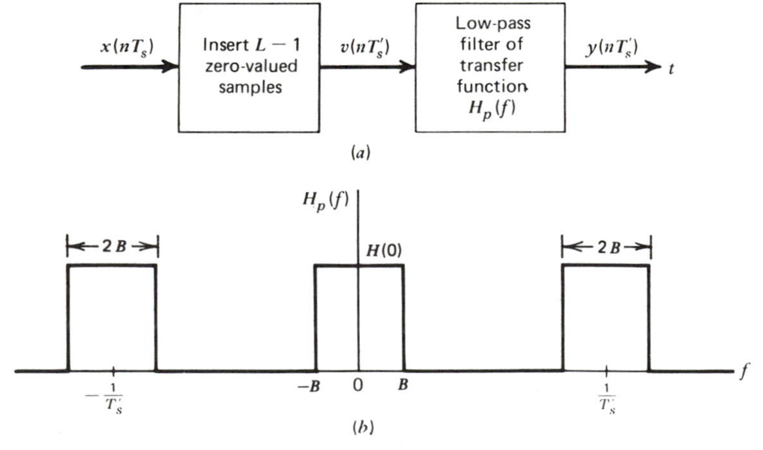

Figure P7.3

Problem 7.12 A band-pass signal $g(t)$ has no frequency components outside the interval $f_1 \leq |f| \leq f_2$, where $f_1 = 0.995$ MHz and $f_2 = 1.0$ MHz. Find the lowest possible sampling rate for this signal, so that there is no distortion due to sampling.

Problem 7.13 A signal, strictly band-limited to 250 Hz, is amplitude-modulated onto a carrier wave of frequency f_c. The resulting band-pass signal is sampled at the rate of 1000 samples per second. Find the permissible values of f_c that would enable sampling at the minimum possible rate.

Problem 7.14 Consider the time functions $\phi_1(t)$ and $\phi_2(t)$, defined as follows, respectively,

$$\phi_1(t) = 4W \,\mathrm{sinc}(2Wt)\cos(2\pi f_c t)$$

and

$$\phi_2(t) = \mathrm{sinc}(2W't)\cos(2\pi f_c' t)$$

where $f_c + W = f_c' + W'$, and $f_c > f_c'$. Show that the convolution of these two time functions is given by

$$\phi_1(t) \otimes \phi_2(t) = \frac{W}{W'} \,\mathrm{sinc}(2Wt)\cos(2\pi f_c t)$$

How is this result affected if $\phi_2(t)$ is delayed by t_0 seconds?

Problem 7.15

(a) Consider a low-pass signal $g(t)$ of finite energy and infinite duration. The highest frequency component of this signal is 1 Hz. Using $g_+(t)$ to denote the pre-envelope of $g(t)$, prove the *complex sampling theorem* which states that the signal $g(t)$ is determined uniquely by specifying the complex samples of the pre-envelope $g_+(t)$ taken at the rate of one complex sample per second throughout the time domain. In particular, show that

$$g_+(t) = \sum_{n=-\infty}^{\infty} g_+(n) e(t-n)$$

where $e(t)$ is a unit-energy extrapolation function defined by

$$e(t) = \mathrm{sinc}(t)\exp(j\pi t)$$

Note that one complex sample is equivalent to two real samples.

(b) How will the result in part (a) be modified if the low-pass signal $g(t)$ has its highest frequency component at W hertz?

(c) Consider next a band-pass signal whose spectrum is limited to the frequency interval $f_0 \leq |f| \leq f_0 + 1$. Here again, prove that the signal $g(t)$ is determined uniquely by specifying the complex samples of the pre-envelope $g_+(t)$ taken at the rate of one complex sample per second throughout the time domain, as shown by

$$g_+(t) = \sum_{n=-\infty}^{\infty} g_+(n)\exp[j2\pi f_0(t-n)] e(t-n)$$

where $e(t)$ is as defined in part (a).

(d) How will the result in part (c) be modified if the band-pass signal has its spectrum limited to the frequency interval $f_0 \leq |f| \leq f_0 + 2W$?

Problem 7.16 A band-pass signal $g(t)$, of duration T, has its spectrum limited essentially to the interval $f_c - W \leq |f| \leq f_c + W$. Show that this signal is uniquely described by specifying $2WT$ samples of its in-phase component and $2WT$ samples of its quadrature component.

Problem 7.17 Six independent message sources of bandwidths W, W, $2W$, $2W$, $3W$, *and* $3W$ hertz are to be transmitted on a time-division multiplexed basis using a common communication channel.

(a) Set up a scheme for accomplishing this multiplexing requirement, with each message signal sampled at its Nyquist rate.
(b) Determine the minimum transmission bandwidth of the channel.

Problem 7.18 Twenty-four voice signals are sampled uniformly and then time-division multiplexed. The sampling operation uses flat-top samples with 1 microsecond duration. The multiplexing operation includes provision for synchronization by adding an extra pulse of sufficient amplitude and also 1 microsecond duration. The highest frequency component of each voice signal is 3.4 kHz.

(a) Assuming a sampling rate of 8 kHz, calculate the spacing between successive pulses of the multiplexed signal.
(b) Repeat your calculation assuming the use of Nyquist rate sampling.

Problem 7.19 Twelve different message signals, each with a bandwidth of 10 kHz, are to be multiplexed and transmitted. Determine the minimum bandwidth required for each method if the multiplexing/modulation method used is:

(a) FDM, SSB.
(b) TDM, PAM.

Problem 7.20 A PAM *telemetry* system involves the multiplexing of four input signals: $s_i(t)$, $i=1, 2, 3, 4$. Two of the signals, $s_1(t)$ and $s_2(t)$, have bandwidths of 80 Hz each, whereas the remaining two signals, $s_3(t)$ and $s_4(t)$ have bandwidths of 1 kHz each. The signals $s_3(t)$ and $s_4(t)$ are each sampled at the rate of 2,400 samples for second. This sampling rate is divided by 2^n (i.e., an integer power of 2) in order to derive the sampling rate for $s_1(t)$ and $s_2(t)$.

(a) Find the maximum value of n.
(b) Using the value of n thus found, design a multiplexing system that first multiplexes $s_1(t)$ and $s_2(t)$ into a new sequence, $s_5(t)$, and then multiplexes $s_3(t)$, $s_4(t)$, and $s_5(t)$.

Problem 7.21 The block diagram of Fig. P7.4 shows the processing of a baseband signal $x_k(t)$, the highest frequency component of which is limited to W hertz. In the upper path the

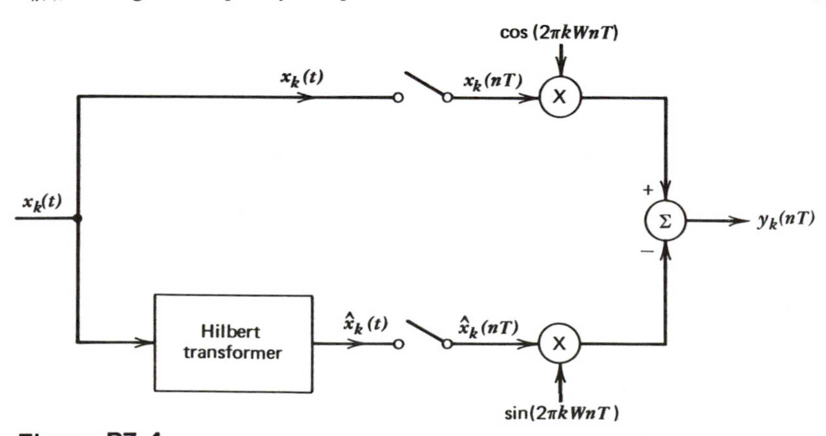

Figure P7.4

signal $x_k(t)$ is sampled at the rate

$$\frac{1}{T} = 2NW$$

where N is an integer. In the lower path, the signal $x_k(t)$ is passed through a Hilbert transformer, after which it is sampled at the same rate $1/T$. The sampled signals $x_k(nT)$ and $\hat{x}_k(nT)$ are then multiplied by $\cos(2\pi kWnT)$ and $\sin(2\pi kWnT)$, respectively, where k is an integer and $0 \leqslant k \leqslant N-1$. The resultant products are subtracted to produce the output $y_k(nT)$.

(a) Sketch the spectrum of the output $y_k(nT)$, and show that it is the sampled form of an SSB modulated signal. Illustrate your answer for $N=4$.
(b) Suppose that we have a set of baseband signals $\{x_k(t)\}$, $k=0, 1, \ldots, N-1$ with the highest frequency of each signal equal to W hertz. Each baseband signal is processed in a manner similar to that described in Fig. P7.4. The resultant individual outputs are summed to produce the overall output

$$y(nT) = \sum_{k=0}^{N-1} y_k(nT)$$

Show that $y(nT)$ represents a sampled SSB–FDM signal of N channels separated by a frequency W hertz.

***Problem* 7.22**

(a) Plot the spectrum of a PAM wave produced by the modulating wave

$$m(t) = A_m \cos(2\pi f_m t)$$

assuming a modulation factor $\mu < 1$, modulation frequency $f_m = 0.25$ Hz, sampling period $T_s = 15$ seconds, and pulse duration $T = 0.45$ seconds.
(b) Using an ideal reconstruction filter, plot the spectrum of the filter output. Compare this result with the output that would be obtained if there was no aperture effect.

***Problem* 7.23** Consider a PAM wave transmitted through a channel with white Gaussian noise and the minimum bandwidth $B_T = 1/2T_s$, where T_s is the sampling period. The noise is of zero mean and power spectral density $N_0/2$. The PAM wave uses a standard pulse $g(t)$ with its Fourier transform defined by

$$G(f) = \begin{cases} \dfrac{1}{2B_T}, & |f| < B_T \\ 0, & |f| > B_T \end{cases}$$

By considering a full-load sinusoidal modulating wave, show that PAM and baseband-signal transmission have equal signal-to-noise ratios for the same average transmitted power.

***Problem* 7.24** Explain why a single-channel PPM system requires the transmission of a synchronizing signal, whereas a single-channel PAM or PDM system does not.

***Problem* 7.25** In a multiplex PPM system with N message sources, each message has bandwidth W and is sampled at its Nyquist rate. The system uses a raised-cosine pulse defined by Eq. (7.60). Assuming a full-load sinusoidal modulating wave, find the output signal-to-noise ratio of the system when the average noise power is small compared with the peak pulse power.

***Problem* 7.26** The unmodulated pulse train in a PPM system is as shown in Fig. P7.5. The slicing level in the receiver is set at $A/2$.

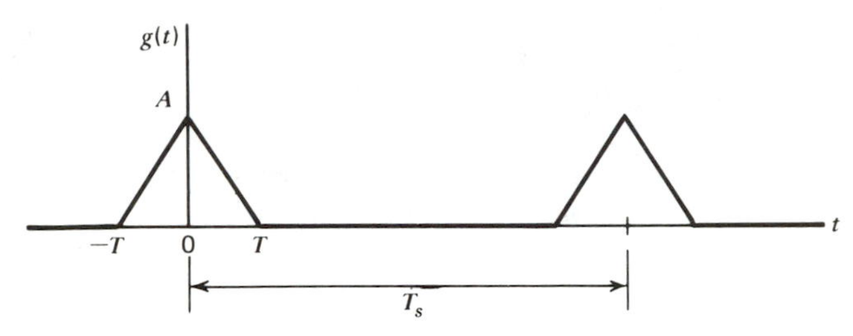

Figure P7.5

(a) Assuming a full-load sinusoidal modulating wave, and front-end receiver noise of zero mean and power spectral density $N_0/2$, determine the output signal-to-noise ratio and figure of merit of the system. Assume a high peak pulse-to-noise ratio.

(b) For the case when the message signal is sampled at its Nyquist rate, find the value of the transmission bandwidth for which the figure of merit of the system is greater than unity.

***Problem* 7.27** Repeat Problem 7.26 for a PDM system using the unmodulated pulse train shown in Fig. P7.6.

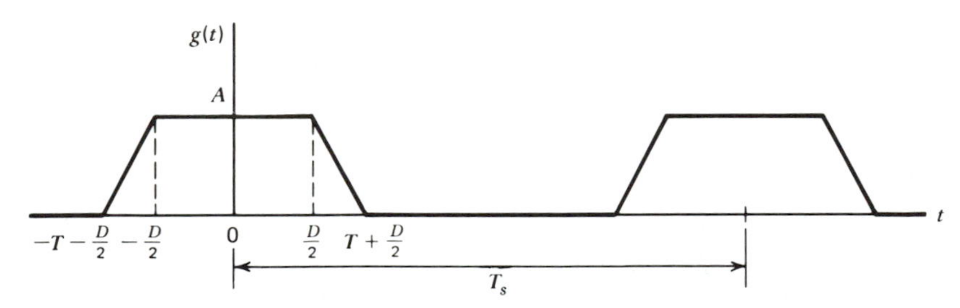

Figure P7.6

chapter 8 PULSE-DIGITAL MODULATION

In pulse-amplitude modulation (PAM), pulse-duration modulation (PDM), and pulse-position modulation (PPM) only time is expressed in discrete form, whereas the respective modulation parameters (namely, pulse amplitude, duration, and position) are varied in a continuous manner in accordance with the message. Thus, in these modulation systems, information transmission is accomplished in analog form at discrete times. On the other hand, in *pulse-code modulation* (PCM),* the message signal is sampled and the amplitude of each sample is rounded off to the nearest one of a finite set of allowable values, so that both time and amplitude are in discrete form. This allows the message to be transmitted by means of coded electrical signals, thereby distinguishing PCM from all other methods of modulation.

The use of digital representation of analog signals (e.g., voice, video) offers the following advantages: (1) *ruggedness* to transmission noise and interference, (2) efficient *regeneration* of the coded signal along the transmission path, and (3) the possibility of a *uniform format* for different kinds of baseband signals. These advantages, however, are attained at the cost of increased transmission bandwidth requirement and increased system complexity. With the increasing availability of wide-band communication channels, coupled with the emergence of the requisite device technology, the use of PCM has become a practical reality.

This chapter is devoted to a study of pulse-code modulation and other related methods of modulation. We begin the chapter by describing the basic operations involved in the design of a PCM system.

8.1 ELEMENTS OF PULSE-CODE MODULATION (PCM)

Pulse-code modulation systems are considerably more complex than PAM, PDM, and PPM systems, in that the message signal is subjected to a greater number of operations. The essential operations in the transmitter of a PCM system are *sampling*, *quantizing*, and *encoding*, as shown in Fig. 8.1. The quantizing and encoding operations are usually performed in the same circuit, which is called an *analog-to-digital converter*. The essential operations in the receiver are *regeneration* of impaired signals, *decoding*, and *demodulation* of the train of quantized samples. Regeneration usually occurs at intermediate points along the transmission route as necessary. When time-division multiplexing is used, it becomes necessary to synchronize the receiver to the transmitter for the overall system to operate satisfactorily.

Sampling

The incoming message wave is sampled with a train of narrow rectangular pulses so as to closely approximate the instantaneous sampling process. In order to ensure perfect reconstruction of the message at the receiver, the sampling rate must be greater than twice the highest frequency component W of the message wave in accordance with the sampling theorem. In practice, a low-pass filter is used at the front end of the sampler in order to exclude frequencies greater than W before

* Pulse-code modulation was invented by Reeves in 1937. For a historical account of this invention, see: A. H. Reeves, "The past, present, and future of PCM," *IEEE Spectrum*, pp. 58–63, May 1975.

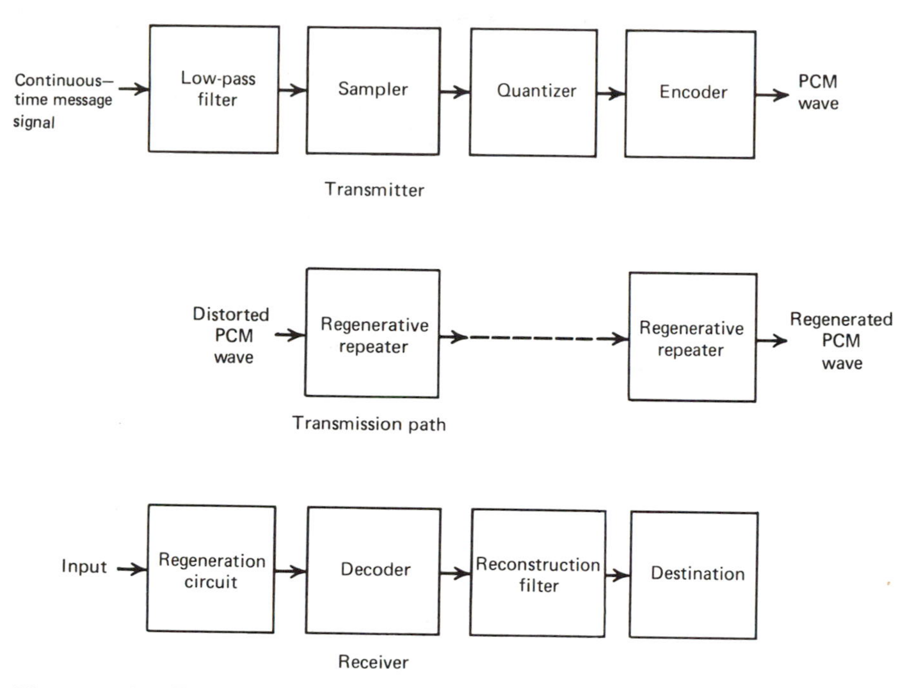

Figure 8.1 The basic elements of a PCM system.

sampling. Thus the application of sampling permits the reduction of the continuously varying message wave to a limited number of discrete values per second.

Quantizing

A continuous signal, such as voice, has a continuous range of amplitudes and therefore its samples have a continuous amplitude range. In other words, within the finite amplitude range of the signal we find an infinite number of amplitude levels. It is not necessary in fact to transmit the exact amplitudes of the samples. Any human sense (the ear or the eye), as ultimate receiver, can only detect finite intensity differences. This means that the original continuous signal may be approximated by a signal constructed of discrete amplitudes selected on a minimum error basis from an available set. The existence of a finite number of discrete amplitude levels is a basic condition of PCM. Clearly, if we assign the discrete amplitude levels with sufficiently close spacing, we may make the approximated signal practically indistinguishable from the original continuous signal.

The conversion of an analog (continuous) sample of the signal into a digital (discrete) form is called the *quantizing* process. Graphically, the quantizing process means that a straight line representing the relation between the input and output of a linear continuous system is replaced by a *staircase* characteristic, as in Fig. 8.2(*a*). The difference between two adjacent discrete values is called a *quantum* or

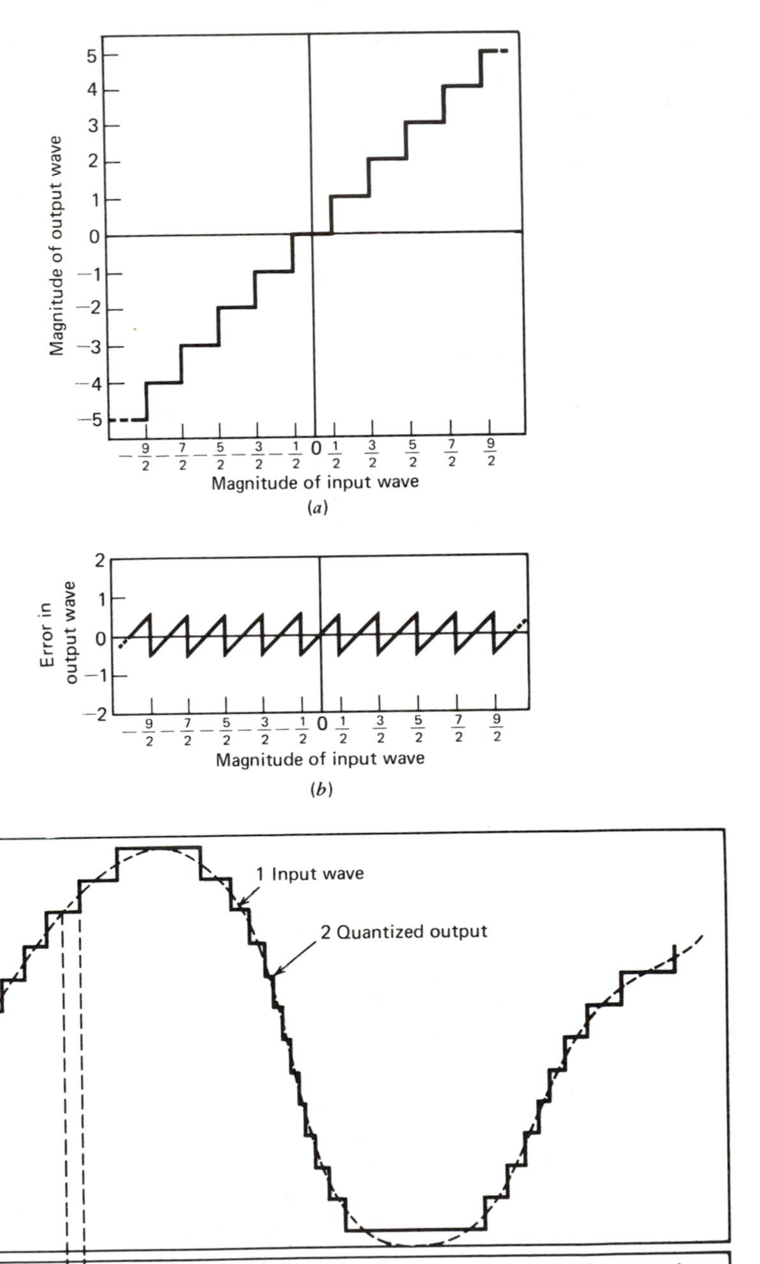

Figure 8.2 Illustration of the quantizing principle. *(a)* Quantizing characteristic. *(b)* Characteristic of errors in quantizing. *(c)* A quantized signal wave and the corresponding error curve.

step size. Signals applied to a *quantizer*, with the input–output characteristic of Fig. 8.2(*a*), are sorted into amplitude slices (the treads of the staircase), and all input signals within plus or minus half a quantum step of the mid-value of a slice are replaced in the output by the mid-value in question.

The *quantizing error* consists of the difference between the input and output signals of the quantizer. It is apparent that the maximum instantaneous value of this error is half of one quantum step, and the total range of variation is from minus half a step to plus half a step. In part (*b*) of Fig. 8.2 the error is shown plotted as a function of the input signal, and in part (*c*) of the figure a typical variation of the error as a function of time is indicated.

According to the staircase-like graph of Fig. 8.2(*a*), the quantizer output may be expressed in the form $H_i\delta$, where $\pm H_i = 0, 1, 2, \ldots$ and δ is the size of a quantum. In this figure, δ is normalized to a value of one. A quantizer having this input-output relation is said to be of the *mid-tread* type, because the origin lies in the middle of a tread of the staircase-like graph.

Another way of designing the quantizer is to define its output in the form $H_i\delta/2$, where $\pm H_i = 1, 3, 5, \ldots$. This quantizer is said to be of the *mid-riser* type, because in this case the origin lies in the middle of a rising part of this staircase-like input-output relation (see Problem 8.1).

The quantizing process based on Fig. 8.2(*a*) uses a uniform separation between the quantizing values. In certain applications, however, it is preferrable to use a variable separation between the quantizing levels. For example, the range of voltages covered by voice signals, from the peaks of loud talk to the weak passages of weak talk, is of the order of 1000 to 1. By using a *nonuniform quantizer* with the feature that the step size increases as the separation from the origin of the input–output amplitude characteristic is increased, the large end step of the quantizer can take care of possible excursions of the voice signal into the large amplitude ranges which occur relatively infrequently. In other words, the weak passages, which need more protection, are favored at the expense of the loud passages. In this way, a nearly uniform percentage precision is achieved through the greater part of the amplitude range of the input signal, with the result that fewer steps are needed than would be the case if a uniform quantizer were used.

The use of a nonuniform quantizer is equivalent to passing the baseband signal through a *compressor* and then applying the compressed signal to a uniform quantizer. A particular form of compression law that is used in practice is the so-called *μ-law* defined by*

$$|v_2| = \frac{\log(1+\mu|v_1|)}{\log(1+\mu)} \tag{8.1}$$

where v_1 and v_2 are normalized input and output voltages, and μ is a positive constant. In Fig. 8.3(*a*), we have plotted the μ-law for varying μ. The case of uniform quantizing corresponds to $\mu = 0$. For a given value of μ, the reciprocal slope of the

* B. Smith, "Instantaneous companding of quantized signals," *Bell System Tech, J.*, vol. 36, pp. 653–709, May 1957.

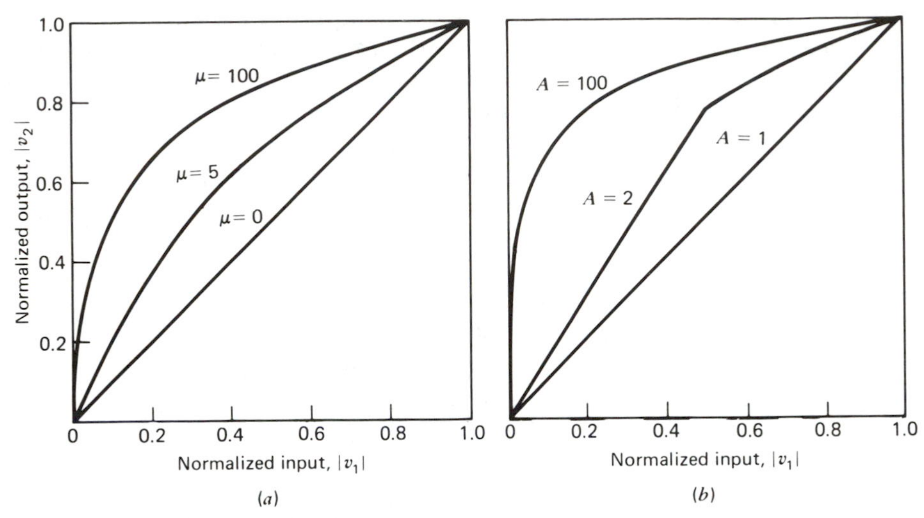

Figure 8.3 Compression laws. *(a)* μ-law. *(b)* A-law.

compression curve, which defines the quantum steps, is

$$\frac{d|v_1|}{d|v_2|} = \frac{\log(1+\mu)}{\mu}(1+\mu|v_1|) \tag{8.2}$$

We see therefore that the μ-law is neither strictly linear nor strictly logarithmic, but it is approximately linear at low input levels corresponding to $\mu|v_1| \ll 1$, and approximately logarithmic at high input levels corresponding to $\mu|v_1| \gg 1$.

Another compression law that is used in practice is the so-called *A-law* defined by*

$$|v_2| = \begin{cases} \dfrac{A|v_1|}{1+\log A}, & 0 \leqslant |v_1| \leqslant \dfrac{1}{A} \\ \dfrac{1+\log(A|v_1|)}{1+\log A}, & \dfrac{1}{A} \leqslant |v_1| \leqslant 1 \end{cases} \tag{8.3}$$

which is shown plotted in Fig. 8.3(*b*). Practical values of A (as of μ in the μ-law) tend to be in the vicinity of 100. The case of uniform quantizing corresponds to $A=1$. The reciprocal slope of this compression curve is

$$\frac{d|v_1|}{d|v_2|} = \begin{cases} \dfrac{1+\log A}{A}, & 0 \leqslant |v_1| \leqslant \dfrac{1}{A} \\ (1+\log A)|v_1|, & \dfrac{1}{A} \leqslant |v_1| \leqslant 1 \end{cases} \tag{8.4}$$

* 1. K. W. Cattermole, *Principles of Pulse-code Modulation*, pp. 133–140 (Iliffe, 1969).
2. H. Kaneko, "A unified formulation of segment companding laws and synthesis of codecs and digital companders," *Bell Syst. Tech. J.*, vol. 49, pp. 1555–1588, Sept. 1970.

Thus the quantum steps over the central linear segment, which have the dominant effect on small signals, are diminished by the factor $A/(1+\log A)$. This is typically about 25 dB in practice, as compared with uniform quantizing.

In order to restore the signal samples to their correct relative level, we must, of course, use a device in the receiver with a characteristic complementary to the compressor. Such a device is called an *expander*. Ideally, the compression and expansion laws are exactly inverse so that, except for the effect of quantization, the expander output is equal to the compressor input. The combination of a *com*pressor and an ex*pander* is called a *compander*.

In actual PCM systems, the companding circuitry does not produce an exact replica of the nonlinear compression curves shown in Fig. 8.3. Rather, it provides a *piecewise linear* approximation to the desired curve. By using a large enough number of linear segments, the approximation can approach the true compression curve very closely. This form of approximation is illustrated in Example 1.

Encoding

In combining the processes of sampling and quantizing, the specification of a continuous baseband signal becomes limited to a discrete set of values, but not in the form best suited to transmission over a line or radio path. To exploit the advantages of sampling and quantizing, we require the use of an *encoding process* to translate the discrete set of sample values to a more appropriate form of signal. Any plan for representing each of this discrete set of values as a particular arrangement of discrete events is called a *code*. One of the discrete events in a code is called a *code element* or *symbol*. For example, the presence or absence of a pulse is a symbol. A particular arrangement of symbols used in a code to represent a single value of the discrete set is called a *code word* or *character*.

In a *binary code*, each symbol may be either of two distinct values or kinds, such as the presence or absence of a pulse. The two symbols of a binary code are customarily denoted as 0 and 1. In a *ternary code*, each symbol may be one of three distinct values or kinds, and so on for other codes. However, *the maximum advantage over the effects of noise in a transmission medium is obtained by using a binary code, because a binary symbol withstands a relatively high level of noise and is easy to regenerate*.

Suppose that, in a binary code, each code word consists of n *bits*; the bit is an acronym for *bi*nary digi*t*. Then, using such a code, we may represent a total of 2^n distinct numbers. For example, a sample quantized into one of 128 levels may be represented by a 7-bit code word. There are several ways of establishing a one-to-one correspondence between quantized levels and code words. A convenient one is to express the ordinal number of the quantized level as a binary number. In the binary number system, each digit has a place-value that is a power of 2, as illustrated in Table 8.1 for the case of $n=4$.

There are several ways by which binary symbols 1 and 0 can be represented by electrical signals:

Table 8.1

Ordinal number of quantized level	Level number expressed as sum of powers of 2	Binary number
0		0 0 0 0
1	2^0	0 0 0 1
2	2^1	0 0 1 0
3	2^1+2^0	0 0 1 1
4	2^2	0 1 0 0
5	2^2+2^0	0 1 0 1
6	2^2+2^1	0 1 1 0
7	$2^2+2^1+2^0$	0 1 1 1
8	2^3	1 0 0 0
9	2^3+2^0	1 0 0 1
10	2^3+2^1	1 0 1 0
11	$2^3+2^1+2^0$	1 0 1 1
12	2^3+2^2	1 1 0 0
13	$2^3+2^2+2^0$	1 1 0 1
14	$2^3+2^2+2^1$	1 1 1 0
15	$2^3+2^2+2^1+2^0$	1 1 1 1

1. Symbol 1 is represented by transmitting a pulse of constant amplitude for the duration of the symbol, and symbol 0 is represented by switching off the pulse, as in Fig. 8.4(*a*). This type of signal is referred to as an *on-off signal.*
2. Symbols 1 and 0 are represented by pulses of equal positive and negative amplitudes, as in Fig. 8.4(*b*). This type of signal is referred to as a *polar signal* or *nonreturn-to-zero* (*NRZ*) *signal.*
3. A rectangular pulse (half-symbol wide) is used for a 1 and no pulse for a 0, as in Fig. 8.4(*c*). This type of signal is called a *return-to-zero* (RZ) *signal.*
4. Positive and negative pulses (of equal amplitude) are used alternately for symbol 1, and no pulse for symbol 0, as in Fig. 8.4(*d*). This type of signal is called a *bipolar signal.* A useful property of this method of signaling is that the power spectrum of the transmitted signal has no dc component and relatively insignificant low-frequency components for the case when symbols 1 and 0 occur with equal probability (see Problem 9.15).
5. Symbol 1 is represented by a positive pulse followed by a negative pulse, with both pulses being of equal amplitude and half-symbol wide; for symbol 0, the polarities of these pulses are reversed, as in Fig. 8.4(*e*). This type of signal is called a *split-phase* or *Manchester code.* It suppresses the dc component and has relatively insignificant low-frequency components, regardless of the signal statistics, which is essential in some applications (see Problem 8.3).
6. It is sometimes desirable to encode the information in a binary PCM wave in terms of signal transitions. For example, a transition in a binary PCM wave may be used to designate symbol 0, while no transition is used to designate symbol 1, as illustrated in Fig. 8.4(*f*). This method of representation is called

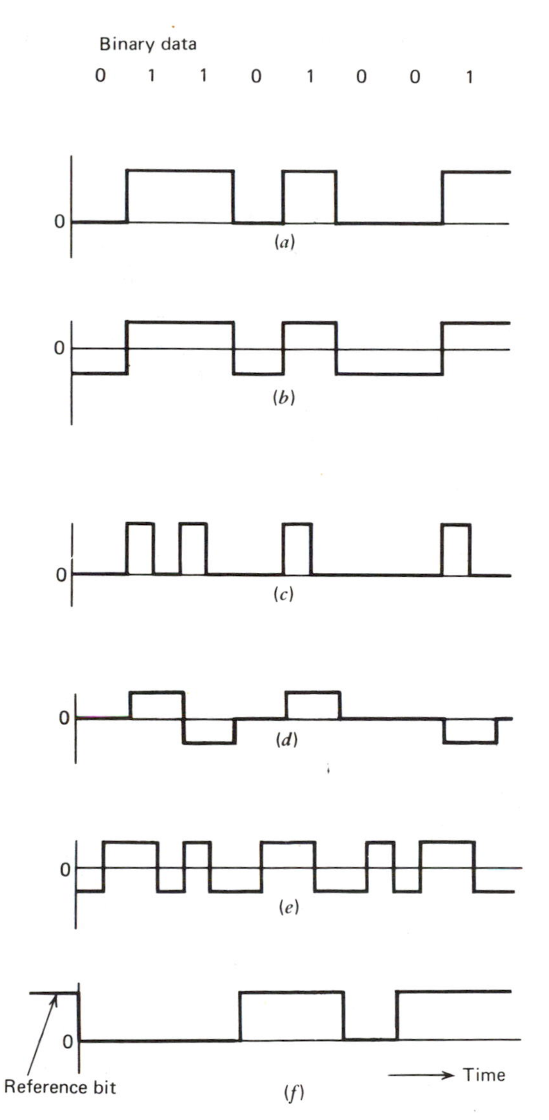

Figure 8.4 Electrical representations of binary data. *(a)* On-off signaling. *(b)* Polar signaling. *(c)* Return-to-zero signaling. *(d)* Bipolar signaling. *(e)* Split phase or Manchester code. *(f)* Differential encoding.

differential encoding. It is apparent that a differentially encoded signal may be inverted without affecting its interpretation. The original binary information is recovered by sampling the received wave and comparing the polarity of adjacent samples to establish whether or not a transition has occurred.

Note that in the waveforms shown in Fig. 8.4 no pulse shaping is used. The issue of pulse shaping and its effect on the transmission of PCM waves are discussed in Chapter 9.

Regeneration

The most important feature of PCM systems lies in the ability to control the effects of distortion and noise produced by transmitting a PCM wave through a channel. This capability is accomplished by reconstructing the PCM wave by means of a chain of *regenerative repeaters* located at sufficiently close spacing along the transmission route. As illustrated in Fig. 8.5, three basic functions are performed by a regenerative repeater, namely, *equalization*, *timing*, and *decision making*. The

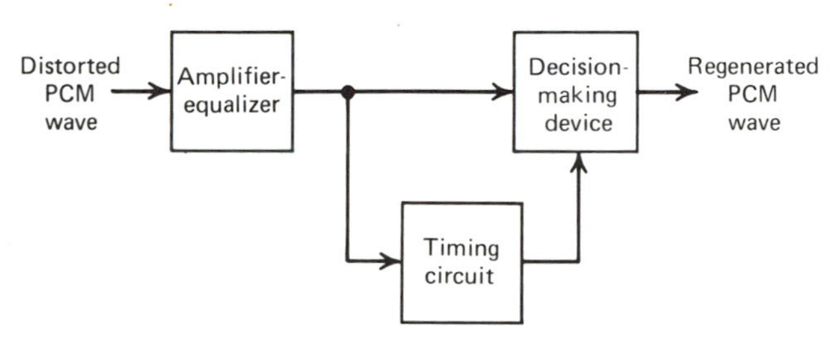

Figure 8.5 Block diagram of a regenerative repeater.

equalizer shapes the received pulses so as to compensate for the effects of amplitude and phase distortions produced by the transmission characteristics of the channel. The timing circuitry provides a periodic pulse train, derived from the received pulses, for sampling the equalized pulses at the instants of time where the signal-to-noise ratio is a maximum. The decision device is enabled, when at the sampling time determined by the timing circuitry, the amplitude of the equalized pulse plus noise exceeds a predetermined voltage level. Thus, for example, in a PCM system with on–off signaling, the repeater makes a decision in each bit interval as to whether or not a pulse is present. If the decision is "yes," a clean new pulse is transmitted to the next repeater. If, on the other hand, the decision is "no," a clean base line is transmitted. In this way, the accumulation of distortion and noise in a repeater span is completely removed, provided that the disturbance is not too large to cause an error in the decision-making process. Ideally, except for delay, the regenerated signal is exactly the same as the signal originally transmitted. In practice, however, the regenerated signal departs from the original signal for two main reasons:

1. The presence of transmission noise and interference causes the repeater to make wrong decisions occasionally, thereby introducing *bit errors* into the regenerated signal; we shall have more to say on this issue in Section 8.2.
2. If the spacing between received pulses deviates from its assigned value, a *jitter* is introduced into the regenerated pulse position, thereby causing distortion.

Decoding

The first operation in the receiver is to regenerate (i.e., reshape and clean up) the received pulses. These clean pulses are then regrouped into code words and decoded (i.e., mapped back) into a quantized PAM signal. The *decoding* process involves generating a pulse the amplitude of which is the linear sum of all the pulses in the code word, with each pulse weighted by its place-value ($2^0, 2^1, 2^2, 2^3, \ldots$) in the code.

Filtering

The final operation in the receiver is to recover the signal wave by passing the decoder output through a low-pass reconstruction filter whose cutoff frequency is equal to the message bandwidth W. Assuming that the transmission path is error-free, the recovered signal includes no noise with the exception of the initial distortion introduced by the quantization process.

Multiplexing

In applications using PCM, it is natural to multiplex different message sources by time-division, whereby each source keeps its individuality throughout the journey from the transmitter to the receiver. This individuality accounts for the comparative ease with which message sources may be dropped or reinserted in a time-division multiplex system. As the number of independent message sources is increased, the time interval that may be allotted to each source has to be reduced, since all of them must be accommodated into a time interval equal to the reciprocal of the sampling rate. This in turn means that the allowable duration of a code word representing a single sample is reduced. However, pulses tend to become more difficult to generate and to transmit as their duration is reduced. Furthermore, if the pulses become too short, impairments in the transmission medium begin to interfere with the proper operation of the system. Accordingly, in practice, it is necessary to restrict the number of independent message sources that can be included within a time-division group.

Synchronization

For a PCM system with time-division multiplexing to operate satisfactorily, it is necessary that the timing operations at the receiver, except for the time lost in transmission and regenerative repeatering, follow closely the corresponding operations at the transmitter. In a general way, this amounts to requiring a local clock at the receiver keep the same time as a distant standard clock at the transmitter, except that the local clock is somewhat slower by an amount corresponding to the time required to transport the message signals from the transmitter to the receiver. One possible procedure to synchronize the transmitter and receiver clocks is to set aside a code element or pulse at the end of a *frame* (consisting of a code word derived from each of the independent message sources in succession) and to transmit this pulse every other frame only. In such a case, the receiver includes a circuit that would search for the pattern of 1's and 0's alternating at half the frame rate, and thereby establish synchronization between the transmitter and receiver.

When the transmission is interrupted, it is highly unlikely that the transmitter and receiver clocks will continue to indicate the same time for long. Accordingly, in carrying out a synchronization process, we must set up an orderly procedure for detecting the synchronizing pulse. The procedure consists of observing the code elements one by one until the synchronizing pulse is detected. That is, after observing a particular code element long enough to establish the absence of the synchronizing pulse, the receiver clock is set back by one code element and the next code element is observed. This searching process is repeated until the synchronizing pulse is detected. Clearly, the time required for synchronization depends on the epoch at which proper transmission is reestablished.

Example 1 The T1 System

In this example we describe the important characteristics of a PCM system known as the *T1 carrier system*, which is designed to accommodate 24 voice channels, primarily for short-distance, heavy usage in metropolitan areas. The T1 system was pioneered by the Bell System in the United States in the early 1960s, and with its introduction the shift to digital communication facilities started.* The T1 system has been adopted for use throughout the United States, Canada, and Japan. It forms the basis for a complete hierarchy of higher order multiplexed systems that are used for either long-distance transmission or transmission in heavily populated urban centers.

A voice signal (male or female) is essentially limited to a band from 300 to 3400 Hz in that frequencies outside this band do not contribute much to articulation efficiency. Indeed, telephone circuits that respond to this range of frequencies give quite satisfactory service. Accordingly, it is customary to pass the voice signal through a low-pass filter with a cutoff frequency of about 3.4 kHz prior to sampling. Hence, with $W = 3.4$ kHz, the nominal value of the Nyquist rate is 6.8 kHz. The filtered voice signal is usually sampled at a slightly higher rate, namely, 8 kHz, which is the *standard* sampling rate in telephone systems.

For companding, the T1 system uses a *piecewise-linear* characteristic (consisting of 15 linear segments) to approximate the logarithmic μ-law of Eq. (8.1) with the constant $\mu = 255$. This approximation is constructed in such a way that the segment-end points lie on the compression curve computed from Eq. (8.1), and their projections onto the vertical axis are spaced uniformly. Table 8.2 gives the projections of the segment-end points onto the horizontal axis, and the step sizes of the individual segments. The table is normalized to 8159, so that all values are represented as integer numbers. Segment 0 of the approximation is a colinear segment, passing through the origin; it contains a total of 30 uniform quantizing levels. Linear segments $1a$, $2a, \ldots, 7a$ lie above the horizontal axis, whereas linear segments $1b, 2b, \ldots, 7b$ lie below the

* For a description of the original version of the T1 PCM system, see K. E. Fultz and D. B. Penick, "T1 carrier system," *Bell System Tech. J.*, vol. 44, pp. 1405–1451, Sept. 1965.
The description given in Example 1 is based on an updated version of this system; see H. H. Henning and J. W. Pan, "D2 channel bank: system aspects," *Bell System Tech. J.*, vol. 51, pp. 1641–1657, Oct. 1972.

Table 8.2 The 15-segment companding characteristic ($\mu = 255$)

Linear segment number	Step size	Projections of segment-end point onto the horizontal axis
0	2	
		± 31
1*a*, 1*b*	4	
		± 95
2*a*, 2*b*	8	
		± 223
3*a*, 3*b*	16	
		± 479
4*a*, 4*b*	32	
		± 991
5*a*, 5*b*	64	
		± 2015
6*a*, 6*b*	128	
		± 4063
7*a*, 7*b*	256	
		± 8159

horizontal axis; each of these 14 segments contains 16 uniform quantizing levels. For colinear segment 0 the quantizing levels at the compressor input are $\pm 1, \pm 3, \ldots, \pm 31$, and the corresponding compressor output levels are $0, \pm 1, \ldots, \pm 15$. For linear segments 1*a* and 1*b*, the quantizing levels at the compressor input are $\pm 31, \pm 35, \ldots, \pm 95$, and the corresponding compressor output levels are $\pm 16, \pm 17, \ldots, \pm 31$, and so on for the other linear segments.

There are a total of $31 + 14 \times 16 = 255$ output levels associated with the 15-segment companding characteristic described above. To accommodate this number of output levels, each of the 24 voice channels uses a binary code with an 8-bit word. The first bit indicates whether the input voice sample is positive or negative; this bit is a 1 if positive and a 0 if negative. The next three bits of the code word identify the particular segment inside which the amplitude of the input voice sample lies, and the last four bits identify the actual quantizing step inside that segment.

With a sampling rate of 8 kHz, each frame of the multiplexed signal occupies a period of 125 μs. In particular, it consists of twenty-four 8-bit words, plus a single bit that is added at the end of the frame for the purpose of synchronization. Hence, each frame consists of a total of $24 \times 8 + 1 = 193$ bits. Correspondingly, the duration of each bit equals 0.647 μs, and the resultant transmission rate is 1.544 megabits per second.

In addition to the voice signal, a telephone system must also pass special supervisory signals to the far end. This *signaling information* is needed to transmit dial pulses, as well as telephone off-hook/on-hook signals. In the T1 system this requirement is accomplished as follows. Every sixth frame, the least significant (that is, the eighth) bit of each voice channel is deleted and a *signaling bit* is inserted in its place, thereby yielding an average $7\frac{5}{6}$-bit operation for each voice input. The sequence of signaling bits is thus transmitted at a rate equal to the sampling rate divided by six, that is, 1.333 kilobits per second.

8.2 NOISE IN PCM SYSTEMS

The performance of a PCM system is influenced by two major sources of noise:

1. *Transmission noise*, which may be introduced anywhere between the transmitter output and the receiver input.
2. *Quantizing noise*, which is introduced in the transmitter and is carried along to the receiver output.

Although these two sources of noise appear simultaneously when the system is operating, we consider them separately, so that we may develop some insight into their individual effects on the receiver performance.

Transmission Noise and Error Probability

The effect of transmission noise is to introduce *bit errors* into the received PCM wave, with the result that, in the case of a binary system, a symbol 1 occasionally is mistaken for a symbol 0, or vice versa. Clearly, the more frequently such errors occur, the more dissimilar the receiver output becomes compared with the original message signal. The fidelity of information transmission by PCM in the presence of transmission noise is conveniently measured in terms of the *error rate* or *probability of error*—the probability that the symbol at the receiver output differs from that transmitted.

Consider a binary-encoded PCM wave $s(t)$ consisting of a sequence of binary digits, in which symbols 0 and 1 are represented by the levels zero and A volts, respectively. This representation corresponds to on-off signaling. The received signal, $x(t)$, consists of the PCM wave $s(t)$ plus a white Gaussian noise, $w(t)$, of zero mean and power spectral density $N_0/2$; that is,

$$x(t) = s(t) + w(t) \tag{8.5}$$

The received signal $x(t)$ is first processed by a low-pass filter of bandwidth B that is large enough to pass the PCM wave essentially unchanged and yet small enough to limit the effects of the noise $w(t)$. Accordingly, we may express the filtered version of the received signal, shown as $y(t)$ in Fig. 8.6, as follows

$$y(t) = s(t) + n(t) \tag{8.6}$$

where the noise $n(t)$ has zero mean and variance $\sigma^2 = N_0 B$.

We are interested in determining, in each bit interval, whether the transmitted symbol was a 0 or a 1. We do this by sampling the filtered signal $y(t)$ every T_b seconds, where T_b is the *bit duration*, and then comparing the value of each sample with some predetermined *threshold*. It is customary to make this decision at the mid-point of each bit interval, for maximum reliability. This assumes that synchronous timing is available in the receiver, so that we have an accurate knowledge of the time interval of occurrence of each symbol in the received waveform.

We assume that binary symbols 0 and 1 occur with equal probability. Intuitively,

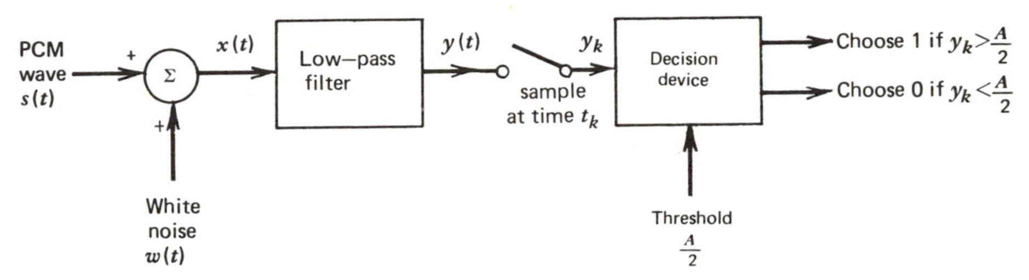

Figure 8.6 Receiver for baseband transmission of binary-encoded PCM wave.

then, it seems reasonable for us to set the threshold at the decision device at $A/2$ volts, so that if the filtered sample value is greater than $A/2$ volts, a decision is made in favor of symbol 1, and if it is less than $A/2$ volts, a decision is made in favor of symbol 0. We adopt the convention that when the sample value equals the threshold $A/2$ volts, the decision as to which symbol was transmitted is made by the flip of a fair coin; clearly, the outcome of such a flip will not alter the average probability of error.

Owing to the presence of the noise $w(t)$ at the front end of the receiver, the decision device will occasionally make errors. Two possible kinds of decision errors can occur in practice:

1. Symbol 1 is chosen when a 0 was actually transmitted.
2. Symbol 0 is chosen when a 1 was actually transmitted.

To determine the average probability of error, we first consider the two possible kinds of error separately. Assume that symbol 0 was transmitted, corresponding to a level of zero volts. The filtered signal $y(t)$ then consists of noise alone, as shown by

$$y(t) = n(t), \qquad \text{symbol 0 was sent} \tag{8.7}$$

Let y_k denote the sample value obtained by observing the sample function $y(t)$ at time $t = t_k$, and let Y_k denote the corresponding random variable. In the case when symbol 0 is transmitted, the probability of error equals the probability of the event $y_k > A/2$ volts. Now from Eq. (8.7) we see that when symbol 0 is transmitted, the random variable Y_k is Gaussian-distributed with zero mean and variance σ^2. Hence, the conditional probability density function of Y_k, given that symbol 0 was transmitted, equals

$$f_{Y_k|0}(y_k|0) = \frac{1}{\sqrt{2\pi}\sigma} \exp\left(-\frac{y_k^2}{2\sigma^2}\right) \tag{8.8}$$

This function is shown sketched in Fig. 8.7(a). Let P_{e0} denote the *conditional probability of error, given that symbol* 0 *was transmitted*. This probability is defined by the area under the $f_{Y_k|0}(y_k|0)$ curve, from $A/2$ to infinity, which corresponds to

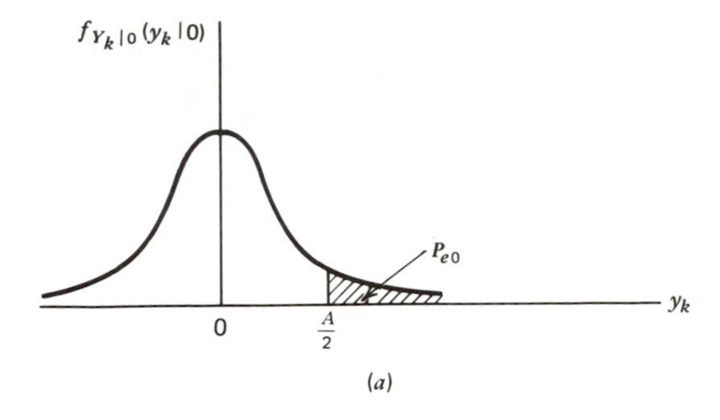

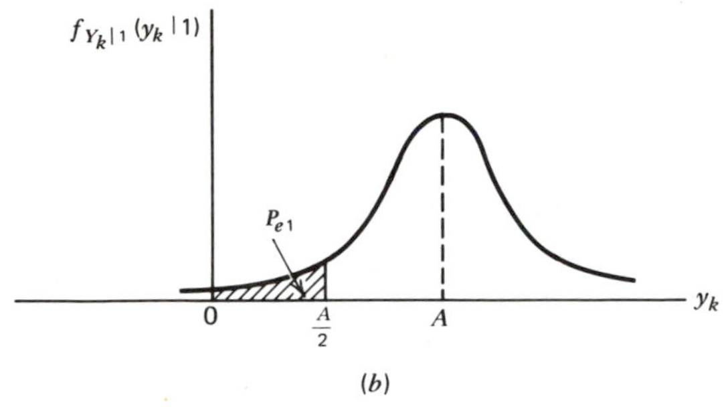

Figure 8.7 Analysis of the effect of transmission noise in a PCM system. *(a)* Probability density function of Y_k when a 0 is transmitted. *(b)* Probability density function of Y_k when a 1 is transmitted.

the shaded area in Fig. 8.7(*a*). That is,

$$P_{e0}=P\left(y_k>\frac{A}{2}\,\middle|\,\text{symbol 0 was transmitted}\right)$$

$$=\int_{A/2}^{\infty} f_{Y_k|0}(y_k|0)\,dy_k$$

$$=\frac{1}{\sqrt{2\pi}\sigma}\int_{A/2}^{\infty}\exp\left(-\frac{y_k^2}{2\sigma^2}\right)dy_k \tag{8.9}$$

Let

$$z=\frac{y_k}{\sqrt{2}\sigma} \tag{8.10}$$

Then we may rewrite Eq. (8.9) as

$$P_{e0}=\frac{1}{\sqrt{\pi}}\int_{A/2\sqrt{2}\sigma}^{\infty}\exp(-z^2)\,dz \tag{8.11}$$

Therefore, using the definition of the *complementary error function* (see Appendix 2):

$$\text{erfc}(u)=\frac{2}{\sqrt{\pi}}\int_u^\infty \exp(-z^2)dz, \tag{8.12}$$

we may express the conditional probability of error P_{e0} as follows:

$$P_{e0}=\frac{1}{2}\,\text{erfc}\left(\frac{A}{2\sqrt{2}\sigma}\right) \tag{8.13}$$

Assume next that symbol 1 was transmitted, corresponding to a level of A volts. The filtered signal equals

$$y(t)=A+n(t), \qquad \text{symbol 1 was sent} \tag{8.14}$$

We now find that the random variable Y_k is Gaussian-distributed with mean A and variance σ^2. Therefore, given that a 1 was transmitted, the conditional probability density function of Y_k is:

$$f_{Y_k|1}(y_{k|1})=\frac{1}{\sqrt{2\pi}\sigma}\exp\left[-\frac{(y_k-A)^2}{2\sigma^2}\right] \tag{8.15}$$

which is sketched in Fig. 8.7(*b*). Let P_{e1} denote the *conditional probability of error, given that symbol* 1 *was transmitted.* This probability is defined by the area under the $f_{Y_k|1}(y_k|1)$ curve from $-\infty$ to $A/2$, as indicated by the shaded area in Fig. 8.7(*b*). That is,

$$\begin{aligned} P_{e1} &= P\left(y_k<\frac{A}{2}\,\middle|\ \text{symbol 1 was transmitted}\right) \\ &= \int_{-\infty}^{A/2} f_{Y_k|1}(y_k|1)dy_k \\ &= \frac{1}{\sqrt{2\pi}\sigma}\int_{-\infty}^{A/2}\exp\left[-\frac{(y_k-A)^2}{2\sigma^2}\right]dy_k \end{aligned} \tag{8.16}$$

Putting $(y_k-A)/\sqrt{2}\sigma=-z$, we find that $P_{e1}=P_{e0}$. This result is a consequence of assuming equally probable symbols and setting the threshold midway between zero and A volts. A channel for which the error probabilities P_{e1} and P_{e0} are equal is said to be *symmetric.*

To determine the average probability of error in the receiver, we note that the two possible kinds of error considered above are mutually exclusive events in that if the receiver, at a particular sampling instant, chooses symbol 1, then symbol 0 is excluded from appearing, and vice versa. Furthermore, P_{e0} and P_{e1} are conditional probabilities with P_{e0} assuming that a 0 was transmitted and P_{e1} assuming that a 1 was transmitted. Thus, assuming that the *a priori probability* of transmitting a 0 is p_0, and the a priori probability of transmitting a 1 is p_1, we find that the *average probability of error* P_e in the receiver is given by

$$P_e=p_0P_{e0}+p_1P_{e1} \tag{8.17}$$

Since $P_{e1}=P_{e0}$ and we have assumed that $p_0=p_1$, we obtain

$$P_e=P_{e1}=P_{e0}$$

or

$$P_e = \frac{1}{2}\,\mathrm{erfc}\left(\frac{A}{2\sqrt{2}\sigma}\right) \tag{8.18}$$

With the transmitted signal alternating between 0 and A volts, we may regard A^2 as the *peak pulse power*. Since σ^2 is the average noise power, we may express the average probability of error P_e in the form

$$P_e = \frac{1}{2}\,\mathrm{erfc}\left(\frac{1}{2}\sqrt{\frac{\gamma}{2}}\right) \tag{8.19}$$

where γ is the *peak pulse-to-noise ratio* at the decoder input, defined by

$$\gamma = \frac{A^2}{\sigma^2} \tag{8.20}$$

We thus see that *the average probability of error in a PCM receiver depends solely on the ratio of the peak pulse power to the average noise power, measured at the decoder input in the receiver*. As shown in Fig. 8.8, the error probability P_e decreases very rapidly as the peak pulse-to-noise ratio γ is increased, so that eventually a very small increase in transmitted pulse power will make the reception of binary pulses almost error free. The nature of this improvement is further emphasized in Table 8.3 where, in the last column, we have assumed a bit rate of 10^5 bits per second.

Clearly, there is an *error threshold* (at about 20 dB, say) below which the receiver performance may involve significant numbers of errors, and above which the effect of transmission noise is practically negligible. In other words, provided that

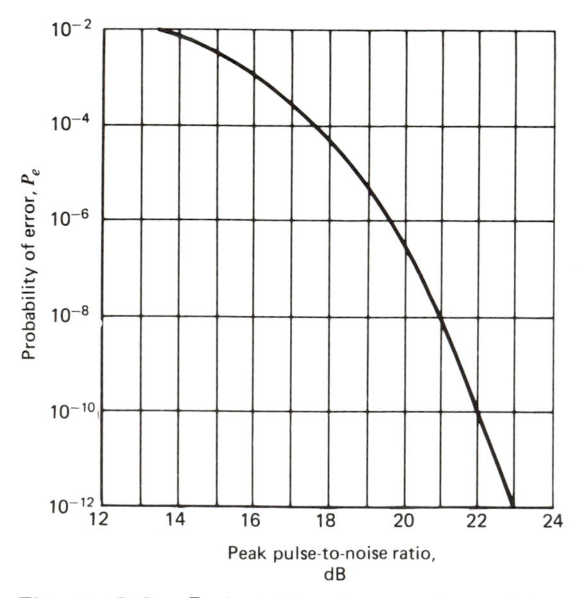

Figure 8.8 Probability of error in a PCM receiver.

Table 8.3

Peak pulse-to-noise ratio γ	Probability of error P_e	For a bit rate of 10^5 bits per second, this is about one error every
13.3 dB	10^{-2}	10^{-3} second
17.4	10^{-4}	10^{-1} second
19.6	10^{-6}	10 seconds
21.0	10^{-8}	20 minutes
22.0	10^{-10}	1 day
23.0	10^{-12}	3 months

the peak pulse-to-noise ratio γ exceeds the error threshold, transmission noise has virtually no effect on the receiver performance, which is precisely the goal of PCM. When, however, γ drops below the error threshold, there is a sharp increase in the rate at which errors occur in the receiver. Because decision errors result in the construction of incorrect code words, we find that when the errors are frequent, the reconstructed message at the receiver output bears little resemblance to the original message.

Comparing the figure of 20 dB for the error threshold in a PCM system with the 60–70 dB required for high-quality transmission of speech using amplitude modulation, we see that PCM requires much less power, even though the average noise power in the PCM system is increased by the n-fold increase in bandwidth, where n is the number of bits in a code word.

In most transmission systems, the effects of noise and distortion from the individual links cumulate. For a given quality of overall transmission, the longer the system, the more severe are the requirements on each link. In a PCM system, however, because the signal can be regenerated as often as necessary, the effects of amplitude, phase, and nonlinear distortions in one link (if not too severe) have practically no effect on the regenerated input signal to the next link. We have also seen that the effect of transmission noise can be made practically negligible by using a peak pulse-to-noise ratio above threshold. For all practical purposes, then, the transmission requirements for a PCM link are almost independent of the total length of the system.

Another important characteristic of a PCM system is its *ruggedness to interference*. We have seen that transmission noise in a PCM system, using on–off signaling, produces no effect unless the peak amplitude is greater than half the pulse height. Similarly, interference caused by stray impulses or crosstalk will produce no effect unless the peak amplitude of this interference plus noise is greater than half the pulse height. Thus the combined presence of transmission noise and interference causes the error threshold necessary for satisfactory operation to increase. If an adequate margin over the error threshold is provided in the first place, however, the system can withstand the presence of relatively large amounts of interference. That is, a PCM system, particularly one that uses a binary code with on–off signaling, is quite *rugged*.

Quantizing Noise

As explained in Section 8.1, quantizing noise is produced in the transmitter end of a PCM system by rounding off the sampled values of a continuous baseband signal to the nearest permitted quantizing level. We assume a quantizing process with a uniform step size denoted by δ volts, so that the quantizing levels are at 0, $\pm\delta$, $\pm 2\delta$, $\pm 3\delta, \ldots$. Consider a particular sample at the quantizer input, with an amplitude that lies in the range $i\delta-(\delta/2)$ to $i\delta+(\delta/2)$, where i is an integer (positive or negative, including zero) and $i\delta$ defines the corresponding quantizer output. We thus have a region of uncertainty of width δ, centered about $i\delta$, as illustrated in Fig. 8.9. Let q_e denote the value of the error produced by the quantizing process. Then the amplitude of the sample at the quantizer input is $i\delta+q_e$. It is apparent that with a random input signal, the quantizing error is a random variable bounded by $-\delta/2 \leqslant q_e \leqslant \delta/2$.

When the quantization is fine enough (say, the number of quantizing levels is greater than 64), the distortion produced by quantizing noise affects the performance of a PCM system as though it were an additive independent source of noise with zero mean and mean-square value determined by the quantizer step size δ. The reason for this is that the power spectral density of the quantizing noise in the receiver output is practically independent of that of the baseband signal over a wide range of input signal amplitudes. Furthermore, for a baseband signal of a root mean-square value that is large compared to a quantum step, it is found that the power spectral density of the quantizing noise has a large bandwidth compared

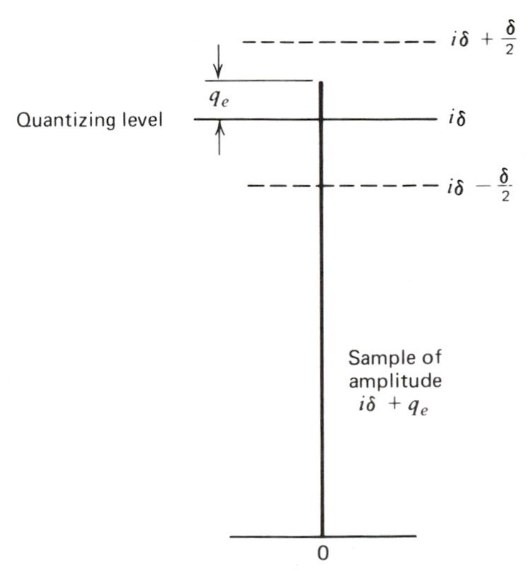

Figure 8.9 Illustrating the quantizing error q_e.

with the signal bandwidth. Thus, with the quantizing noise uniformly distributed throughout the signal band, its interfering effect on a signal is similar to that of thermal noise.*

Let the random variable Q_e denote the quantizing error, and q_e its sample value. Lacking information to the contrary, we assume that the random variable Q_e is uniformly distributed over the possible range $-\delta/2$ to $\delta/2$, as shown by

$$f_{Q_e}(q_e)=\begin{cases}\dfrac{1}{\delta}, & -\dfrac{\delta}{2}\leq q_e\leq\dfrac{\delta}{2}\\ 0, & \text{otherwise}\end{cases} \tag{8.21}$$

For this to be true, we must ensure that the incoming signal does not overload the quantizer. Then the mean of the quantizing error is zero, and its variance is the same as the mean-square value

$$E[Q_e^2]=\int_{-\delta/2}^{\delta/2} q_e^2 f_{Q_e}(q_e)dq_e \tag{8.22}$$

where $f_{Q_e}(q_e)$ is the probability density function of the quantizing error. Substituting Eq. (8.21) in (8.22), we get

$$\begin{aligned}E[Q_e^2]&=\frac{1}{\delta}\int_{-\delta/2}^{\delta/2} q_e^2\, dq_e\\ &=\frac{\delta^2}{12}\end{aligned} \tag{8.23}$$

Let the root mean-square (rms) value of the baseband signal $m(t)$ at the quantizer input be denoted by A_{rms}. The average signal power is A_{rms}^2. When the baseband signal is reconstructed at the receiver output, we obtain the original signal plus quantizing noise. With the average power of the quantizing noise equal to $\delta^2/12$, the output signal-to-noise ratio of the PCM system is therefore

$$(\text{SNR})_O=\frac{A_{\text{rms}}^2}{\delta^2/12} \tag{8.24}$$

Suppose that baseband signal $m(t)$ is modeled as the sample function of a Gaussian random process of zero mean, and that the amplitude range of $m(t)$ at the quantizer input extends from $-4A_{\text{rms}}$ to $4A_{\text{rms}}$. We find that samples of the signal $m(t)$ will fall outside the amplitude range $8A_{\text{rms}}$ with a probability of overload that is less than 1 in 10^4 (see Problem 8.13). If we further assume the use of a binary code with each code word having a length n, so that the number of quantizing levels is 2^n, we find that the resulting quantizer step size is

$$\delta=\frac{8A_{\text{rms}}}{2^n} \tag{8.25}$$

* 1. W. R. Bennett, "Spectra of quantized signals," *Bell System Tech. J.* vol. 27, pp. 446–472, July 1948.
2. H. E. Rowe, *Signals and Noise in Communication Systems*, pp. 311–321 (Van Nostrand, 1965).

Substituting Eq. (8.25) in (8.24), we get

$$(\mathrm{SNR})_{\mathrm{O}}=\tfrac{3}{16}(2^{2n}) \tag{8.26}$$

Expressing the signal-to-noise ratio in decibels:

$$10\log_{10}(\mathrm{SNR})_{\mathrm{O}}=6n-7.2 \tag{8.27}$$

This formula states that each bit in the code word of a PCM system contributes 6dB to the signal-to-noise ratio. It gives a good description of the noise performance of a PCM system, provided that the following conditions are satisfied:

1. The system operates with an average signal power above the error threshold, so that the effect of transmission noise is made negligible, and performance is thereby limited essentially by quantizing noise alone.
2. The quantizing error is uniformly distributed.
3. The quantization is fine enough (say, $n>6$) to prevent signal-correlated patterns in the quantizing error waveform.
4. The quantizer is aligned with the amplitude range from $-4A_{\mathrm{rms}}$ to $4A_{\mathrm{rms}}$.

In general, conditions (1) through (3) are true of toll quality voice signals. However, when demands on voice quality are not severe, we may use a coarse quantizer corresponding to $n \leqslant 6$. In such a case, degradation in system performance is reflected not only by a lower signal-to-noise ratio, but also by an undesirable presence of signal-dependent patterns in the waveform of quantizing error.

Example 2 Sinusoidal modulating wave

Consider the special case of a full-load sinusoidal modulating wave of amplitude A_m, which utilizes all the quantizing levels provided. The average signal power is $A_m^2/2$. The total range of the quantizer input is $2A_m$, because the modulating wave swings between $-A_m$ to A_m. Assuming the use of an n-bit binary code word, the number of quantizing levels L equals 2^n. The quantizer step size is

$$\delta=\frac{2A_m}{L}=A_m 2^{1-n}$$

The average quantizing noise power is therefore

$$E[Q_e^2]=\frac{\delta^2}{12}=\frac{1}{3}A_m^2 2^{-2n}$$

Thus the output signal-to-noise ratio of the PCM system, for a full-load test tone, is

$$(\mathrm{SNR})_{\mathrm{O}}=\frac{A_m^2/2}{A_m^2 2^{-2n}/3}=\frac{3}{2}(2^{2n}) \tag{8.28}$$

Expressing the signal-to-noise ratio in decibels, we get

$$10\log_{10}(\mathrm{SNR})_{\mathrm{O}}=1.8+6n \tag{8.29}$$

Table 8.4

Number of quantizing levels, L	Number of binary digits, n	Signal-to-noise ratio (dB) Sinusoidal modulating wave (Eq. 8.29)	Signal-to-noise ratio (dB) Gaussian modulating wave (Eq. 8.27)
32	5	31.8	22.8
64	6	37.8	28.8
128	7	43.8	34.8
256	8	49.8	40.8

For various values of L and n, the corresponding values of signal-to-noise ratio are as given in Table 8.4. In this table we have also included the corresponding set of values obtained from Eq. (8.27), assuming a Gaussian distribution for the input signal. From Table 8.4 we can make a quick estimate of the number of bits required for a particular PCM system provided that we have some idea of the required signal-to-noise ratio for a full-load test tone or Gaussian input.

We have seen that a message signal of bandwidth W requires a minimum sampling rate of $2W$. With each signal sample represented by an n-bit binary code word, the bit duration T_b equals a maximum value of $1/2nW$. In Section 9.2 it is shown that the bandwidth B_T required to transmit a rectangular pulse of this duration is given by

$$B_T = \kappa n W \tag{8.30}$$

where κ is a constant with a value between 1 and 2. Accordingly, using Eqs. (8.28) and (8.30), we see that a PCM system is capable of improving the output signal-to-noise ratio exponentially with the bandwidth expansion ratio B_T/W, as shown by

$$(\mathrm{SNR})_O = \tfrac{3}{2}(4^{B_T/\kappa W}) \tag{8.31}$$

It is interesting to compare Eq. (8.31) with the corresponding result we obtained in Example 3 of Chapter 6 for the case of frequency modulation. In wide-band FM, the improvement in signal-to-noise ratio produced by increased transmission bandwidth effectively follows a square law. That is, by doubling the bandwidth in an FM system which operates above threshold, the signal-to-noise ratio is improved by 6 dB. In binary PCM, doubling the bandwidth permits twice the number of binary digits n in a code word, and therefore increases the signal-to-noise ratio by $6n$ dB. It follows therefore that FM is much less efficient than PCM in exchanging increased bandwidth for improved signal-to-noise ratio.

8.3 MEASURE OF INFORMATION

In order to assess the capacity of a PCM system to transmit information from source to destination, and compare its performance not only with other modulation systems but also an *ideal* communication system, we need a quantitative measure of the information contained in message signals.* In this section we define a measure of information and in the next section we show how this measure is related to bandwidth and signal-to-noise ratio for an ideal channel.

Entropy of Zero-Memory Information Source

Suppose that we have a source emitting a sequence of symbols from a finite source alphabet $S=\{s_1, s_2, \ldots, s_K\}$. We assume that successive symbols emitted from the source are statistically independent. Such an information source is called a *zero-memory source*, and is completely described by the source alphabet S and the probabilities with which the symbols occur, namely, $P(s_1), P(s_2), \ldots, P(s_K)$.

If we are told that symbol s_k, $k=1, 2, \ldots K$, has occurred, then, by definition, we say that we have received

$$I(s_k)=\log \frac{1}{P(s_k)}= -\log P(s_k) \tag{8.32}$$

units of information.†

The base of the logarithm in Eq. (8.32) is quite arbitrary. Nevertheless, it is standard today to use a logarithm to base 2. The resulting unit of information is called the *bit* (a contraction of *bi*nary uni*t*). We thus write

$$I(s_k)= -\log_2 P(s_k) \text{ bits} \tag{8.33}$$

Note that a bit also refers to a binary digit. Ordinarily, we are able to determine which meaning is applicable from the context.

The probability that symbol s_k will occur is $P(s_k)$, so that the *average* amount of information obtained per symbol from the source equals

$$\sum_{k=1}^{K} P(s_k)I(s_k)= -\sum_{k=1}^{K} P(s_k)\log_2 P(s_k) \text{ bits}$$

* The origins of *information theory* date back to Shannon's publication of a paper in 1948:
C. E. Shannon, "A mathematical theory of communication," *Bell System Tech. J.*, vol. 27, pp. 379–423, July 1948, and pp. 623–656, Oct. 1948.
See also C. E. Shannon, "Communication in the presence of noise," *Proc. IRE*, vol. 37, pp. 10–21, Jan. 1949.
For books on information theory and the closely related subject of coding, see

1. N. Abramson, *Information Theory and Coding* (McGraw-Hill, 1963).
2. J. M. Wozencraft and I. M. Jacobs, *Principles of Communication Engineering*, (Wiley, 1965).
3. F. Jelinek, *Probabilistic Information Theory*, (McGraw-Hill, 1968).
4. R. G. Gallager, *Information Theory and Reliable Communication*, (Wiley, 1968).
5. T. Berger, *Rate Distortion Theory*, (Prentice-Hall, 1971).
6. R. W. Hamming, *Coding and Information Theory*, (Prentice-Hall, 1980).
7. A. J. Viterbi and J. K. Omura, *Principles of Digital Communication and Coding*, (McGraw-Hill, 1979).

† The use of a logarithmic measure of information was first suggested by Hartley in 1928; see: R. V. L. Hartley, "Transmission of information," *Bell System Tech. J.*, vol. 7, pp. 535–563, 1928. Hartley used logarithms to base 10, and so the corresponding unit of information is called the *hartley*.

This quantity, *the average amount of information per source symbol*, is called the *entropy* $H(S)$ of the zero-memory source, as shown by*

$$H(S) = -\sum_{k=1}^{K} P(s_k)\log_2 P(s_k) \text{ bits} \tag{8.34}$$

Example 3 Entropy of Binary Source

Consider a binary source for which symbol 1 occurs with probability p_1 and symbol 0 with probability $p_0 = 1 - p_1$. We assume that the source is a zero-memory source, so that successive symbols emitted by the source are statistically independent.

The entropy of such a source equals

$$\begin{aligned} H(S) &= -p_0 \log_2 p_0 - p_1 \log_2 p_1 \\ &= -(1-p_1)\log_2(1-p_1) - p_1 \log_2 p_1 \end{aligned} \tag{8.35}$$

We note that:

1. When $p_1 = 0$, the entropy $H(S) = 0$. This follows from the fact that $x \log x \to 0$ as $x \to 0$.
2. When $p_1 = 1$, the entropy $H(S) = 0$.
3. The entropy $H(S)$ attains its maximum value, $H_{\max} = 1$ bit, when $p_1 = p_0 = 1/2$, that is, symbols 1 and 0 are equally probable. In this case the transmission of a single binary digit (in the form of symbol 1 or 0) represents one bit of information.

Thus the entropy of a zero-memory binary source varies with p_1 as shown in Fig. 8.10.

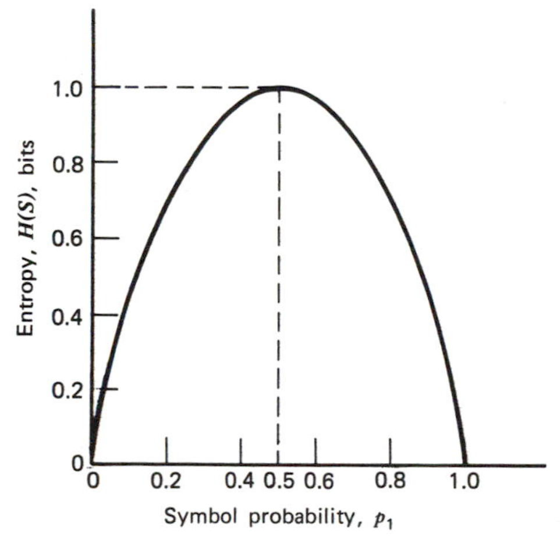

Figure 8.10 Entropy of a zero-memory binary source.

* For a discussion of the relation of the entropy of information theory to the entropy of statistical thermodynamics, see:
L. Brillouin, *Science and Information Theory*, Second Edition (Academic Press, 1962).

Entropy of Continuous Distribution

Equation (8.34) defines the entropy of a set of probabilities $P(s_1), P(s_2), \ldots, P(s_K)$. In an analogous manner we define the entropy of a continuous random variable X with the probability density function $f_X(x)$ by

$$H(X) = -\int_{-\infty}^{\infty} f_X(x)\log_2[f_X(x)]dx \text{ bits} \tag{8.36}$$

When we have a random vector $\mathbf{X}$ consisting of n random variables $X_1, X_2, \ldots, X_n$, we define the entropy of the random vector $\mathbf{X}$ by the n-fold integral

$$H(\mathbf{X}) = -\int_{-\infty}^{\infty} f_{\mathbf{X}}(\mathbf{x})\log_2[f_{\mathbf{X}}(\mathbf{x})]d\mathbf{x} \text{ bits} \tag{8.37}$$

where $f_{\mathbf{X}}(\mathbf{x})$ is the joint probability density function of the random vector $\mathbf{X}$.

Example 4 Maximum entropy for specified mean square value

In this example we use the formula of Eq. (8.36) to determine the form of $f_X(x)$ that gives a maximum entropy subject to two constraints:

1. $$\int_{-\infty}^{\infty} f_X(x)dx = 1 \tag{8.37}$$

   This condition merely states that the total area under the probability density function $f_X(x)$ must equal one.

2. $$\int_{-\infty}^{\infty} x^2 f_X(x)dx = \text{constant} \tag{8.38}$$

   This condition states that the mean-square value of the random variable X is a constant.

The entropy $H(X)$ will attain its maximum value when the integral

$$\int_{-\infty}^{\infty} [-f_X(x)\log_2 f_X(x) + \lambda_1 x^2 f_X(x) + \lambda_2 f_X(x)]dx$$

is stationary, where λ_1 and λ_2 are constants. That is to say, when the derivative of the integrand

$$-f_X(x)\log_2 f_X(x) + \lambda_1 x^2 f_X(x) + \lambda_2 f_X(x)$$

with respect to $f_X(x)$ is zero. This yields the result

$$-\log_2 e + \lambda_1 x^2 + \lambda_2 = \log_2 f_X(x) = \log_2 e \ln f_X(x)$$

where $e = 2.71828$. Solving for $f_X(x)$, we get

$$f_X(x) = \exp\left(-1 + \frac{\lambda_2}{\log_2 e} + \frac{\lambda_1 x^2}{\log_2 e}\right) \tag{8.39}$$

Note that the constant λ_1 has to be negative if the integrals of $f_X(x)$ and $x^2 f_X(x)$ are to converge. Substituting Eq. (8.39) into (8.37) and (8.38), and solving for λ_1 and λ_2, we get

$$\lambda_1 = -\frac{\log_2 e}{2\sigma^2}$$

and

$$\lambda_2 = \log_2\left(\frac{e}{\sqrt{2\pi}\sigma}\right)$$

where

$$\int_{-\infty}^{\infty} x^2 f_X(x)dx = \sigma^2$$

The final result is

$$f_X(x) = \frac{1}{\sqrt{2\pi}\sigma} \exp\left(-\frac{x^2}{2\sigma^2}\right) \tag{8.40}$$

We thus see that a Gaussian distribution yields the maximum entropy for a given mean-square value or variance. Substituting Eq. (8.40) in (8.36) gives the entropy of a Gaussian distribution of zero mean and variance σ^2 as

$$\begin{aligned} H(X) &= \int_{-\infty}^{\infty} f_X(x)\left[\log_2(\sqrt{2\pi}\sigma) + \frac{x^2}{2\sigma^2}\log_2 e\right]dx \\ &= \log_2(\sqrt{2\pi}\sigma) + \frac{\sigma^2}{2\sigma^2}\log_2 e \\ &= \log_2(\sqrt{2\pi e}\sigma) \text{ bits} \end{aligned} \tag{8.41}$$

8.4 CHANNEL CAPACITY

We define *channel capacity* as the maximum rate of reliable information transmission over a channel. The importance of the concept of channel capacity stems from a remarkable theorem called the *capacity theorem* due to Shannon.

The capacity theorem is usually stated for a *discrete memoryless channel.* A discrete channel is one for which the input consists of a symbol belonging to an alphabet of K symbols and the output consists of a symbol belonging to the same alphabet of K input symbols. Due to errors in the channel, the output symbol may be different from the input symbol. We say that the channel is memoryless when the channel output at a given time is a function of the channel input at that time and is not a function of previous channel inputs. Let C be the capacity of a discrete memoryless channel and let H be the entropy of a discrete information source emitting r symbols per second. *The capacity theorem states that if $rH \leqslant C$, then there exists a coding scheme such that the output of the source can be transmitted over the channel with an arbitrarily small probability of error. A converse of this theorem states that it is not possible to transmit messages without error if $rH > C$.*

Thus the capacity theorem predicts essentially error-free transmission in the presence of additive noise. Unfortunately, the theorem tells us only of the existence of codes and says nothing of how to construct them.

Gaussian Channel Capacity

Another important result due to Shannon defines the channel capacity of a *continuous channel* having an average power limitation and perturbed by an additive band-limited white Gaussian noise. In a continuous channel the input and output signals are continuous functions of time.

Consider then a channel of bandwidth B_T hertz, and let $x(t)$ denote the input or transmitted signal and $y(t)$ denote the output or received signal. We assume that the transmitted signal $x(t)$ is noise-like, i.e., the sample function of a zero-mean white Gaussian noise process, band-limited to B_T hertz. Let P denote the average power of the transmitted signal. At the receiving end of the channel, $x(t)$ is perturbed by the sample function of an independent zero-mean white Gaussian noise process that is also band-limited to B_T hertz. Let N denote the average power of this additive noise component. Thus the average power of the received signal $y(t)$ equals $P+N$.

Since both $x(t)$ and $y(t)$ are assumed to have a power spectral density that is constant for $-B_T \leqslant f \leqslant B_T$ and zero elsewhere, we note that:

1. They may be completely described by their respective samples taken at the Nyquist rate of $1/2B_T$ samples per second in accordance with the sampling theorem.
2. The samples are uncorrelated, and being Gaussian with zero mean, they are statistically independent. This ensures that the maximum information transfer will take place.

Let Y denote a sample of the received signal $y(t)$. Since Y has a Gaussian distribution with zero mean and variance (average power) equal to $P+N$, we find from Eq. (8.41) that the entropy of the received signal (measured in bits per sample) equals

$$H(Y)=\log_2\sqrt{2\pi e(P+N)} \text{ bits/sample} \tag{8.42}$$

The noisy channel extracts its toll in terms of errors. Accordingly, we find that the extropy $H(X)$ of a sample X of the transmitted signal differs from $H(Y)$. The difference equals the *conditional entropy* $H(Y|X)$, which is a measure of the average uncertainty of the received sample Y, given that the sample X was transmitted. We may thus write

$$H(X)=H(Y)-H(Y|X) \text{ bits/sample} \tag{8.43}$$

The conditional entropy $H(Y|X)$ is sometimes called the *equivocation*. It is defined by

$$H(Y|X)=-\int_{-\infty}^{\infty}\int_{-\infty}^{\infty} f_{X,Y}(x,y)\log_2\left[f_{Y|X}(y|x)\right]dx\,dy \tag{8.44}$$

where $f_{X,Y}(x, y)$ is the joint probability density function of X and Y, and $f_{Y|X}(y|x)$ is the conditional probability density function of Y, given X. From Eq. (5.20) we have

$$f_{X,Y}(x, y) = f_{Y|X}(y|x) f_X(x) \tag{8.45}$$

Hence, we may rewrite Eq. (8.44) as

$$H(Y|X) = -\int_{-\infty}^{\infty} dx f_X(x) \int_{-\infty}^{\infty} f_{Y|X}(y|x) \log_2 [f_{Y|X}(y|x)] dy \tag{8.46}$$

When the additive noise at the channel output is independent of the transmitted signal, we find that $f_{Y|X}(y|x)$ is a function of only the noise term or difference $(y-x)$, and $f_{Y|X}(y|x)$ does not depend on either x or y except in the combination $(y-x)$. Accordingly, in this case we find from Eq. (8.46) that the equivocation $H(Y|X)$ equals the entropy of the additive noise. Furthermore, since the noise is a Gaussian process of zero mean and variance (average power) equal to N, the entropy of a noise sample equals $\log_2 \sqrt{2\pi e N}$ bits. Hence, we have

$$H(Y|X) = \log_2 \sqrt{2\pi e N} \text{ bits/sample} \tag{8.47}$$

Thus from Eqs. (8.43) and (8.47) we find that the average transmitted information (measured in bits per sample) equals

$$\begin{aligned} I(X; Y) &= H(Y) - H(Y|X) \\ &= \log_2 \sqrt{2\pi e(P+N)} - \log_2 \sqrt{2\pi e N} \\ &= \tfrac{1}{2} \log_2 \left(1 + \frac{P}{N}\right) \text{ bits/sample} \end{aligned} \tag{8.48}$$

Since samples are taken at the maximum rate of $2B_T$ samples per second, it follows that for a white band-limited Gaussian channel the channel capacity is

$$\begin{aligned} C &= 2B_T I(X; Y) \\ &= B_T \log_2 \left(1 + \frac{P}{N}\right) \text{ bits/s} \end{aligned} \tag{8.49}$$

This result is called the *Hartley–Shannon law*. It is so named in recognition of the early work by Hartley on the subject and its rigorous derivation by Shannon.

The Hartley–Shannon law has two important implications:

1. It gives the upper limit for the rate of reliable information transmission over a Gaussian channel.
2. For a specified channel capacity, it defines the way in which transmission bandwidth B_T may be traded off for improved signal-to-noise ratio P/N, and vice versa.

It is important to realize, however, that the result given in Eq. (8.49) is derived for a white band-limited Gaussian channel. Nevertheless, this restriction does not diminish the usefulness of the Hartley–Shannon law for two reasons:

1. In general, most physical channels are at least approximately Gaussian.

2. It has been shown that the result obtained for the Gaussian channel provides a *lower bound* on the performance of a system operating over a non-Gaussian channel.

Example 5 The Shannon bound.

In this example we wish to determine the bound on the channel capacity C as the transmission bandwidth B_T is increased without limit.

For white noise of power spectral density $N_0/2$, limited to the frequency interval $-B_T \leq f \leq B_T$, the average power is

$$N = N_0 B_T$$

Substituting this result in Eq. (8.49) yields

$$C = B_T \log_2\left(1 + \frac{P}{N_0 B_T}\right) \text{ bits/s} \tag{8.50}$$

Before taking the limit, it is convenient to rewrite Eq. (8.50) in the form

$$C = \frac{P}{N_0} \log_2\left[\left(1 + \frac{P}{N_0 B_T}\right)^{N_0 B_T/P}\right]$$

The requirement is therefore to determine the limit

$$\lim_{B_T \to \infty} \frac{P}{N_0} \log_2\left[\left(1 + \frac{P}{N_0 B_T}\right)^{N_0 B_T/P}\right]$$

We may now use the relation

$$\lim_{x \to 0} (1 + x)^{1/x} = e$$

Putting $x = P/N_0 B_T$, we thus get the desired result

$$\lim_{B_T \to \infty} C = \frac{P}{N_0} \log_2 e \tag{8.51}$$

This result is known as the *Shannon bound*. It defines the maximum rate of information transmission for a communication system of given average transmitted power but no bandwidth limitation. In Chapter 10 we describe a method of signaling that approaches the Shannon bound in the limit.

8.5 CHANNEL CAPACITY OF A PCM SYSTEM

Consider a PCM system operating over the error threshold, so that the probability of error due to transmission noise is negligible. The system is used to transmit an analog signal $m(t)$ with W hertz as its highest frequency component. The signal is sampled at the Nyquist rate of $2W$ samples per second. Let L denote the number of amplitude levels used in the quantization process. We assume that the L quantizing levels are equally likely, so that the probability of occurrence of any one of them is

$1/L$. It follows therefore that the amount of information carried by a single sample of the signal is $\log_2 L$ bits. With a maximum sampling rate of $2W$ samples per second, the maximum rate of information transmission or channel capacity of the PCM system is

$$C = 2W \log_2 L \text{ bits/s} \tag{8.52}$$

We assume that the PCM system uses a code word consisting of n code elements, each having one of M possible discrete amplitude values. By using such a code, we have M^n different possible code words. For a unique encoding process, we require

$$L = M^n \tag{8.53}$$

Clearly, the rate of information transmission in the system is unaffected by the use of an encoding process. We may therefore eliminate L between Eqs. (8.52) and (8.53) to obtain

$$C = 2Wn \log_2 M \tag{8.54}$$

In order to provide an adequate noise margin, there must be a certain separation between the M discrete amplitude levels used to represent the elements of the code word. Call this separation $k\sigma$, where k is a constant and σ is the root mean-square amplitude of the noise. The number of amplitude levels M is usually an integer power of two, and transmitted signal power will be least if the amplitude range is symmetrical about zero; that is, if the discrete amplitude levels have the values $\pm k\sigma/2, \pm 3k\sigma/2, \ldots, \pm(M-1)(k\sigma/2)$. Assuming that these M different levels are equally likely, we find that the average transmitted signal power is

$$\begin{aligned} P &= \frac{2}{M}\left[\left(\frac{k\sigma}{2}\right)^2 + \left(\frac{3k\sigma}{2}\right)^2 + \cdots + (M-1)^2\left(\frac{k\sigma}{2}\right)^2\right] \\ &= \frac{k^2\sigma^2}{12}(M^2-1) \\ &= \frac{k^2 N}{12}(M^2-1) \end{aligned} \tag{8.55}$$

where $N = \sigma^2$ is the average noise power. Solving Eq. (8.55) for M, we get

$$M = \left(1 + \frac{12P}{k^2 N}\right)^{1/2} \tag{8.56}$$

Therefore, substituting Eq. (8.56) in (8.54), we obtain

$$C = Wn \log_2\left(1 + \frac{12P}{k^2 N}\right) \tag{8.57}$$

For a message bandwidth W and code word of length n, the minimum transmission bandwidth B_T equals nW, which is obtained by putting the constant κ in Eq. (8.30) equal to unity. We may thus rewrite Eq. (8.57) as:

$$C = B_T \log_2\left(1 + \frac{12P}{k^2 N}\right) \tag{8.58}$$

Comparing Eq. (8.58) with the Hartley–Shannon law, described by Eq. (8.49), we see that they are identical if the transmitted signal power is increased by the factor $k^2/12$, compared with the ideal system. The parameter k determines the separation between the discrete amplitude levels used to represent the elements of a code word, so that as k increases, the error rate of the system decreases. From Fig. 8.8 or Table 8.3, we note that k is about 10 for an average error rate of 1 in 10^6 bits. For this value of k, a binary PCM system requires approximately eight times as much power as the ideal system for the same channel capacity.

Perhaps the most interesting thing to note about Eq. (8.58) is that the form is right. Power and bandwidth are exchanged on a logarithmic basis, and the channel capacity is proportional to the transmission bandwidth B_T.

Example 6

The received signal power P is related to the *signal energy per bit*, E_b, as follows:

$$P = \frac{E_b}{T_b}$$

where T_b is the bit duration. Assuming a noise spectral density of $N_0/2$ (defined for both positive and negative frequencies), the average noise power N in a bandwidth B_T is given by

$$N = N_0 B_T$$

We may therefore express the received signal-to-noise ratio as

$$\begin{aligned} \frac{P}{N} &= \frac{E_b/T_b}{N_0 B_T} \\ &= \frac{E_b/N_0}{T_b B_T} \end{aligned} \tag{8.59}$$

Substituting Eq. (8.59) in (8.49), and assuming communication at channel capacity (i.e., $C = 1/T_b$), we find that for the ideal system,

$$\frac{1}{T_b B_T} = \log_2\left(1 + \frac{E_b/N_0}{T_b B_T}\right) \tag{8.60}$$

The term $1/T_b B_T$ is called *bandwidth efficiency*; it is measured in bits per second per hertz. For the ideal system, the bandwidth efficiency is shown plotted versus the ratio E_b/N_0 in Fig. 8.11. We see that as the ratio E_b/N_0 decreases, the transmission bandwidth B_T necessary to achieve 100 percent efficiency (i.e., communication at the channel capacity C) increases. In the limit when the bandwidth B_T approaches infinity, the ratio E_b/N_0 approaches the limiting value of ln 2 or -1.6 dB.

Similarly, using Eq. (8.58), we find that the corresponding relation between the ratio E_b/N_0

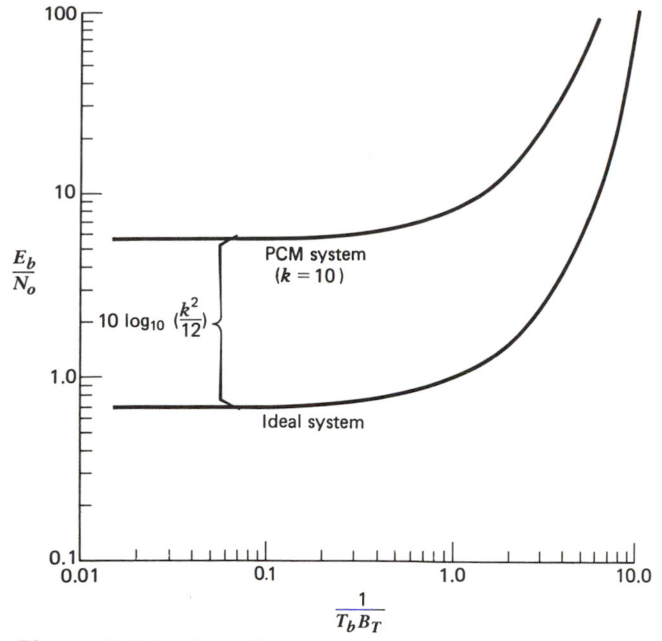

Figure 8.11 The effect of varying the ratio E_b/N_0 on the bandwidth efficiency $1/T_bB_T$ for the ideal transmission system or a PCM system.

and bandwidth efficiency $1/T_bB_T$ for a PCM system is as follows:

$$\frac{1}{T_B B_T} = \log_2\left(1 + \frac{12}{k^2}\frac{E_b/N_0}{T_b B_T}\right) \tag{8.61}$$

This relation is shown plotted in Fig. 8.11 for the case when the parameter k is equal to 10. As indicated earlier, the product $k\sigma$ defines the separation between the discrete amplitude levels used to represent the elements of the code word. Figure 8.11 clearly shows that, for a given value of bandwidth efficiency $1/T_bB_T$, the PCM system requires a value of E_b/N_0 greater than that for the ideal system by an amount equal to 10 $\log_{10}$ $(k^2/12)$ decibels.

8.6 DIFFERENTIAL PULSE-CODE MODULATION (DPCM)

When a voice or video signal is sampled at a rate slightly higher than the Nyquist rate, the resulting sampled signal is found to exhibit a high correlation between adjacent samples. The meaning of this high correlation is that, in an average sense, the signal does not change rapidly from one sample to the next with the result that the difference between adjacent samples has a variance that is smaller than the variance of the signal itself (see Problem 8.21). When these highly correlated samples are encoded, as in a standard PCM system, the resulting encoded signal contains *redundant information.* This means that symbols that are not absolutely essential to the transmission of information are generated as a result of the encoding process. By removing this redundancy before encoding, we obtain a more efficient coded signal.

Now, if we know a sufficient part of a redundant signal, we may infer the rest, or at least make the most probable estimate. In particular, if we know the past behavior of a signal up to a certain point in time, it is possible to make some inference about its future values; such a process is commonly called *prediction*. Suppose then a baseband signal $m(t)$ is sampled at the rate $1/T_s$ to produce a sequence of correlated samples T_s seconds apart; which is denoted by $\{m(nT_s)\}$. The fact that it is possible to predict future values of the signal $m(t)$ provides motivation for the *differential quantization* scheme shown in Fig. 8.12(*a*). In this scheme the input to the quantizer is a signal

$$e(nT_s) = m(nT_s) - \hat{m}(nT_s) \tag{8.62}$$

which is the difference between the unquantized input sample $m(nT_s)$ and a prediction of it, denoted by $\hat{m}(nT_s)$. This predicted value is produced by using a *prediction filter* whose input, as we will see, consists of a quantized version of the input signal

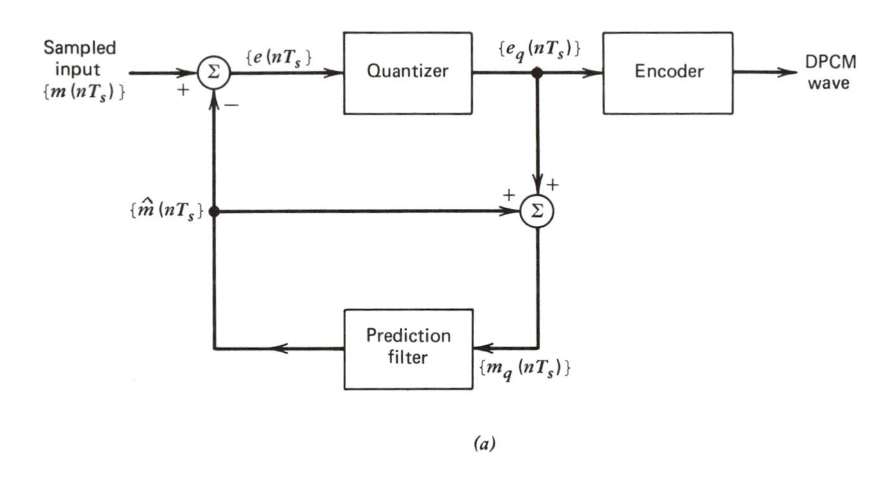

(*a*)

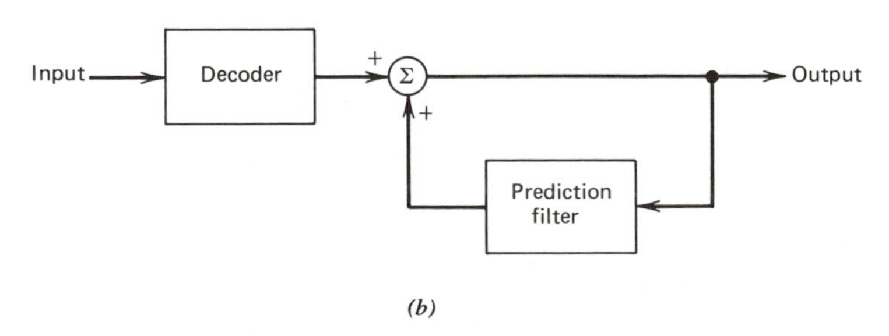

(*b*)

Figure 8.12 DPCM system. (*a*) Transmitter. (*b*) Receiver.

$\{m(nT_s)\}$. The difference signal $\{e(nT_s)\}$ is called a *prediction error*, since it is the amount by which the prediction filter fails to predict the input exactly.

By encoding the quantizer output, as in Fig. 8.12(a), we obtain a variation of PCM, which is known as *differential pulse-code modulation* (DPCM).* It is this encoded signal that is used for transmission.

The quantizer output may be represented as

$$e_q(nT_s) = e(nT_s) + q_e(nT_s) \tag{8.63}$$

where $q_e(nT_s)$ is the quantizing error. According to Fig. 8.12(a), the quantizer output $e_q(nT_s)$ is added to the predicted value $\hat{m}(nT_s)$ to produce the prediction-filter input

$$m_q(nT_s) = \hat{m}(nT_s) + e_q(nT_s) \tag{8.64}$$

Substituting Eq. (8.63) in (8.64), we get

$$m_q(nT_s) = \hat{m}(nT_s) + e(nT_s) + q_e(nT_s) \tag{8.65}$$

However, from Eq. (8.62) we observe that $\hat{m}(nT_s) + e(nT_s)$ is equal to the input signal $m(nT_s)$. Therefore, we may rewrite Eq. (8.65) as follows

$$m_q(nT_s) = m(nT_s) + q_e(nT_s) \tag{8.66}$$

which represents a quantized version of the input signal $m(nT_s)$. That is, irrespective of the properties of the prediction filter, the quantized signal, $m_q(nT_s)$, at the prediction filter input differs from the original input signal $m(nT_s)$ by the quantizing error $q_e(nT_s)$. Accordingly, if the prediction is good, the variance of the prediction error $e(nT_s)$ will be smaller than the variance of $m(nT_s)$, so that a quantizer with a given number of levels can be adjusted to produce a quantizing error with a smaller variance than would be possible if the input signal $m(nT_s)$ were quantized directly as in a standard PCM system.

The receiver for reconstructing the quantized version of the input is shown in Fig. 8.12(b). It consists of a decoder to reconstruct the quantized error signal. The quantized version of the original input is reconstructed from the decoder output using the same prediction filter as used in the transmitter of Fig. 8.12(a). In the absence of transmission noise, we find that the encoded signal at the receiver input is identical to the encoded signal at the transmitter output. Accordingly, the corresponding receiver output is equal to $m_q(nT_s)$, which differs from the original input $m(nT_s)$ only by the quantizing error $q_e(nT_s)$ incurred as a result of quantizing the prediction error $e(nT_s)$.

From the above analysis we observe that, in a noise-free environment, the prediction filters in the transmitter and receiver operate on the same sequence of samples, $m_q(nT_s)$. It is with this purpose in mind that a feedback path is added to the quantizer in the transmitter, as shown in Fig. 8.12(a).

* Differential pulse-code modulation was invented by Cutler; the invention is described in a patent issued in 1952: C. C. Cutler, "Differential Quantization of Communication Signals," United States Patent, No. 2 605 361 (1952).

The output signal-to-noise ratio of the DPCM system shown in Fig. 8.12 is, by definition,

$$(\mathrm{SNR})_O = \frac{\sigma_M^2}{\sigma_Q^2} \tag{8.67}$$

where σ_M^2 is the variance of the original input $m(nT_s)$, assumed to be of zero mean, and σ_Q^2 is the variance of the quantizing error $q_e(nT_s)$. We may rewrite Eq. (8.67) as

$$\begin{aligned}(\mathrm{SNR})_O &= \left(\frac{\sigma_M^2}{\sigma_E^2}\right)\left(\frac{\sigma_E^2}{\sigma_Q^2}\right) \\ &= G_P(\mathrm{SNR})_Q \end{aligned} \tag{8.68}$$

where $(\mathrm{SNR})_Q$ is the *signal-to-quantizing noise ratio* defined by

$$(\mathrm{SNR})_Q = \frac{\sigma_E^2}{\sigma_Q^2} \tag{8.69}$$

and G_p is the *gain* produced by the differential quantization scheme, defined by

$$G_P = \frac{\sigma_M^2}{\sigma_E^2} \tag{8.70}$$

The quantity G_P, when greater than unity, represents the gain in signal-to-noise ratio that is due to the differential quantization scheme of Fig. 8.12. Now, for a given baseband signal, the variance σ_M^2 is fixed, so that G_P is maximized by minimizing the variance σ_E^2 of the prediction error $e(nT_s)$. Accordingly, our objective should be to design the prediction filter so as to minimize σ_E^2.

The Prediction Filter

One approach to specify the nature of the prediction filters in the transmitter and the receiver of the DPCM system shown in Fig. 8.12 is to use a tapped-delay line filter as the basis of the design. An advantage of this approach is that it leads to tractable mathematics, and it is simple to implement. Thus the predicted value $\hat{m}(nT_s)$ is modeled as a linear combination of past values of the quantized input, as

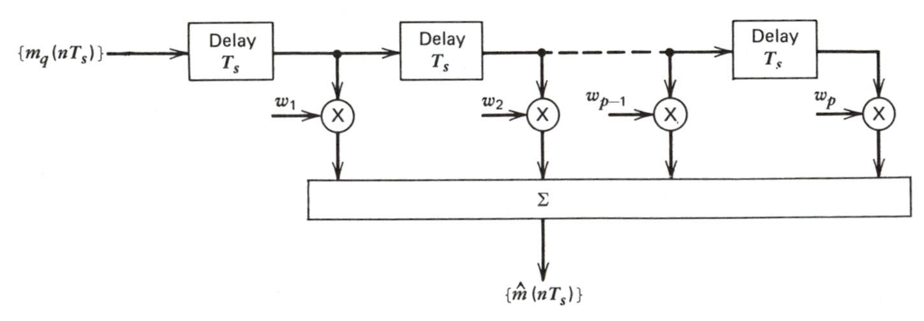

Figure 8.13 Tapped-delay-line filter used as a prediction filter.

shown by (see Fig. 8.13)

$$\hat{m}(nT_s) = \sum_{k=1}^{p} w_k m_q(nT_s - kT_s) \tag{8.71}$$

where the tapped-delay-line weights $w_1, w_2, \ldots, w_p$ define the desired prediction-filter coefficients, and p is *order* of the prediction filter. Substitution of Eq. (8.71) with (8.62) yields the prediction error

$$e(nT_s) = m(nT_s) - \sum_{k=1}^{p} w_k m_q(nT_s - kT_s) \tag{8.72}$$

With the input signal $m(nT_s)$, and therefore the quantized input $m_q(nT_s)$, assumed to have zero mean, the prediction error $e(nT_s)$ has zero mean too. The variance of the prediction error is therefore

$$\begin{aligned}\sigma_E^2 &= E[e^2(nT_s)] \\ &= E\left[\left(m(nT_s) - \sum_{k=1}^{p} w_k m_q(nT_s - kT_s)\right)^2\right]\end{aligned} \tag{8.73}$$

In order to choose a set of weights that minimize the variance σ_E^2, we must differentiate σ_E^2 with respect to each weight and then put the resulting derivatives equal to zero, thereby obtaining the following set of equations:

$$\begin{aligned}\frac{\partial \sigma_E^2}{\partial w_k} &= -2E\left[m_q(nT_s - kT_s)\left(m(nT_s) - \sum_{i=1}^{p} w_i m_q(nT_s - iT_s)\right)\right] \\ &= 0, \qquad 1 \leqslant k \leqslant p\end{aligned} \tag{8.74}$$

Equivalently, we may use Eq. (8.72) to write

$$E[m_q(nT_s - kT_s)e(nT_s)] = 0, \qquad 1 \leqslant k \leqslant p \tag{8.75}$$

Equation (8.75) states that if the prediction filter is optimized by minimizing the variance of the prediction error (error signal), the resulting prediction error is orthogonal to the past values of the prediction-filter input. This result is known as the *principle of orthogonality*.

Substituting Eq. (8.66) in (8.74) and expanding the pertinent terms, we get the following result:

$$\begin{aligned}E[m(nT_s)m(nT_s - kT_s)] &+ E[m(nT_s)q_e(nT_s - kT_s)] \\ &= \sum_{i=1}^{p} w_i E[m(nT_s - iT_s)m(nT_s - kT_s)] \\ &+ \sum_{i=1}^{p} w_i E[m(nT_s - iT_s)q_e(nT_s - kT_s)] \\ &+ \sum_{i=1}^{p} w_i E[q_e(nT_s - iT_s)m(nT_s - kT_s)] \\ &+ \sum_{i=1}^{p} w_i E[q_e(nT_s - iT_s)q_e(nT_s - kT_s)]\end{aligned} \tag{8.76}$$

where $1 \leqslant k \leqslant p$. Now, if the quantization is fine enough, we may assume the following:

1. The quantizing error $q_e(nT_s)$ is orthogonal to the input signal $m(nT_s)$, that is,
$$E[m(nT_s - kT_s)q_e(nT_s - iT_s)] = 0 \qquad \text{for all } n,\ i, \text{ and } k \tag{8.77}$$
2. The quantizing error $q_e(nT_s)$ is a stationary white noise sequence; that is,
$$E[q_e(nT_s - kT_s)q_e(nT_s - iT_s)] = \sigma_Q^2 \delta_{ik} \tag{8.78}$$
where σ_Q^2 is the variance of the quantizing noise, and δ_{ik} is the *Kronecker delta* defined by
$$\delta_{ik} = \begin{cases} 1, & i = k \\ 0, & i \neq k \end{cases} \tag{8.79}$$

Based on these assumptions, we may simplify Eq. (8.76) as follows

$$R_M(kT_s) = \sum_{i=1}^{p} w_i[R_M(kT_s - iT_s) + \sigma_Q^2 \delta_{ik}], \qquad 1 \leqslant k \leqslant p \tag{8.80}$$

where $R_M(kT_s)$ is the autocorrelation function of the input signal $m(nT_s)$ for a lag of kT_s. Divide both sides of Eq. (8.80) by the variance $\sigma_M^2 = R_M(0)$, and define the *normalized autocorrelation* of the input signal as

$$\rho_M(kT_s) = \frac{R_M(kT_s)}{R_M(0)} \tag{8.81}$$

Then, using matrix notation, we may rewrite the set of p simultaneous equations in (8.80) in a compact form as follows

$$\boldsymbol{\rho} = \mathbf{CW} \tag{8.82}$$

where

$$\boldsymbol{\rho} = \begin{bmatrix} \rho(T_s) \\ \rho(2T_s) \\ \vdots \\ \rho(pT_s) \end{bmatrix} \tag{8.83}$$

$$\mathbf{C} = \begin{bmatrix} 1 + \dfrac{1}{(\text{SNR})_O} & \rho(T_s) & \dots & \rho(pT_s - T_s) \\ \rho(T_s) & 1 + \dfrac{1}{(\text{SNR})_O} & \dots & \rho(pT_s - 2T_s) \\ \vdots & \vdots & & \vdots \\ \rho(pT_s - T_s) & \rho(pT_s - 2T_s) & \dots & 1 + \dfrac{1}{(\text{SNR})_O} \end{bmatrix} \tag{8.84}$$

and

$$\mathbf{W} = \begin{bmatrix} w_1 \\ w_2 \\ \vdots \\ w_p \end{bmatrix} \tag{8.85}$$

and

$$(\text{SNR})_O = \sigma_M^2/\sigma_Q^2 \tag{8.86}$$

The terms on the principal diagonal of the matrix $\mathbf{C}$ in Eq. (8.84) depend on the output signal-to-noise $(\text{SNR})_O$. But $(\text{SNR})_O$ depends on the prediction-filter coefficients, which in turn depend on $(\text{SNR})_O$ through Eq. (8.82). It follows therefore that, in general, Eq. (8.82) cannot be solved. One possibility, however, is to neglect the term $1/(\text{SNR})_O$ in Eq. (8.84) in order to obtain a solution for the prediction-filter coefficients. Such a procedure would be justified if the output signal-to-noise ratio is high. The desired solution is then obtained by premultiplying both sides of Eq. (8.82) by the inverse of the matrix $\mathbf{C}$. We thus find that the vector of prediction-filter coefficients is given by

$$\mathbf{W} = \mathbf{C}^{-1}\boldsymbol{\rho} \tag{8.87}$$

where $\mathbf{C}^{-1}$ is the *inverse* of matrix $\mathbf{C}$.

8.7 DELTA MODULATION (DM)

The exploitation of signal correlations in DPCM suggests the further possibility of oversampling a baseband signal (i.e., at a rate much higher than the Nyquist rate) to purposely increase the correlation between adjacent samples of the signal, so as to permit the use of a simple quantizing strategy for constructing the encoded signal. *Delta modulation* (DM), which is the one-bit (or two-level) version of DPCM, is precisely such a scheme.*

In its simple form, DM provides a staircase approximation to the oversampled version of an input baseband signal, as illustrated in Fig. 8.14(a). The difference between the input and the approximation is quantized into only two levels, namely, $\pm\delta$, corresponding to positive and negative differences, respectively. Thus, if the approximation falls below the signal at any sampling epoch, it is increased by δ. If, on the other hand, the approximation lies above the signal, it is diminished by δ. Provided that the signal does not change too rapidly from sample to sample, we find that the staircase approximation remains within $\pm\delta$ of the input signal.

* For the original papers on delta modulation, see

1. J. S. Schouten, F. E. DeJager, and J. A. Greefkes, "Delta modulation, a new modulation system for telecommunications," *Phillips Technical Report*, pp. 237–245, March 1952.
2. F. E. DeJager, "Delta modulation, a method of PCM transmission using a 1-unit code." *Phillips Research Report*, pp. 442–466, Dec. 1952.

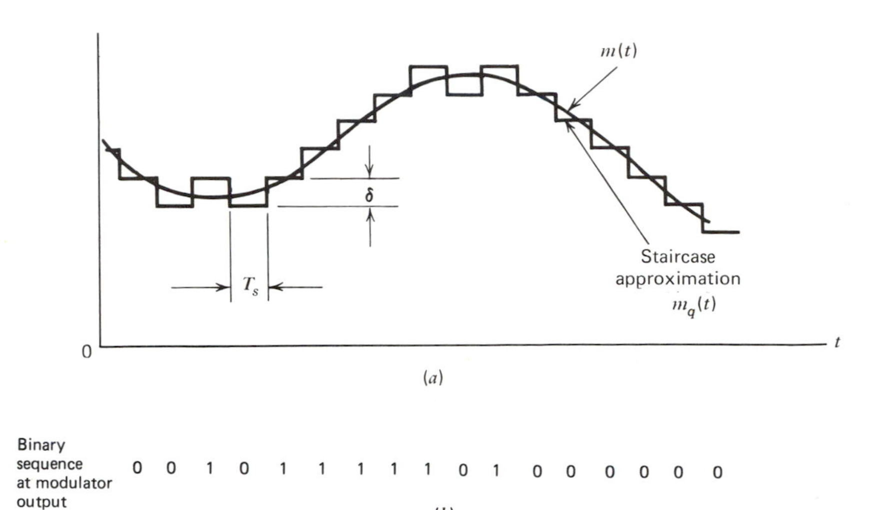

Figure 8.14 Illustration of delta modulation.

Denoting the input signal as $m(t)$ and the staircase approximation as $m_q(t)$, the basic principle of delta modulation may be formalized in the following set of discrete-time relations:

$$e(nT_s) = m(nT_s) - m_q(nT_s - T_s) \tag{8.88}$$

$$e_q(nT_s) = \delta \operatorname{sgn}[e(nT_s)] \tag{8.89}$$

and

$$m_q(nT_s) = m_q(nT_s - T_s) + e_q(nT_s), \tag{8.90}$$

where T_s is the sampling period; $e(nT_s)$ is an error signal representing the difference between the present sample value $m(nT_s)$ of the input signal and the latest approximation to it, namely, $\hat{m}(nT_s) = m_q(nT_s - T_s)$; and $e_q(nT_s)$ is the quantized version of $e(nT_s)$. The quantizer output $e_q(nT_s)$ is finally coded to produce the desired DM wave.

Part (*a*) of Fig. 8.14 illustrates the way in which the staircase approximation $m_q(t)$ follows variations in the input signal $m(t)$ in accordance with Eqs. (8.88) to (8.90), and part (*b*) of the figure displays the corresponding binary sequence at the delta modulator output. It is apparent that in a delta modulation system the rate of information transmission is simply equal to the sampling rate $1/T_s$.

The principal virtue of delta modulation is its simplicity. It may be generated by applying the sampled version of the incoming baseband signal to a modulator that involves a summer, quantizer, and accumulator interconnected as shown in Fig. 8.15(*a*). Details of the modulator follow directly from Eqs. (8.88) to (8.90). In particular, the quantizer consists of a *hard limiter* with an input–output relation defined by Eq. (8.89), which is depicted in Fig. 8.16. The quantizer output is applied

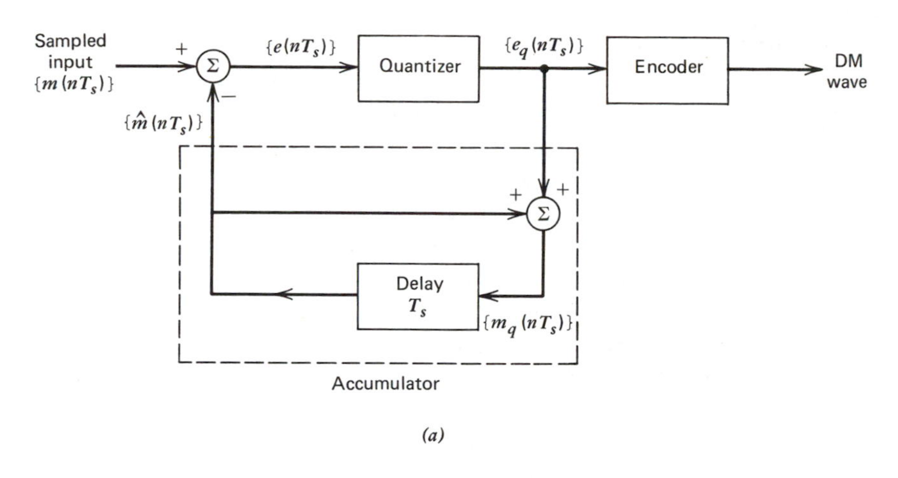

(a)

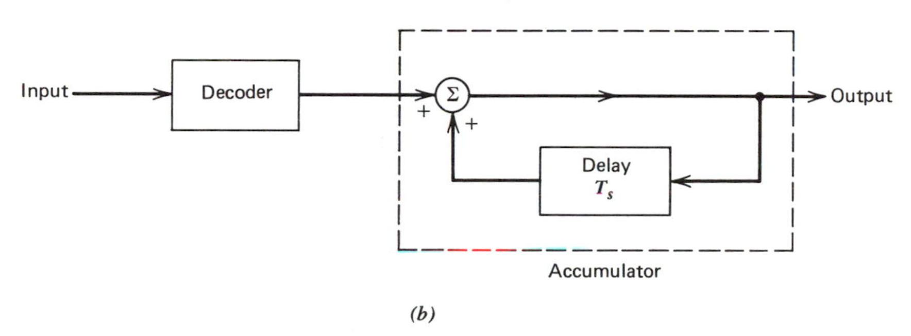

(b)

Figure 8.15 DM system. *(a)* Transmitter. *(b)* Receiver.

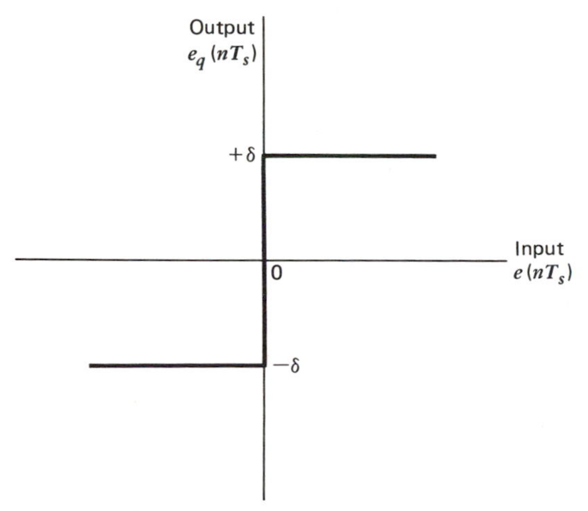

Figure 8.16 Input–output characteristic of quantizer for DM system.

to an *accumulator*, producing the result

$$\begin{aligned} m_q(nT_s) &= \delta \sum_{i=1}^{n} \mathrm{sgn}[e(iT_s)] \\ &= \sum_{i=1}^{n} e_q(iT_s) \end{aligned} \tag{8.91}$$

which is obtained by solving Eqs. (8.89) and (8.90) for $m_q(nT_s)$. Thus, at the sampling instant nT_s, the accumulator increments the approximation by a step δ in a positive or negative direction, depending on the algebraic sign of the error signal $e(nT_s)$. If the input signal $m(nT_s)$ is greater than the most recent approximation $\hat{m}(nT_s)$, a positive increment $+\delta$ is applied to the approximation. If, on the other hand, the input signal is smaller, a negative increment $-\delta$ is applied to the approximation. In this way, the accumulator does the best it can to track the input samples by one step (of amplitude $+\delta$ or $-\delta$) at a time. In the receiver, shown in Fig. 8.15(*b*), the staircase approximation $m_q(t)$ is reconstructed by passing the sequence of positive and negative pulses, produced at the decoder output, through an accumulator in a manner similar to that used in the transmitter. The out-of-band quantizing noise in the high-frequency staircase waveform $m_q(t)$ is rejected by passing it through a low-pass filter with a bandwidth equal to the original signal bandwidth.

In comparing the DPCM and DM networks of Figs. (8.12) and (8.15), we note that they are basically similar, except for two important differences, namely, the use of a one-bit (two-level) quantizer in the delta modulator and the replacement of the prediction filter by a single delay element.

Quantizing Noise

Delta modulation systems are subject to two types of quantizing error: (1) slope overload distortion, and (2) granular noise. We first discuss the cause of slope overload distortion, and then granular noise.

We observe that Eq. (8.90) is the digital equivalent of integration in the sense that it represents the accumulation of positive and negative increments of magnitude δ. Also, denoting the quantizing error by $q_e(nT_s)$, as shown by,

$$m_q(nT_s) = m(nT_s) + q_e(nT_s) \tag{8.92}$$

we observe from Eq. (8.88) that the input to the quantizer is

$$e(nT_s) = m(nT_s) - m(nT_s - T_s) - q_e(nT_s - T_s) \tag{8.93}$$

Thus, except for the quantizing error $q_e(nT_s - T_s)$, the quantizer input is a *first backward difference* of the input signal, which may be viewed as a digital approximation to the derivative of the input signal or, equivalently, as the inverse of the digital integration process. If we consider the maximum slope of the original input waveform $m(t)$, it is clear that in order for the sequence of samples $\{m_q(nT_s)\}$ to increase as fast as the input sequence of samples $\{m(nT_s)\}$ in a region of maximum slope of $m(t)$, we require that the condition

$$\frac{\delta}{T_s} \geqslant \max\left|\frac{dm(t)}{dt}\right| \tag{8.94}$$

be satisfied. Otherwise, we find that the step size is too small for the staircase approximation $m_q(t)$ to follow a steep segment of the input waveform $m(t)$, with the result that $m_q(t)$ falls behind $m(t)$, as illustrated in Fig. 8.17. This condition is called *slope-overload*, and the resulting quantizing error is called *slope-overload distortion* (noise). Note that since the maximum slope of the staircase approximation $m_q(t)$ is fixed by the step size δ, increases and decreases in $m_q(t)$ tend to occur along straight lines. For this reason, a delta modulator using a fixed step size is often referred to as a *linear delta modulator*.

In contrast to slope-overload distortion, *granular noise* occurs when the step size δ is too large relative to the local slope characteristics of the input waveform $m(t)$, thereby causing the staircase approximation $m_q(t)$ to hunt around a relatively flat segment of the input waveform; this phenomenon is also illustrated in Fig. 8.17. The granular noise is analogous to quantizing noise in a PCM system.

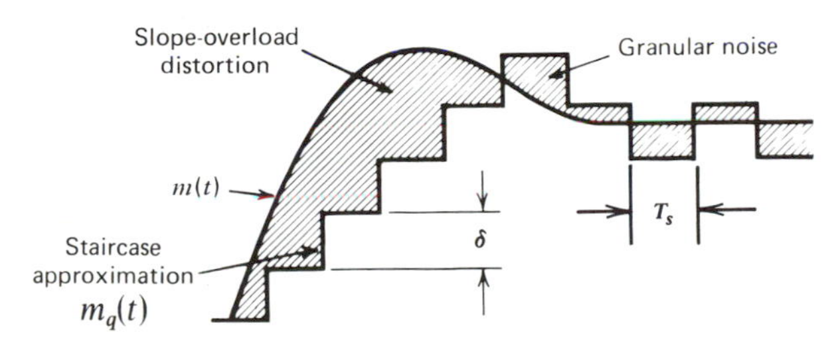

Figure 8.17 Illustration of quantizing error in delta modulation.

We thus see that there is a need to have a large step size to accommodate a wide dynamic range, whereas a small step size is required for the accurate representation of relatively low-level signals. It is therefore clear that the choice of the optimum step size that minimizes the mean-square value of the quantizing error in a linear delta modulator will be the result of a compromise between slope overload distortion and granular noise.*

8.8 DISCUSSION

In this section we discuss the advantages and disadvantages of DPCM and DM, compared with the standard PCM, for the encoding of voice and television signals.

The "relative" behavior of standard PCM and DPCM systems is much the same with either uniform or logarithmic quantizing, because the repertoire of signals

* For details of this evaluation, see J. E. Abate, "Linear and adaptive delta modulation," *Proc. IEEE*, vol. 55, pp. 298–308, 1967.

consists of waveforms similar in character, but differing in mean level. In the case of voice signals, it is found that the optimum signal-to-quantizing noise advantage of DPCM over standard PCM is in the neighborhood of 4–11 dB.* The greatest improvement occurs in going from no prediction to first-order prediction, with some additional gain resulting from increasing the order of the prediction filter up to 4 or 5, after which little additional gain is obtained. Since 6dB of quantizing noise is equivalent to 1 bit per sample, by virtue of Eq. (8.27), the advantage of DPCM may also be expressed in terms of bit rate. For a constant signal-to-quantizing noise ratio, and assuming a sampling rate of 8 kHz, the use of DPCM may provide a saving of about 8–16 kilobits per second (1 to 2 bits per sample) over standard PCM.

In the case of television signals, DPCM provides more of an advantage for high-resolution television systems than for low-resolution systems. For monochrome entertainment television, DPCM provides a signal-to-quantizing noise ratio of approximately 12 dB higher than standard PCM.† For a constant signal-to-quantizing noise ratio, and assuming a sampling rate of 9 MHz, this represents a saving of about 18 megabits per second (2 bits per sample) by DPCM over PCM.

The advantages of DPCM are dependent on the statistics of the signal being encoded, however. In switched telecommunication networks,‡ for example, modulation systems must perform well not only with voice signals but also with various kinds of data signals whose statistics differ markedly from those of voice signals. Thus, DPCM systems optimized for digitized voice signals may not perform well with data. Also a signal encoded into DPCM is more vulnerable to transmission noise, causing more bit errors than one encoded into standard PCM. It is characteristic of DPCM systems that, if they decrease the quantizing noise by k dB over standard PCM, then the noise in the decoded signal at the receiver output caused by bit errors in the digital transmission channel is increased by k dB.

Turning next to delta modulation, subjective voice tests and noise measurements have shown that a DM system operating at 40 kilobits per second is equivalent to a standard PCM system operating with a sampling rate of 8 kHz and 5 bits per

† N. S. Jayant, "Digital coding of speech waveforms; PCM, DPCM, and DM quantizers," *Proc. IEEE*, vol. 62, pp. 611–632, May 1974. See also, L. R. Rabiner and R. W. Schafer, *Digital Processing of Speech Signals*, Chapter 5 (Prentice-Hall, 1978).

† J. B. O'Neal, "Predictive quantizing systems (differential pulse code modulation) for the transmission of television signals," *Bell Syst. Tech. J.*, vol. 45, p. 689, May–June 1966.

‡ Two different types of *switching* are commonly employed in switched telecommunication networks. The first, called *circuit switching*, applies to a system where a continuous two-way path is established in space, time, or frequency between the calling and called station transducers for the duration of a telephone call. The second type of switching applies to data transmission where a continuous path is not established but where memory is used to store the message to be transmitted awaiting the availability of transmission facilities. This form of switching is known generally as *store and forward switching* and includes a currently popular form known as *packet switching*. The switching operation is therefore the heart of a switched telecommunication network; without it, the network is reduced to either a broadcast unit or a set of point-to-point communications links. For further details of switched telecommunication networks, see A. E. Joel, Jr; "What is telecommunications circuit switching?," *Proc. IEEE*, vol. 65, pp. 1237–1253, Sept. 1977.

sample*. At lower bit rates, DM is better than the standard PCM (the latter still using 8 kHz sampling and a reduced number of bits per sample), but at higher bit rates PCM is superior to DM. The quality of 5-bit PCM is low for most purposes in telephony. For telephone quality voice signals, it has become conventional to use a 8-bit PCM. Equivalent voice quality with DM can be obtained only by using bit rates much higher than 64 kilobits per second.

Also, it has been shown† that in a delta modulation system, operating on voice signals under optimum conditions, the SNR is increased by 9 dB by doubling the bit rate. By comparison, in the case of standard PCM, if we double the bit rate by doubling the number of bits per sample, we achieve a 6 dB increase in SNR for each *added* bit. For example, by doubling the bit rate from 40 to 80 kilobits per second, the SNR is increased by 9 dB using DM. On the other hand, if PCM is employed and the bit rate is similarly doubled by increasing the number of bits per sample from 5 to 10 (keeping the sampling rate fixed at 8 kHz), the SNR is improved by 30 dB. Thus the increase of SNR with bit rate is much more dramatic for PCM than for DM.

The use of delta modulation is therefore recommended only in certain special circumstances: (1) if it is necessary to reduce the bit rate below 40 kilobits per second and limited voice quality is tolerable; or (2) if extreme circuit simplicity is of over-riding importance and the accompanying use of a high-bit rate is acceptable.

8.9 ADAPTIVE DIGITAL WAVEFORM CODING SCHEMES

From the discussion presented on PCM using a uniform quantizer with a fixed step size, we see that we have a dilemma in quantizing speech signals. On the one hand, we wish to choose the quantization step size large enough to accommodate the maximum peak-to-peak range of the input signal with the lowest possible number of quantizing levels. On the other hand, we would like to make the quantization step size small enough to minimize the variance of the quantizing noise. This issue is further compounded by the nonstationary nature of the speech signal. The amplitude of the speech signal can vary over a wide range, depending on the speaker, the communication environment, and within a given utterance, from *voiced* to *unvoiced sounds*.‡ One approach to accommodating these conflicting requirements is to use a nonuniform quantizer, as described in Section 8.1. An alternative approach is to use an *adaptive quantizer*, wherein the step size δ is varied automatically so as

* K. W. Cattermole, p. 206, *op. cit.*

† J. E. Abate, *op. cit.* L. R. Rabiner and R. W. Schafer, *op. cit.*

‡ *Voiced sounds* are produced by forcing air through the glottis with the tension of the vocal cords adjusted so that they vibrate in a relaxation oscillation, thereby producing quasi-periodic pulses of air which excite the vocal tract. *Fricative* or *unvoiced sounds* are generated by forming a constriction at some point in the vocal tract (usually toward the mouth end), and forcing air through the constriction at a high enough velocity to produce turbulence. This creates a broad-spectrum noise source to excite the vocal tract.

to match the variance of the input signal. Pulse-code modulation systems using adaptive quantization are called *adaptive PCM* systems.

A flexible means of matching quantizer step size to signal variance is the use of step-size adaptation based on quantizer memory. The idea is to work with a simple uniform quantizer, but to modify its step size (for every new input sample. in general) by a factor depending on the knowledge of which quantizer slots were occupied by the previous samples. In its simplest form, the scheme operates with a one-word memory.* Let the output of an n-bit (uniform) quantizer of the mid-riser type be

$$y_i = H_i \frac{\delta_i}{2} \tag{8.95}$$

where

$$\pm H_i = 1, 3, 5, \ldots, 2^n - 1$$

with $n \geqslant 2$ and $\delta_i > 0$. The step size δ_{i+1} is now chosen to be the previous step size δ_i multiplied by a time-invariant function of the code word magnitude $|H_i|$, as shown by

$$\delta_{i+1} = \delta_i \cdot f(|H_i|) \tag{8.96}$$

When the multiplier function $f(|H_i|)$ is properly designed, the *adaptation rule* defined by Eqs. (8.95) and (8.96) serves to match the step size, at every sample, to an updated estimate of variance of the input signal.

In the case of a DPCM system we have the option of using an adaptive quantizer with a fixed prediction filter, or a fixed quantizer with an *adaptive prediction filter*, or making both of them adaptive. The variation of performance with speakers and speech material, together with variations in signal level inherent in the speech communication process, make the combined use of adaptive quantization and adaptive prediction necessary to achieve best performance over a wide range of speakers and speaking situations. Such systems are called *adaptive DPCM systems*. As with adaptive PCM, we may use the adaptation rule defined by Eqs. (8.95) and (8.96) to realize the adaptive quantization. In adapting the prediction-filter coefficients, it is common to assume that the statistical properties of the speech signal remain fixed over short time intervals. The prediction-filter coefficients are then chosen to minimize the average squared prediction error over a short time interval.

For telephone transmission using standard PCM, it is generally agreed that acceptable speech quality is obtained by using a μ-law quantizer with 6–7 bits per sample. It has been shown† that by using an adaptive DPCM system (in which the quantizer and the prediction filter are both adaptive) with 5 bits per sample, a comparable quality could be obtained to a PCM system with $\mu = 100$ and 7 bits for

* N. S. Jayant, "Adaptive quantization with a one-word memory," *Bell System Tech. J.*, pp. 1119–1144, Sept. 1973.

† P. Noll, "A comparative study of various schemes for speech encoding," *Bell Syst. Tech. J.*, vol. 54, pp. 1597–1614, Nov. 1975.

sample. Also there is strong evidence to suggest that the perceived quality of adaptive DPCM coded speech is better than a comparison of SNR values would suggest. In a study of adaptive DPCM with an adaptive quantizer and a fixed prediction filter, it has been found* that listeners preferred adaptive DPCM-coded speech to coded speech using standard PCM (with logarithmic companding) operating at a higher SNR.

Finally, we should mention that a delta modulator may also be made adaptive, wherein the variable step size increases during a steep segment of the input signal and decreases when the modulator is quantizing an input signal with a slowly varying segment. In this way the step size is adapted to the level of the input signal. The resulting system is called an *adaptive delta modulator*. The problem in adaptive delta modulation, of course, is to specify suitable rules for step-size variation. Note that the simple PCM–DPCM adaptation rule given in Eqs. (8.95) and (8.96) does not apply for $n=1$ (i.e., one-bit code). This is because, with a two-level quantizer, the observation of a single sample of quantizer output does not provide any indication of overload or underload (granularity). In a delta modulation system we need to inspect sequences (of length $\geqslant 2$) of quantizer outputs for meaningful step-size adaptation.

Figure 8.18 illustrates how slope overload and granularity tend to correspond to occurrences of like and unlike (successive) bits, respectively. Let

$$b_i = \text{sgn}[m(iT_s) - m_q(iT_s - T_s)] \tag{8.97}$$

and

$$\delta_i = m_q(iT_s) - m_q(iT_s - T_s) \tag{8.98}$$

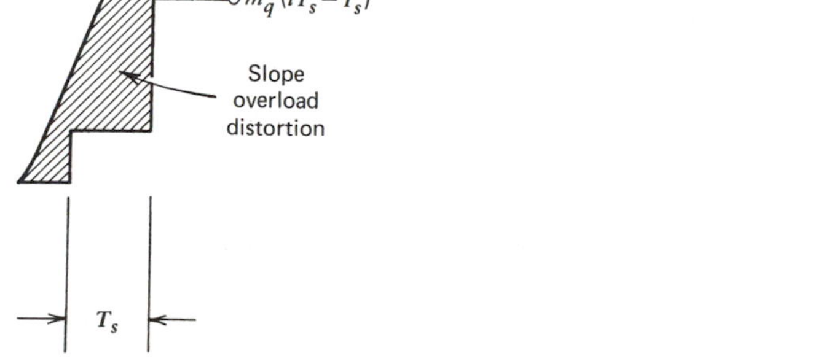

Figure 8.18 Illustrating adaptive delta modulation.

* P. Cummiskey, N. S. Jayant and J. L. Flanagan, "Adaptive quantization in differential PCM coding of speech," *Bell Syst. Tech. J.*, vol. 52, pp. 1105–1118, Sept. 1973.

In a simple realization of adaptive delta modulation, successive bits b_i are compared to detect probable slope overload ($b_i = b_{i-1}$) or probable granularity ($b_i \neq b_{i-1}$). A specific form of adaptation rule suitable for use in delta modulation is the following*

$$\delta_i = \delta_{i-1} \gamma^{b_i b_{i-1}}, \qquad \gamma \geq 1 \tag{8.99}$$

According to this rule, the rate of step-size increase (or decrease) is given by a single factor γ. Note that $\gamma = 1$ represents (nonadaptive) linear delta modulation. Typically, a value of $\gamma_{\text{opt}} = 1.5$ minimizes quantization error power for speech encoding.

In terms of SNR performance, it has been shown† that adaptive delta modulation is superior to linear delta modulation by 8 dB at a bit rate of 20 kilobits per second, and the SNR advantage increases to 14 dB at 60 kilobits per second. Also it is of interest to note that the improved quality of adaptive delta modulation is achieved with only a slight increase in system complexity.

8.10 DIGITAL MULTIPLEXERS

In Section 7.5 we introduced the idea of time-division multiplexing whereby a group of analog signals (e.g., voice signals) are sampled sequentially in time at a *common* sampling rate and then multiplexed for transmission over a common line. In this section we consider the multiplexing of digital signals at different bit rates.‡ This enables us to combine several digital signals, such as computer outputs, digitized voice signals, digitized facsimile and television signals, into a single data stream (at a considerably higher bit rate than any of the inputs). Figure 8.19 shows a conceptual diagram of the digital multiplexing-demultiplexing operation.

The multiplexing of digital signals may be accomplished by using *a bit-by-bit*

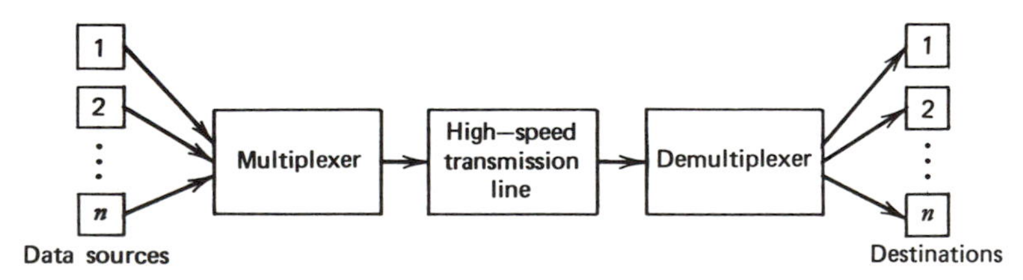

Figure 8.19 Conceptual diagram of multiplexing-demultiplexing.

* N. S. Jayant, "Adaptive delta modulation with a one-bit memory," *Bell System Tech. J.*, vol. 49, pp. 321–342, March 1970.

† 1. N. S. Jayant, *op. cit.*
2. L. R. Rabiner and R. W. Schafer, *op. cit.*

‡ For a more complete description of digital multiplexers, see:
1. Bell Telephone Laboratories, *Transmission Systems for Communications*, Fourth Edition, pp. 608–625 (Western Electric Company, 1970).
2. M. Schwartz, *Information Transmission, Modulation and Noise*, Third Edition, pp. 155–176 (McGraw-Hill, 1980).

interleaving procedure with a selector switch that sequentially takes a bit from each incoming line and then applies it to the high-speed common line. At the receiving end of the system the output of this common line is separated out into its low-speed individual components and then delivered to their respective destinations.

Two major groups of digital multiplexers are used in practice:

1. One group of multiplexers is designed to combine relatively low-speed digital signals, up to a maximum rate of 4800 bits per second, into a higher speed multiplexed signal with a rate of up to 9600 bits per second. These multiplexers are used primarily to transmit data over voice-grade channels of a telephone network. Their implementation requires the use of *modems* in order to convert the digital format into an analog format suitable for transmission over telephone channels. The theory of a modem (*mo*dulator-*dem*odulator) is covered in Chapter 10.
2. The second group of multiplexers, designed to operate at much higher bit rates, forms part of the data transmission service generally provided by communication carriers. For example, Fig. 8.20 shows a block diagram of the digital hierarchy based on the T1 carrier, which has been developed by the Bell System. The T1 carrier, described in Example 1, is designed to operate at 1.544 megabits per second, the T2 at 6.312 megabits per second, the T3 at 44.736 megabits per second, and the T4 at 274.176 megabits per second. The system is thus made up of various combinations of lower order T-carrier subsystems designed to accommodate the transmission of voice signals, Picturephone® service, and television signals by using PCM, as well as (direct) digital signals from data terminal equipment.

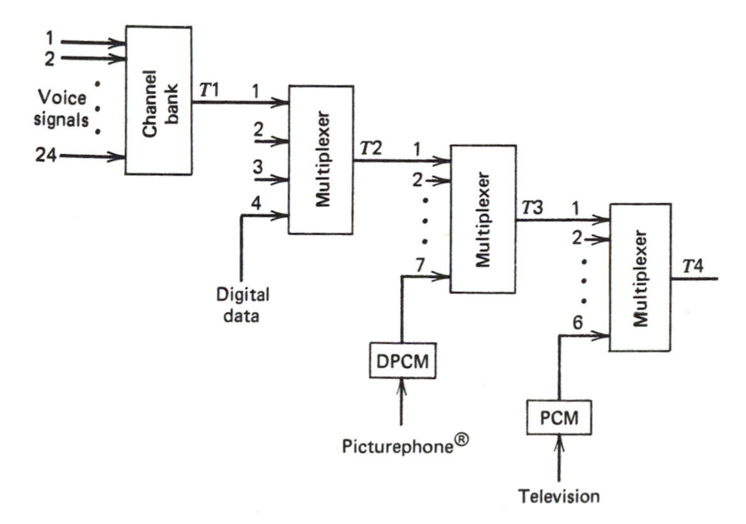

Figure 8.20 Digital hierarchy, Bell system.

There are some basic problems involved in the design of a digital multiplexer, irrespective of its grouping:

1. Digital signals cannot be directly interleaved into a format that allows for their eventual separation unless their bit rates are locked to a common clock. Accordingly, provision has to be made for *synchronization* of the incoming digital signals, so that they can be properly interleaved.
2. The multiplexed signal must include some form of *framing*, so that its individual components can be identified at the receiver.
3. The multiplexer has to handle small variations in the bit rates of the incoming digital signals. For example, a 1000-kilometer coaxial cable carrying 3×10^8 pulses per second will have about one million pulses in transit, with each pulse occupying about one meter of the cable. A 0.01 percent variation in the propagation delay, produced by a 1°F decrease in temperature, will result in 100 fewer pulses in the cable. Clearly, these pulses must be absorbed by the multiplexer.

In order to cater for the requirements of synchronization and rate adjustment to accommodate small variations in the input data rates, we may use a technique known as *bit stuffing*. The idea here is to have the outgoing bit rate of the multiplexer slightly higher than the sum of the maximum expected bit rates of the input channels by stuffing in additional non-information carrying pulses. All incoming digital signals are stuffed with a number of bits sufficient to raise each of their bit rates to equal that of a locally generated clock. To accomplish bit stuffing, each incoming digital signal or bit stream is fed into an *elastic store* at the multiplexer. The elastic store is a device that stores a bit stream in such a manner that the stream may be read out at a rate different from the rate at which it is read in. At the demultiplexer, the stuffed bits must obviously be removed from the multiplexed signal. This requires a method that can be used to identify the stuffed bits. To illustrate one such method, and also show one method of providing frame synchronization, we describe the signal format of the Bell System *M12 multiplexer*, which is designed to combine four T1 bit streams into one T2 bit stream. This corresponds to the second level of the digital hierarchy shown in Fig. 8.20.

Example 7 Signal format of the Bell System M12 multiplexer

Figure 8.21 illustrates the signal format of the M12 multiplexer. Each frame is subdivided into four subframes. The first subframe (first line in Fig. 8.21) is transmitted, then the second, the third, and the fourth, in that order.

Bit-by-bit interleaving of the incoming four T1 bit streams is used to accumulate a total of

M_0	[48]	C_{I}	[48]	F_0	[48]	C_{I}	[48]	C_{I}	[48]	F_1	[48]
M_1	[48]	C_{II}	[48]	F_0	[48]	C_{II}	[48]	C_{II}	[48]	F_1	[48]
M_1	[48]	C_{III}	[48]	F_0	[48]	C_{III}	[48]	C_{III}	[48]	F_1	[48]
M_1	[48]	C_{IV}	[48]	F_0	[48]	C_{IV}	[48]	C_{IV}	[48]	F_1	[48]

Figure 8.21 Signal format of Bell system M12 multiplexer.

48 bits, 12 from each input. A *control bit* is then inserted by the multiplexer. Each frame contains a total of 24 control bits, separated by sequences of 48 data bits. Three types of control bits are used in the M12 multiplexer to provide synchronization and frame indication, and to identify which of the four input signals has been stuffed. These control bits are labelled as F, M, and C in Fig. 8.21. Their functions are as follows:

1. The F-control bits, two per subframe, constitute the *main* framing pulses. The subscripts on the F-control bits denote the actual bit (0 or 1) transmitted. Thus the main framing sequence is $F_0F_1F_0F_1F_0F_1F_0F_1$ or 01010101.
2. The M-control bits, one per subframe, form *secondary* framing pulses to identify the four subframes. Here again the subscripts on the M-control bits denote the actual bit (0 or 1) transmitted. Thus the secondary framing sequence is $M_0M_1M_1M_1$ or 0111.
3. The C-control bits, three per subframe are *stuffing indicators*. In particular, C_{I} refers to input channel I, C_{II} refers to input channel II, and so forth. For example, the three C-control bits in the first subframe following M_0 in the first subframe are stuffing indicators for the first T1 signal. The insertion of a stuffed bit in this T1 signal is indicated by setting all three C-control bits to 1. To indicate no stuffing, all three are set to 0. If the three C-control bits indicate stuffing, the stuffed bit is located in the position of the first information bit associated with the first T1 signal that follows the F_1-control bit in the same subframe. In a similar way, the second, third, and fourth T1 signals may be stuffed, as required. By using *majority logic decoding* in the receiver, a single error in any of the three C-control bits can be detected. This form of decoding means simply that the majority of the C-control bits determine whether an all-one or all-zero sequence was transmitted. Thus three 1's or combinations of two 1's and a 0 indicate that a suffed bit is present in the information sequence, following the control bit F_1 in the pertinent subframe. On the other hand, three 0's or combinations of two 0's and a 1 indicate that no stuffing is used.

The demultiplexer at the receiving M12 unit first searches for the main framing sequence $F_0F_1F_0F_1F_0F_1F_0F_1$. This establishes identity for the four input T1 signals and also for the M- and C- control bits. From the $M_0M_1M_1M_1$ sequence, the correct framing of the C-control bits is verified. Finally, the four T1 signals are properly demultiplexed and destuffed.

The signal format described above has two safeguards:

1. It is possible, although unlikely, that with just the $F_0F_1F_0F_1F_0F_1F_0F_1$ sequence, one of the incoming T1 signals may contain a similar sequence. This could then cause the receiver to look onto the wrong sequence. The presence of the $M_0M_1M_1M_1$ sequence provides verification of the genuine $F_0F_1F_0F_1F_0F_1F_0F_1$ sequence, thereby ensuring that the four T1 signals are properly demultiplexed.
2. The single-error correction capability built into the C-control bits ensures that the four T1 signals are properly destuffed.

The capacity of the M12 multiplexer to accommodate small variations in the input data rates can be calculated from the format of Fig. 8.21. In each M frame, defined as the interval containing one cycle of $M_0M_1M_1M_1$ bits, one bit can be stuffed into each of four input T1 signals. Each such signal has $12 \times 6 \times 4 = 288$ positions in each M frame. Also the T1 signal has a bit rate equal to 1.544 megabits per second. Hence, each input can be incremented by

$$1.544 \times 10^3 \times \frac{1}{288} = 5.4 \text{ kilobits/s}$$

This result is much larger than the expected change in the bit rate of the incoming T1 signal. It follows therefore that the use of only one stuffed bit per input channel in each frame is sufficient to accommodate expected variations in the input signal rate.

The local clock that determines the outgoing bit rate also determines the nominal *stuffing rate S*, defined as the average number of bits stuffed per channel in any frame. The M12 multiplexer is designed for $S = 1/3$. Accordingly, the nominal bit rate of the T2 line is

$$1.544 \times 4 \times \frac{49}{48} \times \frac{288}{288\text{-}S} = 6.312 \text{ megabits/s}$$

This also ensures that the nominal T2 clock frequency is a multiple of 8 kHz (the nominal sampling rate of a voice signal), which is a desirable feature.

Problems

Problem 8.1

(a) A sinusoidal signal, with an amplitude of 3.25 volts, is applied to a uniform quantizer of the mid-tread type whose output takes on the values 0, ± 1, ± 2, ± 3 volts, as in Fig. P8.1(*a*). Sketch the waveform of the resulting quantizer output for one complete cycle of the input.

(b) Repeat this evaluation for the case when the quantizer is of the mid-riser type whose output takes on the values ± 0.5, ± 1.5, ± 2.5, ± 3.5 volts, as in Fig. P8.1(b).

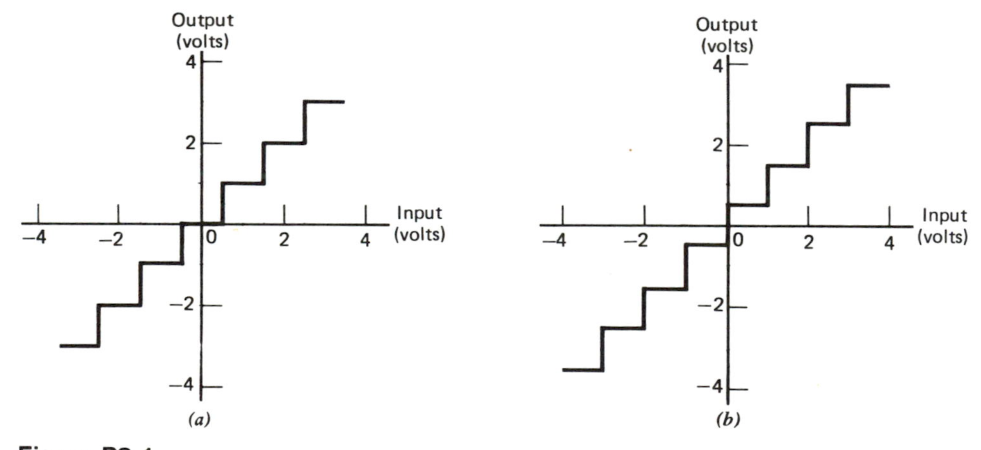

Figure P8.1

Problem 8.2 Consider the following sequences of 1's and 0's:

(a) An alternating sequence of 1's and 0's.
(b) A long sequence of 1's followed by a long sequence of 0's.
(c) A long sequence of 1's followed by a single 0 and then a long sequence of 1's.

Sketch the waveform for each of these sequences using the following methods of representing symbols 1 and 0:

(a) On-off signaling.
(b) Polar signaling.

(c) Return-to-zero signaling.
(d) Bipolar signaling.
(e) Manchester code.

Problem 8.3 This problem is intended to show that the power spectral density of a PCM wave depends on the signaling format used to represent the 1's and 0's. Assuming that the 1's and 0's occur with equal probability, and the symbols in adjacent time slots are statistically independent, determine and plot the power spectral density of the PCM wave for each of the signaling formats:

(a) On-off signaling.
(b) Polar signaling.
(c) Manchester code.

Hint: Use the results of Example 7 of Chapter 5 and Problem 5.9 of the same chapter.

Problem 8.4 The signal

$$m(t)=6\sin(2\pi t)\text{ volts}$$

is transmitted using a 4-bit binary PCM system. The quantizer is of the mid-riser type, with a step size of 1 volt. Sketch the resulting PCM wave for one complete cycle of the input. Assume a sampling rate of four samples per second, with samples taken at $t=\pm 1/8, \pm 3/8, \pm 5/8, \ldots,$ seconds.

Problem 8.5 Figure P8.2 shows a PCM wave in which the amplitude levels of $+1$ volt and -1 volt are used to represent binary symbols 1 and 0, respectively. The code word used consists of three bits. Find the sampled version of an analog signal from which this PCM wave is derived.

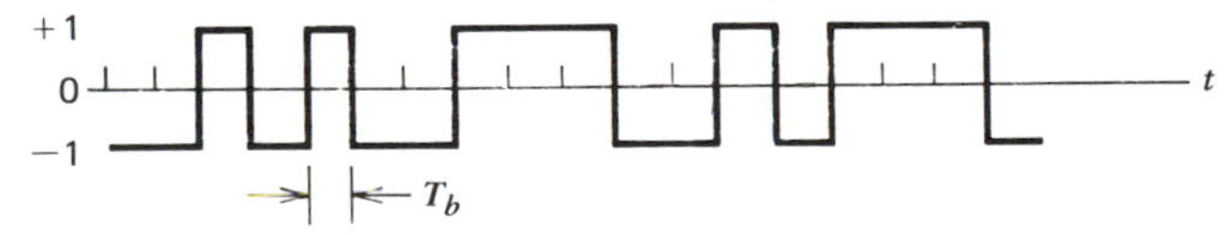

Figure P8.2

Problem 8.6 A block diagram of the overall T carrier TDM/PCM telephone system is shown in Fig. 8.20. It consists of various lower order T-carrier subsystems interconnected so as to accommodate the requirements of voice channels, Picturephone® service, and commercial television network programming. Given that the data rate for one Picturephone® service is 6.312 megabits per second, and that for one television service is 44.736 megabits per second, determine the capacity of each system level measured in terms of the number of (a) voice, (b) picturephone, or (c) television channels, which it can accommodate.

Problem 8.7 The *companding improvement*, C, obtained by using the compressor/expander combination in PCM systems is defined by (in decibels):

$$C=20\log_{10}\left[\frac{\text{uniform (linear) scale of the input}}{\text{companded scale of the input}}\right]$$

(a) Show that for the μ-law, the companding improvement (for small signals) equals

$$C=20\log_{10}\left[\frac{\mu}{\ln(1+\mu)}\right]$$

(b) What is the value of this improvement for the T1 PCM system of Example 1 for which $\mu = 255$?
(c) Find the code-word length for a standard PCM system using a uniform quantizer, which would produce about the same performance as the T1 system of Example 1.

Problem 8.8 A binary PCM system, using on-off signaling, operates just above the error threshold with an average probability of error equal to 10^{-6}. Suppose that the filter bandwidth at the receiver input is doubled. Find the new value of the average probability of error. You may use Table A2.1 of Appendix 2 to evaluate the complementary error function.

Problem 8.9 Consider a binary PCM system that uses polar signaling, with symbols 1 and 0 represented by $+A$ and $-A$ volts, respectively. The receiver is shown in Fig. 8.6, where the additive noise $w(t)$ is white Gaussian with zero mean and power spectral density $N_0/2$. Calculate the average probability of symbol error for this method of signaling.

Problem 8.10 A continuous-time signal is sampled and then transmitted as a PCM wave. The additive noise at the input of the decision device in the receiver has a variance of 0.01 volts^2.

(a) Assuming polar signaling, determine the pulse amplitude that must be transmitted for the average error rate not to exceed 1 bit in 10^8 bits.
(b) If the added presence of interference causes the error rate to increase to 1 bit in 10^6 bits, what is the variance of the interference?

Problem 8.11 In a binary PCM system, symbols 0 and 1 have a priori probabilities p_0 and p_1, respectively. The conditional probability density function of a random variable X (with sample value x) obtained by observing (at some fixed time) the received signal, given that symbol 0 was transmitted, is denoted by $f_{X|0}(x|0)$. Similarly, $f_{X|1}(x|1)$ denotes the conditional probability density function of X, given that symbol 1 was transmitted. Let η denote the threshold used in the receiver, so that if the sample value x exceeds η, the receiver decides in favor of 1; otherwise, it decides in favor of 0.

(a) Show that the optimum threshold η_{opt}, for which the average probability of error is a minimum, is given by the solution of:

$$\frac{f_{X|1}(\eta_{\text{opt}}|1)}{f_{X|0}(\eta_{\text{opt}}|0)} = \frac{p_0}{p_1}$$

(b) For the case of polar signals in additive Gaussian noise of zero mean and variance σ^2, show that the optimum threshold is given by

$$\eta_{\text{opt}} = \frac{\sigma^2}{A} \ln\left(\frac{p_0}{p_1}\right)$$

where $\pm A/2$ define the voltage levels corresponding to symbols 0 and 1.
(c) For the case of on-off signaling, show that the corresponding result is

$$\eta_{\text{opt}} = \frac{A}{2} + \frac{\sigma^2}{A} \ln\left(\frac{p_0}{p_1}\right)$$

where A is the height of the pulse representing symbol 1.

Problem 8.12 Consider a binary PCM system, in which the a priori probabilities of symbols 0 and 1 are p_0 and p_1, respectively. The conditional probability of error, given that a 0 was transmitted, is denoted by P_{e0}, and the conditional probability of error, given that a 1 was transmitted, is denoted by P_{e1}. Let the transmitted symbols 0 and 1 be designated by B_0 and B_1, respectively, and the received symbols 0 and 1 by A_0 and A_1.

(a) Find the four *a posteriori probabilities* of the system, namely, $P(B_j|A_k)$ where j, $k=0$, 1.
(b) What do these a posteriori probabilities reduce to, for the case when $p_0=p_1$ and $P_{e0}=P_{e1}$?

Problem 8.13 Consider a uniform quantizer characterized by the input–output relation illustrated in Fig. 8.2(a). Assume that a Gaussian-distributed random variable with zero mean and unit variance is applied to this quantizer input.

(a) What is the probability that the amplitude of the input lies outside the range -4 to $+4$?
(b) Find the probability density function of the discrete random variable at the quantizer output.

Problem 8.14 A PCM system uses a uniform quantizer followed by a 7-bit binary encoder. The bit rate of the system is equal to 50×10^6 bits per second.

(a) What is the maximum message bandwidth for which the system operates satisfactorily?
(b) Determine the output signal-to-quantizing noise ratio when a full-load sinusoidal modulating wave of frequency 1 MHz is applied to the input.

Problem 8.15 Show that, with a nonuniform quantizer, the mean-square value of the quantizing error is approximately equal to $(1/12)\sum_i \delta_i^2 p_i$, where δ_i is the ith step size and p_i is the probability that the input signal amplitude lies within the ith interval. Assume that the step size δ_i is small compared with the excursion of the input signal.

Problem 8.16 Consider a chain of $(n-1)$ regenerative repeaters, with a total of n sequential decisions made on a binary PCM wave, including the final decision made at the receiver. Assume that any binary symbol transmitted through the system has an independent probability p_1 being inverted by any repeater. Let p_n represent the probability that a binary symbol is in error after transmission through the complete system.

(a) Show that

$$p_n=\tfrac{1}{2}[1-(1-2p_1)^n]$$

(b) If p_1 is very small and n is not too large, what is the corresponding value of p_n?

Problem 8.17 The sample function of a Gaussian process of zero mean and unit variance is uniformly sampled, and then applied to a uniform quantizer of the mid-riser type having the input–output amplitude characteristic shown in Fig. P8.3. Calculate the entropy of the quantizer output.

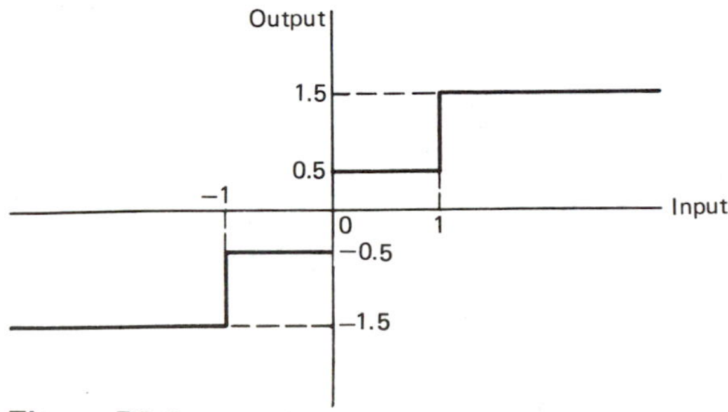

Figure P8.3

Problem 8.18 A voice-grade channel of the telephone network has a bandwidth of 3.4 kHz. Using the Hartley–Shannon law:

(a) Calculate the channel capacity of a telephone channel for a signal-to-noise ratio of 30 dB.
(b) Calculate the minimum signal-to-noise ratio required to support information transmission through a telephone channel at the rate of 4800 bits per second.

Problem 8.19 A black-and-white television picture may be viewed as consisting of approximately 3×10^5 elements, each one of which may occupy one of 10 distinct brightness levels with equal probability. Assume: (a) the rate of transmission is 30 picture frames per second, and (b) the signal-to-noise ratio is 30 dB.

Using the Hartley–Shannon law, calculate the minimum bandwidth required to support the transmission of the resultant video signal.

Note: As a matter of interest, commercial television transmissions actually employ a bandwidth of 4 MHz.

Problem 8.20 In the DPCM system depicted in Fig. P8.4, show that in the absence of transmission noise the transmitting and receiving prediction filters operate on slightly different input signals.

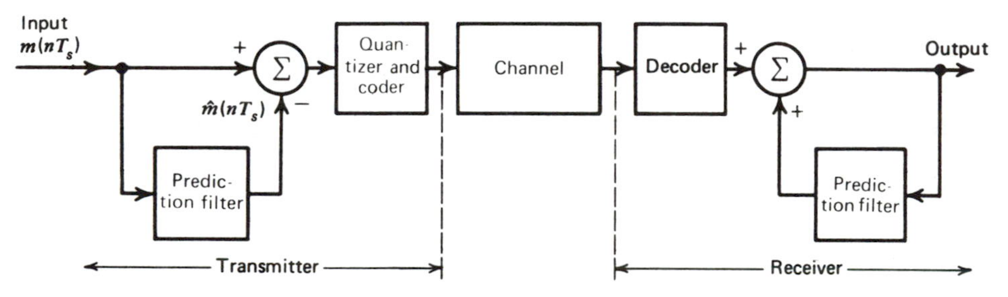

Figure P8.4

Problem 8.21 Consider the first-order prediction defined by

$$\hat{m}(nT_s) = wm(nT_s - T_s)$$

where $m(nT_s)$ is the sample of a stationary signal of zero mean, T_s is the sampling period, and w is a constant.

(a) Show that the prediction error

$$e(nT_s) = m(nT_s) - \hat{m}(nT_s)$$

has variance

$$\sigma_E^2 = \sigma_M^2 \left[1 + w^2 - \frac{2wR_M(T_s)}{\sigma_M^2} \right]$$

where σ_M^2 is the variance of the input signal, and $R_M(T_s)$ is its autocorrelation function for a lag of T_s.

(b) Show that the variance of the prediction error is minimized for

$$w_{\text{opt}} = \frac{R_M(T_s)}{\sigma_M^2}$$

and that the corresponding value of σ_E^2 is

$$\sigma_{E,\min}^2 = \sigma_M^2 - \frac{R_M^2(T_s)}{\sigma_M^2}$$

(c) What are the conditions for which σ_E^2 is less than σ_M^2?

Problem 8.22 Consider a sine-wave of frequency f_m and amplitude A_m, applied to a delta modulator of step size δ. Show that slope-overload distortion will occur if

$$A_m > \frac{\delta}{2\pi f_m T_s}$$

where T_s is the sampling period. What is the maximum power that may be transmitted without slope-overload distortion?

Problem 8.23 The ramp signal $m(t) = at$ is applied to a delta modulator which operates with a sampling period T_s and step size δ.

(a) Show that slope-overload distortion occurs if $\delta < aT_s$.
(b) Sketch the modulator output for the following three values of step size:

(i) $\delta = 0.75aT_s$
(ii) $\delta = aT_s$
(iii) $\delta = 1.25aT_s$

Problem 8.24 Consider a speech signal with maximum frequency of 3.4 kHz and maximum amplitude of 1 volt. This speech signal is applied to a delta modulator whose bit rate is set at 20 kilobits per second. Discuss the choice of an appropriate step size for the modulator.

chapter 9 BASEBAND DATA TRANSMISSION

In the previous chapter, we described various techniques for converting an analog signal into digital form. There is another way in which digital data can arise in practice; the data may represent the output of a source of information that is inherently discrete in nature (e.g., digital computer). In this chapter, we study *discrete pulse modulation* techniques for transmitting digital data (of whatever origin) over a *baseband channel*. In Chapter 10 we study modulation techniques for transmitting digital data over a *band-pass channel*.

In discrete pulse modulation, the amplitude, duration, or position of the transmitted pulses is varied in a discrete manner in accordance with the given digital data. However, for the baseband transmission of digital data, the use of discrete pulse-amplitude modulation (PAM) is the most efficient one in terms of power and bandwidth utilization. Accordingly, in this chapter we restrict our attention to the study of discrete PAM systems.

9.1 ELEMENTS OF BASEBAND BINARY PAM SYSTEMS

The basic elements of a *baseband binary PAM system* are shown in Fig. 9.1. The signal applied to the input of the system consists of a binary data sequence $\{b_k\}$ with a bit duration of T_b seconds; b_k is in the form of 1 or 0. This signal is applied to a pulse generator, producing the pulse waveform

$$x(t) = \sum_{k=-\infty}^{\infty} a_k g(t - kT_b) \tag{9.1}$$

where $g(t)$ denotes the *shaping pulse* that is normalized, so that we may write

$$g(0) = 1 \tag{9.2}$$

The amplitude a_k depends on the identity of the input bit b_k; specifically, we assume that

$$a_k = \begin{cases} +a, & \text{if the input bit } b_k \text{ is represented by symbol 1} \\ -a, & \text{if the input bit } b_k \text{ is represented by symbol 0} \end{cases} \tag{9.3}$$

The PAM signal $x(t)$ passes through a *transmitting filter* of transfer function $H_T(f)$. The resulting filter output defines the transmitted signal, which is modified in a deterministic fashion as a result of transmission through the channel of transfer function $H_C(f)$. In addition, the channel adds random noise to the signal at the receiver input. Then the noisy signal is passed through a *receiving filter* of transfer function $H_R(f)$. This filter output is sampled *synchronously* with the transmitter, with the sampling instants being determined by a *clock* or *timing signal* that is usually extracted from the receiving filter output. Finally, the sequence of samples thus obtained is used to reconstruct the original data sequence by means of a *decision device*. The amplitude of each sample is compared to a *threshold*. If the threshold is exceeded, a decision is made in favor of symbol 1 (say). If the threshold is not exceeded, a decision is made in favor of symbol 0. If the sample amplitude equals the threshold exactly, the flip of a fair coin will determine which symbol was transmitted.

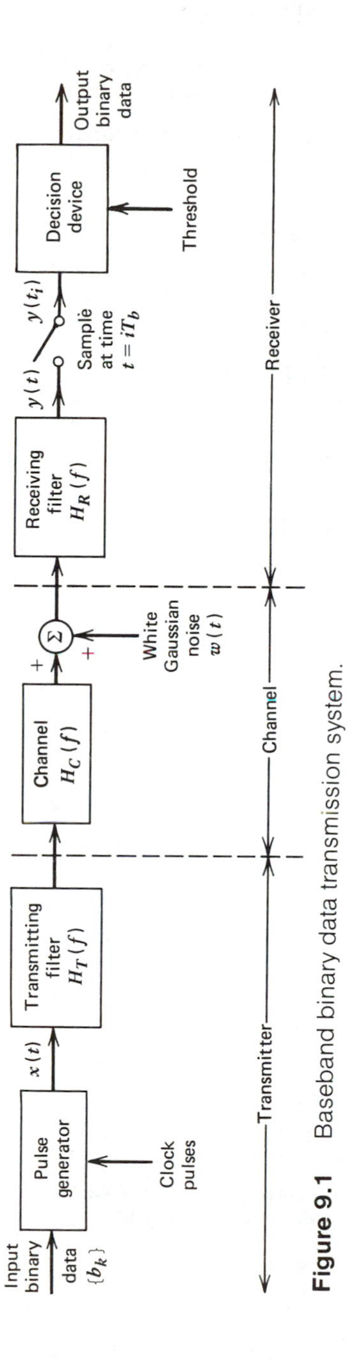

Figure 9.1 Baseband binary data transmission system.

The receiving filter output may be written as*

$$y(t)=\sum_{k=-\infty}^{\infty} A_k p(t-kT_b)+n(t) \tag{9.4}$$

where A_k is the amplitude; the pulse $p(t)$ is normalized such that

$$p(0)=1 \tag{9.5}$$

The pulse $A_k p(t)$ is the response of the cascade connection of the transmitting filter, the channel, and the receiving filter, which is produced by the pulse $a_k g(t)$ applied to the input of this cascade connection. Therefore, we may relate $p(t)$ to $g(t)$ in the frequency domain as follows

$$A_k P(f)=a_k G(f)H_T(f)H_C(f)H_R(f) \tag{9.6}$$

where $P(f)$ and $G(f)$ are the Fourier transforms of $p(t)$ and $g(t)$, respectively. Finally, the term $n(t)$ in Eq. (9.4) is the noise produced at the output of the receiving filter due to the additive noise $w(t)$ at the receiver input. It is customary to model $w(t)$ as a white Gaussian noise of zero mean.

The receiving filter output $y(t)$ is sampled at time $t=iT_b$ (with i taking on integer values), yielding

$$y(t_i)=\sum_{k=-\infty}^{\infty} A_k p[(i-k)T_b)]+n(t_i)$$

$$=A_i+\sum_{\substack{k=-\infty \\ k\neq i}}^{\infty} A_k p[(i-k)T_b)]+n(t_i) \tag{9.7}$$

In Eq. (9.7), the first term A_i represents the ith transmitted bit. The second term represents the residual effect of all other transmitted bits on the decoding of the ith bit; this residual effect is called the *intersymbol interference* (*ISI*). The last term $n(t_i)$ represents the noise sample at time t_i.

In the absence of ISI and noise, we observe from Eq. (9.7) that

$$y(t_i)=A_i$$

which shows that, under these conditions, the ith transmitted bit can be decoded correctly. The unavoidable presence of ISI and noise in the system, however, introduces errors in the decision device at the receiver output. Therefore, in the design of the transmitting and receiving filters, the objective is to minimize the effects of noise and ISI, and thereby deliver the digital data to its destination in an error-free manner as far as possible.

In Section 9.2 we consider the effects of ISI alone; then in Section 9.3 we consider the combined effects of ISI and noise.

* To be precise, an arbitrary time delay t_0 should be included in the argument of the pulse $p(t-kT_b)$ in Eq. (9.4) to represent the effect of transmission delay through the system. For convenience, we have put this delay equal to zero in Eq. (9.4).

9.2 BASEBAND SHAPING

Typically, the transfer function of the channel and the pulse shape are specified, and the problem is to determine the transfer functions of the transmitting and receiving filters so as to enable the receiver to recognize the sequence of values A_i in the received signal wave.

In solving this problem, we have to overcome the intersymbol interference caused by the overlapping tails of other pulses adding to the particular pulse $A_i p(t-iT_b)$, which is examined at the sampling time iT_b. If this form of interference is too strong, it may result in erroneous decisions in the receiver. Clearly, control of intersymbol interference in the system is achieved in the time domain by controlling the function $p(t)$, or in the frequency domain by controlling $P(f)$. One signal waveform that produces *zero* intersymbol interference is defined by the *sinc function*

$$p(t)=\frac{\sin(2\pi B_T t)}{2\pi B_T t}=\operatorname{sinc}(2B_T t) \tag{9.8}$$

where $B_T=1/2T_b$. The corresponding frequency function $P(f)$ is equal to $1/(2B_T)$ for $|f|<B_T$, and zero for $|f|>B_T$. This means that no frequencies of absolute value exceeding half the bit rate are needed. The function $p(t)$ can be regarded as the impulse response of an ideal low-pass filter with passband amplitude response $1/(2B_T)$ and bandwidth B_T. The function $p(t)$ has its peak value at the origin, and goes through zero at integer multiples of the bit duration T_b. It is apparent that if the received waveform $y(t)$ is sampled at the instants of time $t=0, \pm T_b, \pm 2T_b, \ldots,$ then the pulses defined by $A_i p(t-iT_b)$ with arbitrary amplitude A_i and $i=0, \pm 1, \pm 2, \ldots,$ will not interfere with each other.

Although this choice of pulse shape for $p(t)$ achieves economy in bandwidth in that it solves the problem of zero intersymbol interference with the minimum bandwidth possible, there are two practical difficulties that make it an undesirable objective for system design:

1. It requires that the amplitude characteristic of $P(f)$ be flat from $-B_T$ to B_T, and zero elsewhere. This is physically unrealizable, and very difficult to approximate in practice because of the abrupt transitions at $\pm B_T$.
2. The function $p(t)$ decreases as $1/|t|$ for large $|t|$, resulting in a slow rate of decay. This is caused by the discontinuity of $P(f)$ at $\pm B_T$. Accordingly, there is practically no margin of error in sampling times in the receiver. To evaluate the effect of this *timing error*, consider the sample of $y(t)$ at $t=\Delta t$, where Δt is the timing error. To simplify the analysis we have put the correct sampling time t_i equal to zero. We thus obtain, in the absence of noise:

$$\begin{aligned} y(\Delta t) &= \sum_k A_k p(\Delta t - kT_b) \\ &= \sum_k A_k \frac{\sin[2\pi B_T(\Delta t - kT_b)]}{2\pi B_T(\Delta t - kT_b)} \end{aligned} \tag{9.9}$$

Since $2B_T T_b = 1$, we may rewrite Eq. (9.9) as

$$y(\Delta t) = \sum_k A_k \operatorname{sinc}(2B_T \Delta t - k)$$
$$= A_0 \operatorname{sinc}(2B_T \Delta t) + \frac{\sin(2\pi B_T \Delta t)}{\pi} \sum_{\substack{k \\ k \neq 0}} \frac{(-1)^k A_k}{2B_T \Delta t - k} \tag{9.10}$$

The first term on the right-hand side of Eq. (9.10) defines the desired symbol, whereas the remaining series represents the intersymbol interference caused by the timing error Δt in sampling the signal $y(t)$. In certain cases, depending on the values of A_k, it is possible for this series to diverge, thereby causing erroneous decisions in the receiver.

Using the minimum-bandwidth solution described above, it is possible to derive other solutions for $p(t)$ or $P(f)$ with zero intersymbol interference, and yet overcome the two practical difficulties mentioned. We consider one of several such solutions that were first described by Nyquist.* The particular form of $P(f)$, which embodies many desirable features, is constructed by a *raised cosine pulse spectrum*. This frequency characteristic consists of a *flat* portion and a *rolloff* portion that has a sinusoidal form, as follows

$$P(f) = \begin{cases} \dfrac{1}{2B_T}, & |f| < f_1 \\ \dfrac{1}{4B_T}\left\{1 + \cos\left[\dfrac{\pi(|f| - f_1)}{2B_T - 2f_1}\right]\right\}, & f_1 < |f| < 2B_T - f_1 \\ 0, & |f| > 2B_T - f_1 \end{cases} \tag{9.11}$$

The frequency f_1 and bandwidth B_T are related by

$$\rho = 1 - \frac{f_1}{B_T} \tag{9.12}$$

which is called the *rolloff factor*. For $\rho = 0$, that is $f_1 = B_T$, we get the minimum bandwidth solution described earlier.

The frequency response $P(f)$, normalized by multiplying it by $2B_T$, is shown plotted in Fig. 9.2(*a*) for three values of ρ, namely, 0, 0.5, and 1. We see that for $\rho = 0.5$ or 1, the function $P(f)$ cuts off gradually as compared with an ideal low-pass filter (corresponding to $\rho = 0$), and it is therefore easier to realize in practice. Also the function $P(f)$ exhibits odd symmetry about the cutoff frequency B_T of the ideal low-pass filter. The time response $p(t)$, that is, the inverse Fourier transform of $P(f)$, is defined by (see Problem 9.3)

$$p(t) = \operatorname{sinc}(2B_T t)\,\frac{\cos(2\pi\rho B_T t)}{1 - 16\rho^2 B_T^2 t^2} \tag{9.13}$$

* H. Nyquist, "Certain topics in telegraph transmission theory," *Trans. AIEE*, vol. 47, pp. 617–644, Feb. 1928.

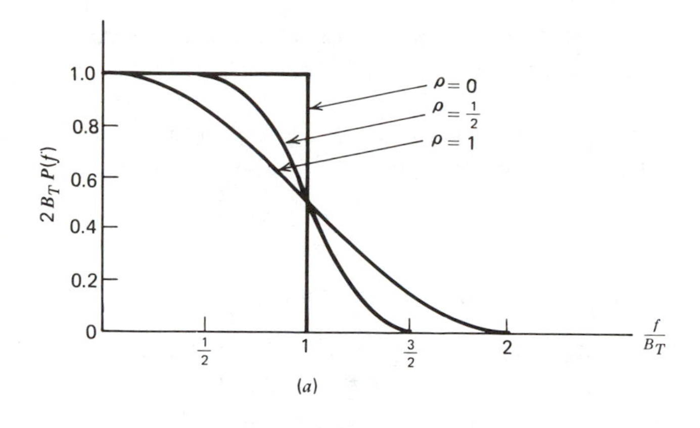

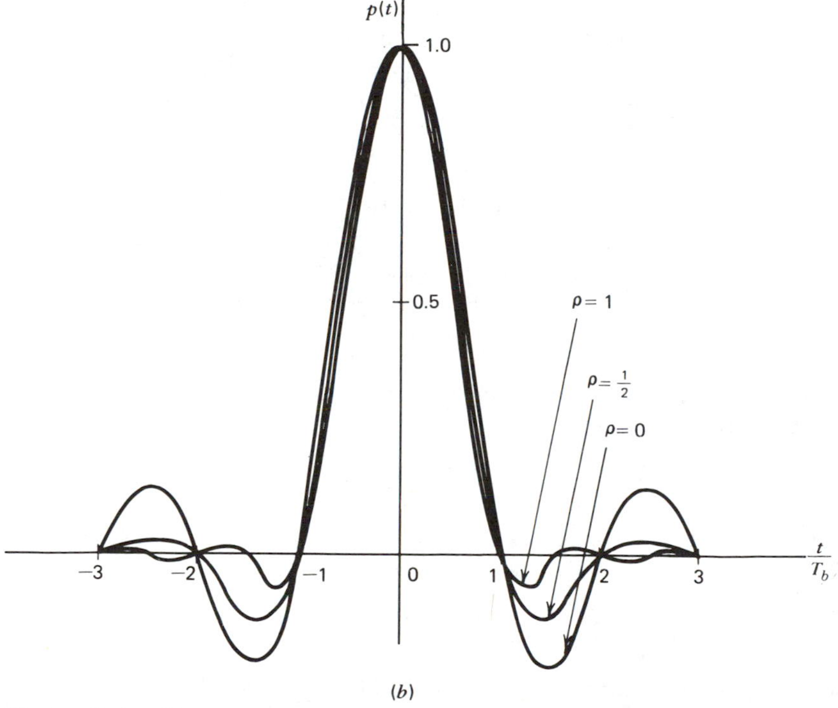

Figure 9.2 Responses for different rolloff factors. *(a)* Frequency response. *(b)* Time response.

This function consists of the product of two factors: the factor $\text{sinc}(2B_T t)$ associated with the ideal filter, and a second factor that decreases as $1/|t|^2$ for large $|t|$. The first factor ensures zero crossings of $p(t)$ at the desired sampling instants of time $t = iT$ with i an integer (positive and negative). The second factor reduces the tails of the pulse considerably below that obtained from the ideal low-pass filter, so that the transmission of binary waves using such pulses is relatively insensitive to sampling time errors. In fact, the amount of intersymbol interference resulting from this timing error decreases as the rolloff factor ρ is increased from zero to unity.

The time response $p(t)$ is shown plotted in Fig. 9.2(b) for $\rho=0$, 0.5 and 1. For the special case of $\rho=1$, the function $p(t)$ simplifies as

$$p(t)=\frac{\text{sinc}(4B_T t)}{1-16B_T^2 t^2} \tag{9.14}$$

This time response exhibits two interesting properties:

1. At $t=\pm T_b/2=\pm 1/4B_T$, we have $p(t)=0.5$; that is, the pulse width measured at half amplitude is exactly equal to the bit duration T_b.
2. There are zero crossings at $t=\pm 3T_b/2, \pm 5T_b/2, \ldots$ in addition to the usual zero crossings at the sampling times $t=\pm T_b, \pm 2T_b, \ldots$.

These two properties are particularly useful in generating a timing signal from the received signal for the purpose of synchronization. However, this requires the use of a transmission bandwidth double that required for the ideal case corresponding to $\rho=0$.

Example 1 Bandwidth Requirements of the T1 System

In Example 1 of Chapter 8 we described the signal format for the T1 carrier system that is used to multiplex 24 independent voice inputs, based on an 8-bit PCM word. It was shown that the bit duration of the resulting time-division multiplexed signal (including a framing bit) is

$$T_b=0.647 \ \mu s$$

Assuming an ideal low-pass characteristic for the channel, it follows that the minimum transmission bandwidth B_T of the T1 system is

$$B_T=\frac{1}{2T_b}=772 \text{ kHz}$$

However, a more realistic value for the necessary transmission bandwidth is obtained by using a cosine-rolloff characteristic with $\rho=1$. In this case, we find that

$$B_T=\frac{1}{T_b}=1.544 \text{ MHz}$$

It is interesting to compare the transmission bandwidth requirement of the T1 system with the minimum bandwidth requirement of a corresponding frequency-division multiplexing (FDM) system. We recall that of all the CW modulation methods, the use of single sideband (SSB) modulation requires the minimum bandwidth possible. Thus, in order to accommodate an FDM system using SSB modulation to transmit 24 independent voice inputs, and assuming a bandwidth of 4 kHz for each voice input, the channel must provide a transmission bandwidth of

$$B_T=24\times 4=96 \text{ kHz}$$

This is an order of magnitude smaller than the minimum bandwidth requirement of the T1 system.

9.3 OPTIMUM TRANSMITTING AND RECEIVING FILTERS FOR NOISE IMMUNITY

With the pulse shape $p(t)$ designed to reduce intersymbol interference at the sampling instants to zero, as described above, the problem that we wish to solve next involves the design of the transmitting and receiving filter such that the *probability of error* due to noise is minimized.

As mentioned previously, the second term in Eq. (9.7) is responsible for the ISI problem. Assuming the use of a pulse shape $p(t)$ that reduces the ISI to zero, at the sampling times, we find that the sample amplitude at the input of the decision device at time $t_i = iT_b$ is given by

$$y(t_i) = A_i + n(t_i) \tag{9.15}$$

where

$$A_i = \begin{cases} +A, & \text{when 1 is transmitted} \\ -A, & \text{when 0 is transmitted} \end{cases} \tag{9.16}$$

Thus, with a threshold of zero volts, if $y(t_i) > 0$, the receiver says symbol 1, and if $y(t_i) < 0$ it says symbol 0.

We assume that the additive noise $w(t)$ at the receiver input is white Gaussian with zero mean and power spectral density $N_0/2$. Therefore, in Eq. (9.15), $n(t_i)$ is the sample value of a Gaussian random variable of zero mean and variance

$$\sigma_N^2 = \frac{N_0}{2} \int_{-\infty}^{\infty} |H_R(f)|^2 \, df \tag{9.17}$$

The average probability of error is given by (see Problem 8.9)

$$P_e = \tfrac{1}{2} \operatorname{erfc}\left(\frac{A}{\sqrt{2}\sigma_N} \right) \tag{9.18}$$

From Eq. (9.18) we observe that P_e decreases as A/σ_N increases. Hence, in order to minimize the average probability of error we have to design the transmitting and receiving filters to as to maximize the ratio A/σ_N or, equivalently, the ratio A^2/σ_N^2; this latter ratio may be viewed as a signal-to-noise ratio. In order to do this maximization, we need to express the ratio A^2/σ_N^2 in terms of the transfer functions $H_T(f)$ and $H_R(f)$ of the transmitting and receiving filters, respectively.

The signal $x(t)$ at the input to the transmitting filter is defined by Eq. (9.1), which is reproduced here for convenience.

$$x(t) = \sum_{k=-\infty}^{\infty} a_k g(t - kT_b) \tag{9.19}$$

where $g(t)$ is a pulse of unit amplitude and duration less than or equal to the bit duraction T_b. We assume that the bits constituting the input signal $x(t)$ are independent and equiprobable. Therefore, the power spectral density of the input signal $x(t)$ is given by (see Example 7 of Chapter 5)*

$$S_X(f)=\frac{a^2\Psi_g(f)}{T_b} \tag{9.20}$$

where $\Psi_g(f)=|G(f)|^2$ is the energy spectral density of the pulse $g(t)$. The signal $x(t)$ is applied to the transmitting filter, producing the output $z(t)$ that constitutes the transmitted signal. Therefore, the power spectral density of $z(t)$ is given by

$$\begin{aligned} S_Z(f)&=|H_T(f)|^2 S_X(f) \\ &=\frac{a^2}{T_b}|H_T(f)|^2|G(f)|^2 \end{aligned} \tag{9.21}$$

and the average transmitted power is

$$\begin{aligned} P&=\int_{-\infty}^{\infty} S_Z(f)df \\ &=\frac{a^2}{T_b}\int_{-\infty}^{\infty} |H_T(f)|^2|G(f)|^2\, df \end{aligned} \tag{9.22}$$

From Eq. (9.6), we observe that for the kth bit,

$$A_k=Ka_k \tag{9.23}$$

or, equivalently,

$$A=Ka \tag{9.24}$$

where K is a scaling factor. Thus, eliminating a between Eqs. (9.22) and (9.24), we get

$$P=\frac{A^2}{K^2T_b}\int_{-\infty}^{\infty} |H_T(f)|^2|G(f)|^2\, df$$

which we can solve for A^2, and so obtain

$$A^2=K^2PT_b\left[\int_{-\infty}^{\infty} |H_T(f)|^2|G(f)|^2\, df\right]^{-1} \tag{9.25}$$

Hence, from Eqs. (9.17) and (9.25), we find that the signal-to-noise ratio to be maximized is

$$\frac{A^2}{\sigma_N^2}=\frac{2K^2PT_b}{N_0}\left[\int_{-\infty}^{\infty} |H_T(f)|^2|G(f)|^2\, df \int_{-\infty}^{\infty} |H_R(f)|^2\, df\right]^{-1} \tag{9.26}$$

However, maximization of the ratio A^2/σ_N^2 is subject to the constraint described by Eq. (9.6). Eliminating $H_T(f)G(f)$ between Eqs. (9.6) and (9.26), and using Eq.

* Note that in Eq. (9.20) the pulse $g(t)$ is defined as having unit amplitude at $t=0$, so that the amplitude-scaled pulse $ag(t)$ used to derive this equation is the same as the pulse $g(t)$ used in Example 7 of Chapter 5.

(9.23), we get

$$\frac{A^2}{\sigma_N^2} = \frac{2PT_b}{N_0}\left[\int_{-\infty}^{\infty} \frac{|P(f)|^2}{|H_C(f)H_R(f)|^2}\,df \int_{-\infty}^{\infty} |H_R(f)|^2\,df\right]^{-1} \tag{9.27}$$

The desired maximization of A^2/σ_N^2 is equivalent to minimizing the quantity

$$\eta = \int_{-\infty}^{\infty} \frac{|P(f)|^2}{|H_C(f)H_R(f)|^2}\,df \int_{-\infty}^{\infty} |H_R(f)|^2\,df \tag{9.28}$$

with respect to $H_R(f)$. This minimization may be performed by using *Schwarz's inequality*, which is discussed in Appendix 5. To proceed with this evaluation, let

$$|H_R(f)| = U(f)$$

and

$$\frac{|P(f)|}{|H_C(f)H_R(f)|} = V(f)$$

so that we may write

$$\eta = \int_{-\infty}^{\infty} V^2(f)df \int_{-\infty}^{\infty} U^2(f)df$$

Then, according to Schwarz's inequality, we have

$$\int_{-\infty}^{\infty} V^2(f)df \int_{-\infty}^{\infty} U^2(f)df \geqslant \left[\int_{-\infty}^{\infty} V(f)U(f)df\right]^2$$

This relation is satisfied with the equality sign (thereby defining the minimum value of the quantity on the left-hand side) when

$$U(f) = CV(f)$$

where C is an arbitrary positive constant. Thus, applying Schwarz's inequality to Eq. (9.28), we may state the following:

1. The minimum value of the quantity η is

$$\eta_{\min} = \left[\int_{-\infty}^{\infty} \frac{|P(f)|}{|H_C(f)|}\,df\right]^2 \tag{9.29}$$

2. The optimum receiving filter that yields this minimum value is defined by

$$|H_{R,\text{opt}}(f)|^2 = \frac{C|P(f)|}{|H_C(f)|} \tag{9.30}$$

There now only remains the problem of deriving the optimum transmitting filter. For this evaluation, we substitute Eqs. (9.23) and (9.30) in (9.6), obtaining

$$|H_{T,\text{opt}}(f)|^2 = \frac{K^2|P(f)|}{C|G(f)|^2|H_C(f)|} \tag{9.31}$$

where C is an arbitrary positive constant.

Equations (9.30) and (9.31) define the squared amplitude responses of the optimum transmitting and receiving filters that maximize the signal-to-noise ratio A^2/σ_N^2 at the sampling points for a constant transmitted power. These two filters may have arbitrary phase responses as long as they compensate one another.

Finally, from Eqs. (9.27), (9.28), and (9.29), we find that the maximum value of the signal-to-noise ratio A^2/σ_N^2 is given by

$$\left(\frac{A^2}{\sigma_n^2}\right)_{\max} = \frac{2PT_b}{N_0}\left[\int_{-\infty}^{\infty} \frac{|P(f)|}{|H_C(f)|}\, df\right]^{-2} \tag{9.32}$$

and from Eq. (9.18), the corresponding minimum value of the average probability of error is

$$P_{e,\min} = \tfrac{1}{2}\,\text{erfc}\left[\frac{1}{\sqrt{2}}\left(\frac{A}{\sigma_N}\right)_{\max}\right] \tag{9.33}$$

A special case of practical significance occurs when the pulse shape $g(t)$ is chosen such that its energy spectral density $\Psi_g(f)$ does not change much over the frequency band of interest. Then, from Eqs. (9.30) and (9.31), we find that except for arbitrary scaling factors, the optimum transmitting and receiving filters have the same amplitude response. Furthermore, if these two filters are chosen to have a linear phase response, their designs become identical, with each one accomplishing "half" of the desired pulse shaping. This is an advantage in system construction, since only one filter design is required. A simple way of ensuring that the energy spectral density $\Psi_g(f)$ is approximately constant over the frequency band of interest is to use a pulse $g(t)$ that consists of a rectangular pulse whose duration is small compared to the bit duration T_b.

9.4 CORRELATIVE CODING

Thus far we have treated intersymbol interference as an undesirable phenomenon that produces a degradation in system performance. Indeed, its very name connotes a nuisance effect. Nevertheless, by adding intersymbol interference to the transmitted signal in a controlled manner, it is possible to achieve a signaling rate of $2B_T$ symbols per second in a channel of bandwidth B_T hertz. Such schemes are called *correlative coding* or *partial-response signaling* schemes.* The design of

* Correlative and partial response are synonomous; both terms are used in the literature. The idea of correlative coding was originated by Lender in 1963: A. Lender, "The duobinary technique for high-speed data transmission," *IEEE Trans. on Communications and Electronics*, vol. 82, pp. 214–218, May 1963. Lender's work was generalized for binary data transmission by Kretzmer: E. R. Kretzmer, "Generalization of a technique for binary data communication," *IEEE Trans. on Communication Technology*, vol. COM-14, pp. 67–68, Feb. 1966.

For further details on correlative techniques, see:

1. P. Kabal and S. Pasupathy, "Partial-response signaling," *IEEE Trans. on Communications*, vol. COM-23, pp. 921–934, Sept. 1975.
2. S. Pasupathy, "Correlative coding—A bandwidth-efficient signaling scheme," *IEEE Communications Society Magazine*, pp. 4–11, July 1977.
3. A. Lender, "Correlative (partial response) techniques and applications to digital radio systems," in K. Feher, *Digital Communications: Microwave Applications*, pp. 144–182 (Prentice-Hall, 1981).

these schemes is based on the premise that since intersymbol interference introduced into the transmitted signal is known, its effect can be interpreted at the receiver. Thus correlative coding may be regarded as a practical means of achieving the theoretical maximum signaling rate of $2B_T$ symbols per second in a bandwidth of B_T hertz, as postulated by Nyquist, using realizable and perturbation-tolerant filters.

Duobinary Signaling

The basic idea of correlative coding will now be illustrated by considering the specific example of *duobinary signaling*, where "duo" implies doubling of the transmission capacity of a straight binary system.

Consider a binary input sequence $\{b_k\}$ consisting of uncorrelated binary digits each having duration T_b seconds, with symbol 1 represented by a pulse of amplitude $+1$ volt, and symbol 0 by a pulse of amplitude -1 volt. When this sequence is applied to a *duobinary encoder*, it is converted into a *three-level output*, namely, -2, 0, and $+2$ volts. To produce this transformation, we may use the scheme shown in Fig. 9.3. The binary sequence $\{b_k\}$ is first passed through a simple filter involving a single delay element. For every unit impulse applied to the input of this filter, we get two unit impulses spread T_b seconds apart at the filter output. We may therefore express the digit c_k at the duobinary coder output as the sum of the present binary digit b_k and its previous value b_{k-1}, as shown by

$$c_k = b_k + b_{k-1} \tag{9.34}$$

One of the effects of the transformation described by Eq. (9.34) is to change the input sequence $\{b_k\}$ of uncorrelated binary digits into a sequence $\{c_k\}$ of correlated digits. This correlation between the adjacent transmitted levels may be viewed as introducing intersymbol interference into the transmitted signal in an artificial manner. However, this intersymbol interference is under the designer's control, which is the basis of correlative coding.

An ideal delay element, producing a delay of T_b seconds, has the transfer function

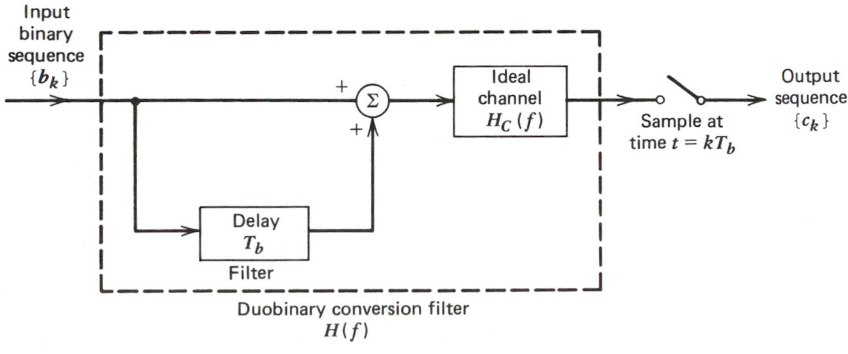

Figure 9.3 Duobinary signaling scheme.

$\exp(-j2\pi f T_b)$, so that the transfer function of the simple filter shown in Fig. 9.3 is $1+\exp(-j2\pi f T_b)$. Hence, the overall transfer function of this filter connected in cascade with the ideal channel $H_C(f)$ is

$$\begin{aligned} H(f) &= H_C(f)[1+\exp(-j2\pi f T_b)] \\ &= H_C(f)[\exp(j\pi f T_b)+\exp(-j\pi f T_b)]\exp(-j\pi f T_b) \\ &= 2H_C(f)\cos(\pi f T_b)\exp(-j\pi f T_b), \end{aligned} \tag{9.35}$$

For an ideal channel of bandwidth $B_T = 1/2T_b$, we have

$$H_C(f) = \begin{cases} 1, & |f| \leqslant 1/2T_b \\ 0, & \text{otherwise} \end{cases} \tag{9.36}$$

Thus the overall frequency response has the form of a half-cycle cosine function, as shown by

$$H(f) = \begin{cases} 2\cos(\pi f T_b)\exp(-j\pi f T_b), & |f| \leqslant 1/2T_b \\ 0, & \text{otherwise} \end{cases} \tag{9.37}$$

for which the amplitude response and phase response are as shown in parts (*a*) and (*b*) of Fig. 9.4, respectively. An advantage of this frequency response is that it can be easily approximated in practice.

The corresponding value of the impulse response consists of two sinc pulses, time-displayed by T_b seconds, as shown by (except for a scaling factor)

$$\begin{aligned} h(t) &= \frac{\sin(\pi t/T_b)}{\pi t/T_b} + \frac{\sin[\pi(t-T_b)/T_b]}{\pi(t-T_b)/T_b} \\ &= \frac{\sin(\pi t/T_b)}{\pi t/T_b} - \frac{\sin(\pi t/T_b)}{\pi(t-T_b)/T_b} \\ &= \frac{T_b^2 \sin(\pi t/T_b)}{\pi t(T_b - t)}. \end{aligned} \tag{9.38}$$

which is shown plotted in Fig. 9.5. We see that the overall impulse response $h(t)$ has only *two* distinguishable values at the sampling instants.

The original data $\{b_k\}$ may be detected from the duobinary-coded sequence $\{c_k\}$ by subtracting the previous decoded binary digit from the currently received digit c_k in accordance with Eq. (9.34). Specifically, letting $\hat{b}_k$ represent the *estimate* of the original binary digit b_k as conceived by the receiver at time $t = kT_b$, we have

$$\hat{b}_k = c_k - \hat{b}_{k-1} \tag{9.39}$$

It is apparent that if c_k is received without error and if also the previous estimate $\hat{b}_{k-1}$ at time $t=(k-1)T_b$ corresponds to a correct decision, then the current estimate $\hat{b}_k$ will be correct too. The technique of using a stored estimate of the previous symbol is called *decision feedback*.

We observe that the detection procedure as described above is essentially an inverse of the operation of the simple filter at the transmitter. However, a major drawback of this detection process is that once errors are made, they tend to

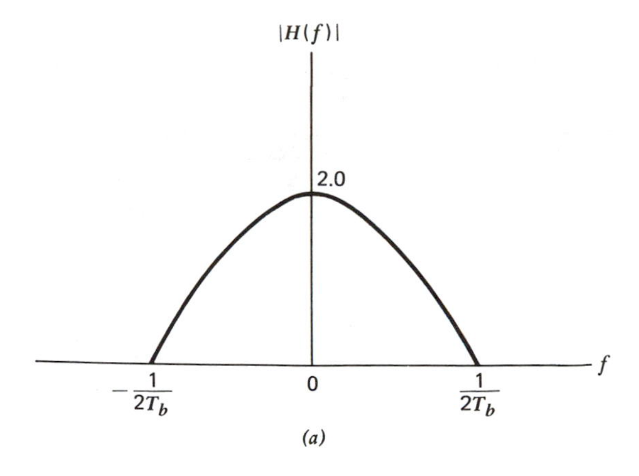

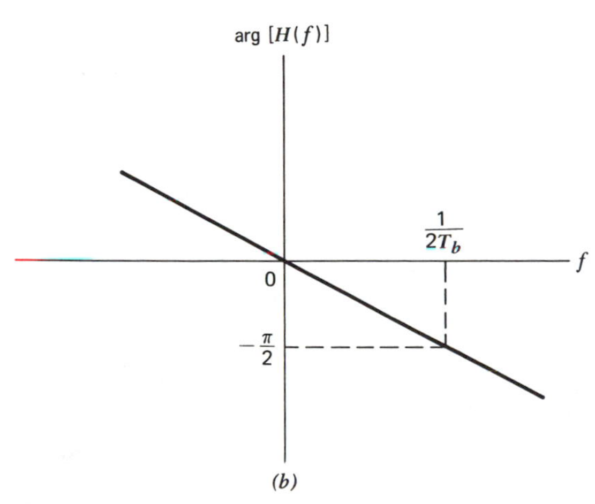

Figure 9.4 Frequency response of the duobinary conversion filter. *(a)* Amplitude response. *(b)* Phase response.

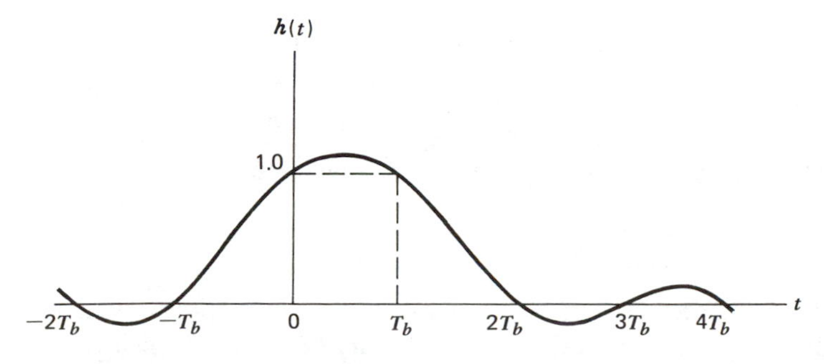

Figure 9.5 Impulse response of duobinary conversion filter.

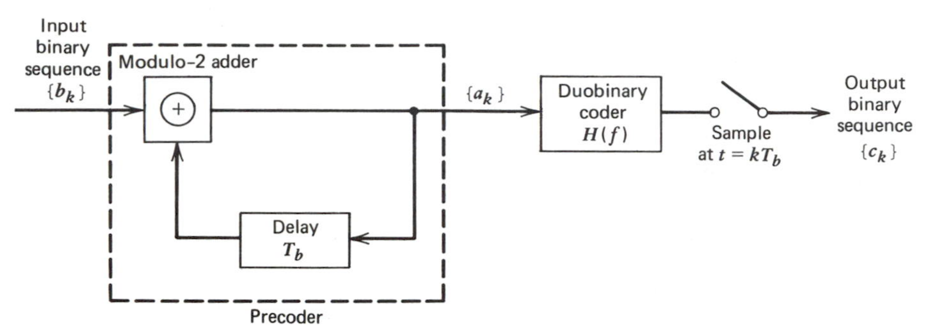

Figure 9.6 A precoded duobinary scheme. Details of the duobinary coder are given in Figure 9.3.

propagate. This is due to the fact that a decision on the current binary digit b_k depends on the correctness of the decision made on the previous binary digit b_{k-1}.

A practical means of avoiding this error propagation is to use *precoding* before the duobinary coding, as shown in Fig. 9.6. The precoding operation performed on the input binary sequence $\{b_k\}$ converts it into another binary sequence $\{a_k\}$ defined by

$$a_k = b_k \oplus a_{k-1} \tag{9.40}$$

where the symbol $\oplus$ denotes *modulo-two addition* of the binary digits b_k and a_{k-1}. This addition is equivalent to the EXCLUSIVE OR operation. An EXCLUSIVE OR gate operates as follows. The output of a two-input EXCLUSIVE OR gate is a 1 if one and only if one input is a 1; otherwise, the output remains a 0. The resulting precoder output $\{a_k\}$ is next applied to the duobinary coder, thereby producing the sequence $\{c_k\}$ that is related to $\{a_k\}$ as follows

$$c_k = a_k + a_{k-1} \tag{9.41}$$

Note that unlike the linear operation of duobinary coding, the precoding is a nonlinear operation.

We assume that symbol 1 at the precoder output in Fig. 9.6 is represented by $+1$ volt and symbol 0 by -1 volt. Therefore, from Eqs. (9.40) and (9.41), we find that

$$c_k = \begin{cases} \pm 2 \text{ volts, if } b_k \text{ is represented by symbol 1} \\ 0 \text{ volts, if } b_k \text{ is represented by symbol 0} \end{cases} \tag{9.42}$$

which is illustrated in Example 2 below. From Eq. (9.42) we deduce the following decision rule for detecting the original input binary sequence $\{b_k\}$ from $\{c_k\}$:

$$b_k = \begin{cases} \text{symbol 0,} & \text{if } |c_k| > 1 \text{ volt} \\ \text{symbol 1,} & \text{if } |c_k| < 1 \text{ volt} \end{cases} \tag{9.43}$$

According to Eq. (9.43), the detector consists of a rectifier, the output of which is compared to a threshold of 1 volt, and the original binary sequence $\{b_k\}$ is thereby detected. A block diagram of the detector is shown in Fig. 9.7. A useful feature of

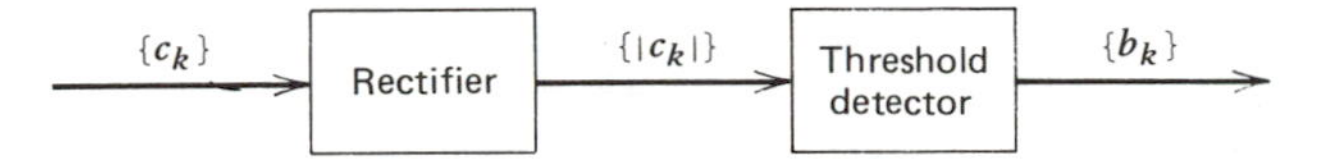

Figure 9.7 Detector for recovering original binary sequence from the precoded duobinary coder output.

this detector is that no knowledge of any input sample other than the present one is required. Hence, error propagation cannot occur in the detector of Fig. 9.7.

Example 2

Consider the input binary sequence 0010110. To proceed with the precoding of this sequence, which involves feeding the precoder output back to the input, we add an extra bit to the precoder output. This extra bit is chosen arbitrarily as a bit 1. Hence, using Eq. (9.40), we find that the sequence $\{a_k\}$ at the precoder output is as shown in row 2 of Table 9.1. We assume that symbol 1 is represented by $+1$ volt and symbol 0 by -1 volt. Accordingly, the precoder output has the amplitudes shown in row 3. Finally, using Eq. (9.34), we find that the duobinary coder output has the amplitudes given in row 4 of Table 9.1.

To detect the original binary sequence, we apply the decision rule of Eq. (9.43), and so obtain the sequence given in row 5 of Table 9.1. This shows that, in the absence of noise, the original binary sequence is detected correctly.

Table 9.1

Binary sequence $\{b_k\}$		0	0	1	0	1	1	0
Binary sequence $\{a_k\}$	1	1	1	0	0	1	0	0
Polar representation of precoder output, a_k (volts)	$+1$	$+1$	$+1$	-1	-1	$+1$	-1	-1
Duobinary coder output, c_k (volts)		2	2	0	-2	0	0	-2
Sequence obtained by applying decision rule of Eq. (9.43)		0	0	1	0	1	1	0

Modified Duobinary Technique

The *modified duobinary* technique involves a correlation span of two binary digits. This is achieved by subtracting input binary digits spaced $2T_b$ seconds apart, as indicated in the block diagram of Fig. 9.8. The output of the modified duobinary conversion filter is related to the sequence $\{a_k\}$ at its input as follows

$$c_k = a_k - a_{k-2} \tag{9.44}$$

Here, again, we find that a three-level signal is generated. If $a_k = \pm 1$ volt, as assumed previously, c_k takes on one of three values: 2, 0, and -2 volts.

The overall transfer function of the tapped-delay-line filter connected in cascade

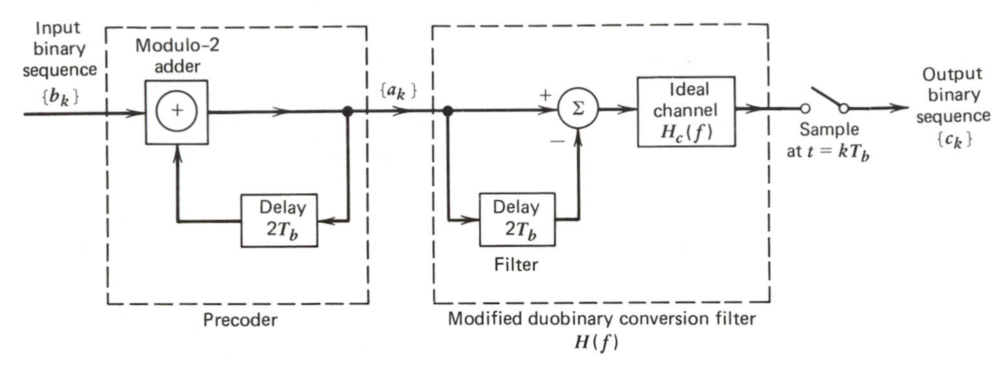

Figure 9.8 Modified duobinary signaling scheme.

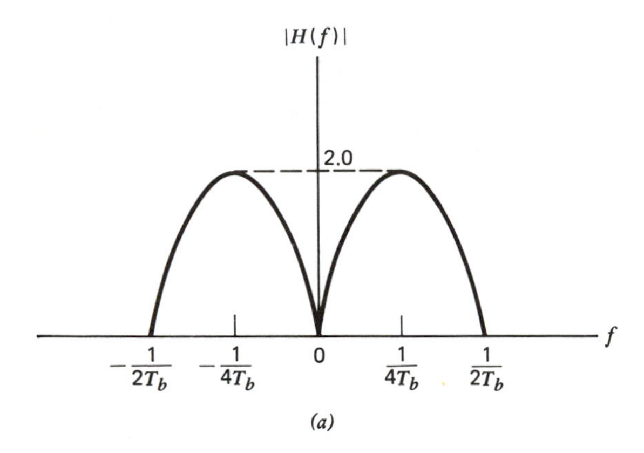

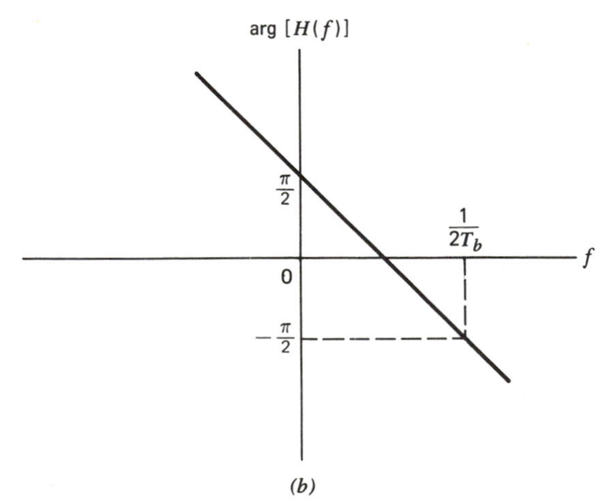

Figure 9.9 Frequency response of modified duobinary conversion filter. *(a)* Amplitude response. *(b)* Phase response.

with the ideal channel $G(f)$, as in Fig. 9.8, is given by

$$\begin{aligned} H(f) &= H_C(f)[1-\exp(-j4\pi f T_b)] \\ &= 2jH_C(f)\sin(2\pi f T_b)\exp(-j2\pi f T_b) \end{aligned} \tag{9.45}$$

where $H_C(f)$ is as defined in Eq. (9.36). We, therefore, have an overall frequency response in the form of a half-cycle sine function, as shown by

$$H(f) = \begin{cases} 2j\sin(2\pi f T_b)\exp(-j2\pi f T_b), & |f| \leq 1/2T_b \\ 0, & \text{elsewhere} \end{cases} \tag{9.46}$$

The corresponding amplitude response and phase response of the modified duobinary-coder as shown in parts (*a*) and (*b*) of Fig. 9.9, respectively. A useful feature of the modified duobinary coder is the fact that its output has no dc component. This property is important since, in practice, many communication channels cannot transmit a dc component.

The impulse response of the modified duobinary coder consists of two sinc pulses that are time-displaced by $2T_b$ seconds, as shown by (except for a scaling factor)

$$\begin{aligned} h(t) &= \frac{\sin(\pi t/T_b)}{\pi t/T_b} - \frac{\sin[\pi(t-2T_b)/T_b)]}{\pi(t-2T_b)/T_b} \\ &= \frac{\sin(\pi t/T_b)}{\pi t/T_b} - \frac{\sin(\pi t/T_b)}{\pi(t-2T_b)/T_b} \\ &= \frac{2T_b^2\sin(\pi t/T_b)}{\pi t(2T_b-t)} \end{aligned} \tag{9.47}$$

This impulse response is plotted in Fig. 9.10, which shows that it has *three* distinguishable levels at the sampling instants.

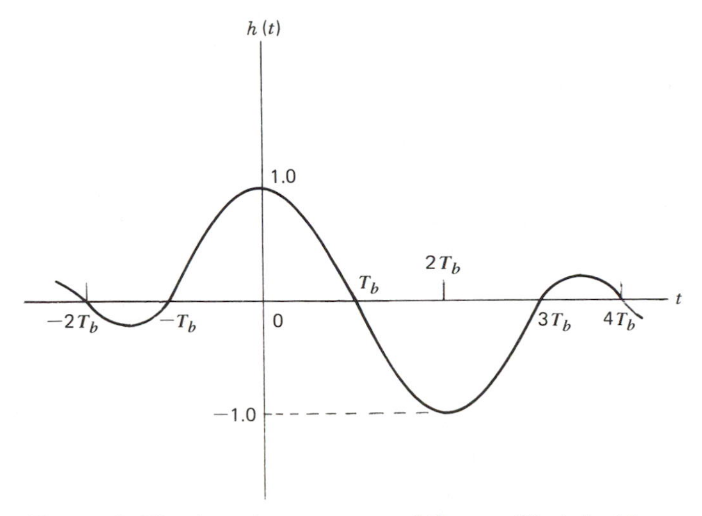

Figure 9.10 Impulse response of the modified duobinary conversion filter.

In order to eliminate the possibility of error propagation in the modified duobinary system, we use a precoding procedure similar to that used for the duobinary case. Specifically, prior to the generation of the modified duobinary signal, a modulo-two logical addition is used on signals $2T_b$ seconds apart, as shown by (see Fig. 9.8)

$$a_k = b_k \oplus a_{k-2} \tag{9.48}$$

where $\{b_k\}$ is the input binary sequence and $\{a_k\}$ is the sequence at the precoder output. Note that modulo-two addition and modulo-two subtraction are the same. The sequence $\{a_k\}$ thus produced is then applied to the modified duobinary conversion filter.

In the case of Fig. 9.8, the output digit c_k equals 0, +1, or −1 volt. Also we find that b_k can be extracted from c_k by disregarding the polarity of c_k. Since plus and minus operations are the same in modulo-two logic, the modulo-two value of c_k also provides the desired value of b_k. Accordingly, at the receiver, we may extract the original sequence $\{b_k\}$ by writing

$$\hat{b}_k = |c_k| = c_k \text{ modulo-two.} \tag{9.49}$$

Generalized Form of Correlative Coding

The duobinary and modified duobinary techniques have correlation spans of 1 binary digit and 2 binary digits, respectively. It is a straightforward matter to generalize these two techniques to other schemes, which are known collectively as *correlative coding schemes.* This generalization is shown in Fig. 9.11, where $H_C(f)$ is defined in Eq. (9.36). It involves the use of a tapped-delay-line filter with tap weights $w_0, w_1, \ldots, w_{N-1}$. Specifically, a correlative sample c_k is obtained from a superposition of N successive input sample values b_k, as shown by

$$c_k = \sum_{n=0}^{N-1} w_n b_{k-n} \tag{9.50}$$

Thus, by choosing various combination of integer values for the w's, we obtain different forms of correlative coding schemes to suit individual applications. For example, in the duobinary case, we have

$$w_0 = +1$$
$$w_1 = +1$$

and $w_n = 0$, for $n \geqslant 2$.
In the modified duobinary case, we have

$$w_0 = +1$$
$$w_1 = 0$$
$$w_2 = -1$$

and $w_n = 0$ for $n \geqslant 3$.

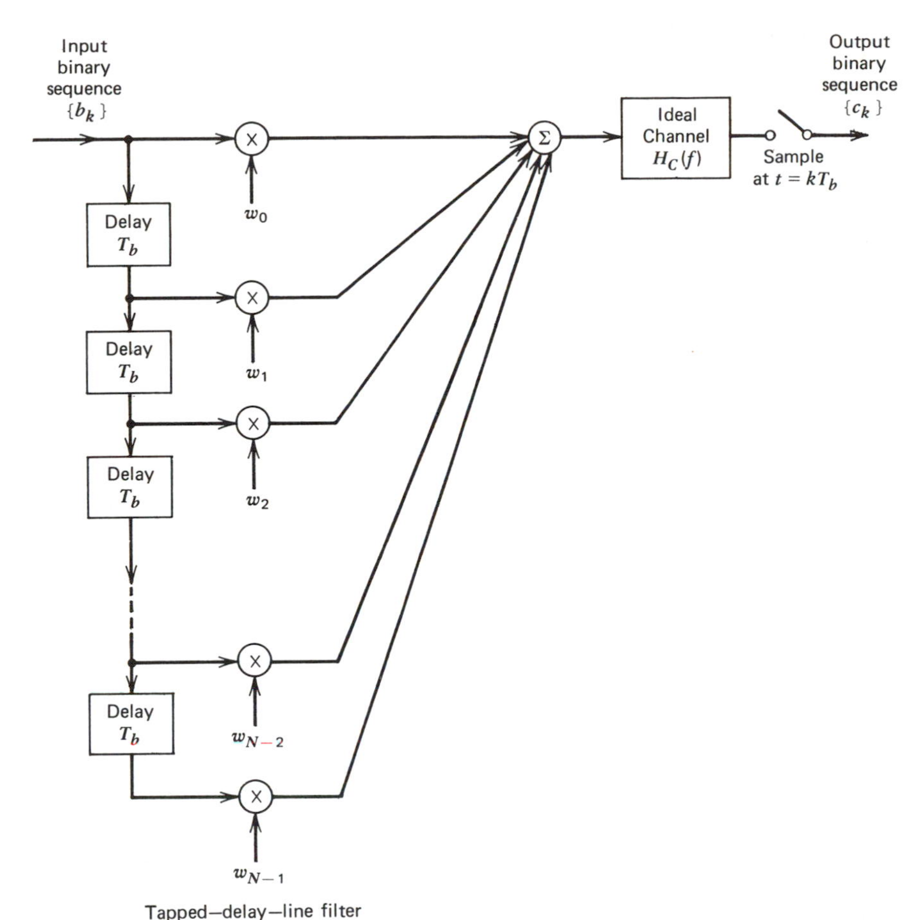

Figure 9.11 Generalized correlative coding scheme.

9.5 BASEBAND M-ARY PAM SYSTEMS

In the baseband binary PAM system of Fig. 9.1, the output of the pulse generator consists of binary pulses, that is, pulses with one of two possible amplitude levels. On the other hand, in a *baseband M-ary PAM system*, the output of the pulse generator takes on one of M possible amplitude levels, with $M > 2$, as illustrated in Fig. 9.12(*a*) for the case of a *quaternary* ($M = 4$) system. The corresponding electrical representation for each of the four possible *dibits* (pairs of bits) is shown in part (*b*) of the figure. In an M-ary system, the information source emits a sequence of symbols from an alphabet that consists of M symbols. Each amplitude level at the pulse generator output corresponds to a distinct symbol, so that there are M distinct amplitude levels to be transmitted. Consider then an *M-ary* PAM system with a signal alphabet that contains M equally likely and statistically independent symbols, with the symbol duration denoted by T seconds. We refer to $1/T$ as the

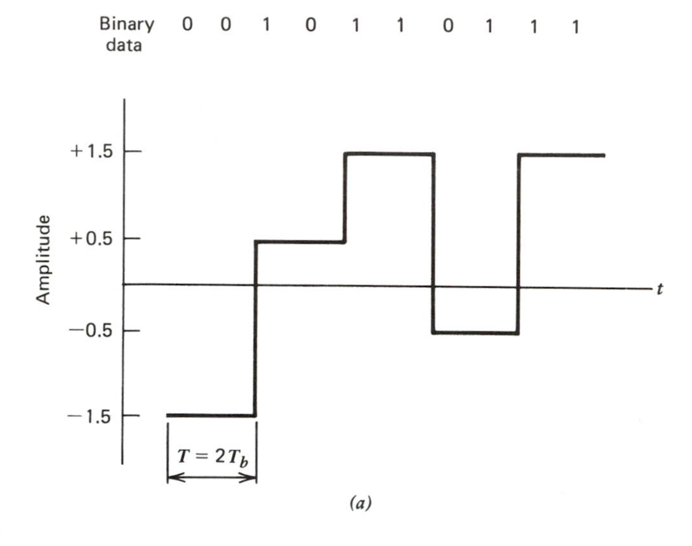

(a)

Dibit	Amplitude
00	−1.5
01	−0.5
10	+0.5
11	+1.5

(b)

Figure 9.12 Output of a quartenary system. *(a)* Waveform. *(b)* Representation of the 4 possible dibits.

signaling rate of the system, which is expressed in *symbols per second* or *bauds.* It is informative to relate the signaling rate of this system to that of an equivalent binary PAM system for which the value of M is 2 and the binary symbols 1 and 0 are equally likely and statistically independent, with the duration of either symbol denoted by T_b seconds. From Example 3 of Chapter 8 we recall that for a zero-memory binary source, with symbols 1 and 0 equally likely and successive symbols statistically independent, the entropy equals one bit. Hence, the binary PAM system produces information at the rate of $1/T_b$ *bits per second.* We also observe that in the case of a *quaternary* PAM system for example, the four possible symbols may be identified with the dibits 00, 01, 10, and 11. We thus see that each symbol represents 2 bits of information, and 1 baud is equal to 2 bits per second. We may generalize this result by stating that in an M-ary PAM system, 1 baud is equal to $\log_2 M$ bits per second, and the symbol duration T of the M-ary PAM system is related to the bit duration T_b of the equivalent binary PAM system as follows:

$$T = T_b \log_2 M \tag{9.51}$$

Therefore, in a given channel bandwidth, we find that by using an M-ary PAM system, we are able to transmit information at a rate that is $\log_2 M$ faster than the corresponding binary PAM system. However, in order to realize the same average probability of symbol error, an M-ary PAM system requires more transmitted power. Specifically, we find that for M much larger than 2 and an average probability of symbol error small compared to 1, the transmitted power must be increased by a factor of $M^2/\log_2 M$, compared to a binary PAM system.*

In a baseband M-ary system, first of all, the sequence of symbols emitted by the information source is converted into an M-level PAM pulse train by a pulse generator at the transmitter input. Next, as with the binary PAM system, this pulse train is shaped by a transmitting filter and then transmitted over the communication channel, which corrupts the signal waveform with both noise and distortion. The signal plus noise is passed through a receiving filter, and then sampled at an appropriate rate in synchronism with the transmitter. Each sample is compared with preset *threshold* values (also called *slicing* levels) and a decision is made as to which symbol was transmitted. We therefore find that the designs of the pulse generator and the decision-making device in an M-ary PAM are more complex than those in a binary PAM system. Intersymbol interference, noise, and imperfect synchronization cause errors to appear at the receiver output. The transmitting and receiving filters are designed to minimize these errors. Procedures used for the design of these filters are similar to those discussed in Sections 9.2 and 9.3 for baseband binary PAM systems.†

9.6 ADAPTIVE EQUALIZATION

An efficient approach to *high-speed transmission* of digital data (e.g., computer data) over a voice-grade telephone channel (which is characterized by a limited bandwidth and high signal-to-noise ratio) involves the use of two basic signal processing operations:

1. Discrete PAM by encoding the amplitudes of successive pulses in a periodic pulse train with a discrete set of possible amplitude levels.
2. A linear modulation scheme that offers bandwidth conservation (e.g., quadrature-amplitude modulation or vestigial sideband modulation) to transmit the encoded pulse train over the telephone channel.

At the receiving end of the system, the received wave is demodulated, and then synchronously sampled and quantized. As a result of dispersion of the pulse shape by the channel, we find that the number of detectable amplitude levels is often

* K. S. Shanmugam, *Digital and Analog Communication Systems*, p. 219 (Wiley, 1979).

† For details of the analysis and design of baseband M-ary PAM systems, see

1. K. S. Shanmugam, pp. 208–222, op. cit.
2. R. W. Lucky, J. Salz, and E. J. Weldon, Jr., *Principles of Data Communication*, pp. 54–58 (McGraw-Hill, 1968).
3. W. R. Bennett and J. R. Davey, *Data Transmission*, pp. 114–118 (McGraw-Hill, 1965).

limited by intersymbol interference rather than by additive noise. In principle, if the channel is known precisely, it is virtually always possible to make the intersymbol interference (at the sampling instants) arbitrarily small by using a suitable pair of transmitting and receiving filters, so as to control the overall pulse shape in the manner described in Section 9.2. The transmitting filter is placed directly before the modulator, whereas the receiving filter is placed directly after the demodulator. Thus, insofar as intersymbol interference is concerned, we may consider the data transmission as being essentially baseband.

However, in a switched telephone network, we find that two factors contribute to the distribution of pulse distortion on different link connections: (1) differences in the transmission characteristics of the individual links that may be switched together, and (2) differences in the number of links in a connection. The result is that the telephone channel is random in the sense of being one of an ensemble of possible channels. Consequently, the use of a fixed pair of transmitting and receiving filters designed on the basis of average channel characteristics may not adequately reduce intersymbol interference. To realize the full transmission capability of a telephone channel, there is need for *adaptive equalization.** By equalization we mean the process of correcting channel-induced distortion. This process is said to be adaptive when it adjusts itself continuously during data transmission by operating on the input signal.

Among the philosophies for adaptive equalization of data transmission systems, we have *prechannel equalization* at the transmitter, and *postchannel equalization* at the receiver. Because the first approach requires a feedback channel, we consider only adaptive equalization at the receiving end of the system. This equalization can be achieved, prior to data transmission, by training the filter with the guidance of a suitable *training sequence* transmitted through the channel so as to adjust the filter parameters to optimum values. The typical telephone channel changes little during an average data call, so that precall equalization with a training sequence is

* Several adaptive equalization schemes have been published which provide equalization for specific synchronous data transmission systems. For some of the earliest references, see

1. R. W. Lucky, "Automatic equalization for digital communication," *Bell Syst. Tech. J.*, vol. 44, pp. 547–588, April 1965.
2. D. C. Coll and D. A. George, "A receiver for time-dispersed pulses," Conference Record, 1965 IEEE Annual Communication Convention, pp. 753–757.
3. M. J. DiToro, "Communication in time-frequency spread media using adaptive equalization," *Proc. IEEE*, vol. 56, pp. 1653–1679, Oct. 1968.
4. R. W. Lucky, J. Salz, and E. J. Weldon, pp. 128–165, *op. cit.*
5. A. Gersho, "Adaptive equalization of highly dispersive channels for data transmission," *Bell Syst. Tech. J.*, vol. 48, pp. 55–70, Jan, 1969.
6. B. Widrow, "Adaptive filters," in *Aspects of Network and System Theory*, edited by R. F. Kalman and N. DeClaris, pp. 563–587 (Holt, Rinehart and Winston, 1971).

For review papers on adaptive equalization, see:

1. J. G. Proakis, "Advances in equalization for intersymbol interference," *Advances in Communication Systems*, vol. 4, pp. 123–198 (Academic Press, 1975).
2. S. Qureshi, "Adaptive equalization," *IEEE Communications Society Magazine*, vol. 20, pp. 9–16, March 1982.

Our treatment is based on the papers by Gersho and Qureshi.

sufficient in most cases encountered in practice. The equalizer is positioned after the receiving filter in the receiver.

In this section, we study an adaptive *synchronous* equalizer based on the tapped-delay-line-filter, the operation of which was described in Example 21, Chapter 2. This equalizer is not only simple to implement but also capable of realizing a relatively satisfactory performance.

Adaptive Tapped-delay-line Filter

Consider a tapped-delay-line filter that consists of a set of delay elements, a set of multipliers connected to the delay-line taps, a corresponding set of adjustable weights, and a summer for adding the multiplier outputs. Let the sequence $\{x(kT)\}$, appearing at the output of the receiving filter, be applied to the input of this tapped-delay-line filter, producing the output (see Fig. 9.13)

$$y(kT) = \sum_{i=0}^{N-1} w_i x(kT - it) \tag{9.52}$$

where w_i is the weight at the ith tap, and N is the number of taps. These N tap weights constitute the adaptive filter coefficients. We assume that the input sequence $x(kT)$ has finite energy. The tap spacing is chosen equal to the symbol duration T of the transmitted signal or the reciprocal of the signaling rate.

The adaptation may be achieved by observing or estimating the error between the actual pulse shape at the filter output and the desired pulse shape, measured at

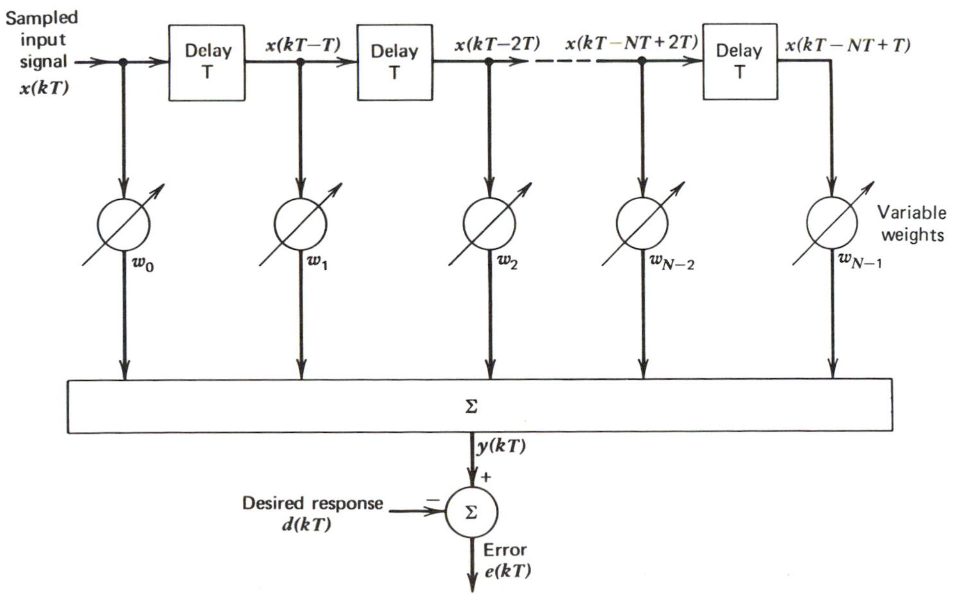

Figure 9.13 Elements of an adaptive filter.

the sampling instants, and then using this error to estimate the direction in which the weight settings of the filter should be changed so as to approach the optimum values. Let $d(kT)$ denote the desired pulse sample at time kT. We assume that all signals of interest are real. Then a criterion of interest is the *mean-square error* defined by

$$\mathscr{E} = \sum_k e^2(kT) \tag{9.53}$$

where $e(kT)$ is the difference between the output and desired pulse samples:

$$e(kT) = y(kT) - d(kT) \tag{9.54}$$

Using Eqs. (9,52) to (9.54), the gradient of the mean-square error $\mathscr{E}$ with respect to the ith tap weight w_i may be expressed as

$$\begin{aligned}
\frac{\partial \mathscr{E}}{\partial w_i} &= 2 \sum_k e(kT) \frac{\partial e(kT)}{\partial w_i} \\
&= 2 \sum_k e(kT) \frac{\partial y(kT)}{\partial w_i} \\
&= 2 \sum_k e(kT) x(kT - iT)
\end{aligned} \tag{9.55}$$

The summation term on the right-hand side of Eq. (9.55) is recognized as the (deterministic) cross-correlation of the output error sequence $e(kT)$ and the input sequence $x(kT)$, that is,

$$R_{ex}(iT) = \sum_k e(kT) x(kT - iT) \tag{9.56}$$

where we have used the discrete-time version of the definition of cross-correlation of energy signals, which was presented in Section 2.9.

We may thus rewrite Eq. (9.55) as

$$\frac{\partial \mathscr{E}}{\partial w_i} = 2R_{ex}(iT) \tag{9.57}$$

The optimality condition for minimum error is

$$\frac{\partial \mathscr{E}}{\partial w_i} = 0, \qquad \text{for} \quad i = 0, 1, 2, \ldots, N-1 \tag{9.58}$$

This is equivalent to the requirement that

$$R_{ex}(iT) = 0, \qquad \text{for} \quad i = 0, 1, 2, \ldots, N-1 \tag{9.59}$$

That is, for minimum error, the cross-correlation between the output error sequence $\{e(kT)\}$ *and the input sequence* $\{x(kT)\}$ *must have zeros for the N components with index values corresponding to the index values of the available tap weights of the filter.* This result is known as the *principle of orthogonality*, another example of which was encountered in Section 8.6 when dealing with prediction filters.

Define the (deterministic) autocorrelation of the input sequence $\{x(kT)\}$ as

$$R_x(iT)=\sum_k x(kT)x(kT-iT), \tag{9.60}$$

and the (deterministic) cross-correlation of the sequence of desired pulse samples $\{d(kT)\}$ and the input sequence $\{x(kT)\}$ as

$$R_{dx}(iT)=\sum_k d(kT)x(kT-iT) \tag{9.61}$$

Then, substituting Eqs. (9.52) and (9.54) in (9.55), and using the definitions of Eqs. (9.60) and (9.61), we may express the gradient $\partial\mathscr{E}/\partial w_i$ in the alternative form

$$\frac{\partial\mathscr{E}}{\partial w_i}=2\sum_{n=0}^{N-1} w_n R_x(iT-nT)-2R_{dx}(iT) \tag{9.62}$$

where $i=0, 1, 2, \ldots, N-1$. Using matrix notation, we may therefore write

$$\mathbf{\nabla}=2\mathbf{R}_x\mathbf{W}-2\mathbf{R}_{dx} \tag{9.63}$$

where $\mathbf{\nabla}$ is the *gradient vector*, $\mathbf{W}$ is the *weight vector* of the tapped-delay-line filter, $\mathbf{R}_x$ is the *correlation matrix* of the sequence of input pulse samples, and $\mathbf{R}_{dx}$ is the *cross-correlation vector* of the sequences of desired and input pulse samples. They are defined as follows, respectively,

$$\mathbf{\nabla}=\begin{bmatrix} \partial\mathscr{E}/\partial w_0 \\ \partial\mathscr{E}/\partial w_1 \\ \partial\mathscr{E}/\partial w_2 \\ \vdots \\ \partial\mathscr{E}/\partial w_{N-1} \end{bmatrix} \tag{9.64}$$

$$\mathbf{W}=\begin{bmatrix} w_0 \\ w_1 \\ w_2 \\ \vdots \\ w_{N-1} \end{bmatrix} \tag{9.65}$$

$$\mathbf{R}_x=\begin{bmatrix} R_x(0) & R_x(-T) & R_x(-2T) & \cdots & R_x(-NT+T) \\ R_x(T) & R_x(0) & R_x(-T) & \cdots & R_x(-NT+2T) \\ R_x(2T) & R_x(T) & R_x(0) & \cdots & R_x(-NT+3T) \\ \vdots & \vdots & \vdots & & \vdots \\ R_x(NT-T) & R_x(NT-2T) & R_x(NT-3T) & \cdots & R_x(0) \end{bmatrix} \tag{9.66}$$

$$\mathbf{R}_{dx}=\begin{bmatrix} R_{dx}(0) \\ R_{dx}(T) \\ R_{dx}(2T) \\ \vdots \\ R_{dx}(NT-T) \end{bmatrix} \tag{9.67}$$

Setting Eq. (9.63) to zero yields the solution for the *optimum tap-weight vector*, $\mathbf{W}_{\text{opt}}$, as shown by

$$\mathbf{W}_{\text{opt}} = \mathbf{R}_x^{-1}\mathbf{R}_{dx} \tag{9.68}$$

where $\mathbf{R}_x^{-1}$ is the *inverse* of the correlation matrix $\mathbf{R}_x$.

Using the definitions of Eqs. (9.60) and (9.61) for the autocorrelation $R_x(iT)$ and cross-correlation $R_{dx}(iT)$, respectively, we may express the mean-square error as follows

$$\mathscr{E} = \sum_{n=0}^{N-1}\sum_{m=0}^{N-1} w_n w_m R_x(mT-nT) - 2\sum_{n=0}^{N-1} w_n R_{dx}(nT) + \sum_k d^2(kT) \tag{9.69}$$

or, in matrix form,

$$\mathscr{E} = \mathbf{W}^t\mathbf{R}_x\mathbf{W} - 2\mathbf{W}^t\mathbf{R}_{dx} + \sum_k d^2(kT) \tag{9.70}$$

where the superscript t denotes the *transpose*. The *minimum-mean-square error* is achieved by choosing the optimum weight vector given by Eq. (9.68); thus

$$\mathscr{E}_{\text{min}} = \mathbf{W}_{\text{opt}}^t\mathbf{R}_x\mathbf{W}_{\text{opt}} - 2\mathbf{W}_{\text{opt}}^t\mathbf{R}_{dx} + \sum_k d^2kT) \tag{9.71}$$

Subtracting Eq. (9.71) from (9.70), and then using Eq. (9.68) to eliminate $\mathbf{R}_{dx}$, we may express the mean-square error $\mathscr{E}$ in the convenient form:

$$\mathscr{E} = \mathscr{E}_{\text{min}} + (\mathbf{W} - \mathbf{W}_{\text{opt}})^t\mathbf{R}_x(\mathbf{W} - \mathbf{W}_{\text{opt}}) \tag{9.72}$$

which shows explicitly the unique optimality of the minimizing weight vector $\mathbf{W}_{\text{opt}}$. It is of interest to note that the residual error $\mathscr{E}_{\text{min}}$ can be made as small as desired for all channels of practical interest by using a sufficiently large number of taps N.

From Eq. (9.69) or (9.72), we observe that the mean-square error $\mathscr{E}$ is precisely a second-order function of the tap weights, the w's. The mean-square-error performance function may be visualized as a bowl-shaped surface that is a parabolic function of the weights. The *adaptation process* has the task of continually seeking the *bottom of the bowl*. It is intuitively reasonable that successive corrections to the tap weights in the direction of steepest descent of the error surface (that is, in a direction opposite to the gradient vector) should lead to the minimum mean-square error $\mathscr{E}_{\text{min}}$, for which $\mathbf{W} = \mathbf{W}_{\text{opt}}$. This is the idea of the *steepest-descent algorithm*:

$$\mathbf{W}(kT+T) = \mathbf{W}(kT) - \tfrac{1}{2}\alpha\nabla(kT) \tag{9.73}$$

The α is a small positive constant called the *adaptation constant or step size parameter*. The $\mathbf{W}(kT)$ is the tap-weight vector and $\nabla(kT)$ is the gradient vector at time kT, and $\mathbf{W}(kT+T)$ is the *updated value* of the tap-weight vector. The adaptation process is started with an *initial guess* $\mathbf{W}(0)$ which is arbitrarily chosen. Equation (9.73) states that the "*next guess is equal to the present guess minus the gradient vector multiplied by a constant.*" Note that in using this algorithm,

$$[\mathbf{W}(kT)]_{\substack{\text{present}\\ \text{iteration}}} = [\mathbf{W}(kT+T)]_{\substack{\text{previous}\\ \text{iteration}}} \tag{9.74}$$

In a practical implementation of the steepest descent algorithm we use an *estimate* of the gradient vector $\mathbf{\nabla}(kT)$. From Eq. (9.55) we see that the *instantaneous value* of the gradient of the mean-square error $\mathscr{E}$ with respect to tap weight w_i at time kT equals $2e(kT)x(kT-iT)$, where $i=0, 1, \ldots, N-1$. Based on this instantaneous value, we define the following estimate for the gradient vector at time kT:

$$\hat{\mathbf{\nabla}}(kT)=\begin{bmatrix} 2e(kT)x(kT) \\ 2e(kT)x(kT-T) \\ 2e(kT)x(kT-2T) \\ \vdots \\ 2e(kT)x\big(kT-(N-1)T\big) \end{bmatrix}$$

$$=2e(kT)\mathbf{x}(kT) \tag{9.75}$$

where $\mathbf{x}(kT)$ is the vector of current and past sample values of the input, defined by

$$\mathbf{x}(kT)=\begin{bmatrix} x(kT) \\ x(kT-T) \\ x(kT-2T) \\ \vdots \\ x\big(kT-(N-1)T\big) \end{bmatrix} \tag{9.76}$$

Using the estimate of Eq. (9.75) for the gradient vector, we may approximate the steepest descent algorithm as follows

$$\begin{aligned}\mathbf{W}(kT+T)&=\mathbf{W}(kT)-\tfrac{1}{2}\alpha\hat{\mathbf{\nabla}}(kT)\\ &=\mathbf{W}(kT)-\alpha e(kT)\mathbf{x}(kT)\end{aligned} \tag{9.77}$$

where the term $\alpha e(kT)\mathbf{x}(kT)$ represents the *correction* applied at time $t=kT$. Equivalently, we may write the recursive formula for updating the tap weights as

$$w_i(kT+T)=w_i(kT)-\alpha e(kT)x(kT-iT), \qquad i=0, 1, \ldots, N-1 \tag{9.78}$$

This algorithm is known as the *least mean square* (*LMS*) *algorithm* or *stochastic gradient algorithm*.

To summarize, we initiate the algorithm with an arbitrary guess $w_i(0)$, $i=0, 1, \ldots, N-1$. A convenient guess is $w_i(0)=0$ for all i. Then we proceed as follows:

1. At time $t=kT$, given the tap weights $w_i(kT)$, $i=0, 1, \ldots, N-1$, and the corresponding set of current and past values of the input signal, we compute the equalizer output

$$y(kT)=\sum_{i=0}^{N-1} w_i(kT)x(kT-iT)$$

2. Given the desired response $d(kT)$, we compute the error signal

$$e(kT)=y(kT)-d(kT)$$

3. For a prescribed step size parameter α, we compute the updated values of the tap

weights in accordance with the relation

$$w_i(kT+T)=w_i(kT)-\alpha e(kT)x(kT-iT), \qquad i=0, 1, \ldots, N-1$$

4. We increment the time index k by one, go back to step 1 and repeat the procedure until steady-state conditions are reached.

Operation of the Equalizer

There are two modes of operating an adaptive equalizer, as shown in Fig. 9.14. During the *training period*, a known sequence is transmitted and a synchronized version of this signal is generated in the receiver where it is applied to the adaptive equalizer as the desired response. The training sequence may, for example, consist of the linear maximal-length or pseudo-noise (PN) sequence described in Example 8 of Chapter 5. The length of this training sequence must be equal to or greater than that of the adaptive equalizer.

When the training period is completed, the adaptive equalizer is switched to its second mode of operation, the *decision-directed mode*. In this mode of operation, the error signal equals

$$e(kT)=y(kT)-\hat{a}(kT)$$

where $y(kT)$ is the equalizer output, and $\hat{a}(kT)$ is the final (not necessarily) correct estimate of the transmitted symbol $a(kT)$. Now, in normal operation the decisions made by the receiver are correct with high probability. This means that the error estimates are correct most of the time, thereby permitting the adaptive equalizer to operate satisfactorily. Furthermore, an adaptive equalizer operating in a decision-directed mode is able to *track* relatively slow variations in channel characteristics.

It turns out that the larger the step size parameter α, the faster the tracking capability of the adaptive equalizer. However, a large step size parameter α may

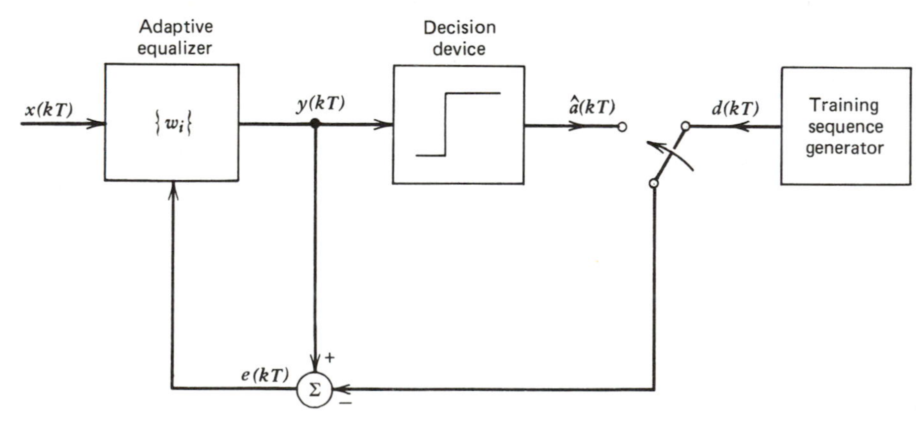

Figure 9.14 Illustrating the two modes of operation of an adaptive equalizer.

result in an unacceptably high *excess mean-square error*, defined as that part of the mean-square value of the error signal in excess of the minimum attainable value $\mathscr{E}_{\min}$ (which results when the tap weights are at their optimum settings). We therefore find that in practice in choosing a suitable value for the step size parameter α, a compromise must be made between fast tracking and excess mean-square error.

Discussion

The convergence behavior of the LMS algorithm is difficult to analyze.* Nevertheless, for a small step size parameter α and a large number of iterations, the behavior of the LMS algorithm is roughly similar to that of the steepest descent algorithm, which uses the actual gradient rather than a *noisy* estimate.

An important advantage of the LMS algorithm is that it is relatively simple to implement. However, this algorithm tends to *converge* to the optimum solution rather slowly. There is another class of algorithms known as the *least-squares algorithms*† which minimize a cost function defined as the sum of error squares. Because these least-squares algorithms make better use of all the past available information than stochastic gradient algorithms, their rate of convergence is faster. But this improvement in start-up time is achieved at the expense of increased complexity.

Another feature of the adaptive synchronous equalizer described above is the fact that the delay-line taps are spaced at the reciprocal of the signaling rate. It is indeed possible to design the equalizer with the taps spaced closer than the reciprocal of the signaling rate. Such equalizers are called *fractionally-spaced equalizers.*‡ For the same total time span, a fractionally spaced equalizer can effectively compensate for more severe delay distortion than the conventional synchronous equalizer. Here again, this improvement in performance is attained at the cost of increased complexity.

* For a detailed analysis of the convergence behavior of the LMS algorithm, see:

1. G. Ungerboeck, "Theory on the speed of convergence in adaptive equalizers for digital communication," *IBM Journal of Research and Development*, vol. 16, pp. 546–555, Nov. 1972.
2. B. Widrow, J. M. McCool, M. G. Larimore, and C. R. Johnson, Jr., "Stationary and nonstationary learning characteristics of the LMS adaptive filter," *Proc. IEEE*, vol. 64, pp. 1151–1162, Aug. 1976.
3. R. D. Gitlin, J. E. Mazo, and M. G. Taylor, "On the design of gradient algorithms for digitally implemented adaptive filters," *IEEE Trans. on Circuit Theory*, vol. CT-20, pp. 125–136, March 1973.

† 1. D. Godard, "Channel equalization using a Kalman filter for fast data transmission," *IBM Journal of Research and Development*, vol. 18, pp. 267–273, May 1974.
2. D. Falconer and L. Ljung, "Application of fast Kalman estimation to adaptive equalization," *IEEE Transactions on Communications*, vol. COM-26 pp. 1439–1446, Oct. 1978.
3. M. S. Mueller, "Least-squares algorithms for adaptive equalizers," *Bell System Tech. J.*, vol. 60, pp. 1905–1925, Oct. 1981.

‡ 1. D. M. Brady, "An adaptive coherent diversity receiver for data transmission through dispersive media," Conference Record, IEEE International Conference on Communications, pp. 21–35—21–40, June 1970.
2. G. Ungerboeck, "Fractional tap-spacing equalizer and consequences for clock recovery in data modems," *IEEE Trans. on Communications*, vol. COM-24, pp. 856–864, Aug. 1976.
3. R. D. Gitlin and S. B. Weinstein, "Fractionally spaced equalization: An improved digital transversal equalizer," *Bell System Tech. J.*, vol. 60, pp. 275–296, Feb. 1981.

Implementation Approaches

The methods of implementing adaptive equalizers may be divided into three broad categories: *analog*, *hardwired digital*, and *programmable digital*, as described below:

1. The analog approach is primarily based on the use of *charge-coupled device* (CCD) technology. The basic circuit realization of the CCD is a row of field-effect transistors with drains and sources connected in series, and the drains capacitively coupled to the gates. The set of adjustable tap weights are stored in digital memory locations, and the multiplications of the analog sample values by the digitized tap weights take place in analog fashion. This approach has significant potential in applications where the symbol rate is too high for digital implementation.
2. In hardwired digital implementation of an adaptive equalizer, the equalizer input is first sampled and then quantized into a form suitable for storage in shift registers. The set of adjustable tap weights are also stored in shift registers. Logic circuits are used to perform the required digital arithmetic (e.g., multiply and accumulate). In this approach the circuitry is hard-wired for the sole purpose of performing equalization. Nonetheless, it is the most widely used method of implementing adaptive equalizers.
3. The use of a programmable digital processor in the form of a *microprocessor*, for example, offers flexibility in that the adaptive equalization is performed as a series of steps or instructions in the microprocessor. An important advantage of this approach is that the same hardware may be time-shared to perform a multiplicity of signal-processing functions such as filtering, modulation and demodulation in a modem (*mo*dulator-*dem*odulator) used to transmit digital data over a telephone channel.

9.7 EYE PATTERN

One way to study intersymbol interference in a PCM or data transmission system experimentally is to apply the received wave to the vertical deflection plates of an oscilloscope and to apply a sawtooth wave at the transmitted symbol rate $1/T$ to the horizontal deflection plates. The waveforms in successive symbol intervals are thereby translated into one interval on the oscilloscope display, as illustrated in Fig. 9.15 for the case of a binary wave for which $T = T_b$. The resulting display is called an *eye pattern* because of its resemblance to the human eye for binary waves. The interior region of the eye pattern is called the *eye opening*.

An eye pattern provides a great deal of information about the performance of the pertinent system, as described below (see Fig. 9.16):

1. The width of the eye opening defines the time interval over which the received wave can be sampled without error from intersymbol interference. It is apparent that the preferred time for sampling is the instant of time at which the eye is open widest.

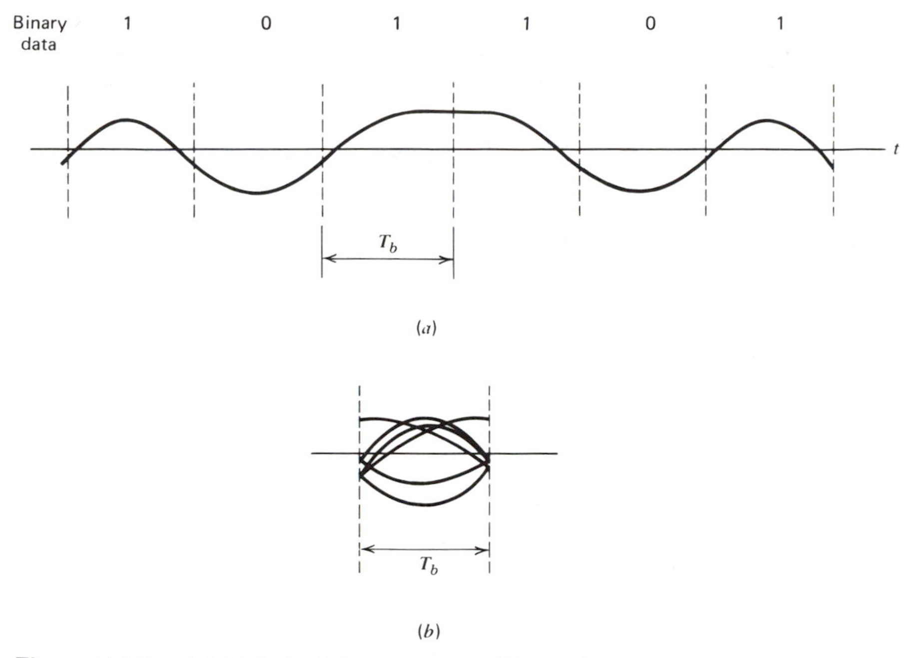

Figure 9.15 *(a)* Distorted binary wave. *(b)* Eye pattern.

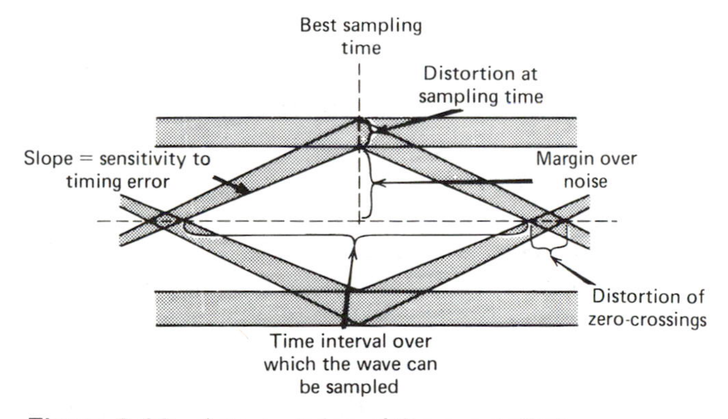

Figure 9.16 Interpretation of the eye pattern.

2. The sensitivity of the system to timing error is determined by the rate of closure of the eye as the sampling time is varied.
3. The height of the eye opening, at a specified sampling time, defines the margin over noise.

When the effect of intersymbol interference is severe, traces from the upper portion of the eye pattern cross traces from the lower portion, with the result that the eye is completely closed. In such a situation, it is impossible to avoid errors due to the combined presence of intersymbol interference and noise in the system.

In the case of an M-ary system, the eye pattern contains $(M-1)$ eye openings stacked up vertically one upon the other, where M is the number of discrete amplitude levels used to construct the transmitted signal. In a strictly linear system with truly random data, all of these eye openings would be identical. In practice, however, it is often possible to discern asymmetries in the eye pattern, which are caused by nonlinearities in the transmission channel.

Problems

Problem 9.1 The overall pulse shape $p(t)$ of a baseband binary PAM system is defined by

$$p(t)=\text{sinc}\left(\frac{t}{T_b}\right)$$

where T_b is the bit duration of the input binary data. The amplitude levels at the pulse generator output are $+1$ volt or -1 volt, depending on whether the binary symbol at the input is 1 or 0, respectively. Sketch the waveform at the output of the receiving filter in response to the input data 001101001.

Problem 9.2 An analog signal is sampled, quantized, and encoded into binary PCM wave. The specifications of the PCM system include the following:

Sampling rate = 8 kHz
Number of quantizing levels = 64

The PCM wave is transmitted over a baseband channel using discrete pulse-amplitude modulation. Determine the minimum bandwidth required for transmitting the PCM wave if each pulse is allowed to take on the following number of amplitude levels:

(a) 2
(b) 4
(c) 8

Problem 9.3 Consider a baseband binary PAM system that is designed to have a raised cosine pulse spectrum $P(f)$, as in Eq. (9.11). Show that the resulting pulse $p(t)$ is as defined in Eq. (9.13). How would this pulse be modified if the system was designed to have a linear phase response?

Problem 9.4 A computer puts out binary data at the rate of 56 kilobits per second. The computer output is transmitted using a baseband binary PAM system that is designed to have a raised cosine pulse spectrum. Determine the transmission bandwidth required for each of the following rolloff factors:

(a) $\rho=0.25$
(b) $\rho=0.5$
(c) $\rho=0.75$
(d) $\rho=1.0$.

Problem 9.5 Repeat Problem 9.4, given that each set of three successive binary digits in the computer output are coded into one of eight possible amplitude levels, and the resulting signal is transmitted by using an 8-level PAM system designed to have a raised cosine pulse spectrum.

Problem 9.6 An analog signal is sampled, quantized, and encoded into a binary PCM wave. The number of quantizing levels used is 128. A synchronizing pulse is added at the end of each code word representing a sample of the analog signal. The resulting PCM wave is transmitted over a channel of bandwidth 12 kHz using a quaternary PAM system with a raised cosine pulse spectrum. The rolloff factor is unity.

(a) Find the rate (in bits per second) at which information is transmitted through the channel.
(b) Find the rate at which the analog signal is sampled. What is the maximum possible value for the highest frequency component of the analog signal?

Problem 9.7 A binary PAM wave is to be transmitted over a low-pass channel with an absolute maximum bandwidth of 75 kHz. The bit duration is 10 microseconds. Find a raised cosine pulse spectrum that satisfies these requirements.

Problem 9.8 Consider a baseband data transmittion system using an idealized channel with the transfer function

$$H_C(f)=\begin{cases} H_C(0), & |f|\leqslant B \\ 0, & \text{elsewhere} \end{cases}$$

The system is designed to have a raised cosine pulse spectrum to reduce the effect of intersymbol interference to zero. Also the transmitting and receiving filters are chosen to minimize the average probability of error for the case when the noise at the input of the receiving filter is white and Gaussian with zero mean and power spectral density $N_0/2$. Show that in this case the variance of the noise at the output of the receiving filter is given by

$$\sigma_N^2=\frac{N_0}{2H_C(0)}$$

Problem 9.9

(a) A baseband binary PAM system is required to transmit data at the rate of 4800 bits per second over an ideal low-pass channel characterized by the transfer function

$$H_C(f)=\begin{cases} 1, & \text{for } |f|\leqslant 4.8 \text{ kHz} \\ 0, & \text{elsewhere} \end{cases}$$

The transmitting and receiving filters are chosen such that each one accomplishes half of the pulse shaping required to minimize the average probability of error. Assuming the use of a raised cosine pulse spectrum, specify the frequency responses of the transmitting and receiving filters.

(b) Given that the additive noise at the input of the receiving filter is white, Gaussian with zero mean and a power spectral density of 10^{-10} watts/Hz, calculate the variance of the resultant noise at the output of this filter.
(c) Using the result of part (b), calculate the peak pulse power required to realize a minimum value of 10^{-5} for the average probability of error.

Problem 9.10 The duobinary, ternary, and bipolar signaling techniques have one common feature: they all employ three amplitude levels. In what ways does the duobinary technique differ from the other two?

Problem 9.11 Suggest and justify a procedure whereby single-sideband modulation may be used to transmit binary data over a microwave radio channel which cannot transmit dc components.

Problem 9.12 The binary data 001101001 is applied to the input of a duobinary system.

(a) Construct the duobinary coder output and corresponding receiver output, without a precoder.
(b) Suppose that due to error during transmission, the level at the receiver input produced by the second digit is reduced to zero. Construct the new receiver output.

Problem 9.13 Repeat Problem 9.12, assuming the use of a precoder in the transmitter.

Problem 9.14 The scheme shown in Fig. P.9.1 may be viewed as a differential encoder (consisting of the modulo-2 adder and the 1-bit delay element) connected in cascade with a special form of correlative coder (consisting of the 1-bit delay element and summer). A single delay element is shown in Fig. P9.1 since it is common to both the differential encoder and the correlative coder. In this differential encoder, a transition is represented by symbol 0 and no transition by symbol 1.

(a) Find the frequency response and inpulse response of the correlative coder part of the scheme shown in Fig. P.9.1.
(b) Show that this scheme may be used to convert the on–off representation of a binary sequence (applied to the input) into the bipolar representation of the sequence at the output. You may do this by considering the sequence 010001101.

Note: For descriptions of on–off, bipolar and differential encoding of binary sequences, see Section 8.1.

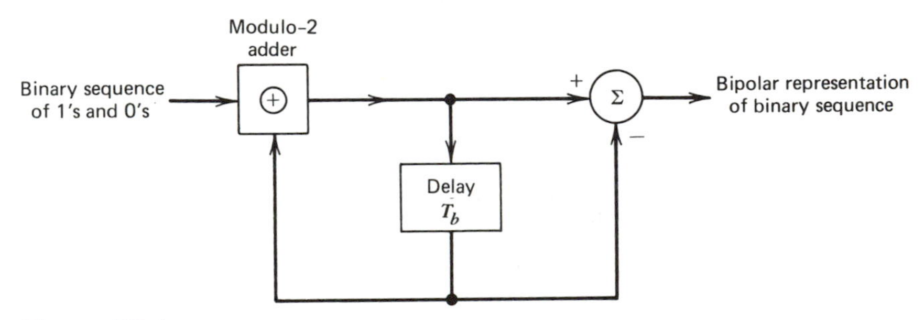

Figure P9.1

Problem 9.15 Consider a random binary wave $x(t)$ in which the 1's and 0's occur with equal probability, the symbols in adjacent time slots are statistically independent, and symbol 1 is represented by A volts and symbol 0 by zero volts. This on–off binary wave is applied to the circuit of Fig. P9.1.

(a) Show that the power spectral density of the bipolar wave $y(t)$ appearing at the output of the circuit equals

$$S_Y(f) = T_b A^2 \sin^2(\pi f T_b)\text{sinc}^2(\pi f T_b)$$

Hint: Use the result of Problem 5.9.
(b) Plot the power spectral densities of the on–off and bipolar binary waves, and compare them.

Problem 9.16 The binary data 011100101 is applied to the input of a modified duobinary system.

(a) Construct the modified duobinary coder output and corresponding receiver output, without a precoder.
(b) Suppose that due to error during transmission, the level produced by the third digit is reduced to zero. Construct the new receiver output.

Problem 9.17 Repeat Problem 9.16 assuming the use of a precoder in the transmitter.

Problem 9.18 Using conventional analog filter design methods, it is difficult to approximate the frequency response of the modified duobinary system defined by Eq. (9.46). To get around this problem, we may use the arrangement shown in Fig. P.9.2. Justify the validity of this scheme.

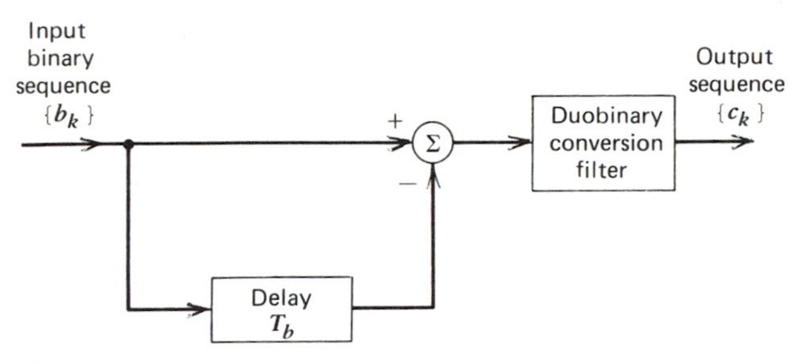

Figure P9.2

Problem 9.19 Consider a baseband M-ary system using M discrete amplitude levels. The receiver model is as shown in Fig. P9.3. Assume that:

(a) The signal component in the received wave is

$$m(t) = \sum_n A_n \operatorname{sinc}\left(\frac{t}{T} - n\right)$$

where $1/T$ is the signaling rate in bauds.
(b) The amplitude levels are $A_n = \pm A/2, \pm 3A/2, \ldots, \pm(M-1)A/2$ if M is even, and $A_n = 0, \pm A, \ldots, \pm(M-1)A/2$ if M is odd.
(c) The M levels are equi-probable, and the symbols transmitted in adjacent time slots are statistically independent.
(d) The noise $w(t)$ at the receiver input is white, Gaussian, and with zero mean and power spectral density $N_0/2$.
(e) The low-pass filter is ideal with bandwidth $B = 1/2T$.
(f) The threshold levels used in the decision device are $0, \pm A, \ldots, \pm(M-3)A/2$ if M is even, and $\pm A/2, \pm 3A/2, \ldots, \pm(M-3)A/2$ if M is odd.

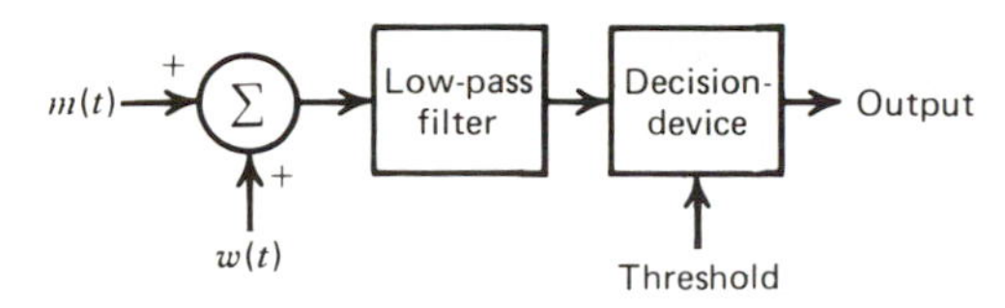

Figure P9.3

The average probability of symbol error in this system is defined by

$$P_e = \left(1 - \frac{1}{M}\right) \operatorname{erfc}\left(\frac{A}{2\sqrt{2}\sigma}\right)$$

where σ is the standard deviation of the noise at the input of the decision device. Demonstrate the validity of this general formula by determining P_e for the following three cases:

(a) $M=2$
(b) $M=3$
(c) $M=4$

Problem 9.20 Suppose that in a baseband M-ary PAM system with M equally likely amplitude levels, as described in Problem 9.19, the average probability of symbol error P_e is less than 10^{-6} so as to make the occurrence of decoding errors negligible. Show that the minimum value of received signal-to-noise ratio in such a system is approximately given by

$$(\text{SNR})_{R,\,\min} \simeq 7.8(M^2 - 1)$$

Problem 9.21 Some radio systems suffer from *multipath distortion* which is caused by the existence of more than one propagation path between the transmitter and the receiver. Consider a channel the output of which, in response to a signal $s(t)$, is defined by

$$x(t) = K_1 s(t - t_{01}) + K_2 s(t - t_{02})$$

where K_1 and K_2 are constants, and t_{01} and t_{02} represent transmission delays. It is proposed to use the three-tap delay-line-filter of Fig. P9.4 to equalize the multipath distortion produced by this channel.

(a) Evaluate the transfer function of the channel.
(b) Evaluate the parameters of the tapped-delay-line-filter in terms of K_1, K_2, t_{01}, and t_{02}, assuming that $K_2 \ll K_1$ and $t_{02} > t_{01}$.

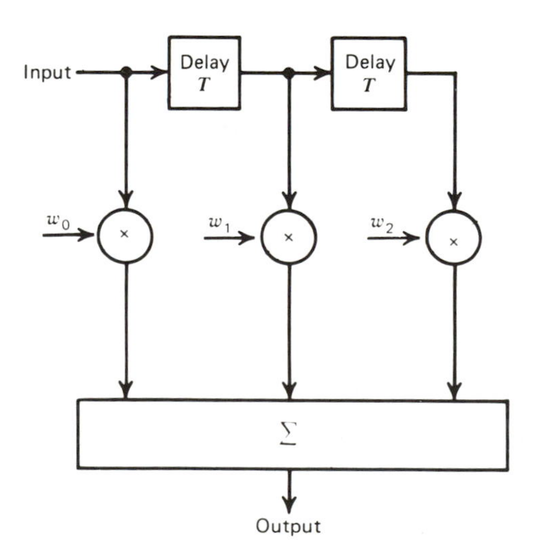

Figure P9.4

Problem 9.22 A binary wave using polar signaling is generated by representing symbol 1 by a pulse of amplitude $+1$ volt and symbol 0 by a pulse of amplitude -1 volt; in both cases the pulse duration equals the bit duration. This signal is applied to a low-pass RC filter with transfer function:

$$H(f)=\frac{1}{1+jf/f_0}$$

Construct the eye pattern for the filter output for the following sequences:

(a) Alternating 1's and 0's.
(b) A long sequence of 1's followed by a long sequence of 0's.
(c) A long sequence of 1's followed by a single 0 and then a long sequence of 1's.

Assume a bit rate of $2f_0$ bits per second.

Problem 9.23 The binary sequence 011010 is transmitted through a channel having a cosine-rolloff characteristic with rolloff factor of unity. Assume the use of polar signaling, with symbols 1 and 0 represented by $+1$ and -1 volt, respectively.

(a) Construct, to scale, the received wave, and indicate the best sampling times for regeneration.
(b) Construct the eye pattern for this received wave and show that it is completely open.
(c) Determine the zero crossings of the received wave.

chapter 10 BAND-PASS DATA TRANSMISSION

When it is required to transmit digital data over a *band-pass channel*, it is necessary to modulate the incoming data onto a carrier wave (usually sinusoidal) with fixed frequency limits imposed by the channel. The data may represent digital computer outputs, or PCM waves generated by digitizing voice or video signals, etc. The channel may be a microwave radio link, or satellite channel, etc. In any event, the modulation process involves switching or keying the amplitude, frequency, or phase of the carrier in accordance with the incoming data. Thus there are three basic signaling techniques known as *amplitude-shift keying* (*ASK*), *frequency-shift keying* (*FSK*), and *phase-shift keying* (*PSK*), which may be viewed as special cases of amplitude modulation, frequency modulation, and phase modulation, respectively.

Ideally, FSK and PSK signals have a constant envelope. This feature makes them impervious to amplitude nonlinearities, as encountered in microwave radio links and satellite channels. Accordingly, we find that, in practice, FSK and PSK signals are much more widely used than ASK signals. This chapter is devoted to a unified study of FSK and PSK systems. In particular, we study two specific issues: (1) the *optimum design* of the receiver in the sense that it will make fewer errors in the long run than any other receiver, and (2) the calculation of its *average error rate* or *probability of error*.

10.1 A MODEL OF BAND-PASS DATA TRANSMISSION SYSTEMS

We may model a band-pass data transmission system as shown in Fig. 10.1. First, there is assumed to exist a *message source* that emits one *symbol* every T seconds, with the symbols belonging to an alphabet of M symbols which we denote by $m_1, m_2, \ldots, m_M$. Consider, for example, the remote connection of two digital computers, with one computer acting as an information source that calculates digital outputs based on observations and inputs fed into it. The resulting computer output is expressed as a sequence of 0's and 1's, which are transmitted to a second computer. In this example, the alphabet consists simply of the two binary symbols 0 and 1. A second example is that of a quaternary PCM encoder with an alphabet consisting of four possible symbols: 00, 01, 10, and 11. In any event, the *a priori probabilities* $P(m_1)$, $P(m_2)$, $\ldots$, $P(m_M)$ specify the message source output. We assume that all M symbols of the alphabet are equally likely. Then we may write

$$\begin{aligned} p_i &= P(m_i) \\ &= \frac{1}{M}, \qquad \text{for all } i \end{aligned} \tag{10.1}$$

The output of the message source is presented to a *vector transmitter*, producing a *vector* of real numbers. In particular, when the input message $m = m_i$, the vector transmitter output takes on the value

$$\mathbf{s}_i = \begin{bmatrix} s_{i1} \\ s_{i2} \\ \vdots \\ s_{iN} \end{bmatrix}, \qquad i = 1, 2, \ldots, M \tag{10.2}$$

where the *dimension* $N \leqslant M$. With this vector as input, the *modulator* then constructs a *distinct* signal $s_i(t)$ of duration T seconds. The signal $s_i(t)$ is necessarily of finite

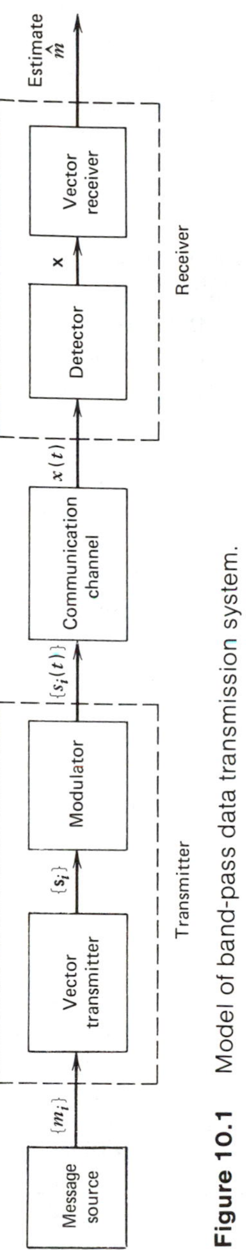

Figure 10.1 Model of band-pass data transmission system.

energy, as shown by

$$E_i = \int_0^T s_i^2(t)dt, \qquad i = 1, 2, \ldots, M \tag{10.3}$$

Note that $s_i(t)$ is real-valued. One such signal is transmitted every T seconds. The particular signal chosen for transmission depends in some fashion on the incoming message and possibly on the signals transmitted in preceding time slots. With a sinusoidal carrier, the feature that is used by the modulator to distinguish one signal from another is a *step* change in the amplitude, frequency, or phase of the carrier.* The result of this modulation process is amplitude-shift keying (ASK), frequency-shift keying (FSK), or phase-shift keying (PSK), respectively, as illustrated in Fig. 10.2 for the special case of a source of binary data for which the *symbol duration* T is the same as the *bit duration* T_b. It is of interest to note that although, in general, it is not easy to distinguish between frequency-modulated and phase-modulated waves (on an oscilloscope, say), this is not so in the case of FSK and PSK signals; for example, compare the waveforms in parts (b) and (c) of Fig. 10.2.

Returning to the model of Fig. 10.1, the band-pass communication channel, coupling the transmitter to the receiver, is assumed to have two characteristics:

1. The channel is linear, with a bandwidth that is wide enough to accommodate the transmission of the modulator output $s_i(t)$ with negligible or no distortion.
2. The transmitted signal $s_i(t)$ is perturbed by an *additive, zero-mean, stationary, white, Gaussian noise* process, denoted by $W(t)$. The reasons for this assumption are that it makes calculations tractable, and also it is a reasonable description of the type of noise present in many communication systems.

We refer to such a channel as an *additive white Gaussian noise* (*AWGN*) *channel*. Accordingly, we may express the received random process, $X(t)$, as

$$X(t) = s_i(t) + W(t), \qquad \begin{matrix} 0 \leqslant t \leqslant T \\ i = 1, 2, \ldots, M \end{matrix} \tag{10.4}$$

We may thus model the channel as in Fig. 10.3.

The receiver has the task of observing a sample function $x(t)$ of the received random process $X(t)$, for a duration of T seconds, and making a best *estimate* of the transmitted signal $s_i(t)$ or, equivalently, the symbol m_i. This task is accomplished in two stages. The first stage is a detector that operates on the received random process $X(t)$ to produce a vector of random variables, $\mathbf{X}$. By using an *observation vector* $\mathbf{x}$ (which is a sample value of $\mathbf{X}$), prior knowledge of the $\mathbf{s}_i$, and the a priori probabilities $P(m_i)$, the second stage of the receiver, called the *vector receiver*, produces the estimate $\hat{m}$. However, owing to the presence of the additive noise at the receiver input, this decision-making process is statistical in nature, with the

* Sometimes, a hybrid form of modulation is used, combining changes in both amplitude and phase, or amplitude and frequency; see Problem 10.31.

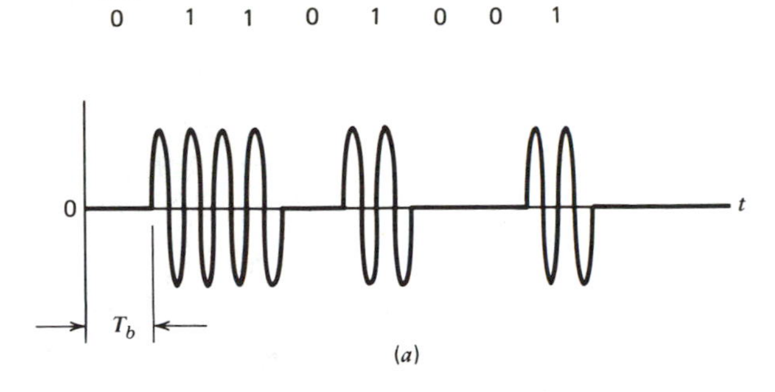

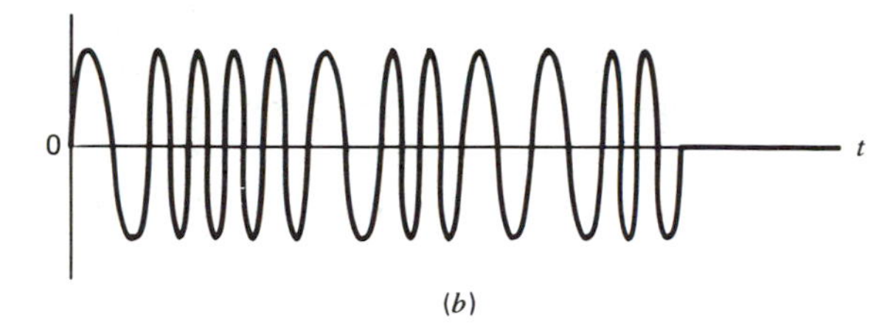

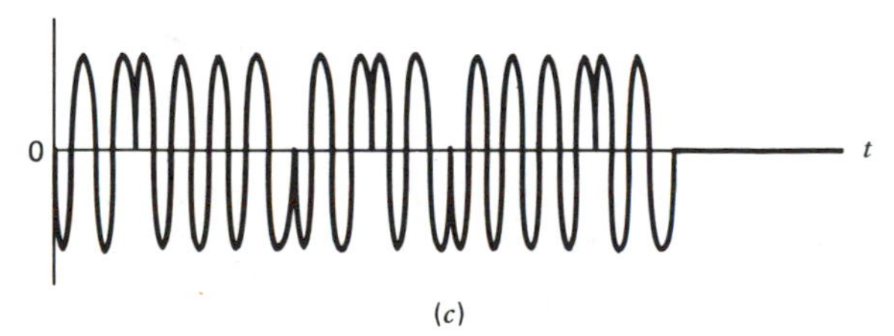

Figure 10.2 Illustrating the three basic forms of signaling binary information. *(a)* Amplitude-shift keying. *(b)* Frequency-shift keying. *(c)* Phase-shift keying.

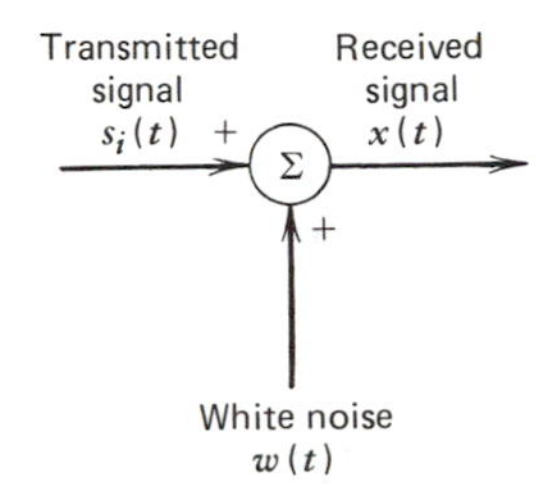

Figure 10.3 Model of additive white Gaussian noise channel.

result that the receiver will make occasional errors. The requirement is to design the vector receiver so as to minimize the *average probability of symbol error* defined as

$$P_e = P(\hat{m} \neq m_i) \tag{10.5}$$

where m_i is the transmitted symbol and $\hat{m}$ is the estimate produced by the vector receiver. The resulting receiver is said to be *optimum in the minimum probability of error sense*.

It is customary to assume that the receiver is *time-synchronized* with the receiver, which means that the receiver knows the instants of time when the modulation changes state. Sometimes, it is also assumed that the receiver is *phase locked* to the transmitter. In such a case, we speak of *coherent detection*, and we refer to the receiver as a *coherent receiver*. On the other hand, there may be no phase synchronism between the transmitter and receiver. In this second case, we speak of *noncoherent detection*, and we refer to the receiver as a *noncoherent receiver*. In this chapter, we shall always assume the existence of time synchronism; however, we shall distinguish between coherent and noncoherent detection.

The model described above provides a basis for the design of the optimum receiver, for which we will use the *geometric representation* of the known set of transmitted signals, $\{s_i(t)\}$. This method provides a great deal of insight, with considerable simplification of detail.*

10.2 GRAM-SCHMIDT ORTHOGONALIZATION PROCEDURE

According to the model of Fig. 10.1, the task of transforming an incoming message m_i, $i = 1, 2, \ldots, M$, into a modulated wave $s_i(t)$ may be divided into separate discrete-time and continuous-time operations. The justification for this separation lies in the *Gram–Schmidt orthogonalization procedure*, which permits the representation

* The geometric representation of signals was first developed by Kotel'nikov in 1947: V. A. Kotel'nikov, *The Theory of Optimum Noise Immunity*, (Dover Publications, 1960), which is a translation of the original doctoral dissertation presented in January 1947 before the Academic Council of the Molotov Energy Institute in Moscow. In particular, see Part II of the book.

This method was subsequently brought to fuller fruition by Wozencraft and Jacobs in 1965: J. M. Wozencraft and I. M. Jacobs, *Principles of Communication Engineering*, pp. 211–284 (Wiley, 1965).

For a detailed tutorial review of different forms of ASK, FSK, and PSK, using a geometric viewpoint, see E. Arthurs and H. Dym, "On the optimum detection of digital signals in the presence of white Gaussian noise—A geometric interpretation and a study of three basic data transmission systems," *IRE Transactions on Communication Systems*, vol. CS-10, pp. 336–372, Dec. 1962.

See also the following list of references:

1. D. J. Sakrison, *Communication Theory: Transmission of Waveforms and Digital Information*, pp. 219–271 (Wiley, 1968).
2. L. E. Franks, *Signal Theory*, pp. 1–65 (Prentice-Hall, 1969).
3. A. J. Viterbi and J. K. Omura, *Principles of Digital Communication and Coding*, pp. 47–127 (McGraw-Hill, 1979).

of any set of M energy signals, $\{s_i(t)\}$, as linear combinations of N *orthonormal basis functions*, where $N \leqslant M$. That is to say, we may represent the given set of real-valued energy signals $s_1(t)$, $s_2(t), \ldots, s_M(t)$, each of duration T seconds, in the form

$$s_i(t) = \sum_{j=1}^{N} s_{ij}\phi_j(t), \qquad \begin{array}{l} 0 \leqslant t \leqslant T \\ i = 1, 2, \ldots, M \end{array} \tag{10.6}$$

where the coefficients of the expansion are defined by

$$s_{ij} = \int_0^T s_i(t)\phi_j(t)dt, \qquad \begin{array}{l} i = 1, 2, \ldots, M \\ j = 1, 2, \ldots, N \end{array} \tag{10.7}$$

The real-valued basis functions $\phi_1(t)$, $\phi_2(t), \ldots, \phi_N(t)$ are orthonormal, by which we mean

$$\int_0^T \phi_i(t)\phi_j(t)dt = \begin{cases} 1, & \text{if } i = j \\ 0, & \text{if } i \neq j \end{cases} \tag{10.8}$$

The first condition of Eq. (10.8) states that each basis function is *normalized* to have unit energy. The second condition states that the basis functions $\phi_1(t)$, $\phi_2(t), \ldots,$ $\phi_N(t)$ are *orthogonal* with respect to each other over the interval $0 \leqslant t \leqslant T$.

Given the set of coefficients $\{s_{ij}\}$, $j = 1, 2, \ldots, N$, operating as input, we may use the scheme shown in Fig. 10.4(*a*) to generate the signal $s_i(t)$, $i = 1, 2, \ldots, M$, which follows directly from Eq. (10.6). It consists of a bank of N multipliers, with each multiplier supplied with its own basis function, followed by a summer. This scheme may be viewed as performing a similar role to that of the second stage or modulator in the transmitter of Fig. 10.1. Conversely, given the set of signals $\{s_i(t)\}$, $i = 1, 2, \ldots,$ M, operating as input, we may use the scheme shown in Fig. 10.4(*b*) to calculate the set of coefficients $\{s_{ij}\}$, $j = 1, 2, \ldots, N$, which follows directly from Eq. (10.7). This second scheme consists of a bank of N *product-integrators* or *correlators* with a common input, and with each one supplied with its own basis function. As will be shown later, such a bank of correlators may be used as the first stage or detector in the receiver of Fig. 10.1.

To prove the Gram–Schmidt orthogonalization procedure, we proceed in two stages, as indicated below:

Stage 1

First, we have to establish whether or not the given set of signals $s_1(t)$, $s_2(t), \ldots,$ $s_M(t)$ is *linearly independent*. If not, then (by definition) there exists a set of coefficients $a_1, a_2, \ldots, a_M$, not all equal to zero, such that we may write.

$$a_1 s_1(t) + a_2 s_2(t) + \cdots + a_M s_M(t) = 0 \tag{10.9}$$

Suppose, in particular, that $a_M \neq 0$. Then, we may express the corresponding signal $s_M(t)$ as

$$s_M(t) = -\left[\frac{a_1}{a_M} s_1(t) + \frac{a_2}{a_M} s_2(t) + \cdots + \frac{a_{M-1}}{a_M} s_{M-1}(t)\right] \tag{10.10}$$

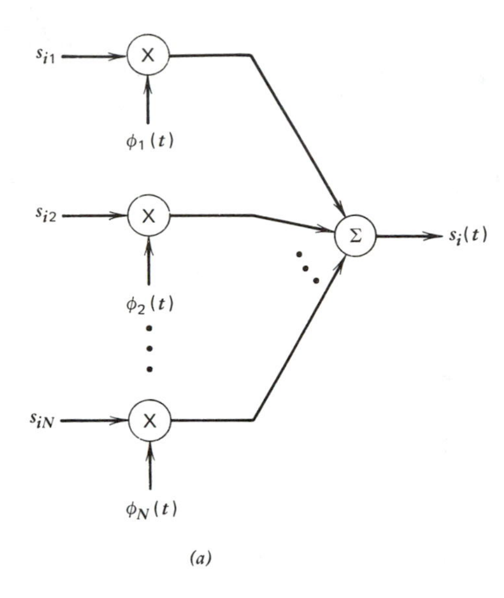

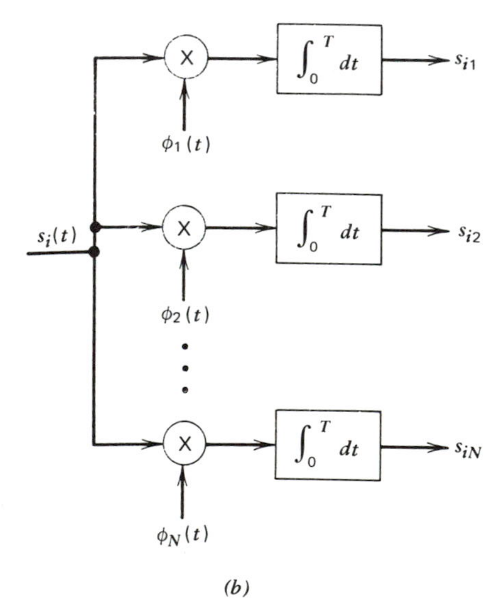

Figure 10.4 *(a)* Scheme for generating the signal $s_i(t)$. *(b)* Scheme for generating the set of coefficients $\{s_i\}$.

which implies that $s_M(t)$ may be expressed in terms of the remaining $(M-1)$ signals.

Consider, next, the set of signals $s_1(t), s_2(t), \ldots, s_{M-1}(t)$. Either this set of signals is linearly independent, or it is not. If not, then there exists a set of numbers $b_1, b_2, \ldots, b_{M-1}$, not all equal to zero, such that

$$b_1 s_1(t) + b_2 s_2(t) + \cdots + b_{M-1} s_{M-1}(t) = 0 \tag{10.11}$$

Suppose that $b_{M-1} \neq 0$. Then, we may express $s_{M-1}(t)$ as a linear combination of the remaining $M-2$ signals, as shown by

$$s_{M-1}(t) = -\left[\frac{b_1}{b_{M-1}} s_1(t) + \frac{b_2}{b_{M-1}} s_2(t) + \cdots + \frac{b_{M-2}}{b_{M-1}} s_{M-2}(t)\right] \tag{10.12}$$

Now, testing the set of signals $s_1(t), s_2(t), \ldots, s_{M-2}(t)$ for linear independence, and continuing in this fashion, it is clear that we will eventually end up with a linearly independent subset of the original set of signals. Let $s_1(t), s_2(t), \ldots, s_N(t)$ denote this subset of linearly independent signals, where $N \leq M$. The important point to note is that each of the original set of signals $s_1(t), s_2(t), \ldots, s_M(t)$ may be expressed as a linear combination of this subset of N signals.

Stage 2

Next, we wish to show that it is possible to construct a set of N orthonormal basis functions $\phi_1(t), \phi_2(t), \ldots, \phi_N(t)$ from the linearly independent signals $s_1(t), s_2(t), \ldots, s_N(t)$. As a starting point, define the first basis function as

$$\phi_1(t) = \frac{s_1(t)}{\sqrt{E_1}} \tag{10.13}$$

where E_1 is the energy of the signal $s_1(t)$. Then, clearly, we have

$$\begin{aligned} s_1(t) &= \sqrt{E_1}\phi_1(t) \\ &= s_{11}\phi_1(t) \end{aligned} \tag{10.14}$$

where the coefficients $s_{11} = \sqrt{E_1}$ and $\phi_1(t)$ has unit energy, as required.

Next, using the signal $s_2(t)$, we define the coefficient s_{21} as

$$s_{21} = \int_0^T s_2(t)\phi_1(t)dt \tag{10.15}$$

We may thus define a new intermediate function

$$g_2(t) = s_2(t) - s_{21}\phi_1(t) \tag{10.16}$$

which is orthogonal to $\phi_1(t)$ over the interval $0 \leq t \leq T$. Now, we are ready to define the second basis function as

$$\phi_2(t) = \frac{g_2(t)}{\sqrt{\int_0^T g_2^2(t)dt}} \tag{10.17}$$

Substituting Eq. (10.16) in (10.17), and simplifying, we get the desired result

$$\phi_2(t) = \frac{s_1(t) - s_{21}\phi_1(t)}{\sqrt{E_2 - s_{21}^2}} \tag{10.18}$$

where E_2 is the energy of the signal $s_2(t)$. It is clear from Eq. (10.17) that,

$$\int_0^T \phi_2^2(t)dt = 1$$

and from Eq. (10.18) that

$$\int_0^T \phi_1(t)\phi_2(t)dt = 0$$

That is to say, $\phi_1(t)$ and $\phi_2(t)$ form an orthonormal set.

Continuing in this fashion, we may define

$$g_i(t) = s_i(t) - \sum_{j=1}^{i-1} s_{ij}\phi_j(t) \tag{10.19}$$

where the coefficients s_{ij}, $j = 1, 2, \ldots, i-1$, are themselves defined by

$$s_{ij} = \int_0^T s_i(t)\phi_j(t)dt \tag{10.20}$$

Then it follows readily that the set of functions

$$\phi_i(t) = \frac{g_i(t)}{\sqrt{\int_0^T g_i^2(t)dt}}, \qquad i = 1, 2, \ldots, N \tag{10.21}$$

form an orthonormal set.

Since we have shown that each one of the derived subset of linearly independent signals $s_1(t), s_2(t), \ldots, s_N(t)$ may be expressed as a linear combination of the orthonormal basis functions $\phi_1(t), \phi_2(t), \ldots, \phi_N(t)$, it follows that each one of the original set of signals $s_1(t), s_2(t), \ldots, s_M(t)$ may be expressed as a linear combination of this set of basis functions, as described in Eq. (10.6). This completes the proof of the Gram–Schmidt orthogonalization procedure.

Note that the conventional Fourier series expansion of a periodic signal is an example of a particular expansion of this type (see Problem 2.4). Also, the representation of a band-limited signal in terms of its samples taken at the Nyquist rate may be viewed as another example of a particular expansion of this type [see Eqs. (7.12) to (7.14)]. There are, however, two important distinctions that should be made:

1. The form of the basis functions $\phi_1(t), \phi_2(t), \ldots, \phi_N(t)$ has not been specified. That is to say, unlike the Fourier series expansion of a periodic signal or the sampled representation of a band-limited signal, we have not restricted the Gram–Schmidt orthogonalization procedure to be in terms of sinusoidal functions or sinc functions of time.
2. The expansion of the signal $s_i(t)$ in terms of a finite number of terms is not an approximation wherein only the first k terms are significant but rather an exact expression where N and only N terms are significant.

Example 1

Consider the signals $s_1(t)$, $s_2(t)$, $s_3(t)$, and $s_4(t)$ shown in Fig. 10.5(a). We wish to use the Gram–Schmidt orthogonalization procedure to find an orthonormal basis for this set of signals.

We observe immediately that $s_4(t) = s_1(t) + s_3(t)$ which means that this set of signals is not

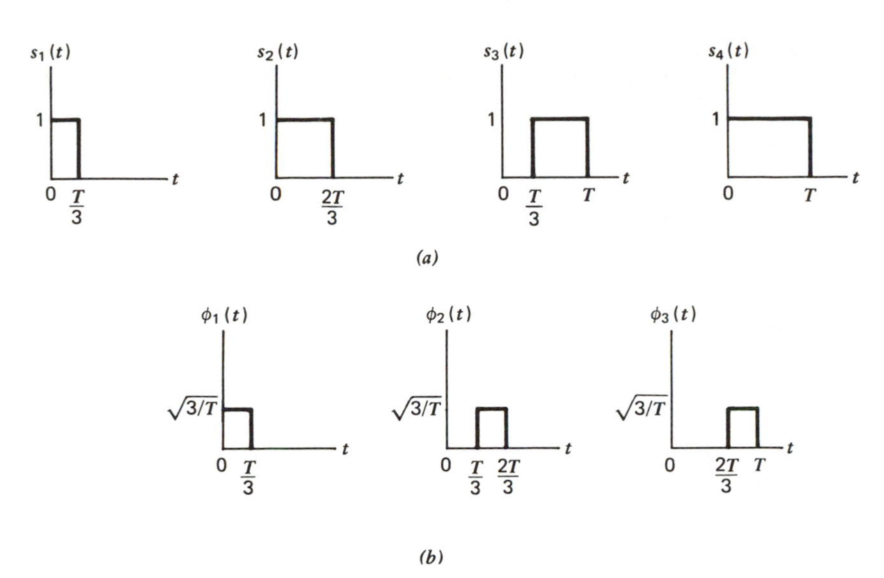

Figure 10.5 *(a)* Set of signals to be orthonormalized. *(b)* The resulting set of orthonormal functions.

linearly independent. Accordingly, we base the Gram–Schmidt orthogonalization procedure on a subset consisting of the signals $s_1(t)$, $s_2(t)$, and $s_3(t)$, which are linearly independent.

Step 1 We note that the energy of signal $s_1(t)$ is

$$\begin{aligned} E_1 &= \int_0^T s_1^2(t)dt \\ &= \int_0^{T/3} (1)^2\, dt \\ &= \frac{T}{3} \end{aligned}$$

The first basis function $\phi_1(t)$ is therefore [see Eq. (10.13)]

$$\begin{aligned} \phi_1(t) &= \frac{s_1(t)}{\sqrt{E_1}} \\ &= \begin{cases} \sqrt{3/T}, & 0 \leqslant t \leqslant T/3 \\ 0, & \text{elsewhere} \end{cases} \end{aligned}$$

Step 2 From Eq. (10.15), we find that the coefficient s_{21} equals

$$\begin{aligned} s_{21} &= \int_0^T s_2(t)\phi_1(t)dt \\ &= \int_0^{T/3} (1)\left(\sqrt{\frac{3}{T}}\right)dt \\ &= \sqrt{\frac{T}{3}} \end{aligned}$$

The energy of signal $s_2(t)$ is

$$E_2 = \int_0^T s_2^2(t)$$

$$= \int_0^{2T/3} (1)^2 \, dt$$

$$= \frac{2T}{3}$$

The second basis function $\phi_2(t)$ is therefore [see Eq. (10.18)]

$$\phi_2(t) = \frac{s_2(t) - s_{21}\phi_1(t)}{\sqrt{E_2 - s_{21}^2}}$$

$$= \begin{cases} \sqrt{3/T}, & (T/3) \leqslant t \leqslant 2T/3 \\ 0, & \text{elsewhere} \end{cases}$$

Step 3 Using Eq. (10.20), we find that the coefficient s_{31} equals

$$s_{31} = \int_0^T s_3(t)\phi_1(t) dt$$

$$= 0$$

and the coefficient s_{32} equals

$$s_{32} = \int_0^T s_3(t)\phi_2(t) dt$$

$$= \int_{T/3}^{2T/3} (1)\left(\sqrt{\frac{3}{T}}\right) dt$$

$$= \sqrt{\frac{T}{3}}$$

The pertinent value of the intermediate function $g_i(t)$, with $i = 3$, is therefore [see Eq. (10.19)]

$$g_3(t) = s_3(t) - s_{31}\phi_1(t) - s_{32}\phi_2(t)$$

$$= \begin{cases} 1, & (2T/3) \leqslant t \leqslant T \\ 0, & \text{elsewhere} \end{cases}$$

Finally, using Eq. (10.21), we find that the third basis function $\phi_3(t)$ is

$$\phi_3(t) = \frac{g_3(t)}{\sqrt{\int_0^T g_3^2(t) dt}}$$

$$= \begin{cases} \sqrt{T/3}, & (2T/3) \leqslant t \leqslant T \\ 0 & \text{elsewhere} \end{cases}$$

The resulting basis functions $\phi_1(t)$, $\phi_2(t)$, and $\phi_3(t)$ are shown in Fig. 10.5(*b*). It is clear that these three basis functions are orthonormal, and that any of the original signals $s_1(t)$, $s_2(t)$, $s_3(t)$, and $s_4(t)$ may be expressed as a linear combination of them.

10.3 GEOMETRIC INTERPRETATION OF SIGNALS

Once we have adopted a convenient set of orthonormal basis functions $\{\phi_j(t)\}$, $j=1, 2, \ldots, N$, then each signal in the set $\{s_i(t)\}$, $i=1, 2, \ldots, M$ may be expanded as in Eq. (10.6), reproduced here for convenience:

$$s_i(t)=\sum_{j=1}^{N} s_{ij}\phi_j(t), \qquad \begin{array}{l} 0\leqslant t\leqslant T \\ i=1, 2, \ldots, M \end{array} \tag{10.22}$$

The coefficients of the expansion, s_{ij}, are themselves defined by Eq. (10.7), also reproduced here for convenience:

$$s_{ij}=\int_0^T s_i(t)\phi_j(t)dt, \qquad \begin{array}{l} i=1, 2, \ldots, M \\ j=1, 2, \ldots, N \end{array} \tag{10.23}$$

Accordingly, we may state that each signal in the set $\{s_i(t)\}$ is completely determined by the *vector* of its coefficients

$$\mathbf{s}_i=\begin{bmatrix} s_{i1} \\ s_{i2} \\ \vdots \\ s_{iN} \end{bmatrix}, \qquad i=1, 2, \ldots, M \tag{10.24}$$

The vector $\mathbf{s}_i$ is called the *signal vector*. Furthermore, if we conceptually extend our conventional notion of two- and three-dimensional Euclidean spaces to an *N-dimensional Euclidean space*, we may visualize the set of signal vectors $\{\mathbf{s}_i\}$, $i=1, 2, \ldots, M$, as defining a corresponding set of M points in an N-dimensional Euclidean space, with N mutually perpendicular axes labeled $\phi_1, \phi_2, \ldots, \phi_N$. This N-dimensional Euclidean space is called the *signal space*.

The idea of visualizing a set of energy signals geometrically, as described above, is of fundamental importance. For example, Fig. 10.6 illustrates the case of a two-dimensional signal space with three signals, that is, $N=2$ and $M=3$.

In an N-dimensional Euclidean space, we may define *lengths* of vectors and *angles* between vectors.* However, for our purpose here, only length is of interest. It is customary to denote the length (also called the *absolute value* or *norm*) of a signal vector $\mathbf{s}_i$ by the symbol $\|\mathbf{s}_i\|$. The squared-length of any signal vector $\mathbf{s}_i$ is defined to be the *inner product* or *dot product* of $\mathbf{s}_i$ with itself. In the familiar case of $N=2$ or 3, we have

$$\begin{aligned} \|\mathbf{s}_i\|^2 &= (\mathbf{s}_i, \mathbf{s}_i) \\ &= \sum_{j=1}^{N} s_{ij}^2 \end{aligned} \tag{10.25}$$

* For further details on signal spaces, see:

1. L. E. Franks, *op cit*.
2. G. Birkhoff and S. MacLane, *A Survey of Modern Algebra*, Chapter 7 (Macmillan, 1965).

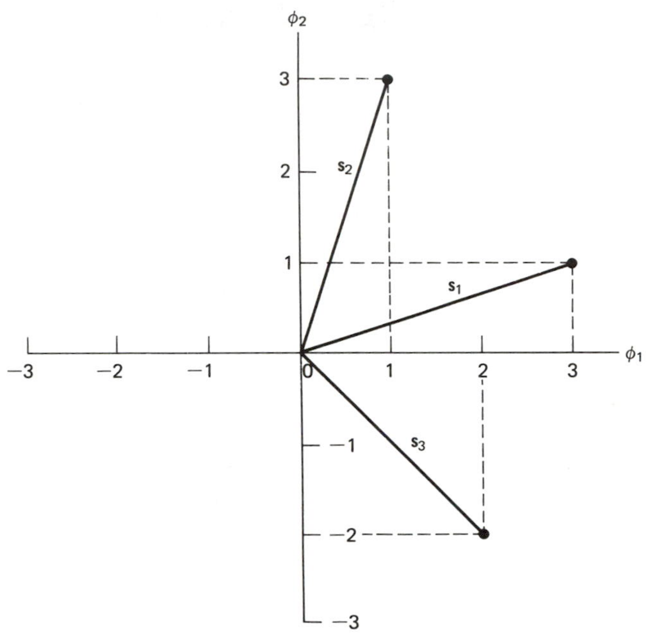

Figure 10.6 Illustrating the geometric representation of signals for the case when $N = 2$ and $M = 3$.

where the s_{ij} are the elements of $\mathbf{s}_i$. For larger values of N, length is defined in the same way, and Eq. (10.25) remains valid.

There is an interesting relationship between the energy content of a signal and its representation as a vector. By definition, the energy of a signal $s_i(t)$ of duration T seconds is equal to

$$E_i = \int_0^T s_i^2(t)dt \tag{10.26}$$

Therefore, substituting Eq. (10.22) in (10.26), we get

$$E_i = \int_0^T \left[\sum_{j=1}^{N} s_{ij}\phi_j(t) \right] \left[\sum_{k=1}^{N} s_{ik}\phi_k(t) \right] dt \tag{10.27}$$

Interchanging the order of summation and integration:

$$E_i = \sum_{j=1}^{N} \sum_{k=1}^{N} s_{ij}s_{ik} \int_0^T \phi_j(t)\phi_k(t)dt \tag{10.28}$$

But, since the $\phi_j(t)$ form an orthonormal set, then, in accordance with the two conditions of Eq. (10.8), we find that Eq. (10.28) reduces simply to

$$E_i = \sum_{j=1}^{N} s_{ij}^2 \tag{10.29}$$

Thus Eqs. (10.25) and (10.29) show that the energy of a signal $s_i(t)$ is equal to the squared-length of the signal vector $\mathbf{s}_i$ representing it.

In the case of a pair of signals $s_i(t)$ and $s_k(t)$, represented by the signal vectors $\mathbf{s}_i$ and $\mathbf{s}_k$, respectively, we may similarly show that

$$\begin{aligned} \|\mathbf{s}_i - \mathbf{s}_k\|^2 &= \sum_{j=1}^{N} (s_{ij} - s_{kj})^2 \\ &= \int_0^T [s_i(t) - s_k(t)]^2 dt \end{aligned} \tag{10.30}$$

where $\|\mathbf{s}_i - \mathbf{s}_k\|$ is the distance between the points represented by the signal vectors $\mathbf{s}_i$ and $\mathbf{s}_k$.

10.4 RESPONSE OF BANK OF CORRELATORS TO NOISY INPUT

Suppose that the input to the bank of N product-integrators or correlators in Fig. 10.4(*b*) is not the transmitted signal $s_i(t)$ but rather the received random process $X(t)$ of the idealized AWGN channel of Fig. 10.3. That is to say,

$$X(t) = s_i(t) + W(t), \qquad \begin{matrix} 0 \leqslant t \leqslant T \\ i = 1, 2, \ldots, M \end{matrix} \tag{10.31}$$

where $W(t)$ is a white Gaussian noise process of zero mean and power spectral density $N_0/2$. Correspondingly, we find that the output of each correlator is a random variable defined by

$$\begin{aligned} X_j &= \int_0^T X(t)\phi_j(t)dt \\ &= s_{ij} + W_j, \qquad j = 1, 2, \ldots, N \end{aligned} \tag{10.32}$$

The first component, s_{ij}, is a deterministic quantity contributed by the transmitted signal $s_i(t)$; it is defined by

$$s_{ij} = \int_0^T s_i(t)\phi_j(t)dt \tag{10.33}$$

The second component, W_j, is a random variable that arises because of the presence of noise at the input; it is defined by

$$W_j = \int_0^T W(t)\phi_j(t)dt \tag{10.34}$$

Consider next a new random process, $X'(t)$, that is related to the received random process $X(t)$ as follows:

$$X'(t) = X(t) - \sum_{j=1}^{N} X_j\phi_j(t) \tag{10.35}$$

Substituting Eqs. (10.31) and (10.32) in (10.35), we get

$$\begin{aligned} X'(t) &= s_i(t) + W(t) - \sum_{j=1}^{N} (s_{ij} + W_j)\phi_j(t) \\ &= W(t) - \sum_{j=1}^{N} W_j\phi_j(t) \\ &= W'(t) \end{aligned} \tag{10.36}$$

which depends only on the noise $W(t)$ at the front end of the receiver and not at all on the transmitted signal $s_i(t)$. Thus we may express the received random process as

$$\begin{aligned} X(t) &= \sum_{j=1}^{N} X_j\phi_j(t) + X'(t) \\ &= \sum_{j=1}^{N} X_j\phi_j(t) + W'(t) \end{aligned} \tag{10.37}$$

Accordingly, we may view $W'(t)$ as a sort of remainder term which must be included on the right in order to preserve the equality in Eq. (10.37). It is of interest to contrast the expansion of the received random process $X(t)$, as in Eq. (10.37), with the corresponding expansion of the transmitted deterministic signal $s_i(t)$, as in Eq. (10.6).

We now wish to characterize the set of correlator outputs, $\{X_j\}$, $j = 1, 2, \ldots, N$. Since the received random process $X(t)$ is Gaussian, we deduce that each X_j is a Gaussian random variable (see Property 1 of a Gaussian process, Section 5.9). Hence, each X_j is characterized completely by its mean value and variance.

The noise process $W(t)$ has zero mean. Hence, the random variable W_j extracted from $W(t)$ in accordance with Eq. (10.34) has a zero mean too. This implies that the mean value of the jth correlator output X_j depends only on s_{ij}, as shown by

$$\begin{aligned} m_{X_j} &= E[X_j] \\ &= E[s_{ij} + W_j] \\ &= s_{ij} + E[W_j] \\ &= s_{ij} \end{aligned} \tag{10.38}$$

To find the variance of X_j, we note that

$$\begin{aligned} \sigma_{X_j}^2 &= \text{Var}[X_j] \\ &= E[(X_j - s_{ij})^2] \\ &= E[W_j^2] \end{aligned} \tag{10.39}$$

Substituting Eq. (10.34) in (10.39), we get

$$\begin{aligned} \sigma_{X_j}^2 &= E\left[\int_0^T W(t)\phi_j(t)dt \int_0^T W(u)\phi_j(u)du\right] \\ &= E\left[\int_0^T \int_0^T \phi_j(t)\phi_j(u)W(t)W(u)dt\,du\right] \end{aligned} \tag{10.40}$$

Interchanging the order of integration and expectation:

$$\sigma_{X_j}^2 = \int_0^T \int_0^T \phi_j(t)\phi_j(u)E[W(t)W(u)]dt\,du$$

$$= \int_0^T \int_0^T \phi_j(t)\phi_j(u)R_W(t, u)dt\,du \qquad (10.41)$$

where $R_W(t, u)$ is the autocorrelation function of the noise process $W(t)$. Since this noise is stationary, $R_W(t, u)$ depends only on the time difference $t-u$. Furthermore, since the noise $W(t)$ is white with a power spectral density $N_0/2$, we may express $R_W(t, u)$ as follows [see Eq. (5.142)]

$$R_W(t, u) = \frac{N_0}{2}\delta(t-u) \qquad (10.42)$$

Therefore, substituting Eq. (10.42) in (10.41), we get

$$\sigma_{X_j}^2 = \frac{N_0}{2}\int_0^T \int_0^T \phi_j(t)\phi_j(u)\delta(t-u)dt\,du$$

$$= \frac{N_0}{2}\int_0^T \phi_j^2(t)dt \qquad (10.43)$$

Since the $\phi_j(t)$ have unit energy, we finally get the simple result

$$\sigma_{X_j}^2 = \frac{N_0}{2}, \qquad \text{for all } j \qquad (10.44)$$

This shows that all the correlator outputs, $X_j, j = 1, 2, \ldots, N$, have a variance equal to the power spectral density $N_0/2$ of the additive noise $W(t)$.

Similarly, we find that since the $\phi_j(t)$ form an orthogonal set, then the X_j are mutually uncorrelated, as shown by

$$\begin{aligned}
\text{Cov}[X_jX_k] &= E[(X_j - m_{X_j})(X_k - m_{X_k})] \\
&= E[(X_j - s_{ij})(X_k - s_{ik})] \\
&= E[W_jW_k] \\
&= E\left[\int_0^T W(t)\phi_j(t)dt \int_0^T W(u)\phi_k(u)du\right] \\
&= \int_0^T \int_0^T \phi_j(t)\phi_k(u)R_W(t, u)dt\,du \\
&= \frac{N_0}{2}\int_0^T \int_0^T \phi_j(t)\phi_k(u)\delta(t-u)dt\,du \\
&= \frac{N_0}{2}\int_0^T \phi_j(t)\phi_k(t)dt \\
&= 0, \qquad j \neq k
\end{aligned} \qquad (10.45)$$

Since the X_j are Gaussian random variables, Eq. (10.45) implies that they are also statistically independent. (See Property 4 of Gaussian Processes, Section 5.9).

Define the vector of N random variables at the correlator outputs as

$$\mathbf{X}=\begin{bmatrix} X_1 \\ X_2 \\ \vdots \\ X_N \end{bmatrix} \tag{10.46}$$

whose elements are independent Gaussian random variables with mean values equal to s_{ij} and variances equal to $N_0/2$. Since the elements of the vector $\mathbf{X}$ are statistically independent, we may express the conditional probability density function of the vector $\mathbf{X}$, given that the signal $s_i(t)$ or correspondingly the symbol m_i was transmitted, as the product of the conditional probability density functions of its individual elements, as shown by

$$f_{\mathbf{X}|M_i}(\mathbf{x}|m_i)=\prod_{j=1}^{N} f_{X_j|M_i}(x_j|m_i), \qquad i=1, 2, \ldots, M \tag{10.47}$$

where the vector $\mathbf{x}$ and scalar x_j are sample values of the random vector $\mathbf{X}$ and random variable X_j, respectively. The conditional probability density functions, $f_{\mathbf{X}|M_i}(\mathbf{x}|m_i)$, for each transmitted message m_i, $i=1, 2, \ldots, M$ are called *likelihood functions*. These likelihood functions, which are in fact the channel characterization, are also called *channel transition probabilities*. Any channel whose likelihood functions satisfy Eq. (10.47) is called a *memoryless* channel.

Since each X_j is a Gaussian random variable with mean s_{ij} and variance $N_0/2$, we have

$$f_{X_j|M_i}(x_j|m_i)=\frac{1}{\sqrt{\pi N_0}}\exp\left[-\frac{1}{N_0}(x_j-s_{ij})^2\right], \qquad \begin{matrix} j=1, 2, \ldots, N \\ i=1, 2, \ldots, M \end{matrix} \tag{10.48}$$

Therefore, substituting Eq. (10.48) in (10.47), we find that the likelihood functions of an AWGN channel are defined by

$$f_{\mathbf{X}|M_i}(\mathbf{x}|m_i)=(\pi N_0)^{-N/2}\exp\left[-\frac{1}{N_0}\sum_{j=1}^{N}(x_j-s_{ij})^2\right], \qquad i=1, 2, \ldots, M \tag{10.49}$$

Returning to Eq. (10.37), while it is now clear that the elements of the random vector $\mathbf{X}$ completely characterize the summation term $\Sigma X_j\phi_j(t)$, there remains the term $W'(t)$ which depends only on the original noise $W(t)$. Since the noise $W(t)$ is Gaussian with zero mean, it follows that $W'(t)$ is also a zero-mean Gaussian process. Finally, we note that any random variable $W'(t_k)$, say, derived from the noise process $W'(t)$ by sampling it at time t_k, is in fact independent of the set of random variables $\{X_j\}$. That is to say (see Problem 10.4)

$$E[W'(t_k)X_j]=0, \qquad \begin{matrix} j=1, 2, \ldots, N \\ 0\leq t_k\leq T \end{matrix} \tag{10.50}$$

Since any random variable based on the remainder noise term $W'(t)$ is independent of the set of random variables $\{X_j\}$ and the set of transmitted signals $\{s_i(t)\}$, we conclude that such a random variable is *irrelevant* to the decision as to which signal was transmitted. In other words, the set of random variables $\{X_j\}$ at the correlator outputs, based on the received random process $X(t)$, are the only data that are useful for the decision-making process and, hence, represent *sufficient statistics*.

10.5 COHERENT DETECTION OF SIGNALS IN THE PRESENCE OF NOISE

Assume that, in each time slot of duration T seconds, one of the M possible signals $s_1(t), s_2(t), \ldots, s_M(t)$ is transmitted with equal probability, namely, $1/M$. Then, for an AWGN channel, a possible realization or sample function, $x(t)$, of the received random process $X(t)$ may be described by

$$x(t) = s_i(t) + w(t), \qquad \begin{array}{l} 0 \leqslant t \leqslant T \\ i = 1, 2, \ldots, M \end{array} \tag{10.51}$$

where $w(t)$ is a sample function of the white Gaussian noise process $W(t)$, assumed to be of zero mean and power spectral density $N_0/2$. The receiver has to observe the signal $x(t)$ and make a "best estimate" of the transmitted signal $s_i(t)$ or equivalently the symbol m_i.

We note that when the transmitted signal $s_i(t)$, $i = 1, 2, \ldots, M$, is applied to a bank of correlators, with a common input and supplied with an appropriate set of N orthonormal basis functions, the resulting correlator outputs define the *signal vector* $\mathbf{s}_i$ [see Eq. (10.7)]. Since knowledge of the signal vector $\mathbf{s}_i$ is as good as knowing the transmitted signal $s_i(t)$ itself, and vice versa, we may represent $s_i(t)$ by a point in a Euclidean space of dimension $N \leqslant M$. We refer to this point as the *transmitted signal point* or *message point*.

However, the representation of the received signal $x(t)$ is complicated by the presence of the additive noise $w(t)$. We note that when the received signal $x(t)$ is applied to the same bank of N correlators as above, the correlator outputs define a new vector, $\mathbf{x}$, called the *observation vector*. The vector $\mathbf{x}$ differs from the signal vector $\mathbf{s}_i$ by the *noise vector* $\mathbf{w}$ whose orientation is completely random. That is to say,

$$\mathbf{x} = \mathbf{s}_i + \mathbf{w}, \qquad i = 1, 2, \ldots, M \tag{10.52}$$

The vectors $\mathbf{x}$ and $\mathbf{w}$ are sample values of random vectors $\mathbf{X}$ and $\mathbf{W}$, respectively, that were described in Section 10.4. Note also that the noise vector $\mathbf{w}$ is completely characterized by the noise $w(t)$; the converse of this, however, is not true. The noise vector $\mathbf{w}$ represents that portion of the noise $w(t)$ which will interfere with the detection process. The remaining portion of this noise may be thought of as being tuned out by the bank of correlators.

Now, based on the observation vector $\mathbf{x}$, we may represent the received signal $x(t)$ by a point in the same Euclidean space used to represent the transmitted

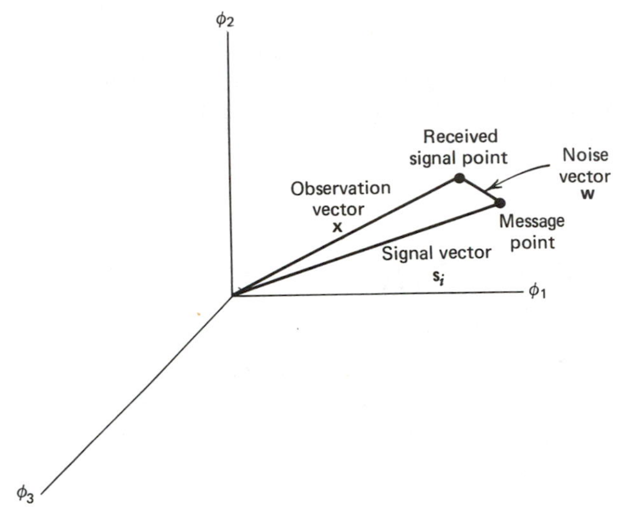

Figure 10.7 Illustrating the effect of noise perturbation on the location of the received signal point.

signal. We refer to this second point as the *received signal point*. The relationship between the observation vector, $\mathbf{x}$, the signal vector, $\mathbf{s}_i$, and the noise vector, $\mathbf{w}$, is illustrated in Fig. 10.7 for the case of $N=3$.

We are now ready to state the detection problem. Given the observation vector $\mathbf{x}$, we have to perform a *mapping* from $\mathbf{x}$ to an estimate $\hat{m}$ of the transmitted symbol, m_i, in a way that would minimize the probability of error in the decision. The so-called maximum-likelihood decoder provides the solution to this basic problem.

Maximum-Likelihood Decoder

Suppose that, when the observation vector has the value $\mathbf{x}$, we make the decision $\hat{m}=m_i$. The probability of error in this decision, which we denote by $P_e(m_i, \mathbf{x})$, is simply

$$\begin{aligned} P_e(m_i, \mathbf{x}) &= P(m_i \text{ not sent} \mid \mathbf{x}) \\ &= 1 - P(m_i \text{ sent} \mid \mathbf{x}) \end{aligned} \tag{10.53}$$

Now, since our criterion is to minimize the probability of error in mapping each given observation vector, $\mathbf{x}$, into a decision, we deduce from Eq. (10.53) that the *optimum decision rule* is as follows:

$$\begin{aligned} &\text{set } \hat{m}=m_i \text{ if} \\ &P(m_i \text{ sent}|\mathbf{x}) \geq P(m_k \text{ sent}|\mathbf{x}), \text{for all } k \neq i \end{aligned} \tag{10.54}$$

where $k=1, 2, \ldots, M$. This decision rule is referred to as *maximum a posteriori probability*.

The condition of Eq. (10.54) may be expressed more explicitly in terms of the a priori probabilities of the transmitted signals and in terms of the likelihood functions. Applying Bayes rule to Eq. (10.54), and for the moment ignoring possible ties in the decision-making process, we may restate this decision rule as:

$$\text{set } \hat{m}=m_i \text{ if}$$

$$\frac{p_k f_{\mathbf{X}|M_k}(\mathbf{x}|m_k)}{f_{\mathbf{X}}(\mathbf{x})} \text{ is maximum for } k=i \tag{10.55}$$

where p_k is the a priori probability of occurrence of symbol m_k, $f_{\mathbf{X}|M_k}(\mathbf{x}|m_k)$ is the likelihood function that results when symbol m_k is transmitted, and $f_{\mathbf{X}}(\mathbf{x})$ is the unconditional joint probability density function of the random vector $\mathbf{X}$. However, in Eq. (10.55), we note that: (1) the denominator term, $f_{\mathbf{X}}(\mathbf{x})$, is independent of the transmitted signal, and (2) the a priori probability $p_k=p_i$ when all the signals are transmitted with equal probability. Therefore, we may simplify the decision rule of Eq. (10.55) as:

$$\text{set } \hat{m}=m_i \text{ if}$$

$$f_{\mathbf{X}|M_k}(\mathbf{x}|m_k) \text{ is maximum for } k=i \tag{10.56}$$

Ordinarily, we find it more convenient to work with the natural logarithm of the likelihood function rather than the likelihood function itself. For a memoryless channel, the logarithm of the likelihood function is commonly called the *metric*. Since the likelihood function $f_{\mathbf{X}|M_k}(\mathbf{x}|m_k)$ is always nonnegative, and since if $A>B>0$ then $\ln A>\ln B$, we may restate the decision rule of Eq. (10.56) in terms of the metric as:

$$\text{set } \hat{m}=m_i \text{ if}$$

$$\ln[f_{\mathbf{X}|M_k}(\mathbf{x}|m_k)] \text{ is maximum for } k=i \tag{10.57}$$

This decision rule is referred to as *maximum likelihood*, and the device for its implementation is correspondingly referred to as the *maximum-likelihood decoder*. According to Eq. (10.57), a maximum-likelihood decoder computes the metric for each transmitted message, compares them, and then decides in favor of the maximum.

It is useful to have a graphical interpretation of the maximum-likelihood decision rule. Let Z denote the N-dimensional space of all possibly observed vectors $\mathbf{x}$. We refer to this space as the *observation space*. Because we have assumed that the decision rule must say $\hat{m}=m_i$, where $i=1, 2, \ldots, M$, the total observation space Z is correspondingly partitioned into M *decision regions*, denoted by $Z_1, Z_2, \ldots, Z_M$. Accordingly, we may restate the decision rule of Eq. (10.57) as follows:

$$\text{vector } \mathbf{x} \text{ lies inside region } Z_i \text{ if}$$

$$\ln[f_{\mathbf{X}|M_k}(\mathbf{x}|m_k)] \text{ is maximum for } k=i \tag{10.58}$$

Aside from the boundaries between the decision regions $Z_1, Z_2, \ldots, Z_M$, it is clear that this set of regions covers the entire space of possibly observed vectors $\mathbf{x}$. We adopt the convention that all ties are resolved at random. Specifically, if the observation vector $\mathbf{x}$ falls on the boundary between any two decision regions, Z_i and Z_k, say, the choice between the two possible decisions $\hat{m}=m_i$ and $\hat{m}=m_k$ is resolved a priori by the flip of a fair coin. Clearly, the outcome of such an event does not affect the ultimate value of the probability of error since, on this boundary, the condition of Eq. (10.54) is satisfied with the equality sign.

To illustrate the above concept, consider an AWGN channel for which the likelihood function is defined by Eq. (10.49). We thus have

$$f_{\mathbf{X}|M_k}(\mathbf{x}|m_k)=(\pi N_0)^{-N/2}\exp\left[-\frac{1}{N_0}\sum_{j=1}^{N}(x_j-s_{kj})^2\right], \qquad k=1,2,\ldots,M \qquad (10.59)$$

The corresponding value of the metric is therefore

$$\ln[f_{\mathbf{X}|M_k}(\mathbf{x}|m_k)]=-\frac{N}{2}\ln(\pi N_0)-\frac{1}{N_0}\sum_{j=1}^{N}(x_j-s_{kj})^2, \qquad k=1,2,\ldots,M \qquad (10.60)$$

The constant term $-(N/2)\ln(\pi N_0)$ on the right-hand side of Eq. (10.60) is of no consequence in so far as application of the decision rule is concerned. Therefore, ignoring this term, and then substituting the remainder of Eq. (10.60) into (10.58), we may formulate the maximum-likelihood decision rule for an AWGN channel as follows:

$$\text{vector } \mathbf{x} \text{ lies inside region } Z_i \text{ if}$$
$$-\frac{1}{N_0}\sum_{j=1}^{N}(x_j-s_{kj})^2 \text{ is maximum for } k=i \qquad (10.61)$$

Equivalently, we may state:

$$\text{vector } \mathbf{x} \text{ lies inside region } Z_i \text{ if}$$
$$\sum_{j=1}^{N}(x_j-s_{kj})^2 \text{ is minimum for } k=i \qquad (10.62)$$

However, we note that

$$\sum_{j=1}^{N}(x_j-s_{kj})^2=\|\mathbf{x}-\mathbf{s}_k\|^2 \qquad (10.63)$$

where $\|\mathbf{x}-\mathbf{s}_k\|$ is the distance between the received signal point and message point, represented by the vectors $\mathbf{x}$ and $\mathbf{s}_k$, respectively. Accordingly, we may rewrite the decision rule of Eq. (10.62) as follows:

$$\text{vector } \mathbf{x} \text{ lies inside region } Z_i \text{ if}$$
$$\|\mathbf{x}-\mathbf{s}_k\| \text{ is minimum for } k=i \qquad (10.64)$$

This shows that the maximum likelihood decision rule is simply to choose the message point closest to the received signal point.

In practice the need for squarers, as in the decision rule of Eq. (10.64), is avoided by recognizing that

$$\sum_{j=1}^{N}(x_j-s_{kj})^2=\sum_{j=1}^{N}x_j^2-2\sum_{j=1}^{N}x_js_{kj}+\sum_{j=1}^{N}s_{kj}^2 \qquad (10.65)$$

The first summation term is independent of the index k and may therefore be ignored. The second summation term is the inner product of the observation vector $\mathbf{x}$ and signal vector $\mathbf{s}_k$. The third summation term is the energy of the transmitted signal $s_k(t)$. Accordingly, a decision rule equivalent to that of Eq. (10.64) is:

$$\text{vector } \mathbf{x} \text{ lies inside region } Z_i \text{ if}$$

$$\sum_{j=1}^{N} x_j s_{kj} - \tfrac{1}{2}E_k \text{ is maximum for } k=i \tag{10.66}$$

where E_k is the energy of the transmitted signal $s_k(t)$:

$$E_k = \sum_{j=1}^{N} s_{kj}^2 \tag{10.67}$$

From Eq. (10.66) we deduce that, for an AWGN channel, the decision regions are regions of the N-dimensional observation space Z, bounded by linear [$(N-1)$-dimensional hyperplane] boundaries. Figure 10.8 shows the example of decision regions for $M=4$ signals and $N=2$ dimensions, assuming that the signals are transmitted with equal energy, E, and equal probability.

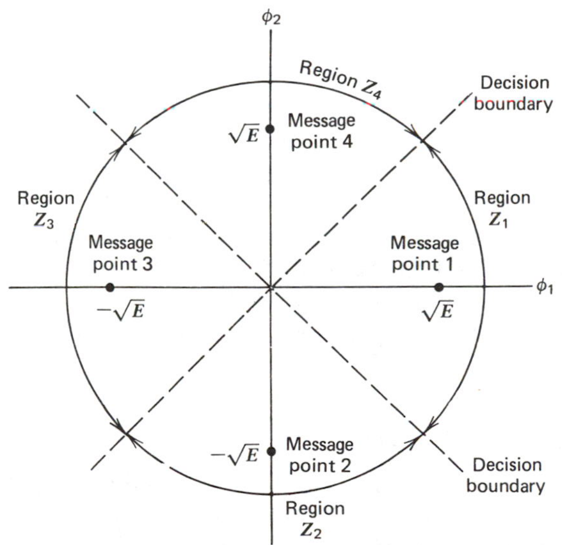

Figure 10.8 Illustrating the partitioning of the observation space into decision regions for the case when $N=2$ and $M=4$.

Probability of Error

Suppose that the observation space Z is partitioned, in accordance with the maximum-likelihood decision rule, into a set of M regions $\{Z_i\}$, $i=1, 2, \ldots, M$. Suppose also that symbol m_i (or, equivalently, signal vector $\mathbf{s}_i$) is transmitted, and an observation vector $\mathbf{x}$ is received. Then an error occurs whenever the received signal point represented by $\mathbf{x}$ does not fall inside region Z_i associated with the message point represented by $\mathbf{s}_i$. Averaging over all possibly transmitted symbols, we readily see that the *average probability of symbol error*, P_e, equals

$$\begin{aligned}P_e &= \sum_{i=1}^{M} P(m_i \text{ sent})P(\mathbf{x} \text{ does not lie inside } Z_i|m_i \text{ sent})\\ &= \frac{1}{M}\sum_{i=1}^{M} P(\mathbf{x} \text{ does not lie inside } Z_i|m_i \text{ sent})\\ &= 1 - \frac{1}{M}\sum_{i=1}^{M} P(\mathbf{x} \text{ lies inside } Z_i|m_i \text{ sent}) \qquad (10.68)\end{aligned}$$

where we have used standard notation to denote the probability of an event and the conditional probability of an event (see Section 5.1). Since $\mathbf{x}$ is the sample value of random vector $\mathbf{X}$, we may rewrite Eq. (10.68) in terms of the likelihood function (when m_i is sent) as follows

$$P_e = 1 - \frac{1}{M}\sum_{i=1}^{M} \int_{Z_i} f_{\mathbf{X}|M_i}(\mathbf{x}|m_i)\, d\mathbf{x} \qquad (10.69)$$

For an N-dimensional observation vector $\mathbf{x}$, the integral in Eq. (10.69) is likewise N-dimensional.

For AWGN channels, it is possible to obtain exact expressions for the average probability of error in integral form by substituting Eq. (10.49) into (10.69). Unfortunately, however, the integrals in question are not integrable, except in a few simple (but important) cases. In such difficult situations, it is sometimes possible to derive *upper* and *lower bounds* which are usually adequate to predict the signal-to-noise ratio (within a decibel or so) required to maintain a prescribed error rate.* We will not discuss these bounds here because, for the PSK and FSK systems that are of major interest to us (within the scope of this book), we will be able to derive exact expressions for the average probability of error.

10.6 CORRELATION RECEIVER

We note that for an AWGN channel and for the case when the transmitted signals $s_1(t), s_2(t), \ldots, s_M(t)$ are equally likely, the optimum receiver consists of two subsystems, detailed in Fig. 10.9 and described below:

1. The detector part of the receiver is shown in part (*a*) of Fig. 10.9. It consists of a bank of M *product-integrators* or *correlators* supplied with a corresponding set of coherent reference signals or orthonormal basis functions $\phi_1(t), \phi_2(t), \ldots,$

* For details of bounds on probability of error. see:
1. E. Arthurs and H. Dym, *op. cit.*
2. A. J. Viterbi and J. K. Omura, pp. 58–65, *op. cit.*

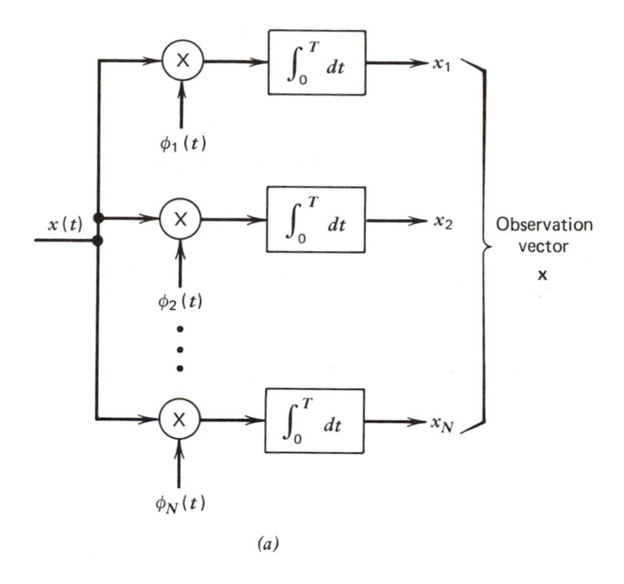

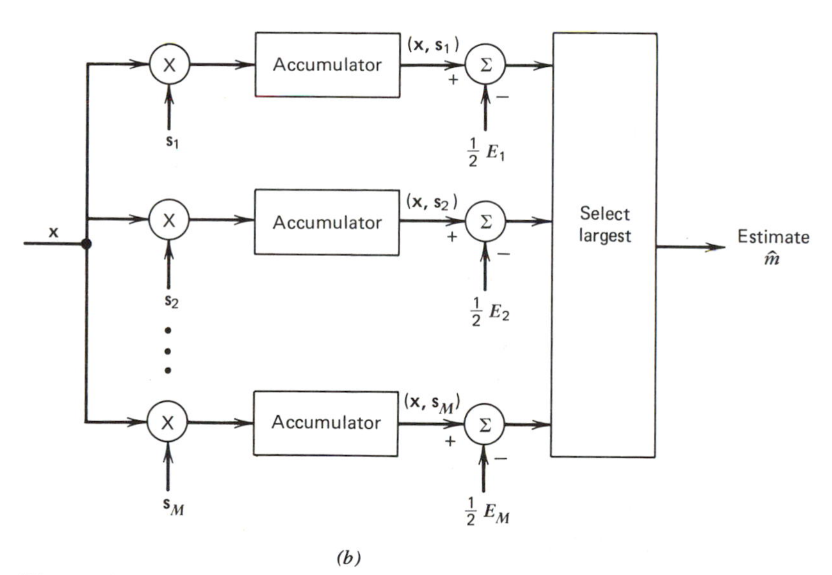

Figure 10.9 *(a)* Detector or demodulator. *(b)* Vector receiver.

$\phi_N(t)$ that are generated locally. This bank of correlators operate on the received signal $x(t)$, $0 \leqslant t \leqslant T$, to produce the observation vector $\mathbf{x}$.

2. The second part of the receiver, namely, the vector receiver is shown in part (*b*) of Fig. 10.9. It is implemented in the form of a maximum-likelihood decoder that operates on the observation vector $\mathbf{x}$ to produce an estimate, $\hat{m}$, of the transmitted symbol m_i, $i = 1, 2, \ldots, M$, in a way that would minimize the average probability of symbol error. In accordance with Eq. (10.66), the N elements of

the observation vector $\mathbf{x}$ are first multiplied by the corresponding N elements of each of the M signal vectors $\mathbf{s}_1, \mathbf{s}_2, \ldots, \mathbf{s}_M$, and the resulting products are successively summed in accumulators to form the corresponding set of inner products $\{(\mathbf{x}, \mathbf{s}_k)\}$, $k=1, 2, \ldots, M$. Next, the inner products are corrected for the fact that the transmitted signal energies may be unequal. Finally, the largest in the resulting set of numbers is selected, and a corresponding decision on the transmitted message is made.

The optimum receiver of Fig. 10.9 is commonly referred to as a *correlation receiver*.

10.7 MATCHED FILTER RECEIVER

Since each of the orthonormal basis functions $\phi_1(t), \phi_2(t), \ldots, \phi_N(t)$ is assumed to be zero outside the interval $0 \leqslant t \leqslant T$, the use of multipliers shown in Fig. 10.9(*a*) may be avoided. This is desirable because analog multipliers are usually hard to build. Consider, for example, a linear filter with impulse response $h_j(t)$. With the received signal $x(t)$ used as the filter input, as in Fig. 10.10. The resulting filter output, $y_j(t)$, is defined by the convolution integral:

$$y_j(t) = \int_{-\infty}^{\infty} x(\tau) h_j(t-\tau) d\tau \qquad (10.70)$$

Suppose we now set the impulse response

$$h_j(t) = \phi_j(T-t) \qquad (10.71)$$

Then the resulting filter output is

$$y_j(t) = \int_{-\infty}^{\infty} x(\tau) \phi_j(T-t+\tau) d\tau \qquad (10.72)$$

Sampling this output at time $t=T$, we get

$$y_j(T) = \int_{-\infty}^{\infty} x(\tau) \phi_j(\tau) d\tau \qquad (10.73)$$

and since $\phi_j(T)$ is zero outside the interval $0 \leqslant t \leqslant T$, we finally get

$$y_j(T) = \int_0^T x(\tau) \phi_j(\tau) d\tau \qquad (10.74)$$

We note that $y_j(T) = x_j$, where x_j is the jth correlator output produced by the received signal $x(t)$ in Fig. 10.9(*a*). Thus the detector part of the optimum receiver may also be implemented as in Fig. 10.11.

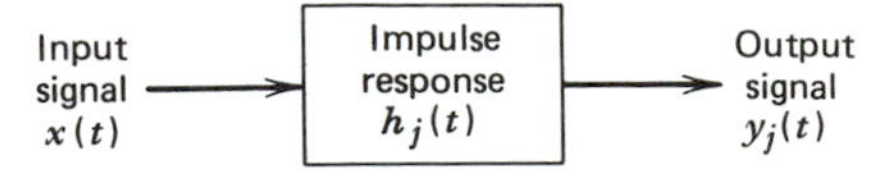

Figure 10.10 Linear filter.

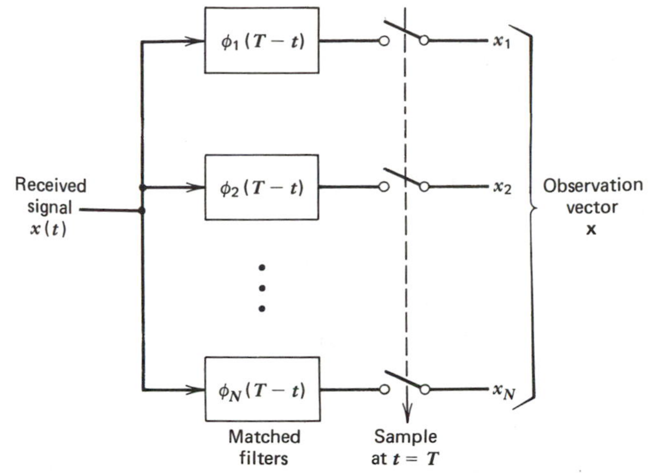

Figure 10.11 Detector part of matched filter receiver; the vector receiver part is as shown in Fig. 10.9*(b)*.

A filter whose impulse response is a time-reversed and delayed version of some signal $\phi_j(t)$, as in Eq. (10.71), is said to be *matched* to $\phi_j(t)$. Correspondingly, the optimum receiver based on the detector of Fig. 10.11 is referred to as the *matched filter receiver.**

For a matched filter operating in real time to be physically realizable, it must be causal. That is to say, its impulse response must be zero for negative time, as shown by

$$h_j(t)=0, \qquad t<0$$

With $h_j(t)$ defined in terms of $\phi_j(t)$ as in Eq. (10.71), we see that the causality condition is satisfied provided that the signal $\phi_j(t)$ is zero outside the interval $0 \leqslant t \leqslant T$.

Maximization of Output Signal-to-Noise Ratio

We may gain further insight into the operation of a matched filter by using output signal-to-noise ratio as the optimality criterion.

* The characterization of a matched filter was first derived by North in a classified report (RCA Laboratories Report PTR-6C, June 1943), which was published 20 years later: D. O. North, "An analysis of the factors which determine signal/noise discrimination in pulsed-carrier systems," *Proc. IEEE*, vol. 51, pp. 1016–1027, July 1963.

A similar result was obtained independently by Van Vleck and Middleton, who coined the term "matched filter": J. H. Van Vleck and D. Middleton, "A theoretical comparison of the visual, aural and meter reception of pulsed signals in the presence of noise," *J. Appl. Phys.*, vol. 17, pp. 940–971, Nov. 1946.

For review papers on the matched filter and its properties see:

1. G. L. Turin, "An introduction to matched filters," *IRE Trans., Information Theory*, vol. IT-6, pp. 311–329, June 1960.
2. G. L. Turin, "An introduction to digital matched filters," *Proc. IEEE*, vol. 64, pp. 1092–1112, July 1976.

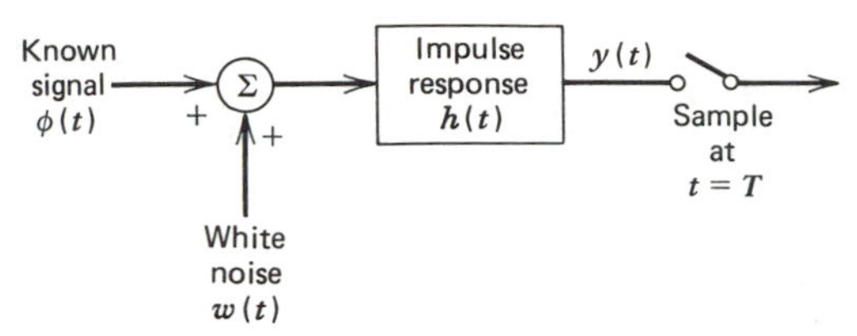

Figure 10.12 Illustrating the condition for derivation of the matched filter.

Consider then a linear filter of impulse response $h(t)$, with an input that consists of a known signal, $\phi(t)$, and an additive noise component, $w(t)$, as shown in Fig. 10.12. We may thus write

$$x(t)=\phi(t)+w(t), \qquad 0\leqslant t\leqslant T \tag{10.75}$$

where T is an arbitrary observation instant. In particular, we may choose $\phi(t)$ to be one of the orthonormal basis functions. The $w(t)$ is the sample function of a white Gaussian noise process of zero mean and power spectral density $N_0/2$. Since the filter is linear, the resulting output, $y(t)$, may be expressed as

$$y(t)=\phi_o(t)+n(t) \tag{10.76}$$

where $\phi_o(t)$ and $n(t)$ are produced by the signal and noise components of the input $x(t)$, respectively. A simple way of describing the requirement that the output signal component $\phi_o(t)$ be considerably greater than the output noise component $n(t)$ is to have the filter make the instantaneous power in the output signal $\phi_o(t)$, measured at time $t=T$, as large as possible compared with the average power of the output noise $n(t)$. This is equivalent to maximizing the *output signal-to-noise ratio* defined as

$$(\mathrm{SNR})_{\mathrm{O}}=\frac{|\phi_o(T)|^2}{E[n^2(t)]} \tag{10.77}$$

We now show that this maximization occurs when the filter is matched to the known signal $\phi(t)$ at the input.

Let $\Phi(f)$ denote the Fourier transform of the known signal $\phi(t)$, and $H(f)$ denote the transfer function of the filter. Then the Fourier transform of the output signal $\phi_o(t)$ is equal to $H(f)\Phi(f)$, and $\phi_o(t)$ is itself given by the inverse Fourier transform

$$\phi_o(t)=\int_{-\infty}^{\infty} H(f)\Phi(f)\exp(j2\pi ft)df \tag{10.78}$$

Hence, when the filter output is sampled at time $t=T$, we may write

$$|\phi_o(T)|^2=\left|\int_{-\infty}^{\infty} H(f)\Phi(f)\exp(j2\pi fT)df\right|^2 \tag{10.79}$$

Consider next the effect of the noise $w(t)$ alone on the filter output. The power spectral density $S_N(f)$ of the output noise $n(t)$ is equal to the power spectral density of the input noise $w(t)$ times the squared magnitude of the transfer function $H(f)$

(see Section 5.8). Since $w(t)$ is white with constant power spectral density $N_0/2$, it follows that

$$S_N(f)=\frac{N_0}{2}|H(f)|^2 \tag{10.80}$$

The average power of the output noise $n(t)$ is therefore

$$E[n^2(t)]=\int_{-\infty}^{\infty} S_N(f)df$$
$$=\frac{N_0}{2}\int_{-\infty}^{\infty} |H(f)|^2 df \tag{10.81}$$

Thus, substituting Eqs. (10.79) and (10.81) into (10.77), we may rewrite the expression for the output signal-to-noise ratio as

$$(\mathrm{SNR})_O=\frac{\left|\int_{-\infty}^{\infty} H(f)\Phi(f)\exp(j2\pi fT)df\right|^2}{\frac{N_0}{2}\int_{-\infty}^{\infty} |H(f)|^2 df} \tag{10.82}$$

Our problem is to find, while holding the Fourier transform $\Phi(f)$ of the input signal fixed, the form of the transfer function $H(f)$ of the filter which makes $(\mathrm{SNR})_O$ a maximum. To find the solution to this optimization problem, we may apply a mathematical result known as Schwarz's inequality to the numerator of Eq. (10.82); a derivation of Schwarz's inequality is given in Appendix 5. We may thus write

$$\left|\int_{-\infty}^{\infty} H(f)\Phi(f)\exp(j2\pi fT)df\right|^2 \leqslant \int_{-\infty}^{\infty} |H(f)|^2 df \int_{-\infty}^{\infty} |\Phi(f)|^2 df \tag{10.83}$$

Using this relation in Eq. (10.82), we may rewrite it as

$$(\mathrm{SNR})_O \leqslant \frac{2}{N_0}\int_{-\infty}^{\infty} |\Phi(f)|^2 df \tag{10.84}$$

The right-hand side of this relation does not depend on the transfer function $H(f)$ of the filter but only on the signal energy and the noise power spectral density. Consequently, the output signal-to-noise ratio will be a maximum when $H(f)$ is chosen so that the equality holds; that is,

$$(\mathrm{SNR})_{O,\max}=\frac{2}{N_0}\int_{-\infty}^{\infty} |\Phi(f)|^2 df \tag{10.85}$$

For this case, $H(f)$ assumes its optimum value denoted as $H_{opt}(f)$. From Schwarz's inequality, we also find that, except for a scaling factor, the optimum value of this transfer function is defined by

$$H_{opt}(f)=\Phi^*(f)\exp(-j2\pi fT) \tag{10.86}$$

where $\Phi^*(f)$ is the complex conjugate of the Fourier transform of the input signal

$\phi(t)$. This relation states that, except for the necessary time delay factor $\exp(-j2\pi fT)$, the transfer of the optimum filter is the same as the complex conjugate of the spectrum of the input signal.

Equation (10.86) specifies the matched filter in the frequency domain. To characterize it in the time domain, we take the inverse Fourier transform of $H_{opt}(f)$ in Eq. (10.86) to obtain the impulse response of the matched filter as

$$h_{opt}(t)=\int_{-\infty}^{\infty} \Phi^*(f)\exp[-j2\pi f(T-t)]df$$

and, since for a real-valued signal $\phi(t)$ we have $\Phi^*(f)=\Phi(-f)$,

$$\begin{aligned} h_{opt}(t) &= \int_{-\infty}^{\infty} \Phi(-f)\exp[-j2\pi f(T-t)]df \\ &= \phi(T-t) \end{aligned} \tag{10.87}$$

Equation (10.87) shows that the impulse response of the optimum filter is a time-reversed and delayed version of the input signal $\phi(t)$; that is, it is matched to the input signal. Note that the only assumption we have made about the input noise $w(t)$ is that it is stationary and white with zero mean and power spectral density $N_0/2$.

Example 2 Matched filter for RF pulse

Consider a rectangular RF pulse of duration T seconds and unit energy, as shown by [see Fig. 10.13(a)]

$$\phi(t)=\begin{cases} \sqrt{\dfrac{2}{T}}\cos(2\pi f_c t), & 0 \leqslant t \leqslant T \\ 0, & \text{elsewhere} \end{cases} \tag{10.88}$$

where f_c is an integral multiple of $1/T$. The impulse response of a filter matched to $\phi(t)$ is therefore

$$\begin{aligned} h_{opt}(t) &= \phi(T-t) \\ &= \begin{cases} \sqrt{\dfrac{2}{T}}\cos(2\pi f_c t), & 0 \leqslant t \leqslant T \\ 0, & \text{elsewhere} \end{cases} \end{aligned} \tag{10.89}$$

which, in this example, works out to be the same as the signal $\phi(t)$ itself. The corresponding filter output, $\phi_o(t)$, is determined by convolving $h_{opt}(t)$ with $\phi(t)$; we thus obtain (see Problem 2.65):

$$\begin{aligned} \phi_o(t) &= h_{opt}(t) \otimes \phi(t) \\ &= \begin{cases} \dfrac{2}{T}\left[1-\left|\left(\dfrac{t}{T}\right)-1\right|\cos(2\pi f_c|t-T|)\right], & 0 \leqslant t \leqslant 2T \\ 0, & \text{elsewhere} \end{cases} \end{aligned} \tag{10.90}$$

which is shown sketched in Fig. 10.13(b). As expected, the filter output attains its maximum value at time $t=T$.

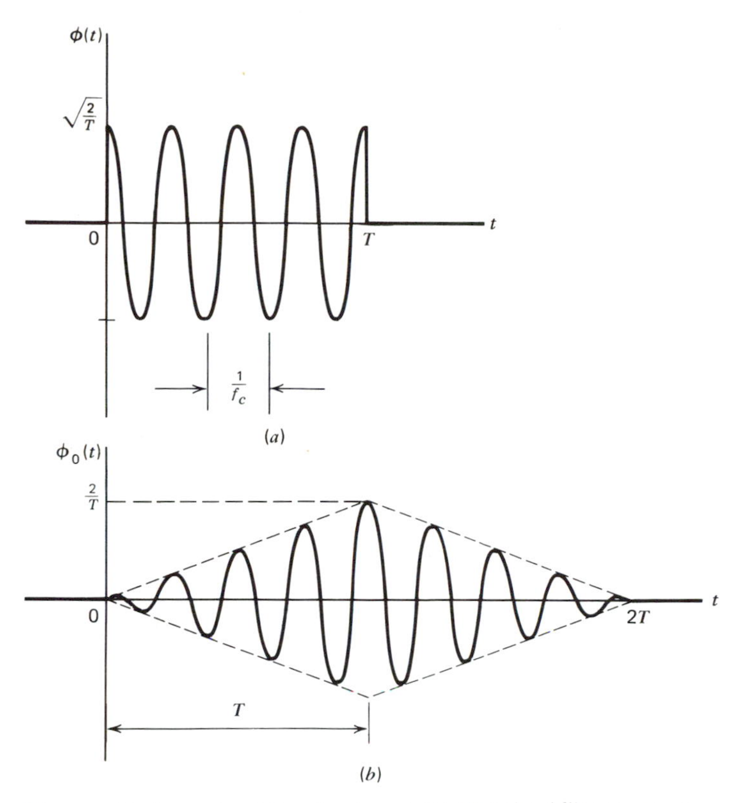

Figure 10.13 *(a)* RF pulse input. *(b)* Matched filter output.

When a unit impulse of current is applied to the infinite $-Q$ parallel tuned circuit shown in Fig. 10.14(*a*), the resulting voltage response is equal to the impulse response of the circuit, as shown by

$$h(t)=\frac{1}{C}\cos\left(\frac{t}{\sqrt{LC}}\right), \qquad 0\leqslant t\leqslant\infty \tag{10.91}$$

where it is assumed that the initial energy in both L and C at time $t=0$ is zero. Suppose we choose the circuit parameters L and C such that

$$\frac{1}{\sqrt{LC}}=2\pi f_c$$

and

$$\frac{1}{C}=\sqrt{\frac{2}{T}}$$

We then find that the response $h(t)$ of the tuned circuit coincides with $h_{opt}(t)$ over the interval $0\leqslant t\leqslant T$. However, this is not so for $t>T$. Thus, the matched filtering operation for the RF pulse $\phi(t)$ of Fig. 10.13(*a*) may be implemented by using the modification shown in Fig. 10.14(*b*). The parallel switch closes briefly at time $t=0$, dumping any residual energy in the filter. This ensures that signal energy received before $t=0$ does not contribute to the output

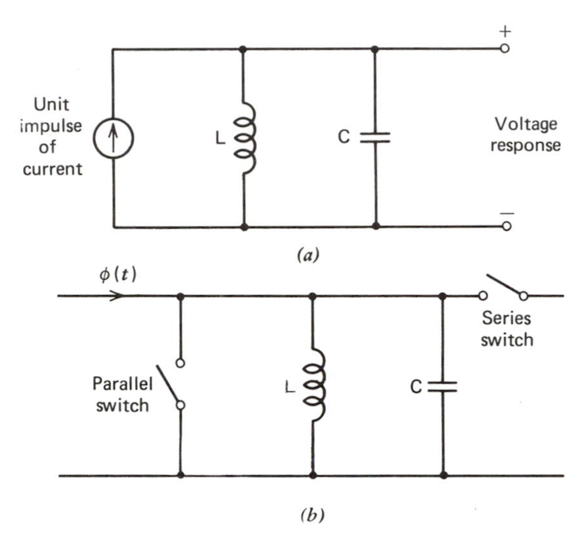

Figure 10.14 *(a)* Parallel tuned LC circuit. *(b)* Integrate-and-dump circuit.

at time $t=T$. The series switch then closes briefly at time $t=T$, thereby sampling the filter output at the right time. This form of a matched filter is called an *integrate-and-dump filter*. Note that such a filter is not time-invariant; however, it does exhibit the desired impulse response as long as the timing of the two switches is properly synchronized with respect to the input RF pulse $\phi(t)$.

Properties of Matched Filters

We note that a filter, which is matched to a known signal $\phi(t)$ of duration T seconds, is characterized by an impulse response that is a time-reversed and delayed version of the input $\phi(t)$, as shown by

$$h_{opt}(t)=\phi(T-t) \tag{10.92}$$

In the frequency domain, the matched filter is characterized by a transfer function that is, except for a delay factor, the complex conjugate of the Fourier transform of the input $\phi(t)$, as shown by

$$H_{opt}(f)=\Phi^*(f)\exp(-j2\pi fT) \tag{10.93}$$

Based on this fundamental pair of relations, we may derive some important properties of matched filters, which should help the reader develop an intuitive grasp of how a matched filter operates.

Property 1

The spectrum of the output signal of a matched filter with the matched signal as input is, except for a delay factor, proportional to the energy spectral density of the input signal.

Let $\Phi_o(f)$ denote the Fourier transform of the filter output $\phi_o(t)$. Then,

$$\begin{aligned}\Phi_o(f) &= H_{opt}(f)\Phi(f) \\ &= \Phi^*(f)\Phi(f)\exp(-j2\pi f T) \\ &= |\Phi(f)|^2 \exp(-j2\pi f T)\end{aligned} \tag{10.94}$$

which is the desired result.

Property 2

The output signal of a matched filter is proportional to a shifted version of the autocorrelation function of the input signal to which the filter is matched.

This property follows directly from Property 1, recognizing that the autocorrelation function and energy spectral density of a signal form a Fourier transform pair (see Section 2.8). Thus, taking the inverse Fourier transform of Eq. (10.94), we may express the matched-filter output as

$$\phi_o(t) = R_\phi(t - T) \tag{10.95}$$

where $R_\phi(\tau)$ is the autocorrelation function of the input $\phi(t)$. Equation (10.95) is the desired result. Note that at time $t = T$, we have

$$\phi_o(T) = R_\phi(0) = E \tag{10.96}$$

where E is the signal energy. That is, in the absence of noise, the maximum value of the matched-filter output, attained at time $t = T$, is proportional to the signal energy.

Example 3

A possible exploitation of this property of a matched filter is illustrated in Fig. 10.15. Let us suppose that we have a signal lasting from $t=0$ to $t=T$, which has the appearance and character of a sample function of a random process with a broad power spectrum, so that its autocorrelation function approximates a delta function. This signal may be generated by applying, at $t=0$, a short pulse (short enough to approximate a delta function) to a linear filter with impulse response $\phi(t)$. The impulse-like input signal has components occupying a very wide frequency band, but their amplitudes and phases are such that they add constructively only at and near $t=0$ and cancel each other out elsewhere. We may therefore view the signal-generating filter

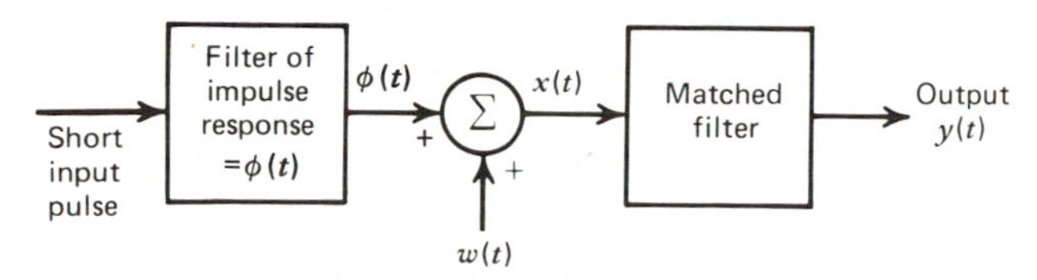

Figure 10.15 Viewing the matched filtering operation as an encoding-decoding process.

as an *encoder*, whereby the amplitudes and phases of the frequency components of the impulse-like input signal are coded in such a way that the filter output becomes noise-like in character, lasting from $t=0$ to $t=T$, as in Fig. 10.16(*a*). The signal $\phi(t)$ generated in this way is to be transmitted to a receiver via a distortionless but noisy channel. The requirement is to reconstruct at the receiver output a signal that closely approximates the original impulse-like signal at the transmitter input.

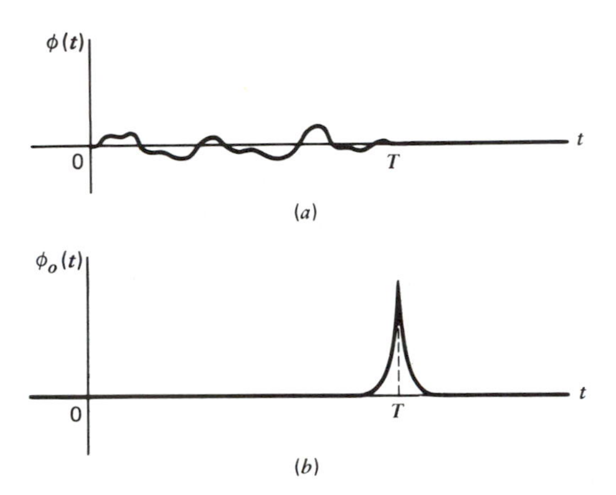

Figure 10.16 *(a)* Noise-like input signal. *(b)* Matched filter output.

The optimum solution to such a requirement, in the presence of additive white Gaussian noise, is to employ a matched filter in the receiver, as in Fig. 10.15. We may view this matched filter as a *decoder*, whereby the useful signal component $\phi(t)$ of the receiver input is decoded in such a way that all frequency components at the filter output have zero phase at time $t=T$, and add constructively to produce a large pulse of nonzero width, as in Fig. 10.16(*b*). Thus, in coding the impulse-like signal at the transmitter input we have spread the signal energy out over a duration T, and in decoding the noise-like signal at the receiver input we are able to concentrate this energy into a relatively narrow pulse. The extent to which the receiver output $\phi_o(t)$ approximates the original impulse-like signal is simply a reflection of the extent to which the autocorrelation function of the transmitted signal $\phi(t)$ approximates a delta function. The signal generating and reconstruction filters in Fig. 10.15 are said to constitute a *matched-filter pair*.*

Property 3

The output signal-to-noise ratio of a matched filter depends only on the ratio of the signal energy to the power spectral density of the white noise at the filter input.

To demonstrate this property, consider a filter matched to an input signal $\phi(t)$.

* The use of a matched-filter pair forms the basis of *pulse compression radar* which is used to obtain fine-range resolution and high-signal detectability. For more details, see: A. W. Rihaczek, *Principles of High-Resolution Radar* (McGraw-Hill, 1969).

From Property 2, the maximum value of the filter output, at time $t=T$, is proportional to the signal energy E. Substituting Eq. (10.86) in (10.81) gives the average output noise power as

$$E[n^2(t)]=\frac{N_0}{2}\int_{-\infty}^{\infty}|\Phi(f)|^2 df=\frac{N_0}{2}E \tag{10.97}$$

where we have made use of Rayleigh's energy theorem. Therefore, the output signal-to-noise ratio has the maximum value

$$(\mathrm{SNR})_{\mathrm{O,max}}=\frac{E^2}{N_0E/2}=\frac{2E}{N_0} \tag{10.98}$$

This result is perhaps the most important parameter in the calculation of the performance of signal processing systems using matched filters. From Eq. (10.98) we see that dependence on the waveform of the input $\phi(t)$ has been completely removed by the matched filter. Accordingly, in evaluating the ability of a matched-filter receiver to combat additive white Gaussian noise, we find that all signals that have the same energy are equally effective. Note that the signal energy E is in joules and the noise spectral density $N_0/2$ is in watts per hertz, so that the ratio $2E/N_0$ is dimensionless; however, the two quantities have different physical meaning. We refer to E/N_0 as the *signal energy-to-noise density ratio.*

Property 4

The matched-filtering operation may be separated into two matching conditions; namely, spectral phase matching that produces the desired output peak at time T, and spectral amplitude matching that gives this peak value its optimum signal-to-noise ratio.

In polar form, the spectrum of the signal $\phi(t)$ being matched may be expressed as

$$\Phi(f)=|\Phi(f)|\exp[j\theta(f)]$$

where $|\Phi(f)|$ is the amplitude spectrum and $\theta(f)$ is the phase spectrum of the signal. The filter is said to be *spectral phase matched* to the signal $\phi(t)$ if the transfer function of the filter is defined by*

$$H(f)=|H(f)|\exp[-j\theta(f)-j2\pi fT]$$

where $|H(f)|$ is real and nonnegative and T is a positive constant. The output of such a filter is

$$\begin{aligned}\phi_o'(t)&=\int_{-\infty}^{\infty}H(f)\Phi(f)\exp(j2\pi ft)df\\&=\int_{-\infty}^{\infty}|H(f)||\Phi(f)|\exp[j2\pi f(t-T)]df\end{aligned}$$

* T. G. Birdsall, "On understanding the matched filter in the frequency domain," *IEEE Trans, on Education*, vol. E-19, pp. 168–169, Nov. 1976.

where the product $|H(f)||\Phi(f)|$ is real and nonnegative. The spectral phase matching ensures that all the spectral components of the output $\phi_o'(t)$ add constructively at time $t = T$, thereby causing the output to attain its maximum value, as shown by

$$\phi_o'(t) \leq \phi_o'(T) = \int_{-\infty}^{\infty} |\Phi(f)||H(f)| df$$

For *spectral amplitude matching*, we choose the amplitude response $|H(f)|$ of the filter to shape the output for best signal-to-noise ratio at $t = T$ by using

$$|H(f)| = |\Phi(f)|$$

and the standard matched filter is the result.

10.8 COHERENT BINARY SIGNALING TECHNIQUES

We begin our study of digital modulation by considering the simple and yet important class of *coherent binary signaling techniques*, for which the modulation process corresponds to switching or keying the amplitude, frequency, or phase of the carrier between either of *two* possible values corresponding to binary symbols 0 and 1, with fixed frequency limits set by the channel. We may thus distinguish three basic signaling techniques, namely, amplitude-shift keying (ASK), frequency-shift keying (FSK), and phase-shift keying (PSK), as described below:

1. In an ASK system, binary symbol 1 is represented by transmitting a sinusoidal carrier wave of fixed amplitude and fixed frequency for the bit duration T_b seconds, whereas binary symbol 0 is represented by switching off the carrier for T_b seconds, as illustrated in Fig. 10.2(*a*).
2. In an FSK system, two sinusoidal waves of the same amplitude but different frequencies are used to represent binary symbols 1 and 0, as illustrated in Fig. 10.2(*b*).
3. In a PSK system, a sinusoidal carrier wave of fixed amplitude and fixed frequency is used to represent both symbols 1 and 0, except that whenever symbol 0 is transmitted the carrier phase is shifted by 180 degrees, as illustrated in Fig. 10.2(*c*).

It is apparent, therefore, that ASK, FSK, and PSK signals are special cases of amplitude-modulated, frequency-modulated, and phase-modulated waves, respectively.

In the next two sections we evaluate the error performances of coherent binary PSK and FSK receivers for an AWGN channel. For the analysis of an ASK system, the reader is referred to Problem 10.25.

10.9 COHERENT BINARY PSK

In a coherent binary PSK system, the pair of signals, $s_1(t)$ and $s_2(t)$, used to represent binary symbols 1 and 0, respectively, are defined by

$$s_1(t)=\sqrt{\frac{2E_b}{T_b}}\cos(2\pi f_c t) \tag{10.99}$$

$$s_2(t)=\sqrt{\frac{2E_b}{T_b}}\cos(2\pi f_c t+\pi)=-\sqrt{\frac{2E_b}{T_b}}\cos(2\pi f_c t) \tag{10.100}$$

where $0 \leqslant t \leqslant T_b$, and E_b is the *transmitted signal energy per bit*. In order to ensure that each transmitted bit contains an integral number of cycles of the carrier wave, the carrier frequency f_c is chosen equal to n_c/T_b for some fixed integer n_c. A pair of sinusoidal waves differing only in phase by 180 degrees, as defined above, are referred to as *antipodal signals*.

From Eqs. (10.99) and (10.100), it is clear that there is only one basis function of unit energy, namely

$$\phi_1(t)=\sqrt{\frac{2}{T_b}}\cos(2\pi f_c t), \qquad 0 \leqslant t \leqslant T_b \tag{10.101}$$

Then we may express the transmitted signals $s_1(t)$ and $s_2(t)$ in terms of $\phi_1(t)$ as follows

$$s_1(t)=\sqrt{E_b}\phi_1(t) \tag{10.102}$$

and

$$s_2(t)=-\sqrt{E_b}\phi_1(t) \tag{10.103}$$

A coherent binary PSK system is therefore characterized by having a signal space that is one-dimensional (i.e., $N=1$), and with two message points (i.e., $M=2$), as shown in Fig. 10.17. The coordinates of the message points equal

$$\begin{aligned} s_{11} &= \int_0^{T_b} s_1(t)\phi_1(t)\,dt \\ &= +\sqrt{E_b} \end{aligned} \tag{10.104}$$

and

$$\begin{aligned} s_{21} &= \int_0^{T_b} s_2(t)\phi_1(t)\,dt \\ &= -\sqrt{E_b} \end{aligned} \tag{10.105}$$

The message point corresponding to $s_1(t)$ is located at $s_{11}=+\sqrt{E_b}$, and the message point corresponding to $s_2(t)$ is located at $s_{21}=-\sqrt{E_b}$.

To realize the decision rule, we must partition this signal space into two regions, as described below:

1. The set of points closest to the message point at $+\sqrt{E_b}$.

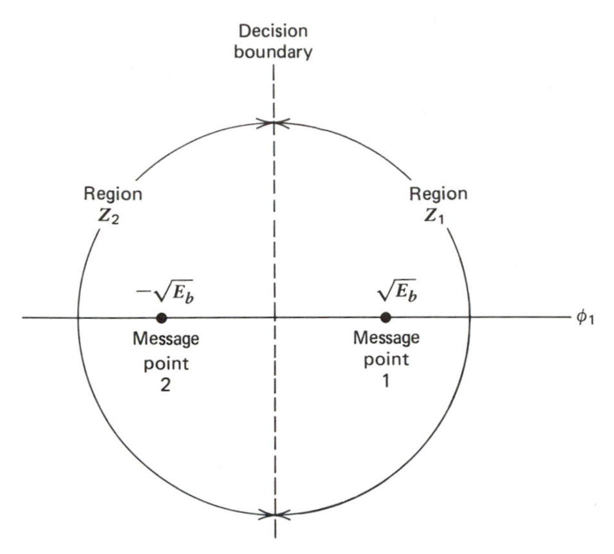

Figure 10.17 Signal space diagram for coherent binary PSK system.

2. The set of points closest to the message point at $-\sqrt{E_b}$.

This is accomplished by constructing the perpendicular bisector of the line joining these two message points, and then marking off the appropriate decision regions. In Fig. 10.17, these decision regions are marked Z_1 and Z_2, according to the message point around which they are constructed.

The decision rule is now simply to guess signal $s_1(t)$ or binary symbol 1 was transmitted if the received signal point falls in region Z_1, and guess signal $s_2(t)$ or binary symbol 0 was transmitted if the received signal point falls in region Z_2. Two kinds of erroneous decisions will, however, be made. Signal $s_1(t)$ is transmitted, but the noise is such that the received signal point falls inside region Z_2 and so the receiver decides in favor of signal $s_2(t)$. Alternatively, signal $s_2(t)$ is transmitted, but the noise is such that the received signal point falls inside region Z_1 and so the receiver decides in favor of signal $s_1(t)$.

To calculate the probability of making an error of the first kind, we note from Fig. 10.17 that the decision region associated with symbol 1 or signal $s_1(t)$ is described by

$$Z_1: \qquad 0 < x_1 < \infty$$

where

$$x_1 = \int_0^{T_b} x(t)\phi_1(t)dt \tag{10.106}$$

From Eq. (10.49) we deduce that the likelihood function, when symbol 0 or signal $s_2(t)$ is transmitted, is defined by

$$f_{X_1|0}(x_1|0)=\frac{1}{\sqrt{\pi N_0}}\exp\left[-\frac{1}{N_0}(x_1-s_{21})^2\right]$$
$$=\frac{1}{\sqrt{\pi N_0}}\exp\left[-\frac{1}{N_0}(x_1+\sqrt{E_b})^2\right] \tag{10.107}$$

The conditional probability of the receiver deciding in favor of symbol 1, given that symbol 0 was transmitted, is therefore

$$P_{e0}=\int_0^\infty f_{X_1|0}(x_1|0)dx_1$$
$$=\frac{1}{\sqrt{\pi N_0}}\int_0^\infty \exp\left[-\frac{1}{N_0}(x_1+\sqrt{E_b})^2\right]dx_1 \tag{10.108}$$

Putting

$$z=\frac{1}{\sqrt{N_0}}(x_1+\sqrt{E_b}) \tag{10.109}$$

and changing the variable of integration from x_1 to z, we may rewrite Eq. (10.108) in the form

$$P_{e0}=\frac{1}{\sqrt{\pi}}\int_{\sqrt{E_b/N_0}}^\infty \exp(-z^2)dz$$
$$=\tfrac{1}{2}\operatorname{erfc}\left(\sqrt{\frac{E_b}{N_0}}\right) \tag{10.110}$$

where $\operatorname{erfc}(\sqrt{E_b/N_0})$ is the complementary error function (see Appendix 2).

Similarly, we may show that P_{e1}, the conditional probability of the receiver deciding in favor of symbol 0, given that symbol 1 was transmitted, also has the same value as in Eq. (10.110). Thus, averaging the conditional error probabilities P_{e0} and P_{e1}, we find that the average probability of symbol error for coherent binary PSK equals

$$P_e=\tfrac{1}{2}\operatorname{erfc}\left(\sqrt{\frac{E_b}{N_0}}\right) \tag{10.111}$$

It is of interest to note that, as a rule, whenever the observation space is partitioned in a symmetric manner (as in Fig. 10.17, for example), then the conditional symbol error probabilities and the average probability of symbol error will all have the same value.

Binary PSK Transmitter and Receiver

To generate a binary PSK wave, we see from Eqs. (10.101) to (10.103) that we have to represent the input binary sequence in polar form with symbols 1 and 0 represented by constant amplitude levels of $+\sqrt{E_b}$ and $-\sqrt{E_b}$, respectively. This binary

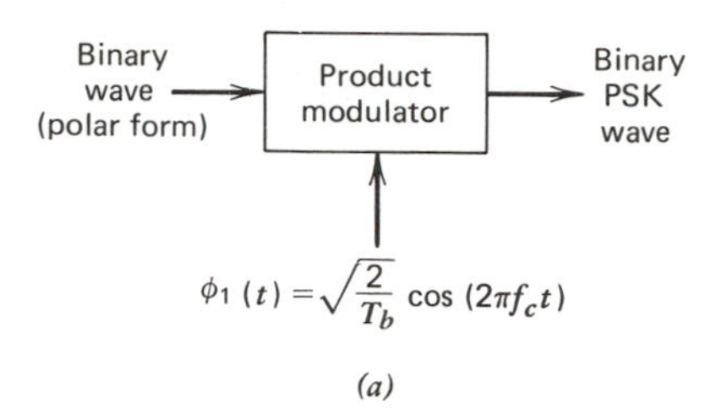

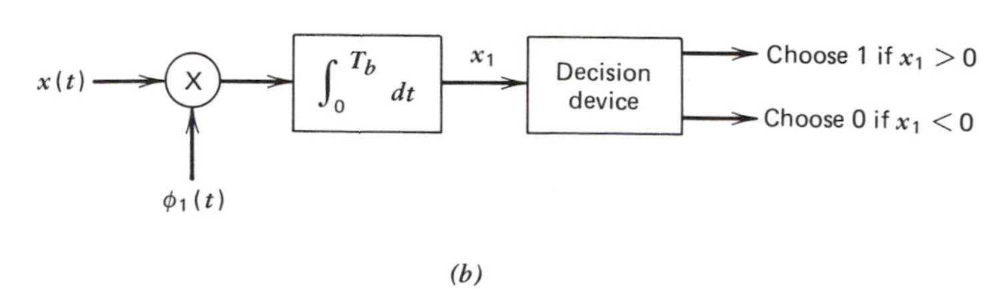

Figure 10.18 Block diagrams for *(a)* binary PSK transmitter and *(b)* coherent binary PSK receiver.

wave and a sinusoidal carrier wave $\phi_1(t)$ (whose frequency $f_c = n_c/T_b$ for some fixed integer n_c) are applied to a product modulator, as in Fig. 10.18(*a*). The carrier and the timing pulses used to generate the binary wave are usually extracted from a common master clock. The desired PSK wave is obtained at the modulator output. Thus a binary PSK wave may also be viewed as a special form of double-sideband suppressed-carrier (DSBSC) modulated wave.

To reconstruct the original binary sequence of 1's and 0's, we apply the noisy PSK wave $x(t)$ (at the channel output) to a correlator, which is also supplied with a locally generated coherent reference signal $\phi_1(t)$. The correlator output, x_1, is compared with a threshold of zero volts. If $x_1 > 0$, the receiver decides in favor of symbol 1. On the other hand, if $x_1 < 0$, it decides in favor of symbol 0.

The carrier $\phi_1(t)$ may be extracted from the received signal $x(t)$ by using a *Costas loop* or *squaring loop*, which were described in Section 3.6. These synchronization devices operate just as well with the binary PSK wave since this modulated wave, as mentioned above, is a special form of DSBSC wave. Note, however, that both of these carrier-tracking loops exhibit a 180-degree phase ambiguity. One way of overcoming this difficulty is to differentially encode the incoming binary data at the transmitter input, and correspondingly decode the signal at the receiver output.

10.10 COHERENT BINARY FSK

In a binary FSK system, symbols 1 and 0 are distinguished from each other by transmitting one of two sinusoidal waves that differ in frequency by a fixed amount. A typical pair of sinusoidal waves is described by

$$s_i(t)=\begin{cases}\sqrt{\dfrac{2E_b}{T_b}}\cos(2\pi f_i t), & 0\leqslant t\leqslant T_b\\ 0, & \text{elsewhere}\end{cases} \tag{10.112}$$

where $i=1, 2$, and E_b is the transmitted signal energy per bit, and the transmitted frequency equals

$$f_i=\frac{n_c+i}{T_b}, \qquad \text{for some fixed integer } n_c \text{ and } i=1, 2 \tag{10.113}$$

Thus symbol 1 is represented by $s_1(t)$, and symbol 0 by $s_2(t)$.

From Eqs. (10.112) we observe directly that the signals $s_1(t)$ and $s_2(t)$ are orthogonal, but not normalized to have unit energy. We therefore deduce that the most useful form for the set of orthonormal basis functions is:

$$\phi_i(t)=\begin{cases}\sqrt{\dfrac{2}{T_b}}\cos(2\pi f_i t), & 0\leqslant t\leqslant T_b\\ 0, & \text{elsewhere}\end{cases} \tag{10.114}$$

where $i=1, 2$. Correspondingly, the coefficient s_{ij} for $i=1, 2$, and $j=1, 2$, is defined by

$$\begin{aligned} s_{ij} &= \int_0^{T_b} s_i(t)\phi_j(t)dt \\ &= \int_0^{T_b} \sqrt{\frac{2E_b}{T_b}}\cos(2\pi f_i t)\sqrt{\frac{2}{T_b}}\cos(2\pi f_j t)dt \\ &= \begin{cases}\sqrt{E_b}, & i=j\\ 0, & i\neq j\end{cases} \end{aligned} \tag{10.115}$$

Thus a coherent binary FSK system is characterized by having a signal space that is two-dimensional (i.e., $N=2$) with two message points (i.e., $M=2$), as in Fig. 10.19. The two message points are defined by the signal vectors:

$$\mathbf{s}_1=\begin{bmatrix}\sqrt{E_b}\\ 0\end{bmatrix} \tag{10.116}$$

and

$$\mathbf{s}_2=\begin{bmatrix}0\\ \sqrt{E_b}\end{bmatrix} \tag{10.117}$$

Note that the distance between the two message points is equal to $\sqrt{2E_b}$.

The observation vector $\mathbf{x}$ has two elements, x_1 and x_2, that are defined by, respectively,

$$x_1=\int_0^{T_b} x(t)\phi_1(t)dt \tag{10.118}$$

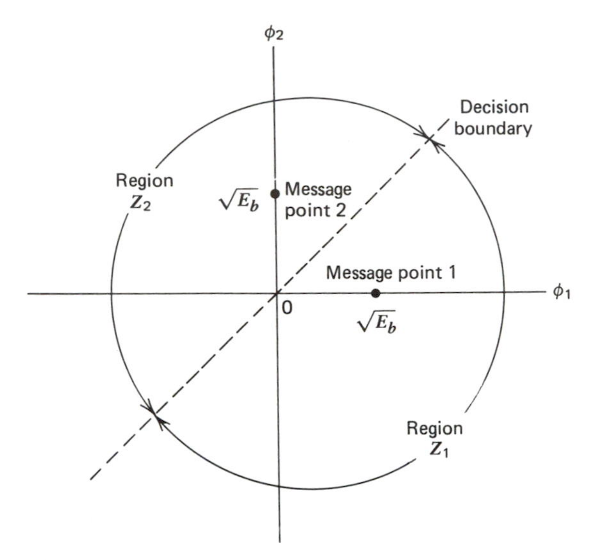

Figure 10.19 Signal space diagram for coherent binary FSK system.

and

$$x_2 = \int_0^{T_b} x(t)\phi_2(t)dt \tag{10.119}$$

where $x(t)$ is the received signal, the form of which depends on which symbol was transmitted. Given that symbol 1 was transmitted, $x(t)$ equals $s_1(t) + w(t)$, where $w(t)$ is the sample function of a white Gaussian noise process of zero mean and power spectral density $N_0/2$. If, on the other hand, symbol 0 was transmitted, $x(t)$ equals $s_2(t) + w(t)$.

Now, applying the decision rule of Eq. (10.64), we find that the observation space is partitioned into two decision regions, labeled as Z_1 and Z_2 in Fig. 10.19. Accordingly, the receiver decides in favor of symbol 1 if the received signal point represented by the observation vector $\mathbf{x}$ falls inside region Z_1. This occurs when $x_1 > x_2$. If, on the other hand, we have $x_1 < x_2$, the received signal point falls inside region Z_2, and the receiver decides in favor of symbol 0. The decision boundary, separating region Z_1 from region Z_2, is defined by $x_1 = x_2$.

Define a new Gaussian random variable L whose sample value l is equal to the difference between x_1 and x_2; thus

$$l = x_1 - x_2 \tag{10.120}$$

The mean value of the random variable L depends on which binary symbol was transmitted. Given that symbol 1 was transmitted, the Gaussian random variables X_1 and X_2, whose sample values are denoted by x_1 and x_2, have mean values equal to $\sqrt{E_b}$ and zero, respectively. Correspondingly, the conditional mean of the random variable L, given that symbol 1 was transmitted, is given by

$$E[L|1] = E[X_1|1] - E[X_2|1]$$
$$= +\sqrt{E_b} \tag{10.121}$$

On the other hand, given that symbol 0 was transmitted, the random variables X_1 and X_2 have mean values equal to zero and $\sqrt{E_b}$, respectively. Correspondingly, the conditional mean of the random variable L, given that symbol 0 was transmitted, is given by

$$E[L|0] = E[X_1|0] - E[X_2|0]$$
$$= -\sqrt{E_b} \tag{10.122}$$

The variance of the random variable L is independent of which binary symbol was transmitted. Since the random variables X_1 and X_2 are statistically independent, each with a variance equal to $N_0/2$, it follows that

$$\text{Var}[L] = \text{Var}[X_1] + \text{Var}[X_2]$$
$$= N_0 \tag{10.123}$$

Suppose, now, we know that symbol 0 was transmitted. Then the corresponding value of the conditional probability density function of the random variable L equals

$$f_{L|0}(l|0) = \frac{1}{\sqrt{2\pi N_0}} \exp\left[-\frac{(l+\sqrt{E_b})^2}{2N_0}\right] \tag{10.124}$$

Since the condition $x_1 > x_2$, or, equivalently, $l > 0$, corresponds to the receiver making a decision in favor of symbol 1, we deduce that the conditional probability of error, given that symbol 0 was transmitted, is given by

$$P_{e0} = P(l > 0|\text{symbol 0 was sent})$$
$$= \int_0^\infty f_{L|0}(l|0)dl$$
$$= \frac{1}{\sqrt{2\pi N_0}} \int_0^\infty \exp\left[-\frac{(l+\sqrt{E_b})^2}{2N_0}\right]dl \tag{10.125}$$

Put

$$\frac{l+\sqrt{E_b}}{\sqrt{2N_0}} = z \tag{10.126}$$

Then, changing the variable of integration from l to z, we may rewrite Eq. (10.125) as follows

$$P_{e0} = \frac{1}{\sqrt{\pi}} \int_{\sqrt{E_b/2N_0}}^\infty \exp(-z^2)dz$$
$$= \tfrac{1}{2}\text{erfc}\left(\sqrt{\frac{E_b}{2N_0}}\right) \tag{10.127}$$

Similarly, we may show that P_{e1}, the conditional probability of error, given that symbol 1 was transmitted, has the same value as in Eq. (10.127). Accordingly, averaging P_{e0} and P_{e1}, we find that the average probability of symbol error for coherent binary FSK is

$$P_e = \frac{1}{2}\operatorname{erfc}\left(\sqrt{\frac{E_b}{2N_0}}\right) \tag{10.128}$$

Comparing Eqs. (10.111) and (10.128), we see that in a coherent binary FSK system we have to double the *bit energy-to-noise density ratio*, E_b/N_0, in order to maintain the same average error rate as in a coherent binary PSK system. This result is in perfect accord with the signal space diagrams of Fig. 10.17 and 10.19, where we see that in a binary PSK system the distance between the two message points is equal to $2\sqrt{E_b}$, whereas in a binary FSK system the corresponding distance is $\sqrt{2E_b}$. This shows that, in an AWGN channel, the detection performance of equal energy binary signals depends only on the "distance" between the two pertinent message points in the signal space. In particular, the larger we make this distance, the smaller will the average probability of error be. This is intuitively appealing, since the larger the distance between the message points, the less will be the probability of mistaking one signal for the other.

Coherent Binary FSK Transmitter and Receiver

In order to generate a binary FSK signal, we may use the scheme shown in Fig. 10.20(*a*). The input binary sequence is represented in its on–off form, with symbol 1 represented by a constant amplitude of $\sqrt{E_b}$ volts and symbol 0 represented by zero volts. By using an *inverter* in the lower channel in Fig. 10.20(*a*), we in effect make sure that when we have symbol 1 at the input, the oscillator with frequency f_1 in the upper channel is switched on while the oscillator with frequency f_2 in the lower channel is switched off, with the result that frequency f_1 is transmitted. Conversely, when we have symbol 0 at the input, the oscillator in the upper channel is switched off, and the oscillator in the lower channel is switched on, with the result that frequency f_2 is transmitted. The two frequencies f_1 and f_2 are chosen to equal integer multiples of the bit rate $1/T_b$, as in Eq. (10.113).

In the transmitter of Fig. 10.20(*a*), we assume that the two oscillators are synchronized, so that their outputs satisfy the requirements of the two orthonormal basis functions $\phi_1(t)$ and $\phi_2(t)$, as in Eq. (10.114). Alternatively, we may use a single keyed (voltage-controlled) oscillator. In either case, the frequency of the modulated wave is shifted with a *continuous phase*, in accordance with the input binary wave. That is to say, phase continuity is always maintained, including the inter-bit switching times. We refer to this form of digital modulation as *continuous-phase frequency-shift keying* (CPFSK).

In order to recover the original binary sequence from the noisy received wave $x(t)$, we may use the receiver shown in Fig. 10.20(*b*). It consists of two correlators with a common input, and which are supplied with locally generated coherent referent signals $\phi_1(t)$ and $\phi_2(t)$. The correlator outputs are then subtracted, one

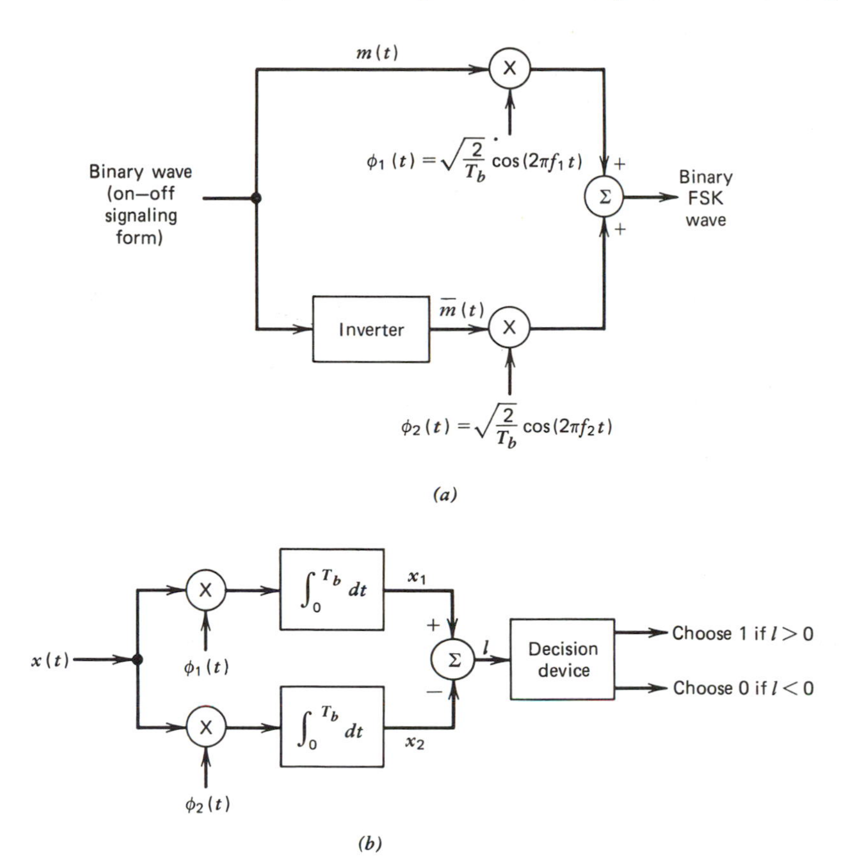

Figure 10.20 Block diagram for *(a)* binary FSK transmitter and *(b)* coherent binary FSK receiver.

from the other, and the resulting difference, l, is compared with a threshold of zero volts. If $l > 0$, the receiver decides in favor of 1. On the other hand, if $l < 0$, it decides in favor of 0.

The transmitted signal is rich in both frequencies f_1 and f_2 (see the spectral properties of binary FSK signals described below). Therefore, we may extract $\phi_1(t)$ and $\phi_2(t)$ from the received signal $x(t)$ by applying it to a pair of narrow-band filters, one tuned to f_1 and the other to f_2.

10.11 SPECTRAL PROPERTIES OF BINARY PSK AND FSK SIGNALS

Phase-shift keying and frequency-shift keying signals are examples of a band-pass signal. Using the complex notation described in Section 2.14, we may express a band-pass signal $s(t)$ in terms of its complex envelope $\tilde{s}(t)$ as the real part of the product $\tilde{s}(t)\exp(j2\pi f_c t)$, where f_c is the carrier frequency. Based on this representation, we define the *baseband power spectral density*, $S_B(f)$, of a band-pass signal

$s(t)$ to be the average power of the complex envelope $\tilde{s}(t)$ as a function of frequency. The power spectral density, $S_S(f)$, of the band-pass signal $s(t)$ itself is, except for a scaling factor, a frequency-shifted version of $S_B(f)$, as shown by

$$S_S(f)=\tfrac{1}{4}[S_B(f-f_c)+S_B(f+f_c)] \tag{10.129}$$

It is therefore sufficient to evaluate the baseband power spectral density $S_B(f)$. This information is useful in system design, because it indicates an estimate of bandwidth requirements, and the interference which may result in other systems.

Consider first the case of a binary PSK wave. From the modulator of Fig. 10.18(*a*), we see that the complex envelope of a binary PSK wave consists of an in-phase component only. Furthermore, depending on whether we have a symbol 1 or a symbol 0 at the modulator input during the signaling interval $0 \leqslant t \leqslant T_b$, we find that this in-phase component equals $+g(t)$ or $-g(t)$, respectively, where $g(t)$ is the *symbol shaping function* defined by

$$g(t)=\begin{cases}\sqrt{\dfrac{2E_b}{T_b}}, & 0 \leqslant t \leqslant T_b \\ 0, & \text{elsewhere}\end{cases} \tag{10.130}$$

We assume that the input binary wave is random, with symbols 1 and 0 equally likely, and the symbols transmitted during the different time slots being statistically independent. Then the baseband power spectral density of a binary PSK wave equals (see Example 7, Chapter 5)

$$\begin{aligned} S_B(f) &= 2E_b \operatorname{sinc}^2(T_b f) \\ &= \frac{2E_b \sin^2(\pi T_b f)}{(\pi T_b f)^2} \end{aligned} \tag{10.131}$$

This power spectrum falls off as the inverse square of frequency.

Consider next the case of a binary FSK wave, for which the two transmitted frequencies f_1 and f_2 differ by an amount equal to the bit rate $1/T_b$, and their arithmetic mean equals the nominal carrier frequency f_c. We also assume that phase continuity is always maintained, including inter-bit switching times. We may thus express the binary FSK signal as follows*

$$s(t)=\sqrt{\frac{2E_b}{T_b}}\cos\left(2\pi f_c t \pm \frac{\pi t}{T_b}\right)$$

and using a well-known trigonometric identity, we get

$$\begin{aligned} s(t) &= \sqrt{\frac{2E_b}{T_b}}\cos\left(\pm\frac{\pi t}{T_b}\right)\cos(2\pi f_c t) - \sqrt{\frac{2E_b}{T_b}}\sin\left(\pm\frac{\pi t}{T_b}\right)\sin(2\pi f_c t) \\ &= \sqrt{\frac{2E_b}{T_b}}\cos\left(\frac{\pi t}{T_b}\right)\cos(2\pi f_c t) \mp \sqrt{\frac{2E_b}{T_b}}\sin\left(\frac{\pi t}{T_b}\right)\sin(2\pi f_c t) \end{aligned} \tag{10.132}$$

* This special form of binary FSK signal is sometimes referred to as *Sunde's FSK*; see E. D. Sunde, "Ideal pulses transmitted by AM and FM," *Bell System Tech. J.*, vol. 38, pp. 1357–1426, Nov. 1959.

In the last line of Eq. (10.132), the plus sign corresponds to transmitting symbol 0, and the minus sign corresponds to transmitting symbol 1. As before, we assume that the symbols 1 and 0 in the random binary wave at the modulator input are equally likely, and that the symbols transmitted in adjacent time slots are statistically independent. Then, based on the representation of Eq. (10.132), we may make the following observations pertaining to the in-phase and quadrature components of a binary FSK signal with continuous phase:

1. The in-phase component is completely independent of the input binary wave. It equals $\sqrt{2E_b/T_b}\cos(\pi t/T_b)$ for all values of time t. The power spectral density of this component therefore consists of two delta functions, weighted by the factor $E_b/2T_b$, and occurring at $f=\pm 1/2T_b$.
2. The quadrature component is directly related to the input binary wave. During the signaling interval $0 \leqslant t \leqslant T_b$, it equals $+g(t)$ when we have symbol 1, and $-g(t)$ when we have symbol 0. The symbol shaping function $g(t)$ is now defined by

$$g(t)=\begin{cases}\sqrt{\dfrac{2E_b}{T_b}}\sin\left(\dfrac{\pi t}{T_b}\right), & 0 \leqslant t \leqslant T_b \\ 0, & \text{elsewhere}\end{cases} \tag{10.133}$$

The energy spectral density of this symbol shaping function equals (see Problem 2.27)

$$\Psi_g(f)=\frac{8E_bT_b\cos^2(\pi T_b f)}{\pi^2(4T_b^2f^2-1)^2} \tag{10.134}$$

The power spectral density of the quadrature component equals $\Psi_g(f)/T_b$. It is also apparent that the in-phase and quadrature components of the binary FSK wave are independent of each other. Accordingly, the baseband power spectral density of the binary FSK wave equals the sum of the power spectral densities of these two components, as shown by

$$S_B(f)=\frac{E_b}{2T_b}\left[\delta\left(f-\frac{1}{2T_b}\right)+\delta\left(f+\frac{1}{2T_b}\right)\right]+\frac{8E_b\cos^2(\pi T_b f)}{\pi^2(4T_b^2f^2-1)^2} \tag{10.135}$$

Substituting Eq. (10.135) in (10.129), we find that the power spectrum of the binary FSK signal contains two discrete frequency components at $(f_c+1/2T_b)=f_1$ and $(f_c-1/2T_b)=f_2$, with their average powers adding up to one-half the total power of the binary FSK signal. The presence of these two discrete frequency components provides a means of synchronizing the receiver with the transmitter.

Note also that the baseband power spectral density of a binary FSK signal with continuous phase ultimately falls off as the inverse fourth power of frequency. This is readily established by taking the limit in Eq. (10.135) as f approaches infinity. If, however, the FSK signal exhibits phase discontinuity at the inter-bit switching instants (this arises when the two oscillators supplying frequencies f_1 and f_2 operate independently of each other), the power spectral density ultimately falls

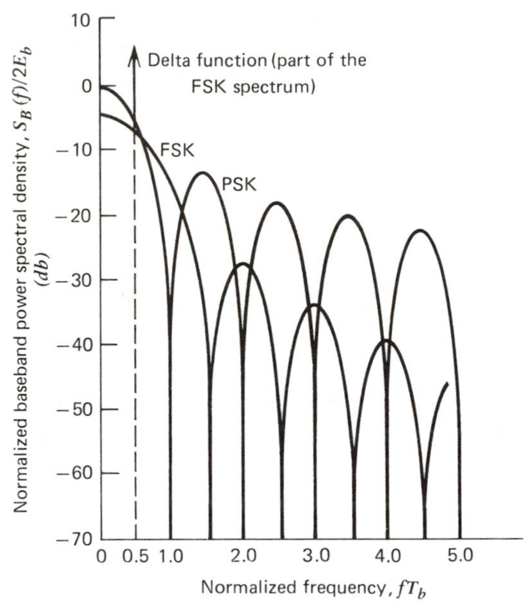

Figure 10.21 Normalized baseband power spectra of binary PSK and FSK signals. (Note that for negative frequencies the spectra are the mirror images of those shown for positive frequencies)

off as the inverse square of frequency* (see Problem 10.23). Accordingly, an FSK signal with continuous phase does not produce as much interference outside the signal band of interest as an FSK signal with discontinuous phase.

In Fig. 10.21 we have plotted the baseband power spectra of Eqs. (10.131) and (10.135). In both cases, $S_B(f)$ is shown normalized with respect to $2E_b$, and the frequency is normalized with respect to the bit rate $1/T_b$. The difference in the rates of falloff of these spectra can be explained on the basis of the pulse shape $g(t)$. The smoother the pulse, the faster is the drop of spectral tails to zero. Thus, with binary FSK (with continuous phase) having a smoother pulse shape, it has lower sidelobes than binary PSK.

10.12 COHERENT QUADRATURE — SIGNALING TECHNIQUES

Channel bandwidth and transmitted power constitute two primary "communication resources," the efficient utilization of which provides the motivation for the

* 1. W. R. Bennett and S. O. Rice, "Spectral density and autocorrelation functions associated with binary frequency shift keying," *Bell System Tech. J.*, vol. 42, pp. 2355–2385, Sept. 1963.

2. R. R. Anderson and J. Salz, "Spectra of digital FM," *Bell System Tech. J.*, vol. 44 pp. 1165–1189, July–Aug. 1965.

search for *spectrally efficient* modulation schemes.* The primary objective of spectrally efficient modulation is to maximize the *bandwidth efficiency*, defined as the ratio of data rate to channel bandwidth (in units of bits per second per hertz) for a specified probability of symbol error. A secondary objective is to achieve this bandwidth efficiency at a minimum practical expenditure of average signal power or, equivalently, in a channel perturbed by additive white Gaussian noise, a minimum practical expenditure of average signal-to-noise ratio.

In the next two sections we study two spectrally efficient modulation schemes for the transmission of binary data, which are examples of the quadrature-carrier multiplexing system described in Section 3.7. We first consider a quadrature signaling technique known as quadriphase-shift keying, which is an extension of the binary PSK. Next we consider the so-called minimum-shift keying, which is a special form of continuous-phase frequency-shift keying (CPFSK), with the detection in the receiver being performed in two successive bit intervals. In particular, we show that, for an AWGN channel, these two coherent quadrature-signaling techniques have an identical error performance.

10.13 QUADRIPHASE — SHIFT KEYING (QPSK)

As with the binary PSK, this modulation scheme is characterized by the fact that the information carried by the transmitted wave is contained in the phase. In particular, in a *quadriphase-shift keying* (QPSK) wave, the phase of the carrier takes on one of four possible values, such as $\pi/4$, $3\pi/4$, $5\pi/4$, and $7\pi/4$, as shown by

$$s_i(t)=\begin{cases}\sqrt{\dfrac{2E}{T}}\cos\left[2\pi f_c t+(2i-1)\dfrac{\pi}{4}\right], & 0\leqslant t\leqslant T\\ 0, & \text{elsewhere}\end{cases} \tag{10.136}$$

where $i=1, 2, 3, 4$, and E is the transmitted signal energy per symbol, T is the symbol duration, and the carrier frequency f_c equals n_c/T for some fixed integer n_c. Each possible value of the phase corresponds to a unique pair of bits called a *dibit*. Thus, for example, we may choose the above set of phase values to represent the following set of dibits: 10, 00, 01, and 11.

Using a well-known trigonometric identity, we may rewrite Eq. (10.136) in the equivalent form:

$$s_i(t)=\begin{cases}\sqrt{\dfrac{2E}{T}}\cos\left[(2i-1)\dfrac{\pi}{4}\right]\cos(2\pi f_c t)-\sqrt{\dfrac{2E}{T}}\sin\left[(2i-1)\dfrac{\pi}{4}\right]\sin(2\pi f_c t), & 0\leqslant t\leqslant T\\ 0, & \text{elsewhere}\end{cases} \tag{10.137}$$

where $i=1$, 2, 3, 4. Based on this representation, we may make the following observations:

* J. G. Smith, "Spectrally efficient modulation," IEEE International Conference on Communications, pp. 3.1-37–3.1-41, Chicago, Illinois, June 1977.

1. There are only two orthonormal basis functions, $\phi_1(t)$ and $\phi_2(t)$, contained in the expansion of $s_i(t)$, and the appropriate form for $\phi_1(t)$ and $\phi_2(t)$ is defined by

$$\phi_1(t) = \sqrt{\frac{2}{T}} \cos(2\pi f_c t), \qquad 0 \leqslant t \leqslant T \tag{10.138}$$

and

$$\phi_2(t) = \sqrt{\frac{2}{T}} \sin(2\pi f_c t), \qquad 0 \leqslant t \leqslant T \tag{10.139}$$

2. There are four message points, and the associated signal vectors are defined by

$$\mathbf{s}_i = \begin{bmatrix} \sqrt{E} \cos\left((2i-1)\frac{\pi}{4}\right) \\ -\sqrt{E} \sin\left((2i-1)\frac{\pi}{4}\right) \end{bmatrix}, \qquad i = 1, 2, 3, 4 \tag{10.140}$$

Table 10.1

Input dibit $0 \leqslant t \leqslant T$	Phase of QPSK signal (*radians*)	Coordinates of message points	
		s_{i1}	s_{i2}
10	$\pi/4$	$+\sqrt{E/2}$	$-\sqrt{E/2}$
00	$3\pi/4$	$-\sqrt{E/2}$	$-\sqrt{E/2}$
01	$5\pi/4$	$-\sqrt{E/2}$	$+\sqrt{E/2}$
11	$7\pi/4$	$+\sqrt{E/2}$	$+\sqrt{E/2}$

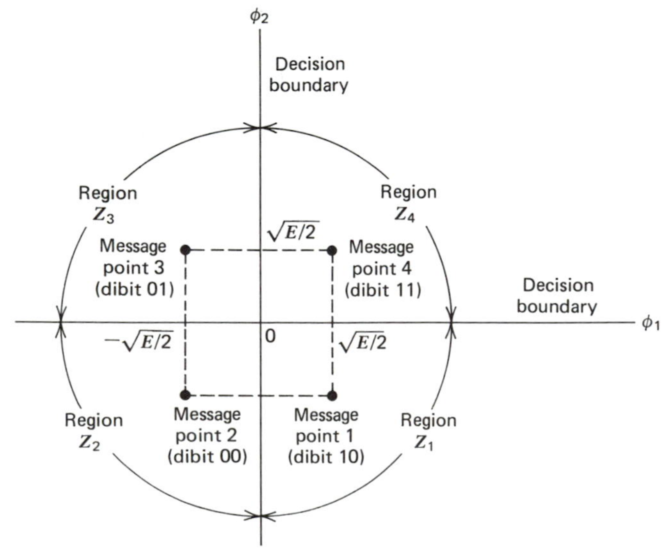

Figure 10.22 Signal space diagram for coherent QPSK system.

The elements of the signal vectors, namely, s_{i1} and s_{i2}, have their values summarized in Table 10.1. The first two columns of this table give the associated dibits and phase of the QPSK signal.

Accordingly, a QPSK signal is characterized by having a two-dimensional signal space (i.e., $N=2$) and four message points (i.e., $M=4$), as illustrated in Fig. 10.22.

Example 4

Figure 10.23 shows the waveforms involved in the generation of a QPSK signal for the input binary sequence 01101000. Part (*a*) of the figure shows the input binary wave $m(t)$ represented in its polar form, with symbol 1 represented by $+\sqrt{E_b}$ volts and symbol 0 by $-\sqrt{E_b}$ volts. The binary wave $m(t)$ is divided into two separate binary waves, $m_1(t)$ and $m_2(t)$, consisting of the odd- and even-numbered input bits, respectively, as illustrated in part (*b*) of the figure. Note that in any signaling interval, the amplitudes of $m_1(t)$ and $m_2(t)$ are equal to s_{i1} and s_{i2}, respectively, depending on the dibit being transmitted. Note also that $E=2E_b$. We may thus express the QPSK wave, for all values of time t, in the following form

$$s(t)=m_1(t)\phi_1(t)+m_2(t)\phi_2(t) \tag{10.141}$$

where $m_1(t)\phi_1(t)$ and $m_2(t)\phi_2(t)$ represent two binary PSK waves. These modulated waves are shown in part (*c*) of Fig. 10.23. Adding these two modulated waves, we get the QPSK wave shown in part (*d*) of the figure.

To realize the decision rule, we must partition the signal space into four regions, as described below:

1. The set of points closest to the message point associated with signal vector $\mathbf{s}_1$.
2. The set of points closest to the message point associated with signal vector $\mathbf{s}_2$.
3. The set of points closest to the message point associated with signal vector $\mathbf{s}_3$.
4. The set of points closest to the message point associated with signal vector $\mathbf{s}_4$.

This is accomplished by constructing the perpendicular bisectors of the four-sided polygon formed by joining the four message points, and then marking off the appropriate regions. We thus find that the decision regions are cones whose vertices coincide with the origin. These regions are marked as Z_1, Z_2, Z_3, and Z_4, in Fig. 10.22, according to the message point about which they are constructed.

The received signal, $x(t)$, is defined by

$$x(t)=s_i(t)+w(t), \qquad \begin{array}{l} 0\leqslant t\leqslant T \\ i=1, 2, 3, 4 \end{array} \tag{10.142}$$

where $w(t)$ is the sample function of a white Gaussian noise process of zero mean and power spectral density $N_0/2$. The observation vector, $\mathbf{x}$, of a coherent QPSK receiver has two elements, x_1 and x_2, that are defined by

$$\begin{aligned} x_1 &= \int_0^T x(t)\phi_1(t)dt \\ &= \sqrt{E}\cos\left[(2i-1)\frac{\pi}{4}\right]+w_1 \end{aligned} \tag{10.143}$$

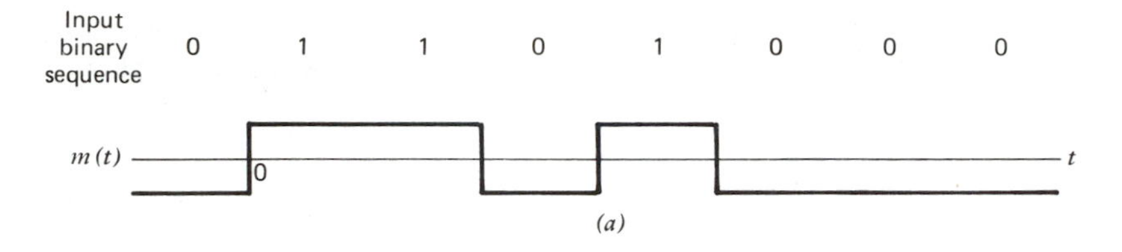

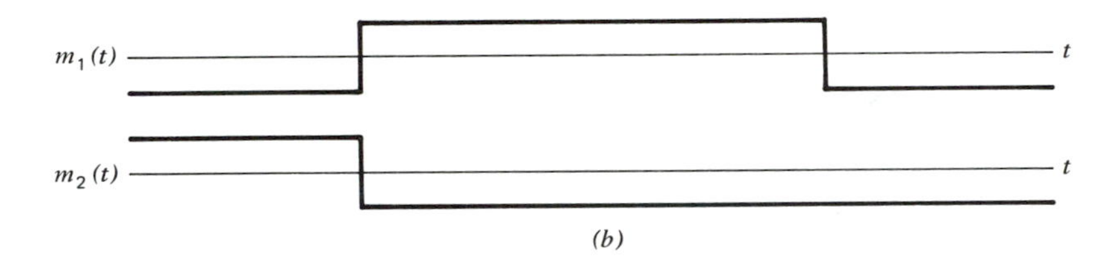

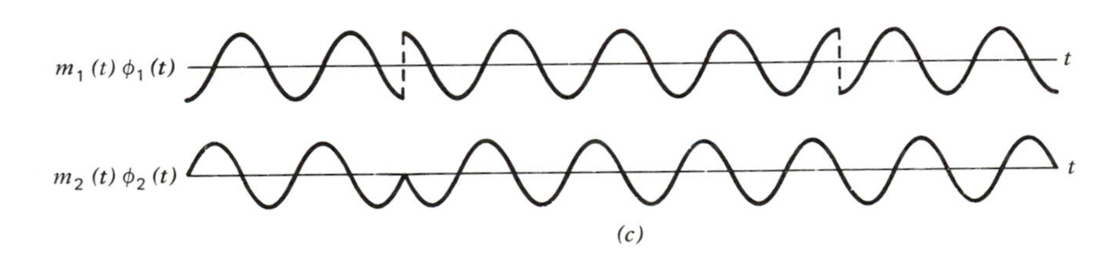

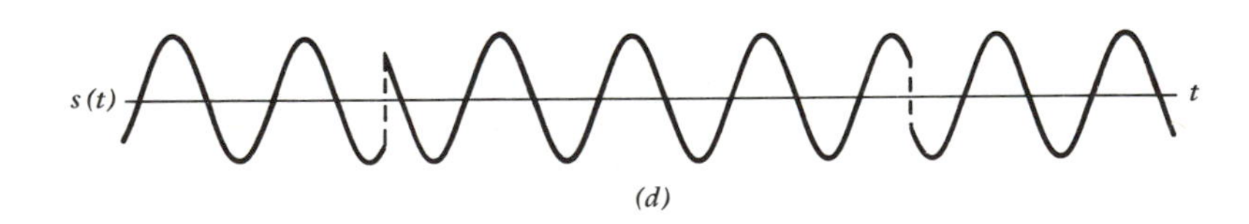

Figure 10.23 *(a)* Input binary wave $m(t)$ in polar form. *(b)* Decomposition of $m(t)$ into two new binary wave components $m_1(t)$ and $m_2(t)$. *(c)* PSK waves $m_1(t)\phi_1(t)$ and $m_2(t)\phi_2(t)$. *(d)* QPSK wave $s(t)$.

and

$$x_2 = \int_0^T x(t)\phi_2(t)dt$$
$$= -\sqrt{E}\sin\left[(2i-1)\frac{\pi}{4}\right] + w_2 \tag{10.144}$$

where $i = 1, 2, 3, 4$.

Thus x_1 and x_2 are sample values of independent Gaussian random variables with mean values equal to $\sqrt{E}\cos[(2i-1)\pi/4]$ and $-\sqrt{E}\,\sin[(2i-1)\pi/4]$, respectively, and with a common variance equal to $N_0/2$.

The decision rule is now simply to guess $s_1(t)$ was transmitted if the received signal point associated with the observation vector **x** falls inside region Z_1, guess $s_2(t)$ was transmitted if the received signal point falls inside region Z_2, and so on. An erroneous decision will be made if, for example, signal $s_4(t)$ is transmitted but the noise $w(t)$ is such that the received signal point falls outside region Z_4.

We note that, owing to the symmetry of the decision regions, the probability of interpreting the received signal point correctly is the same regardless of which particular signal was actually transmitted. Suppose, for example, we know that signal $s_4(t)$ was transmitted. The receiver will then make a correct decision provided that the received signal point represented by the observation vector **x** lies inside region Z_4 of the signal space diagram in Fig. 10.22. Accordingly, for a correct decision when signal $s_4(t)$ is transmitted, the elements x_1 and x_2 of the observation vector **x** must be both positive, as illustrated in Fig. 10.24. This means that the *probability of a correct decision*, P_c, equals the conditional probability of the joint event $x_1 > 0$ and $x_2 > 0$, given that signal $s_4(t)$ was transmitted. Since the random variables X_1 and X_2 (with sample values x_1 and x_2, respectively) are independent, P_c also equals the product of the conditional probabilities of the events $x_1 > 0$ and $x_2 > 0$, both given that signal $s_4(t)$ was transmitted. Furthermore, both X_1 and X_2 are Gaussian random variables with a conditional mean equal to $\sqrt{E/2}$ and a variance equal to $N_0/2$. Hence, we may write

$$P_c = \int_0^\infty \frac{1}{\sqrt{\pi N_0}} \exp\left[-\frac{(x_1-\sqrt{E/2})^2}{N_0}\right]dx_1 \int_0^\infty \frac{1}{\sqrt{\pi N_0}} \exp\left[-\frac{(x_2-\sqrt{E/2})^2}{N_0}\right]dx_2 \tag{10.145}$$

where the first integral on the right-hand side is the conditional probability of the event $x_1 > 0$ and the second integral is the conditional probability of the event $x_2 > 0$, both given that signal $s_4(t)$ was transmitted. Let

$$\frac{x_1-\sqrt{E/2}}{\sqrt{N_0}} = \frac{x_2-\sqrt{E/2}}{\sqrt{N_0}} = z \tag{10.146}$$

Then, changing the variables of integration from x_1 and x_2 to z, we may rewrite Eq. (10.145) in the form

$$P_c = \left[\frac{1}{\sqrt{\pi}} \int_{-\sqrt{E/2N_0}}^{\infty} \exp(-z^2)dz\right]^2 \tag{10.147}$$

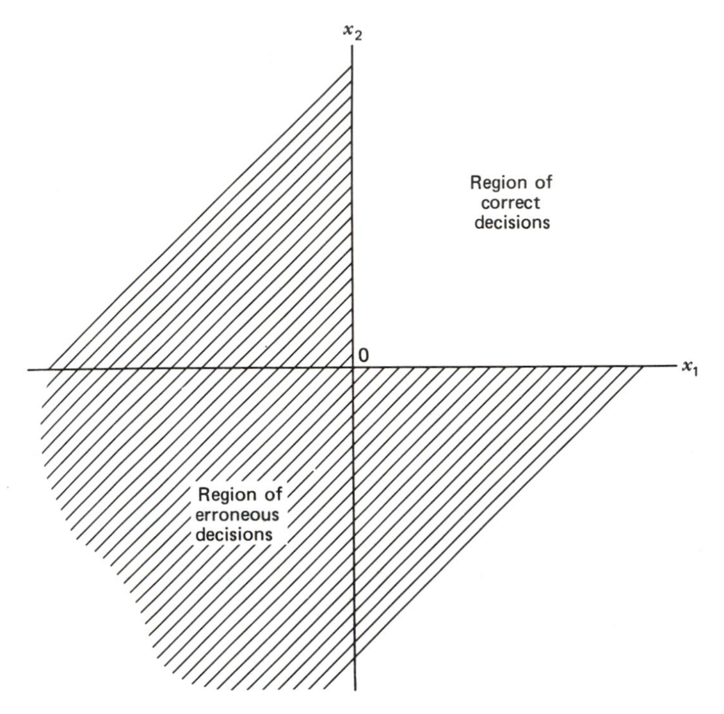

Figure 10.24 Illustrating the region of correct decisions and the region of erroneous decisions, given that signal $s_4(t)$ was transmitted.

However, from the definition of the complementary error function, we find that (see Appendix 2)

$$\frac{1}{\sqrt{\pi}} \int_{-\sqrt{E/2N_0}}^{\infty} \exp(-z^2) dz = 1 - \tfrac{1}{2} \operatorname{erfc}\left(\sqrt{\frac{E}{2N_0}}\right) \tag{10.148}$$

Accordingly, we have

$$\begin{aligned} P_c &= \left[1 - \tfrac{1}{2} \operatorname{erfc}\left(\sqrt{\frac{E}{2N_0}}\right)\right]^2 \\ &= 1 - \operatorname{erfc}\left(\sqrt{\frac{E}{2N_0}}\right) + \tfrac{1}{4} \operatorname{erfc}^2\left(\sqrt{\frac{E}{2N_0}}\right) \end{aligned} \tag{10.149}$$

The average probability of symbol error for coherent QPSK is therefore

$$\begin{aligned} P_e &= 1 - P_c \\ &= \operatorname{erfc}\left(\sqrt{\frac{E}{2N_0}}\right) - \tfrac{1}{4} \operatorname{erfc}^2\left(\sqrt{\frac{E}{2N_0}}\right) \end{aligned} \tag{10.150}$$

In the region where $(E/2N_0) \gg 1$, we may ignore the second term on the right-hand side of Eq. (10.150), and so approximate the formula for the average probability of symbol error for coherent QPSK as

$$P_e \simeq \text{erfc}\left(\sqrt{\frac{E}{2N_0}}\right) \tag{10.151}$$

In a QPSK system, we note that there are two bits per symbol. This means that the transmitted signal energy per symbol is twice the signal energy per bit; that is,

$$E = 2E_b \tag{10.152}$$

Thus, expressing the average probability of symbol error in terms of the ratio E_b/N_0, we may write

$$P_e \simeq \text{erfc}\left(\sqrt{\frac{E_b}{N_0}}\right) \tag{10.153}$$

Coherent QPSK Transmitter and Receiver

Figure 10.25(*a*) shows the block diagram of a typical QPSK transmitter. The input binary sequence is represented in polar form, with symbols 1 and 0 represented by $+\sqrt{E_b}$ and $-\sqrt{E_b}$ volts, respectively. This binary wave is divided by means of a *demultiplexer* into two separate binary waves consisting of the odd- and even-numbered input bits. These two binary waves are denoted by $m_1(t)$ and $m_2(t)$. We note that in any signaling interval, the amplitudes of $m_1(t)$ and $m_2(t)$ equal s_{i1} and s_{i2}, respectively, depending on the particular dibit that is being transmitted. The two binary waves $m_1(t)$ and $m_2(t)$ are used to modulate a pair of quadrature carriers or orthonormal basis functions $\phi_1(t) = \sqrt{2/T}\cos(2\pi f_c t)$ and $\phi_2(t) = \sqrt{2/T}\sin(2\pi f_c t)$. The result is a pair of binary PSK waves, which may be detected independently due to the orthogonality of $\phi_1(t)$ and $\phi_2(t)$. Finally, the two binary PSK waves are added to produce the desired QPSK wave. Note that the symbol duration, T, of a QPSK wave is twice as long as the bit duration, T_b, of the input binary wave. That is, for a given bit rate $1/T_b$, a QPSK wave requires half the transmission bandwidth of the corresponding binary PSK wave. Equivalently, for a given transmission bandwidth, a QPSK wave carries twice as many bits of information as the corresponding binary PSK wave.

The QPSK receiver consists of a pair of correlators with a common input and supplied with a locally generated pair of coherent reference signals $\phi_1(t)$ and $\phi_2(t)$, as in Fig. 10.25(*b*). The correlator outputs, x_1 and x_2, are each compared with a threshold of zero volts. If $x_1 > 0$, a decision is made in favor of symbol 1 for the upper or in-phase channel output, but if $x_1 < 0$ a decision is made in favor of symbol 0. Similarly, if $x_2 > 0$, a decision is made in favor of symbol 1 for the lower or quadrature channel output, but if $x_2 < 0$, a decision is made in favor of symbol 0. Finally, these two binary sequences at the in-phase and quadrature channel outputs are combined in a *multiplexer* to reproduce the original binary sequence at the transmitter input with the minimum probability of symbol error.

For the receiver of Fig. 10.25(*b*) to operate successfully, we must provide an efficient and accurate method for establishing a pair of coherent reference signals $\phi_1(t)$ and $\phi_2(t)$ that are independent of the modulation. The significance of this

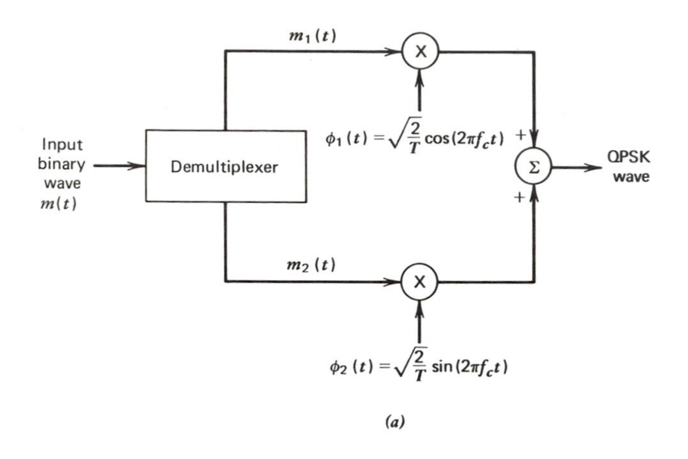

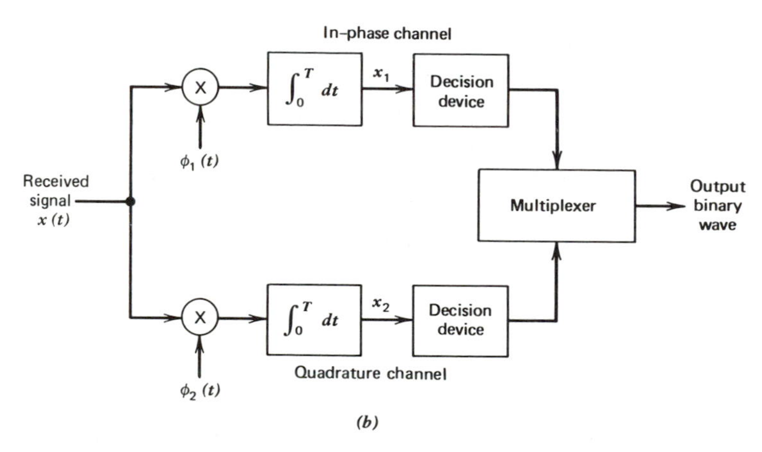

Figure 10.25 Block diagrams for *(a)* QPSK transmitter and *(b)* QPSK receiver.

statement is that the *synchronizing loop* (used to generate such a pair of reference signals) must track the randomly varying phase of the incoming QPSK wave without concern for which of the four possible dibits phase-modulates the carrier. A simple synchronizing loop that satisfies this requirement is the *fourth power loop* shown in block diagrametic form in Fig. 10.26, which is an extension of the squaring loop described in Section 3.6. The received signal $x(t)$, after being band-pass filtered to minimize the effects of noise, is raised to the fourth power, and the fourth harmonic of the carrier so produced is tracked by a phase-locked loop. The resulting sinusoidal output is next applied to a frequency divide-by-four circuit which yields $\phi_1(t)$. By applying this output to a 90-degree phase shifter, the second reference signal, $\phi_2(t)$, is obtained. Note that the frequency divide-by-four circuit in Fig. 10.26 produces four phase ambiguities in the interval $(0, 2\pi)$. We may overcome this minor difficulty by differentially encoding the incoming binary wave at the transmitter input and correspondingly decoding the receiver output.

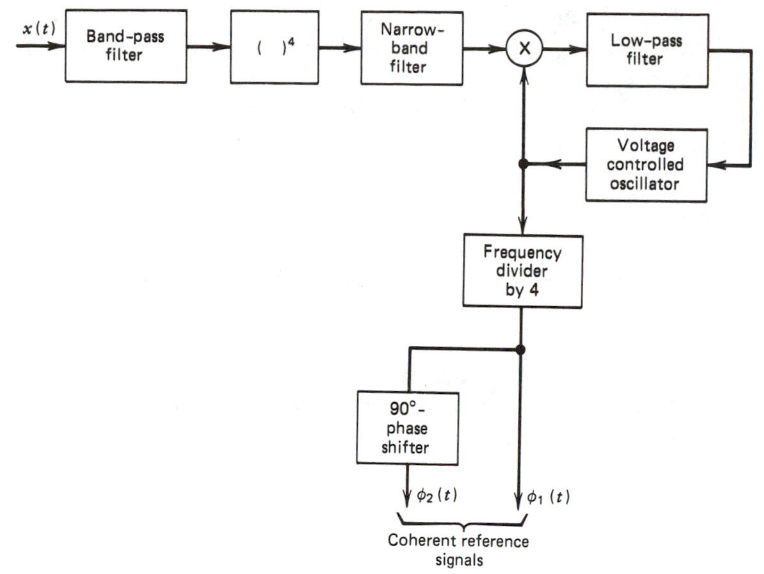

Figure 10.26 Synchronization circuit for QPSK using fourth power loop.

10.14 MINIMUM SHIFT KEYING (MSK)

In the coherent detection of binary FSK signals described in Section 10.10, the phase information contained in the received signal was not fully exploited, other than to provide for synchronization of the receiver to the transmitter. We now show that by proper utilization of the phase when performing detection, it is possible to improve the noise performance of the receiver significantly. This improvement is, however, achieved at the expense of increased receiver complexity.

Consider a continuous-phase frequency-shift keying (CPFSK) signal, which, for the interval $0 \leqslant t \leqslant T_b$, is defined by

$$s(t)=\begin{cases}\sqrt{\dfrac{2E_b}{T_b}}\cos[2\pi f_1 t+\theta(0)], & \text{for symbol 1}\\[2ex] \sqrt{\dfrac{2E_b}{T_b}}\cos[2\pi f_2 t+\theta(0)], & \text{for symbol 0}\end{cases} \tag{10.154}$$

where E_b is the transmitted signal energy per bit, and T_b is the bit duration. The phase $\theta(0)$, denoting the value of the phase at time $t=0$, depends on the past history of the modulation process. The frequencies f_1 and f_2 are transmitted in response to binary symbols 1 and 0 appearing at the modulator input, respectively.

Another useful way of representing the CPFSK signal $s(t)$ is to express it in the conventional form of an angle-modulated wave as follows

$$s(t)=\sqrt{\frac{2E_b}{T_b}}\cos[2\pi f_c t+\theta(t)] \tag{10.155}$$

where $\theta(t)$ is the phase of $s(t)$. When the phase $\theta(t)$ is a continuous function of time, we find that the modulated wave $s(t)$ itself is also continuous at all times, including the inter-bit switching times.

The *nominal* carrier frequency f_c in Eq. (10.155) is equal to the arithmetic mean of the two frequencies f_1 and f_2 transmitted to represent symbols 1 and 0; that is,

$$f_c = \tfrac{1}{2}(f_1 + f_2) \tag{10.156}$$

Moreover, we choose the carrier frequency f_c equal to an integral multiple of $1/4T_b$, one-fourth of the bit rate, in order to make the phase $\theta(t)$ continuous at the bit transition instants.

Comparing Eqs. (10.154) and (10.155), and using (10.156), we deduce that the phase $\theta(t)$ of a CPFSK signal increases or decreases linearly with time during each bit period of T_b seconds, as shown by

$$\theta(t) = \theta(0) \pm \frac{\pi h}{T_b} t \tag{10.157}$$

where the plus sign corresponds to transmitting symbol 1, and the minus sign corresponds to transmitting symbol 0. The parameter h is defined by

$$h = T_b(f_1 - f_2) \tag{10.158}$$

We refer to h as the *deviation ratio*, measured with respect to the bit rate $1/T_b$.

From Eq. (10.157), we find that at time $t = T_b$,

$$\theta(T_b) - \theta(0) = \begin{cases} \pi h, & \text{for symbol 1} \\ -\pi h, & \text{for symbol 0} \end{cases} \tag{10.159}$$

That is to say, the transmission of symbol 1 increases the phase of the CPFSK signal $s(t)$ by πh radians, whereas the transmission of symbol 0 reduces it by an equal amount. The possible values of $\theta(t)$ are shown in Fig. 10.27. It is, therefore, evident that the phase of the CPFSK signal is an odd or even multiple of πh radians at odd or even multiples of the bit duration T_b, respectively. Since all phase shifts are modulo 2π, the case of $h = 1/2$ is of special interest, because then the phase can take on only the two values $\pm\pi/2$ at odd multiples of T_b, and only the two values 0 and π at even multiples of T_b. Each path from left to right through the trellis of Fig. 10.27 corresponds to a specific binary sequence input. For example, the path shown in Fig. 10.28, for $h = \frac{1}{2}$, corresponds to the binary sequence 01101000 with $\theta(-T_b) - \theta(0) = \pi/2$. From here on we assume that $h = \frac{1}{2}$.

Using a well-known trigonometric identity in Eq. (10.155), we may express the CPFSK signal $s(t)$ in terms of its in-phase and quadrature components as follows:

$$s(t) = \sqrt{\frac{2E_b}{T_b}} \cos[\theta(t)] \cos(2\pi f_c t) - \sqrt{\frac{2E_b}{T_b}} \sin[\theta(t)] \sin(2\pi f_c t) \tag{10.160}$$

Consider first the in-phase component $\sqrt{2E_b/T_b} \cos[\theta(t)]$. With the deviation ratio

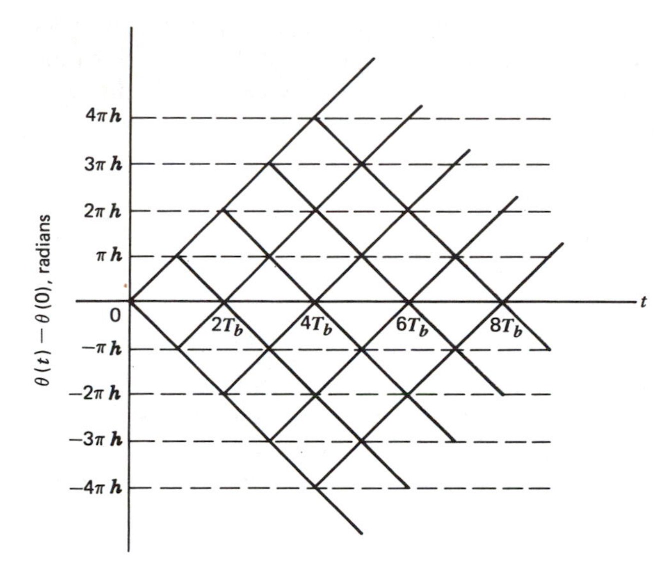

Figure 10.27 Possible values of the phase $\theta(t) - \theta(0)$.

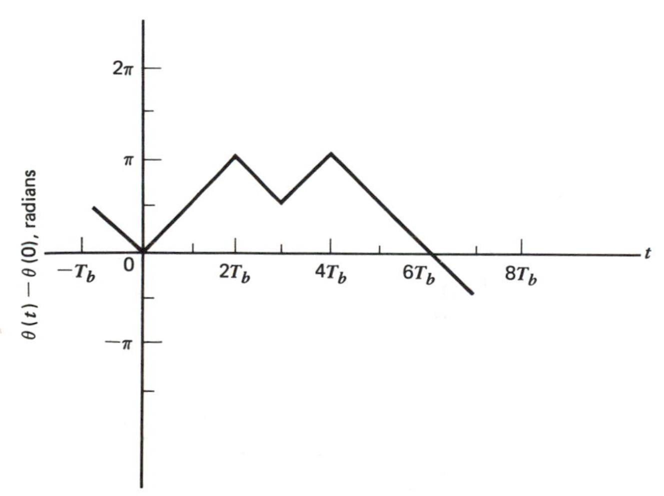

Figure 10.28 Excess phase $\theta(t) - \theta(0)$ for $h = 1/2$ and the input binary sequence 01101000.

$h = 1/2$, we have from Eq. (10.157) that

$$\theta(t) = \theta(0) \pm \frac{\pi}{2T_b} t, \qquad 0 \leqslant t \leqslant T_b \tag{10.161}$$

where the plus sign corresponds to symbol 1 and the minus sign corresponds to symbol 0. A similar result holds for $\theta(t)$ in the interval $-T_b \leqslant t \leqslant 0$, except that the algebraic sign is not necessarily the same in both intervals. Since the phase $\theta(0)$ is 0 or π, depending on the past history of the modulation process, we find that, in the interval $-T_b \leq t \leq T_b$, the polarity of $\cos[\theta(t)]$ depends only on $\theta(0)$, regardless of the sequence of 1's and 0's transmitted before or after $t=0$. Thus, for this time interval, the in-phase component, $s_c(t)$, consists of a *half-cosine pulse* defined as follows:

$$\begin{aligned} s_c(t) &= \sqrt{\frac{2E_b}{T_b}} \cos[\theta(t)] \\ &= \sqrt{\frac{2E_b}{T_b}} \cos[\theta(0)] \cos\left(\frac{\pi}{2T_b} t\right) \\ &= \pm \sqrt{\frac{2E_b}{T_b}} \cos\left(\frac{\pi}{2T_b} t\right), \qquad -T_b \leq t \leq T_b \end{aligned} \tag{10.162}$$

where the plus sign corresponds to $\theta(0)=0$ and the minus sign corresponds to $\theta(0)=\pi$. In a similar way, we may show that, in the interval $0 \leq t \leq 2T_b$, the quadrature component, $s_s(t)$, consists of a *half-sine pulse*, whose polarity depends only on $\theta(T_b)$, as shown by

$$\begin{aligned} s_s(t) &= \sqrt{\frac{2E_b}{T_b}} \sin[\theta(t)] \\ &= \sqrt{\frac{2E_b}{T_b}} \sin[\theta(T_b)] \sin\left(\frac{\pi}{2T_b} t\right) \\ &= \pm \sqrt{\frac{2E_b}{T_b}} \sin\left(\frac{\pi}{2T_b} t\right) \end{aligned} \tag{10.163}$$

where the plus sign corresponds to $\theta(T_b)=\pi/2$ and the minus sign corresponds to $\theta(T_b)=-\pi/2$.

With $h=1/2$, we find from Eq. (10.158) that the frequency deviation (i.e., the difference between the two signaling frequencies f_1 and f_2) equals half the bit rate. This is the minimum frequency spacing that allows the two FSK signals in Eq. (10.154) to be coherently orthogonal in the sense that they do not interfere with one another in the process of detection. It is for this reason, a CPFSK signal with a deviation ratio of one-half is referred to as *minimum-shift keying* (*MSK*).* Since the frequency spacing is only half as much as the conventional spacing of $1/T_b$ that is used in the coherent detection of binary FSK signals (see Section 10.10), it is also referred to as *fast FSK*.†

* 1. M. L. Dolez and E. H. Heald, "Minimum-Shift Data Communication system," U.S. Patent 2977417, March 28, 1961.
2. W. A. Sullivan, "High-capacity microwave system for digital data transmission," *IEEE Trans. on Communications*, vol. COM-20, pp. 466–470, June 1972.
3. For a tutorial review of MSK and comparison with QPSK, see S. Pasupathy "Minimum shift keying: A spectrally efficient modulation," *IEEE Communications Society Magazine*, vol. 17, 11, 14–22, July 1979.

† R. de Buda, "Coherent demodulation of frequency-shift keying with low deviation ratio," *IEEE Trans. on Communications*, vol. COM-20, pp. 419–439, June 1972.

From the above discussion we see that since the phase states $\theta(0)$ and $\theta(T_b)$ can each assume one of two possible values, any one of four possibilities can arise, as described below:

1. The phase $\theta(0)=0$ and $\theta(T_b)=\pi/2$, corresponding to the transmission of symbol 1.
2. The phase $\theta(0)=\pi$ and $\theta(T_b)=\pi/2$, corresponding to the transmission of symbol 0.
3. The phase $\theta(0)=\pi$ and $\theta(T_b)=-\pi/2$ (or, equivalently, $3\pi/2$, modulo 2π), corresponding to the transmission of symbol 1.
4. The phase $\theta(0)=0$ and $\theta(T_b)=-\pi/2$, corresponding to the transmission of symbol 0.

This, in turn, means that the MSK signal itself may assume any one of four possible forms, determined by

$$s(t)=\sqrt{\frac{2E_b}{T_b}}\cos[\theta(0)]\cos\left(\frac{\pi}{2T_b}t\right)\cos(2\pi f_c t)-\sqrt{\frac{2E_b}{T_b}}\sin[\theta(T_b)]\sin\left(\frac{\pi}{2T_b}t\right)\sin(2\pi f_c t) \tag{10.164}$$

where $0\leqslant t\leqslant T_b$.

From the expansion of Eq. (10.164), we deduce that in the case of an MSK signal the appropriate form for the orthonormal basis functions $\phi_1(t)$ and $\phi_2(t)$ is as follows:

$$\phi_1(t)=\sqrt{\frac{2}{T_b}}\cos\left(\frac{\pi}{2T_b}t\right)\cos(2\pi f_c t),\qquad 0\leqslant t\leqslant 2T_b \tag{10.165}$$

and

$$\phi_2(t)=\sqrt{\frac{2}{T_b}}\sin\left(\frac{\pi}{2T_b}t\right)\sin(2\pi f_c t),\qquad 0\leqslant t\leqslant 2T_b \tag{10.166}$$

Note that both $\phi_1(t)$ and $\phi_2(t)$ are defined for a period equal to twice the bit duration. This is necessary, so as to ensure that they satisfy the condition of orthogonality.

Correspondingly, we may express the MSK signal in the form

$$s(t)=s_1\phi_1(t)+s_2\phi_2(t),\qquad 0\leqslant t\leqslant T_b \tag{10.167}$$

where the coefficients s_1 and s_2 are related to the phase states $\theta(0)$ and $\theta(T_b)$, respectively. To evaluate s_1 we integrate the product $s(t)\phi_1(t)$ between the limits $-T_b$ and T_b [in accordance with the observation interval $-T_b\leqslant t\leqslant T_b$ for the in-phase component of $s(t)$, as in Eq. (10.162)], obtaining

$$\begin{aligned} s_1 &= \int_{-T_b}^{T_b} s(t)\phi_1(t)dt \\ &= \sqrt{E_b}\cos[\theta(0)] \end{aligned} \tag{10.168}$$

Similarly, to evaluate s_2 we integrate the product $s(t)\phi_2(t)$ between the limits 0 and $2T_b$ [in accordance with the observation interval $0 \leqslant t \leqslant 2T_b$ for the quadrature component of $s(t)$, as in Eq. (10.163)], obtaining

$$\begin{aligned} s_2 &= \int_0^{2T_b} s(t)\phi_2(t)dt \\ &= -\sqrt{E_b}\sin[\theta(T_b)] \end{aligned} \tag{10.169}$$

Note that in Eqs. (10.168) and (10.169):

1. Both integrals are evaluated for a time interval twice the bit duration, for which $\phi_1(t)$ and $\phi_2(t)$ are orthogonal.
2. Both the lower and upper limits of the product integration used to evaluate the coefficient s_1 are shifted by T_b seconds with respect to those used to evaluate the coefficient s_2.
3. The time interval $0 \leqslant t \leqslant T_b$, for which the phase states $\theta(0)$ and $\theta(T_b)$ are defined, is common to both integrals.

We thus see that the signal space for an MSK signal is two-dimensional (i.e., $N=2$), with four message points (i.e., $M=4$), as illustrated in Fig. 10.29. The coordinates of the message points are as follows: $(+\sqrt{E_b}, -\sqrt{E_b})$, $(-\sqrt{E_b}, -\sqrt{E_b})$, $(-\sqrt{E_b}, +\sqrt{E_b})$, and $(+\sqrt{E_b}, +\sqrt{E_b})$. The possible values of $\theta(0)$ and $\theta(T_b)$, corresponding to these four message points, are also included in Fig. 10.29.

Comparing Figs. 10.22 and 10.29 we see that the signal space diagrams for QPSK and MSK signals have an identical format. Note, however, that the coordinates of the message points for the QPSK signal in Fig. 10.22 are expressed in terms of the signal energy per symbol, E, whereas for the MSK signal in Fig. 10.29 they are expressed in terms of the signal energy per bit, E_b, with $E_b = E/2$. The basic difference between QPSK and MSK signals is in the choice of the orthonormal signals $\phi_1(t)$ and $\phi_2(t)$. For a QPSK signal, $\phi_1(t)$ and $\phi_2(t)$ are represented by a pair of quadrature carriers, as in Eqs. (10.138) and (10.139), whereas for an MSK signal, they are represented by a pair of sinusoidally modulated quadrature carriers, as in Eqs. (10.165) and (10.166).

Table 10.2 presents a summary of the values of $\theta(0)$ and $\theta(T_b)$, as well as the corresponding values of s_1 and s_2 that are calculated for $-T_b \leqslant t \leqslant T_b$ and $0 \leqslant t \leqslant 2T_b$,

Table 10.2

Transmitted binary symbol, $0 \leqslant t \leqslant T_b$	Phase states (*radians*)		Coordinates of message points	
	$\theta(0)$	$\theta(T_b)$	s_1	s_2
1	0	$+\pi/2$	$+\sqrt{E_b}$	$-\sqrt{E_b}$
0	π	$+\pi/2$	$-\sqrt{E_b}$	$-\sqrt{E_b}$
1	π	$-\pi/2$	$-\sqrt{E_b}$	$+\sqrt{E_b}$
0	0	$-\pi/2$	$+\sqrt{E_b}$	$+\sqrt{E_b}$

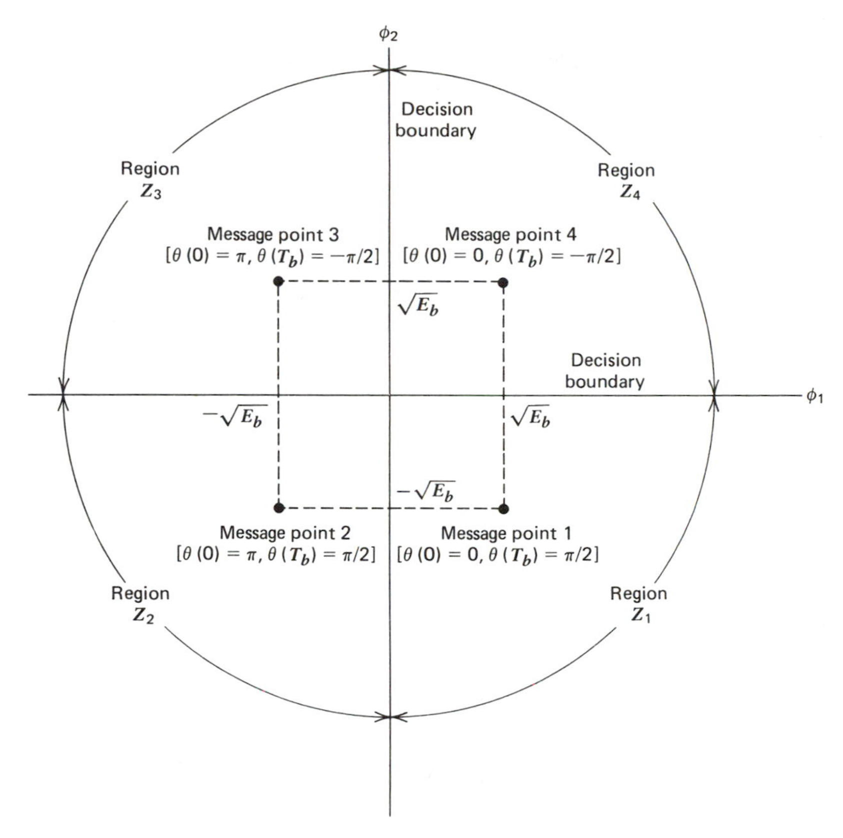

Figure 10.29 Signal space diagram for MSK system.

respectively. The first column of this table indicates whether symbol 1 or symbol 0 was transmitted in the interval $0 \leq t \leq T_b$. Note that the coordinates of the message points, s_1 and s_2, have opposite signs when symbol 1 is transmitted, but the same sign when symbol 0 is transmitted.

Example 5

Figure 10.30 shows the sets of numbers and waveforms involved in the generation of an MSK signal for the input binary sequence 01101000. This input is represented by the binary wave $m(t)$ shown in part (a) of the figure. Assuming the initial value $\theta(-T_b) = \pi/2$ radians, we find that the phase state $\theta(kT_b)$, $k = 0, 1, 2, \ldots$, is as shown in part (b) of the figure.

With $\theta(0) = 0$ and $\theta(T_b) = \pi/2$ radians, we find from Eqs. (10.168) and (10.169) that, respectively,

$$s_1 = \sqrt{E_b}$$

and

$$s_2 = -\sqrt{E_b}$$

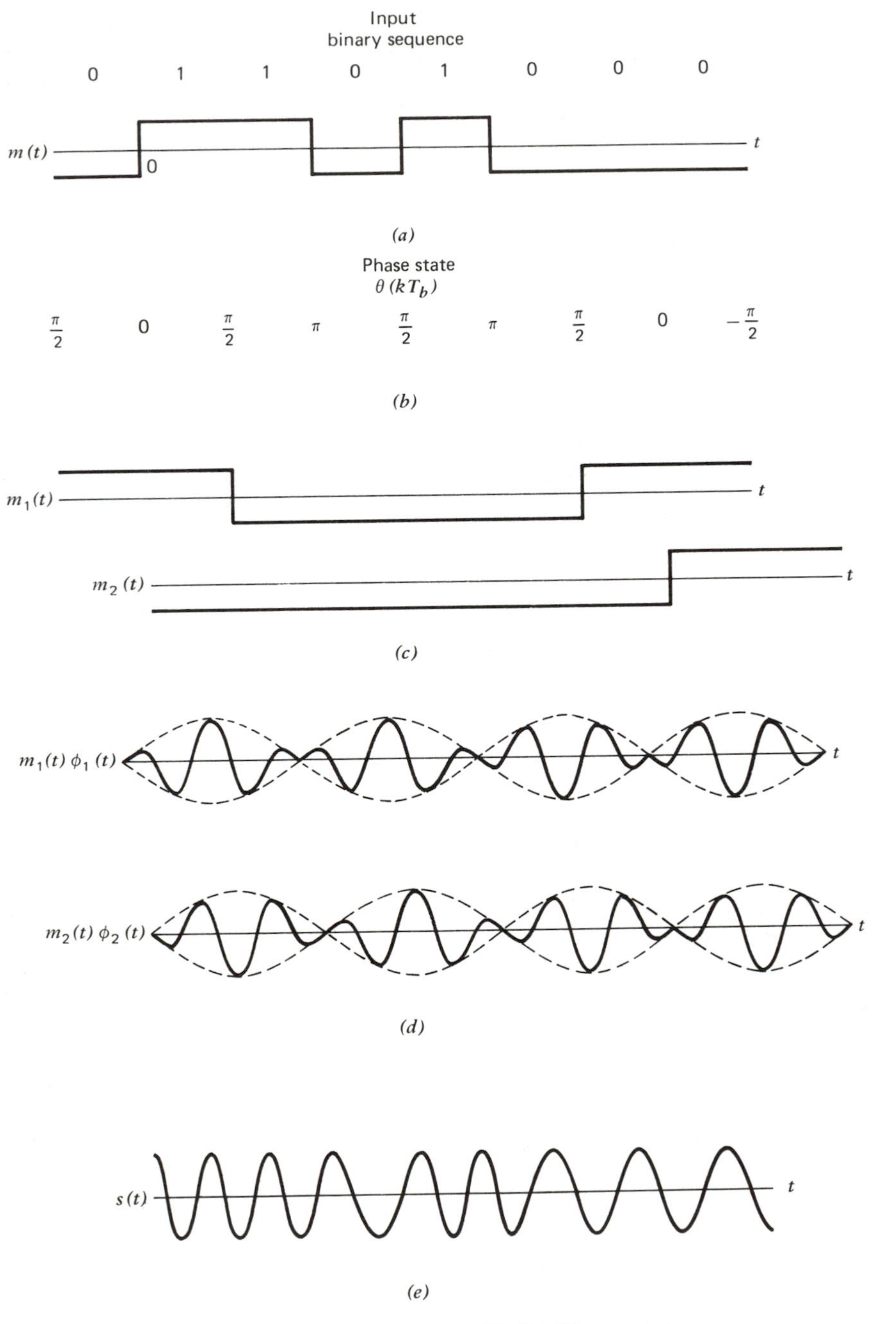

Figure 10.30 *(a)* Input binary wave $m(t)$. *(b)* Phase state $\theta(t)$ at $t = kT_b$, where k is an integer. *(c)* Decomposition of $m(t)$ into two new binary waves $m_1(t)$ and $m_2(t)$. *(d)* Modulated waves $m_1(t)\phi_1(t)$ and $m_2(t)\phi_2(t)$. *(e)* MSK wave $s(t)$.

Define two new binary waves $m_1(t)$ and $m_2(t)$ such that

$$m_1(t) = s_1, \qquad -T_b \leqslant t \leqslant T_b$$

and

$$m_2(t) = s_2, \qquad 0 \leqslant t \leqslant 2T_b$$

We note that by transmitting binary symbol 1, the phase of the MSK signal $s(t)$ is increased by $\pi/2$ radians, whereas by transmitting binary symbol 0 it is decreased by $\pi/2$ radians. This means that by transmitting the dibit 01 or 10, the net change in the phase of the MSK signal $s(t)$ is zero. On the other hand, by transmitting the dibit 11 or 00 the phase of the MSK signal is changed by $+\pi$ or $-\pi$ radians (which are equivalent, modulo 2π). Accordingly, for the interval $T_b \leqslant t \leqslant 3T_b$, we may write

$$m_1(t) = \begin{cases} s_1, & \text{if, for } 0 \leqslant t \leqslant 2T_b, \text{ dibit 10 or 01 is sent} \\ -s_1, & \text{if, for } 0 \leqslant t \leqslant 2T_b, \text{ dibit 11 or 00 is sent} \end{cases}$$

and so on for subsequent time intervals. Similarly, for the interval $2T_b \leqslant t \leqslant 4T_b$, we may write

$$m_2(t) = \begin{cases} s_2, & \text{if, for } T_b \leqslant t \leqslant 3T_b, \text{ dibit 10 or 01 is sent} \\ -s_2, & \text{if, for } T_b \leqslant t \leqslant 3T_b, \text{ dibit 11 or 00 is sent} \end{cases}$$

and so on for subsequence time intervals. We thus find that for the binary sequence given above, the binary waves $m_1(t)$ and $m_2(t)$ are as shown in part (*c*) of Fig. 10.30.

The MSK signal $s(t)$, for all values of time t, is defined by

$$s(t) = m_1(t)\phi_1(t) + m_2(t)\phi_2(t) \tag{10.170}$$

The modulated waves $m_1(t)\phi_1(t)$ and $m_2(t)\phi_2(t)$ are as shown in part (*d*) of Fig. 10.30. Finally, adding these two modulated waves, we get the desired MSK signal $s(t)$ shown in part (*e*) of the figure.

In the case of an AWGN channel, the received signal is given by

$$x(t) = s(t) + w(t) \tag{10.171}$$

where $s(t)$ is the transmitted MSK signal, and $w(t)$ is the sample function of a white, Gaussian noise of zero mean and power spectral density $N_0/2$. In order to decide whether symbol 1 or symbol 0 was transmitted in the interval $0 \leqslant t \leqslant T_b$, say, we have to establish a procedure for the use of $x(t)$ to detect the phase states $\theta(0)$ and $\theta(T_b)$. For the optimum detection of $\theta(0)$, we first determine the projection of the received signal $x(t)$ onto the reference signal $\phi_1(t)$, obtaining

$$\begin{aligned} x_1 &= \int_{-T_b}^{T_b} x(t)\phi_1(t)dt \\ &= s_1 + w_1 \end{aligned} \tag{10.172}$$

where s_1 is as defined by Eq. (10.168), and w_1 is the sample value of a Gaussian random variable of zero mean and variance $N_0/2$. From the signal space diagram of Fig. 10.29 we observe that if $x_1 > 0$, the receiver chooses the estimate $\hat{\theta}(0) = 0$. On the other hand, if $x_1 < 0$, it chooses the estimate $\hat{\theta}(0) = \pi$.

Similarly, for the optimum detection of $\theta(T_b)$, we determine the projection of the received signal $x(t)$ onto the second reference signal $\phi_2(t)$, obtaining

$$x_2 = \int_0^{2T_b} x(t)\phi_2(t)dt$$

$$= s_2 + w_2 \qquad (10.173)$$

where s_2 is as defined by Eq. (10.169), and w_2 is the sample value of another independent Gaussian random variable of zero mean and variance $N_0/2$. Referring again to the signal space diagram of Fig. 10.29, we observe that if $x_2 > 0$, the receiver chooses the estimate $\hat{\theta}(T_b) = -\pi/2$. If, on the other hand, $x_2 < 0$, it chooses the estimate $\hat{\theta}(T_b) = \pi/2$.

To reconstruct the original binary sequence, we interleave the above two sets of phase decisions, as described below (see Table 10.2):

1. If we have the estimates $\hat{\theta}(0) = 0$ and $\hat{\theta}(T_b) = -\pi/2$, or alternatively $\hat{\theta}(0) = \pi$ and $\hat{\theta}(T_b) = \pi/2$, the receiver makes a final decision in favor of symbol 0.
2. If we have the estimates $\hat{\theta}(0) = \pi$ and $\hat{\theta}(T_b) = -\pi/2$, or alternatively $\hat{\theta}(0) = 0$ and $\hat{\theta}(T_b) = \pi/2$, the receiver makes a final decision in favor of symbol 1.

Earlier we remarked that the MSK and QPSK signals have similar signal space diagrams. It follows, therefore, that for the case of an AWGN channel, they will have the same formula for their average probability of symbol error. Accordingly, the average probability of symbol error for the MSK is given by

$$P_e = \text{erfc}\left(\sqrt{\frac{E_b}{N_0}}\right) - \tfrac{1}{4}\text{erfc}^2\left(\sqrt{\frac{E_b}{N_0}}\right) \qquad (10.174)$$

Here again, we may ignore the second term on the right-hand side of Eq. (10.174) in the region where $E_b/N_0 \gg 1$, and so approximate the formula for the average probability of symbol error as

$$P_e \simeq \text{erfc}\left(\sqrt{\frac{E_b}{N_0}}\right) \qquad (10.175)$$

Thus, comparing Eqs. (10.111) and (10.175), we find that for high values of E_b/N_0 the average probability of error for an MSK system is approximately the same as that for a coherent binary PSK system.

MSK Transmitter and Receiver

Figure 10.31(*a*) shows the block diagram of a typical MSK transmitter. The advantage of this method of generating MSK signals is that the signal coherence and deviation ratio are largely unaffected by variations in the input data rate. Two input sine-waves, one of frequency $f_c = n_c/4T_b$ for some fixed integer n_c, and the other of frequency $1/4T_b$, are first applied to a product modulator. This produces two phase-coherent sine-waves at frequencies f_1 and f_2, which are related to f_c and the bit rate $1/T_b$ by Eqs. (10.156) and (10.158) for $h = 1/2$. These two sine waves

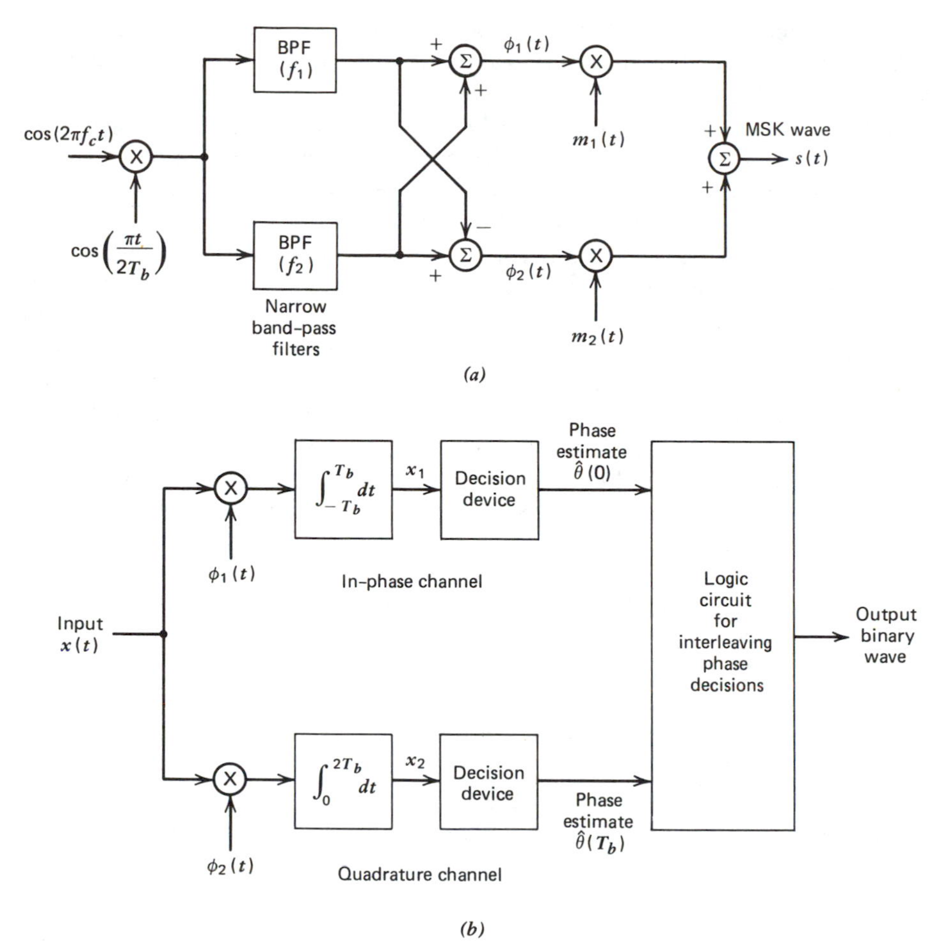

Figure 10.31 Block diagrams for *(a)* MSK transmitter and *(b)* MSK receiver

are separated from each other by two narrow-band filters, one centered at f_1 and the other at f_2. The resulting filter outputs are next summed to produce the pair of quadrature carriers or orthonormal basis functions $\phi_1(t)$ and $\phi_2(t)$. Finally, $\phi_1(t)$ and $\phi_2(t)$ are multiplied with two binary waves $m_1(t)$ and $m_2(t)$, both of which have a bit rate equal to $1/2T_b$. These two binary waves are extracted from the incoming binary wave $m(t)$ in the manner described in Example 5.

Figure 10.31(*b*) shows the block diagram of a typical MSK receiver. The received signal $x(t)$ is correlated with locally generated replicas of the coherent reference signals, $\phi_1(t)$ and $\phi_2(t)$. Note that in both cases the integration interval is $2T_b$ seconds, and that the integration in the quadrature channel is delayed by T_b seconds with respect to that in the in-phase channel. The resulting in-phase and quadrature channel correlator outputs, x_1 and x_2, are next compared with a threshold of zero volts, and estimates of the phase $\theta(0)$ and $\theta(T_b)$ are derived in the manner described previously. Finally, these phase decisions are interleaved so as to reconstruct the original input binary wave $m(t)$ with the minimum average probability of symbol error.

The pair of coherent reference signals, $\phi_1(t)$ and $\phi_2(t)$, and the clock signal at one-half the bit rate may be recovered from the received signal $x(t)$ by using the scheme shown in Fig. 10.32. Now, although the MSK signal itself does not contain discrete frequency components that can be used for the purpose of synchronization, nevertheless, after passing it through a frequency doubler we do obtain strong discrete frequency components at $2f_1$ and $2f_2$. In effect, the frequency doubler produces a new FSK signal with a deviation ratio equal to one. This special form of FSK has the useful property that one-half of its total power is contained in the two discrete frequency components $2f_1$ and $2f_2$ (see Section 10.11). These two frequency components are extracted by means of narrow-band filters (which usually, in practice, take the form of phase-locked loops). The resulting outputs are next applied to a pair of frequency dividers to produce two sinusoidal waves, $c_1(t)$ and $c_2(t)$, whose frequencies are equal to f_1 and f_2, respectively. Next, the sum and difference, namely, $c_1(t)+c_2(t)$ and $c_1(t)-c_2(t)$, produce the desired pair of orthonormal basis functions $\phi_1(t)$ and $\phi_2(t)$. Finally, by multiplying $c_1(t)$ and $c_2(t)$ together and then low-pass filtering the product, we obtain a sine-wave at one-half the bit rate, thereby providing the desired timing wave or clock signal. Thus the MSK format lends itself readily to self-synchronization.

However, the frequency divide-by-two circuits in Fig. 10.32 result in a 180-degree phase ambiguity, as it is not possible to decide whether the outputs are $\pm c_1(t)$ and $\pm c_2(t)$. This phase ambiguity may be removed by first differentially encoding the incoming binary wave, and then applying the encoded wave to the modulator. At the receiver, a differential decoder is used to generate the original binary wave.

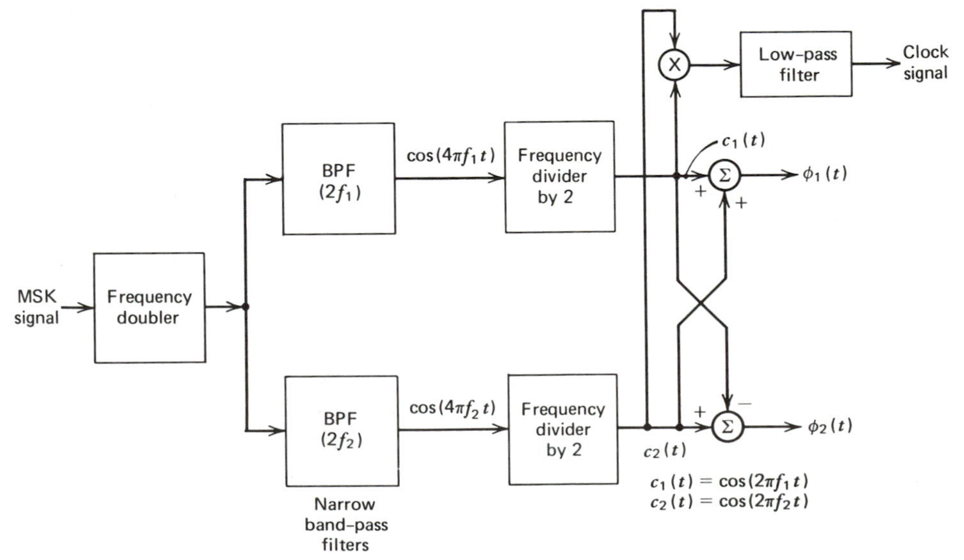

Figure 10.32 Synchronization circuit for MSK signals.

10.15 SPECTRAL PROPERTIES OF QPSK AND MSK SIGNALS

We note that QPSK and MSK signals have exactly the same noise performance, and they both require about the same order of circuit complexity for their implementation. In practice, the final decision to adopt the one method or the other is usually determined by spectral considerations. We now evaluate and compare their spectral properties.

Consider first the case of QPSK. We assume that the binary wave at the modulator input is random, with symbols 1 and 0 being equally likely, and with the symbols transmitted during adjacent time slots being statistically independent. We may make the following observations pertaining to the in-phase and quadrature components of a QPSK signal:

1. Depending on the dibit transmitted during the signaling interval $-T_b \leqslant t \leqslant T_b$, the in-phase component equals $+g(t)$ or $-g(t)$, and similarly for the quadrature component. The $g(t)$ denotes the symbol shaping function, defined by

$$g(t)=\begin{cases}\sqrt{\dfrac{E}{T}}, & 0\leqslant t\leqslant T\\ 0, & \text{elsewhere}\end{cases} \tag{10.176}$$

   Hence, using the result of Example 7, Chapter 5, we find that the in-phase and quadrature components have the same power spectral density, namely, $E\,\text{sinc}^2(Tf)$.
2. The in-phase and quadrature components are statistically independent. Accordingly, the baseband power spectral density of the QPSK signal equals the sum of the individual power spectral densities of the in-phase and quadrature components, and so we may write

$$\begin{aligned}S_B(f)&=2E\,\text{sinc}^2(Tf)\\&=4E_b\,\text{sinc}^2(2T_bf)\end{aligned} \tag{10.177}$$

Consider next the MSK signal. Here again we assume that the input binary wave is random, with symbols 1 and 0 equally likely, and the symbols transmitted during different time slots being statistically independent. In this case, we may make the following observations:

1. Depending on the value of phase state $\theta(0)$, the in-phase component equals $+g(t)$ or $-g(t)$, where

$$g(t)=\begin{cases}\sqrt{\dfrac{2E_b}{T_b}}\cos\left(\dfrac{\pi t}{2T_b}\right), & -T_b\leqslant t\leqslant T_b\\ 0, & \text{elsewhere}\end{cases} \tag{10.178}$$

The energy spectral density of this symbol shaping function equals [see Problem 2.27]:

$$\Psi_g(f)=\frac{32E_bT_b}{\pi^2}\left[\frac{\cos(2\pi T_bf)}{16T_b^2f^2-1}\right]^2 \tag{10.179}$$

Hence, using the result of Example 7, Chapter 5, we find that the power spectral density of the in-phase component equals $\Psi_g(f)/2T_b$.

2. Depending on the value of the phase state $\theta(T_b)$, the quadrature component equals $+g(t)$ or $-g(t)$, where

$$g(t)=\begin{cases}\sqrt{\dfrac{2E_b}{T_b}}\sin\left(\dfrac{\pi t}{2T_b}\right), & 0\leqslant t\leqslant 2T_b\\ 0, & \text{elsewhere}\end{cases} \tag{10.180}$$

The energy spectral density of this second symbol shaping function is also given by Eq. (10.179). Hence, the in-phase and quadrature components have the same power spectral density.

3. As with the QPSK signal, the in-phase and quadrature component of the MSK signal are also statistically independent. Hence, the baseband power spectral density of the MSK signal equals

$$\begin{aligned}S_B(f)&=2\left[\frac{\Psi_g(f)}{2T_b}\right]\\&=\frac{32E_b}{\pi^2}\left[\frac{\cos(2\pi T_bf)}{16T_b^2f^2-1}\right]^2\end{aligned} \tag{10.181}$$

The baseband power spectra of Eqs. (10.177) and (10.181) for the QPSK and MSK signals, respectively, are shown plotted in Fig. 10.33. The power spectral density is normalized with respect to $4E_b$, and the frequency is normalized with respect to the bit rate $1/T_b$. Note that for $f\gg 1/T_b$, the baseband power spectral density of the MSK signal falls off as the inverse fourth power of frequency, whereas in the case of the QPSK signal, it falls off as the inverse square of frequency. Accordingly, the MSK does not produce as much interference outside the signal band of interest as the QPSK. This is a desirable characteristic of MSK when operating with a bandwidth limitation.

A measure of the compactness of a modulation waveform's spectrum is the transmission bandwidth B_T which contains 99 percent of the total power. For MSK, we find that B_T is approximately $1.17/T_b$, while for QPSK it is approximately $8/T_b$.* This suggests that in relatively wide-band communication channels, MSK may be spectrally more efficient than QPSK. However, we see from Fig. 10.33 that the power spectrum of MSK has a wider mainlobe than QPSK, which would

* 1. V. K. Prabhu, "Spectral occupancy of digital angle-modulated signals," *Bell System Tech. J.*, vol. 55, pp. 429–453, April 1976.

2. S. A. Gronemeyer and A. L. McBride, "MSK and offset QPSK modulation," *IEEE Trans. on Communications*, vol. COM-24, pp. 809–820, Aug. 1976.

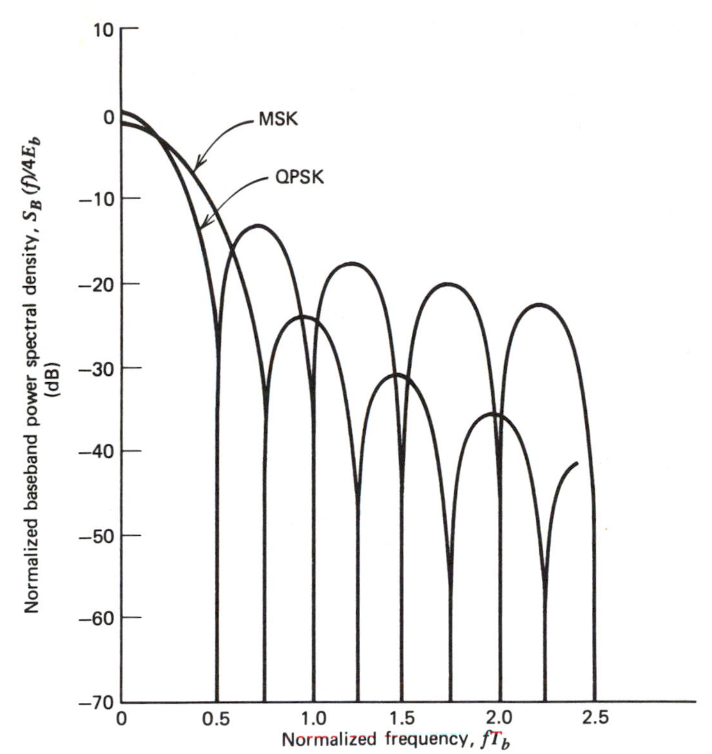

Figure 10.33 Normalized baseband power spectra of QPSK and MSK signals. (Note that for negative frequencies the spectra are the mirror images of those shown for positive frequencies).

indicate that in narrow-band communication channels, QPSK may be the preferred method.

Also, comparing the power spectra of Fig. 10.33 with those of Fig. 10.21, we see that QPSK and MSK signals both occupy bandwidths that are one-half those of the binary PSK and FSK signals, respectively. This further confirms the observation we made earlier that, in a given bandwidth, QPSK and MSK signals transmit information at twice the rate attainable by binary PSK and FSK signals, respectively.

10.16 M-ARY SIGNALING TECHNIQUES

In an *M-ary signaling scheme*, we may send any one of M possible signals, $s_1(t)$, $s_2(t), \ldots, s_M(t)$, during each signaling interval of duration T. For almost all applications, the number of possible signals $M = 2^n$, where n is an integer. The symbol duration $T = nT_b$, where T_b is the bit duration. These signals are generated by changing the amplitude, phase, or frequency of a carrier in M discrete steps. Thus, we have M-ary ASK, M-ary PSK, and M-ary FSK digital modulation schemes.

The QPSK system considered in Section 10.13 is an example of M-ary PSK with $M=4$.

M-ary signaling schemes are preferred over binary signaling schemes for transmitting digital information over bandpass channels when the requirement is to conserve bandwidth at the expense of increasing power. In practice, we rarely find a communication channel that has the exact bandwidth required for transmitting the output of an information source by means of binary signaling schemes. Thus, when the bandwidth of the channel is less than the required value, we may use M-ary signaling schemes so as to utilize the channel efficiently.

In this section we briefly highlight some important features of two kinds of coherent M-ary digital modulation schemes: (1) M-ary PSK, and (2) M-ary FSK.*

Coherent M-ary PSK

In M-ary PSK systems, the phase of the carrier takes on one of M possible values $\theta_i = 2i\pi/M$, where $i = 1, 2, \ldots, M$. Accordingly, during each signaling interval of duration T, one of the M possible signals:

$$s_i(t) = \sqrt{\frac{2E}{T}} \cos\left(2\pi f_c t + \frac{2i\pi}{M}\right), \qquad i = 1, 2, \ldots, M \tag{10.182}$$

is transmitted, where E is the signal energy per symbol. The carrier frequency $f_c = n_c/T$ for some fixed integer n_c.

If the information to be transmitted consists of a binary sequence with a bit duration T_b, then the bandwidth required for transmitting this information by means of binary PSK is inversely proportional to T_b. Now, if we take blocks of n bits and use an M-ary PSK scheme with $M = 2^n$ and $T = nT_b$, the bandwidth required is inversely proportional to nT_b. This shows that the use of M-ary PSK enables a reduction in transmission bandwidth by a factor of n over binary PSK signaling.

The M-ary PSK receiver consists of a phase discriminator whose output is directly proportional to the phase of the incoming carrier plus noise as measured over a signaling interval of duration T. The phase of the signal component at the receiver input in the signaling interval $0 \leqslant t \leqslant T$ is determined as θ_i if the phase discriminator output at time $t = T$ is within $\pm\pi/M$ of θ_i. Accordingly, the receiver makes an error when the magnitude of the noise-induced phase perturbation at the phase discriminator output exceeds π/M.

In Table 10.3 we have summarized typical values of power-bandwidth requirements for coherent binary and M-ary PSK schemes, assuming an average probability of symbol error equal to 10^{-4} and that the systems operate in identical

* For more betails on M-ary digital modulation schemes, see

1. W. C. Lindsey and M. K. Simon, Chapters 5 and 10, *Telecommunication Systems Engineering* (Prentice-Hall, 1973).
2. A. J. Viterbi, *Principles of Coherent Communication*, Chapter 8 (McGraw-Hill, 1966).
3. E. Arthurs and H. Dym, *op. cit.*

Table 10.3 Comparison of power-bandwidth requirements for M-ary PSK with binary PSK. Probability of symbol error = 10^{-4}.

Value of M	$\dfrac{(\text{Bandwidth})_{\text{M-ary}}}{(\text{Bandwidth})_{\text{Binary}}}$	$\dfrac{(\text{Average power})_{\text{M-ary}}}{(\text{Average power})_{\text{Binary}}}$
4	0.5	0.34 dB
8	0.333	3.91 dB
16	0.25	8.52 dB
32	0.2	13.52 dB

noise environments.* This table shows that the QPSK (corresponding to $M=4$) offers the best trade-off between power and bandwidth requirements. It is for this reason that we find QPSK is widely used in practice. For $M>8$, power requirements become excessive; accordingly, M-ary PSK schemes with $M>8$ are rarely used in practice. Also, it should be noted that coherent M-ary PSK schemes require considerably more complex equipment than coherent binary PSK schemes for signal generation or detection.

Coherent M-ary FSK

Consider next an FSK scheme in which the M transmitted signals are defined by

$$s_i(t)=\begin{cases}\sqrt{\dfrac{2E}{T}}\cos\left[\dfrac{\pi}{T}(n_c+i)t\right], & 0\leqslant t\leqslant T \\ 0, & \text{elsewhere}\end{cases} \tag{10.183}$$

where $i=1, 2, \ldots, M$. The carrier frequency f_c is equal to $n_c/2T$ for some fixed integer n_c. These signals are of equal duration, T, and have equal energy, E. Since the signal frequencies are separated by $1/2T$ hertz, the signals in Eq. (10.183) are also orthogonal.

An M-ary FSK system based on Eq. (10.183) has the following properties:†

1. For fixed values of bit rate $1/T_b$, noise power spectral density $N_0/2$, and probability of symbol error P_e, an increase in M results in a reduced power requirement. However, this reduction in transmitted power is achieved at the cost of increased bandwidth.
2. In the limiting case as $M\rightarrow\infty$, the average probability of symbol error P_e

* K. S. Shanmugam, *Digital and Analog Communication Systems*, p. 424 (Wiley, 1979).
† W. C. Lindsey and M. K. Simon, *op. cit.*

satisfies the condition

$$P_e = \begin{cases} 1, & \text{if } \dfrac{1}{T_b} > \dfrac{P}{N_0}\log_2 e \\ 0, & \text{if } \dfrac{1}{T_b} < \dfrac{P}{N_0}\log_2 e \end{cases} \tag{10.184}$$

where P is the average signal power at the receiver input, and T_b is the bit duration. Equation (10.184) indicates that the *maximum errorless rate* at which data can be transmitted using an M-ary orthogonal FSK signaling scheme is

$$\frac{1}{T_b} = \frac{P}{N_0}\log_2 e \tag{10.185}$$

Of course, the bandwidth required for such a transmission is infinite, since the bandwidth of the signal set approaches infinity as M approaches infinity. From Example 5 of Chapter 8, we recall that the capacity C of a channel with additive white Gaussian noise is $(P/N_0)\log_2 e$ when the bandwidth is infinite. We conclude therefore that if the bit rate $1/T_b$ is less than the channel capacity C, the probability of error can be made arbitrarily small. Thus an M-ary FSK signaling system is capable of transmitting data at a rate up to channel capacity with an arbitrarily small probability of error.

10.17 DETECTION OF SIGNALS WITH RANDOM PHASE IN THE PRESENCE OF NOISE

Up to this point, in our discussion of band-pass data transmission techniques, we have assumed that the information-bearing signal is completely known at the receiver. In practice, however, it is often found that in addition to the uncertainty due to the additive noise of a receiver, there is an additional uncertainty due to the randomness of certain signal parameters. The usual cause of this uncertainty is distortion in the transmission medium. Perhaps the most common random signal parameter is the phase, which is especially true for narrow-band signals. For example, transmission over a multiplicity of paths of different and variable lengths, or rapidly varying delays in the propagating medium from transmitter to receiver, may cause the phase of the received signal to change in a way that the receiver cannot follow. Synchronization with the phase of the transmitted carrier may then be too costly, and the designer may simply choose to disregard the phase information in the received signal at the expense of some degradation in the noise performance of the system.

Consider then a digital communication system in which the transmitted signal equals

$$s_i(t) = \sqrt{\frac{2E}{T}}\cos(2\pi f_i t), \qquad 0 \leq t \leq T \tag{10.186}$$

where E is the signal energy, T is the duration of the signaling interval, and the frequency f_i is an integral multiple of $1/2T$. When no provision is made to phase synchronize the receiver with the transmitter, the received signal will, for an AWGN channel, be of the form:

$$x(t)=\sqrt{\frac{2E}{T}}\cos(2\pi f_i t+\theta)+w(t), \qquad 0\leqslant t\leqslant T \tag{10.187}$$

where $w(t)$ is the sample function of a white Gaussian noise process of zero mean and power spectral density $N_0/2$. The phase θ is unknown, and it is usually considered to be the sample value of a random variable uniformly distributed between 0 and 2π radians. This implies a complete lack of knowledge of the phase. A digital communication system characterized in this way is said to be *noncoherent*.

We readily see that the detection schemes presented previously are inadequate for dealing with noncoherent systems, because if the received signal has the form described by Eq. (10.187), the output of the associated correlator in the receiver will be a function of the unknown phase θ. We shall now discuss, in a rather intuitive manner, the necessary modifications which may be introduced into the receiver in order to deal with this new situation.

Using a well-known trigonometric identity, we may rewrite Eq. (10.187) in the expanded form

$$x(t)=\sqrt{\frac{2E}{T}}\cos(\theta)\cos(2\pi f_i t)-\sqrt{\frac{2E}{T}}\sin\theta\sin(2\pi f_i t)+w(t), \qquad 0\leqslant t\leqslant T \tag{10.188}$$

Suppose, now, the received signal $x(t)$ is applied to a pair of correlators; we assume that one correlator is supplied with the reference signal $\sqrt{2/T}\cos(2\pi f_i t)$, and the other is supplied with the reference signal $\sqrt{2/T}\sin(2\pi f_i t)$. For both correlators, the observation interval is $0\leqslant t\leqslant T$. Then in the absence of noise, we find that the first correlator output equals $\sqrt{E}\cos\theta$ and the second correlator output equals $-\sqrt{E}\sin\theta$. The dependence on the unknown phase θ may be removed by summing the squares of the two correlator outputs, and then taking the square root of the sum. Thus, when the noise $w(t)$ is zero, the result of these operations is simply $\sqrt{E}$, which is independent of the unknown phase θ. This suggests that for the detection of a sinusoidal signal of arbitrary phase, and which is corrupted by an additive white Gaussian noise, as in the model of Eq. (10.187), we may use the so-called *quadrature receiver* shown in Fig. 10.34(a). Indeed, this receiver is optimum in the sense that it realizes this detection with the minimum probability of error.*

We next derive two equivalent forms of the quadrature receiver. The first form is easily obtained by replacing each correlator in Fig. 10.34(a) with a corresponding equivalent matched filter. We thus obtain the alternative form of quadrature receiver shown in Fig. 10.34(b). In one branch of this receiver, we have a filter matched to the signal $\sqrt{2/T}\cos(2\pi f_i t)$, and in the other branch we have a filter

* For mathematical details of this derivation, see A. D. Whalen, *Detection of Signals in Noise*, pp. 196–205 (Academic Press, 1971).

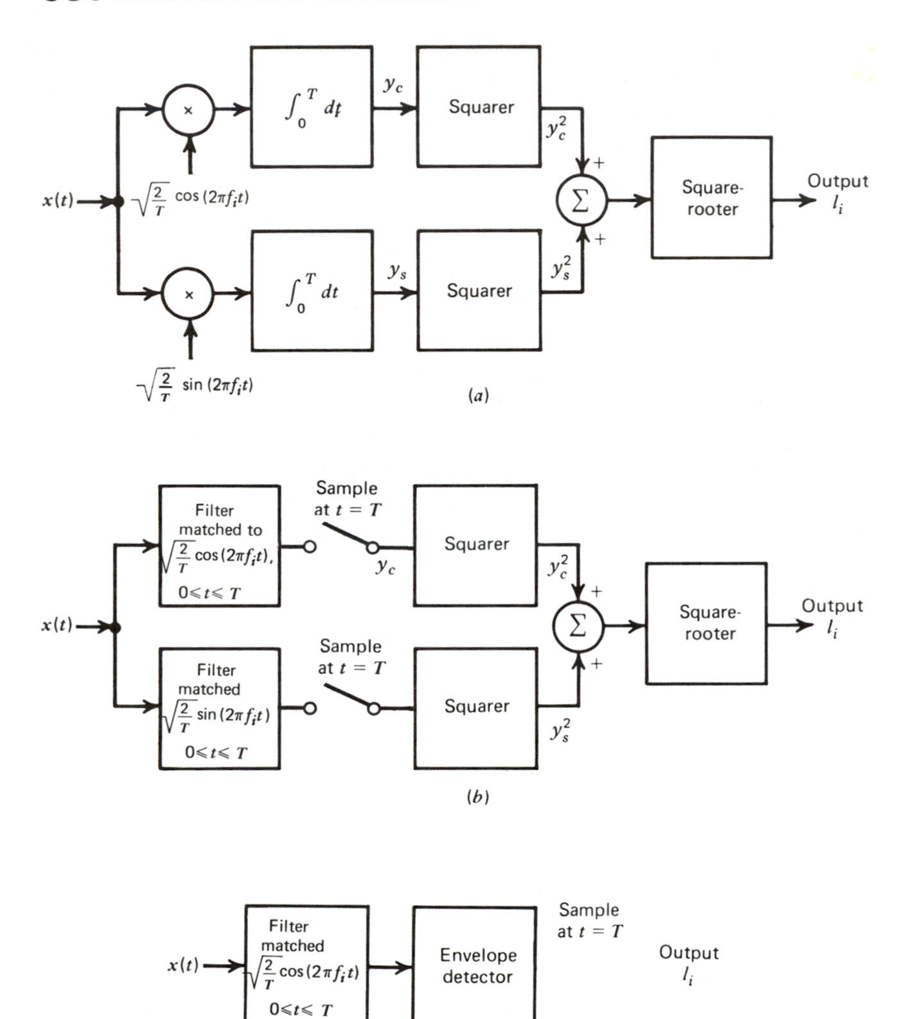

Figure 10.34 Noncoherent receivers. *(a)* Quadrature receiver using correlators. *(b)* Quadrature receiver using matched filters. *(c)* Another equivalent form of the quadrature receiver based on envelope detection.

matched to $\sqrt{2/T}\sin(2\pi f_i t)$, both defined for the time interval $0 \leq t \leq T$. The filter outputs are sampled at time $t = T$, squared, and added together.

To obtain the second equivalent form of the quadrature receiver, suppose we have a filter that is matched to $s(t) = \sqrt{2/T}\cos(2\pi f_i t + \theta)$, for $0 \leq t \leq T$. The envelope of the matched filter output is obviously unaffected by the value of phase θ. Therefore, for convenience, we will choose a matched filter with impulse response $\sqrt{2/T}\cos[2\pi f_i(T - t)]$, corresponding to $\theta = 0$. The output of such a filter in response to the received signal $x(t)$ is given by

$$y(t)=\sqrt{\frac{2}{T}}\int_0^T x(\tau)\cos[2\pi f_i(T-t+\tau)]d\tau$$
$$=\sqrt{\frac{2}{T}}\cos[2\pi f_i(T-t)]\int_0^T x(\tau)\cos(2\pi f_i\tau)d\tau$$
$$-\sqrt{\frac{2}{T}}\sin[2\pi f_i(T-t)]\int_0^T x(\tau)\sin(2\pi f_i\tau)d\tau \qquad (10.189)$$

The envelope of the matched filter output is proportional to the square root of the sum of the squares of the integrals in Eq. (10.189). The envelope, evaluated at time $t=T$, will therefore equal

$$l_i=\left\{\left[\int_0^T x(\tau)\sqrt{\frac{2}{T}}\cos(2\pi f_i\tau)d\tau\right]^2+\left[\int_0^T x(\tau)\sqrt{\frac{2}{T}}\sin(2\pi f_i\tau)d\tau\right]^2\right\}^{1/2} \qquad (10.190)$$

But, this is just the output of the quadrature receiver. Therefore, the output (at time T) of a filter matched to the signal $\sqrt{2/T}\cos(2\pi f_i t+\theta)$, of arbitrary phase θ, followed by an envelope detector is the same as the corresponding output of the quadrature receiver of Fig. 10.34(*a*). This form of receiver is shown in Fig. 10.34(*c*).

The need for an envelope detector following the matched filter in Fig. 10.34(*c*) may also be justified intuitively as follows. The output of a filter matched to a rectangular RF wave reaches a positive peak at the sampling instant $t=T$ (see Example 2). If, however, the phase of the filter is not matched to that of the signal, it is apparent that the peak will occur at a time different from the sampling instant. In actual fact, if the phases differ by π radians, we get a negative peak at the sampling instant. To avoid such poor sampling in the absence of prior information about the phase θ, it is reasonable to retain only the envelope of the matched filter output, since this is completely independent of the phase mismatch.

We illustrate the noncoherent detection theory described above by considering the specific problem of noncoherently detecting binary FSK signals in an AWGN channel. By disregarding the phase information in the received signal, hardware implementation of the binary FSK receiver is greatly simplified. However, this simplification is achieved at the expense of some degradation in the noise performance of the receiver.

10.18 NONCOHERENT DETECTION OF BINARY FSK SIGNALS

In the binary FSK case, the transmitted signal is defined by

$$s_i(t)=\begin{cases}\sqrt{\dfrac{2E_b}{T_b}}\cos(2\pi f_i t), & 0\le t\le T_b\\ 0, & \text{elsewhere}\end{cases} \qquad (10.191)$$

where the carrier frequency f_i equals one of two possible values f_1 and f_2. The transmission of frequency f_1 represents symbol 1, and the transmission of frequency f_2

represents symbol 0. For the noncoherent detection of this frequency-modulated wave, the receiver consists of a pair of matched filters followed by envelope detectors, as in Fig. 10.35. The filter in the upper channel of the receiver is matched to $\sqrt{2/T_b}\cos(2\pi f_1 t)$, and the filter in the lower channel is matched to $\sqrt{2/T_b}\cos(2\pi f_2 t)$, $0 \leqslant t \leqslant T_b$. The resulting envelope detector outputs are sampled at $t = T_b$, and their values are compared. Let l_1 and l_2 denote the envelope samples of the upper and lower channels, respectively. Then, if $l_1 > l_2$, the receiver decides in favor of symbol 1, and if $l_1 < l_2$ it decides in favor of symbol 0.

Suppose symbol 1 or frequency f_1 is transmitted. Then a correct decision will be made by the receiver if $l_1 > l_2$. If, however, the noise is such that $l_1 < l_2$, the receiver decides in favor of symbol 0, and an erroneous decision will have been made. To calculate the probability of error, we must have the probability density functions of the random variables L_1 and L_2 whose sample values are denoted by l_1 and l_2, respectively.

When frequency f_1 is transmitted, and there is no synchronism between the receiver and transmitter, the received signal $x(t)$ is of the form

$$\begin{aligned} x(t) &= \sqrt{\frac{2E_b}{T_b}}\cos(2\pi f_1 t + \theta) + w(t) \\ &= \sqrt{\frac{2E_b}{T_b}}\cos\theta\cos(2\pi f_1 t) - \sqrt{\frac{2E_b}{T_b}}\sin\theta\sin(2\pi f_1 t) + w(t), \qquad 0 \leqslant t \leqslant T_b \end{aligned} \tag{10.192}$$

Let

$$x_{ci} = \int_0^{T_b} x(t)\sqrt{\frac{2}{T_b}}\cos(2\pi f_i t)dt, \qquad i = 1, 2 \tag{10.193}$$

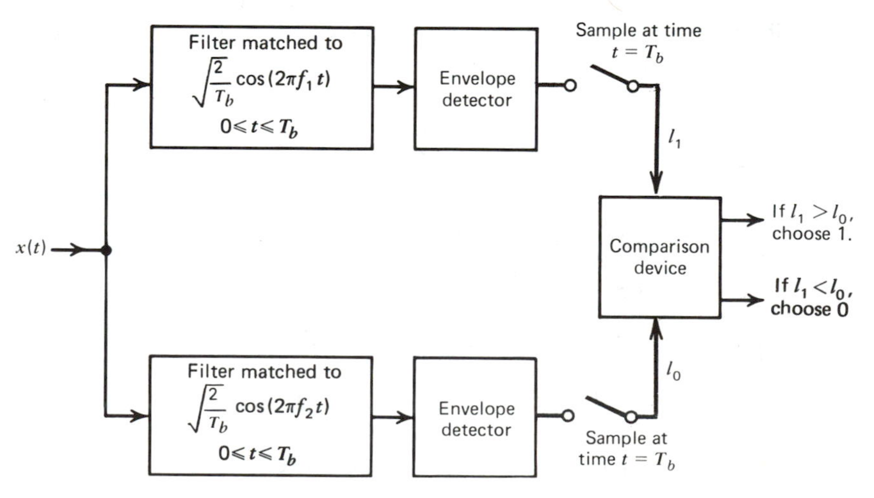

Figure 10.35 Noncoherent receiver for the detection of binary FSK signals.

and

$$x_{si} = \int_0^{T_b} x(t) \sqrt{\frac{2}{T_b}} \sin(2\pi f_i t) dt, \qquad i = 1, 2 \tag{10.194}$$

The x_{ci} and x_{si}, $i=1, 2$, define the coordinates of the received signal point. Note that, although each transmitted signal $s_i(t)$, $i=1, 2$, is represented by a point in a two-dimensional space, the presence of the unknown phase θ makes it necessary to use four orthonormal basis functions in order to resolve the received signal $x(t)$. With the received signal $x(t)$ having the form shown in Eq. (10.192), we find that the output of the upper channel in the receiver of Fig. 10.35 equals

$$l_1 = \sqrt{x_{c1}^2 + x_{s1}^2} \tag{10.195}$$

where

$$x_{c1} = \sqrt{E_b} \cos \theta + w_{c1} \tag{10.196}$$

and

$$x_{s1} = -\sqrt{E_b} \sin \theta + w_{s1} \tag{10.197}$$

On the other hand, the corresponding value of the lower channel-output is

$$l_2 = \sqrt{x_{c2}^2 + x_{s2}^2} \tag{10.198}$$

where

$$x_{c2} = w_{c2} \tag{10.199}$$

and

$$x_{s2} = w_{s2} \tag{10.200}$$

The w_{ci} and w_{si}, $i=1, 2$, are related to the noise $w(t)$ as follows:

$$w_{ci} = \int_0^{T_b} w(t) \sqrt{\frac{2}{T_b}} \cos(2\pi f_i t) dt, \qquad i = 1, 2 \tag{10.201}$$

and

$$w_{si} = \int_0^{T_b} w(t) \sqrt{\frac{2}{T_b}} \sin(2\pi f_i t) dt, \qquad i = 1, 2 \tag{10.202}$$

Accordingly, w_{ci} and w_{si}, $i=1, 2$, are sample values of independent Gaussian random variables of zero mean and variance $N_0/2$.

When symbol 1 or frequency f_1 is transmitted, we see from Eqs. (10.199) and (10.200) that x_{c2} and x_{s2} are sample values of two Gaussian and statistically independent random variables, X_{c2} and X_{s2}, with zero mean and variance $N_0/2$. Accordingly, the lower channel output l_2, related to x_{c2} and x_{s2} by Eq. (10.198), is the sample value of a Rayleigh-distributed random variable L_2 (see Section 5.11). We may thus express the conditional probability density function of L_2, given that symbol 1 was transmitted, as follows:

$$f_{L_2|1}(l_2|1)=\frac{2l_2}{N_0}\exp\left(-\frac{l_2^2}{N_0}\right), \qquad l_2 \geq 0 \tag{10.203}$$

Again under the condition that symbol 1 or frequency f_1 is transmitted, we see from Eqs. (10.196) and (10.197) that x_{c1} and x_{s1} are sample values of two Gaussian and statistically independent random variables, X_{c1} and X_{s1}, with mean values equal to $\sqrt{E_b}\cos\theta$ and $-\sqrt{E_b}\sin\theta$, respectively, and variance $N_0/2$. Therefore, the joint probability density function of X_{c1} and X_{s1}, given that symbol 1 was transmitted and that the random phase $\Theta=\theta$, may be expressed as follows

$$f_{X_{c1},X_{s1}|1,\Theta}(x_{c1}, x_{s1}|1, \theta)=\frac{1}{\pi N_0}\exp\left\{-\frac{1}{N_0}\left[(x_{c1}-\sqrt{E_b}\cos\theta)^2+(x_{s1}+\sqrt{E_b}\sin\theta)^2\right]\right\} \tag{10.204}$$

Define the transformations

$$x_{c1}=l_1\cos\psi_1 \tag{10.205}$$

and

$$x_{s1}=l_1\sin\psi_1 \tag{10.206}$$

where $\psi_1=\tan^{-1}(x_{s1}/x_{c1})$, with $0\leq\psi_1\leq 2\pi$. Then, applying this transformation and following a procedure similar to that described in Section 5.12, we find that the upper channel output l_1 is the sample value of a Rician-distributed random variable L_1. Hence, the conditional probability density function of L_1, given that symbol 1 was transmitted and that the random phase $\Theta=\theta$, is given by the Rician distribution

$$\begin{aligned} f_{L_1|1,\Theta}(l_1|1,\theta) &= \int_0^{2\pi} f_{L_1,\Psi|1,\Theta}(l_1,\psi|1,\theta)\,d\psi \\ &= \frac{2l_1}{N_0}\exp\left(-\frac{l_1^2+E_b}{N_0}\right)I_0\left(\frac{2l_1\sqrt{E_b}}{N_0}\right) \end{aligned} \tag{10.207}$$

where $I_0(2l_1\sqrt{E_b}/N_0)$ is the modified Bessel function of the first kind of zero order (see Appendix 4). Since Eq. (10.207) does not depend on θ, which is to be expected, it follows that the conditional probability density function of L_1, given that symbol 1 was transmitted, is

$$f_{L_1|1}(l_1|1)=\frac{2l_1}{N_0}\exp\left(-\frac{l_1^2+E_b}{N_0}\right)I_0\left(\frac{2l_1\sqrt{E_b}}{N_0}\right), \qquad l_1\geq 0 \tag{10.208}$$

Note that by putting $E_b=0$ and recognizing that $I_0(0)=1$, Eq. (10.208) reduces to a Rayleigh distribution.

When symbol 1 is transmitted, the receiver makes an error whenever the envelope sample l_2 obtained from the lower channel (due to noise alone) exceeds the envelope sample l_1 obtained from the upper channel (due to signal plus noise), for all possible values of l_1. Consequently, the probability of this error is obtained by integrating

$f_{L_2|1}(l_2|1)$ with respect to l_2 from l_1 to infinity, and then averaging over all possible values of l_1. That is to say,

$$
\begin{aligned}
P_{e1} &= P(l_2 > l_1 | \text{symbol 1 was sent}) \\
&= \int_0^\infty dl_1 f_{L_1|1}(l_1|1) \int_{l_1}^\infty dl_2 f_{L_2|1}(l_2|1)
\end{aligned}
\tag{10.209}
$$

where the inner integral is the conditional probability of error for a *fixed* value of l_1, given that symbol 1 was transmitted, and the outer integral is the average of this conditional probability for all possible values of l_1. Since the random variable L_2 is Rayleigh-distributed when symbol 1 is transmitted, the inner integral in Eq. (10.209) is equal to $\exp(-l_1^2/N_0)$. Thus, using Eq. (10.208) in (10.209), we get

$$
P_{e1} = \int_0^\infty \frac{2l_1}{N_0} \exp\left(-\frac{2l_1^2 + E_b}{N_0}\right) I_0\left(\frac{2l_1\sqrt{E_b}}{N_0}\right) dl_1 \tag{10.210}
$$

Define a new variable v related to l_1 by

$$
v = \frac{2l_1}{\sqrt{N_0}} \tag{10.211}
$$

Then, changing the variable of integration from l_1 to v, we may rewrite Eq. (10.210) in the form

$$
P_{e1} = \tfrac{1}{2} \exp\left(-\frac{E_b}{2N_0}\right) \int_0^\infty v \exp\left(-\frac{v^2 + a^2}{2}\right) I_0(av)\, dv \tag{10.212}
$$

where $a = \sqrt{E_b/N_0}$. The integral in Eq. (10.212) represents the total area under the normalized form of the Rician distribution (see Section 5.12). Since this integral must be equal to one, we may simplify Eq. (10.212) as

$$
P_{e1} = \tfrac{1}{2} \exp\left(-\frac{E_b}{2N_0}\right) \tag{10.213}
$$

Similarly, when symbol 0 or frequency f_2 is transmitted, we may show that P_{e0}, the probability that $l_1 > l_2$ and therefore the probability that the receiver makes an error by deciding in favor of symbol 1, has the same value as in Eq. (10.213). Thus, averaging P_{e0} and P_{e1}, we find that the average probability of symbol error for the noncoherent binary FSK equals

$$
P_e = \tfrac{1}{2} \exp\left(-\frac{E_b}{2N_0}\right) \tag{10.214}
$$

10.19 DIFFERENTIAL PHASE-SHIFT KEYING (DPSK)

This digital modulation scheme may be viewed as the noncoherent version of the PSK. It eliminates the need for a coherent reference signal at the receiver by combining two basic operations at the transmitter: (1) *differential encoding* of the input binary wave, and (2) *phase-shift keying*—hence, the name, *differential phase-shift keying* (DPSK). In effect, to send symbol 0 we phase advance the current signal waveform by 180 degrees, and to send symbol 1 we leave the phase of the current signal waveform unchanged. The receiver is equipped with a *storage* capability, so that it can measure the *relative phase difference* between the waveforms received during two successive bit intervals. Provided that the unknown phase θ contained in the received wave varies slowly (that is, slow enough for it to be considered essentially constant over two bit intervals), then the phase difference between waveforms received in two successive bit intervals will be independent of θ.

In conceptual terms, the task of the receiver is three-fold:

1. It calculates the coordinates of the received signal by correlating it with two locally generated reference signals:

$$\phi_1(t)=\sqrt{\frac{2}{T_b}}\cos(2\pi f_c t), \qquad 0\leqslant t\leqslant T_b \tag{10.215}$$

and

$$\phi_2(t)=\sqrt{\frac{2}{T_b}}\sin(2\pi f_c t), \qquad 0\leqslant t\leqslant T_b \tag{10.216}$$

where the carrier frequency f_c is an integral multiple of $1/T_b$.
2. It plots the received signal points, and thereby measures the phase difference between the currently received signal point and the previously received signal point, which has been stored.
3. It decides in favor of symbol 1 if this phase difference lies inside the range $-\pi/2$ to $\pi/2$ radians; if the phase difference lies outside this range, modulo 2π, it decides in favor of symbol 0.

Thus, a distinguishing feature of DPSK is the fact that there are no preassigned decision regions in the observation space, with each region associated with a particular message point. Instead, the decision is based on the phase difference between successively received signals. This suggests that in order to calculate the probability of error for the DPSK,* it is sufficient to consider the particular case where the transmitter keeps on sending the same waveform, so that a pair of successively received signals have the form

$$x_1(t)=\sqrt{\frac{2E_b}{T_b}}\cos(2\pi f_c t+\theta)+w_1(t) \tag{10.217}$$

and

$$x_2(t-T_b)=\sqrt{\frac{2E_b}{T_b}}\cos(2\pi f_c t+\theta)+w_2(t) \tag{10.218}$$

* The procedure presented here for calculating the probability of error is based on the paper by E. Arthurs and H. Dym, *op. cit.*

where θ is the unknown phase; $w_1(t)$ and $w_2(t)$ are sample functions of a white Gaussian noise process of zero mean and power spectral density $N_0/2$. Correspondingly, the phase difference measurement will (at least, conceptually) be based on a pair of observation vectors: (1) the stored vector $\mathbf{x}_1$ with elements x_{11} and x_{12}, and (2) the currently received vector $\mathbf{x}_2$ with elements x_{21} and x_{22}, where

$$\begin{aligned} x_{11} &= \int_0^{T_b} x_1(t)\phi_1(t)dt \\ &= \sqrt{E_b}\cos\theta + w_{11} \end{aligned} \tag{10.219}$$

$$\begin{aligned} x_{12} &= \int_0^{T_b} x_1(t)\phi_2(t)dt \\ &= -\sqrt{E_b}\sin\theta + w_{12} \end{aligned} \tag{10.220}$$

$$\begin{aligned} x_{21} &= \int_0^{T_b} x_2(t-T_b)\phi_1(t)dt \\ &= \sqrt{E_b}\cos\theta + w_{21} \end{aligned} \tag{10.221}$$

and

$$\begin{aligned} x_{22} &= \int_0^{T_b} x_2(t-T_b)\phi_2(t)dt \\ &= -\sqrt{E_b}\sin\theta + w_{22} \end{aligned} \tag{10.222}$$

The w_{11}, w_{12}, w_{21}, and w_{22} are sample values of independent Gaussian random variables, each with zero mean and variance $N_0/2$. The unknown phase θ is the sample value of a uniformly distributed random variable, Θ, over a 2π interval. A possible set of received signal points are shown in Fig. 10.36. Each vector is represented as the sum of a signal vector and a noise vector.

Let ψ_1 denote the phase perturbation produced by the noise $w_1(t)$, so that $\psi_1+\theta$ equals the total phase of the received signal $x_1(t)$. Similarly, let ψ_2 denote the phase perturbation produced by the noise $w_2(t)$, so that $\psi_2+\theta$ equals the total phase of the received signal $x_2(t-T_b)$. The angles ψ_1 and ψ_2 are shown defined in Fig. 10.36. The decision will be based on the phase difference $\psi_2-\psi_1$. An erroneous decision will be made, if and only if $|\psi_2-\psi_1|>\pi/2$.

Assume, for the moment, that the angle ψ_1 is known. Also, let v denote the noise component of the vector, $\mathbf{x}_2$, which lies in a direction parallel to the orientation of the stored vector, $\mathbf{x}_1$. Then, from Fig. 10.36 we deduce that an error will occur, if and only if v exceeds the value $\sqrt{E_b}\cos\psi_1$, for then the phase difference $\psi_2-\psi_1$ will have an absolute value that exceeds $\pi/2$. We note that v is the sample value of a Gaussian random variable, V, of zero mean and variance $N_0/2$. Hence, the probability density function of this random variable is

$$f_V(v) = \frac{1}{\sqrt{\pi N_0}} \exp\left(-\frac{v^2}{N_0}\right) \tag{10.223}$$

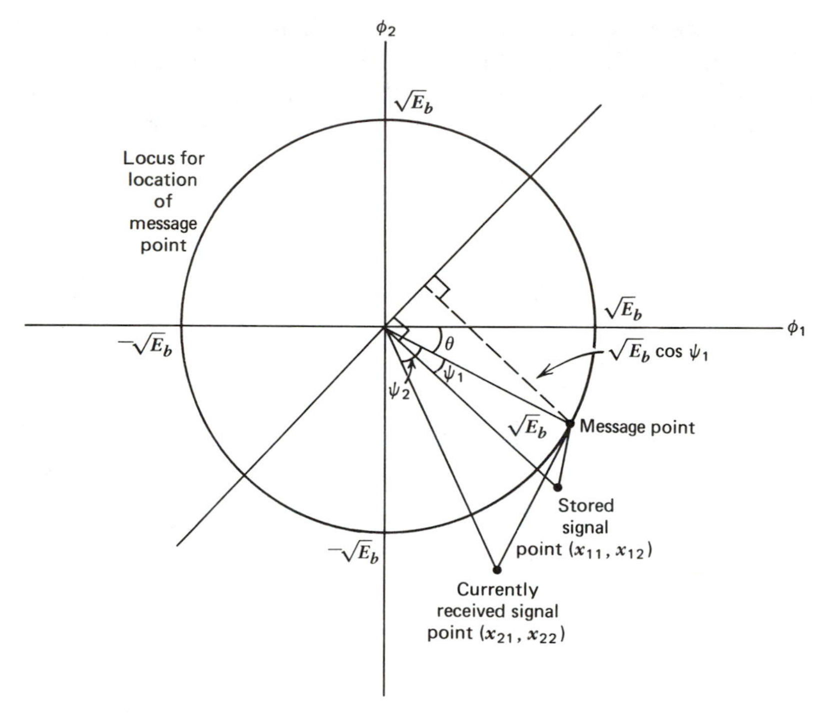

Figure 10.36 Signal space diagram for DPSK.

The conditional probability of error, given the value ψ_1, is therefore

$$
\begin{aligned}
P(\text{error}|\psi_1) &= \int_{\sqrt{E_b}\cos\psi_1}^{\infty} f_V(v)dv \\
&= \frac{1}{\sqrt{\pi N_0}} \int_{\sqrt{E_b}\cos\psi_1}^{\infty} \exp\left(-\frac{v^2}{N_0}\right)dv
\end{aligned}
\tag{10.224}
$$

Define the *error function* (see Appendix 2):

$$
\operatorname{erf}\left(\sqrt{\frac{E_b}{N_0}}\cos\psi_1\right) = \frac{2}{\sqrt{\pi}} \int_0^{\sqrt{E_b/N_0}\cos\psi_1} \exp(-z^2)dz \tag{10.225}
$$

Then putting $z = v/\sqrt{N_0}$ in Eq. (10.224), we may express the conditional probability of error, given ψ_1, in terms of this error function as follows

$$
P(\text{error}|\psi_1) = \tfrac{1}{2} - \tfrac{1}{2}\operatorname{erf}\left(\sqrt{\frac{E_b}{N_0}}\cos\psi_1\right) \tag{10.226}
$$

The reason for using the error function rather than the complementary error function will become apparent later.

To calculate the average probability of error, P_e, we have to average the conditional probability of error in Eq. (10.226) over all possible values which the angle ψ_1 may assume, as shown by

$$P_e = \int_{-\pi}^{\pi} P(\text{error}|\psi_1) f_{\Psi_1}(\psi_1) d\psi_1$$

$$= \frac{1}{2} \int_{-\pi}^{\pi} \left[1 - \text{erf}\left(\sqrt{\frac{E_b}{N_0}} \cos \psi_1 \right) \right] f_{\Psi_1}(\psi_1) d\psi_1 \qquad (10.227)$$

where $f_{\Psi}(\psi_1)$ is the probability density function of random variable Ψ_1 with sample value ψ_1.

To calculate $f_{\Psi_1}(\psi_1)$, it is convenient to first define a new coordinate system represented by the axes ϕ'_1 and ϕ'_2, which are parallel and perpendicular to the signal component of the stored vector $\mathbf{x}_1$, as illustrated in Fig. 10.37. Designate the new elements of the stored vector as z_1 and z_2. We note that z_1 and z_2 are sample values of two independent Gaussian random variables, Z_1 and Z_2, with mean values equal to $\sqrt{E_b}$ and zero, respectively, and variance $N_0/2$. Hence, the joint probability density function of the random variables Z_1 and Z_2 is

$$f_{Z_1,Z_2}(z_1, z_2) = \frac{1}{\pi N_0} \exp\left[-\frac{(z_1 - \sqrt{E_b})^2 + z_2^2}{N_0} \right] \qquad (10.228)$$

Next, we transform to polar coordinates by means of the relationships (see Fig. 10.37)

$$z_1 = r \cos \psi_1 \qquad (10.229)$$

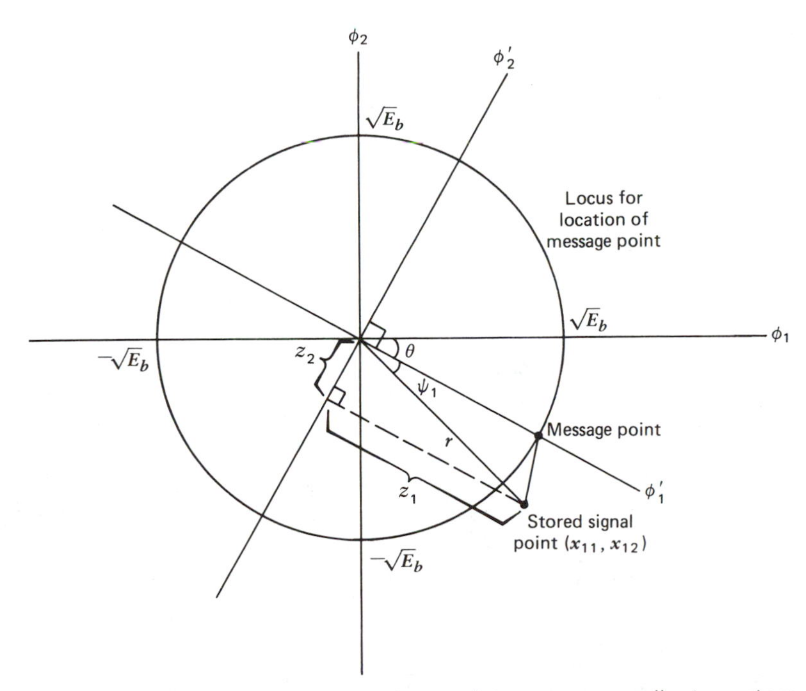

Figure 10.37 Illustrating the change into a new coordinate system represented by the axes ϕ'_1 and ϕ'_2.

and

$$z_2 = r \sin \psi_1 \tag{10.230}$$

We note that (see the derivation of the Rayleigh distribution, Section 5.11)

$$dz_1\, dz_2 = r\, dr\, d\psi_1 \tag{10.231}$$

and

$$f_{Z_1,Z_2}(z_1, z_2)dz_1\, dz_2 = f_{R,\Psi_1}(r, \psi_1)dr\, d\psi_1 \tag{10.232}$$

Thus, using Eqs. (10.228) to (10.232), we find that the joint probability density function $f_{R,\Psi_1}(r, \psi_1)$ is given by

$$\begin{aligned} f_{R,\Psi_1}(r, \psi_1) &= \frac{r}{\pi N_0} \exp\left[-\frac{(r \cos \psi_1 - \sqrt{E_b})^2 + (r \sin \psi_1)^2}{N_0} \right] \\ &= \frac{r}{\pi N_0} \exp\left(-\frac{r^2 - 2\sqrt{E_b}\, r \cos \psi_1 + E_b}{N_0} \right) \end{aligned} \tag{10.233}$$

If we complete the square in the exponent of Eq. (10.233) by writing

$$r^2 - 2\sqrt{E_b} r \cos \psi_1 + E_b = (r - \sqrt{E_b} \cos \psi_1)^2 + E_b \sin^2 \psi_1 \tag{10.234}$$

we may rewrite Eq. (10.233) in the form

$$f_{R,\Psi_1}(r, \psi_1) = \frac{r}{\pi N_0} \exp\left[-\frac{(r - \sqrt{E_b} \cos \psi_1)^2 + E_b \sin^2 \psi_1}{N_0} \right] \tag{10.235}$$

Integrating out the dependence on r, we get the marginal density:

$$\begin{aligned} f_{\Psi_1}(\psi_1) &= \int_0^\infty f_{R,\Psi_1}(r, \psi_1)dr \\ &= \frac{1}{\pi N_0} \exp\left(-\frac{E_b \sin^2 \psi_1}{N_0} \right) \int_0^\infty r \exp\left[-\frac{(r - \sqrt{E_b} \cos \psi_1)^2}{N_0} \right] dr \end{aligned} \tag{10.236}$$

Put

$$\frac{r - \sqrt{E_b} \cos \psi_1}{\sqrt{N_0}} = y \tag{10.237}$$

and

$$\frac{dr}{\sqrt{N_0}} = dy \tag{10.238}$$

Then, changing the variable of integration from r to y, we may rewrite Eq. (10.236) in the form

$$f_{\Psi_1}(\psi_1) = \frac{1}{\pi} \exp\left(-\frac{E_b \sin^2 \psi_1}{N_0} \right) \int_{-\sqrt{E_b/N_0} \cos \psi_1}^\infty \left(y + \sqrt{\frac{E_b}{N_0}} \cos \psi_1 \right) \exp(-y^2) dy \tag{10.239}$$

However, we note that

$$\int_{-\sqrt{E_b/N_0}\cos\psi_1}^{\infty} y\exp(-y^2)dy=\tfrac{1}{2}\exp\left(-\frac{E_b\cos^2\psi_1}{N_0}\right) \tag{10.240}$$

and

$$\int_{-\sqrt{E_b/N_0}\cos\psi_1}^{\infty} \exp(-y^2)dy=\frac{\sqrt{\pi}}{2}\left[1+\operatorname{erf}\left(\sqrt{\frac{E_b}{N_0}}\cos\psi_1\right)\right] \tag{10.241}$$

where the error function is as defined in Eq. (10.225). Therefore, substituting Eqs. (10.240) and (10.241) into (10.239), we get

$$f_{\Psi_1}(\psi_1)=\frac{1}{2\pi}\exp\left(-\frac{E_b}{N_0}\right) +\frac{1}{2}\sqrt{\frac{E_b}{\pi N_0}}\cos\psi_1\exp\left(-\frac{E_b\sin^2\psi_1}{N_0}\right)\left[1+\operatorname{erf}\left(\sqrt{\frac{E_b}{N_0}}\cos\psi_1\right)\right] \tag{10.242}$$

Next, substituting Eq. (10.242) into (10.227), we find that the average probability of error equals

$$\begin{aligned}P_e=&\tfrac{1}{2}\exp\left(-\frac{E_b}{N_0}\right)\\ &-\frac{1}{4\pi}\exp\left(-\frac{E_b}{N_0}\right)\int_{-\pi}^{\pi}\operatorname{erf}\left(\sqrt{\frac{E_b}{N_0}}\cos\psi_1\right)d\psi_1\\ &+\frac{1}{4}\sqrt{\frac{E_b}{\pi N_0}}\int_{-\pi}^{\pi}\cos\psi_1\exp\left(-\frac{E_b\sin^2\psi_1}{N_0}\right)d\psi_1\\ &-\frac{1}{4}\sqrt{\frac{E_b}{\pi N_0}}\int_{-\pi}^{\pi}\cos\psi_1\exp\left(-\frac{E_b\sin^2\psi_1}{N_0}\right)\operatorname{erf}^2\left(\sqrt{\frac{E_b}{N_0}}\cos\psi_1\right)d\psi_1\end{aligned} \tag{10.243}$$

The three integrands on the right-hand side of Eq. (10.243) have two common features:

1. They are periodic functions of ψ_1 with period 2π.
2. They alternate each half-period; that is, their functional dependence on ψ_1 satisfies the condition $g(\psi_1+\pi)=-g(\psi_1)$ for all ψ_1.

Now, the average value of any such function over one complete period is zero. Hence, we may write

$$\int_{-\pi}^{\pi}\operatorname{erf}\left(\sqrt{\frac{E_b}{N_0}}\cos\psi_1\right)d\psi_1=0 \tag{10.244}$$

$$\int_{-\pi}^{\pi}\cos\psi_1\exp\left(-\frac{E_b\sin^2\psi_1}{N_0}\right)d\psi_1=0 \tag{10.245}$$

$$\int_{-\pi}^{\pi}\cos\psi_1\exp\left(-\frac{E_b\sin^2\psi_1}{N_0}\right)\operatorname{erf}^2\left(\sqrt{\frac{E_b}{N_0}}\cos\psi_1\right)d\psi_1=0 \tag{10.246}$$

Accordingly, the expression for the average probability of symbol error for the DPSK, as defined in Eq. (10.243), reduces to the simple result

$$P_e = \frac{1}{2}\exp\left(-\frac{E_b}{N_0}\right) \tag{10.247}$$

It is of interest to note that this result may also be derived by viewing the detection of a DPSK signal as a special form of noncoherent detection of a pair of orthogonal signals, each having a duration equal to $2T_b$ seconds (see Problem 10.40).

DPSK Transmitter and Receiver

The differential encoding process at the transmitter input starts with an arbitrary first bit, serving as a reference, and thereafter the differentially encoded sequence $\{d_k\}$ is generated by using the *logical* equation

$$d_k = d_{k-1} b_k \oplus \bar{d}_{k-1} \bar{b}_k \tag{10.248}$$

where b_k is the input binary digit at time kT_b, and d_{k-1} is the previous value of the differentially encoded digit. The symbol $\oplus$ denotes *modulo-two-addition*, and the use of an overbar denotes logical *inversion*. Table 10.4 illustrates the logical operations involved in the use of Eq. (10.248), assuming that the reference bit added to the differentially encoded sequence $\{d_k\}$ is a 1. The differentially encoded sequence $\{d_k\}$ thus generated is used to phase-shift key a carrier with the phase angles 0 and π radians, as illustrated in the last row of Table 10.4.

The block diagram of a DPSK transmitter is shown in part (*a*) of Fig. 10.38. It consists, in part, of a logic network and a one-bit delay element interconnected so as to convert an input binary sequence $\{b_k\}$ into a differentially encoded sequence $\{d_k\}$ in accordance with Eq. (10.248). This sequence is amplitude-level

Table 10.4 Illustrating the generation of a DPSK signal

$\{b_k\}$		1	0	0	1	0	0	1	1
$\{\bar{b}_k\}$		0	1	1	0	1	1	0	0
$\{d_{k-1}\}$		1	1	0	1	1	0	1	1
$\{\bar{d}_{k-1}\}$		0	0	1	0	0	1	0	0
$\{b_k d_{k-1}\}$		1	0	0	1	0	0	1	1
$\{\bar{b}_k \bar{d}_{k-1}\}$		0	0	1	0	0	1	0	0
Differentially encoded sequence $\{d_k\}$	1	1	0	1	1	0	1	1	1
Transmitted phase (*radians*)	0	0	π	0	0	π	0	0	0

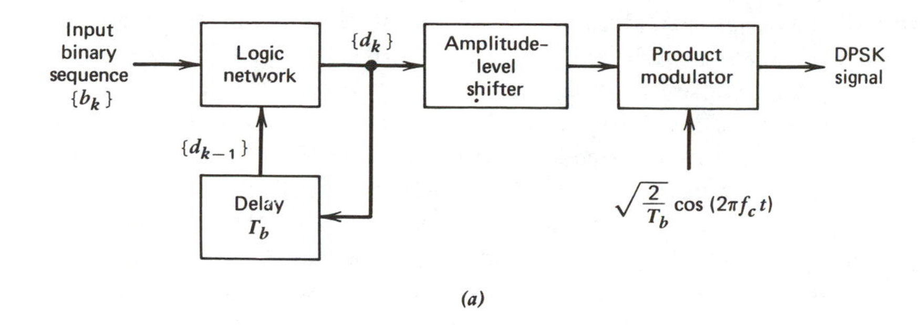

(a)

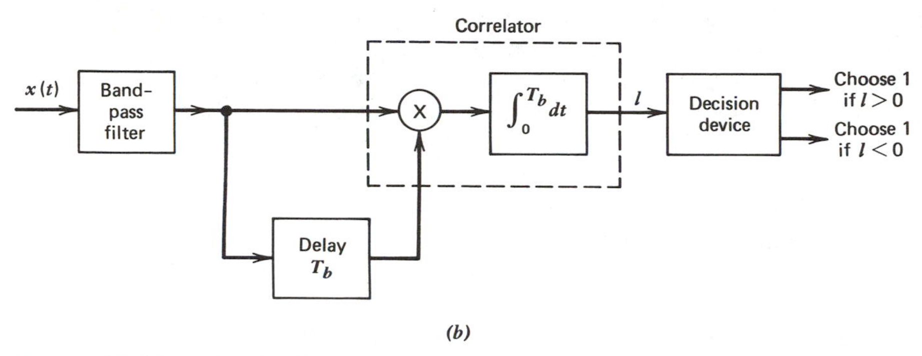

(b)

Figure 10.38 Block diagrams for *(a)* DPSK transmitter and *(b)* DPSK receiver.

shifted and then used to modulate a carrier wave of frequency f_c, thereby producing the desired DPSK wave.

At the receiver input, the received DPSK signal plus noise is passed through a band-pass filter centered at the carrier frequency f_c, so as to limit the noise power. The filter output and a delayed version of it, with the delay equal to the bit duration T_b, are applied to a correlator, as depicted in Fig. 10.38(*b*). The resulting correlator output is proportional to the cosine of the difference between the carrier phase angles in the two correlator inputs. The correlator output is finally compared with a threshold of zero volts, and a decision is thereby made in favor of symbol 1 or symbol 0. If the correlator output is positive, the phase difference between the waveforms received during the pertinent pair of bit intervals lies inside the range $-\pi/2$ to $\pi/2$, and the receiver decides in favor of symbol 1. If, on the other hand, the correlator output is negative, the phase difference lies outside the range $-\pi/2$ to $\pi/2$, modulo 2π, and the receiver decides in favor of symbol 0.

10.20 DISCUSSION

Throughout this chapter, we have used the average probability of committing a symbol error as the figure of merit for evaluating the noise performance of a digital communication system. It should be realized, however, that even if two systems yield the same average probability of symbol error, their performances, from the users' viewpoint, may be quite different. In particular, the greater the number of bits per symbol, the more the bit errors will cluster together. For example, if the average probability of symbol error is 10^{-3}, the expected number of symbols occurring between any two erroneous symbols is 1000. If each symbol represents one bit of information (as in a binary PSK or binary FSK system), the expected number of bits separating two erroneous bits is 1000. If, on the other hand, there are 2 bits per symbol (as in a QPSK system), the expected separation is 2000 bits. Of course, a symbol error generally creates more bit errors in the second case, so that the percentage of bit errors tends to be the same. Nevertheless, this clustering effect may make one system more attractive than another, even at the same symbol error rate. In the final analysis, which system is preferable will depend on the particular situation.

Two systems having an unequal number of symbols may be compared in a meaningful way only if they use the same amount of energy to transmit each bit of information. It is the total amount of energy needed to transmit the complete message that represents the cost of the transmission, not the amount of energy needed to transmit a particular symbol satisfactorily. Accordingly, in comparing the different data transmission systems considered above, we will use, as the basis of our comparison, the average probability of symbol error expressed as a function of the bit energy-to-noise density ratio E_b/N_0.

In Table 10.5, we have summarized the expressions for the average probability of symbol error P_e for the coherent PSK, conventional coherent FSK with one-bit decoding, DPSK, noncoherent FSK, QPSK and MSK, when operating over an AWGN channel. In Fig. 10.39 we have used these expressions to plot P_e as a function of E_b/N_0.

Table 10.5 Summary of Formulas for the Symbol Error Probability for Different Data Transmission Systems.

	Error probability, P_e
Coherent Binary Signaling:	
(a) Coherent PSK	$\frac{1}{2}\operatorname{erfc}(\sqrt{E_b/N_0})$
(b) Coherent FSK	$\frac{1}{2}\operatorname{erfc}(\sqrt{E_b/2N_0})$
Noncoherent Binary Signaling:	
(a) DPSK	$\frac{1}{2}\exp(-E_b/N_0)$
(b) Noncoherent FSK	$\frac{1}{2}\exp(-E_b/2N_0)$
Coherent Quadrature Signaling:	
(a) QPSK (b) MSK	$\operatorname{erfc}(\sqrt{E_b/N_0})-\frac{1}{4}\operatorname{erfc}^2(\sqrt{E_b/N_0})$

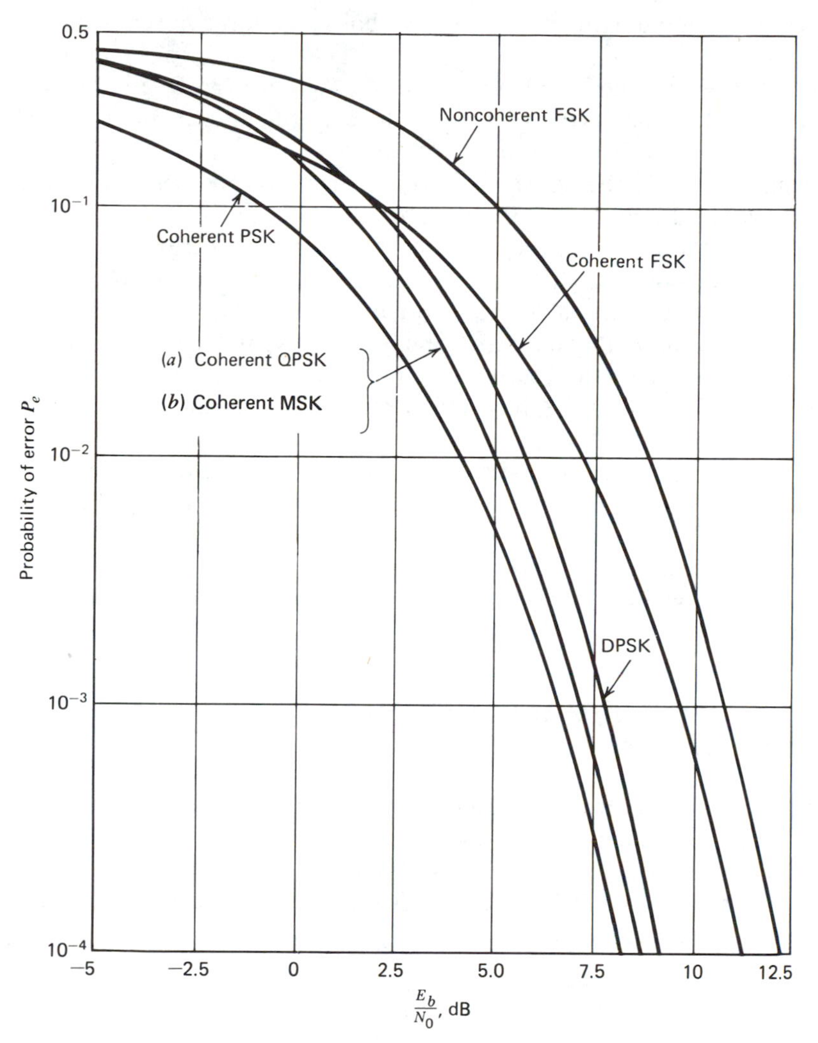

Figure 10.39 Comparison of the noise performances of different PSK and FSK systems.

In summary, we may thus state the following:

1. The error rates for all the systems decrease monotonically with increasing values of E_b/N_0.
2. For any value of E_b/N_0, the coherent PSK produces a smaller error rate than any of the other systems. Indeed, it may be shown that in the case of systems restricted to one-bit decoding, perturbed by additive white Gaussian noise, the coherent PSK system is the optimum system for transmitting binary data in the sense that

it achieves the minimum probability of symbol error for a given value of E_b/N_0.*

3. The coherent PSK and the DPSK require an E_b/N_0 that is 3 dB less than the corresponding values for the conventional coherent FSK and the noncoherent FSK, respectively, to realize the same error rate.
4. At high values of E_b/N_0, the DPSK and the noncoherent FSK perform almost as well (to within about 1 dB) as the coherent PSK and the conventional coherent FSK, respectively, for the same bit rate and signal energy per bit.
5. In QPSK two orthogonal carriers $\sqrt{2/T}\cos(2\pi f_c t)$ and $\sqrt{2/T}\sin(2\pi f_c t)$ are used, where the carrier frequency f_c is an integral multiple of the symbol rate $1/T$, with the result that two independent bit streams can be transmitted and subsequently detected in the receiver. At high values of E_b/N_0, coherently detected binary PSK and QPSK have the same error rate performances for the same value of E_b/N_0.
6. In MSK the two orthogonal carriers $\sqrt{2/T_b}\cos(2\pi f_c t)$ and $\sqrt{2/T_b}\sin(2\pi f_c t)$ are modulated by antipodal symbol shaping pulses $\cos(\pi t/2T_b)$ and $\sin(\pi t/2T_b)$, respectively, over $2T_b$ intervals, where T_b is the bit duration. Correspondingly, the receiver uses a coherent phase decoding process over two successive bit intervals to recover the original bit stream.† We thus find that MSK has exactly the same error rate performance as the QPSK.

10.21 EFFECT OF INTERSYMBOL INTERFERENCE

From the above discussion we conclude that the performance analysis of band-pass data transmission systems in the presence of additive white Gaussian noise is well-understood for both coherent and noncoherent reception. In practice, however, we find that because bandwidth occupancy is a major factor in the design of these systems, there is indeed a second source of interference, namely, intersymbol

* S. Stein, "Unified analysis of certain coherent and noncoherent binary communication systems," *IRE Trans. Information Theory*, vol. IT-10, pp. 43–51, Jan. 1964.

† In continuous-phase frequency-shift keying (CPFSK), we may use *multi-bit phase decoding*, with appropriate values of the deviation ratio h of the CPFSK signal, to realize a further improvement in the noise performance of the receiver. For example, by using $h = 2/3$ and extending the observation to at least 4-bit intervals, a marginal improvement of 0.8 dB over the case of MSK with $h = \frac{1}{2}$ is obtained. However, the complexity of the circuits required to implement multi-h signaling schemes does not seem to favor these schemes over the relatively simple and yet efficient MSK. For further details on multi-h signaling, see

1. W. P. Osborne and M. B. Lutz, "Coherent and noncoherent detection of CPFSK," *IEEE Trans. on Communications*, vol. COM-22, pp. 1023–1036, Aug. 1974.
2. R. deBuda, "About optimal properties of fast frequency-shift keying," *IEEE Trans. on Communications*, vol. COM-22, pp. 1726–1727, Oct. 1974.
3. J. B. Anderson and D. P. Taylor, "A bandwidth-efficient class of signal-space codes," *IEEE Trans. Information Theory*, vol. IT-24, pp. 703–712, Nov. 1978.
4. V. K. Bhargava, D. Haccoun, R. Matyas, and P. Nuspl, *Digital Communications by Satellite*, pp. 167–197 (Wiley, 1981).

interference (ISI), which must be accounted for in error rate calculations. As explained in Section 9.1, intersymbol interference is generated by the use of band-limiting filters at the transmitter output, in the transmission medium, and at the receiver input, or combinations thereof. When ISI is present, we find that (for the case of coherent reception) the correlation receiver (described in Section 10.6) or the matched filter receiver (described in Section 10.7) is no longer optimum, with the result that there is degradation in the actual error rate of the receiver. This also applies to noncoherent receivers.

Owing to the linear nature of the detection process in coherent PSK systems, we find that the effect of ISI on the performance of these systems has been treated in great detail in the literature. In particular, numerical methods have been developed for calculating the average probability of symbol error in coherent M-ary PSK systems in the combined presence of additive white Gaussian noise and intersymbol interference.* On the other hand, in the case of noncoherent reception (as, for example, DPSK), the detection process is inherently nonlinear. Accordingly, we find that under these conditions the performance analysis of noncoherent receivers is very difficult.†

Digital Computer Simulation

When explicit performance analysis of a band-pass data transmission system defies a satisfactory solution, the use of *digital computer simulation* provides the only alternative approach to actual hardware evaluation. The speed and flexibility usually associated with digital computers are compelling reasons for adopting this approach.

In order to simulate a given band-pass data-transmission system, we first develop its baseband (low-pass) equivalent model using the complex notation described in detail in Sections 2.14 and 2.15. To do this, we proceed as follows:

1. The transmitted signal $s(t)$ is represented by its complex envelope $\tilde{s}(t)$. These two signals are related by

$$s(t) = \text{Re}[\tilde{s}(t)\exp(j2\pi f_c t)]$$

where f_c is the carrier frequency. The complex envelope $\tilde{s}(t)$ equals

$$\tilde{s}(t) = s_c(t) + js_s(t)$$

where $s_c(t)$ and $s_s(t)$ are the in-phase and quadrature components of the trans-

* 1. O. Shimbo, R. J. Fang, and M. Celebiler, "Performance of M-ary PSK systems in Gaussian noise and intersymbol interference," *IEEE Trans. Information Theory*, vol. IT-19, pp. 44–58, Jan. 1973.
2. V. K. Prabhu, "Error probability performance of M-ary CPSK systems with intersymbol interference," *IEEE Trans. Communications*, vol. COM-21, pp. 97–109, Feb. 1973.

† For a rigorous method for calculating the error rate of M-ary DPSK in the combined presence of additive white Gaussian noise and intersymbol interference, see V. K. Prabhu and J. Salz, "On the performance of phase-shift keying systems," *Bell System Tech. J.*, vol. 60, pp. 2307–2343, Dec. 1981.

mitted signal $s(t)$, respectively. The characteristics of the input data waveform and the particular signaling method used to transmit the data are completely contained in the description of the complex envelope $\tilde{s}(t)$. Furthermore, by using the complex envelope $\tilde{s}(t)$, we eliminate the need for simulating the high-frequency carrier component.

2. To simulate the effect of noise at the front end of the receiver, we add to $s(t)$ the complex envelope

$$\tilde{w}(t) = w_c(t) + jw_s(t)$$

where the in-phase component $w_c(t)$ and the quadrature component $w_s(t)$ are modeled as sample functions of independent, zero-mean, white Gaussian noise processes, and with each one having a power spectral density of $N_0/2$ watts per hertz. Thus the received signal is represented by the complex envelope

$$\begin{aligned}\tilde{x}(t) &= \tilde{s}(t) + \tilde{w}(t)\\ &= [s_c(t) + w_c(t)] + j[s_s(t) + w_s(t)]\end{aligned}$$

3. Correspondingly, the matched filter (or correlation) receiver and any band-limiting filter in the system are replaced by their respective complex equivalent low-pass versions. Here, we use the fact that the actual impulse response, $h(t)$, of a narrow-band filter and the complex impulse response, $\tilde{h}(t)$, of its low-pass equivalent are related by

$$h(t) = 2\mathrm{Re}[\tilde{h}(t)\exp(j2\pi f_c t)]$$

In carrying out the simulation, it is customary to assume that the symbols of the alphabet used in the particular system under study are equally likely, and that the symbols transmitted in adjacent time slots are statistically independent. One way of accomplishing this requirement is to use linear maximal-length or PN sequences of sufficient length (see Example 8 of Chapter 5). Suppose that in a particular simulation run a total of K such symbols are transmitted and, say, L of them are misinterpreted by the receiver. Then, with K assumed to be large enough, the average probability of error will (almost always) be approximately equal to

$$P_e \simeq \frac{L}{K}$$

and the approximation becomes better as K approaches infinity. This suggests that in order to measure (with some degree of confidence) an average probability of symbol error as low as 10^{-5}, for example, the number K will have to be at least as large as 10^7; that is, 100 times the reciprocal of the error rate. This results in a standard deviation or root mean-square measurement error of no more than 10 percent.

In the case of linear systems, we may avoid the use of such long simulation runs on a computer (which can be quite expensive) by applying an *indirect* procedure to evaluate the effects of the transmitted signal and noise separately. We illustrate the procedure by considering a quadrature signaling system. Using the baseband equivalent model of the receiver, we first compute the amplitude of the correlator output (or decision device input) in the in-phase channel of the receiver in response to the in-phase component, $s_c(t)$, of the transmitted signal. This computation is done for each transmitted symbol. Let $A_{i,k}$ denote the value of this amplitude for symbol m_k, say. The subscript i refers to the in-phase channel. Next, we compute the variance of the noise at the same point in the receiver, in response to the in-phase noise component, $w_c(t)$. Let σ_i^2 denote this variance. Thus, for symbol m_k, we compute the conditional probability of error $P_{ei,k}$ by using the formula (see Problem 8.9):

$$P_{ei,k} = \frac{1}{2}\,\text{erfc}\left(\sqrt{\frac{\gamma_{i,k}}{2}}\right), \qquad k = 1, 2, \ldots, K$$

where $\gamma_{i,k} = A_{i,k}^2/\sigma_i^2$, and K is the total number of symbols transmitted.

The above computation is next repeated for the quadrature channel of the receiver. Let $P_{eq,k}$ denote this probability of error, conditional on the transmission of symbol m_k. The subscript q refers to the quadrature channel. Since the in-phase and quadrature channels are independent, the corresponding value of the conditional probability of symbol error for the complete receiver equals

$$P_{e,k} = P_{ei,k} + P_{eq,k} - P_{ei,k}P_{eq,k}, \qquad k = 1, 2, \ldots, K$$

By averaging this result over the number of transmitted symbols, K, we get the average probability of symbol error for the complete receiver as

$$P_e \simeq \frac{1}{K}\sum_{k=1}^{K} P_{e,k}$$

In the case of a binary signaling system, we only have an in-phase channel, so that $P_{eq,k}$ is zero, and the average probability of symbol error reduces to

$$P_e \simeq \frac{1}{K}\sum_{k=1}^{K} P_{ei,k}$$

By using the indirect procedure described above, the length of a simulation run is reduced by several orders of magnitude compared to the direct procedure. The only requirement is that the PN sequence used to generate the in-phase and quadrature components of the transmitted signal should have a length that is in excess of 100, say, and that all symbols of the alphabet for the system under study occur with approximately equal frequency.

10.22 MAXIMUM LIKELIHOOD ESTIMATION

In the preceding sections, our primary concern was with the detection of signals in additive white Gaussian noise. Specifically, the requirement was to determine whether or not a signal of interest was present in a received waveform. In the final section of this chapter, we wish to consider the related problem of *estimating* the signal parameters themselves.

The model for the received waveform in the parameter estimation problem is

$$x(t) = s(t, \alpha) + w(t), \qquad 0 \leqslant t \leqslant T \tag{10.249}$$

where α represents the unknown signal parameter to be estimated, and $w(t)$ denotes white Gaussian noise (of zero mean and power spectral density $N_0/2$) at the front end of the receiver. For example, the parameter α may represent the unknown phase or amplitude of a sinusoidal wave. There are various procedures available for carrying out this estimation, depending on the available information.* We consider a particular method called the *maximum likelihood estimation* procedure, which is well-suited for estimating a real nonrandom parameter when neither a priori knowledge of the parameter nor any meaningful cost function associated with the estimation error is available.

We approach the problem by choosing a set of orthonormal basis functions $\{\phi_i(t)\}$, $i = 1, 2, \ldots, N$, and approximate the received signal $x(t)$ in terms of a corresponding finite set of numbers $\{x_i\}$, $i = 1, 2, \ldots, N$, as shown by

$$x(t) \simeq \sum_{i=1}^{N} x_i \phi_i(t), \qquad 0 \leqslant t \leqslant T \tag{10.250}$$

where

$$x_i = \int_0^T x(t)\phi_i(t)dt, \qquad i = 1, 2, \ldots, N \tag{10.251}$$

To represent $x(t)$ completely, we require an infinite set of numbers, due to the presence of the noise term $w(t)$. Substituting Eq. (10.249) into (10.251), we may thus write

$$x_i = s_i(\alpha) + w_i \tag{10.252}$$

where

$$s_i(\alpha) = \int_0^T s(t, \alpha)\phi_i(t)dt, \qquad i = 1, 2, \ldots, N \tag{10.253}$$

and

$$w_i = \int_0^T w(t)\phi_i(t)dt, \qquad i = 1, 2, \ldots, N \tag{10.254}$$

We note that the w_i are sample values of a set of independent Gaussian random variables of zero mean and variance $N_0/2$. Correspondingly, the x_i are sample values of a set of independent Gaussian random variables of mean values equal to $s_i(\alpha)$, $i = 1, 2, \ldots, N$, and variance $N_0/2$. Therefore, the likelihood function, given α,

* For a detailed treatment of estimation theory, see

1. H. L. Van Trees, *Detection, Estimation, and Modulation Theory*, Part I, (Wiley, 1968).
2. C. W. Helstrom, *Statistical Theory of Signal Detection* (Pergamon Press, 1968).
3. A. D. Whalen, *op. cit*.

is defined by

$$f_{\mathbf{X}|\alpha}(\mathbf{x}|\alpha)=(\pi N_0)^{-N/2}\prod_{i=1}^{N}\exp\left\{-\frac{[x_i-s_i(\alpha)]^2}{N_0}\right\} \tag{10.255}$$

Now, if we let $N\to\infty$, this likelihood function is not well-defined. Noting that we may divide a likelihood function by anything that does not depend on α and still have a likelihood function, we may avoid the convergence problem by dividing by

$$f_{\mathbf{X}}(\mathbf{x})=(\pi N_0)^{-N/2}\prod_{i=1}^{N}\exp\left(-\frac{x_i^2}{N_0}\right) \tag{10.256}$$

before letting $N\to\infty$. Clearly, we may do this, because this function does not depend on α. Define

$$\Lambda[x(t),\,\alpha]=\lim_{N\to\infty}\frac{f_{\mathbf{X}|\alpha}(\mathbf{x}|\alpha)}{f_{\mathbf{X}}(\mathbf{x})} \tag{10.257}$$

Then, substituting Eqs. (10.255) and (10.256) into (10.257), cancelling common terms, and taking the logarithm, we get

$$\ln\Lambda[x(t),\,\alpha]=\frac{2}{N_0}\int_0^T x(t)s(t,\,\alpha)dt-\frac{1}{N_0}\int_0^T s^2(t,\,\alpha)dt \tag{10.258}$$

The *maximum likelihood estimate* $\hat{\alpha}$ is defined as that estimate which maximizes the likelihood function. Therefore, differentiating Eq. (10.258) with respect to α and setting the result equal to zero, we find that the maximum likelihood estimate $\hat{\alpha}$ is the solution of the equation:

$$\int_0^T [x(t)-s(t,\,\hat{\alpha})]\,\frac{\partial s(t,\,\hat{\alpha})}{\partial\hat{\alpha}}\,dt=0 \tag{10.259}$$

It should be noted that the estimate $\hat{\alpha}$ obtained in this way depends on the received wave $x(t)$. It is therefore a random variable with a mean and a variance.*

Example 6 Estimation of phase

Consider a signal of known form

$$s(t,\,\theta)=A_c\sin(2\pi f_c t+\theta),\qquad 0\leqslant t\leqslant T \tag{10.260}$$

* An estimator is said to be *unbiased* if the expected value of the estimate is exactly the same as the true value of the parameter in question. It is said to be a *minimum-variance estimator* if its variance is less than or equal to the variance of any other estimator. Furthermore, the minimum-variance-unbiased estimator is unique. In this regard, it may be shown, subject to certain conditions, that any estimator has a variance which cannot be less than a particular lower bound. This bound is called the *Cramer-Rao bound*. For further details, see:

1. A. D. Whalen, pp. 325–331, *op. cit.*
2. H. L. Van Trees, pp. 66–72, *op. cit.*
3. C. W. Helstrom, pp. 260–261, *op. cit.*

where the amplitude A_c and frequency f_c are known. The problem is to estimate the unknown phase θ. From Eq. (10.259) we find that the maximum likelihood estimate $\hat{\theta}$ of the phase θ is the solution of

$$\int_0^T [x(t) - A_c \sin(2\pi f_c t + \hat{\theta})]\cos(2\pi f_c t + \hat{\theta})dt = 0 \tag{10.261}$$

We assume that $2f_cT = k$ where k is an integer. Then the integral of the second term in Eq. (10.261) is zero, so that the phase estimate $\hat{\theta}$ is the solution of

$$\int_0^T x(t)\cos(2\pi f_c t + \hat{\theta})dt = 0 \tag{10.262}$$

Expanding the cosine term in the integrand in Eq. (10.262), and rearranging terms, we get

$$\cos\hat{\theta}\int_0^T x(t)\cos(2\pi f_c t)dt = \sin\hat{\theta}\int_0^T x(t)\sin(2\pi f_c t)dt$$

Solving for $\hat{\theta}$ yields the desired estimate:

$$\hat{\theta} = \tan^{-1}\left[\frac{\int_0^T x(t)\cos(2\pi f_c t)dt}{\int_0^T x(t)\sin(2\pi f_c t)dt}\right] \tag{10.263}$$

The operations in Eq. (10.263) may be performed by using a pair of correlators or filters matched to $\cos(2\pi f_c t)$ and $\sin(2\pi f_c t)$, as indicated in Fig. 10.40.

From Eq. (10.262) we may also deduce the phase-locked loop realization shown in Fig. 10.41, consisting of a multiplier, an integrator, and a voltage-controlled oscillator (VCO),

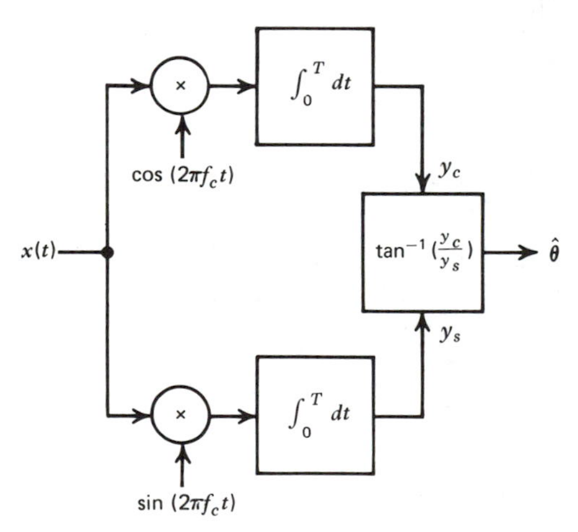

Figure 10.40 Estimator of phase of a sinusoidal wave of known amplitude and frequency.

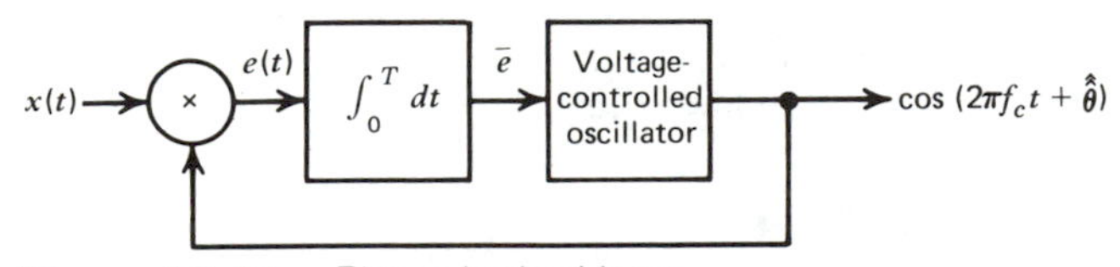

Figure 10.41 Phase-locked loop.

interconnected in the form of a feedback loop. The phase of the VCO output is denoted by $\hat{\hat{\theta}}$ to distinguish it from the maximum likelihood estimate $\hat{\theta}$. When the incoming signal $x(t)$ is noise-free, we have

$$x(t) = A_c \sin(2\pi f_c t + \theta)$$

The corresponding multiplier output is

$$\begin{aligned} e(t) &= A_c \sin(2\pi f_c t + \theta)\cos(2\pi f_c t + \hat{\hat{\theta}}) \\ &= \frac{A_c}{2}\sin(\theta - \hat{\hat{\theta}}) + \frac{A_c}{2}\sin(4\pi f_c t + \theta + \hat{\hat{\theta}}) \end{aligned}$$

The integrator averages $e(t)$ with respect to time, thereby yielding an output $\bar{e}$ that is proportional to $(\sin \theta - \hat{\hat{\theta}})$. For a small phase difference, the integrator output $\bar{e}$ is therefore proportional to $\theta - \hat{\hat{\theta}}$. Also, the integrator acts to smooth the variations of $e(t)$ due to noise. When the integrator output $\bar{e}$ is applied to the VCO, the phase $\hat{\hat{\theta}}$ of the VCO output changes in such a way as to reduce $\bar{e}$ to zero. Then $\hat{\hat{\theta}}$ approaches the maximum likelihood estimate $\hat{\theta}$. This is the desired condition for the phase-locked loop and the condition suggested by Eq. (10.262). Thus a phase-locked loop may be viewed as a realization of the maximum likelihood estimation procedure.

Problems

Problem 10.1

(a) Using the Gram–Schmidt orthogonalization procedure, find a set of orthonormal basis functions to represent the three signals $s_1(t)$, $s_2(t)$, and $s_3(t)$ shown in Fig. P10.1.
(b) Express each of these signals in terms of the set of basis functions found in part (a).

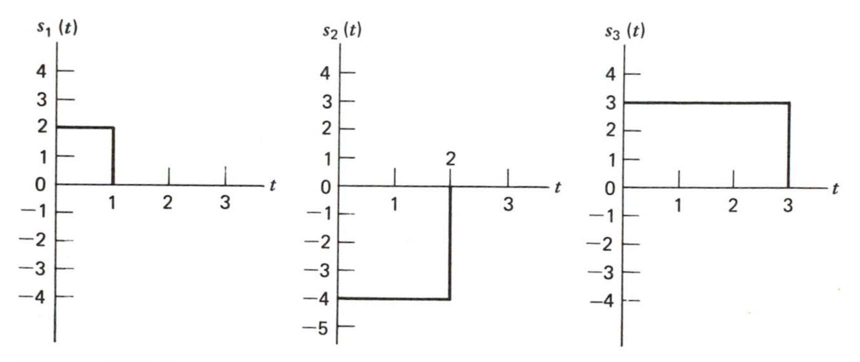

Figure P10.1

Problem 10.2 Consider the set of signals

$$s_i(t) = \begin{cases} \sqrt{\dfrac{2E}{T}} \cos\left(2\pi f_c t + i\dfrac{\pi}{4}\right), & 0 \leqslant t \leqslant T \\ 0, & \text{elsewhere} \end{cases}$$

where $i = 1, 2, 3, 4$, and $f_c = n_c/T$ for some fixed integer n_c.

(a) What is the dimensionality, N, of the space spanned by this set of signals?
(b) Find a set of orthonormal basis functions to represent this set of signals.
(c) Using the expansion,

$$s_i(t)=\sum_{j=1}^{N} s_{ij}\phi_j(t), \qquad i=1, 2, 3, 4$$

find the coefficients s_{ij}.
(d) Plot the locations of $s_i(t)$, $i=1, 2, 3, 4$, in the signal space, using the results of parts (b) and (c).

Problem 10.3 Given two real-valued vectors $\mathbf{x}$ and $\mathbf{y}$, prove Schwarz's inequality, which, using vector notation, states that

$$|(\mathbf{x}, \mathbf{y})| \leqslant \|\mathbf{x}\| \, \|\mathbf{y}\|$$

where $|(\mathbf{x}, \mathbf{y})|$ is the absolute value of the inner product of $\mathbf{x}$ and $\mathbf{y}$, and $\|\mathbf{x}\|$ and $\|\mathbf{y}\|$ are the lengths of $\mathbf{x}$ and $\mathbf{y}$, respectively.

Hint: Consider the inequality $\|\mathbf{x} \pm a\mathbf{y}\|^2 \geqslant 0$, and find an appropriate choice for the scalar a.

Problem 10.4 Consider Eq. (10.37), which shows that the received random process $X(t)$ may be expressed as

$$X(t)=\sum_{j=1}^{N} X_j\phi_j(t)+W'(t), \qquad 0\leqslant t\leqslant T$$

where $W'(t)$ is a remainder noise term. The $\{\phi_j(t)\}$, $j=1, 2, \ldots, N$ form an orthonormal set, and the X_j are defined by

$$X_j=\int_0^T X(t)\phi_j(t)dt$$

Let $W'(t_k)$ denote a random variable obtained by observing $W'(t)$ at time $t=t_k$. Show that

$$E[X_j W'(t_k)]=0, \qquad \begin{array}{l} j=1, 2, \ldots, N \\ 0\leqslant t\leqslant T \end{array}$$

Problem 10.5 In the *Bayes test*, applied to a binary hypothesis testing problem where we have to choose one of two possible hypotheses H_0 and H_1, we minimize the *risk* R defined by:

$$\begin{aligned} R = {} & C_{00}p_0(\text{say } H_0|H_0 \text{ is true}) \\ & + C_{10}p_0P(\text{say } H_1|H_0 \text{ is true}) \\ & + C_{11}p_1P(\text{say } H_1|H_1 \text{ is true}) \\ & + C_{01}p_1P(\text{say } H_0|H_1 \text{ is true}) \end{aligned}$$

The C_{00}, C_{10}, C_{11}, and C_{01} denote the costs assigned to the four possible outcomes of the experiment; the first subscript indicates the hypothesis chosen and the second the hypothesis that was true. Assume that $C_{10}>C_{00}$ and $C_{01}>C_{11}$. The p_0 and p_1 denote the a priori probabilities of hypotheses H_0 and H_1, respectively.

(a) Given the observation vector $\mathbf{x}$, show that the partitioning of the observation space so as to minimize the risk R leads to the *likelihood ratio test*:

$$\text{say } H_0 \text{ if } \Lambda(\mathbf{x})<\eta$$
$$\text{say } H_1 \text{ if } \Lambda(\mathbf{x})>\eta$$

where $\Lambda(\mathbf{x})$ is the *likelihood ratio*

$$\Lambda(\mathbf{x})=\frac{f_{\mathbf{X}|H_1}(\mathbf{x}|H_1)}{f_{\mathbf{X}|H_0}(\mathbf{x}|H_0)}$$

and η is the *threshold* of the test defined by

$$\eta=\frac{p_0(C_{10}-C_{00})}{p_1(C_{01}-C_{11})}$$

(b) What are the cost values for which the Bayes' criterion reduces to the minimum probability of error criterion?

Problem 10.6 Consider the signal $s(t)$ shown in Fig. P10.2

(a) Determine the impulse response of a filter matched to this signal and sketch it as a function of time.
(b) Plot the matched filter output as a function of time.
(c) What is the peak value of the output?

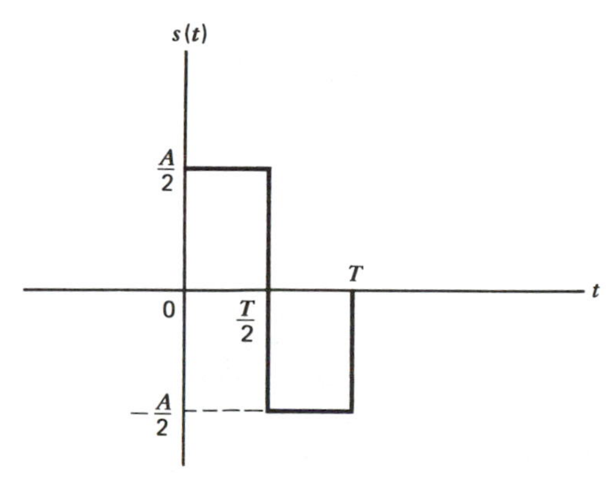

Figure P10.2

Problem 10.7

(a) Specify a matched filter for the signal $s_1(t)$ shown in Fig. P10.3(*a*), and sketch the resulting filter output as a function of time.
(b) Repeat the problem for the signal $s_2(t)$ shown in Fig. 10.3(*b*).
(c) Sketch the output of the filter matched to the signal $s_2(t)$ when the signal $s_1(t)$ is applied to the filter input.

Problem 10.8 It is proposed to implement a matched filter in the form of a tapped-delay-line filter with a set of tap weights $\{w_k\}$, $k=0, 1, \ldots, K$. Given a signal $s(t)$ of duration T seconds to which the filter is matched, find the value of w_k.

Problem 10.9 Show that the *input-to-output SNR gain* of a matched filter depends on the product of the input signal duration and the noise bandwidth.

Note: The input SNR is defined as the ratio of input signal power to input noise power measured in the noise equivalent bandwidth of the matched filter. The output SNR is defined in Section 10.7. The input-to-output SNR gain is defined as the ratio of the output SNR to input SNR.

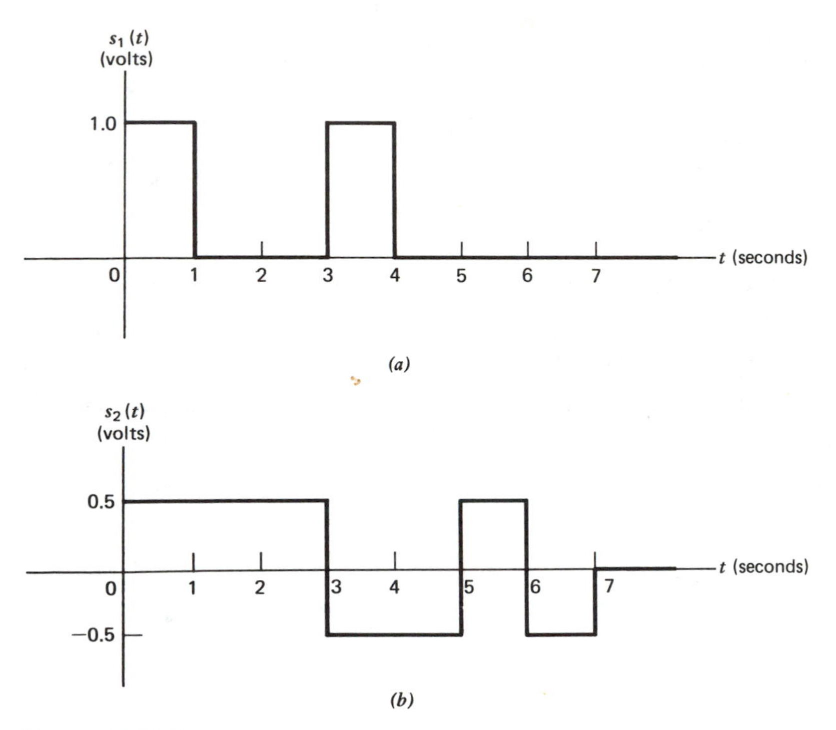

Figure P10.3

Problem 10.10 An *inverse filter*, with respect to a signal $s(t)$ of duration T and Fourier transform $S(f)$, is defined as having a transfer function equal to $\exp(-j2\pi fT)/S(f)$. Explain why an inverse filter is inferior to a matched filter when detecting the signal $s(t)$ in the presence of additive white noise.

Problem 10.11 Consider a pulse $s(t)$ defined by

$$s(t)=\begin{cases} A, & 0\leqslant t\leqslant T \\ 0, & \text{elsewhere} \end{cases}$$

It is proposed to approximate the matched filter for this pulse by a low-pass RC filter defined by the transfer function

$$H(f)=\frac{1}{1+jf/f_0}$$

where $f_0=1/2\pi RC$ is the 3-dB bandwidth of the filter.

(a) Determine the optimum value of f_0 for which the RC filter provides the best approximation to the matched filter.
(b) Assuming an additive white noise of zero mean and power spectral density $N_0/2$, what is the peak output signal-to-noise ratio?
(c) By how many decibels must the transmitted energy be increased so as to realize the same performance as the perfectly matched filter?

Problem 10.12 In a binary PCM system using on–off signaling, symbol 1 is represented by the pulse

$$s(t)=\begin{cases} A, & 0\leqslant t\leqslant T_b \\ 0, & \text{elsewhere} \end{cases}$$

and symbol 0 is represented by switching off the pulse. For predetection filtering, the receiver uses a matched filter, the maximum output of which is sampled and then applied to a decision device. Assume that the receiver noise is white, Gaussian, with zero mean and power spectral density $N_0/2$. Determine the average probability of error when symbols 1 and 0 occur with equal probability.

Problem 10.13 A binary PCM wave uses the Manchester code to describe symbols 1 and 0, as illustrated in Fig. P10.4. The additive noise at the receiver input is white, Gaussian, with zero mean and power spectral density $N_0/2$. Assuming that symbols 1 and 0 occur with equal probability, find an expression for the average probability of error at the receiver output, using a matched filter.

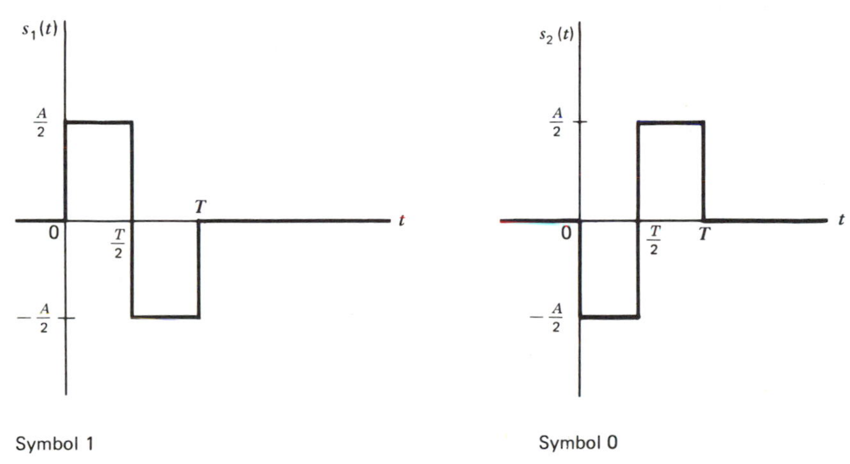

Figure P10.4

Problem 10.14

(a) The two signals

$$s_1(t)=-s_2(t)=\begin{cases} \exp(-t), & t\geqslant 0 \\ 0, & \text{elsewhere} \end{cases}$$

are transmitted with equal probability over a channel with additive white Gaussian noise of zero mean and power spectral density $N_0/2$. The receiver bases its decision on the received signal over the interval $0\leqslant t\leqslant 2$. Determine the minimum attainable probability of error at the receiver output.

(b) Compare the result obtained in part (a) with the performance of an optimum receiver which observes the received signal over the entire interval $-\infty<t<\infty$.

Problem 10.15 Consider a signal $s(t)$ corrupted by additive *colored* noise $n(t)$ of power spectral density $S_N(f)$. To maximize the detection of the signal $s(t)$, the input is passed through a structure consisting of two components: (1) a *prewhitening filter* to transform the input noise $n(t)$

into a white noise, and (2) a filter matched to the signal component at the prewhitening filter output. Determine the overall transfer function of this structure.

Problem 10.16 Consider a phase-locked loop consisting of a multiplier, loop filter, and voltage-controlled oscillator (VCO). Let the signal applied to the multiplier input be a PSK signal defined by

$$s(t) = A_c \cos[2\pi f_c t + k_p m(t)]$$

where k_p is the phase sensitivity, and the data signal $m(t)$ takes on the value $+1$ volt for binary symbol 1 and -1 volt for binary symbol 0. The VCO output is

$$r(t) = A_c \sin[2\pi f_c t + \theta(t)]$$

(a) Evaluate the loop filter output, assuming that this filter removes only modulated components with carrier frequency $2f_c$.
(b) Show that this output is proportional to the data signal $m(t)$ when the loop is phase-locked, that is, $\theta(t) = 0$.

Problem 10.17 The signal component of a coherent PSK system is defined by

$$s(t) = A_c k \sin(2\pi f_c t) \pm A_c \sqrt{1-k^2} \cos(2\pi f_c t)$$

where $0 \leqslant t \leqslant T_b$, and the plus sign corresponds to symbol 1 and the minus sign to symbol 0. The first term represents a carrier component included for the purpose of synchronizing the receiver to the transmitter.

(a) Show that, in the presence of additive white Gaussian noise of zero mean and power spectral density $N_0/2$, the average probability of error is

$$P_e = \frac{1}{2} \operatorname{erfc}\left(\sqrt{\frac{E_b}{N_0}(1-k^2)}\right)$$

where

$$E_b = \tfrac{1}{2} A_c^2 T_b$$

(b) Suppose that 10 percent of the transmitted signal power is allocated to the carrier component. Determine the E_b/N_0 required to realize a probability of error equal to 10^{-4}.
(c) Compare this value of E_b/N_0 with that required for a conventional PSK system with the same probability of error.

Problem 10.18 A PSK signal is applied to a correlator supplied with a phase reference that lies within ϕ radians of the exact carrier phase. Determine the effect of the phase error ϕ on the average probability of error of the system.

Problem 10.19 Figure P10.5 shows the block diagram of a *spread-spectrum communication system*. Foremost among the applications of this system is the suppression of interference which may be intentional (hostile) or unintentional. Binary information, consisting of a random sequence of 1's and 0's, is modulated onto a carrier as a PSK wave. This modulated wave is then passed to a second product modulator where further phase reversals are applied by means of a *spreading code sequence*. Assume that this sequence consists of a linear maximal length (or PN) sequence whose bit rate is much greater than that of the original binary wave. The result is a transmitted signal with a noise-like spectrum. For a description of the maximal length sequence, see Example 8, Chapter 5. Note that the original binary sequence and the spreading code sequence are represented in their polar form. At the receiver, the received signal (consisting

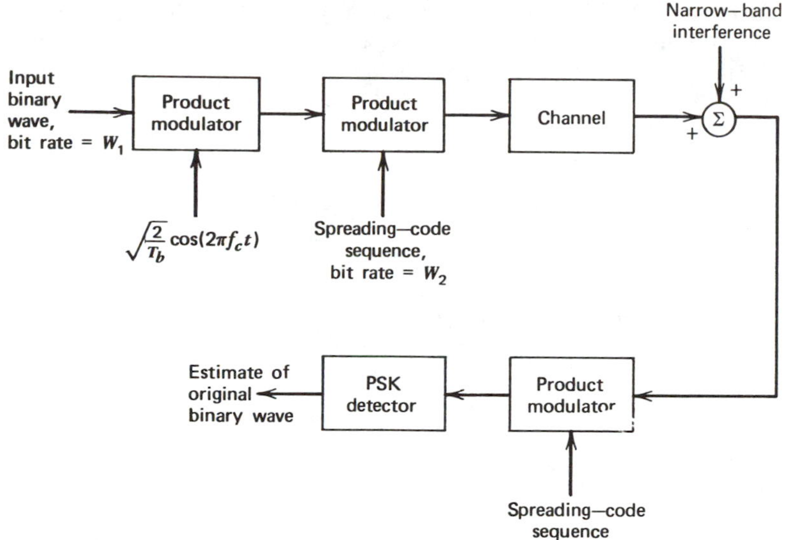

Figure P10.5

of the transmitted signal plus interference) is applied to a product modulator where it is further phase-reversed by an exact replica of the spreading code sequence. The original PSK wave will have thus been subjected to phase-reversals twice by the same binary code, and thereby restored to its initial form. The resulting signal is finally applied to a conventional coherent PSK detector, and the original binary information is thus recovered.

(a) Explain the reason for the property that the bandwidth of the transmitted signal is roughly equal to the bit rate of the spreading code sequence, and thus essentially independent of the input binary information.

(b) Suppose that a narrow-band undesired (jamming) interference enters the receiver input with an average power of J watts. If the desired component of the received signal has an average power of P watts, show that the *jamming power-to-signal power ratio* is

$$\frac{J}{P} \simeq \frac{W_2/W_1}{E_b/N_0}$$

where E_b/N_0 is the equivalent bit energy-to-noise density ratio needed for a coherent binary PSK to support the same error rate as that produced by the presence of the interference alone, and W_2/W_1 is the ratio of the spread bandwidth (bit rate of the spreading code sequence) to the original binary data bandwidth. The ratio W_2/W_1 is referred to as the *processing gain* of the system.

Problem 10.20 The signal vectors $\mathbf{s}_1$ and $\mathbf{s}_2$ are used to represent binary symbols 1 and 0, respectively, in a coherent binary FSK system. The receiver decides in favor of symbol 1 when

$$(\mathbf{x}, \mathbf{s}_1) < (\mathbf{x}, \mathbf{s}_2)$$

where $(\mathbf{x}, \mathbf{s}_1)$ is the inner product of the observation vector $\mathbf{x}$ and the signal vector $\mathbf{s}_1$, and similarly for $(\mathbf{x}, \mathbf{s}_2)$. Show that this decision rule is equivalent to the condition $x_1 > x_2$, where x_1 and x_2 are the elements of the observation vector $\mathbf{x}$ [see Eqs. (10.118) and (10.119)]. Assume that the signal vectors $\mathbf{s}_1$ and $\mathbf{s}_2$ have equal energy.

Problem 10.21 An FSK system transmits binary data at the rate of 2.5×10^6 bits per second. During the course of transmission, white Gaussian noise of zero mean and power spectral density 10^{-20} watts per hertz is added to the signal. In the absence of noise, the amplitude of the received sinusoidal wave for digit 1 or 0 is 1 microvolt. Determine the average probability of symbol error for the following system configurations:

(a) Coherent binary FSK.
(b) MSK.
(c) Noncoherent binary FSK.

Problem 10.22

(a) In a coherent FSK system, the signals $s_1(t)$ and $s_2(t)$ representing symbols 1 and 0, respectively, are defined by

$$s_1(t),\, s_2(t) = A_c \cos\left[2\pi\left(f_c \pm \frac{\Delta f}{2}\right)t\right], \qquad 0 \leqslant t \leqslant T_b$$

Assuming that $f_c > \Delta f$, show that the correlation coefficient of the signals $s_1(t)$ and $s_2(t)$ is approximately given by

$$\rho = \frac{\int_0^{T_b} s_1(t)s_2(t)dt}{\int_0^{T_b} s_1^2(t)dt} \simeq \text{sinc}(2\,\Delta f\,T_b)$$

(b) What is the minimum value of frequency shift Δf for which the signals $s_1(t)$ and $s_2(t)$ are orthogonal?
(c) What is the value of Δf that minimizes the average probability of symbol error?
(d) For the value of Δf obtained in part (c), determine the increase in E_b/N_0 required so that this coherent FSK system has the same noise performance as a coherent binary PSK system.

Problem 10.23 A binary FSK signal with *discontinuous phase* is defined by

$$s(t) = \begin{cases} \sqrt{\dfrac{2E_b}{T_b}} \cos\left[2\pi\left(f_c + \dfrac{\Delta f}{2}\right)t + \theta_1\right], & \text{for symbol 1} \\ \sqrt{\dfrac{2E_b}{T_b}} \cos\left[2\pi\left(f_c - \dfrac{\Delta f}{2}\right)t + \theta_2\right], & \text{for symbol 0} \end{cases}$$

where E_b is the signal energy per bit, T_b is the bit duration, and θ_1 and θ_2 are sample values of uniformly distributed random variables over the interval 0 to 2π. In effect, the two oscillators supplying the transmitted frequencies $f_c \pm \Delta f/2$ operate independently of each other. Assume that $f_c \gg \Delta f$.

(a) Evaluate the power spectral density of the FSK signal.
(b) Show that for frequencies far removed from the carrier frequency f_c, the power spectral density falls off as the inverse square of frequency.

Problem 10.24 Set up a block diagram for the generation of a binary FSK signal $s(t)$ with continuous phase by using the representation given in Eq. (10.132), which is reproduced here for convenience:

$$s(t) = \sqrt{\frac{2E_b}{T_b}} \cos\left(\frac{\pi t}{T_b}\right)\cos(2\pi f_c t) \mp \sqrt{\frac{2E_b}{T_b}} \sin\left(\frac{\pi t}{T_b}\right)\sin(2\pi f_c t)$$

Problem 10.25 In the on–off keying version of an ASK system, symbol 1 is represented by transmitting a sinusoidal carrier of amplitude $\sqrt{2E_b/T_b}$, where E_b is the bit energy and T_b is the bit duration. Symbol 0 is represented by switching off the carrier. Assume that symbols 1 and 0 occur with equal probability.

For an AWGN channel, determine the average probability of error for this ASK system, assuming:

(a) Coherent reception.
(b) Noncoherent reception, operating with a large value of bit energy-to-noise density ratio E_b/N_0.

Note: When x is large, the modified Bessel function of the first kind of zero order may be approximated as follows (see Appendix 4)

$$I_0(x) \simeq \frac{\exp(x)}{\sqrt{2\pi x}}$$

Problem 10.26 The purpose of a *radar system* is basically to detect the presence of a target, and to extract useful information about the target. Suppose that in such a system, hypotheses H_0 is that there is no target present, so that the received signal $x(t) = w(t)$, where $w(t)$ is white Gaussian noise of zero mean and power spectral density $N_0/2$. For hypothesis H_1, a target is present, and $x(t) = w(t) + s(t)$, where $s(t)$ is an echo produced by the target. Assume that $s(t)$ is completely known. Evaluate:

(a) The *probability of false alarm* defined as the probability that the receiver decides a target is present when it is not,
(b) The *probability of detection* defined as the probability that the receiver decides a target is present when it is.

Problem 10.27 Binary data is transmitted over a microwave link at the rate of 10^6 bits per second and the power spectral density of the noise at the receiver input is 10^{-10} watts per hertz. Find the average carrier power required to maintain an average probability of error $P_e \leqslant 10^{-4}$ for (a) coherent binary PSK, and (b) DPSK.

Problem 10.28 The values of E_b/N_0 required to realize an average probability of symbol error $P_e = 10^{-4}$ using coherent binary PSK and coherent FSK (conventional) systems are equal to 7.2 and 13.5, respectively. Using the approximation (see Appendix 2)

$$\text{erfc}(u) \simeq \frac{1}{\sqrt{\pi u}} \exp(-u^2),$$

determine the separation in the values of E_b/N_0 for $P_e = 10^{-4}$, using

(a) Coherent binary PSK and DPSK.
(b) Coherent binary PSK and QPSK.
(c) Coherent binary FSK (conventional) and noncoherent binary FSK.
(d) Coherent binary FSK (conventional) and MSK.

Problem 10.29

(a) Given the input binary sequence 1100100010, sketch the waveforms of the in-phase and quadrature components of a modulated wave obtained by using the QPSK based on the signal set of Fig. 10.22.
(b) Sketch the QPSK waveform itself for the input binary sequence specified in part (a).

Problem 10.30 Figure P10.6 shows the signal-space diagram for a QPSK signal, where E is the signal energy per symbol.

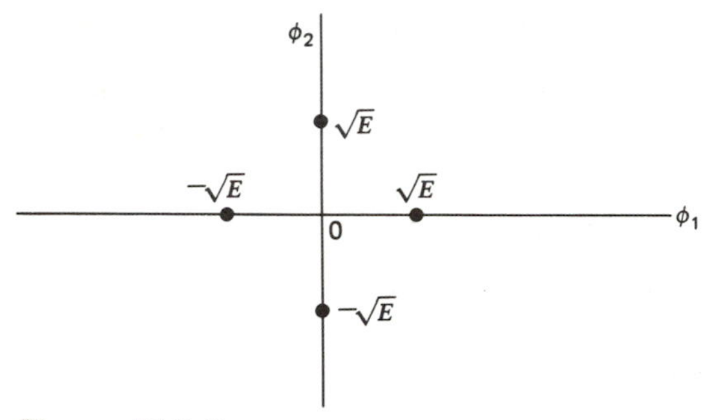

Figure P10.6

(a) Explain why the average probability of symbol error for this signal set is the same as that for the signal set shown in Fig. 10.22, when operating in an AWGN channel.

(b) Sketch the waveforms for the in-phase and quadrature components of this QPSK signal produced by the input binary sequence 1100100010.

(c) Sketch the waveform of the QPSK signal for the binary sequence in part (b), assuming that the carrier frequency is an integral multiple of the symbol rate $1/T$

Problem 10.31 Figure P10.7 shows the signal-space diagram for a *16-level quadrature-amplitude modulation* (*QAM*) *scheme* that represents a hybrid form of amplitude and phase modulation. This method of digital modulation provides a high bandwidth efficiency, which makes it attractive for use in applications where the requirement is to facilitate efficient use of the radio frequency spectrum. Assume that the 16 message points of Fig. P.10.7 are equally likely, with each one being represented by a corresponding signal of duration T seconds. During each signaling interval, one of these signals is transmitted over an additive white Gaussian noise channel of zero mean and power spectral density $N_0/2$.

(a) Suppose that binary data is transmitted with this modulation scheme at a rate of 200 megabits per second. What is the corresponding value of the symbol rate?

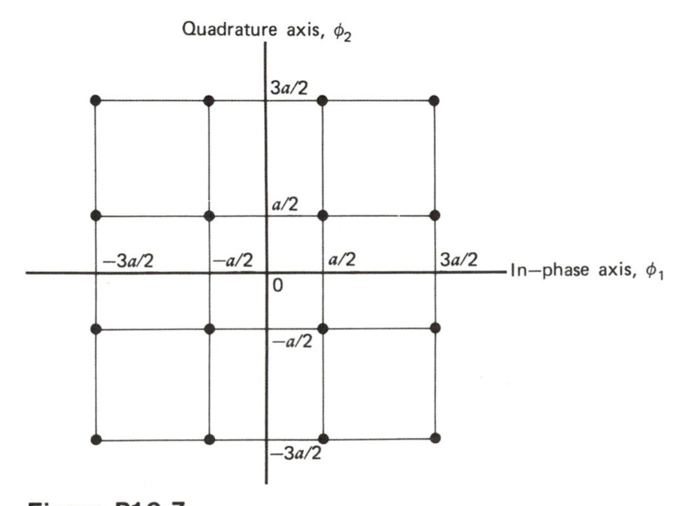

Figure P10.7

(b) By how many decibels do the highest and lowest transmitted amplitude levels differ?
(c) The 16-level QAM system transmits information at a rate equal to that achievable by a combination of two QPSK systems and an 8-level PSK system. Identify the message points for these three PSK systems.
(d) Each message point in Fig. P10.7 may be considered to correspond to a specific four-bit code word. Assuming that the encoding rule is such that the words corresponding to adjacent message points (along horizontal or vertical lines) differ by only one bit, find a set of four-bit code words for the 16 message points that satisfy this requirement.
(e) Construct the optimum decision regions.
(f) The 16-level QAM wave may be considered to consist of two components, an in-phase component and a quadrature component, with each component taking on one of four possible amplitude levels. Construct the signal-space diagram for each of these two components.
(g) Express the average probability of symbol error in terms of the *average* signal-to-noise density ratio.
(h) Develop an expression for the 16-level QAM wave in terms of its in-phase and quadrature components. Hence, develop block diagrams for the 16-level QAM transmitter and receiver.

Problem 10.32

(a) Sketch the waveforms of the in-phase and quadrature components of the MSK signal in response to the input binary sequence 1100100010.
(b) Sketch the MSK waveform itself for the binary sequence specified in part (a).

Problem 10.33 An M-ary digital communication system is used to transmit binary data, with each symbol consisting of n bits, so that the number of possible symbols $M=2^n$. Show that the *average probability of symbol error*, P_e, and the *average probability of bit error*, P_{eb}, are related by

$$P_{eb}=\frac{M}{2(M-1)}P_e$$

Problem 10.34 In a special form of quadriphase-shift keying known as the *offset QPSK*, the in-phase data stream is delayed relative to the quadrature data stream by half a symbol period $T/2$.

(a) What is the average probability of symbol error for an offset QPSK system?
(b) What is the power spectral density of an offset QPSK signal produced by a random binary sequence in which symbols 1 and 0 (represented by ± 1 volt) are equally likely, and the symbols in different time slots are statistically independent and identically distributed?
(c) What are the similarities between the offset QPSK and MSK, and what features distinguish them?

Problem 10.35 Assuming an average probability of symbol error equal to 10^{-4}, compare the performances of binary coherent PSK, coherent FSK, QPSK, and MSK with the ideal system as defined by the Hartley-Shannon law. For this evaluation, plot the bandwidth efficiency versus E_b/N_0. Which method of modulation comes closest to the ideal system?

Note: For the Hartley–Shannow law, see Section 8.4.

Problem 10.36 The *noise equivalent bandwidth* of a band-pass signal is defined as the value of bandwidth which satisfies the relation

$$4BS(f_c)=P$$

where $2B$ = noise equivalent bandwidth centered around the mid-band frequency f_c.
$S(f_c)$ = maximum value of the power spectral density of the signal at $f=f_c$.
P = average power of the signal.
Show that the noise equivalent bandwidths of binary PSK, QPSK, and MSK are as follows:

TYPE OF MODULATION	NOISE BANDWIDTH/BIT RATE
Binary PSK	1.0
QPSK	0.5
MSK	0.62

Note: You may use the definite integrals in Table A6.6 of Appendix 6.

Problem 10.37 There are two ways of detecting an MSK signal. One way is to use a coherent receiver to take full account of the phase information content of the MSK signal. Another way is to use a noncoherent receiver and disregard the phase information. The second method offers the advantage of simplicity of implementation, at the expense of a degraded noise performance. By how many decibels do we have to increase the bit energy-to-noise density ratio, E_b/N_0, in the second case so as to realize an average probability of symbol error equal to 10^{-5} in both cases?

Problem 10.38 Figure P10.8(*a*) shows a noncoherent receiver using a matched filter for the detection of a sinusoidal signal of known frequency but random phase, in the presence of additive white Gaussian noise. An alternative implementation of this receiver is its mechanization in the frequency domain as a *spectrum analyzer receiver*, as in Fig. P10.8(*b*), where the correlator computes the finite time autocorrelation function $R_X(\tau)$ defined by

$$R_X(\tau)=\int_0^{T-\tau} x(t)x(t+\tau)dt, \qquad 0\leqslant\tau\leqslant T$$

Show that the square-law envelope detector output sampled at time $t=T$ in Fig. P10.8(*a*) is twice the spectral output of the Fourier transformer sampled at frequency $f=f_c$ in Fig. P10.8(*b*)

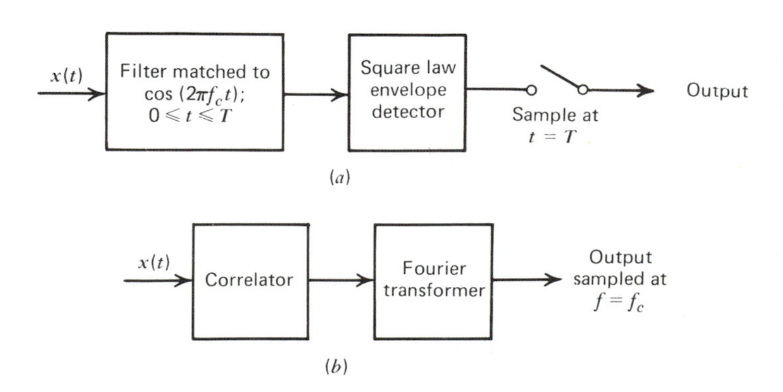

Figure P10.8

Problem 10.39 The binary sequence 1100100010 is applied to the DPSK transmitter of Fig. 10.38(*a*)

(a) Sketch the resulting waveform at the transmitter output.
(b) Applying this waveform to the DPSK receiver of Fig. 10.38(*b*), show that, in the absence of noise, the original binary sequence is reconstructed at the receiver output.

Problem 10.40 Let $s_1(t)$ denote the transmitted DPSK signal, for $0 \leqslant t \leqslant 2T_b$, for the case when we have symbol 1 at the transmitter input for $T_b \leqslant t \leqslant 2T_b$. Similarly, let $s_2(t)$ denote the transmitted DPSK signal, for $0 \leqslant t \leqslant 2T_b$, for the case when we have symbol 0 at the transmitter input for $T_b \leqslant t \leqslant 2T_b$.

(a) By examining all possible phase combinations for the interval $0 \leqslant t \leqslant 2T_b$, show that $s_1(t)$ and $s_2(t)$ are always orthogonal over this interval.
(b) Hence, by using the formula for the average probability of error for a noncoherent binary FSK system, find the average probability of error for a DPSK system.

Problem 10.41 Consider a signal of the form

$$s(t, A) = \begin{cases} As(t), & 0 \leqslant t \leqslant T \\ 0, & \text{elsewhere} \end{cases}$$

where $s(t)$ is completely known and the amplitude A is unknown. Find the maximum likelihood estimate of A in the presence of white Gaussian noise of zero mean and power spectral density $N_0/2$. What are the mean and variance of this estimate?

Appendix 1

CONTINUOUS PROBABILITY DISTRIBUTIONS

In this appendix we list the probability density function, distribution function, mean, and variance of some continuous random variables that are frequently encountered. In each case, the parameter a is a positive constant. Also, the probability density function in each case is illustrated.

1 UNIFORM DISTRIBUTION

$$f_X(x)=\begin{cases} \dfrac{1}{2a}, & -a\leqslant x\leqslant a \\ 0, & \text{elsewhere} \end{cases}$$

$$F_X(x)=\begin{cases} 0, & x<-a \\ \dfrac{1}{2a}(x+a), & -a\leqslant x\leqslant a \\ 0, & x>a \end{cases}$$

$$\text{Mean}=0$$

$$\text{Variance}=\frac{a^2}{3}$$

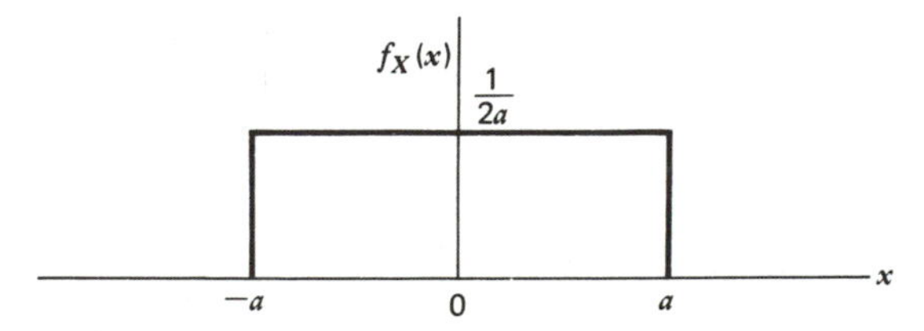

Figure A1.1 Uniform distribution.

2 GAUSSIAN DISTRIBUTION

$$f_X(x)=\frac{1}{\sqrt{2\pi}a}\exp\left(-\frac{x^2}{2a^2}\right), \qquad -\infty<x<\infty$$

$$F_X(x)=\frac{1}{2}\left[1+\operatorname{erf}\left(\frac{x}{\sqrt{2}a}\right)\right], \qquad -\infty<x<\infty$$

Mean $=0$
Variance $=a^2$

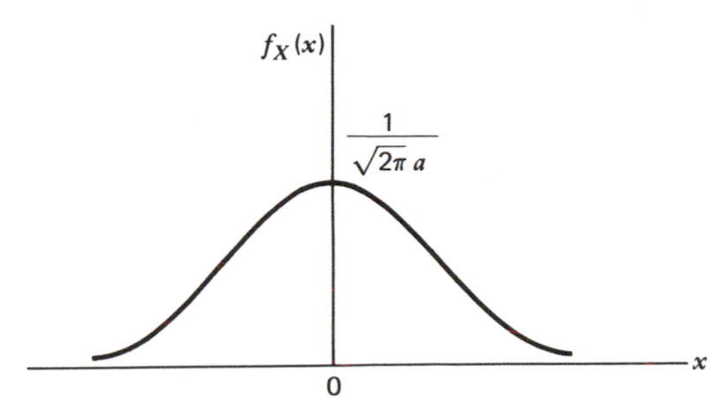

Figure A1.2 Gaussian distribution.

3 RAYLEIGH DISTRIBUTION

$$f_X(x)=\begin{cases}\dfrac{x}{a^2}\exp\left(-\dfrac{x^2}{2a^2}\right), & x\geqslant 0\\[2ex] 0, & x<0\end{cases}$$

$$F_X(x)=\begin{cases}0, & x<0\\[2ex] 1-\exp\left(-\dfrac{x^2}{2a^2}\right), & x\geqslant 0\end{cases}$$

Mean $=\sqrt{\dfrac{a^2\pi}{2}}$

Variance $=\left(2-\dfrac{\pi}{2}\right)a^2$

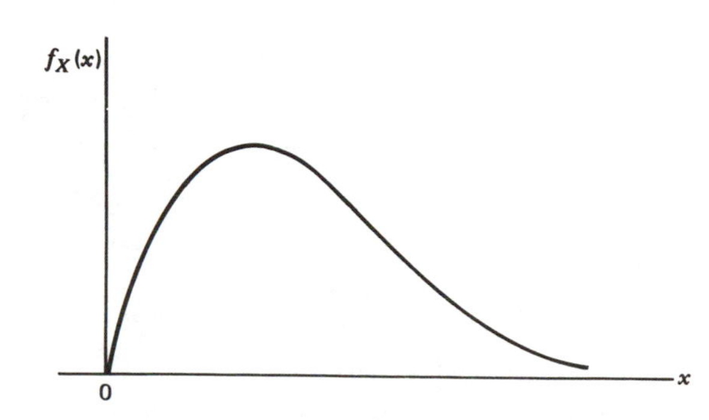

Figure A1.3 Rayleigh distribution.

4 EXPONENTIAL DISTRIBUTION

$$f_X(x) = \begin{cases} \dfrac{1}{a}\exp\left(-\dfrac{x}{a}\right), & x \geqslant 0 \\ 0, & x < 0 \end{cases}$$

$$F_X(x) = \begin{cases} 0, & x < 0 \\ 1 - \exp\left(-\dfrac{x}{a}\right), & x \geqslant 0 \end{cases}$$

Mean $= a$
Variance $= a^2$

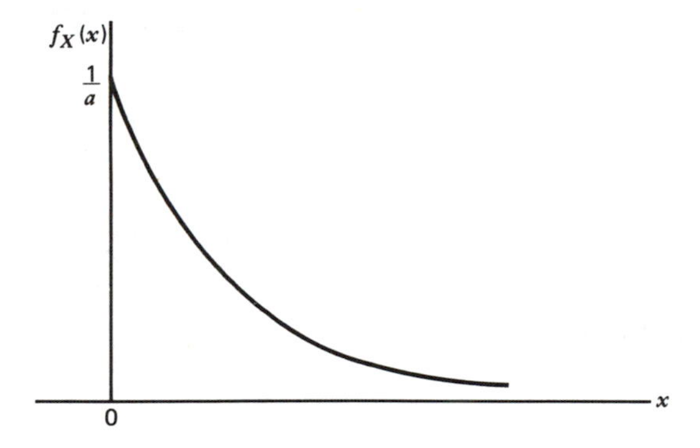

Figure A1.4 Exponential distribution.

Appendix 2
ERROR FUNCTION

The *error function*, denoted by erf(u), is defined in a number of different ways in the literature. We shall use the following definition*

$$\operatorname{erf}(u)=\frac{2}{\sqrt{\pi}}\int_0^u \exp(-z^2)dz \tag{A2.1}$$

The error function has two useful properties:

1. $\operatorname{erf}(-u)=-\operatorname{erf}(u)$ (A2.2)

   This is known as the *symmetry relation.*

2. As u approaches infinity, erf(u) approaches unity; that is

$$\frac{2}{\sqrt{\pi}}\int_0^\infty \exp(-z^2)dz=1 \tag{A2.3}$$

The *complementary error function* is defined by

$$\operatorname{erfc}(u)=\frac{2}{\sqrt{\pi}}\int_u^\infty \exp(-z^2)dz \tag{A2.4}$$

It is related to the error function as follows

$$\operatorname{erfc}(u)=1-\operatorname{erf}(u) \tag{A2.5}$$

Table A2.1 gives values of the error function erf(u) for u in the range 0 to 3.

* The error function is tabulated extensively in several references; see, for example, M. Abramowitz and I. A. Stegun, *Handbook of Mathematical Functions*, pp. 297–316 (Dover Publications, 1965).

Table A2.1 The Error Function

u	erf (u)	u	erf (u)
0.00	0.00000	1.10	0.88021
0.05	0.05637	1.15	0.89612
0.10	0.11246	1.20	0.91031
0.15	0.16800	1.25	0.92290
0.20	0.22270	1.30	0.93401
0.25	0.27633	1.35	0.94376
0.30	0.32863	1.40	0.95229
0.35	0.37938	1.45	0.95970
0.40	0.42839	1.50	0.96611
0.45	0.47548	1.55	0.97162
0.50	0.52050	1.60	0.97635
0.55	0.56332	1.65	0.98038
0.60	0.60386	1.70	0.98379
0.65	0.64203	1.75	0.98667
0.70	0.67780	1.80	0.98909
0.75	0.71116	1.85	0.99111
0.80	0.74210	1.90	0.99279
0.85	0.77067	1.95	0.99418
0.90	0.79691	2.00	0.99532
0.95	0.82089	2.50	0.99959
1.00	0.84270	3.00	0.99998
1.05	0.86244		

BOUNDS OF THE COMPLEMENTARY ERROR FUNCTION*

Substituting $u-x$ for z in Eq. (A2.4), we get

$$\operatorname{erfc}(u)=\frac{2}{\sqrt{\pi}}\exp(-u^2)\int_{-\infty}^{0}\exp(2ux)\exp(-x^2)dx$$

For any real x, the value of $\exp(-x^2)$ lies between the successive partial sums of the power series

$$1-\frac{x^2}{1!}+\frac{(x^2)^2}{2!}-\frac{(x^2)^3}{3!}+\cdots$$

Therefore, for $u>0$, we find, on using $(n+1)$ terms of this series, that $\operatorname{erfc}(u)$ lies between the values taken by

$$\frac{2}{\sqrt{\pi}}\exp(-u^2)\int_{-\infty}^{0}\left(1-x^2+\frac{x^4}{2}-\cdots\pm\frac{x^{2n}}{n!}\right)\exp(2ux)dx$$

for even n and for odd n. Putting $2ux=-v$, and using the integral

$$\int_{0}^{\infty}v^n\exp(-v)dv=n!$$

we obtain the following *asymptotic expansion* for $\operatorname{erfc}(u)$, assuming $u>0$,

$$\operatorname{erfc}(u)\simeq\frac{\exp(-u^2)}{\sqrt{\pi}u}\left[1-\frac{1}{2u^2}+\frac{1\cdot 3}{2^2u^4}-\cdots\pm\frac{1\cdot 3\cdot 5\cdots(2n-1)}{2^n u^{2n}}\right] \tag{A2.6}$$

For large positive values of u, the successive terms of the series on the right-hand side of Eq. (A2.6) decrease very rapidly. We thus deduce two simple bounds on $\operatorname{erfc}(u)$, as shown by

$$\frac{\exp(-u^2)}{\sqrt{\pi}u}\left(1-\frac{1}{2u^2}\right)<\operatorname{erfc}(u)<\frac{\exp(-u^2)}{\sqrt{\pi}u} \tag{A2.7}$$

For large positive u, a second bound on the complementary error function $\operatorname{erfc}(u)$ is obtained by omitting the multiplying factor $1/u$ in the upper bound of Eq. (A2.7):

$$\operatorname{erfc}(u)<\frac{\exp(-u^2)}{\sqrt{\pi}} \tag{A2.8}$$

In Fig. A2.1, we have plotted $\operatorname{erfc}(u)$, the two bounds defined by Eq. (A2.7), and the upper bound of Eq. (A2.8). We see that for $u\geqslant 1.5$ the bounds on $\operatorname{erfc}(u)$, defined by Eq. (A2.7), become increasingly tight.

*N. M. Blachman, *Noise and Its Effect on Communication*, pp. 5–6 (McGraw-Hill, 1966).

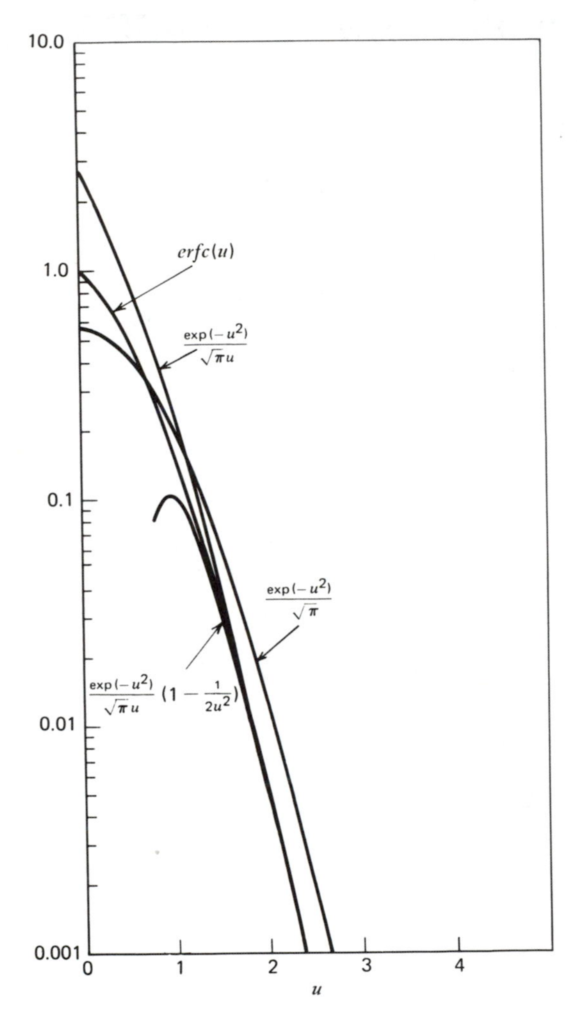

Figure A2.1 The complementary error function and its bounds.

Appendix 3

NOISE FIGURE

A convenient measure of the noise performance of a linear two-port device is furnished by the so-called *noise figure*.* Consider a linear two-port device connected to a signal source of internal impedance $Z(f)=R(f)+jX(f)$ at the input, as in Fig. A3.1. The noise voltage $v(t)$ represents the thermal noise associated with the internal resistance $R(f)$ of the source. The output noise of the device is made up of two contributions, one due to the source and the other due to the device itself. *We define the available output noise power in a band of width Δf centered at frequency f as the maximum average noise power in this band, obtainable at the output of the device.* The maximum noise power which the two-port device can deliver to an external load is obtained when the load impedance is the complex conjugate of the output impedance of the device, that is, when the resistance is matched and the reactance is tuned out. *We define the noise figure of the two-port device as the ratio of the total available output noise power (due to the device and the source) per unit bandwidth to the portion thereof due solely to the source.*

Let the spectral density of the total available noise power of the device output be $S_{NO}(f)$, and the spectral density of the available noise power due to the source at the device input be $S_{NS}(f)$. Also let $G(f)$ denote the *available power gain of the two-port device, defined as the ratio of the available signal power at the output of the device to the available signal power of the source when the signal is a sinusoidal wave of frequency f.* Then we may express the noise figure F of the device as

$$F=\frac{S_{NO}(f)}{G(f)S_{NS}(f)} \tag{A3.1}$$

If the device were noise free, $S_{NO}(f)=G(f)S_{NS}(f)$, and the noise figure would then be unity. In a physical device, however, $S_{NO}(f)$ is larger than $G(f)S_{NS}(f)$, so that the noise figure is always larger than unity. The noise figure is commonly expressed in decibels, that is, as $10 \log_{10} F$.

The noise figure may also be expressed in an alternative form. Let $P_S(f)$ denote the available signal power from the source, which is the maximum average signal power that can be obtained. For the case of a source providing a single-frequency signal component with open-circuit voltage $V_0 \cos(2\pi ft)$, the available signal power

* H. T. Friis, "Noise figures in radio receivers," *Proc. IRE*, vol. 32, pp. 419–422, July 1944.

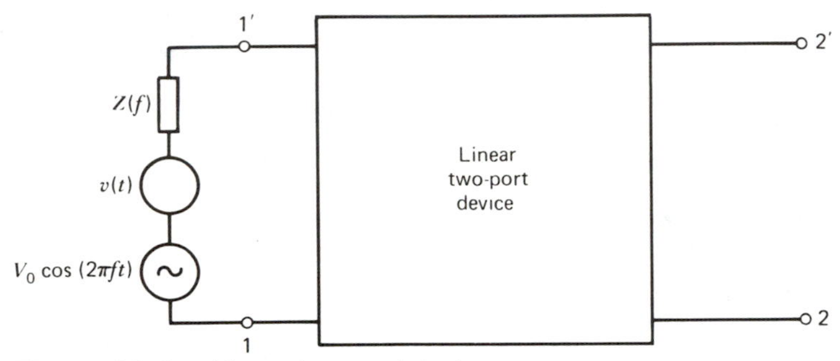

Figure A3.1 Linear two-port device.

is obtained when the load connected to the source is $Z^*(f)=R(f)-jX(f)$, yielding the value

$$P_S(f)=\left[\frac{V_0}{2R(f)}\right]^2 R(f)$$

$$=\frac{V_0^2}{4R(f)} \tag{A3.2}$$

The available signal power at the output of the device is therefore,

$$P_O(f)=G(f)P_S(f) \tag{A3.3}$$

Then, multiplying both the numerator and denominator of the right-hand side of Eq. (A3.1) by $P_S(f)\,\Delta(f)$, we have

$$F=\frac{P_S(f)S_{NO}(f)\,\Delta f}{G(f)P_S(f)S_{NS}(f)\,\Delta f}$$

$$=\frac{P_S(f)S_{NO}(f)\,\Delta f}{P_O(f)S_{NS}(f)\,\Delta f}$$

$$=\frac{\rho_S(f)}{\rho_O(f)} \tag{A3.4}$$

where

$$\rho_S(f)=\frac{P_S(f)}{S_{NS}(f)\Delta f} \tag{A3.5}$$

$$\rho_O(f)=\frac{P_O(f)}{S_{NO}(f)\,\Delta f} \tag{A3.6}$$

We refer to $\rho_S(f)$ as the *available signal-to-noise ratio of the source* and to $\rho_O(f)$ as the *available signal-to-noise ratio at the device output*, both measured in a narrow band of width Δf centered at f. Since the noise figure is always greater than unity, it follows from Eq. (A3.4) that the signal-to-noise ratio always decreases with amplification, which is a significant result.

The noise figure F, as defined above, is a function of the operating frequency and hence is referred to as the *spot noise figure*. In contrast, we may define an *average noise figure* F_0 of a two-port device as the ratio of the total noise power at the device output to the output noise power due solely to the source. That is,

$$F_0 = \frac{\int_{-\infty}^{\infty} S_{NO}(f)\,df}{\int_{-\infty}^{\infty} G(f) S_{NS}(f)\,df} \tag{A3.7}$$

It is apparent that in the case of thermal noise in the input circuit with $R(f)$ constant, and constant gain throughout a fixed band with zero gain at other frequencies, the spot noise figure F and the average noise figure F_0 are identical.

EQUIVALENT NOISE TEMPERATURE

A disadvantage of the noise figure F is that when it is used to compare low-noise devices, the values obtained are all close to unity, which makes the comparison rather difficult. In such cases, it is preferable to use the *equivalent noise temperature*. Consider a linear two-port device whose input resistance is matched to the internal resistance of the source as shown in Fig. A3.2. In this diagram, we have also included the noise voltage generator associated with the internal resistance R_s of the source. The mean-square value of this noise voltage is $4kTR_s\Delta f$. Hence, the available noise power at the device input is

$$N_1 = kT\,\Delta f \tag{A3.8}$$

Let N_d denote the noise power contributed by the two-port device to the total available output noise power N_2. We define N_d as

$$N_d = GkT_e\,\Delta f \tag{A3.9}$$

where G is the available power gain of the device and T_e is its equivalent noise

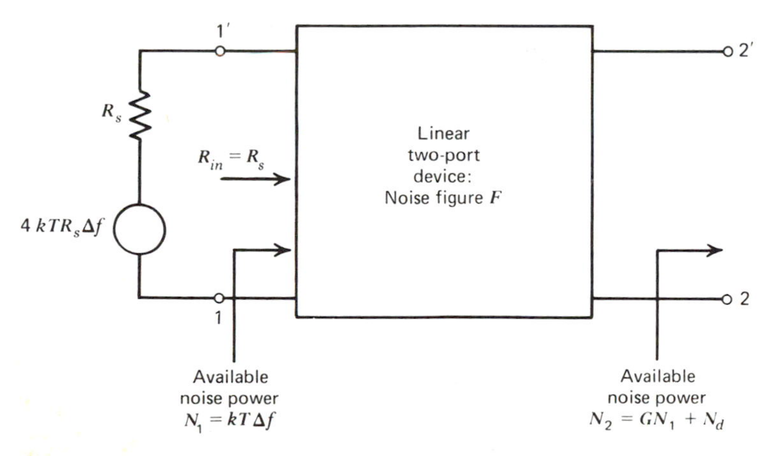

Figure A3.2 Linear two-port device matched to the internal resistance of a source connected to the input.

temperature. Then it follows that the total output noise power is

$$\begin{aligned} N_2 &= GN_1 + N_d \\ &= Gk(T + T_e)\,\Delta f \end{aligned} \tag{A3.10}$$

The noise figure of the device is therefore

$$F = \frac{T + T_e}{T} \tag{A3.11}$$

Solving for the equivalent noise temperature:

$$T_e = T(F - 1) \tag{A3.12}$$

where F is the noise figure of the device measured under matched input conditions, and with the noise source at temperature T.

NOISE FIGURE OF COMPOSITE TWO-PORT NETWORKS

It is often necessary to evaluate the noise figure of a cascade connection of two-port networks whose individual noise figures are known. Consider Fig. A3.3 consisting of a pair of two-port networks of noise figures F_1 and F_2 and power gains G_1 and G_2, connected in cascade. It is assumed that the devices are matched, and that the noise figure F_2 of the second network is defined assuming an input noise power N_1.

At the input of the first network we have a noise power N_1 contributed by the source, plus an equivalent noise power $(F_1 - 1)N_1$ contributed by the network itself. The output noise power from the first network is therefore $F_1 N_1 G_1$. Added to this noise power at the input of the second network, we have the equivalent extra power $(F_2 - 1)N_1$ contributed by the second network itself. The output noise power from this network is therefore equal to $F_1 G_1 N_1 G_2 + (F_2 - 1)N_1 G_2$. We may consider the noise figure F as the ratio of the actual output noise power to the output noise power assuming the networks to be noiseless. We may therefore express the overall noise figure of the cascade connection of Fig. A3.3 as

$$\begin{aligned} F &= \frac{F_1 G_1 N_1 G_2 + (F_2 - 1)N_1 G_2}{N_1 G_1 G_2} \\ &= F_1 + \frac{F_2 - 1}{G_1} \end{aligned} \tag{A3.13}$$

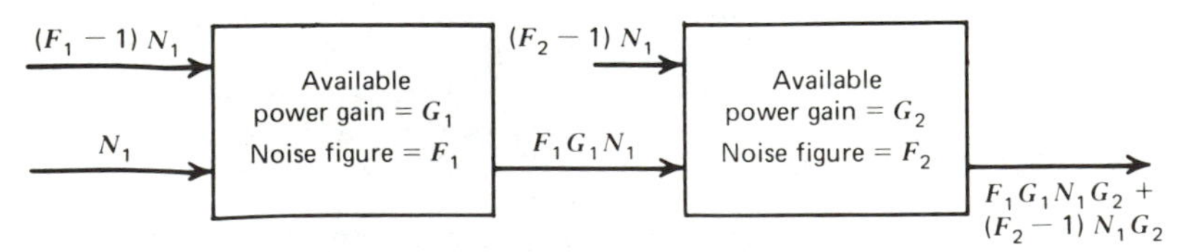

Figure A3.3 A cascade of two noisy two-port networks.

The result may be readily extended to the cascade connection of any number of two-port networks, as shown by

$$F = F_1 + \frac{F_2 - 1}{G_1} + \frac{F_3 - 1}{G_1 G_2} + \frac{F_4 - 1}{G_1 G_2 G_3} + \cdots \tag{A3.14}$$

where F_1, F_2, $F_3, \ldots$ are the individual noise figures, and G_1, G_2, $G_3, \ldots$ are the individual available power gains. Equation (A3.14) shows that if the first stage of the cascade connection in Fig. A3.3 has a high gain, the overall noise figure F is practically the same as that of the first stage.

Appendix 4
BESSEL FUNCTIONS

A Bessel function of the first kind of order n and argument x, commonly denoted by $J_n(x)$, is defined by*

$$J_n(x) = \frac{1}{2\pi} \int_{-\pi}^{\pi} \exp(jx \sin\theta - jn\theta)\, d\theta \tag{A4.1}$$

or, equivalently,

$$J_n(x) = \frac{1}{\pi} \int_{0}^{\pi} \cos(x \sin\theta - n\theta)\, d\theta \tag{A4.2}$$

Just as the trigonometric functions can be expanded in power series, so can the Bessel function $J_n(x)$ be expanded in a power series:

$$J_n(x) = \sum_{m=0}^{\infty} \frac{(-1)^m (\frac{1}{2}x)^{n+2m}}{m!(n+m)!} \tag{A4.3}$$

In particular, for $n=0$, we have

$$J_0(x) = 1 - \frac{x^2}{2^2} + \frac{x^4}{2^2 \cdot 4^2} - \frac{x^6}{2^2 \cdot 4^2 \cdot 6^2} + \cdots \tag{A4.4}$$

for $n=1$,

$$J_1(x) = \frac{x}{2} - \frac{x^3}{2^2 \cdot 4} + \frac{x^5}{2^2 \cdot 4^2 \cdot 6} - \cdots \tag{A4.5}$$

and for $n=2$,

$$J_2(x) = \frac{x^2}{2 \cdot 4} - \frac{x^4}{2^2 \cdot 4 \cdot 6} + \frac{x^6}{2^2 \cdot 4^2 \cdot 6 \cdot 8} - \cdots \tag{A4.6}$$

and so on for higher values of n.

* 1. G. N. Watson, *A Treatise on the Theory of Bessel Functions*, Second Edition (Cambridge University Press, 1966).
  2. L. A. Pipes, *Applied Mathematics for Engineers and Physicists*, Second Edition (McGraw-Hill, 1958).

The Bessel function $J_n(x)$ has the following properties:

1.
$$J_n(x)=(-1)^n J_{-n}(x) \tag{A4.7}$$

To prove this relation, we replace θ by $(\pi-\theta)$ in Eq. (A4.2). Then, noting that $\sin(\pi-\theta)=\sin\theta$, we get

$$J_n(x)=\frac{1}{\pi}\int_0^\pi \cos(x\sin\theta+n\theta-n\pi)d\theta$$
$$=\frac{1}{\pi}\int_0^\pi [\cos(n\pi)\cos(x\sin\theta+n\theta)+\sin(n\pi)\sin(x\sin\theta+n\theta)]d\theta$$

For integer values of n, we have

$$\cos(n\pi)=(-1)^n$$
$$\sin(n\pi)=0$$

Therefore,

$$J_n(x)=\frac{(-1)^n}{\pi}\int_0^\pi \cos(x\sin\theta+n\theta)d\theta \tag{A4.8}$$

From Eq. (A4.2), we also find that by replacing n with $-n$

$$J_{-n}(x)=\frac{1}{\pi}\int_0^\pi \cos(x\sin\theta+n\theta)d\theta \tag{A4.9}$$

The desired result follows immediately from Eqs. (A4.8) and (A4.9).

2.
$$J_n(x)=(-1)^n J_n(-x) \tag{A4.10}$$

This relation is obtained by replacing x with $-x$ in Eq. (A4.2), and then using Eq. (A4.8)

3.
$$J_{n-1}(x)+J_{n+1}(x)=\frac{2n}{x}J_n(x) \tag{A4.11}$$

This *recurrence formula* is useful in constructing tables of Bessel coefficients.

4. For small values of x, we have

$$J_n(x)\simeq\frac{x^n}{2^n n!} \tag{A4.12}$$

This relation is obtained simply by retaining the first term in the power series of Eq. (A4.3) and ignoring the higher-order terms. Thus, when x is small, we have

$$J_0(x)\simeq 1$$
$$J_1(x)\simeq\frac{x}{2} \tag{A4.13}$$
$$J_n(x)\simeq 0, \qquad \text{for } n>1$$

5. For large values of x, we have

$$J_n(x) \simeq \sqrt{\frac{2}{\pi x}} \cos\left(x - \frac{\pi}{4} - \frac{n\pi}{2}\right) \tag{A4.14}$$

This shows that for large values of x, the Bessel function $J_n(x)$ behaves like a sine wave with progressively decreasing amplitude.

6. With x real and fixed, $J_n(x)$ approaches zero as the order n goes to infinity.

7.
$$\sum_{n=-\infty}^{\infty} J_n(x)\exp(jn\phi) = \exp(jx \sin \phi) \tag{A4.15}$$

To prove this property, consider the sum $\sum_{n=-\infty}^{\infty} J_n(x)\exp(jn\phi)$ and use the formula of Eq. (A4.1) for $J_n(x)$ to obtain

$$\sum_{n=-\infty}^{\infty} J_n(x)\exp(jn\phi) = \frac{1}{2\pi} \sum_{n=-\infty}^{\infty} \exp(jn\phi) \int_{-\pi}^{\pi} \exp(jx \sin \theta - jn\theta) d\theta$$

Interchanging the order of integration and summation:

$$\sum_{n=-\infty}^{\infty} J_n(x)\exp(jn\phi) = \frac{1}{2\pi} \int_{-\pi}^{\pi} d\theta \exp(jx \sin \theta) \sum_{n=-\infty}^{\infty} \exp[jn(\phi - \theta)] \tag{A4.16}$$

From Example 19 of Chapter 2, we note that

$$\delta(\phi - \theta) = \frac{1}{2\pi} \sum_{n=-\infty}^{\infty} \exp[jn(\phi - \theta)], \qquad -\pi \leqslant \phi - \theta \leqslant \pi \tag{A4.17}$$

Therefore, substituting Eq. (A4.17) in (A4.16), and using the sifting property of a delta function, we get

$$\sum_{n=-\infty}^{\infty} J_n(x)\exp(jn\phi) = \int_{-\pi}^{\pi} \exp(jx \sin \theta)\delta(\phi - \theta) d\theta$$

$$= \exp(jx \sin \phi)$$

which is the desired result.

8.
$$\sum_{n=-\infty}^{\infty} J_n^2(x) = 1, \qquad \text{for all } x \tag{A4.18}$$

To prove this property, we may proceed as follows. We observe that $J_n(x)$ is real. Hence, multiplying Eq. (A4.1) by its complex conjugate, and summing over all possible values of n, we get

$$\sum_{n=-\infty}^{\infty} J_n^2(x) = \frac{1}{(2\pi)^2} \sum_{n=-\infty}^{\infty} \int_{-\pi}^{\pi} \int_{-\pi}^{\pi} \exp(jx \sin \theta - jn\theta - jx \sin \phi + jn\phi) d\theta \, d\phi$$

Interchanging the order of double integration and summation:

$$\sum_{n=-\infty}^{\infty} J_n^2(x) = \frac{1}{(2\pi)^2} \int_{-\pi}^{\pi} \int_{-\pi}^{\pi} d\theta \, d\phi \, exp[jx(\sin \theta - \sin \phi)] \sum_{n=-\infty}^{\infty} \exp[jn(\phi - \theta)] \tag{A4.19}$$

Substituting Eq. (A4.17) in (A4.19), and using the sifting property of a delta function, we finally get

$$\sum_{n=-\infty}^{\infty} J_n^2(x) = \frac{1}{2\pi} \int_{-\pi}^{\pi} d\theta = 1$$

which is the desired result.

Many of these properties of the Bessel function $J_n(x)$ may also be verified by referring to the plots of Fig. 4.5 or Table A4.1.* Note that the β in Fig. 4.5 corresponds to the variable x as used in this appendix.

MODIFIED BESSEL FUNCTION OF THE FIRST KIND

Suppose in Eq. (A4.15) we replace jx with x, the angle ϕ with $\theta - \pi/2$, and define a new function $I_n(x)$ that is related to $J_n(x)$ as follows:

$$I_n(x) = j^{-n} J_n(jx) \tag{A4.20}$$

We thus obtain

$$\sum_{n=-\infty}^{\infty} I_n(x) \exp(jn\theta) = \exp(x \cos\theta) \tag{A4.21}$$

From this relation it follows that

$$I_n(x) = \frac{1}{2\pi} \int_{-\pi}^{\pi} \exp(x\cos\theta)\cos(n\theta)\, d\theta \tag{A4.22}$$

The function $I_n(x)$ defined in this manner is a real function of x, and it is known as the *modified Bessel function of the first kind of order n* and argument x.†

In contrast to the Bessel function $J_n(x)$, the function $I_n(x)$ is not of the oscillatory type, but rather its behavior is similar to that of an exponential function. The function $I_n(x)$ can be expressed in the form of a power series as:

$$I_n(x) = \sum_{m=0}^{\infty} \frac{(\frac{1}{2}x)^{n+2m}}{m!(m+n)!} \tag{A4.23}$$

For the special case of $n = 0$, we find from Eq. (A4.22) that

$$I_0(x) = \frac{1}{2\pi} \int_{-\pi}^{\pi} \exp(x\cos\theta)\, d\theta \tag{A4.24}$$

* For more extensive tables of Bessel functions, see:
1. G. N. Watson, pp. 666–697, *op. cit.*
2. M. Abramowitz and I. A. Stegun, *Handbook of Mathematical Functions*, pp. 358–406 (Dover Publications, 1965).

† For tabulations of the modified Bessel function of the first kind, see:
1. G. N. Watson, pp. 698–713, *op. cit.*
2. M. Abramowitz and I. A. Stegun, pp. 416–422, *op. cit.*

For small values of x,

$$I_0(x) \simeq 1, \tag{A4.25}$$

while for large values of x,

$$I_0(x) \simeq \frac{\exp(x)}{\sqrt{2\pi x}} \tag{A4.26}$$

Table A4.1
Table of Bessel Functions

$J_n(x)$

n \ x	0.5	1	2	3	4	6	8	10	12
0	0.9385	0.7652	0.2239	−0.2601	−0.3971	0.1506	0.1717	−0.2459	0.0477
1	0.2423	0.4401	0.5767	0.3391	−0.0660	−0.2767	0.2346	0.0435	−0.2234
2	0.0306	0.1149	0.3528	0.4861	0.3641	−0.2429	−0.1130	0.2546	−0.0849
3	0.0026	0.0196	0.1289	0.3091	0.4302	0.1148	−0.2911	0.0584	0.1951
4	0.0002	0.0025	0.0340	0.1320	0.2811	0.3576	−0.1054	−0.2196	0.1825
5	—	0.0002	0.0070	0.0430	0.1321	0.3621	0.1858	−0.2341	−0.0735
6		–	0.0012	0.0114	0.0491	0.2458	0.3376	−0.0145	−0.2437
7			0.0002	0.0025	0.0152	0.1296	0.3206	0.2167	−0.1703
8			—	0.0005	0.0040	0.0565	0.2235	0.3179	0.0451
9				0.0001	0.0009	0.0212	0.1263	0.2919	0.2304
10				—	0.0002	0.0070	0.0608	0.2075	0.3005
11					—	0.0020	0.0256	0.1231	0.2704
12						0.0005	0.0096	0.0634	0.1953
13						0.0001	0.0033	0.0290	0.1201
14						—	0.0010	0.0120	0.0650

Appendix 5
SCHWARZ'S INEQUALITY

Let $g_1(t)$ and $g_2(t)$ be functions of the real variable t in the interval $a \leqslant t \leqslant b$. We assume that $g_1(t)$ and $g_2(t)$ satisfy the conditions

$$\int_a^b |g_1(t)|^2 \, dt < \infty \tag{A5.1}$$

$$\int_a^b |g_2(t)|^2 \, dt < \infty \tag{A5.2}$$

Then, according to *Schwarz's inequality*, we have

$$\left| \int_a^b g_1(t) g_2(t) dt \right|^2 \leqslant \int_a^b |g_1(t)|^2 \, dt \int_a^b |g_2(t)|^2 \, dt \tag{A5.3}$$

To prove this inequality, form the integral*

$$\int_a^b [\lambda g_1^*(t) + g_2^*(t)][\lambda g_1(t) + g_2(t)] dt = \lambda^2 A + \lambda(B + B^*) + C \tag{A5.4}$$

where λ is a real variable, the asterisk signifies complex conjugate, and

$$A = \int_a^b |g_1(t)|^2 \, dt \geqslant 0 \tag{A5.5}$$

$$B = \int_a^b g_1^*(t) g_2(t) dt \tag{A5.6}$$

$$C = \int_a^b |g_2(t)|^2 \, dt \geqslant 0 \tag{A5.7}$$

The integral of Eq. (A5.4) exists, is real, and is a nonnegative function of λ, say $f(\lambda)$. Since $f(\lambda)$ is nonnegative, it must have no real roots except possibly a double root. From the quadratic formula, we must then have

$$(B + B^*)^2 \leqslant 4AC \tag{A5.8}$$

Note that $(B + B^*)/2$ is equal to the real part of B. On substituting Eqs. (A5.5)

* Our proof follows: J. B. Thomas, *An Introduction to Statistical Communication Theory*, pp. 619–620 (Wiley, 1969).

to (A5.7) in (A5.8), we get

$$\left\{\int_a^b [g_1^*(t)g_2(t)+g_1(t)g_2^*(t)]dt\right\}^2 \leqslant 4\int_a^b |g_1(t)|^2\,dt \int_a^b |g_2(t)|^2\,dt \tag{A5.9}$$

This is the form of Schwarz's inequality that is appropriate for complex functions $g_1(t)$ and $g_2(t)$. For the case when both $g_1(t)$ and $g_2(t)$ are real, we have

$$g_1^*(t)g_2(t)+g_1(t)g_2^*(t)=2g_1(t)g_2(t) \tag{A5.10}$$

and Eq. (A5.3) follows immediately.

Note that equality is obtained [aside from the trivial case where both $g_1(t)$ and $g_2(t)$ are zero] when the double root exists in Eq. (A5.4); that is, when

$$\lambda g_1(t)+g_2(t)=\lambda g_1^*(t)+g_2^*(t)=0 \tag{A5.11}$$

Since λ is real, then $g_1(t)$ and $g_2(t)$ are linearly related. Looking at the problem from a slightly different viewpoint, we see that there is a real value of λ for which Eq. (A5.4) is zero and for which its first derivative with respect to λ vanishes; that is,

$$2\lambda A+(B+B^*)=0 \tag{A5.12}$$

or

$$\begin{aligned}\lambda &= -\frac{B+B^*}{2A}\\ &= -\frac{\int_a^b [g_1^*(t)g_2(t)+g_1(t)g_2^*(t)]dt}{2\int_a^b |g_1(t)|^2\,dt}\end{aligned} \tag{A5.13}$$

This relation holds if and only if

$$g_1(t)=-\lambda g_2(t) \tag{A5.14}$$

This last relationship is equivalent to Eq. (A5.11).

SPECIAL CASE

Consider the special case for which the functions $g_1(t)$ and $g_2(t)$ are equal, as shown by

$$g_1(t)=g_2(t)=\sqrt{g(t)} \tag{A5.15}$$

Then, substituting Eq. (A5.15) in (A5.3), we get

$$\left|\int_a^b g(t)dt\right|^2 \leqslant \left[\int_a^b |g(t)|dt\right]^2$$

or, equivalently,

$$\left|\int_a^b g(t)dt\right| \leqslant \int_a^b |g(t)|dt \tag{A5.16}$$

That is, the absolute value of the total area under a time function $g(t)$ is less than or equal to the absolute integral of $g(t)$.

Appendix 6
MATHEMATICAL TABLES

This appendix presents. (1) a summary of properties of the Fourier transform, (2) a short table of Fourier transform pairs, (3) a short table of Hilbert transform pairs, (4) a list of trigonometric identities, (5) a selected list of series expansions, and (6) a selected list of integrals.

Table A6.1 Summary of Properties of the Fourier Transform

Property	Mathematical Description
1. Linearity	$ag_1(t)+bg_2(t)\rightleftharpoons aG_1(f)+bG_2(f)$ where a and b are constants
2. Time scaling	$g(at)\rightleftharpoons\frac{1}{\lvert a\rvert}G\left(\frac{f}{a}\right)$ where a is a constant
3. Duality	If $g(t)\rightleftharpoons G(f)$, then $G(t)\rightleftharpoons g(-f)$
4. Time shifting	$g(t-t_0)\rightleftharpoons G(f)\exp(-j2\pi f t_0)$
5. Frequency shifting	$\exp(j2\pi f_c t)g(t)\rightleftharpoons G(f-f_c)$
6. Area under $g(t)$	$\int_{-\infty}^{\infty} g(t)dt=G(0)$
7. Area under $G(f)$	$g(0)=\int_{-\infty}^{\infty} G(f)df$
8. Differentiation in the time domain	$\frac{d}{dt}g(t)\rightleftharpoons j2\pi fG(f)$
9. Integration in the time domain	$\int_{-\infty}^{t} g(\tau)d\tau\rightleftharpoons\frac{1}{j2\pi f}G(f)+\frac{G(0)}{2}\delta(f)$
10. Conjugate functions	If $g(t)\rightleftharpoons G(f)$, then $g^*(t)\rightleftharpoons G^*(-f)$
11. Multiplication in the time domain	$g_1(t)g_2(t)\rightleftharpoons\int_{-\infty}^{\infty} G_1(\lambda)G_2(f-\lambda)d\lambda$
12. Convolution in the time domain	$\int_{-\infty}^{\infty} g_1(\tau)g_2(t-\tau)d\tau\rightleftharpoons G_1(f)G_2(f)$.

Table A6.2 Fourier-Transform Pairs.

Time Function	Fourier Transform
$\text{rect}\left(\frac{t}{T}\right)$	$T\,\text{sinc}(fT)$
$\text{sinc}(2Wt)$	$\frac{1}{2W}\,\text{rect}\left(\frac{f}{2W}\right)$
$\exp(-at)u(t), \quad a>0$	$\frac{1}{a+j2\pi f}$
$\exp(-a\|t\|), \quad a>0$	$\frac{2a}{a^2+(2\pi f)^2}$
$\exp(-\pi t^2)$	$\exp(-\pi f^2)$
$\begin{cases} 1-\frac{\|t\|}{T}, & \|t\|<T \\ 0, & \|t\|\geqslant T \end{cases}$	$T\,\text{sinc}^2(fT)$
$\delta(t)$	1
1	$\delta(f)$
$\delta(t-t_0)$	$\exp(-j2\pi f t_0)$
$\exp(j2\pi f_c t)$	$\delta(f-f_c)$
$\cos(2\pi f_c t)$	$\frac{1}{2}[\delta(f-f_c)+\delta(f+f_c)]$
$\sin(2\pi f_c t)$	$\frac{1}{2j}[\delta(f-f_c)-\delta(f+f_c)]$
$\text{sgn}(t)$	$\frac{1}{j\pi f}$
$\frac{1}{\pi t}$	$-j\,\text{sgn}(f)$
$u(t)$	$\frac{1}{2}\delta(f)+\frac{1}{j2\pi f}$
$\sum_{i=-\infty}^{\infty}\delta(t-iT_0)$	$\frac{1}{T_0}\sum_{n=-\infty}^{\infty}\delta\left(f-\frac{n}{T_0}\right)$

Table A6.3 Hilbert-Transform Pairs*

Time Function	Hilbert Transform
$m(t)\cos(2\pi f_c t)$	$m(t)\sin(2\pi f_c t)$
$m(t)\sin(2\pi f_c t)$	$-m(t)\cos(2\pi f_c t)$
$\cos(2\pi f_c t)$	$\sin(2\pi f_c t)$
$\sin(2\pi f_c t)$	$-\cos(2\pi f_c t)$
$\dfrac{\sin t}{t}$	$\dfrac{1-\cos t}{t}$
$\text{rect}(t)$	$-\dfrac{1}{\pi}\ln\left\lvert\dfrac{t-\frac{1}{2}}{t+\frac{1}{2}}\right\rvert$
$\delta(t)$	$\dfrac{1}{\pi t}$
$\dfrac{1}{1+t^2}$	$\dfrac{t}{1+t^2}$
$\dfrac{1}{t}$	$-\pi\delta(t)$

*In the first two pairs, it is assumed that $m(t)$ is band-limited to the interval $-W \leqslant f \leqslant W$, where $W < f_c$.

Table A6.4 Trigonometric Identities

$\exp(\pm j\theta)=\cos\theta\pm j\sin\theta$

$\cos\theta=\frac{1}{2}[\exp(j\theta)+\exp(-j\theta)]$

$\sin\theta=\frac{1}{2j}[\exp(j\theta)-\exp(-j\theta)]$

$\sin^2\theta+\cos^2\theta=1$

$\cos^2\theta-\sin^2\theta=\cos(2\theta)$

$\cos^2\theta=\frac{1}{2}[1+\cos(2\theta)]$

$\sin^2\theta=\frac{1}{2}[1-\cos(2\theta)]$

$2\sin\theta\cos\theta=\sin(2\theta)$

$\sin(\alpha\pm\beta)=\sin\alpha\cos\beta\pm\cos\alpha\sin\beta$

$\cos(\alpha\pm\beta)=\cos\alpha\cos\beta\mp\sin\alpha\sin\beta$

$\tan(\alpha\pm\beta)=\frac{\tan\alpha\pm\tan\beta}{1\mp\tan\alpha\tan\beta}$

$\sin\alpha\sin\beta=\frac{1}{2}[\cos(\alpha-\beta)-\cos(\alpha+\beta)]$

$\cos\alpha\cos\beta=\frac{1}{2}[\cos(\alpha-\beta)+\cos(\alpha+\beta)]$

$\sin\alpha\cos\beta=\frac{1}{2}[\sin(\alpha-\beta)+\sin(\alpha+\beta)]$

Table A6.5 Series Expansions

Taylor series

$$f(x)=f(a)+\frac{f'(a)}{1!}(x-a)+\frac{f''(a)}{2!}(x-a)^2+\cdots+\frac{f^{(n)}(a)}{n!}(x-a)^n+\cdots$$

where

$$f^{(n)}(a)=\left.\frac{d^n f(x)}{dx^n}\right|_{x=a}$$

MacLaurin series

$$f(x)=f(0)+\frac{f'(0)}{1!}x+\frac{f''(0)}{2!}x^2+\cdots+\frac{f^{(n)}(0)}{n!}x^n+\cdots$$

where

$$f^{(n)}(0)=\left.\frac{d^n f(x)}{dx^n}\right|_{x=0}$$

Binomial series

$$(1+x)^n=1+nx+\frac{n(n-1)}{2!}x^2+\cdots, \qquad |nx|<1$$

Exponential series

$$\exp x=1+x+\frac{1}{2!}x^2+\cdots$$

Logarithmic series

$$\ln(1+x)=x-\tfrac{1}{2}x^2+\tfrac{1}{3}x^3-\cdots$$

Trigonometric series

$$\sin x=x-\frac{1}{3!}x^3+\frac{1}{5!}x^5-\cdots$$

$$\cos x=1-\frac{1}{2!}x^2+\frac{1}{4!}x^4-\cdots$$

$$\tan x=x+\tfrac{1}{3}x^3+\tfrac{2}{15}x^5+\cdots$$

$$\sin^{-1} x=x+\tfrac{1}{6}x^3+\tfrac{3}{40}x^5+\cdots$$

$$\tan^{-1} x=x-\tfrac{1}{3}x^3+\tfrac{1}{5}x^5-\cdots, \qquad |x|<1$$

$$\operatorname{sinc} x=1-\frac{1}{3!}(\pi x)^2+\frac{1}{5!}(\pi x)^4-\cdots$$

Table A6.6 Integrals

Indefinite integrals

$$\int x \sin(ax)dx = \frac{1}{a^2}[\sin(ax) - ax\cos(ax)]$$

$$\int x \cos(ax)dx = \frac{1}{a^2}[\cos(ax) + ax\sin(ax)]$$

$$\int x \exp(ax)dx = \frac{1}{a^2}\exp(ax)(ax-1)$$

$$\int x \exp(ax^2)dx = \frac{1}{2a}\exp(ax^2)$$

$$\int \exp(ax)\sin(bx)dx = \frac{1}{a^2+b^2}\exp(ax)[a\sin(bx) - b\cos(bx)]$$

$$\int \exp(ax)\cos(bx)dx = \frac{1}{a^2+b^2}\exp(ax)[a\cos(bx) + b\sin(bx)]$$

$$\int \frac{dx}{a^3+b^2x^2} = \frac{1}{ab}\tan^{-1}\left(\frac{bx}{a}\right)$$

$$\int \frac{x^2\,dx}{a^2+b^2x^2} = \frac{x}{b^2} - \frac{a}{b^3}\tan^{-1}\left(\frac{bx}{a}\right)$$

Definite integrals

$$\int_0^\infty \frac{x\sin(ax)}{b^2+x^2}dx = \frac{\pi}{2}\exp(-ab), \qquad a>0,\ b>0$$

$$\int_0^\infty \frac{\cos(ax)}{b^2+x^2}dx = \frac{\pi}{2b}\exp(-ab), \qquad a>0,\ b>0$$

$$\int_0^\infty \frac{\cos(ax)}{(b^2-x^2)^2}dx = \frac{\pi}{4b^3}[\sin(ab) - ab\cos(ab)], \qquad a>0,\ b>0$$

$$\int_0^\infty \operatorname{sinc} x\,dx = \int_0^\infty \operatorname{sinc}^2 x\,dx = \tfrac{1}{2}$$

$$\int_0^\infty \exp(-ax^2)dx = \frac{1}{2}\sqrt{\frac{\pi}{a}}, \qquad a>0$$

$$\int_0^\infty x^2\exp(-ax^2)dx = \frac{1}{4a}\sqrt{\frac{\pi}{a}}, \qquad a>0$$

Glossary

CONVENTIONS AND NOTATIONS

1. The symbol $|\ |$ means the magnitude of the complex quantity contained within.
2. The symbol arg( ) means the phase angle of the complex quantity contained within.
3. The symbol Re[ ] means the "real part of," and Im[ ] means the "imaginary part of."
4. The symbol ln( ) denotes the natural logarithm of the quantity contained within, whereas the logarithm to the base a is denoted by $\log_a(\ )$.
5. The use of an asterisk as superscript denotes complex conjugate, e.g., x^* is the complex conjugate of x.
6. The symbol $\rightleftharpoons$ indicates a Fourier transform pair, e.g., $g(t) \rightleftharpoons G(f)$, where a lowercase letter denotes the time function and a corresponding uppercase letter denotes the frequency function.
7. The symbol $F[\ ]$ indicates the Fourier transform operation, e.g., $F[g(t)] = G(f)$, and the symbol $F^{-1}[\ ]$ indicates the inverse Fourier transform operation, e.g.,

$$F^{-1}[G(f)] = g(t).$$

8. The symbol $\otimes$ denotes convolution, e.g.,

$$x(t) \otimes h(t) = \int_{-\infty}^{\infty} x(\tau)h(t-\tau)d\tau$$

9. The symbol $\oplus$ denotes modulo-two addition.
10. The use of subscript p indicates that the pertinent function is periodic, e.g., the function $g_p(t)$ is a periodic function of time t.
11. The use of a hat over a function indicates one of two things:
    (a) the Hilbert transform of a function, e.g., the function $\hat{g}(t)$ is the Hilbert transform of $g(t)$, or
    (b) the estimate of an unknown parameter, e.g., the quantity $\hat{\alpha}(\mathbf{x})$ is an estimate of the unknown parameter α, based on the observation vector $\mathbf{x}$.
12. The use of a tilde over a function indicates the complex envelope of a narrow-band signal, e.g., the function $\tilde{g}(t)$ is the complex envelope of the narrow-band signal $g(t)$.
13. The use of subscript + indicates the pre-envelope of a signal, e.g., the function $g_+(t)$ is the pre-envelope of the signal $g(t)$. We may thus write $g_+(t) = g(t) + j\hat{g}(t)$, where $\hat{g}(t)$ is the Hilbert transform of $g(t)$.
14. The use of subscripts c and s indicates the in-phase and quadrature components of a narrow-band signal, a narrow-band random process, or the impulse response of a narrow-band filter, with respect to the carrier $\cos(2\pi f_c t)$.

15. For a low-pass message signal the highest frequency component or message bandwidth is denoted by W. The spectrum of this signal occupies the frequency interval $-W \leqslant f \leqslant W$ and is zero elsewhere. For a band-pass signal with carrier frequency f_c, the spectrum occupies the frequency intervals $f_c - W \leqslant f \leqslant f_c + W$ and $-f_c - W \leqslant f \leqslant -f_c + W$, and so $2W$ denotes the bandwidth of the signal. The complex envelope (low-pass) of this band-pass signal has a spectrum that occupies the frequency interval $-W \leqslant f \leqslant W$.

    For a low-pass filter the bandwidth is denoted by B. A common definition of filter bandwidth is the frequency at which the amplitude response of the filter drops by 3 dB below the zero-frequency value. For a band-pass filter of mid-band frequency f_c the bandwidth is denoted by $2B$, centered on f_c. The complex low-pass equivalent of this band-pass filter has a bandwidth equal to B.

    The transmission bandwidth of a communication channel, required to transmit a modulated wave, is denoted by B_T.
16. Random variables are uppercase (e.g., X or $\mathbf{X}$), whereas sample values of random variables are lowercase (e.g., x or $\mathbf{x}$).
17. A vertical bar in an expression means "given that," e.g., $f_{X|H_0}(x|H_0)$ is the probability density function of the random variable X, given that hypothesis H_0 is true.
18. The symbol $E[\ \ ]$ means the expected value of the random variable enclosed within.
19. The symbol Var[ ] means the variance of the random variable enclosed within.
20. The symbol Cov[ ] means the covariance of the two random variables enclosed within.
21. The average probability of symbol error is denoted by P_e.

    In the case of binary signaling techniques, P_{e0} denotes the conditional probability of error given that symbol 0 was transmitted, and P_{e1} denotes the conditional probability of error given that symbol 1 was transmitted. The a priori probabilities of symbols 0 and 1 are denoted by p_0 and p_1, respectively.
22. The symbol $\langle\ \ \rangle$ denotes the time average of the sample function enclosed within.
23. Boldface letter denotes a vector or matrix. The inverse of a square matrix $\mathbf{R}_x$ is denoted by $\mathbf{R}_x^{-1}$. The transpose of a vector $\mathbf{W}$ is denoted by $\mathbf{W}^t$.
24. The length of a vector $\mathbf{x}$ is denoted by $\|\mathbf{x}\|$. The inner product or dot product of two vectors $\mathbf{x}$ and $\mathbf{y}$ is denoted by $(\mathbf{x}, \mathbf{y})$.

Functions

1. Rectangular function:

$$\text{rect}(t) = \begin{cases} 1, & -\frac{1}{2} < t < \frac{1}{2} \\ 0, & |t| > \frac{1}{2} \end{cases}$$

2. Unit step function:

$$u(t) = \begin{cases} 1, & t > 0 \\ 0, & t < 0 \end{cases}$$

3. Signum function:

$$\text{sgn}(t) = \begin{cases} 1, & t > 0 \\ -1, & t < 0 \end{cases}$$

4. Dirac delta function:

$$\delta(t) = 0, \qquad t \neq 0$$

$$\int_{-\infty}^{\infty} \delta(t)\,dt = 1$$

or equivalently

$$\int_{-\infty}^{\infty} g(t)\delta(t - t_0)\,dt = g(t_0)$$

5. Sinc function: $\operatorname{sinc}(x) = \dfrac{\sin(\pi x)}{\pi x}$

6. Sine integral: $\operatorname{Si}(u) = \displaystyle\int_0^u \frac{\sin x}{x}\,dx$

7. Error function: $\operatorname{erf}(u) = \dfrac{2}{\sqrt{\pi}} \displaystyle\int_0^u \exp(-z^2)\,dz$

   Complementary error function: $\operatorname{erfc}(u) = 1 - \operatorname{erf}(u)$.

8. Bessel function of the first kind of order n: $J_n(x) = \dfrac{1}{2\pi} \displaystyle\int_{-\pi}^{\pi} \exp(jx \sin\theta - jn\theta)\,d\theta$

   Modified Bessel function of the first kind of zero order: $I_0(x) = \dfrac{1}{2\pi} \displaystyle\int_{-\pi}^{\pi} \exp(x \cos\theta)\,d\theta$

9. Binomial coefficient $\dbinom{n}{k} = \dfrac{n!}{(n-k)!k!}$

ABBREVIATIONS

ac: alternating current
AWGN: additive white Gaussian noise
AM: amplitude modulation
ASK: amplitude shift-keying
BPF: band-pass filter
CCD: charge-coupled device
CPFSK: continuous-phase frequency-shift keying
CW: continuous wave
dB: decibel
dc: direct current
DEM: demodulator
DM: delta modulation
DPCM: differential pulse-code modulation
DPSK: differential phase-shift keying
DSBSC: double-sideband suppressed-carrier
FDM: frequency-division multiplexing
FFT: fast Fourier transform
FMFB: frequency modulator with feedback
FSK: frequency shift-keying
Hz: hertz
IF: intermediate frequency
ISI: intersymbol interference
kHz: kilohertz
LMS: least mean-square
LPF: low-pass filter
MHz: megahertz
ms: millisecond

μs: microsecond
MOD: modulator
modem: modulator-demodulator
MSK: minimum shift keying
NRZ: Non-return-to-zero
PAM: pulse-amplitude modulation
PCM: pulse-code modulation
PDM: pulse-duration modulation
PLL: phase-locked loop
PN: pseudo-noise
PPM: pulse-position modulation
PSK: phase-shift keying
QAM: quadrature-amplitude modulation
QPSK: quadriphase-shift keying
RF: radio frequency
rms: root mean-square
RZ: return-to-zero
s: second
SNR: signal-to-noise ratio
$(SNR)_C$: channel signal-to-noise ratio
$(SNR)_O$: output signal-to-noise ratio
SSB: single sideband
TDM: time-division multiplexing
VCO: voltage-controlled oscillator
VSB: vestigial sideband

Index

Adaptive encoders, 451
Adaptive equalizer, 487
  decision-directed mode, 494
  fractionally spaced, 495
  implementation approaches, 496
  training mode, 494
Aliasing, 370
Amplitude distortion, 69
Amplitude modulation, 114
  channel signal-to-noise ratio, 328
  figure of merit, 330
  output signal-to-noise ratio, 329
  phasor representation, 187
  spectrum, 116
  threshold effect, 330
  time-domain description, 114
  transmission bandwidth, 117
Amplitude response, 66
Amplitude sensitivity, 114
Amplitude spectrum, bounds on, 98
  continuous, 20
  discrete, 14
AM-to-PM conversion, 218
Analytic signal, *see* Pre-envelope
Analog signals, 6
Angle modulation, 179
Antipodal signals, 541
Aperture effect, 381
a posteriori probability, 461
a priori probability, 423
Area, under frequency function, 31
  under time function, 30
Autocorrelation function, periodic signals, 56
  energy signals, 54
  measurement of, 254
  random processes, 244, 250
Autocovariance function, 245
Average power, 5, 53

Balanced modulator, 127
Band-pass filter, 70, 85
  complex impulse response, 85
Band-pass signals, 80
  envelope and phase, 82
  in-phase and quadrature components, 82
Band-stop filter, 70
Bandwidth, 66
Bandwidth efficiency, 438, 553
Baseband shaping, 469
Baseband signals, 113
Bayes rule, 232
Bayes test, 604
Baud, 486
Bessel function of first kind, 188, 628
Binary amplitude-shift keying, 540
  probability of error, 610
Binary code, 413
Binary data transmission, band-pass, 540
  baseband, 466
Binary frequency-shift keying, coherent, 544
  probability of error, 548
  signal space diagram, 546
  spectral properties, 551
  transmitter and receiver, 548
Binary frequency-shift keying, noncoherent, 581
  probability of error, 585
Binary phase-shift keying, coherent, 541
  effect of phase error, 608
  probability of error, 543
  signal space diagram, 542
  spectral properties, 550
  transmitter and receiver, 543
Binary symbols, 413
  electrical representation of, 413
Bipolar signaling, 414
Bit, 413
Bit duration, 419
Bit energy-to-noise density ratio, 548
Bit rate, 486, 508
Bit stuffing, 456
Broadcast systems, 156

Capacity theorem, 433
Capture effect in FM, 341
Carrier delay, *see* Phase delay
Carrier-to-noise ratio, 343

Carson's rule, 195
Cauchy's principal value, 74
Causality, criterion for, 63
Central limit theorem, 272
Central moments, 237
Channel, 8
  additive white Gaussian noise, 508
  band-limited, 8
  dispersive, 69
  memoryless, 522
  power-limited, 8
  symmetric, 423
Channel capacity, 433
  Gaussian channel, 434
  PCM system, 436
  television system, 462
Channel signal-to-noise ratio, 321
Characteristic function, 237
Chebyshev inequalilty, 237
Code word, 413
Coherent detection, DSBSC waves, 130
  SSB waves, 146
  VSB waves, 151
Coherent receiver, for band-pass data transmission, 510
Color television, 168
Colored noise, 607
Comb filter, 265
Commutator, 384
Compander, 413
Companding improvement, 459
Complementary error function, 619
Complex envelope, 80
Complex exponential Fourier series, 13
Complex impulse response, 85
Compression, A-law, 412
  μ-law, 411
Communication, 4
Conditional probability, 232
Conjugate functions, 35
Continuous-phase frequence-shift keying, 548, 561
Continuous-wave modulation, 114
Convolution theorem, 39
Correlation coefficient, 238
Correlation matrix, 249
Correlation receiver, 528
Correlative coding, 476
  generalized scheme for, 484
Cosine function, Fourier transform of, 46
Costas receiver, 132
Cramer-Rao bound, 601
Cross-correlation, periodic signals, 60
  energy signals, 59
  random processes, 249
Cross-spectral density, 268
Cumulative distribution function, *see* Distribution function

DC signal, 45
Decibel, 66
Decimation, 400
Decision feedback, 478
Decision regions, 525
Decorrelation time, 246
Delay distortion, 70
Delta function, *see* Dirac delta function
Delta modulation, 445
  adaptive, 453
  granular noise, 449
  linear, 449
  slope-overload distortion, 449
Demodulation, 8, 114
Detection of signals in noise, coherent, 523
  noncoherent, 578
Deterministic signals, 5, 229
Deviation ratio, CPFSK signal, 562
  FM signal, 196
Dibit, 485
Differential encoding, 415
Differential phase-shift keying, 585
  probability of error, 592, 615
  signal-space diagram, 588
  transmitter and receiver, 592
Differential pulse-code modulation, 439
  adaptive, 452
  gain, 442
Differentiation in time domain, 31
Digital computer simulation, 217, 597
Digital filter, 63
Digital multiplexers, 454
Digital signals, 6
Dirac comb, *see* Ideal sampling function
Dirac delta function, 42
  even function property, 100
  replication property, 43
  sifting property, 43
Direct method of generating FM waves, 200
Dirichlet's conditions, 13, 19
Discrete Fourier transform, 40
  properties of, 100
Distribution function, 234
Distortionless transmission, 68
  conditions for, 69
Dolby systems, 352
Donald Duck voice effect, 149
Double-sideband suppressed-carrier modulation, 125

channel-signal-to-noise ratio, 323
coherent detection, 130
figure of merit, 325
output signal-to-noise ratio, 324
spectrum, 127
time-domain description, 126
transmission bandwidth, 127
Duality (time-frequency), 25
Duobinary signaling, 477
precoder for, 480

Elastic store, 456
Encoding, 413
Energy, 5, 51
signals, 6
spectral density, 51
Entropy, 431
conditional, 434
Envelope delay, *see* Group relay
Envelope detector, 124
Envelope distortion, 116
Equivalent noise temperature, 278, 625
Ergodic process, 251
ergodicity of the autocorrelation function, 252
erogodicity of the mean, 251
Error function, 619
Error rate, *see* Probability of error
Expectation operator, 236
Expected value, *see* Mean value
Exponential distribution, 618
Exponential function, Fourier transform of, 21
Eye pattern, 496

Fading channel, 313
Fast Fourier transform algorithm, 41
Fast frequency-shift keying, *see* Minimum-shift keying
Figure of merit, 321
Filters, 70
Flat-top samples, 378
FMFB demodulator, 346
Fourier series, 12
properties of, 94
Fourier transform, 17
numerical computation of, 40
properties of, 23, 637
Fourier-transform pairs, 638
Fourth power loop, 560
Frame, 417
Frequency deviation, 184
Frequency discrimination method of generating SSB waves, 141
Frequency discriminator, 202
Frequency distortion in demodulating SSB waves, 147
Frequency-division multiplexing, 160
Frequency-modulated radar, 219
Frequency modulation, 181
capture effect, 341
channel signal-to-noise ratio, 339
clicks, 341
direct method of generation, 200
effect of filtering, 216
effect of limiting, 226
figure of merit, 339
indirect method of generation, 197
multitone, 190
narrow-band, 185
nonlinear effects, 217
output signal-to-noise ratio, 339
phasor representation, 187
pre-emphasis and de-emphasis, 348
root mean-square bandwidth, 308
single-sideband version, 226
single-tone, 183
threshold effect, 341
threshold reduction, 346
transmission bandwidth, 194
wide-band, 187
Frequency sensitivity, 181
Frequency shifting, 28
Frequency synthesizer, 176
Frequency translation, 158

Gain, 66
Gaussian distribution, 271, 617
Gaussian process, 270
properties of, 272
Gaussian pulse, 32
Gibb's phenomenon, 38
Gradient vector, 491
Gram-Schmidt orthogonalization procedure, 510
Group delay, 92

Half-cosine pulse, 96, 564
Half-sine pulse, 96, 564
Hartley modulator, 144
Hartley oscillator, 200
Hartley-Shannon law, 435
Heterodyning, *see* Mixing
High-pass filter, 70
Hilbert transform, 74
properties of, 77
Hilbert-transform pairs, 639

Ideal low-pass filter, 70
Ideal sampling function, 50
Image signal, 319

Impulse response, 61
Indirect method of generating FM waves, 197
Information, measure of, 429
Inner product, 517
Instantaneous frequency, 180
Instantaneous sampling, 365
Integrate-and-dump filter, 536
Integration in time domain, 33
Intermediate frequency, 319
Interpolation, 400
Intersymbol interference, 468, 596
Inverse discrete Fourier transform, 41
Inverse filter, 606
Inverse Fourier transform, 19
Inverse Hilbert transform, 74

Joint distribution function, 235
Jointly Gaussian random variables, 274
Joint probability, 231
Joint probability density function, 235

Kronecker delta, 444

Least-mean-square algorithm, 493
Least-squares algorithm, 495
Leibnitz's rule, 308
Likelihood function, 522
Likelihood ratio, 605
Linear maximal sequences, *see* Pseudo-noise sequences
Linear modulation, 156
Linear systems, 60
Low-pass equivalent model of band-pass filter, 88
Low-pass filter, 70
Low-pass signal, 80

M-12 multiplexer, 456
MacLaurin series, 641
Majority logic decoding, 457
Manchester code, 414
M-ary data transmission, band-pass, 575
  baseband, 485
Marginal density, 236
Matched filter, 531
  maximization of output signal-to-noise ratio, 531
  properties of, 536
Matched-filter pair, 538
Matched-filter receiver, 530
Maximum *a posteriori* probability decoder, 524
Maximum likelihood decoder, 525
Maximum likelihood estimation, 599
Maximum-power transfer theorem, 277
Mean-square value, 237
Mean value, 236, 244, 250
Message bandwidth, 116
Message point, 523
Minimum probability of error criterion, 510
Minimum-shift keying, 561
  probability of error, 570
  signal space diagram, 566
  spectral properties, 574
  transmitter and receiver, 570
Minimum-variance estimator, 601
Mixer, 159
Modem, 455
Modified Bessel function of first kind, 298, 631
Modified duobinary signaling, 481
  precoder for, 484
Modulation, 8, 114
Modulation factor of AM wave, 117
Modulation index of FM wave, 184
Multiplication theorem, 36

Narrow-band noise, 283
  envelope and phase components, 294
  in-phase and quadrature components, 285
Narrow-band signal, 80
  complex envelope, 80
  envelope and phase, 82
  in-phase and quadrature components, 82
Natural sampling, 389
Noise equivalent, bandwidth, 283, 614
Noise figure, 623
Noncoherent receiver for band-pass data transmission, 510
Nonperiodic signals, 5
Nonreturn-to-zero signaling, *see* Polar signaling
Normal distribution, *see* Gaussian distribution
Normalized transmission bandwidth, 352
Norton equivalent circuit of noisy resistor, 277
Nyquist rate, 370

Observation space, 525
Observation vector, 523
Offset QPSK, 613
On-off signaling, 414
Output signal-to-noise ratio, 320
Orthogonal random variables, 239
Orthogonal signals, 59
Overmodulation, 116

Packet-switching, 450
Paley-Wiener criterion, 68
Parseval's power theorem, 53
Partial response signaling, *see* Correlative coding
Peak pulse-to-noise ratio, 424
Percentage modulation, 116

Periodic pulse train, 15
  duty cycle, 17
  spectrum, 16
Periodic signals, 4
  complex exponential Fourier series, 13
  Fourier series, 12
  Fourier transform, 49
  mean value, 12
Periodogram, 267
Phase, 66
Phase ambiguity, 135, 544, 560, 572
Phase delay, 92
Phase-discrimination method of generating SSB waves, 143
Phase distortion, *see* Delay distortion
Phase distortion in demodulating SSB waves, 148
Phase modulation, 181
  pre-emphasis and de-emphasis, 360
  signal-to-noise ratios, 359
  spectrum, 220
  transmission bandwidth, 222
Phase response, 66
Phase sensitivity, 181
Phase spectrum, continuous, 20
  discrete, 14
Phase-locked loop, 207
  first-order, 211
  FM threshold reduction, 348
  linearized model, 209
  nonlinear model, 209
  open-loop transfer function, 210
  phase error, 208
  phase lock, 209
  second-order, 213
  use in estimating phase, 601
Pilot carrier, 162
Plancherel's theorem, 19
Point-to-point communication, 156
Poisson's sum formula, 50
Polar signaling, 414
Power signals, 5
Power spectral density, 53, 256
Prediction filter, 442
Pre-envelope, 79
Prewhitening filter, 607
Principle of orthogonality, 443, 490
Principle of stationary phase, 111
Principle of superposition, 7
Probability, 230
Probability density function, 234
Probability of detection, 611
Probability of error, 527
  bounds of, 528
Probability of false alarm, 611
Product modulator, 127
Pseudo-noise sequence, 262
Pulse-amplitude modulation, 385
  signal-to-noise ratio, 403
  spectrum, 403
  synchronization, 387
Pulse-compression radar, 538
Pulse-code modulation, 408
  adaptive, 452
  bandwidth efficiency, 438
  channel capacity, 436
  decoding, 417
  encoding, 413
  error threshold, 424
  multiplexing, 417
  output signal-to-noise ratio, 427, 428
  probability of error, 424
  quantizing, 409
  regeneration, 416
  ruggedness to interference, 425
  sampling, 408
  synchronization, 417
  transmission bandwidth, 429
Pulse-duration modulation, 389
  channel signal-to-noise ratio, 396
  figure of merit, 396
  output signal-to-noise ratio, 396
  spectrum, 391
  synchronization, 397
Pulse-length modulation, *see* Pulse-duration modulation
Pulse modulation, 363
Pulse-position modulation, 389
  channel signal-to-noise ratio, 395
  figure of merit, 395
  output signal-to-noise ratio, 394
  spectrum, 391
  synchronization, 397
Pulse-time modulation, 389
Pulse-width modulation, *see* Pulse-duration modulation

Quadrature-carrier multiplexing, 135, 553
Quadrature null effect, 131
Quadrature receiver, 579
Quadriphase-shift keying, coherent, 553
  probability of error, 558
  signal space diagram, 555
  spectral properties, 573
  transmitter and receiver, 559
Quantizer, adaptive, 451
  mid-riser type, 411, 458
  mid-tread type, 411, 458
  nonuniform, 411

uniform, 411
Quantizing noise (error), 411, 426

Radar detection, 611
Radio frequency pulse, 28
Raised-cosine pulse, 392
Raised-cosine pulse spectrum, 470
Random binary wave, 247, 260
Random processes, 240
filtering of, 254
mixing of, 264
nonstationary, 243
sample function of, 241
sampling of, 382
strictly stationary, 243
wide-sense stationary, 245
Random telegraph signal, 305
Random variables, 233
correlation, 238
covariance, 238
mean value, 236
probability density function, 234
statistical independence of, 236
transformation of, 239
Random vector, 242
Rayleigh distribution, 296, 617
Rayleigh's energy theorem, 52
Received signal point, 524
Rectangular function, 20
Regenerative repeaters, 416
Repeaters, 156
Return-to-zero signaling, 414
Rician distribution, 298
Ring modulator, 128
Rolloff factor, 470
Root mean-square bandwidth, 102
Root mean-square duration, 102

Sample point, 233
Sample space, 233
Sampling process, 364
error, 399
interpolation function, 368
practical aspects, 376
rate, 365
Sampling theorem, 369
band-pass signals, 371
complex signals, 401
random processes, 382
Schottky formula, 276
Schwarz's inequality, 634
vector interpretation, 604
Scrambler, 173
Shannon bound, 436
Shot noise, 275
Sidebands, 117
Signal energy-to-noise density ratio, 539
Signal space, 517
Signal vector, 517
Signaling rate, 486
Signum function, 48
Sine function, 15
Fourier transform of, 47
Sine integral, 38
Sine-wave with random phase, 246
Single-sideband modulation, 137
channel signal-to-noise ratio, 326
figure of merit, 328
output signal-to-noise ratio, 328
spectrum, 137
time-domain description, 140
transmission bandwidth, 137
16-level quadrature amplitude modulation, 612
Slope circuit, 202
Spectral amplitude matching, 540
Spectral density, 51
Spectral phase matching, 539
Spectrum, 7
Spectrum analyzer, 175, 222
Spectrum analyzer receiver, 614
Split-phase code, *see* Manchester code
Spread-spectrum communications, 608
Square-law detector, 123
Square-law modulator, 119
Squaring loop, 133
Stability, criterion for, 64
Stabilization of frequency modulator, 201
Standard deviation, 237
Statistically independent events, 233
Statistically independent random variables, 236
Steepest descent algorithm, 492
Step size parameter, 492
Stereophonic FM waves, 223
Stochastic gradient algorithm, *see* Least-mean-square algorithm
Stochastic processes, *see* Random processes
Sunde's FSK, 550
Supergroup, 162
Superheterodyne receivers, 318
Switching modulator, 121
Switched telecommunication network, 450
Symbol, 413
Synchronous detection, *see* Coherent detection

T1 carrier, 418, 472
Tapped-delay-line filter, 62
Taylor series, 641
Thermal noise, 276

Thevenin-equivalent circuit of noisy resistor, 277
Threshold effect, AM, 330
  FM, 341
  PCM, 424
  PDM, 397
  PPM, 397
Time averages, 250
Time-bandwidth product, 102
Time-division multiplexing, 384
Time scaling, 24
Time shifting, 27
Transfer function, 65
Transformer, ideal, 74
Transmultiplexer, 385

Unbiased estimator, 251, 601
Uncorrelated random variables, 239
Uniform distribution, 616
Unit step function, 47
Unvoiced sounds, 451

Variance, 237
Vector receiver, 508
Vector transmitter, 506
Vestigial sideband modulation, 149
  envelope detection, 155
  signal-to-noise ratio, 359
  spectrum, 150
  time-domain description, 153
  transmission bandwidth, 150
Voiced sounds, 451

Weaver's method of generating SSB waves, 145, 171
Wiener-Khintchine relations, 58, 258
White noise, 277

Zero-crossing detector, 225
Zero crossings, 181
Zero-memory source, 430